Aleksey Ber, Vladimir Chilin, Galina Levitina, Fedor Sukochev, Dmitriy Zanin

Algebras of Unbounded Operators

Also of Interest

Algebraic Topology
Constructions, Retractions, and Fixed Point Theory
Smail Djebali, 2024
ISBN 978-3-11-151736-0, e-ISBN (PDF) 978-3-11-151738-4

Algebraic Topology
A Toolkit
Kevin P. Knudson, 2024
ISBN 978-3-11-101481-4, e-ISBN (PDF) 978-3-11-101485-2

Abstract Algebra
With Applications to Galois Theory, Algebraic Geometry, Representation Theory and Cryptography
Gerhard Rosenberger, Annika Schürenberg, Leonard Wienke, 2024
ISBN 978-3-11-113951-7, e-ISBN (PDF) 978-3-11-114252-4

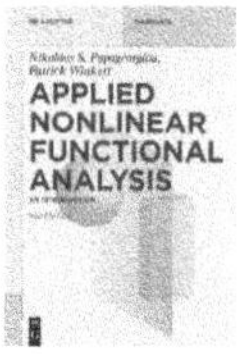

Applied Nonlinear Functional Analysis
An Introduction
Nikolaos S. Papageorgiou, Patrick Winkert, 2024
ISBN 978-3-11-128421-7, e-ISBN (PDF) 978-3-11-128695-2

Topics in Complex Analysis
Dan Romik, 2023
ISBN 978-3-11-079678-0, e-ISBN (PDF) 978-3-11-079681-0

Aleksey Ber, Vladimir Chilin, Galina Levitina,
Fedor Sukochev, Dmitriy Zanin

Algebras of Unbounded Operators

Algebraic and Topological Aspects of
Murray–von Neumann Algebras

DE GRUYTER

Mathematics Subject Classification 2020
Primary: 47B47, 47L60; Secondary: 46H05, 46H35, 46L10, 46L57, 46L52

Authors
Dr. Aleksey Ber
V.I.Romanovskiy Institute of Mathematics
Uzbekistan Academy of Sciences,
National University of Uzbekistan
Department of Mathematics
Uzbekistan
aber1960@mail.ru

Prof. Vladimir Chilin
V.I.Romanovskiy Institute of Mathematics
Uzbekistan Academy of Sciences,
National University of Uzbekistan
Department of Mathematics
Uzbekistan
vladimirchil@gmail.com

Dr. Galina Levitina
Australian National University
Mathematical Sciences Institute
Australia
galina.levitina@anu.edu.au

Prof. Fedor Sukochev
University of New South Wales
School of Mathematics & Statistics
Australia
f.sukochev@unsw.edu.au

Dr. Dmitriy Zanin
University of New South Wales
School of Mathematics & Statistics
Australia
d.zanin@unsw.edu.au

ISBN 978-3-11-159791-1
e-ISBN (PDF) 978-3-11-159968-7
e-ISBN (EPUB) 978-3-11-160088-8

Library of Congress Control Number: 2024948923

Bibliographic information published by the Deutsche Nationalbibliothek
The Deutsche Nationalbibliothek lists this publication in the Deutsche Nationalbibliografie; detailed bibliographic data are available on the Internet at http://dnb.dnb.de.

Cover image: dianaarturovna / iStock / Getty Images Plus
Typesetting: VTeX UAB, Lithuania

www.degruyter.com
Questions about General Product Safety Regulation:
productsafety@degruyterbrill.com

Preface

The purpose of this monograph is to present the general theory of algebras of unbounded operators affiliated with general von Neumann algebras. Our motivation to write this book has two rather different sources. The first source is the study of derivations on algebras of (unbounded) operators, which is a classical theme in those parts of quantum mechanics which find its inspiration in W^*- and C^*-algebras as evidenced in the books by Bratelli–Robinson [49, 50] and Sakai [134, 135]. The second source of inspiration is linked with an old question in mathematical analysis: "Which functions are differentiable?" and its more recent noncommutative counterpart. For the classical roots of this question, we refer the reader to a survey by A. Khinčin [99] and for its noncommutative version to Sh. Ayupov [9, 12] and Kadison–Liu [90]. We address in more detail these sources of inspiration in the introduction below.

This book is targeted at graduate students and researchers working in the area of operator algebras and derivations on them as well as in noncommutative integration theory. While writing the book, the authors made the choice to work with von Neumann algebras acting on separable Hilbert spaces only. While the absolute majority of results hold in the case of nonseparable Hilbert spaces, many proofs become significantly harder and may require a different approach altogether.

The authors would like to express their deepest gratitude to Professors Shavkat Ayupov and Karimbergen Kudaybergenov for their inspriruational work and decisive contributions to the field of derivations on algebras of unbounded operators. G. L. thanks Alan Carey and Adam Rennie for their continuous support and encouragement.

The authors also would like to thank Hongyin Zhao and Angus Alexander, as well as Eva-Maria Hekkelman, Yerlan Nessipbaev, and Thomas Scheckter. They diligently and unwaveringly helped with the final editing of the manuscript.

https://doi.org/10.1515/9783111599687-201

Contents

General notation

Here, we collect general notation that we shall use throughout the book. For specific notation, we refer to the Notation Index.

$\mathbb{N}_0$	set of all natural numbers including zero,
$\mathbb{N}$	set of all natural numbers excluding zero,
$\mathbb{Z}$	set of all integers,
$\mathbb{R}$	set of all real numbers,
$\mathbb{C}$	set of all complex numbers,
χ_M	characteristic function of a set M.

For a sequence of some elements, we shall use notation with braces such as $\{x_n\}_{n\in\mathbb{N}}$. The notation $\{x_i\}_{i\in I}$ stands for a net of elements x_i indexed by an index set I.

For an algebra $\mathcal{A}$ and fixed $n \in \mathbb{N}$, we denote by $M_{n\times n}(\mathcal{A})$ the algebra of all $n \times n$ matrices with elements from $\mathcal{A}$. For a fixed $n \in \mathbb{N}$, we use round brackets $(x_{ij})_{i,j=1}^{n}$ to denote an $n \times n$-matrix with ij-th element x_{ij}. For a fixed $n \in \mathbb{N}$, the matrix unit e_{ij} is an $n \times n$ matrix with 1 on the ij-th place and 0 everywhere else.

A pair of elements is denoted by (x, y). The direct product of two sets X and Y is denoted by $X \times Y$.

The symbol H stands for a Hilbert space with scalar product $\langle\cdot,\cdot\rangle$ and corresponding norm $\|\cdot\|$. Scalar products are always denoted by angle brackets $\langle\cdot,\cdot\rangle$. Occasionally, subscripts such as $\langle\cdot,\cdot\rangle_H$ and $\|\cdot\|_H$ are added. The vectors in the Hilbert space will usually be denoted by small Greek letters $\xi, \eta, \ldots$. We always assume that H is a separable Hilbert space.

By an operator in a Hilbert space H, we always mean linear operator (not necessarily everywhere defined). Operators in a Hilbert space H will be denoted by small Latin letters $x, y, u, v, a, b, \ldots$. The identity operator on H is denoted by $\mathbf{1} = \mathbf{1}_H$. By a projection in H, we mean an orthogonal projection. For a subspace $K \subset H$, we denote by $K^\perp$ its orthogonal complement (in H).

For an operator x in a Hilbert space H, we denote by

$\mathrm{dom}(x)$	domain of x,
$\ker(x)$	kernel of x,
$\mathrm{ran}(x)$	range of x,
$\sigma(x)$	spectrum of x,
$\rho(x)$	resolvent set of x,
E_x	spectral measure of x.

We shall typically use italic capital letters such as $\mathcal{A}$, $\mathcal{B}$, and $\mathcal{M}$ to denote algebras. The notation $\mathcal{M}$ is reserved for a von Neumann algebra acting on a Hilbert space H.

For a bounded linear mapping T from a (quasi)normed space $(X, \|\cdot\|_X)$ to a (quasi)normed space $(Y, \|\cdot\|_Y)$ we shall denote by $\|T\|_{(X,\|\cdot\|_X)\to(Y,\|\cdot\|_Y)}$ (or $\|T\|_{\|\cdot\|_X\to\|\cdot\|_Y}$, or $\|T\|_{X\to Y}$) its operator norm.

https://doi.org/10.1515/9783111599687-202

Introduction

The theory of algebras of operators acting on a Hilbert space began in the 1930s with a series of papers by Murray and von Neumann (see [111–113, 158]), motivated by the theory of group representations and certain aspects of the quantum mechanical formalism. They analyzed the structure of the family of $*$-algebras, which are referred to nowadays as von Neumann algebras or W^*-algebras and which have the distinct property of being closed in the weak operator topology.

Nowadays, the theory of operator algebras is an intensively developed field characterized by the interplay between pure mathematical and applied aspects. This theory plays an important role in mathematical physics, motivated by the fact that operator algebras, their states, representations, groups of automorphisms, and derivations, can describe and analyze the properties of model systems in quantum field theory and statistical physics.

Let $\mathcal{M}$ be an associative algebra over the field $\mathbb{C}$ of complex numbers and let $\mathcal{A}$ be an $\mathcal{M}$-bimodule. A linear map δ from $\mathcal{M}$ into $\mathcal{A}$ is called a derivation if $\delta(xy) = \delta(x)y + x\delta(y)$ for all $x, y \in \mathcal{M}$. It is clear that each element $a \in \mathcal{A}$ defines a derivation δ_a from $\mathcal{M}$ into $\mathcal{A}$ by $\delta_a(x) = [a, x] = ax - xa$, $x \in \mathcal{M}$. Such derivations δ_a are called *inner derivations*.

One of the classic problems in operator algebra theory is the following.

Derivation problem. *Let $\mathcal{M}$ be an algebra and $\mathcal{A}$ be an $\mathcal{M}$-bimodule. For what classes of $\mathcal{M}$-bimodules $\mathcal{A}$ is every derivation $\delta : \mathcal{M} \to \mathcal{A}$ inner?*

In the case when $\mathcal{A} = \mathcal{M}$ is a von Neumann algebra, loosely speaking, a derivation $\delta : \mathcal{M} \to \mathcal{M}$ acts as the generator of a one-parameter automorphism group of the operator algebra $\mathcal{M}$. The automorphism groups provide the mathematical description of the time (dynamical)-evolution of a physical system. In the case of classical mechanics, time-differentiation of the dynamical variable is Poisson bracketing with the (classical) Hamiltonian (the total energy). In quantum mechanics, differentiation of the "evolving observable" is Lie bracketing with the (quantum) Hamiltonian. This bracketing with the Hamiltonian represents an (inner) derivation of the system, and similar to other generators of the one-parameter automorphism groups of the operator algebras, it describes the physical system and its symmetries. This explains the significant interest attached to the study of such derivations. Now, the Hamiltonian will, in general, correspond to an unbounded operator on a Hilbert space H. Of course, these unbounded operators will not lie in a von Neumann algebra, but they may be affiliated with the von Neumann algebra corresponding to our quantum system. This makes it very desirable to study derivations of algebras into the bimodules that include such unbounded operators.

The study of derivations of operator algebras is a large, complicated topic that underwent a vast development in the 1960s and 1970s. In 1953, I. Kaplansky asked I. Singer if he had an idea of what the derivations of $C(X)$, the algebra of complex-valued, continuous functions on a compact Hausdorff space X, might be. The following day, Singer showed Kaplansky the elegant proof that all such derivations are zero (see [89] for an

https://doi.org/10.1515/9783111599687-001

account of this). Kaplansky's paper [97] and the strong interest in derivations of operator algebras grew out of Singer's result. Kaplansky went on to prove that each derivation of a type I von Neumann algebra $\mathcal{M}$ is inner [97]. In the course of the proof, Kaplansky proved that a derivation $\delta : \mathcal{M} \to \mathcal{M}$ of such an algebra is (norm) continuous (this was one of the earliest of the so-called "automatic continuity" results) and conjectured that this "automatic" continuity holds for derivations of an arbitrary C^*-algebra. This was proved by S. Sakai in [132] and extended to derivations of a C^*-algebra $\mathcal{M}$ into a Banach bimodule $\mathcal{E}$ by J. R. Ringrose in [128]. In [88] and [133], Kadison and Sakai proved that each derivation of a C^*-algebra acting on a Hilbert space H extends to a derivation of the strong-operator closure of that algebra, a von Neumann algebra, and that each derivation of a von Neumann algebra is inner. Surprisingly enough, this theorem is an extension of Singer's result in the sense that it tells us that each such derivation is a null element as an element of the 1-cohomology group of the von Neumann algebra (see [91]). In particular, this means that in commutative von Neumann algebras each derivation is identically equal to zero. Further development of the theory of derivations on operator algebras is expounded in detail in [49] and [135].

At the same time, there is a fundamental example of a nontrivial (in particular, non-inner) derivation on a commutative algebra, which is known to any first-year calculus student.

Example (Noninner derivations). *Let $C^\infty[0,1]$ be the commutative algebra of all infinitely differentiable functions defined on the interval $[0,1]$. The derivative operator $\partial : C^\infty[0,1] \to C^\infty[0,1]$, where $(\partial f)(t) = f'(t)$, $t \in \mathbb{R}$ is a derivation on $C^\infty[0,1]$, which is not inner.*

The algebra $C^\infty[0,1]$ is a subalgebra in the commutative von Neumann algebra $L_\infty[0,1]$ of all (classes of a. e. equal) essentially bounded Lebesgue measurable functions on $[0,1]$. Since there are no nonzero derivations on $L_\infty[0,1]$, the classical derivation ∂ cannot be extended up to a derivation $\delta : L_\infty[0,1] \to L_\infty[0,1]$. This is where the theory of derivations on operator algebras intersects with the theory of possible extensions of the classical derivations.

In the early 20th century, several ways to extend the class of differentiable functions were proposed. For a measurable function f on $\mathbb{R}$, one can define several types of differentiability depending on whether the limit of the fraction $\frac{f(t+h)-f(t)}{h}$, $h \to 0$ exists in a certain sense. In particular, Borel differentiable, approximately differentiable, and Riesz differentiable functions can be defined. Each of these classes extends the notion of the classical differentiation and forms a proper subset in the set of all measurable functions.

In [99], Aleksandr Khinčin came to the following conclusion:

> The question of finding such a method of differentiation, which would yield that every continuous function has a derivation remains open. It appears that by remaining in the realm of real variables it is extremely hard to determine the structural foundation of such a method, if it is possible at all.

In essence, Khinčin sought to find nontrivial derivations δ from the algebra $C[0,1]$ of continuous functions on the interval $[0,1]$ to the algebra $L_0[0,1]$ of all (Lebesgue) measurable functions on $[0,1]$.

The algebra $L_0[0,1]$ can be identified with the algebra of all (closed and densely defined) operators affiliated with the von Neumann algebra $L_\infty[0,1]$. These kinds of algebras were first introduced by John von Neumann in [156], who was originally motivated by problems in harmonic analysis, dynamical systems, and the newly discovered field of quantum mechanics. Von Neumann termed these algebras "Rings of Operators" and showed that Rings of Operators may be treated analogously to functions and the integration theory can be generalized by replacing the integral with the trace. Despite this tacit understanding, it was only through the landmark work of Irving Segal in 1953 [141] that it was formally shown that such a theory is not simply analogous to, but is a genuine extension of the classical theory of Lebesgue integration. This broader understanding that von Neumann algebras form a genuine extension of classical measure theory to the noncommutative case establishes our interest in studying their structures through derivations.

In his paper [141], Segal considered new classes of (not necessarily Banach) algebras of unbounded operators, in particular, the algebra $S(\mathcal{M})$ of measurable operators affiliated with a von Neumann algebra $\mathcal{M}$. The notion of measurable operators allowed defining on the set $S(\mathcal{M})$ the algebraic operations of the strong sum and strong product $x+y, x\cdot y, \lambda x, \lambda \in \mathbb{C}$, with respect to which the set $S(\mathcal{M})$ becomes a $*$-algebra over the field $\mathbb{C}$ of complex numbers, where the involution x^* is defined as the standard adjoint operation for closed densely defined operators $x \in S(\mathcal{M})$. The von Neumann algebra $\mathcal{M}$ itself is a $*$-subalgebra of $S(\mathcal{M})$. In the special case when $\mathcal{M}$ is a commutative von Neumann algebra $L_\infty(\Omega,\Sigma,\mu)$ of all (classes of a.e. equal) essentially bounded measurable functions on a σ-finite measure space (Ω,Σ,μ), the $*$-algebra $S(\mathcal{M})$ is the $*$-algebra $L_0(\Omega,\Sigma,\mu)$ of all classes of almost everywhere equal complex-valued measurable functions on (Ω,Σ,μ).

Since Segal's fundamental paper, other types of algebras of unbounded operators affiliated with a given von Neumann algebra were introduced. For the theory of noncommutative integration, two of these algebras are the most important.

One is the algebra $S(\mathcal{M},\tau)$ of all τ-measurable operators affiliated with a semifinite von Neumann algebra $\mathcal{M}$ equipped with a faithful normal semifinite trace τ. This algebra was introduced by E. Nelson [116], who showed, in particular, that $S(\mathcal{M},\tau)$ can be viewed as the completion of the von Neumann algebra $\mathcal{M}$ with respect to the so-called measure topology. In particular, $S(\mathcal{M},\tau)$ itself is a complete metrizable topological $*$-algebra when equipped with the measure topology. We refer the reader to [64] for detailed treatment of these algebras as well as other aspects of noncommutative integration theory.

The other important algebra in the noncommutative integration is the algebra $LS(\mathcal{M})$ of all locally measurable operators affiliated with a (not necessarily semifinite) von Neumann algebra $\mathcal{M}$. This algebra was first introduced by S. Sankaran [137] for $\mathcal{M}$

with a countably decomposable center. In its full generality this algebra was introduced by Yeadon in [162]. The algebra $LS(\mathcal{M})$ can be viewed as the completion of the von Neumann algebra $\mathcal{M}$ with respect to the so-called local measure topology, so that the algebra $LS(\mathcal{M})$ equipped with the local measure topology is also a complete metrizable topological $*$-algebra.

The algebra $LS(\mathcal{M})$ of all locally measurable operators has a few significant properties, which distinguish it drastically from both the algebras $S(\mathcal{M})$ and $S(\mathcal{M},\tau)$. First, Yeadon in [163] proved that $LS(\mathcal{M})$ is the largest possible operator bimodule over a given von Neumann algebra $\mathcal{M}$. Second, as shown by M. Lennon [105], the algebra $LS(\mathcal{M})$ is closed with respect to direct integrals of von Neumann algebras. Namely, if $\mathcal{M} = \int_Z^{\oplus} \mathcal{M}_\zeta \, dw(\zeta)$ is a decomposable von Neumann algebra, then $LS(\mathcal{M}) = \int_Z^{\oplus} LS(\mathcal{M}_\zeta) \, dw(\zeta)$. In general, neither of these properties hold for the algebras $S(\mathcal{M})$ and $S(\mathcal{M},\tau)$.

The development of the theory of noncommutative integration, as well as further research on derivations through the second half of the 20th century, led Shavkat Ayupov to pose a problem [9] concerning the description of derivations in these algebras of unbounded operators affiliated with a von Neumann algebra. In 2014, Richard Kadison and his student, Zhe Liu, restated this problem [90]. Ayupov, Kadison, and Liu posed the following questions.

Problem (Ayupov–Kadison–Liu).

- *Is any derivation on the algebra of measurable operators affiliated with a commutative von Neumann algebra identically zero?*
- *Is any derivation on the algebra of measurable (resp., locally measurable, τ-measurable) operators affiliated with an arbitrary von Neumann algebra necessarily continuous with respect to the measure topology?*
- *Is any derivation on the algebra of measurable (resp., locally measurable, τ-measurable) operators affiliated with an arbitrary von Neumann algebra inner?*

Not only are these questions fundamental, but they are extremely natural. The third part of this problem is the most general and its affirmative answer implies an affirmative answer to the previous two parts. Furthermore, in the particular case of a commutative von Neumann algebra, the third question is intrinsically related to the question posed by Khinčin.

As noted by Kadison and Liu in [90], "the complete cohomological result would say that each derivation of (the algebra of measurable operators) is inner.... The authors *strongly* feel that this is true; but it is still open." These questions were the focus of Kadison's research during the last years of his life. In this book, we fully resolve the Ayupov–Kadison–Liu problem and give necessary and sufficient conditions on the algebra $\mathcal{M}$, for the algebra $LS(\mathcal{M})$ (resp., $S(\mathcal{M},\tau)$) to have noninner derivations (see Theorem 7.6.2).

It turns out that when a von Neumann algebra $\mathcal{M}$ has no finite type I direct summand, any derivation on $LS(\mathcal{M})$ (resp., $S(\mathcal{M},\tau)$) is necessarily inner. For a finite type I

von Neumann algebra, the existence of a noninner derivation on $LS(\mathcal{M})$ depends on the center $\mathcal{Z}(\mathcal{M})$ of the von Neumann algebra $\mathcal{M}$. Namely, there are noninner derivations on $LS(\mathcal{M})$ if and only if the center $\mathcal{Z}(\mathcal{M})$ of $\mathcal{M}$ is not atomic. In particular, for the commutative von Neumann algebra $L_\infty[0,1]$, this result relates back to Khinčin's questions and shows that the classical derivation ∂ can be extended up to infinitely many distinct noninner derivations on the algebra $L_0[0,1]$. In Section 6.5, we show also that the maximal subalgebra of $L_0[0,1]$, which has a unique extension of the classical derivation ∂, is the algebra of all (classes of a.e. equal) approximately differentiable functions on the interval $[0,1]$.

Structure of the book

In the first chapter, we introduce the necessary background material such as spectral theory of self-adjoint operators, von Neumann algebras, classification of projections in a von Neumann algebra, and traces on a von Neumann algebra. Special attention is paid to the dimension function on a von Neumann algebra and direct integrals of bounded as well as unbounded operators.

In the second chapter, we introduce several classes of measurable operators associated with a given von Neumann algebra, which extend the classical notion of a measurable function. The most important and the most studied are locally measurable and τ-measurable operators. This chapter establishes that both the locally measurable and τ-measurable operators form $*$-algebras and discusses basic properties of these algebras.

In Chapter 3, we study properties of locally measurable operators, including their functional calculi and order properties. It is established that the algebra of all locally measurable operators is the maximal possible bimodule over a given von Neumann algebra. In this chapter, we also develop a form of commutator estimates for locally measurable operators, which shall be used in the subsequent chapters.

Chapter 4 introduces the natural topologies of the local measure convergence on the algebra of all locally measurable operators and the measure topology on the algebra of all τ-measurable operators. These topologies are natural noncommutative counterparts of the classical local measure and measure topologies on measurable functions.

In Chapter 5, using the techniques developed in the previous chapters, we present the essential properties of derivations on the algebras of locally measurable and τ-measurable operators. The fundamental result of this chapter is that a derivation of the algebra of locally measurable (resp., τ-measurable) operators is inner if and only if it is continuous in the respective topology on this algebra.

In Chapter 6, we study derivations on the algebra of all measurable functions $L_0(\Omega)$ on a σ-finite measure space (Ω, Σ, μ) (which can be viewed as the algebra of all (locally) measurable operators affiliated with a commutative von Neumann algebra). This chapter establishes the fundamental criterion that a noninner (and so, nonzero) derivation on $L_0(\Omega)$ exists if and only if (Ω, Σ, μ) is not atomic. This provides a complete resolution to the question posed by Khinčin. A similar criterion is then proved for derivations on the algebra of measurable operators affiliated with a finite type I von Neumann algebra.

In the final Chapter 7, we prove that if a von Neumann algebra has no finite type I summand, then any derivation on the algebra of all locally measurable (resp., τ-measurable) operators is necessarily inner. With the criteria of Chapter 6 at hand, this then gives a criterion for the existence of noninner derivations on these algebras of operators affiliated with an arbitrary von Neumann algebra and provides a complete answer to the Ayupov–Kadison–Liu problem.

1 Preliminaries on the theory of von Neumann algebras and general theory of linear operators in a Hilbert space

The primary aim of this book is to introduce algebras of unbounded operators affiliated with a von Neumann algebra $\mathcal{M}$ and investigate derivations on these algebras. In particular, we shall deal with unbounded operators and we need to recall some basic facts from the theory of such operators. In this chapter, we recall well-known notions and results related to the theory of von Neumann algebras and the general theory of closed linear operators in a Hilbert space.

We give a full exposition of the notions of the dimension function on a von Neumann algebra and direct integrals of fields of unbounded closed operators.

Writing this chapter, we used materials from [49, 60, 61, 92, 93, 114, 127, 129, 134, 140, 147, 149], and the majority of results are presented without proofs.

1.1 Bounded linear operators on a Hilbert space

In this section, we collect some notation and terminology concerning bounded linear operators on a Hilbert space.

Let H be a Hilbert space over the field $\mathbb{C}$ of complex numbers equipped with the inner product $\langle \cdot, \cdot \rangle$ and the corresponding norm $\| \cdot \|$. We always assume that H is a separable Hilbert space. The identity operator on H is denoted by $\mathbf{1} = \mathbf{1}_H$.

We denote the algebra of all bounded linear operators in H by $\mathcal{B}(H)$. When equipped with the operator norm $\| \cdot \|_{\mathcal{B}(H)}$, defined by

$$\|x\|_{\mathcal{B}(H)} = \sup_{\|\xi\|_H \leq 1} \|x\xi\|_H, \quad x \in \mathcal{B}(H),$$

the space $\mathcal{B}(H)$ is a Banach space.

For any $x \in \mathcal{B}(H)$, there exists a unique *adjoint* operator $x^* \in \mathcal{B}(H)$, such that

$$\langle x\xi, \eta \rangle = \langle \xi, x^*\eta \rangle, \quad \forall \xi, \eta \in H.$$

The mapping $x \longmapsto x^*$ is a conjugate linear involution in $\mathcal{B}(H)$ satisfying $\|x^*\|_{\mathcal{B}(H)} = \|x\|_{\mathcal{B}(H)}$ and $\|x\|^2_{\mathcal{B}(H)} = \|x^*x\|_{\mathcal{B}(H)}$ for all $x \in \mathcal{B}(H)$ (so $\mathcal{B}(H)$ is an example of a C^*-algebra, which we will discuss in Section 1.7).

An operator $x \in \mathcal{B}(H)$ satisfying $x^* = x$ is called *self-adjoint*. The collection of all self-adjoint operators is denoted by $\mathcal{B}_h(H)$, which is a real linear subspace of $\mathcal{B}(H)$. An operator $x \in \mathcal{B}(H)$ is called *normal* if $xx^* = x^*x$. Furthermore, if $u \in \mathcal{B}(H)$ satisfies $u^*u = uu^* = \mathbf{1}$ (equivalently, $u^{-1} = u^*$), then u is called *unitary*.

https://doi.org/10.1515/9783111599687-002

A self-adjoint operator $x \in \mathcal{B}(H)$ is called *positive* if $\langle x\xi, \xi\rangle \geq 0$ for all $\xi \in H$. The collection of all positive elements of $\mathcal{B}(H)$ is denoted by $\mathcal{B}_+(H)$. This set is a proper closed cone in $\mathcal{B}_h(H)$, and it induces a partial ordering in $\mathcal{B}_h(H)$ by defining $x \leq y$ whenever $y - x \in \mathcal{B}_+(H)$, which turns $\mathcal{B}_h(H)$ into a partially ordered vector space.

In what follows, we use the following notation from the theory of partially ordered sets.

Let $(X, \leq)$ be an arbitrary partially ordered set, and let $\{x_i\}_{i\in I}$ be a net of elements of X. The net $\{x_i\}_{i\in I}$ is called *increasing* (resp., *decreasing*) if $x_i \leq x_j$ (resp., $x_j \leq x_i$) for all $i \leq j$, $i, j \in I$.

For every subset $Y \subseteq X$, denote by $\bigvee Y$ (or $\sup Y$) (resp., $\bigwedge Y$ (or $\inf Y$)) the least upper bound (resp., greatest lower bound) of the set Y of X, assuming that this bound exists.

The notation $x_i \uparrow x$ (resp., $x_i \downarrow x$) means that the net $\{x_i\}_{i\in I}$ is increasing (resp., decreasing) and $\bigvee_{i\in I} x_i = x$ (resp., $\bigwedge_{i\in I} x_i = x$).

Let K be a closed subspace in H. The operator p_K defined by $p_K(\xi + \eta) = \xi$ for $\xi \in K$ and $\eta \in K^\perp$, where $K^\perp$ is the orthogonal complement of K in H, is called the (orthogonal) *projection* of H on K. An operator $p \in \mathcal{B}(H)$ is a projection if and only if $p^2 = p = p^*$; in this case, p is the projection on $p(H)$. The set of all projections in H is denoted by $P(\mathcal{B}(H))$. If $p \in P(\mathcal{B}(H))$, then it is clear that $\mathbf{1} - p \in P(\mathcal{B}(H))$, which is called the *complement* of p and this projection will also be denoted by $p^\perp$.

The set $P(\mathcal{B}(H))$ of all projections in H is a subset of $\mathcal{B}_h(H)$, so we may equip $P(\mathcal{B}(H))$ with the partial ordering inherited from $\mathcal{B}_h(H)$, i. e., if $p, q \in P(\mathcal{B}(H))$, then $p \leq q$ if and only if $\langle p\xi, \xi\rangle \leq \langle q\xi, \xi\rangle$ for all $\xi \in H$. The order $p \leq q$ for $p, q \in P(\mathcal{B}(H))$ can be characterized as follows.

Proposition 1.1.1 ([92, Proposition 2.5.2]). *Let p, q be projections on closed subspaces K and L of H, respectively. The following conditions are equivalent:*
(i) $p \leq q$;
(ii) $pq = qp = p$;
(iii) $\|p\xi\| \leq \|q\xi\|$ *for all* $\xi \in H$;
(iv) $K \subset L$.

Proposition 1.1.1 implies, in particular, that the partial ordering of projections in $\mathcal{B}_h(H)$ corresponds to the partial ordering of closed subspaces by the inclusion relation $\subseteq$. Consequently, each family $\{p_i\}_{i\in I}$ of projections acting on H has a greatest lower bound $\bigwedge_{i\in I} p_i = \inf_{i\in I} p_i$ and a least upper bound $\bigvee_{i\in I} p_i = \sup_{i\in I} p_i$, which are projections onto $\bigcap_{i\in I} p_i(H)$ and $\overline{\bigcup_{i\in I} p_i(H)}$, respectively. Thus, $P(\mathcal{B}(H))$ is a complete lattice with smallest element 0 and largest element $\mathbf{1}$. For every $p \in P(\mathcal{B}(H))$, the projection $p^\perp = \mathbf{1} - p$ is a complement of p, i. e., $p \vee p^\perp = \mathbf{1}$ and $p \wedge p^\perp = 0$. The mapping $p \longmapsto p^\perp$ reverses the ordering in $P(\mathcal{B}(H))$, i. e., $p \leq q$ if and only if $q^\perp \leq p^\perp$, and so, $(p \vee q)^\perp = p^\perp \wedge q^\perp$ and $(p \wedge q)^\perp = p^\perp \vee q^\perp$ for all $p, q \in P(\mathcal{B}(H))$. If $p, q \in P(\mathcal{B}(H))$ commute, then $p \wedge q = pq$ and $p \vee q = p + q - pq$.

The supremum of an increasing system in $P(\mathcal{B}(H))$ can be characterized as follows.

Proposition 1.1.2 ([92, Proposition 2.5.6]). *If $\{p_i\}_{i\in I}$ is an increasing net in $P(\mathcal{B}(H))$ and $p \in P(\mathcal{B}(H))$, then $p = \sup_{i\in I} p_i$ if and only if $p_i\xi \to p\xi$ for all $\xi \in H$.*

Two projections $p, q \in P(\mathcal{B}(H))$ are called mutually *orthogonal* if $pq = 0$ (equivalently, $p(H)$ and $q(H)$ are mutually orthogonal subspaces). Suppose that $\{p_i\}_{i\in I}$ is a collection of pairwise orthogonal projections in $P(\mathcal{B}(H))$ (i. e., $p_ip_j = 0$ whenever $i \neq j$ in I). For each finite subset F of I, we may define the projection $p_F = \sum_{i\in F} p_i$. It is clear that $\{p_F\}$ is an increasing net (with respect to the inclusion ordering of the finite subsets of I). Hence, there exists $p \in P(\mathcal{B}(H))$ such that $p_F \uparrow p$ with respect to the directed set of finite subsets of $\mathbb{N}$ ordered by inclusion. This projection p is denoted by $\sum_{i\in I} p_i$. It follows from Proposition 1.1.2 that $p\xi = \sum_{i\in I} p_i\xi$ for all $\xi \in H$.

1.2 Topologies on the algebra of bounded linear operators

In addition to the norm topology, generated by the operator norm $\|\cdot\|_{\mathcal{B}(H)}$, there are a number of other important topologies on $\mathcal{B}(H)$. In this section, we recall these topologies. For the proofs, we refer the reader to, e. g., [49, 92].

For every $\xi \in H$, we define the seminorm ρ_ξ on $\mathcal{B}(H)$ by $\rho_\xi(x) = \|x\xi\|_H, x \in \mathcal{B}(H)$. The locally convex Hausdorff topology on $\mathcal{B}(H)$ generated by the family of seminorms $\{\rho_\xi : \xi \in H\}$ is called the *strong operator topology* (briefly, *so-topology*). A net $\{x_i\}_{i\in I}$ in $\mathcal{B}(H)$ *so*-converges to an operator $x \in \mathcal{B}(H)$, denoted by $x_i \xrightarrow{so} x$, if and only if $\|x_i\xi - x\xi\|_H \to 0$ for all $\xi \in H$. Clearly, the *so*-topology is weaker than the topology generated by the norm $\|\cdot\|_{\mathcal{B}(H)}$. Multiplication in $\mathcal{B}(H)$ is continuous with respect to the *so*-topology in each factor separately, but in general not jointly *so*-continuous in both factors (however, multiplication is jointly *so*-continuous when restricted to norm bounded sets). The mapping $x \longmapsto x^*$ is not *so*-continuous (unless H is finite-dimensional).

The following theorem emphasizes some important properties of the *so*-topology, related to the partial order in $\mathcal{B}_h(H)$.

Proposition 1.2.1 ([147, Proposition 2.16]). *If $\{x_i\}_{i\in I}$ is an increasing (resp., decreasing) net in $\mathcal{B}_h(H)$ and $x_i \leq y$ (resp., $x_i \geq y$) for all $i \in I$, where $y \in \mathcal{B}_h(H)$, then there exists an operator $x \in \mathcal{B}_h(H)$ such that $\{x_i\}_{i\in I}$ (so)-converges to x and $x = \sup_{i\in I} x_i$ (resp., $x = \inf_{i\in I} x_i$) with respect to the natural partial order on $\mathcal{B}(H)$, in addition $a^*x_ia \uparrow a^*xa$ (resp., $a^*x_ia \downarrow a^*xa$) for every operator $a \in \mathcal{B}(H)$.*

For $\xi, \eta \in H$, we define the seminorm $\rho_{\xi,\eta}$ by $\rho_{\xi,\eta}(x) = |\langle x\xi, \eta\rangle|, x \in \mathcal{B}(H)$. The locally convex Hausdorff topology on $\mathcal{B}(H)$ generated by the family of seminorms $\{\rho_{\xi,\eta} : \xi, \eta \in H\}$ is called the *weak operator topology* (briefly, *wo-topology*). A net $\{x_i\}_{i\in I}$ in $\mathcal{B}(H)$ *wo*-converges to an operator $x \in \mathcal{B}(H)$, denoted by $x_i \xrightarrow{wo} x$, if and only if $\langle x_i\xi, \eta\rangle \to \langle x\xi, \eta\rangle$ for all $\xi, \eta \in H$. Obviously, the *wo*-topology is weaker than the *so*-topology and coincides with the latter only if H is finite-dimensional. At the same time on the set $P(\mathcal{B}(H))$, these topologies coincide. Multiplication is *wo*-continuous in each factor separately, but is not jointly *wo*-continuous (unless H is finite-dimensional). The mapping $x \longmapsto x^*$ is evidently *wo*-continuous.

Next, we consider the locally convex Hausdorff topology on $\mathcal{B}(H)$ generated by the family of seminorms given by $\rho_{\{\xi_n\},\{\eta_n\}}(x) = |\sum_{n=1}^{\infty}\langle x\xi_n, \eta_n\rangle|$, where $\{\xi_n\}_{n\in\mathbb{N}}$ are $\{\eta_n\}_{n\in\mathbb{N}}$ sequences in H satisfying $\sum_{n=1}^{\infty}\|\xi_n\|_H^2 < \infty$ and $\sum_{n=1}^{\infty}\|\eta_n\|_H^2 < \infty$. This topology is called the *ultraweak operator topology* (briefly, *uwo-topology*). The ultraweak operator topology is stronger than the weak operator topology. On norm bounded subsets of $\mathcal{B}(H)$, the ultraweak operator topology and weak operator topology coincide. Convergence of a net $\{x_i\}_{i\in I}$ to an element x in $\mathcal{B}(H)$ with respect to the *uwo*-topology is denoted by $x_i \overset{uwo}{\longrightarrow} x$.

Given a sequence $\{\xi_n\}_{n\in\mathbb{N}}$ in H satisfying $\sum_{n=1}^{\infty}\|\xi_n\|_H^2 < \infty$, the seminorm $\rho_{\{\xi_n\}}$ on $\mathcal{B}(H)$ is defined by $\rho_{\{\xi_n\}}(x) = (\sum_{n=1}^{\infty}\|x\xi_n\|_H^2)^{1/2}$. The Hausdorff locally convex topology on $\mathcal{B}(H)$ generated by the family of these seminorms $\rho_{\{\xi_n\}}$ is called the *ultrastrong operator topology* (briefly, *uso-topology*). The ultrastrong operator topology is stronger than the strong operator and ultraweak operator topologies and is weaker than the norm topology. On norm bounded subsets of $\mathcal{B}(H)$, the ultrastrong operator and strong operator topologies coincide. Convergence of a net $\{x_i\}_{i\in I}$ to an element x in $\mathcal{B}(H)$ with respect to the ultrastrong topology is denoted by $x_i \overset{uso}{\longrightarrow} x$.

1.3 Closed linear operators

In this section, we recall the basic theory of closed densely defined operators in a Hilbert space H (see, e. g., [126, 127, 140]).

A linear operator x in H is a linear mapping from its *domain* $\mathrm{dom}(x)$, which is a linear subspace in H, into the space H itself. The *range* and *kernel* of a linear operator x are defined as $\mathrm{ran}(x) = \{x\xi : \xi \in \mathrm{dom}(x)\}$ and $\ker(x) = \{\xi \in \mathrm{dom}(x) : x\xi = 0\}$, respectively.

A linear operator y is called an *extension* of the operator x (or x is a *restriction* of the operator y) (notation: $x \subseteq y$), if $\mathrm{dom}(x) \subseteq \mathrm{dom}(y)$ and $x\xi = y\xi$ for all $\xi \in \mathrm{dom}(x)$.

The equality $x = y$ means that both $x \subseteq y$ and $y \subseteq x$, i. e., $\mathrm{dom}(x) = \mathrm{dom}(y)$ and $x\xi = y\xi$ for all $\xi \in \mathrm{dom}(x) = \mathrm{dom}(y)$.

In the collection of linear operators, we may introduce the algebraic operations of scalar multiplication, addition, and multiplication as follows. Given linear operators x, y in H with their respective domains $\mathrm{dom}(x)$ and $\mathrm{dom}(y)$, we define

- *scalar multiplication* λx of the operator x by a scalar $\lambda \in \mathbb{C}$, where $\mathrm{dom}(\lambda x) = \mathrm{dom}(x)$ and $(\lambda x)\xi = \lambda(x\xi)$ for all $\xi \in \mathrm{dom}(\lambda x)$;
- (algebraic) *sum* $x+y$, where $\mathrm{dom}(x+y) = \mathrm{dom}(x)\cap\mathrm{dom}(y)$ and $(x+y)(\xi) = x(\xi)+y(\xi)$ for all $\xi \in \mathrm{dom}(x+y)$;
- (algebraic) *product* xy, where $\mathrm{dom}(xy) = \{\xi \in \mathrm{dom}(y) : y(\xi) \in \mathrm{dom}(x)\}$ and $(xy)(\xi) = x(y(\xi))$ for all $\xi \in \mathrm{dom}(xy)$;

- *inverse operator* x^{-1} (in the case when x is injective), where $\mathrm{dom}(x^{-1}) = \mathrm{ran}(x) = x(\mathrm{dom}(x))$ and $x^{-1}(\eta) = \xi$ if and only if $x(\xi) = \eta$, $\eta \in \mathrm{dom}(x^{-1})$.

We note that in general it may happen that $\mathrm{dom}(x + y) = \{0\}$ or $\mathrm{dom}(xy) = \{0\}$. Furthermore, with respect to these algebraic operations, the set of all linear operators is not a vector space. However, we do have the following relations for arbitrary linear operators x, y, and z:

(a) $(x + y) + z = x + (y + z)$;
(b) $(xy)z = x(yz)$;
(c) $(x + y)z = xz + yz$;
(d) $zx + zy \subseteq z(x + y)$.

It follows from (a) and (b) above that we can form, without ambiguity, sums and products of an arbitrary number of linear operators. In particular, polynomials in linear operators are well-defined.

Definition 1.3.1. Let x be a linear operator with domain $\mathrm{dom}(x)$. We say that a bounded operator y *commutes* with x if $yx \subset xy$, i. e., if $y(\mathrm{dom}(x)) \subset \mathrm{dom}(x)$ and $yx\xi = xy\xi$ for all $\xi \in \mathrm{dom}(x)$.

Let x be a linear operator in H. A linear subspace $\Gamma(x)$ of the direct product $H \times H$, defined by

$$\Gamma(x) = \{(\xi, x\xi); \xi \in \mathrm{dom}(x)\},$$

is called the *graph* of the operator x. It is clear that $x \subseteq y$ if and only if $\Gamma(x) \subseteq \Gamma(y)$, in particular, $x = y$ if and only if $\Gamma(x) = \Gamma(y)$.

A linear operator x in H is called *closed* if $\Gamma(x)$ is a closed subspace in $H \times H$. In other words, x is closed if and only if it follows from

$$\{\xi_n\}_{n\in\mathbb{N}} \subseteq \mathrm{dom}(x), \quad \xi_n \to \xi_0, \quad x(\xi_n) \to \eta_0$$

that

$$\xi_0 \in \mathrm{dom}(x) \quad \text{and} \quad x(\xi_0) = \eta_0.$$

If x is closed, then the kernel $\ker(x)$ is a closed subspace in H.

It is clear that every bounded linear operator $x \in \mathcal{B}(H)$ is closed. Furthermore, if a linear operator x in H is closed and there exists the inverse operator x^{-1}, then the operator x^{-1} is also closed.

Let x be a closed linear operator with domain $\mathrm{dom}(x)$. The *resolvent set* $\rho(x)$ is the set of all complex numbers λ, such that the operator $x - \lambda\mathbf{1}$ has bounded everywhere defined inverse $(x - \lambda\mathbf{1})^{-1}$. The *spectrum* $\sigma(x)$ of x is the complement of $\rho(x)$ in $\mathbb{C}$.

Let x be a linear operator in H. The operator x is called *closable* if the closure $\overline{\Gamma(x)}$ of its graph $\Gamma(x)$ in $H \times H$ is the graph of some operator in H. Note that $\overline{\Gamma(x)}$ is the graph of a linear operator if and only if $(0, \eta) \in \overline{\Gamma(x)}$ implies that $\eta = 0$. In particular, the operator x is closable if and only if the following condition holds: if $\{\xi_n\}_{n\in\mathbb{N}} \subseteq \mathrm{dom}(x)$, $\xi_n \to 0$, and $x(\xi_n)$ converges in H, then $x(\xi_n) \to 0$.

If x is a closable operator, then the closed linear operator whose graph is $\overline{\Gamma(x)}$ is denoted by $\overline{x}$ and is called the *closure* of the operator x.

Suppose that x is a linear operator in H and that $\mathfrak{D}$ is a linear subspace of $\mathrm{dom}(x)$. Then $\mathfrak{D}$ is called a *core* of x if $\Gamma(x) \subseteq \overline{\Gamma(x_{|\mathfrak{D}})}$, where $x_{|\mathfrak{D}}$ denotes the restriction of x to $\mathfrak{D}$. In other words, $\mathfrak{D}$ is a core of x if and only if for every $\xi \in \mathrm{dom}(x)$ there exists a sequence $\{\xi_n\}_{n\in\mathbb{N}} \subseteq \mathfrak{D}$ such that $\xi_n \to \xi$ and $x\xi_n \to x\xi$ as $n \to \infty$. It is easily verified that two closable operators, which coincide on a common core, have identical closures.

Note that if x, y are closed operators in H, then the sum $x + y$ and the product xy may not be even closable. We now introduce the notions of the strong sum and strong product, which will be used in the sequel.

Definition 1.3.2. Let x, y be closed linear operators in H.
(i) If the operator $x + y$ is closable, then the *strong sum* $x \dot{+} y$ is defined as $x \dot{+} y = \overline{x + y}$.
(ii) If the operator xy is closable, then the *strong product* $x \cdot y$ is defined as $x \cdot y = \overline{xy}$.

Note that if x is bounded and y is closed, then $x + y$ and yx are necessarily closed, too. However, the operator xy may be not closable.

A linear operator x in H is called *densely defined* if $\mathrm{dom}(x)$ is a dense subspace of H. Now suppose that x is a densely defined operator in H and consider the linear subspace $\mathfrak{D}$ of H given by

$$\mathfrak{D} = \{\eta \in H : \exists \zeta \in H \text{ such that } \langle x\xi, \eta\rangle = \langle \xi, \zeta\rangle \ \forall \xi \in \mathrm{dom}(x)\}.$$

If $\eta \in \mathfrak{D}$, then the element $\zeta \in H$ satisfying $\langle x\xi, \eta\rangle = \langle \xi, \zeta\rangle$ for all $\xi \in \mathrm{dom}(x)$ is uniquely determined by η, as $\mathrm{dom}(x)$ is dense in H. Therefore, we may define the mapping $x^* : \eta \longmapsto \zeta$ from $\mathfrak{D}$ into H. It is readily verified that x^* is linear. Hence, x^* is a linear operator in H with domain $\mathrm{dom}(x^*) = \mathfrak{D}$. The operator x^* is called the *adjoint* of x. Note that, by definition, we have

$$\langle x\xi, \eta\rangle = \langle \xi, x^*\eta\rangle, \quad \xi \in \mathrm{dom}(x), \eta \in \mathrm{dom}(x^*). \tag{1.1}$$

It is straightforward that the graph $\Gamma(x^*)$ of the operator x^* is the orthogonal complement in the Hilbert space $H \times H$ of the set

$$\{(ix(\xi), -i\xi),\ \xi \in \mathrm{dom}(x)\},$$

where i is the imaginary unit, i. e., $i^2 = -1$. It means that $\Gamma(x^*)$ is a closed linear subspace in $H \times H$ and, therefore, x^* is a closed operator. If x is densely defined and closable, then $\overline{x}^* = x^*$. The following theorem lists some elementary properties of adjoint operators.

Theorem 1.3.3. *Let x and y be densely defined linear operators in H.*
(i) *If $x \subseteq y$, then $y^* \subseteq x^*$;*
(ii) $(\lambda x)^* = \bar{\lambda} x^*$ *for all* $\lambda \in \mathbb{C}$, $\lambda \neq 0$;
(iii) *If $x + y$ is densely defined, then $x^* + y^* \subseteq (x + y)^*$;*
(iv) *If xy is densely defined, then $y^* x^* \subseteq (xy)^*$;*
(v) *If $y \in \mathcal{B}(H)$, then $(x + y)^* = x^* + y^*$ and $(yx)^* = x^* y^*$;*
(vi) *If $u \in \mathcal{B}(H)$ is unitary, then $(ux)^* = x^* u^*$, $(xu)^* = u^* x^*$, and $(uxu^*)^* = ux^* u^*$;*
(vii) *If x is injective and* $\mathrm{ran}(x)$ *is dense, then $(x^{-1})^* = (x^*)^{-1}$.*

Given a densely defined linear operator x in H, the adjoint $x^* : \mathrm{dom}(x^*) \to H$ is closed. If x^* is densely defined, then we may consider $(x^*)^* = x^{**}$ and it is easy to see that $x \subseteq x^{**}$. Hence, x is closable. Conversely, if x is closable, then it can be shown that x^* is densely defined. This is contained in the next theorem.

Theorem 1.3.4. *If x is a densely defined linear operator in H, then x^* is densely defined if and only if x is closable. Moreover, if x is closable, then $\overline{x} = x^{**}$. In particular, for every densely defined closed operator x we have $x^{**} = x$.*

A densely defined operator x is said to be *symmetric* if $x \subseteq x^*$, or equivalently, if $\langle x(\xi), \eta \rangle = \langle \xi, x(\eta) \rangle$ for all $\xi, \eta \in \mathrm{dom}(x)$. Evidently, a symmetric operator is closable and $\langle x(\xi), \xi \rangle$ is a real number for every $\xi \in \mathrm{dom}(x)$. For every symmetric operator x, the inclusions $x \subseteq \overline{x} = x^{**} \subseteq x^*$ hold, and if x is a closed symmetric operator, then $x = x^{**} \subseteq x^*$.

A closed and densely defined linear operator x in H is called *normal* when $x^* x = xx^*$. A densely defined operator x is called *self-adjoint* if $x = x^*$. If $\overline{x}$ is self-adjoint, then the operator x is said to be *essentially self-adjoint*. It is clear that every self-adjoint operator is closed and symmetric, however, not every closed symmetric operator is self-adjoint. The following theorem contains some necessary and sufficient conditions for a given symmetric operator to be self-adjoint.

Theorem 1.3.5. *Let x be a symmetric operator in H. The following conditions are equivalent:*
(i) $x = x^*$;
(ii) *x is closed and* $\ker(x^* \pm i\mathbf{1}) = \{0\}$;
(iii) $\mathrm{ran}(x \pm i\mathbf{1}) = H$.

For self-adjoint operators, we have the following important result.

Theorem 1.3.6. *If x is a self-adjoint operator in H, then $\sigma(x) \subseteq \mathbb{R}$.*

There are a number of important projections associated with a closed densely defined operator in H, which will be used extensively throughout the book. First, we recall the following simple result.

Proposition 1.3.7. *If x is a densely defined closed linear operator in H, then $\operatorname{ran}(x)^\perp = \ker(x^*)$ and $\ker(x)^\perp = \overline{\operatorname{ran}(x^*)}$. In particular, if x is a self-adjoint operator in H, then $\operatorname{ran}(x)^\perp = \ker(x)$ and $\ker(x)^\perp = \overline{\operatorname{ran}(x)}$.*

Definition 1.3.8. Given a closed densely defined linear operator x in H, we define:
(i) the *null projection* of x, denoted by $n(x)$, as the projection onto $\ker(x)$;
(ii) the *left support projection* of x, denoted by $s_l(x)$, as the projection onto $\overline{\operatorname{ran}(x)}$;
(iii) the *right support projection* of x, denoted by $s_r(x)$ as the projection $\mathbf{1} - n(x)$, which is the projection onto $\overline{\operatorname{ran}(x^*)}$.

Observe that $n(x^*) = \mathbf{1} - s_l(x)$, $s_r(x^*) = s_l(x)$, and $s_l(x^*) = s_r(x)$. In particular, if x is self-adjoint, then $s_r(x) = s_l(x)$. In this case, we denote $s(x) := s_r(x) = s_l(x)$ and the projection $s(x)$ is referred to as the *support projection*.

It is easy to see that $x = s_l(x)x = xs_r(x)$, as the following theorem shows the left and right support projections are smallest projections, which satisfy this property.

Theorem 1.3.9. *Let x be a closed and densely defined linear operator in H.*
(i) *$s_r(x)$ is the smallest projection $p \in P(\mathcal{B}(H))$ satisfying $x = xp$;*
(ii) *$s_l(x)$ is the smallest projection $p \in P(\mathcal{B}(H))$ satisfying $x = px$.*

Next, we discuss the notion of the characteristic matrix of a closed operator.

Every operator x in $\mathcal{B}(H \times H)$ is uniquely identified with a 2×2 matrix $(x_{ij})_{i,j=1}^2$ of bounded linear operators in H, through the relation

$$x : (\xi_1, \xi_2) \longrightarrow (x_{11}\xi_1 + x_{12}\xi_2, x_{21}\xi_1 + x_{22}\xi_2), \quad \xi_i \in H, i = 1, 2.$$

Note that $p \in \mathcal{B}(H \times H)$ is a projection if and only if for the matrix representation $p = (p_{ij})_{i,j=1,2}$ the following relations hold:

$$p_{ij}^* = p_{ji}, \quad \sum_{k=1}^{2} p_{ik}p_{kj} = p_{ij}, \quad i, j = 1, 2. \tag{1.2}$$

Definition 1.3.10. Let x be a closed operator in H. The matrix $(p_{ij})_{i,j=1}^2$ of the projection p of $H \times H$ onto the graph $\Gamma(x)$ is called the *characteristic matrix* of the operator x.

Since a closed subspace H_0 in $H \times H$ is the graph of a linear operator if and only if $(0, \eta) \in H_0$ implies that $\eta = 0$, we have the following necessary and sufficient condition for a projection in $H \times H$ to be a characteristic matrix.

Proposition 1.3.11. *Suppose that $p = (p_{ij})_{i,j=1,2}$ is the projection of $H \times H$ onto a closed subspace $H_0 \subset H \times H$. Then p is the characteristic matrix of some closed operator x if and* *only if* $\ker(\mathbf{1} - p_{22}) = \{0\}$, *i. e., if and only if the inverse* $(\mathbf{1} - p_{22})^{-1}$ *exists. In this case, the operator* x *is uniquely determined by the equation*

$$x(p_{11}\xi_1 + p_{12}\xi_2) = p_{21}\xi_1 + p_{22}\xi_2, \quad \xi_i \in H, i = 1, 2.$$

In particular, $p_{21} = xp_{11}$ *and* $p_{22} = xp_{12}$.

Proof. Note that the subspace $H_0 \subset H \times H$ is a graph of a closed linear operator if and only if $(0, \eta) \in H_0$ for some $\eta \in H$ implies that $\eta = 0$. Since p is the projection onto H_0, it follows that $(0, \eta) = p(0, \eta) = (p_{12}\eta, p_{22}\eta)$. Hence, H_0 is a graph if and only if $p_{22}\eta = \eta$. Therefore, H_0 is a graph if and only if $\ker(\mathbf{1} - p_{22}) = \{0\}$, as required.

Assume now that p is the characteristic matrix of some closed operator x. For any $\xi_1, \xi_2 \in H$, we have that $p(\xi_1, \xi_2) = (p_{11}\xi_1 + p_{12}\xi_2, p_{21}\xi_1 + p_{22}\xi_2) \in H_0$. Since H_0 is a closed graph, we can define a closed operator by setting $x(p_{11}\xi_1 + p_{12}\xi_2) = p_{21}\xi_1 + p_{22}\xi_2$ for all $\xi_i \in H$, $i = 1, 2$. The equalities $p_{21} = xp_{11}$, $p_{22} = xp_{12}$ follow by taking $\xi_1 = 0$ and $\xi_2 = 0$, respectively. □

To avoid ambiguity, we note that the equalities $p_{21} = xp_{11}$ and $p_{22} = xp_{12}$, established in the proposition above, mean that $\mathrm{dom}(xp_{11}) = \mathrm{dom}(p_{21}) = H$, $\mathrm{dom}(xp_{12}) = \mathrm{dom}(p_{22}) = H$, and for any $\xi \in H$, $p_{21}\xi = xp_{11}\xi$ and $p_{22}\xi = xp_{12}\xi$.

A requirement for a closed operator to be densely defined can also be described in terms of its characteristic matrix.

Proposition 1.3.12. *Let* $p = (p_{ij})_{i,j=1}^2$ *be the characteristic matrix of a closed operator* x *in* H. *Then the domain* $\mathrm{dom}(x)$ *of* x *is dense in* H *if and only if* $\ker(p_{11}) = 0$, *i. e., if and only if* p_{11} *has an inverse.*

Proof. It follows from Proposition 1.3.11 that $\mathrm{dom}(x)$ is dense in H if and only if the equality $\langle p_{11}\xi_1 + p_{12}\xi_2, \eta\rangle = 0$ for all $\xi_1, \xi_2 \in H$ implies that $\eta = 0$. Hence, referring to (1.2) we obtain that $\mathrm{dom}(x)$ is dense in H if and only $p_{11}^*\eta = p_{11}\eta = 0$ and $p_{12}^*\eta = p_{21}\eta = 0$ imply that $\eta = 0$. Thus, $\mathrm{dom}(x)$ is dense in H if and only if $\ker(p_{11}) \cap \ker(p_{21}) = 0$.

It remains to show that $\ker(p_{11}) \subset \ker(p_{21})$. Suppose that $\eta \in \ker(p_{11})$. Then using (1.2), we have that $p_{11} = p_{11}p_{11} + p_{12}p_{21}$ and, therefore,

$$\begin{aligned}\|p_{21}\eta\|^2 &= \langle p_{21}\eta, p_{21}\eta\rangle = \langle p_{21}^*p_{21}\eta, \eta\rangle = \langle p_{12}p_{21}\eta, \eta\rangle \\ &= \langle p_{11}\eta, \eta\rangle - \langle (p_{11} - p_{11}p_{11})\eta, \eta\rangle = 0,\end{aligned}$$

proving that $\eta \in \ker(p_{21})$. □

Combining Proposition 1.3.11 and Proposition 1.3.12, we obtain the following necessary and sufficient conditions for a matrix $(p_{ij})_{i,j=1,2}$ of bounded operators in H to be a characteristic matrix of a closed densely defined operator.

Corollary 1.3.13. *A 2×2-matrix $p = (p_{ij})_{i,j=1}^2$ of bounded operators in H is the characteristic matrix of a closed densely defined operator x in H if and only if it satisfies the relations:*
(i) $p_{ij}^* = p_{ji}$, $i, j = 1, 2$;
(ii) $\sum_{k=1}^2 p_{ik}p_{kj} = p_{ij}$, $i, j = 1, 2$;
(iii) $\ker(p_{11}) = \ker(\mathbf{1} - p_{22}) = \{0\}$.

In this case, x is uniquely determined by the equation

$$x(p_{11}\xi_1 + p_{12}\xi_2) = p_{21}\xi_1 + p_{22}\xi_2, \quad \xi_i \in H, i = 1, 2.$$

In particular, $p_{21} = xp_{11}$ and $p_{22} = xp_{12}$.

Next, we describe the adjoint x^* of a closed densely defined operator x in terms of characteristic matrices.

Proposition 1.3.14. *Let x be a closed densely defined operator with the characteristic matrix $(p_{ij})_{i,j=1}^2$. Then the characteristic matrix $(q_{ij})_{i,j=1}^2$ of the adjoint x^* satisfies*

$$q_{11} = \mathbf{1} - p_{22}, \quad q_{12} = p_{21}, \quad q_{21} = p_{12}, \quad q_{22} = \mathbf{1} - p_{11}.$$

In particular, the operator x^* can be described as the mapping

$$x^*((-p_{22})\xi_1 + p_{21}\xi_2) = p_{12}\xi_1 + (-p_{11})\xi_2, \quad \xi_1, \xi_2 \in H.$$

Proof. It is well known (see, e. g., [140, Lemma 1.10]) that $\Gamma(x^*) = V(\Gamma(x)^\perp)$, where $V : H \times H \to H \times H$ is a unitary operator defined by $V(\xi, \eta) = (-\eta, \xi)$, $\xi, \eta \in H$. Writing

$$\begin{aligned}\Gamma(x)^\perp &= (\mathbf{1} - p)(H \times H)\\ &= \{((\mathbf{1} - p_{11})\xi_1 - p_{12}\xi_2, -p_{21}\xi_1 + (\mathbf{1} - p_{22})\xi_2), \xi_1, \xi_2 \in H\},\end{aligned}$$

we obtain

$$\begin{aligned}\Gamma(x^*) &= V((\mathbf{1} - p)(H))\\ &= \{(p_{21}\xi_1 - (\mathbf{1} - p_{22})\xi_2, (\mathbf{1} - p_{11})\xi_1 - p_{12}\xi_2), \xi_1, \xi_2 \in H\}\\ &= \{((\mathbf{1} - p_{22})\eta_1 + p_{21}\eta_2, p_{12}\eta_1 + (\mathbf{1} - p_{11})\eta_2), \eta_1, \eta_2 \in H\}.\end{aligned}$$

Therefore,

$$q_{11} = \mathbf{1} - p_{22}, \quad q_{12} = p_{21}, \quad q_{21} = p_{12}, \quad q_{22} = \mathbf{1} - p_{11}.$$

The second part of the assertion follows from Proposition 1.3.11. □

We now give a precise description of the characteristic matrix of a closed densely defined operator.

Proposition 1.3.15. *For a closed densely defined operator x with the characteristic matrix $p = (p_{ij})_{i,j=1,2}$, we have*

$$\begin{aligned} p_{11} &= (x^*x+\mathbf{1})^{-1},\\ p_{12} &= x^*(xx^*+\mathbf{1})^{-1},\\ p_{21} &= x(x^*x+\mathbf{1})^{-1} = p_{12}^*,\\ p_{22} &= xx^*(xx^*+\mathbf{1})^{-1} = \mathbf{1}-(xx^*+\mathbf{1})^{-1}. \end{aligned}$$

Proof. By Proposition 1.3.12, the matrix

$$\begin{pmatrix} \mathbf{1}-p_{22} & p_{21} \\ p_{12} & \mathbf{1}-p_{11} \end{pmatrix}$$

is the characteristic matrix of x^*.

Since x is densely defined, Proposition 1.3.12 implies that p_{11} is invertible. By Corollary 1.3.13 and Proposition 1.3.14, we have that $p_{21} = xp_{11}$ and $\mathbf{1} - p_{11} = x^*p_{21} = x^*xp_{11}$, showing that $(\mathbf{1} + x^*x)p_{11} = \mathbf{1}$. Therefore, $(\mathbf{1} + x^*x)$ is an extension of p_{11}^{-1}. Since both $(\mathbf{1}+x^*x)$ and p_{11}^{-1} are self-adjoint, it follows that $(\mathbf{1}+x^*x) = p_{11}^{-1}$, and so $p_{11} = (\mathbf{1}+x^*x)^{-1}$. Since $p_{21} = xp_{11}$, it immediately follows that $p_{21} = x(\mathbf{1}+x^*x)^{-1}$.

Similarly, $p_{22} = xp_{12} = xx^*(\mathbf{1}-p_{22})$, and hence, an argument as above implies that $\mathbf{1}-p_{22} = (\mathbf{1}+xx^*)^{-1}$. Since $p_{12} = x^*(\mathbf{1}-p_{22})$, the assertion follows. □

The characteristic matrix gives also a necessary and sufficient condition for invertibility of a closed operator, as the following proposition shows.

Proposition 1.3.16. *Let $(p_{ij})_{i,j=1}^2$ be the characteristic matrix of a closed linear operator x. Then x has an inverse if and only if $\mathbf{1}-p_{11}$ has an inverse. If x^{-1} exists, it is a closed operator and its characteristic matrix $(q_{ij})_{i,j=1}^2$ satisfies the relations*

$$q_{11} = p_{22}, \quad q_{12} = p_{21}, \quad q_{21} = p_{12}, \quad q_{22} = p_{11}.$$

Proof. Let $\xi \in H$. Then x is invertible if and only if $(\xi, 0) \in \Gamma(x)$ implies that $\xi = 0$. Since p is the projection onto $\Gamma(x)$, it follows that x is invertible if and only if $p_{11}\xi = \xi$ and $p_{21}\xi = 0$, i. e., if and only if $\ker(\mathbf{1}-p_{11}) \cap \ker(p_{21}) = 0$. Assume that $\xi \in \ker(\mathbf{1}-p_{11})$. By (1.2), we have that

$$\|p_{21}\xi\|^2 = \langle p_{21}\xi, p_{21}\xi\rangle = \langle p_{12}p_{21}\xi, \xi\rangle = \langle p_{11}(\mathbf{1}-p_{11})\xi, \xi\rangle = 0.$$

Therefore, $\xi \in \ker(p_{21})$, i. e., $\ker(\mathbf{1}-p_{11}) \subset \ker(p_{21})$. Thus, x is invertible if and only if $\ker(\mathbf{1}-p_{11})$ is trivial.

Assume now that x is invertible. Since $\Gamma(x^{-1}) = \{(x\xi, \xi), \xi \in \operatorname{dom}(x)\}$, it follows that

$$\Gamma(x^{-1}) = \{(p_{21}\xi_1 + p_{22}\xi_2, p_{11}\xi_1 + p_{12}\xi_2), \xi_1, \xi_2 \in H\}.$$

Since the characteristic matrix q of x^{-1} has to satisfy (1.2), the assertion follows. □

In conclusion of this section, we present a sufficient condition for the characteristic matrix of a closed operator to commute with a given bounded operator.

Proposition 1.3.17. *If y is a bounded operator on H and x is a closed operator in H, such that both y and y^* commute with x, then y commutes with the components of the characteristic matrix of x.*

Proof. Define a bounded operator $z(y) = (z(y)_{ij})_{i,j=1,2}$ on $H \times H$ by setting $z(y)_{11} = z(y)_{22} = y$ and $z(y)_{12} = z(y)_{21} = 0$. Since y commutes with x, it follows that $z(y)$ maps $\Gamma(x)$ into itself. Since the characteristic matrix p is the projection on $\Gamma(x)$, it follows that $pz(y)p = z(y)p$. Similarly, $z(y^*) = z(y)^*$ maps $\Gamma(x)$ into itself and so $pz(y)^*p = z(y)^*p$. Hence,

$$pz(y) = p^* z(y)^{**} = (z(y)^* p)^* = (pz(y)^* p)^* = pz(y)p = z(y)p.$$

Therefore, y commutes with the components of the characteristic matrix p. □

1.4 The spectral theorem

In this section, we recall the spectral theory of self-adjoint operators on a Hilbert space. We refer the reader to [43, 127, 140] for the proofs.

As before, we assume that $(H, \langle\cdot,\cdot\rangle)$ is a complex Hilbert space and denote by $P(\mathcal{B}(H))$ the lattice of all projections on H. Suppose that Ω is a nonempty set and that Σ is a σ-algebra of subsets of Ω, so (Ω, Σ) is a measurable space.

Definition 1.4.1. A *spectral measure* on (Ω, Σ) is a mapping $E : \Sigma \to P(\mathcal{B}(H))$ such that:
(i) $E(\emptyset) = 0$ and $E(\Omega) = \mathbf{1}$;
(ii) if $M_j \in \Sigma, j \in \mathbb{N}$, are pairwise disjoint, then $E(\bigcup_{j=1}^{\infty} M_j) = \sum_{j=1}^{\infty} E(M_j)$, where the series is *so*-convergent in $\mathcal{B}(H)$.

Note that any spectral measure E necessarily satisfies the equality

$$E(M_1 \cap M_2) = E(M_1)E(M_2)$$

for all $M_1, M_2 \in \Sigma$. In particular, $E(M_1)$ and $E(M_2)$ commute for any two $M_1, M_2 \in \Sigma$.

A set $M \in \Sigma$ is called an *E-null set* if $E(M) = 0$. As usual, we say that a property P holds E-almost everywhere (or, E-a. e. shortly) if there exists an E-null set $N \in \Sigma$, such that property P holds for all $x \in \Omega \setminus N$.

The support of a spectral measure E is the complement in Ω of the union of all open E-null sets. The support of a spectral measure is denoted by $\operatorname{supp}(E)$.

Given a spectral measure E on (Ω, Σ) and $\xi, \eta \in H$, we define the σ-additive complex measure $E_{\xi,\eta} : \Sigma \to \mathbb{C}$ by $E_{\xi,\eta}(M) = \langle E(M)\xi, \eta\rangle$ for all $M \in \Sigma$. The total variation $\|E_{\xi,\eta}\|$ of the measure satisfies $\|E_{\xi,\eta}\| \le \|\xi\|\|\eta\|$.

We denote by $\mathcal{B}(\Omega,\Sigma)$ the set of all complex-valued Σ-measurable functions f, which are E-a. e. finite, i. e., $E(\{t \in \Omega : f(t) = \infty\}) = 0$. Clearly, $\mathcal{B}(\Omega,\Sigma)$ is an algebra with respect to the pointwise operations. The Banach algebra of all bounded Σ-measurable functions on Ω equipped with the norm

$$\|f\|_\infty = \operatorname{esssup}\{|f(t)| : t \in \Omega\}$$

is denoted by $\mathcal{B}_b(\Omega,\Sigma)$. Finally, we denote by $\mathcal{B}_s(\Omega,\Sigma)$ the subalgebra of all simple functions in $\mathcal{B}(\Omega,\Sigma)$ taking on finitely many values. That is, $f \in \mathcal{B}_s(\Omega,\Sigma)$ if it can be written in the form

$$f = \sum_{j=1}^{n} \alpha_j \chi_{M_j} : \quad \alpha_j \in \mathbb{C}, M_j \in \Sigma, n \in \mathbb{N},$$

where χ_A is the characteristic function of a set A.

For $f \in \mathcal{B}(\Omega,\Sigma)$, we write that $f \geq 0$, if $f(t) \geq 0$ for E-a. e. $t \in \Omega$. If $f, g \in \mathcal{B}(\Omega,\Sigma)$ are such that $f - g \geq 0$, we write $f \geq g$.

If Ω is a topological space and Σ is the Borel σ-algebra on Ω, then we shall simply write $\mathcal{B}(\Omega)$ (resp., $\mathcal{B}_b(\Omega)$), instead of $\mathcal{B}(\Omega,\Sigma)$ (resp., $\mathcal{B}_b(\Omega)$).

Given a spectral measure $E : \Sigma \to \mathcal{B}(H)$ and a simple function $f = \sum_{j=1}^{n} \alpha_j \chi_{M_j}$ in $\mathcal{B}_s(\Omega,\Sigma)$, we define

$$\int_\Omega f dE = \sum_{j=1}^{n} \alpha_j E(M_j).$$

The integration mapping $f \longmapsto \int_\Omega f dE$ is an algebra homomorphism from $\mathcal{B}_s(\Omega,\Sigma)$ into $\mathcal{B}(H)$, which satisfies

$$\left\| \int_\Omega f dE \right\|_{\mathcal{B}(H)} \leq \|f\|_\infty, \quad f \in \mathcal{B}_s(\Omega,\Sigma). \tag{1.3}$$

Since the space $\mathcal{B}_s(\Omega,\Sigma)$ is dense in $\mathcal{B}_b(\Omega,\Sigma)$, for any $f \in \mathcal{B}_b(\Omega,\Sigma)$ one can find a sequence $\{f_n\}_{n\in\mathbb{N}} \subset \mathcal{B}_s(\Omega,\Sigma)$, such that $\|f - f_n\|_\infty \to 0$ as $n \to \infty$. Then inequality (1.3) implies that the sequence $\{\int_\Omega f_n dE\}_{n\in\mathbb{N}}$ is a Cauchy sequence in $\mathcal{B}(H)$. Consequently, this sequence is convergent in $\mathcal{B}(H)$ and it is readily verified that its limit depends only on the function f and not on the choice of the particular sequence $\{f_n\}_{n\in\mathbb{N}}$, which approximates f. Hence, one can define the spectral integral for $f \in \mathcal{B}_b(\Omega,\Sigma)$ by setting

$$\int_\Omega f dE = \lim_{n\to\infty} \int_\Omega f_n dE,$$

where convergence is taken in the norm of $\mathcal{B}(H)$ and $\{f_n\}_{n\in\mathbb{N}}$ is any sequence in $\mathcal{B}_s(\Omega,\Sigma)$, such that $\|f - f_n\|_\infty \to 0$ as $n \to \infty$.

In the next theorem, we collect the basic properties of the spectral integral of bounded functions.

Theorem 1.4.2. *If $E : \Sigma \to P(\mathcal{B}(H))$ is a spectral measure on the measurable space (Ω, Σ), then the following statements hold:*

(i) *The spectral integration mapping $f \longmapsto \int_\Omega f dE$ is an algebra homomorphism from $\mathcal{B}_b(\Omega, \Sigma)$ into $\mathcal{B}(H)$;*

(ii) *$\int_\Omega \bar{f} dE = (\int_\Omega f dE)^*$ for all $f \in \mathcal{B}_b(\Omega, \Sigma)$;*

(iii)

$$\left\langle \left(\int_\Omega f dE \right) \xi, \eta \right\rangle = \int_\Omega f dE_{\xi,\eta}$$

for all $\xi, \eta \in H$ and all $f \in \mathcal{B}_b(\Omega, \Sigma)$;

(iv) *$\| \int_\Omega f dE \|_{\mathcal{B}(H)} \leq \|f\|_\infty$ for all $f \in \mathcal{B}_b(\Omega, \Sigma)$;*

(v) *For every $f \in \mathcal{B}_b(\Omega, \Sigma)$, the operator $\int_\Omega f dE$ is normal. If f is real-valued, then $\int_\Omega f dE$ is self-adjoint;*

(vi) *If $f \in \mathcal{B}_b(\Omega, \Sigma)$ is such that $f \geq 0$ E-a. e., then $\int_\Omega f dE \geq 0$ in $\mathcal{B}(H)$;*

(vii) *If $f, g \in \mathcal{B}_b(\Omega, \Sigma)$ are such that $|f| \leq |g|$ E-a. e., then*

$$\left\| \left(\int_\Omega f dE \right) \xi \right\|_H \leq \left\| \left(\int_\Omega g dE \right) \xi \right\|_H, \qquad \xi \in H.$$

In particular, $\| \int_\Omega f dE \|_{\mathcal{B}(H)} \leq \| \int_\Omega g dE \|_{\mathcal{B}(H)}$;

(viii) *If $\{f_n\}_{n \in \mathbb{N}}$ is a uniformly bounded sequence in $\mathcal{B}_b(\Omega, \Sigma)$ and if $f \in \mathcal{B}_b(\Omega, \Sigma)$, such that $f_n(\omega) \to f(\omega)$ as $n \to \infty$ for all $\omega \in \Omega$, then $\int_\Omega f_n dE \to \int_\Omega f dE$ in the strong operator topology as $n \to \infty$.*

As usual for a function $f \in \mathcal{B}_b(\Omega, \Sigma)$ and $\Delta \in \Sigma$, the integral $\int_\Delta f dE$ is defined by setting

$$\int_\Delta f dE := \int_\Omega f \chi_\Delta dE.$$

Our next objective is to extend the spectral integrals to all measurable functions on (Ω, Σ). If $f \in \mathcal{B}(\Omega, \Sigma)$, then there exists a *bounding sequence* $\{M_n\}_{n \in \mathbb{N}} \subset \Sigma$, such that $M_n \subset M_{n+1}$, $E(\bigcup_{n \in \mathbb{N}} M_n) = E(\Omega) = \mathbf{1}$ and $f \chi_{M_n} \in \mathcal{B}_b(\Omega, \Sigma)$. Since $f \chi_{M_n} \in \mathcal{B}_b(\Omega, \Sigma)$, it follows that the spectral integral $\int_\Omega f \chi_{M_n} dE$ is well-defined.

We now define

$$\operatorname{dom}\left(\int_\Omega f dE \right) = \left\{ \xi \in H : \int_\Omega |f|^2 dE_{\xi,\xi} < \infty \right\}.$$

The subspace $\operatorname{dom}(\int_\Omega f dE)$ can be equivalently described as

$$\operatorname{dom}\Big(\int_\Omega f dE\Big) = \Big\{\xi \in H : \Big(\int_\Omega f\chi_{M_n} dE\Big)\xi \text{ converges in } H\Big\}$$
$$= \Big\{\xi \in H : \sup_{n\in\mathbb{N}}\Big\|\Big(\int_\Omega f\chi_{M_n} dE\Big)\xi\Big\| < \infty\Big\}.$$

It is easy to show that for any $\xi \in \operatorname{dom}(\int_\Omega f dE)$ the limit of the sequences $\{(\int_\Omega f\chi_{M_n} dE)\xi\}_{n\in\mathbb{N}}$ as $n \to \infty$ does not depend on the choice of the bounding sequence $\{M_n\}_{n\in\mathbb{N}}$. Hence, the following definition makes sense.

Definition 1.4.3. Let $E : \Sigma \to \mathcal{B}(H)$ be a spectral measure on (Ω, Σ). Let $\{M_n\}_{n\in\mathbb{N}}$ be a bounding sequence for a function $f \in \mathcal{B}(\Omega, \Sigma)$. The integral of f with respect to E is defined by

$$\Big(\int_\Omega f dE\Big)\xi := \lim_{n\to\infty}\Big(\int_\Omega f\chi_{M_n} dE\Big)\xi,$$
$$\xi \in \operatorname{dom}\Big(\int_\Omega f dE\Big) = \Big\{\xi \in H : \int_\Omega |f|^2 dE_{\xi,\xi} < \infty\Big\}.$$

In the next theorem, we collect the basic properties of the spectral integration mapping. Recall (see Definition 1.3.2) that for closed densely defined operators x, y the strong sum $x \dotplus y$ and strong product $x \cdot y$ are defined as closures of the algebraic sum and product whenever these operators are closable.

Theorem 1.4.4. *Let $E : \Sigma \to P(\mathcal{B}(H))$ be a spectral measure on a measurable space (Ω, Σ). For all $f, g \in \mathcal{B}(\Omega, \Sigma)$, the following statements hold:*

(i) $\| \int_\Omega f dE_{\xi,\xi}\|^2 = \int_\Omega |f|^2 dE_{\xi,\xi}$ *for all* $\xi \in \operatorname{dom}(\int_\Omega f dE)$;
(ii) *if $\{M_n\}_{n\in\mathbb{N}}$ is a bounding sequence for f, then $\bigcup_{n\in\mathbb{N}} E(M_n)H$ is a core for the operator $\int_\Omega f dE$;*
(iii) $\int_\Omega \bar{f} dE = (\int_\Omega f dE)^*$;
(iv) *If $f = g$ E-a. e., then $\int_\Omega f dE = \int_\Omega g dE$;*
(v) *if f is real-valued E-a. e., then $\int_\Omega f dE$ is self-adjoint and if $f \geq 0$ E-a. e., then $\int_\Omega f dE \geq 0$;*
(vi) *if f, g are real-valued functions such that $f \leq g$ E-a. e., then $\int_\Omega f dE \leq \int_\Omega g dE$;*
(vii) $\int_\Omega (f + g) dE = \int_\Omega f dE \dotplus \int_\Omega g dE$;
(viii) $\int_\Omega fg dE = \int_\Omega f dE \cdot \int_\Omega g dE$;
(ix) $\int_\Omega p(f) dE = p(\int_\Omega f dE)$ *for all polynomials p. In particular, $p(\int_\Omega f dE)$ is closed for any polynomial p;*
(x) *if $\Delta \in \Sigma$ such that $f\chi_\Delta \in \mathcal{B}_b(\Omega, \Sigma)$, then $E(\Delta)H \subseteq \operatorname{dom}(\int_\Omega f dE)$ and $\int_\Omega f\chi_\Delta dE = (\int_\Omega f dE)E(\Delta)$.*

The spectral theorem for (unbounded) normal operators is the following result, which establishes that any normal operator x is of the form $\int_{\mathbb{R}} \lambda dE_x(\lambda)$ for a uniquely determined spectral measure E_x. We denote by $B(\mathbb{C})$ (resp., $B(\mathbb{R})$) the Borel σ-algebra on $\mathbb{C}$ (resp., on $\mathbb{R}$).

Theorem 1.4.5 (Spectral theorem). *If $x : \mathrm{dom}(x) \to H$ is a normal operator, then there exists a uniquely determined spectral measure*

$$E_x : B(\mathbb{C}) \to P(\mathcal{B}(H)),$$

called the spectral measure of x, *such that $x = \int_{\mathbb{C}} \lambda dE_x(\lambda)$. The support of E_x coincides with $\sigma(x)$.*

The spectral measure E_x of a normal operator x in H allows us to define (Borel) functional calculus of x.

Definition 1.4.6 (Functional calculus). Given a normal operator $x : \mathrm{dom}(x) \to H$ with the spectral measure $E_x : B(\mathbb{C}) \to P(\mathcal{B}(H))$, we define

$$f(x) = \int_{\mathbb{C}} f(\lambda) dE_x(\lambda)$$

for any $f \in \mathcal{B}(\mathbb{C})$. In particular,

$$\mathrm{dom}(f(x)) = \left\{ \xi \in H : \int_{\mathbb{C}} |f(t)|^2 d\langle E_x(t)\xi, \xi\rangle < \infty \right\}.$$

The mapping $f \longmapsto f(x)$ is called the *Borel functional calculus of the operator x* and it satisfies the properties listed in Theorem 1.4.4.

For any normal operator x and any $f \in \mathcal{B}(\mathbb{C})$, the operator $f(x)$ is also normal. As stated in the following proposition, the spectral measures of x and $f(x)$ are related to one another.

Proposition 1.4.7. *Let $x : \mathrm{dom}(x) \to H$ be a normal operator and let $f \in \mathcal{B}(\mathbb{C})$. Then $E_{f(x)}(M) = E_x(f^{-1}(M \cap f(\mathbb{C})))$ for any $M \in B(\mathbb{C})$.*

Here, $f^{-1}(M)$ denotes the preimage of a set M, i. e., $f^{-1}(B) = \{t \in \Omega : f(t) \in B\}$.

The spectral measure E_x of a normal operator x gives a necessary and sufficient condition for a bounded operator y to commute with x.

Theorem 1.4.8. *For a normal operator x and a bounded operator y, the following conditions are equivalent:*

(i) $yx \subset xy$;

(ii) $yE_x(M) = E_x(M)y$ *for any* $M \in B(\mathbb{C})$.

Let $x : \mathrm{dom}(x) \to H$ be an operator in H. Recall that x is called *positive*, denoted by $x \geq 0$, if $\langle x\xi, \xi \rangle \geq 0$ for all $\xi \in \mathrm{dom}(x)$. A self-adjoint operator x is positive if and only if $\sigma(x) \subseteq [0, \infty)$.

As an application of the functional calculus, we obtain the following result.

Proposition 1.4.9. *Suppose that x is a positive self-adjoint operator in the Hilbert space H. For each $n \in \mathbb{N}$, there exists a unique positive self-adjoint operator y in H such that $y^n = x$.*

If x is a positive self-adjoint operator in H, then the unique positive self-adjoint operator y satisfying $y^2 = x$ is denoted by $x^{1/2}$. Thus, x has a unique *positive square root* $x^{1/2}$. Note, furthermore, as it follows from Theorem 1.4.4, we have that $\mathrm{dom}(x) \subseteq \mathrm{dom}(x^{1/2})$ and that $\mathrm{dom}(x)$ is a core of $x^{1/2}$.

The *positive part* x_+ and *negative part* x_- of a self-adjoint operator x are defined by

$$x_+ = \int_{\mathbb{R}} \lambda_+ dE_x(\lambda) \quad \text{and} \quad x_- = \int_{\mathbb{R}} \lambda_- dE_x(\lambda),$$

respectively, where $\lambda_+ = \max\{\lambda, 0\}$ and $\lambda_- = \max\{-\lambda, 0\}$. By Theorem 1.4.4(v), $x_+, x_- \geq 0$. Furthermore, $\mathrm{dom}(x) = \mathrm{dom}(x_+) \cap \mathrm{dom}(x_-)$ and so, referring to Theorem 1.4.4(vii), we have that $x = x_+ - x_-$.

Note that the positive and negative parts of a self-adjoint operator are defined uniquely in the following sense.

Corollary 1.4.10. *Suppose that x is a self-adjoint operator in the Hilbert space H. If y and z are self-adjoint positive operators on H, such that $x = y - z$ and $s(y)s(z) = 0$, then $y = x_+$ and $z = x_-$.*

1.5 Polar decomposition

In this section, we recall the polar decomposition for any closed densely defined operator x on a Hilbert space H (see, e. g., [126, 127, 140]). The polar decomposition and its properties will be used repeatedly in the future chapters.

We start this section by stating the following important result.

Theorem 1.5.1. *If x is a closed densely defined operator in H, then the operator x^*x is self-adjoint and positive. Moreover,* $\mathrm{dom}(x^*x)$ *is a core for the operator x.*

Since the operator x^*x is positive and self-adjoint, it has unique positive square root $(x^*x)^{1/2}$.

Definition 1.5.2. For every closed densely defined operator x in H the *absolute value* $|x|$ of x is defined by $|x| = (x^*x)^{1/2}$.

It is clear that for a self-adjoint operator x, we have that $|x| = f(x)$ for $f(t) = |t|$. In particular, $|x| = x_+ + x_-$.

Note that $\mathrm{dom}(x^*x)$ is a core for the operators $|x|$ and x. Furthermore, $\mathrm{dom}(x) = \mathrm{dom}(|x|)$ and $\|x\xi\| = \||x|\xi\|$ for all $\xi \in \mathrm{dom}(x)$. In particular, $\ker(x) = \ker(|x|)$ and $s_r(x) = s(|x|)$.

Next, we recall the notion of partial isometries. An operator $v \in \mathcal{B}(H)$ is called a *partial isometry* if $\|v\xi\| = \|\xi\|$ for all $\xi \in \ker(v)^\perp$. In this case, $\mathrm{ran}(v)$ is a closed subspace in H. The projections $p = v^*v$ and $q = vv^*$ are projections on subspaces $\ker(v)^\perp$ and $\mathrm{ran}(v)$, respectively, and called *initial* and *final* projections of the partial isometry v. It is easy to see that $p = v^*qv$, $q = vqv^*$, and $v = vp = qv$.

The characterization of partial isometries follows.

Proposition 1.5.3 ([114, Chapter 1, Section 5]). *For an operator $v \in \mathcal{B}(H)$, the following conditions are equivalent:*
(i) *v is a partial isometry;*
(ii) *v^*v is a projection in $\mathcal{B}(H)$;*
(iii) *vv^* is a projection in $\mathcal{B}(H)$;*
(iv) *$vv^*v = v$;*
(v) *$v^*vv^* = v^*$.*

Theorem 1.5.4 (Polar decomposition). *Let $x : \mathrm{dom}(x) \to H$ be a closed and densely defined operator. There exists a partial isometry v, with initial space $s_r(x) = s(|x|)$ and final projection $s_l(x)$, such that $x = v|x|$. Moreover, if $x = wa$, where a is a positive self-adjoint operator and w is a partial isometry with initial projection $s(a)$, then $a = |x|$ and $w = v$.*

The factorization $x = v|x|$ in the above theorem is called the *polar decomposition* of x. The last statement in this theorem is usually referred to as *the uniqueness of the polar decomposition*. Since $v^*v = s(|x|)$, it is also clear that $|x| = v^*x$.

We note that if $x = v|x|$ is the polar decomposition of a closed densely defined operator, then

$$x^* = v^*vx^*, \quad |x^*| = vx^* = xv^*, \quad x = |x^*|v, \quad x^* = v^*|x^*| = |x|v^*. \tag{1.4}$$

Another application of the uniqueness of the polar decomposition is given in the following proposition.

Proposition 1.5.5. *Suppose that $x : \mathrm{dom}(x) \to H$ is a closed densely defined operator with polar decomposition $x = v|x|$ and let $u \in \mathcal{B}(H)$ be unitary. Then uxu^* is closed and densely defined with polar decomposition given by*

$$uxu^* = (uvu^*)(u|x|u^*).$$

In particular, $uxu^* = x$ if and only if $uvu^* = v$ and $u|x|u^* = |x|$.

1.6 Involutive algebras

We now move onto the theory of abstract operator algebras. In this section, we start with the notion of a $*$-algebras and in the following sections cover C^*-algebras and von Neumann algebras.

Let $\mathcal{A}$ be an associative algebra over the field $\mathbb{C}$ of complex numbers. A mapping $x \mapsto x^*$ from $\mathcal{A}$ into $\mathcal{A}$ is called an *involution*, if for all $x, y \in \mathcal{A}$, and $\lambda \in \mathbb{C}$, the following properties hold:
(i) $(x+y)^* = x^* + y^*$;
(ii) $(\lambda x)^* = \bar{\lambda} x^*$;
(iii) $(xy)^* = y^* x^*$;
(iv) $(x^*)^* = x$.

An algebra $\mathcal{A}$ with involution is called a *$*$-algebra.*

Recall that an algebra $\mathcal{A}$ is called *commutative* (or *abelian*) if $xy = yx$ for all $x, y \in \mathcal{A}$. We say that two elements $x, y \in \mathcal{A}$ are *disjoint*, if $xy = yx = 0$. A family $\{x_i\}_{i \in I}$ is said to be pairwise disjoint if $x_i x_j = x_j x_i = 0$ for all $i \neq j$.

An algebra $\mathcal{A}$ is called *unital* if there exists a multiplicative identity (a *unit*), which is necessarily unique and which we denote by $\mathbf{1}$ or $\mathbf{1}_{\mathcal{A}}$.

If $\mathcal{A}$ is a $*$-algebra, then an element $x \in \mathcal{A}$ is called *self-adjoint* if $x = x^*$, and *normal* if $x^* x = x x^*$. It is clear that if $\mathcal{A}$ is unital, then $\mathbf{1}$ is necessarily a self-adjoint element in $\mathcal{A}$. The set $\{x \in \mathcal{A} : x = x^*\}$ of all self-adjoint elements from $\mathcal{A}$ is denoted by $\mathcal{A}_h$. It is clear that the set $\mathcal{A}_h$ is a real linear subspace in $\mathcal{A}$ and, moreover, $x^* x, x x^* \in \mathcal{A}_h$ for any $x \in \mathcal{A}$. For every element $x \in \mathcal{A}$, we set

$$\operatorname{Re}(x) = \frac{x + x^*}{2}, \quad \operatorname{Im}(x) = \frac{x - x^*}{2i}.$$

Clearly, $\operatorname{Re}(x), \operatorname{Im}(x) \in \mathcal{A}_h$, in addition, $x = \operatorname{Re}(x) + i \operatorname{Im}(x)$. Conversely, if $x = x_1 + i x_2$ for some $x_1, x_2 \in \mathcal{A}_h$, then $x_1 = \operatorname{Re}(x)$ and $x_2 = \operatorname{Im}(x)$. The elements $\operatorname{Re}(x)$ and $\operatorname{Im}(x)$ are called the *real* and *imaginary* part of the element $x \in \mathcal{A}$.

An element x in a unital algebra $\mathcal{A}$ is called *invertible* if there exists an element $y \in \mathcal{A}$, such that $xy = yx = \mathbf{1}$. The element y, satisfying this condition is uniquely defined and is denoted by x^{-1}. It is easy to see that an element $x \in \mathcal{A}$ in a $*$-algebra $\mathcal{A}$ is invertible if and only if the element x^* is invertible. In this case, the equality

$$(x^{-1})^* = (x^*)^{-1}$$

holds.

An element $u \in \mathcal{A}$ is called *unitary*, if

$$u^* u = u u^* = \mathbf{1}.$$

It is clear that an element $u \in \mathcal{A}$ is unitary if and only if it is invertible and $u^* = u^{-1}$. The set of all unitary elements of the $*$-algebra $\mathcal{A}$ forms a (multiplicative) group, which is denoted by $U(\mathcal{A})$.

An element p of $\mathcal{A}$ is called an *idempotent* if $p^2 = p$. A self-adjoint idempotent p from the $*$-algebra $\mathcal{A}$ is called a *projection* (or *orthoprojection*). The set of all projections of the algebra $\mathcal{A}$ is denoted by $P(\mathcal{A})$.

A subalgebra $\mathcal{B}$ of a $*$-algebra $\mathcal{A}$ is called *$*$-subalgebra*, if $x^* \in \mathcal{B}$ for all $x \in \mathcal{B}$. In this case, $\mathcal{B}$ itself is a $*$-algebra with respect to the algebraic operations and involution inherited from $\mathcal{A}$.

For a subalgebra $\mathcal{B}$ of $\mathcal{A}$ and $x, y \in \mathcal{A}$, we denote

$$x\mathcal{B}y = \{xay, a \in \mathcal{B}\}.$$

A linear subspace $\mathcal{B}$ of a $*$-algebra is a *(two-sided) ideal* of $\mathcal{A}$ if $xy, yx \in \mathcal{B}$ for any $x \in \mathcal{B}$ and $y \in \mathcal{A}$. An ideal $\mathcal{B}$ of $\mathcal{A}$ is said to be *$*$-ideal* if $x^* \in \mathcal{B}$ for any $x \in \mathcal{B}$.

Let $\mathcal{A}$ and $\mathcal{B}$ be $*$-algebras. A mapping $\varphi : \mathcal{A} \to \mathcal{B}$ is called a *$*$-homomorphism*, if for all $x, y \in \mathcal{A}$, $\lambda \in \mathbb{C}$ the following equalities hold:

(i) $\varphi(x + y) = \varphi(x) + \varphi(y)$;
(ii) $\varphi(\lambda x) = \lambda\varphi(x)$;
(iii) $\varphi(xy) = \varphi(x)\varphi(y)$;
(iv) $\varphi(x^*) = \varphi(x)^*$.

If, in addition, the mapping φ is bijective, then φ is called a *$*$-isomorphism*. In this case, the algebras $\mathcal{A}$ and $\mathcal{B}$ are called *$*$-isomorphic*. If $\mathcal{A}$ and $\mathcal{B}$ are two $*$-algebras, which are $*$-isomorphic, then we shall write $\mathcal{A} \cong \mathcal{B}$.

1.7 C^*-algebras

If an associative $*$-algebra $\mathcal{A}$ is a Banach space equipped with a norm $\|\cdot\|$ satisfying for all $x, y \in \mathcal{A}$, the conditions

(i) $\|xy\| \le \|x\| \cdot \|y\|$;
(ii) $\|x\| = \|x^*\|$;

then $\mathcal{A}$ is called a *Banach $*$-algebra*. If, in addition, the algebra $\mathcal{A}$ has unit $\mathbf{1}$, such that $\|\mathbf{1}\| = 1$, then $\mathcal{A}$ is called a *unital Banach $*$-algebra*.

A Banach $*$-algebra $\mathcal{A}$ is called a *C^*-algebra* if $\|x^*x\| = \|x\|^2$ for all $x \in \mathcal{A}$. It is clear that every closed $*$-subalgebra of a C^*-algebra $\mathcal{A}$ is a C^*-algebra itself with respect to the same algebraic operations and involution as in the algebra $\mathcal{A}$ (these $*$-algebras are called *C^*-subalgebras* of $\mathcal{A}$).

If a C^*-algebra $\mathcal{A}$ has unit $\mathbf{1}$, then the condition $\|\mathbf{1}\| = 1$ follows immediately from the equalities:

$$\|\mathbf{1}\| = \|\mathbf{1}^*\mathbf{1}\| = \|\mathbf{1}\|^2.$$

We list below some important examples of C^*-algebras.

Example 1.7.1. *Let X be an arbitrary nonempty set. The set $\ell_\infty(X)$ of all bounded complex-valued functions f on X with pointwise operations:*
(i) $(f+g)(x) = f(x) + g(x)$;
(ii) $(fg)(x) = f(x)g(x)$;
(iii) $(\lambda f)(x) = \lambda f(x)$;

and involution given by complex conjugation $f \mapsto \bar{f}$ and the norm

$$\|f\|_\infty = \sup_{x\in X} |f(x)|,$$

is a unital commutative C^*-algebra.

Example 1.7.2. *Let K be a topological space. The set $C_b(K)$ of all bounded continuous complex-valued functions on K is a closed subalgebra of $\ell_\infty(K)$ and, therefore, $C_b(K)$ is a unital commutative C^*-algebra. In the case when K is compact, the C^*-algebra $C_b(K)$ coincides with the set $C(K)$ of all continuous functions on K.*

Example 1.7.3. *Let (Ω, Σ, μ) be a σ-finite measure space, i. e., (Ω, Σ, μ) is a measure space such that there exists at most countable family $\{\Omega_n\}_{n\in\mathbb{N}}$ of pairwise disjoint measurable sets, such that $\mu(\Omega_n) < \infty$ for all $n \in \mathbb{N}$ and $\Omega = \bigcup_{n\in\mathbb{N}} \Omega_n$.*

Let $L_\infty(\Omega, \Sigma, \mu)$ be the algebra of all classes $[f]$ of almost everywhere equal essentially bounded complex-valued measurable functions f on (Ω, Σ, μ). Equipped with the norm

$$\|[f]\|_\infty = \operatorname*{ess\,sup}_{x\in\Omega} |f(x)|,$$

the algebra $L_\infty(\Omega, \Sigma, \mu)$ is a commutative unital C^*-algebra with respect to involution given by complex conjugation $[f] \mapsto [\bar{f}]$. Whenever the measure space (Ω, Σ, μ) is clear from the context, we shall use notation $L_\infty(\Omega)$ instead of $L_\infty(\Omega, \Sigma, \mu)$. For the equivalence class $[f] \in L_\infty(\Omega)$, the support projection $s([f])$ is defined as the equivalence class $[\chi_{G(f)}]$, where $G(f) = \{\omega \in \Omega : f(\omega) \neq 0\} \in \Sigma$.

Example 1.7.4. *Let H be a Hilbert space and let $\mathcal{B}(H)$ be the algebra of all bounded linear operators on H equipped with the operator norm $\|\cdot\|_{\mathcal{B}(H)}$. Then $\mathcal{B}(H)$ is a C^*-algebra when equipped with the involution $x \mapsto x^*$. The algebra $\mathcal{B}(H)$ is commutative if and only if H is a one-dimensional Hilbert space.*

By the Gelfand–Naimark–Segal theorem (see e. g., [60, Section 2.6]), every C^*-algebra $\mathcal{A}$ is $*$-isomorphic to a C^*-subalgebra of the C^*-algebra $\mathcal{B}(H)$ for some Hilbert space H.

Let $\mathcal{A}$ be an arbitrary unital C^*-algebra. The *spectrum* of an element $x \in \mathcal{A}$ is the set

$$\sigma_{\mathcal{A}}(x) = \{\lambda \in \mathbb{C} : (x - \lambda\mathbf{1}) \text{ is not invertible in } \mathcal{A}\}.$$

In general, if $\mathcal{B}$ is a Banach subalgebra of the Banach algebra $\mathcal{A}$ and $x \in \mathcal{B}$, then $\sigma_{\mathcal{A}}(x) \subseteq \sigma_{\mathcal{B}}(x)$. However, if $\mathcal{B}$ is a C^*-subalgebra of the C^*-algebra $\mathcal{A}$, then $\sigma_{\mathcal{A}}(x) = \sigma_{\mathcal{B}}(x)$ for all $x \in \mathcal{B}$. In particular, when no confusion may arise, we denote $\sigma_{\mathcal{A}}(x)$ by $\sigma(x)$.

Let $\mathcal{A}$ be a C^*-algebra. A consequence of the Gelfand–Naimark–Segal theorem (see, e. g., [60, Section 2.6]) is that any element $x \in \mathcal{A}$ is a bounded linear operator on some Hilbert space H and $\sigma_{\mathcal{A}}(x) = \sigma_{\mathcal{B}(H)}(x)$, and so the theory developed in the previous sections is applicable for all $x \in \mathcal{A}$. In particular, for any self-adjoint x the (Borel) functional calculus $f \mapsto f(x)$ is well-defined.

An element $x \in \mathcal{A}$ is called *positive* if $x = x^*$ and $\sigma(x) \subseteq [0, \infty)$. The set of all positive elements in $\mathcal{A}$ is denoted by $\mathcal{A}_+$. It is clear that

$$\mathcal{A}_+ \cap (-\mathcal{A}_+) = \{0\}, \quad \lambda\mathcal{A}_+ \subseteq \mathcal{A}_+$$

for all $\lambda \geq 0$. In the following theorem, the principal properties of positive elements in a C^*-algebra $\mathcal{A}$ are collected.

Theorem 1.7.5 ([110, Section 2.2]).

(i) *For every $x \in \mathcal{A}_+$, there exists a unique element $y \in \mathcal{A}_+$, such that $y^2 = x$ (in this case, the element y is denoted by $x^{1/2}$ and called the* square root *of the elements $x \in \mathcal{A}_+$);*

(ii) $\mathcal{A}_+ + \mathcal{A}_+ \subseteq \mathcal{A}_+$;

(iii) $\mathcal{A}_+ = \{x^*x : x \in \mathcal{A}\} = \{y^2 : y \in \mathcal{A}_h\}$;

(iv) *If $\mathcal{A}$ is a C^*-subalgebra of $\mathcal{B}(H)$, then*

$$\mathcal{A}_+ = \{x \in \mathcal{A} : \langle x\xi, \xi\rangle \geq 0 \text{ for all } \xi \in H\}.$$

We define a partial order on $\mathcal{A}_h$ by setting

$$x \leq y \Longleftrightarrow (y - x) \in \mathcal{A}_+.$$

Theorem 1.7.5(iv) implies, in particular, that the partial order on $\mathcal{A}_h$ coincides with the partial order inherited from the partial order on $\mathcal{B}_h(H)$.

This partial order has the following properties.

Proposition 1.7.6 ([110, Section 2.2]).

(i) *If $x \leq y$, then $x + z \leq y + z$ and $\lambda x \leq \lambda y$ for all $z \in \mathcal{A}_h$ and all $\lambda \in \mathbb{R}^+$;*

(ii) *If $x \leq y$, then $z^*xz \leq z^*yz$ for all $z \in \mathcal{A}$;*

(iii) *If $x, y \in \mathcal{A}_+$ and $xy = yx$, then $xy \in \mathcal{A}_+$;*

(iv) *If $0 \leq x \leq y$, then $\|x\| \leq \|y\|$ and $x^{1/2} \leq y^{1/2}$; moreover, if x and y are invertible, then $0 \leq y^{-1} \leq x^{-1}$.*

Note that by functional calculus for every $x \in \mathcal{A}_h$ with $\|x\| \leq 1$ the element $\mathbf{1} - x^2$ is in $\mathcal{A}_+$, and the elements

$$u = x + i(\mathbf{1} - x^2)^{1/2} \quad \text{and} \quad v = x - i(\mathbf{1} - x^2)^{1/2}$$

are unitary, in addition,

$$x = \frac{u + v}{2}.$$

In particular, every C^*-algebra $\mathcal{A}$ coincides with the linear span of all its unitary elements.

A linear functional $f : \mathcal{A} \to \mathbb{C}$ on a unital C^*-algebra $\mathcal{A}$ is called *positive* if $f(x) \geq 0$ for all $x \in \mathcal{A}_+$. This is denoted by $f \geq 0$. A positive linear function $f : \mathcal{A} \to \mathbb{C}$ is said to be *state* if $\|f\| = 1$, where $\|f\|$ denotes the norm of functional f. Every positive linear functional f is continuous, in addition $\|f\| = f(\mathbf{1})$, and for every $x \in \mathcal{A}$ the following assertions hold:

(i) $f(x^*) = \overline{f(x)}$;
(ii) $|f(x)|^2 \leq \|f\| \cdot f(x^*x)$.

It can be shown that every continuous linear functional on $\mathcal{A}$ can be written as a linear combination positive linear functionals.

1.8 Von Neumann algebras

In this section, we recall the basic theory of von Neumann algebras, which is a fundamental notion for this book.

Let $\mathcal{A}$ be an arbitrary algebra over the field of complex numbers. For an arbitrary nonempty subset $\mathcal{B}$ of $\mathcal{A}$, its *commutant* $\mathcal{B}'$ is defined by the equality

$$\mathcal{B}' = \{y \in \mathcal{A} : xy = yx \text{ for all } x \in \mathcal{B}\}.$$

It is straightforward that $\mathcal{B}'$ is a subalgebra of $\mathcal{A}$ and the *bicommutant* $\mathcal{B}'' = (\mathcal{B}')'$ contains $\mathcal{B}$, moreover, if $\mathcal{A}$ has unit $\mathbf{1}$, then $\mathbf{1} \in \mathcal{B}'$. In addition, the inclusion $\mathcal{B}_1 \subset \mathcal{B}_2$ for two subalgebras $\mathcal{B}_1$ and $\mathcal{B}_2$ implies the inclusion $\mathcal{B}_2' \subset \mathcal{B}_1'$.

For a $*$-subalgebra $\mathcal{B}$ of $\mathcal{A}$, the $*$-subalgebra

$$\mathcal{Z}(\mathcal{B}) = \mathcal{B} \cap \mathcal{B}'$$

is called the *center* of the algebra $\mathcal{B}$. The algebra $\mathcal{B}$ is called a *factor* if $\mathcal{Z}(B) = \mathbb{C}\mathbf{1}$. If the algebra $\mathcal{B}$ is commutative, then $\mathcal{Z}(\mathcal{B}) = \mathcal{B}$. We have $\mathcal{Z}(\mathcal{A})'' = \mathcal{Z}(\mathcal{A})$.

Definition 1.8.1. A $*$-subalgebra $\mathcal{M} \subseteq \mathcal{B}(H)$ is called a *von Neumann algebra* if $\mathcal{M} = \mathcal{M}''$. In this case, the von Neumann algebra $\mathcal{M}$ is said to be *acting* on H (or simply $\mathcal{M}$ is a von Neumann algebra on H).

Elementary examples of von Neumann algebras are the algebra $\mathcal{B}(H)$ and the algebra

$$\mathbb{C}\mathbf{1} = \{\lambda\mathbf{1} : \lambda \in \mathbb{C}\}$$

of all scalar multiples of the identity element of $\mathcal{B}(H)$.

It is clear that every von Neumann algebra $\mathcal{M}$ is a unital C^*-subalgebra of $\mathcal{B}(H)$ and the identity in $\mathcal{M}$ coincides with the identity in $\mathcal{B}(H)$. For any nonempty subset $\mathcal{S}$ of $\mathcal{B}(H)$, which is invariant under the operations of taking adjoints, the double commutant $\mathcal{S}''$ is a von Neumann algebra generated by $\mathcal{S}$, i. e., the smallest von Neumann algebra on H containing $\mathcal{S}$.

Let $\mathcal{M}$ be a von Neumann algebra. As noted in Section 1.7, every element of $\mathcal{M}'$ is a linear combination of unitary elements of $\mathcal{M}'$, and so $x \in \mathcal{B}(H)$ belongs to $\mathcal{M}$ if and only if $xu = ux$ (equivalently, $x = uxu^*$) for all $U(\mathcal{M}')$. Furthermore, if $x \in \mathcal{B}(H)$ is normal, it follows from Theorem 1.4.8 that the following statements are equivalent:

(i) $x \in \mathcal{M}$;
(ii) $E_x(B) \in \mathcal{M}$ for all Borel sets $B \subseteq \mathbb{C}$;
(iii) $f(x) \in \mathcal{M}$ for all bounded Borel functions $f : \sigma(x) \to \mathbb{C}$.

In particular, if $x \in \mathcal{M}$, then $x^*x \in \mathcal{M}$, and so $|x| = (x^*x)^{1/2} \in \mathcal{M}$. Furthermore, if $x \in \mathcal{M}$, with the polar decomposition $x = v|x|$, then it follows from Proposition 1.5.5 that $v = uvu^*$ for all $u \in U(\mathcal{M}')$, and hence, $v \in \mathcal{M}$. This also implies that the left support projection $s_l(x) = v^*v$ and the right support projection $s_r(x) = vv^*$ both belong to $\mathcal{M}$.

Denote by $\|\cdot\|_{\mathcal{M}}$ the C^*-norm on a von Neumann algebra $\mathcal{M}$, which coincides with the operator norm $\|\cdot\|_{\mathcal{B}(H)}$. A von Neumann algebra inherits partial order from $\mathcal{B}(H)$. In particular, we denote by $\mathcal{M}_+$ the cone of all positive operators from $\mathcal{M}$. Recall that $\mathcal{M}_1 = \{x \in \mathcal{M} : \|x\|_{\mathcal{M}} \leq 1\}$ denotes the closed unit ball in $\mathcal{M}$.

As was proved by J. von Neumann, any von Neumann algebra is necessarily closed with respect to the topologies introduced in Section 1.2.

Theorem 1.8.2 (Double commutant theorem, [61, Part 1, Chapter 3]). *Let $\mathcal{M}$ be a unital $*$-subalgebra in $\mathcal{B}(H)$. The following conditions are equivalent:*

(i) $\mathcal{M} = \mathcal{M}''$;
(ii) *$\mathcal{M}$ is closed in the weak operator topology;*
(iii) *$\mathcal{M}$ is closed in the strong operator topology;*
(iv) *$\mathcal{M}$ is closed in the ultraweak operator topology;*
(v) *$\mathcal{M}$ is closed in the ultrastrong operator topology;*
(vi) *$\mathcal{M}_1$ is closed in the weak operator topology;*
(vii) *$\mathcal{M}_1$ is closed in the strong operator topology;*

(viii) $\mathcal{M}_1$ *is closed in the ultraweak operator topology;*
(ix) $\mathcal{M}_1$ *is closed in the ultrastrong operator topology.*

Proposition 1.2.1 and Theorem 1.8.2 imply the following theorem.

Theorem 1.8.3. *If* $\{x_i\}_{i\in I}$ *is an increasing (decreasing) and bounded net of operators from the self-adjoint part* $\mathcal{M}_h$ *of a von Neumann algebra* $\mathcal{M}$*, then there exists an operator* $x \in \mathcal{M}_h$ *such that* $x = \sup_{i\in I} x_i$ *(resp.,* $x = \inf_{i\in I} x_i$*).*

By Theorem 1.8.2, the intersection of an arbitrary family of von Neumann algebras acting on H is a von Neumann algebra acting on H. In particular, the center $\mathcal{Z}(\mathcal{M}) = \mathcal{M} \cap \mathcal{M}'$ of a von Neumann algebra $\mathcal{M}$ is a commutative von Neumann algebra. Furthermore, in this case the following equalities hold:

$$\mathcal{Z}(\mathcal{M}) = \mathcal{M} \cap \mathcal{M}' = \mathcal{M}'' \cap \mathcal{M}' = \mathcal{Z}(\mathcal{M}').$$

Since $\mathcal{B}(H)' = \mathbb{C}\mathbf{1}$ (see, e. g., [114, Chapter YII, Section 34]), it follows that $\mathcal{B}(H)$ is a factor.

The following theorem singles out the class of von Neumann algebras among all C^*-algebras.

Theorem 1.8.4 ([134, Section 1.16]). *For a* C^**-algebra* $\mathcal{A}$*, the following are equivalent:*
(i) *The algebra* $\mathcal{A}$ *(considered as a Banach space) has a* predual space, *i. e., there exists a normed space* $\mathcal{A}_*$ *with the Banach dual* $(\mathcal{A}_*)^*$ *coinciding with* $\mathcal{A}$*;*
(ii) *The algebra* $\mathcal{A}$ *is* $*$*-isomorphic to a von Neumann algebra.*

A C^*-algebra fulfilling condition (i) of Theorem 1.8.4 is called a *W^*-algebra*.

Let $\mathcal{M}$ be a von Neumann algebra, acting on H, and let $\mathcal{M}^*$ be the dual space of $\mathcal{M}$ with respect to the norm topology. We denote by $\mathcal{M}_*$ the set of all *uwo*-continuous linear functionals on $\mathcal{M}$. It is clear that $\mathcal{M}_*$ is a linear subspace in $\mathcal{M}^*$.

Theorem 1.8.5. *The subspace* $\mathcal{M}_*$ *is norm closed in* $\mathcal{M}^*$ *and* $(\mathcal{M}_*)^* = \mathcal{M}$ *with respect to the* canonical bilinear form

$$\langle f, T\rangle = f(T), \quad f \in \mathcal{M}_*, T \in \mathcal{M};$$

in addition, the weak topology $\sigma(\mathcal{M}, \mathcal{M}_*)$ *on* $\mathcal{M}$ *coincides with the ultraweak operator topology.*

We now turn to a characterization of positive functionals in $\mathcal{M}_*$.

Definition 1.8.6. Let ψ be a positive linear functional on a von Neumann algebra $\mathcal{M}$.
(i) ψ is said to be *normal* if $x_i \uparrow x$ in $\mathcal{M}_+$ implies that $\psi(x_i) \uparrow \psi(x)$;
(ii) ψ is called *completely additive* if $\psi(\sum_{i\in I} p_i) = \sum_{i\in I} \psi(p_i)$ for every system $\{p_i\}_{i\in I}$ of pairwise orthogonal projections in $\mathcal{M}$.

Theorem 1.8.7. *For a positive functional ψ on $\mathcal{M}$, the following statements are equivalent:*
(i) *ψ is normal;*
(ii) *ψ is completely additive;*
(iii) *ψ is uwo-continuous (equivalently, $\psi \in \mathcal{M}_*$).*

The notions of positivity and normality can also be introduced for linear mappings between von Neumann algebras.

Definition 1.8.8. Let $\mathcal{M}$ and $\mathcal{N}$ be von Neumann algebras, acting on the Hilbert spaces H and K, respectively, and let ψ be a $*$-homomorphism from $\mathcal{M}$ into $\mathcal{N}$.
(i) ψ is called positive if $\psi(x) \geq 0$ in $\mathcal{N}$ whenever $x \geq 0$ in $\mathcal{M}$;
(ii) If ψ is positive, then ψ is said to be normal if $\psi(x_i) \uparrow \psi(x)$ in $\mathcal{N}$ whenever $x_i \uparrow x$ in $\mathcal{M}$.

Note that any $*$-homomorphism $\pi : \mathcal{M} \to \mathcal{N}$ is positive. Furthermore, a positive linear mapping $\psi : \mathcal{M} \to \mathcal{N}$ is normal if and only if it is continuous with respect to the ultraweak operator topologies. By [44, Section III.2.2] any positive $*$-isomorphism between two von Neumann algebras is necessarily normal.

Next, we describe commutative von Neumann algebras.

Example 1.8.9. *Let (Ω, Σ, μ) be a σ-finite measure space and let $L_\infty(\Omega, \Sigma, \mu)$ be the C^*-algebra of all (classes of a. e. equal) essentially bounded functions on (Ω, Σ, μ). Let $L_2(\Omega, \Sigma, \mu)$ be the Hilbert space of all (classes of a. e. equal) square-integrable functions on (Ω, Σ, μ). For $f \in L_\infty(\Omega, \Sigma, \mu)$, denote by M_f the bounded linear operator on $L_2(\Omega, \Sigma, \mu)$ given by multiplication with the function f, i. e., $(M_f g)(\omega) = f(\omega)g(\omega)$, $\omega \in \Omega$. The set*

$$\mathcal{M} = \{M_f : f \in L_\infty(\Omega, \Sigma, \mu)\}$$

is a commutative von Neumann algebra acting on $L_2(\Omega, \Sigma, \mu)$. Moreover, the mapping $\Phi : L_\infty(\Omega, \Sigma, \mu) \to \mathcal{M}$ defined by the equality $\Phi(f) = M_f$ is a $*$-isomorphism from $L_\infty(\Omega, \Sigma, \mu)$ onto $\mathcal{M}$. In this case, for any $f \in L_\infty(\Omega, \Sigma, \mu)$, the right and left support projections $s_l(M_f), s_r(M_f)$ coincide and are equal to the projection $s(M_f) := M_{[\chi_A]}$, where $A = \{\omega \in \Omega : f(\omega) \neq 0\} \in \Sigma$.

In fact, any commutative von Neumann algebra is $*$-isomorphic to the von Neumann algebra $L_\infty(\Omega, \Sigma, \mu)$ for some measure space (Ω, Σ, μ).

Theorem 1.8.10 ([149, Chapter III, Section 1]). *Let $\mathcal{M}$ be a commutative von Neumann algebra acting on a separable Hilbert space. Then there exists a σ-finite measure space (Ω, Σ, μ) and a positive $*$-isomorphism from the algebra $\mathcal{M}$ onto the von Neumann algebra $L_\infty(\Omega, \Sigma, \mu)$, i. e., we can assume that $\mathcal{M}$ acts on the Hilbert space $L_2(\Omega, \Sigma, \mu)$ and $\mathcal{M}$ coincides with the von Neumann algebra $\mathcal{M} = \{M_f : f \in L_\infty(\Omega, \Sigma, \mu)\}$.*

Remark 1.8.11. *Let $\mathcal{M}$ be an arbitrary von Neumann algebra (acting on a separable Hilbert space H). Then $\mathcal{Z}(\mathcal{M})$ is a commutative von Neumann algebra acting on H. Therefore, by Theorem 1.8.10 there exists a σ-finite measure space (Ω, Σ, μ), such that $\mathcal{Z}(\mathcal{M})$ is $*$-isomorphic to $L_\infty(\Omega, \Sigma, \mu)$. Since (Ω, Σ, μ) is σ-finite, there exists a family $\{\Omega_i\}_{i\in\mathbb{N}}$ of pairwise disjoint measurable sets of finite measure μ, such that $\bigcup_{i\in\mathbb{N}} \Omega_i = \Omega$. We define a finite measure (in fact, probability measure) ν on (Ω, Σ) by setting*

$$\nu(A) = \sum_{i=1}^{\infty} \frac{\mu(A \cap \Omega_i)}{2^i \mu(\Omega_i)}, \quad A \in \Sigma.$$

The measures μ and ν are equivalent (i. e., μ and ν have the same null sets). In particular, the $*$-algebras $L_\infty(\Omega, \Sigma, \mu)$ and $L_\infty(\Omega, \Sigma, \nu)$ coincide.

Thus, in what follows, without loss of generality, for a noncommutative von Neumann algebra $\mathcal{M}$, we shall assume that $\mathcal{Z}(\mathcal{M})$ is $*$-isomorphic to the algebra $L_\infty(\Omega, \Sigma, \mu)$ associated with a probability space (Ω, Σ, μ). Everywhere below, we reserve the notation φ for a positive $*$-isomorphism between $\mathcal{Z}(\mathcal{M})$ and $L_\infty(\Omega, \Sigma, \mu)$ for a probability space (Ω, Σ, μ). We note that the $*$-isomorphism φ is necessarily normal.

Next, we discuss reduced von Neumann algebras.

Let $\mathcal{M}$ be a von Neumann algebra acting on the Hilbert space H, and let $e \in P(\mathcal{M})$ and $L = e(H)$. For every operator $x \in \mathcal{M}$, define an operator $x_e \in \mathcal{B}(L)$ by setting

$$x_e \xi = ex\xi, \quad \xi \in L.$$

Consider the sets

$$\mathcal{M}_e = \{x_e : x \in \mathcal{M}\}$$

and

$$(\mathcal{M}')_e = \{x'_e : x' \in \mathcal{M}'\}.$$

Theorem 1.8.12 ([147, Sections 3.13–3.15]). *Let $\mathcal{M}$ be a von Neumann algebra in $\mathcal{B}(H)$, $e \in P(\mathcal{M})$. Then the following assertions hold:*

(i) *The $*$-algebras $\mathcal{M}_e$ and $(\mathcal{M}')_e$ are von Neumann algebras acting on the Hilbert space $L = e(H)$;*

(ii) $(\mathcal{M}_e)' = (\mathcal{M}')_e$;

(iii) *The mapping*

$$e\mathcal{M}e \ni x \mapsto x_e \in \mathcal{M}_e$$

is a $*$-isomorphism;

(iv) $\mathcal{Z}(\mathcal{M}_e) = (\mathcal{Z}(\mathcal{M}))_e$, *in particular, if $\mathcal{M}$ is a factor, then $\mathcal{M}_e$ and $\mathcal{M}_{e'}$ are factors, too.*

The von Neumann algebra $\mathcal{M}_e$ is called the *reduced* von Neumann algebra for $\mathcal{M}$ with respect to $e \in P(\mathcal{M})$. We shall frequently identify the $*$-algebra $e\mathcal{M}e$ with the von Neumann algebra $\mathcal{M}_e$ on L. Note that if z is a central projection, then $z\mathcal{M}z = z\mathcal{M} = \mathcal{M}z$.

In conclusion of this section, we recall the construction of the direct sum of von Neumann algebras. For a countable family $\{H_n\}_{n\in\mathbb{N}}$ of Hilbert spaces, consider the direct sum $H = \bigoplus_{n\in\mathbb{N}} H_n$, i. e., the Hilbert space H, defined by setting

$$H = \left\{\{\xi_n\}_{n\in\mathbb{N}} : \xi_n \in H_n \text{ for all } n \in \mathbb{N}, \sum_{n=1}^{\infty} \|\xi_n\|_{H_n}^2 < \infty\right\},$$

with pointwise algebraic operations and inner product given by

$$\langle \{\xi_n\}_{n\in\mathbb{N}}, \{\eta_n\}_{n\in\mathbb{N}} \rangle = \sum_{n=1}^{\infty} \langle \xi_n, \eta_n \rangle_{H_n}.$$

For any $n \in \mathbb{N}$, let $\mathcal{M}_n$ be a von Neumann algebra acting on the Hilbert space H_n. For every family $\{x_n\}_{n\in\mathbb{N}}$ with $x_n \in \mathcal{M}_n$ and $\sup_{n\in\mathbb{N}} \|x_n\|_{\mathcal{M}_n} \leq \infty$, we define a bounded operator x on the Hilbert space $\bigoplus_{n\in\mathbb{N}} H_n$ by setting:

$$x(\{\xi_n\}_{n\in\mathbb{N}}) = \{x_n \xi_n\}_{n\in\mathbb{N}}.$$

The set of all such operators x is a von Neumann algebra in $\mathcal{B}(\bigoplus_{n\in\mathbb{N}} H_n)$, which is called the *direct sum* of von Neumann algebras $\mathcal{M}_n$, $n \in \mathbb{N}$, and denoted by

$$\mathcal{M} = \bigoplus_{n\in\mathbb{N}} \mathcal{M}_n.$$

Each von Neumann algebra $\mathcal{M}_n$ shall be referred to as *direct summand* of $\mathcal{M}$. In the case, when the indexing set I is finite (e. g., $I = \{1, 2, \ldots, n\}$), we shall simply write

$$\mathcal{M}_1 \oplus \mathcal{M}_2 \oplus \cdots \oplus \mathcal{M}_n = \oplus_{i=1}^{n} \mathcal{M}_i.$$

A particularly important type of direct sums of von Neumann algebras is given by direct sums over a central partition of unity.

For any von Neumann algebra $\mathcal{M}$, we say that $\{p_n\}_{n\in\mathbb{N}}$ is a *partition* of a projection $p \in P(\mathcal{M})$ in $\mathcal{M}$, if $p_n p_m = 0$ for $n \neq m$ and $\sup_{n\in\mathbb{N}} p_n = \sum_{n=1}^{\infty} p_n = p$. If $\{z_n\}_{n\in\mathbb{N}}$, a partition of a central projection $z \in P(\mathcal{Z}(\mathcal{M}))$ and $z_n \in P(\mathcal{Z}(\mathcal{M}))$ for all $n \in \mathbb{N}$, then we say $\{z_n\}_{n\in\mathbb{N}}$ is a *central partition* of $z \in P(\mathcal{Z}(\mathcal{M}))$ in $\mathcal{M}$. If $\{z_n\}$ is a central partition of the projection $z = \mathbf{1}$, then we say that $\{z_n\}_{n\in\mathbb{N}}$ is a *central partition of unity* in $\mathcal{M}$. Let $\{z_n\}_{n\in\mathbb{N}}$ be a central partition of unity. For every $n \in \mathbb{N}$, we set $\mathcal{M}_n = \mathcal{M}_{z_n}$ and consider the direct sum $\bigoplus_{n\in\mathbb{N}} \mathcal{M}_n$ of the von Neumann algebras $\mathcal{M}_n$. Define a mapping Ψ from $\mathcal{M}$ into $\bigoplus_{n\in\mathbb{N}} \mathcal{M}_n$ by setting $\Psi(x) = \{z_n x_n\}_{n\in\mathbb{N}}$.

Proposition 1.8.13. *The mapping Ψ is a $*$-isomorphism of the von Neumann algebra $\mathcal{M}$ onto the von Neumann algebra $\bigoplus_{n\in\mathbb{N}} \mathcal{M}_n$.*

1.9 Comparison of projections in a von Neumann algebra

As before, we assume that $\mathcal{M}$ is a von Neumann algebra acting on a Hilbert space H and denote by $P(\mathcal{M})$ the collection of all (orthogonal) projections from $\mathcal{M}$, i. e.,

$$P(\mathcal{M}) = \{p \in \mathcal{M} : p = p^* = p^2\}.$$

Clearly, $P(\mathcal{M}) \subset P(\mathcal{B}(H))$, and furthermore, $P(\mathcal{M})$ is a complete sublattice in $P(\mathcal{B}(H))$, i. e., $P(\mathcal{M})$ is a complete lattice with respect to the lattice operations inherited from $P(\mathcal{B}(H))$.

If the von Neumann algebra $\mathcal{M}$ is commutative, then $p \wedge q = pq$ and $p \vee q = p + q - pq$ for all $p, q \in P(\mathcal{M})$. In this case, $P(\mathcal{M})$ is a complete Boolean algebra, where the complement of each $p \in P(\mathcal{M})$ is given by $p^\perp = \mathbf{1} - p$. In particular, for any von Neumann algebra $\mathcal{M}$, $P(\mathcal{Z}(\mathcal{M}))$ is a complete Boolean algebra and a complete sublattice of $P(\mathcal{M})$.

Projections belonging to the center $\mathcal{Z}(\mathcal{M})$ of $\mathcal{M}$ are called *central projections.*

Definition 1.9.1. For $x \in \mathcal{M}$, the *central support* $s_c(x) \in P(\mathcal{M})$ is defined by

$$s_c(x) = \inf\{z \in P(\mathcal{Z}(\mathcal{M})) : x = xz\}.$$

The following proposition gathers the main properties of central supports.

Proposition 1.9.2. *For any $x \in \mathcal{M}$ and $p, \{p_i\}_{i\in I} \subset P(\mathcal{M})$ we have:*

(i) $s_c(x) = s_c(x^*)$;

(ii) $s_l(x), s_r(x) \le s_c(x)$ *and*

$$\begin{aligned} s_c(x) &= \inf\{z \in P(\mathcal{Z}(\mathcal{M})) : s_r(x) \le z\} \\ &= \inf\{z \in P(\mathcal{Z}(\mathcal{M})) : s_l(x) \le z\}; \end{aligned}$$

(iii)

$$s_c(p) = \inf\{z \in P(\mathcal{Z}(\mathcal{M})) : p \le z\};$$

(iv) $s_c(x) = s_c(s_r(x)) = s_c(s_l(x))$;

(v) $s_c(\bigvee_{i\in I} p_i) = \bigvee_{i\in I} s_c(p_i)$;

(vi) $s_c(p) = \bigvee_{u\in U(\mathcal{M})} upu^*$.

Proof. We prove part (vi) only, as all other properties are straightforward.

We show first that $\bigvee_{u\in U(\mathcal{M})} upu^* \in P(\mathcal{Z}(\mathcal{M}))$. For any $v \in U(\mathcal{M})$, we have that

$$v\Big(\bigvee_{u\in U(\mathcal{M})} upu^*\Big)v^* = \bigvee_{u\in U(\mathcal{M})} (vu)p(vu)^* = \bigvee_{u\in U(\mathcal{M})} upu^*.$$

Since any element in $\mathcal{M}$ can be written as a linear combination of at most four unitaries from $\mathcal{M}$, it follows that $\bigvee_{u\in U(\mathcal{M})} upu^* \in P(\mathcal{Z}(\mathcal{M}))$.

On one hand, the inequality $p \le \bigvee_{u\in U(\mathcal{M})} upu^*$ implies that $s_c(p) \le \bigvee_{u\in U(\mathcal{M})} upu^*$. On the other hand,

$$upu^* \le us_c(p)u^* = s_c(p)uu^* = s_c(p)$$

for any $u \in U(\mathcal{M})$ and, therefore, $\bigvee_{u\in U(\mathcal{M})} upu^* \le s_c(p)$. Thus, $s_c(p) = \bigvee_{u\in U(\mathcal{M})} upu^*$, as required. □

Next, we recall the notions of equivalence and majorization of two projections from $\mathcal{M}$.

Definition 1.9.3. Let $p, q \in P(\mathcal{M})$.

(i) Projections p and q from $P(\mathcal{M})$ are called *equivalent* (relative to the von Neumann algebra $\mathcal{M}$) (notation $p \sim q$), if there exists a partial isometry $v \in \mathcal{M}$ such that p is the initial projection for v and q is the final projection for v, i. e., $v^*v = p$, $vv^* = q$;

(ii) The projection p is said to be *majorized* by q (or, *weaker* than q) (notation $p \precsim q$), if there exists a projection $q_1 \le q$ such that $p \sim q_1$;

(iii) The projection p is said to be *strictly weaker* than q (notation $p \prec q$), if $p \precsim q$ and p is not equivalent to q (written, $p \nsim q$).

We note that if $p \sim q$ with a partial isometry $v \in \mathcal{M}$, such that $v^*v = p$, $vv^* = q$, then

$$vp = v = qv, \quad pv^* = v^* = v^*q, \quad vpv^* = vv^* = q, \quad v^*qv = v^*v = p.$$

The introduced relation "$\sim$" is an equivalence relation on the lattice $P(\mathcal{M})$. If p and q are central projections in $P(\mathcal{M})$ and $v^*v = p$, $vv^* = q$, then

$$q = vpv^* = pvv^* = pq, \quad p = v^*qv = qv^*v = qp,$$

i. e., $p = q$. Consequently, if the von Neumann algebra $\mathcal{M}$ is commutative, then projections $p, q \in P(\mathcal{M})$ are equivalent if and only if $p = q$.

Below we list the main properties of relation $\precsim$ and equivalence relation $\sim$ in the lattice $P(\mathcal{M})$.

Theorem 1.9.4. *Let $\mathcal{M}$ be a von Neumann algebra. Let $p, q \in P(\mathcal{M})$, $z \in P(\mathcal{Z}(\mathcal{M}))$, $x \in \mathcal{M}$. Then:*

(i) *If $p \precsim q$ and $q \precsim f$, then $p \sim q$;*

(ii) *If $p \sim q$, then the central supports $s_c(p)$ and $s_c(q)$ coincide;*

(iii) *If $p \precsim q$, then $pz \precsim qz$, in particular, if $p \sim q$, then $pz \sim qz$;*

(iv) *$s_l(x) \sim s_r(x)$;*

(v) *[Kaplansky's identity]*

$$(p \vee q - q) \sim (p - p \wedge q),$$

in particular, if $p \wedge q = 0$, then $p \precsim q^\perp$;

(vi) *If* $\{p_i\}_{i\in I}$, $\{q_i\}_{i\in I} \subset P(\mathcal{M})$, $p_ip_j = 0$, $q_iq_j = 0$ *for* $i \neq j$, $i,j \in I$, *and* $p_i \precsim q_i$ *for all* $i \in I$, *then* $\sup_{i\in I} p_i \precsim \sup_{i\in I} q_i$. *In particular, if* $p_i \sim q_i$ *then* $\sup_{i\in I} p_i \sim \sup_{i\in I} q_i$;

(vii) $p\mathcal{M}q \neq \{0\}$ *if and only if there exist non-zero projections* $p_1, q_1 \in P(\mathcal{M})$, *such that* $p_1 \leq p$, $q_1 \leq q$ *and* $p_1 \sim q_1$;

(viii) *If* $\mathcal{M}$ *is a factor, then either* $p \precsim q$ *or* $q \precsim p$.

We also recall the following important comparison theorem.

Theorem 1.9.5 (Comparison theorem, [93, Chapter 6, Section 2, Theorem 6.2.7]). *Let* $p, q \in P(\mathcal{M})$. *Then there exist unique orthogonal central projections* $z_1, z_2 \in P(\mathcal{Z}(\mathcal{M}))$ *maximal with respect to the properties* $z_2p \sim z_2q$, *and if* z_0 *is a nonzero central subprojection of* z_1, *then* $z_0p \prec z_0q$. *If* r_0 *is a nonzero central subprojection of* $\mathbf{1} - z_1 - z_2$, *then* $r_0q \prec r_0p$.

Let $\mathcal{M}$ be a von Neumann algebra and let e be a projection from $P(\mathcal{M})$. Then p is called *minimal* (or an *atom*) if the condition $0 \neq q \leq p$, $q \in P(\mathcal{M})$ implies that $q = p$. For every atom $p \in P(\mathcal{M})$, the equality $p\mathcal{M}p = \mathbb{C}p$ always holds.

A von Neumann algebra $\mathcal{M}$ is called *atomic* if for every nonzero projection q from $\mathcal{M}$ there exists an atom $p \leq q$. An example of an atomic von Neumann algebra is the algebra $\mathcal{B}(H)$.

A commutative von Neumann algebra $\mathcal{M}$ is atomic if and only if the Boolean algebra $P(\mathcal{M})$ of all projections from $\mathcal{M}$ is an atomic Boolean algebra. If (Ω, Σ, μ) is a σ-finite measure space (Ω, Σ, μ), such that $\mathcal{M}$ is $*$-isomorphic to $L_\infty(\Omega, \Sigma, \mu)$ (see Theorem 1.8.10), then $\mathcal{M}$ is atomic if and only if (Ω, Σ, μ) is an atomic measure space.

Let $\mathcal{M}$ be an arbitrary von Neumann algebra, Δ be the set of all atoms of the lattice $P(\mathcal{M})$ (the set Δ can be empty). Then the projection $z_0 = \bigvee_{p\in\Delta} p$ is central. Indeed, if $e \in \Delta$ then $u^*eu \in \Delta$ for any $u \in U(\mathcal{M})$. Therefore, by Proposition 1.9.2(vi) we have that $z_0 = \bigvee_{e\in\Delta} \bigvee_{u\in U(\mathcal{M})} u^*eu = \bigvee_{e\in\Delta} s_c(e)$, showing that z_0 is a central projection. Furthermore, we have that $\mathcal{M} = z_0\mathcal{M} \oplus (\mathbf{1} - z_0)\mathcal{M}$. Clearly, the von Neumann algebra $z_0\mathcal{M}$ is atomic and the von Neumann algebra $(\mathbf{1} - z_0)\mathcal{M}$ has no atoms. Thus, we have the following.

Proposition 1.9.6. *Every von Neumann algebra* $\mathcal{M}$ *is direct sum of an atomic von Neumann algebra and a von Neumann algebra with no atoms (one of the summands may vanish).*

A projection $p \in P(\mathcal{M})$ is called *abelian* if the reduced algebra $p\mathcal{M}p$ is commutative. It is trivial that the zero projection is abelian. Every atom in $P(\mathcal{M})$ is an abelian projection. In a commutative von Neumann algebra, all projections are abelian.

The majorization relation $\precsim$ allows us to distinguish finite and infinite projections in a way similar to finite and infinite sets.

Definition 1.9.7.

(i) A projection $p \in P(\mathcal{M})$ is called *finite* (relative to $\mathcal{M}$) if $q \leq p$, $q \in P(\mathcal{M})$, $p \sim q$ implies that $p = q$;

(ii) A projection $p \in P(\mathcal{M})$ is called *infinite* if p is not finite;

(iii) A projection $p \in P(\mathcal{M})$ is called *properly infinite* if $p \neq 0$ and for every $z \in P(\mathcal{Z}(\mathcal{M}))$ the projection zp is either infinite or zero.

Every abelian projection is finite.

Definition 1.9.8.
(i) A von Neumann algebra $\mathcal{M}$ is said to be *finite* or *infinite* or *properly infinite* if the identity projection $\mathbf{1}$ is finite, respectively, infinite or properly infinite;
(ii) A von Neumann algebra is said to be *semifinite* if every nonzero projection $q \in \mathcal{M}$ there exists a nonzero finite projection $p \in P(\mathcal{M})$, such that $p \leq q$.

It is said that a projection $p \in P(\mathcal{M})$ is of *countable type* if every family of nonzero pairwise orthogonal projections from the reduced algebra $p\mathcal{M}p$ is at most countable. If the unit $\mathbf{1}$ of the von Neumann algebra, $\mathcal{M}$ is of countable type, then the algebra $\mathcal{M}$ is called a von Neumann algebra of *countable type* or *σ-finite* von Neumann algebra. Clearly, any von Neumann algebra $\mathcal{M}$ acting on a separable Hilbert space H is σ-finite. Since we always assume that the Hilbert space H is separable, $\mathcal{M}$ is necessarily σ-finite.

Theorem 1.9.9. *Let $\mathcal{M}$ be a von Neumann algebra acting on a separable Hilbert space H and let $p, q \in P(\mathcal{M})$.*
(i) *If $p \precsim q$ and q is finite, then p is finite, too;*
(ii) *If projections p, q are finite, then the projection $p \vee q$ is also finite;*
(iii) *If p is properly infinite and $q \sim p$, then q is properly infinite, too;*
(iv) *If $\mathbf{1}$ is properly infinite, then there exists $p_0 \in P(\mathcal{M})$, such that $p_0 \sim (\mathbf{1} - p_0) \sim \mathbf{1}$;*
(v) *A projection p is properly infinite if and only if there exists a sequence $\{p_n\}_{n\in\mathbb{N}} \subseteq P(\mathcal{M})$ of pairwise orthogonal projections, such that $p = \sup_{n\in\mathbb{N}} p_n$ and $p_n \sim p$ for all $n = 1, 2, \ldots$;*
(vi) *If p is infinite, then there exists a unique central projection $z \in P(\mathcal{Z}(\mathcal{M}))$, such that $z \leq s_c(p)$, pz is properly infinite, and $p(\mathbf{1} - z)$ is finite;*
(vii) *Let $\{z_i\}_{i\in I}$ be a collection of central projections and $z = \sup_{i\in I} z_i$. If $z_i p$ is finite for all $i \in I$, then zp is finite;*
(viii) *If q is a properly infinite projection, such that $s_c(p) \precsim s_c(q)$, then $p \precsim q$.*

We denote

$$P_{\text{fin}}(\mathcal{M}) := \{p \in P(\mathcal{M}) : p \text{ is finite projection (in } \mathcal{M})\}.$$

By Theorem 1.9.9(ii), $P_{\text{fin}}(\mathcal{M})$ is a sublattice in $P(\mathcal{M})$.

We mention a particular result, which follows from Theorem 1.9.9(iv).

Proposition 1.9.10. *Let $\mathcal{M}$ be a properly infinite von Neumann algebra.*
(i) *If $p \in P_{\text{fin}}(\mathcal{M})$, then $\mathbf{1} - p$ is a properly infinite projection and the central support of $\mathbf{1} - p$ is $\mathbf{1}$;*

(ii) *For every projection $p \in \mathcal{M}$, there exists a central projection $z \in \mathcal{M}$, such that $pz \sim z$ and $(\mathbf{1} - p)(\mathbf{1} - z) \sim (\mathbf{1} - z)$.*

Proof. (i) Suppose, by contradiction, that the projection $\mathbf{1} - p$ is not properly infinite. Then there exists $z \in P(\mathcal{Z}(\mathcal{M}))$, such that $0 \neq z - zp = z(\mathbf{1} - p) \in P_{\text{fin}}(\mathcal{M})$. In this case, $0 \neq z = zp + (z - zp) \in P_{\text{fin}}(\mathcal{M})$, which contradicts with the assumption that $\mathcal{M}$ is a properly infinite von Neumann algebra.

Suppose that the central support $s_c(\mathbf{1} - p)$ of $\mathbf{1} - p$ is not $\mathbf{1}$. Setting $z = \mathbf{1} - s_c(\mathbf{1} - p)$, we have that $z(\mathbf{1} - p) = 0$, $z \neq 0$. Therefore, $z = zp + (z - zp) \in P_{\text{fin}}(\mathcal{M})$, which again contradicts with the assumption that $\mathcal{M}$ is a properly infinite von Neumann algebra.

(ii) By Theorem 1.9.5, there exists a central projection z such that $(\mathbf{1} - p)z \precsim pz$ and $p(\mathbf{1}-z) \precsim (\mathbf{1}-p)(\mathbf{1}-z)$. We claim that $z \leq s_c(p)$. We define the projection $z' := z - zs_c(p) \in P(\mathcal{Z}(\mathcal{M}))$. Since $(\mathbf{1} - p)z \precsim pz$, Theorem 1.9.4(iii) and definition of the central support projection (see Definition 1.9.1) implies that

$$(\mathbf{1} - p)z' \precsim pz' = pz - ps_c(p)z = pz - pz = 0.$$

Thus, $(\mathbf{1} - p)z' = 0$, or equivalently $z' = pz'$. By the definition of the projection z', the latter equality implies that

$$z - zs_c(p) = z' = p(z - s_c(p)z) = pz - ps_c(p)z = pz - pz = 0,$$

i. e., $z = zs_c(p)$, which proves the claim.

Next, we claim that pz is either a properly infinite projection or $z = 0$. Let $0 \leq z_1 \leq z = s_c(pz)$, $z_1 \in P(\mathcal{Z}(\mathcal{M}))$ be such that pz_1 is finite. Since $\mathcal{M}$ is a properly infinite algebra, it follows that $z_1\mathcal{M}$ is also properly infinite. Therefore, by part (i) we have that the projection $z_1 - pz_1 = (\mathbf{1} - p)z_1$ is properly infinite. Since $(\mathbf{1} - p)z_1 \precsim pz_1$ and pz_1 is finite, this is a contradiction (see Theorem 1.9.9(i)). Therefore, the projection pz is either properly infinite or $pz = 0$. Since $z \leq s_c(p)$, the equality $pz = 0$ implies that $z = 0$. Thus, either the projection pz is properly infinite or $z = 0$, as claimed.

Similarly, the projection $(\mathbf{1} - p)(\mathbf{1} - z)$ is properly infinite or $\mathbf{1} - z = 0$.

By Theorem 1.9.9(iv) applied to the algebra $pz\mathcal{M}pz$, there exist projections $p_1, p_2 \leq pz$ such that $p_1 \sim p_2 \sim pz$, $p_1p_2 = 0$, and $p_1 + p_2 = pz$. We have $(\mathbf{1} - p)z \precsim pz \sim p_1$ and $pz \sim p_2$. By Theorem 1.9.4(vii), we have

$$(\mathbf{1} - p)z + pz \precsim p_1 + p_2,$$

which implies that $z \precsim pz$. Since $pz \leq z$, Theorem 1.9.4(i) implies that $pz \sim z$.

Similarly, one can prove that $(\mathbf{1} - p)(\mathbf{1} - z) \sim \mathbf{1} - z$. □

Lemma 1.9.11. *Suppose that $\mathcal{M}$ is a semifinite von Neumann algebra and let $z \in P(\mathcal{Z}(\mathcal{M}))$ be nonzero. There exists a finite projection $p \in P(\mathcal{M})$ such that $s_c(p) = z$.*

Proof. By Zorn's lemma, there exists a maximal system of finite projections $\Gamma = \{g_i : i \in I\}$, such that $s_c(g_i) \le z$ and $s_c(g_i)$ are pairwise orthogonal. Suppose that $z - \bigvee_{i \in I} s_c(g_i) \neq 0$. Since $\mathcal{M}$ is semifinite, there exists a nonzero finite projection q, such that $q \le z - \bigvee_{i \in I} s_c(g_i)$. Then $s_c(q) \le z$ and $s_c(q)s_c(q_i) = 0$ for all $i \in I$, i. e., $q \in \Gamma$. This contradicts the maximality of Γ. Hence, $z = \bigvee_{i \in I} s_c(g_i)$. We set $p = \bigvee_{i \in I} g_i$. Then, by Theorem 1.9.9(vii) p is finite and, by Proposition 1.9.2(v) we have that $s_c(g) = \bigvee_{i \in I} s_c(q_i) = z$, as required. □

Lemma 1.9.12. *Let p be a finite projection and q be a properly infinite projection. If the central support of q is $\mathbf{1}$, then $p \prec q$.*

Proof. By Theorem 1.9.5, there exists a central projection z such that $pz \prec qz$ and $q(\mathbf{1} - z) \prec p(\mathbf{1} - z)$. Since p is a finite projection, then so are $p(\mathbf{1} - z)$ and $q(\mathbf{1} - z)$. Since q is properly infinite, it follows that $q(\mathbf{1} - z) = 0$. Since the central support of q is $\mathbf{1}$, it follows that $z = \mathbf{1}$. Thus, $p \prec q$. □

Now we move onto types of von Neumann algebras.

Definition 1.9.13. A von Neumann algebra $\mathcal{M}$ is said to be:
- (i) *of type I* (or *discrete*) if for every nonzero central projection $z \in \mathcal{M}$ there exists a nonzero abelian projection $p \in P(\mathcal{M})$, such that $p \le z$;
- (ii) *of type II* if $\mathcal{M}$ is a semifinite algebra containing no nonzero abelian projections;
- (iii) *of type III* (or *purely infinite*) if $\mathcal{M}$ does not contain nonzero finite projections;
- (iv) *of type I_{fin}* if $\mathcal{M}$ is a finite algebra of type I;
- (v) *of type I_∞* if $\mathcal{M}$ is a properly infinite algebra of type I;
- (vi) *of type II_1* if $\mathcal{M}$ is a finite algebra of type II;
- (vii) *of type II_∞* if $\mathcal{M}$ is a properly infinite algebra of type II (equivalently, if $\mathcal{M}$ is a type II algebra with no nonzero central finite projections).

It is clear that von Neumann algebras of type II and III have no atoms, and are sometimes referred to as *continuous* von Neumann algebras.

A fundamental result in the theory of von Neumann algebras is that any von Neumann algebra can be decomposed into a direct sum of algebras of respective type.

Theorem 1.9.14. *Every von Neumann algebra $\mathcal{M}$ contains uniquely defined pairwise orthogonal central projections z_i, $i = 1, \ldots, 5$, such that:*
- (i) $\sum_{i=1}^{5} z_i = \mathbf{1}$;
- (ii) *$z_1\mathcal{M}$ is a von Neumann algebra of type I_{fin};*
- (iii) *$z_2\mathcal{M}$ is a von Neumann algebra of type I_∞;*
- (iv) *$z_3\mathcal{M}$ is a von Neumann algebra of type II_1;*
- (v) *$z_4\mathcal{M}$ is a von Neumann algebra of type II_∞;*
- (vi) *$z_5\mathcal{M}$ is a von Neumann algebra of type III.*

By Theorem 1.9.14, every von Neumann algebra $\mathcal{M}$ is of the form

$$\mathcal{M} = \bigoplus_{i=1}^{5} \mathcal{M}_i, \quad \mathcal{M}_i = z_i\mathcal{M}, i = 1, \dots, 5,$$

where every algebra $\mathcal{M}_1, \mathcal{M}_2, \mathcal{M}_3, \mathcal{M}_4, \mathcal{M}_5$ is a von Neumann algebra of respective type $I_{\text{fin}}, I_\infty, II_1, II_\infty, III$ (some summand may vanish). In this case, it is said that the algebra $\mathcal{M}_1$ (resp., $\mathcal{M}_2, \mathcal{M}_3, \mathcal{M}_4, \mathcal{M}_5$) is a direct summand of type I_{fin} (resp., $I_\infty, II_1, II_\infty, III$) of the algebra $\mathcal{M}$.

Corollary 1.9.15.
(i) *If a von Neumann algebra $\mathcal{M}$ is factor, then it is only one of the following types: I_{fin}, I_∞, II_1, II_∞, III;*
(ii) *If $\mathcal{M}$ is an atomic von Neumann algebra, then $\mathcal{M}$ is $*$-isomorphic to the factor $\mathcal{B}(H)$ for some Hilbert space H;*
(iii) *If a von Neumann algebra $\mathcal{M}$ is finite, then it is of the form*

$$\mathcal{M} = \mathcal{M}_1 \oplus \mathcal{M}_2,$$

where $\mathcal{M}_1$ is a von Neumann algebra of type I_{fin} and $\mathcal{M}_2$ is a von Neumann algebra of type II_1 (one of the summands may vanish);
(iv) *If a von Neumann algebra $\mathcal{M}$ is properly infinite, then it is of the form*

$$\mathcal{M} = \mathcal{M}_1 \oplus \mathcal{M}_2 \oplus \mathcal{M}_3,$$

where $\mathcal{M}_1$ is a von Neumann algebra of type I_∞, $\mathcal{M}_2$ is a von Neumann algebra of type II_∞ and $\mathcal{M}_3$ is a von Neumann algebra of type III (some summands may vanish).

We now give a further description of type I_{fin} von Neumann algebras.

A von Neumann algebra $\mathcal{M}$ is called *type I_n*, where n is a finite or infinite cardinal number, if there exists an orthogonal family $\{p_i\}_{i=1}^n$ of abelian projections of $\mathcal{M}$, such that $p_i \sim p_j$ and $s_c(p_j) = \mathbf{1}$ for all $i, j = 1, \dots, n$, and $\sup_{i \in I} p_i = \mathbf{1}$. By [149, Chapter V, Section 1, Lemma 1.26], the definition of type I_n von Neumann algebra is independent of the choice of $\{p_i\}_{i=1}^n$.

Let $\mathcal{A}$ be an arbitrary commutative von Neumann algebra. Denote by $M_{n\times n}(\mathcal{A})$ the $*$-algebra of all $n \times n$-matrices $(x_{ij})_{i,j=1}^n$ with $x_{ij} \in \mathcal{A}$, $i, j = 1, \dots, n$. As shown in [149, Chapter V, Section 1], the $*$-algebra $M_{n\times n}(\mathcal{A})$ is a von Neumann algebra of type I_n. In fact, any type I_n von Neumann algebra is of this form.

Proposition 1.9.16 ([149, Chapter V, Section 1]).
(i) *If $\mathcal{M}$ is a von Neumann algebra of type I_n, then $\mathcal{M}$ is of type I;*
(ii) *Any von Neumann algebra $\mathcal{M}$ of type I_n, $n \in \mathbb{N}$ is $*$-isomorphic to the von Neumann algebra $M_{n\times n}(\mathcal{Z}(\mathcal{M}))$.*

The following theorem presents the full description of von Neumann algebras of type I_{fin}.

Theorem 1.9.17 ([149, Chapter V, Section 1]). *Let $\mathcal{M}$ be a von Neumann algebra of type I_{fin}. Then there exist a unique family E of natural numbers and a central partition of unity $\{z_n\}_{n\in E}$, such that $z_n\mathcal{M}$ is a von Neumann algebra of type I_n for every $n \in E$, i. e., the von Neumann algebra $\mathcal{M}$ is $*$-isomorphic to the direct sum*

$$\bigoplus_{n\in E} z_n\mathcal{M} \cong \bigoplus_{n\in E} (z_n M_{n\times n}(z_n \mathcal{Z}(\mathcal{M}))).$$

1.10 Traces on a von Neumann algebra

In this section, we recall the notions of center-valued and scalar-valued traces on a von Neumann algebra. We start with center-valued traces. As before, we let $\mathcal{M}$ be a von Neumann algebra, with center $\mathcal{Z}(\mathcal{M})$, acting on a separable Hilbert space H.

Definition 1.10.1. A center-valued trace on $\mathcal{M}$ is a linear mapping $\mathcal{T} : \mathcal{M} \to \mathcal{Z}(\mathcal{M})$, such that:

(i) $\mathcal{T}(x^*x) = \mathcal{T}(xx^*) \geq 0$ for all $x \in \mathcal{M}$;
(ii) $\mathcal{T}(zx) = z\mathcal{T}(x)$ for all $z \in \mathcal{Z}(\mathcal{M})$ and $x \in \mathcal{M}$;
(iii) $\mathcal{T}(\mathbf{1}) = \mathbf{1}$;
(iv) If $x \in \mathcal{M}$, $x \neq 0$, then $\mathcal{T}(x^*x) \neq 0$.

As the following theorem shows, existence of a center-valued trace $\mathcal{T}$ on a von Neumann algebra $\mathcal{M}$ is a necessary and sufficient condition for $\mathcal{M}$ to be finite.

Theorem 1.10.2 ([149, Chapter V, Section 2]). *Let $\mathcal{M}$ be a von Neumann algebra. The following conditions are equivalent:*

(i) *$\mathcal{M}$ is finite;*
(ii) *There exists a center-valued trace $\mathcal{T} : \mathcal{M} \to \mathcal{Z}(\mathcal{M})$.*

In this case, the center-valued trace $\mathcal{T}$ is unique and continuous in the ultraweak operator topology and normal.

Since the center-valued trace on a finite von Neumann algebra $\mathcal{M}$ is unique it is called the *canonical center-valued trace.*

Proposition 1.10.3 ([149, Chapter V, Section 2]). *For the canonical center- valued trace $\mathcal{T}$ on a finite von Neumann algebra $\mathcal{M}$ and any projections $p, q \in \mathcal{M}$, the following are equivalent:*

(i) *$p \precsim q$;*
(ii) *$\mathcal{T}(p) \leq \mathcal{T}(q)$.*

By Theorem 1.10.2, a nonfinite von Neumann algebra $\mathcal{M}$ does not admit a center-valued trace. However, when $\mathcal{M}$ is a semifinite von Neumann algebra it admits a so-called extended center-valued trace. Before we introduce this notion, we recall some additional terminology and notation.

Let $\mathcal{M}$ be a von Neumann algebra acting on a separable Hilbert space H. By Remark 1.8.11, there exists a probability space (Ω, Σ, μ) and a $*$-isomorphism φ from the commutative von Neumann algebra $\mathcal{Z}(\mathcal{M})$ onto $L_\infty(\Omega, \Sigma, \mu)$. Denote by $L^+(\Omega, \Sigma, \mu)$ the set of all measurable real-valued functions defined on (Ω, Σ, μ) and taking values in the extended half-line $[0, \infty]$ (functions that are equal almost everywhere are identified).

Let $\mathrm{Sp}(\mathcal{Z}(\mathcal{M}))$ be the spectrum of the commutative von Neumann algebra $\mathcal{Z}(\mathcal{M})$ (see, e. g., [149, Chapter I]) and let $C(\mathrm{Sp}(\mathcal{Z}(\mathcal{M})))$ be the algebra of all continuous functions on $\mathrm{Sp}(\mathcal{Z}(\mathcal{M}))$. As noted [149, Chapter III, Section 1.11], the measure space (Ω, Σ, μ) can be considered as an open dense subset of the spectrum $\mathrm{Sp}(\mathcal{Z}(\mathcal{M}))$ with a Radon measure. In this case, the complement of any open dense subset of Ω has measure zero [149, Chapter III, Proposition 1.11].

Denoting by γ the isomorphism of the positive cones $L_\infty(\Omega, \Sigma, \mu)_+$ and $C(\mathrm{Sp}(\mathcal{Z}(\mathcal{M})))_+$, we can extend γ up to an isomorphism of the positive cone $L^+(\Omega, \Sigma, \mu)$ and the positive cone of all positive continuous $[0, +\infty]$-valued functions on $\mathrm{Sp}(\mathcal{Z}(\mathcal{M}))$. We will denote this extended isomorphism by γ, too.

Lemma 1.10.4. *Let $f \in L^+(\Omega, \Sigma, \mu)$. Then f has a. e. finite values if and only if $\gamma(f)$ takes finite values on an everywhere dense set.*

Proof. Assume first that f has a. e. finite values. We define $A_n = \{\omega \in \Omega : f(\omega) < n\}$. Then $\bigcup_n A_n$ is a set of full measure in Ω, i. e., $\bigcup_n \chi_{A_n} = \mathbf{1}$ in $L_\infty(\Omega, \Sigma, \mu)$. Hence, $\bigvee_n \gamma(\chi_{A_n}) = \mathbf{1}$ in $C(\mathrm{Sp}(\mathcal{Z}(\mathcal{M})))$. Let $U_n \subset \mathrm{Sp}(\mathcal{Z}(\mathcal{M}))$ be a clopen set such that $\chi_{U_n} = \gamma(\chi_{A_n})$. Since $\bigvee_n \chi_{U_n} = \mathbf{1}$, then $\bigcup_n \chi_{U_n}$ is an open dense set in $\mathrm{Sp}(\mathcal{Z}(\mathcal{M}))$. Since $\gamma(f)\gamma(\chi_{A_n}) \le n$, we also have that $\gamma(f)|_{U_n} \le n$ for any n. Therefore, $\gamma(f)$ takes finite values on an everywhere dense set, as required.

Conversely, assume that $\gamma(f)$ takes finite values on an everywhere dense set. Setting $V_n = \{x \in \mathrm{Sp}(\mathcal{Z}(\mathcal{M})) : \gamma(f)(x) < n\}$, $n \in \mathbb{N}$, we have that V_n is open. Since $\mathrm{Sp}(\mathcal{Z}(\mathcal{M}))$ is a Stonean space, it follows that $U_n := \overline{V_n}$ is open and $\gamma(f)|_{U_n} \le n$. Since $\bigcup_n U_n$ is open and dense in $\mathrm{Sp}(\mathcal{Z}(\mathcal{M}))$, it follows that $\bigvee_n \chi_{U_n} = \mathbf{1}$ in $C(\mathrm{Sp}(\mathcal{Z}(\mathcal{M})))$ and, therefore, $\bigvee_n \gamma^{-1}(\chi_{U_n}) = \mathbf{1}$ in $L_\infty(\Omega, \Sigma, \mu)$. Noting that $f(\gamma^{-1}(\chi_{U_n})) \le n$ for any $n \in \mathbb{N}$, we conclude that f has a. e. finite values. □

Definition 1.10.5. Let $\mathcal{M}$ be a von Neumann algebra with a fixed $*$-isomorphism φ between $\mathcal{Z}(\mathcal{M})$ and $L_\infty(\Omega, \Sigma, \mu)$. A map $\mathcal{T} : \mathcal{M}_+ \to L^+(\Omega, \Sigma, \mu)$ is called an *extended center-valued trace* if T satisfies:

(i) $\mathcal{T}(x + y) = \mathcal{T}(x) + \mathcal{T}(y)$, $x, y \in \mathcal{M}_+$;

(ii) $\mathcal{T}(ax) = \varphi(a)\mathcal{T}(x)$, $a \in \mathcal{Z}(\mathcal{M})_+$, $x \in \mathcal{M}_+$;

(iii) $\mathcal{T}(x^*x) = \mathcal{T}(xx^*)$, $x \in \mathcal{M}$.

Furthermore, an extended center-valued trace $\mathcal{T} : \mathcal{M}_+ \to L^+(\Omega, \Sigma, \mu)$ is said to be
- *normal* if $\mathcal{T}(\sup x_i) = \sup \mathcal{T}(x_i)$ for any bounded increasing net $\{x_i\}_{i \in I} \subset \mathcal{M}_+$;
- *faithful* if $\mathcal{T}(x^*x) = 0$ for some $x \in \mathcal{M}$ implies that $x = 0$;
- *semifinite* if $\{x \in \mathcal{M} : \mathcal{T}(x^*x) \in L_\infty(\Omega, \Sigma, \mu)\}$ is an ultraweakly dense ideal $\mathcal{M}$.

While the existence of a center-valued trace on a given von Neumann algebra $\mathcal{M}$ is equivalent to the assumption that $\mathcal{M}$ is finite, as the following result shows the existence of a faithful normal semifinite extended center-valued trace characterizes semifinite von Neumann algebras. Furthermore, if an extended center-valued trace exists, then it is essentially unique.

Theorem 1.10.6 ([149, Chapter V, Theorem 2.34]). *For a von Neumann algebra $\mathcal{M}$, the following statements are equivalent:*
(i) *$\mathcal{M}$ is semifinite;*
(ii) *$\mathcal{M}$ admits a faithful normal semifinite extended center-valued trace.*

If this is the case, then a faithful normal semifinite extended center-valued trace is unique up to a multiplication by $L^+(\Omega, \Sigma, \mu)$ in the sense that if $\mathcal{T}_1$ and $\mathcal{T}_2$ are two such extended center- valued traces then there exists $c \in L^+(\Omega, \Sigma, \mu)$, such that

$$\mathcal{T}_1(x) = c\mathcal{T}_2(x), \quad x \in \mathcal{M}_+,$$
$$0 < c(w) < +\infty \quad a.\,e.$$

Combining Lemma 1.10.4 and [149, Chapter V, Proposition 2.35], we obtain the following criteria for a projection in a semifinite von Neumann algebra to be finite.

Proposition 1.10.7. *Let $\mathcal{M}$ be a semifinite von Neumann algebra. Let $\mathcal{T}$ be a faithful semifinite normal extended center-valued trace of $\mathcal{M}$. A projection $e \in \mathcal{M}$ is finite if and only if $\mathcal{T}(e)$ takes finite values a. e.*

Next, we recall (scalar valued) traces on a von Neumann algebra. As before, let $\mathcal{M}$ be a von Neumann algebra and let $\mathcal{M}_+$ be the set of all positive elements from $\mathcal{M}$ and $U(\mathcal{M})$ be the set of all unitary operators from $\mathcal{M}$.

Definition 1.10.8. A functional $\tau : \mathcal{M}_+ \to [0, \infty]$ is called a (scalar valued) *trace* on $\mathcal{M}_+$, if:
(i) $\tau(x + y) = \tau(x) + \tau(y)$ for all $x, y \in \mathcal{M}_+$;
(ii) $\tau(\lambda x) = \lambda\tau(x)$ for every $x \in \mathcal{M}_+$ and $\lambda \geq 0$ (here $0 \cdot (+\infty) = 0$);
(iii) $\tau(u^*xu) = \tau(x)$ for all $x \in \mathcal{M}_+$ and $u \in U(\mathcal{M})$.

If $\tau : \mathcal{M}_+ \to [0, \infty]$ is a trace, then it follows immediately from (i) in the above definition that $\tau(x) \leq \tau(y)$ whenever $x \leq y$ in $\mathcal{M}_+$. Furthermore, assumption (iii) in the definition of the trace can be replaced by the requirement that $\tau(x^*x) = \tau(xx^*)$ for all $x \in \mathcal{M}$. In particular, for $p, q \in P(\mathcal{M})$ we have that

$$p \precsim q \Rightarrow \tau(p) \le \tau(q).$$

The latter inequality combined with the Kaplansky's identity (see Theorem 1.9.4(v)) implies that

$$\tau(p \vee q) \le \tau(p) + \tau(q), \quad p, q \in P(\mathcal{M}). \tag{1.5}$$

Definition 1.10.9. A trace $\tau : \mathcal{M}_+ \to [0, \infty]$ is said to be
- *finite* if $\tau(x) < \infty$ for every $x \in \mathcal{M}_+$;
- a *tracial state* if $\tau(\mathbf{1}) = 1$;
- *semifinite* if

$$\tau(x) = \sup\{\tau(y) : y \le x, \tau(y) < \infty\}$$

  for every $x \in \mathcal{M}_+$;
- *faithful* if $\tau(x) = 0$, $x \in \mathcal{M}_+$, implies that $x = 0$;
- *normal* if the conditions $x_i \uparrow x$, $x, x_i \in \mathcal{M}_+$, $i \in I$, imply that $\tau(x_i) \uparrow \tau(x)$.

Note that a normal trace $\tau : \mathcal{M}_+ \to [0, \infty]$ is semifinite if and only if for every $0 < x \in \mathcal{M}_+$ there exists $y \in \mathcal{M}_+$ such that $0 < y \le x$ and $\tau(y) < \infty$.

We now recall the simplest examples of normal faithful semifinite traces.

Example 1.10.10 (Canonical trace on $\mathcal{B}(H)$). *As before, let H be a separable Hilbert space and let $\{\xi_n\}_{n \in \mathbb{N}}$ be an orthonormal basis of the Hilbert space H. For every $x \in \mathcal{B}_+(H)$, we set*

$$\mathrm{tr}(x) = \sum_{n=1}^{\infty} \langle x\xi_n, \xi_n \rangle.$$

The functional tr *is a faithful normal semifinite trace on $\mathcal{B}_+(H)$ and does not depend on a choice of orthonormal basis $\{\xi_n\}_{n \in \mathbb{N}}$. This trace is called the* canonical trace *on $\mathcal{B}(H)$.*

Example 1.10.11 (Commutative von Neumann algebra). *Let (Ω, Σ, μ) be a σ-finite measure space and let $\mathcal{M} = \{M_f : f \in L_\infty(\Omega, \Sigma, \mu)\}$ be a commutative von Neumann algebra acting on the Hilbert space $L_2(\Omega, \Sigma, \mu)$ by multiplication on essentially bounded functions.*

A linear functional $\tau : \mathcal{M}_+ \to [0, \infty]$ defined by

$$\tau(M_f) = \int_\Omega f d\mu,$$

is a faithful normal semifinite trace on $\mathcal{M}_+$. The trace τ is finite if and only if the measure μ is finite. The trace τ is a (tracial) state if and only if $\mu(\Omega) = 1$.

Note that since (Ω, Σ, μ) is a σ-finite measure space, there always exists a state on $\mathcal{M}_+$. Indeed, as noted in Remark 1.8.11 *there exists a probability measure ν equivalent to μ. Then $\tau : \mathcal{M}_+ \to [0, \infty)$ defined by*

$$\tau(M_f) = \int_{\Omega} f dv,$$

is a tracial state.

The traces on $L_\infty(\Omega, \Sigma, \mu)$ given by integration give rise to traces on $\mathcal{Z}(\mathcal{M})$ for any von Neumann algebra $\mathcal{M}$.

Example 1.10.12. *Let $\mathcal{M}$ be an arbitrary von Neumann algebra (acting on a separable Hilbert space). Let φ be a $*$-isomorphism of the algebra $\mathcal{Z}(\mathcal{M})$ onto the algebra $L_\infty(\Omega, \Sigma, \mu)$ for an appropriate probability space (Ω, Σ, μ) (see Remark* 1.8.11). *Then $\psi : \mathcal{Z}(\mathcal{M})_+ \to [0, \infty]$ defined by*

$$\psi(x) = \int_{\Omega} \varphi(x) d\mu,$$

is a (tracial) state on $\mathcal{Z}(\mathcal{M})_+$.

The following theorem classifies finite and semifinite von Neumann algebras in terms of (scalar valued) traces.

Theorem 1.10.13 ([149, Chapter V, Section 2]). *Let $\mathcal{M}$ be a von Neumann algebra.*
(i) *$\mathcal{M}$ is finite if and only if for every nonzero operator $x \in \mathcal{M}_+$ there exists a finite trace τ such that $\tau(x) \neq 0$;*
(ii) *$\mathcal{M}$ is semifinite if and only if there exists a faithful normal semifinite trace τ on $\mathcal{M}_+$.*

Let $\mathcal{M}$ be a semifinite von Neumann algebra and let τ be a faithful normal semifinite trace on $\mathcal{M}_+$. The sets

$$\mathfrak{N}_\tau = \{x \in \mathcal{M} : \tau(x^* x) < \infty\}$$

and

$$\mathfrak{M}_\tau = \left\{ \sum_{i=1}^{n} x_i y_i : x_i, y_i \in \mathfrak{N}_\tau, i = 1, 2, \ldots, n, n \in \mathbb{N} \right\}$$

are two-sided $*$-ideals in $\mathcal{M}$ [149, Chapter V, Section 2]. Furthermore the following theorem holds.

Theorem 1.10.14 ([149, Chapter V, Section 2]).
(i) *$\mathfrak{M}_\tau \cap \mathcal{M}_+ = \{x \in \mathcal{M}_+ : \tau(x) < \infty\}$;*
(ii) *$\mathfrak{M}_\tau$ is a linear span of the set $\mathfrak{M}_\tau \cap \mathcal{M}_+$;*
(iii) *The trace τ extends from $\mathfrak{M}_\tau \cap \mathcal{M}_+$ up to a linear functional $\tilde{\tau}$ on $\mathfrak{M}_\tau$, which satisfies the following conditions:*
  (1) *$\tilde{\tau}(x^*) = \overline{\tilde{\tau}(x)}$ for every $x \in \mathfrak{M}_\tau$;*
  (2) *$\tilde{\tau}(ax) = \tilde{\tau}(xa)$ for every $a \in \mathcal{M}$ and $x \in \mathfrak{M}_\tau$;*

(3) $\tilde{\tau}(xy) = \tilde{\tau}(yx)$ *for every* $x, y \in \mathfrak{N}_\tau$;
(4) $\tilde{\tau}(|x|) = \sup\{|\tilde{\tau}(ax)| : a \in \mathcal{M}, \|a\|_{\mathcal{M}} \leq 1\}$ *for all* $x \in \mathfrak{M}_\tau$;
(5) $|\tilde{\tau}(ax)| \leq \tilde{\tau}(|ax|) \leq \|a\|_{\mathcal{M}} \tilde{\tau}(|x|)$ *for all* $a \in \mathcal{M}$ *and* $x \in \mathfrak{M}_\tau$.

The ideal $\mathfrak{M}_\tau$ is called the ideal of definition of the trace $\tilde{\tau}$. In the case when the trace τ is finite, the ideal $\mathfrak{M}_\tau$ coincides with the von Neumann algebra $\mathcal{M}$ and the trace τ is equal to the extended trace $\tilde{\tau}$. This does not hold in general. In what follows, the extension $\tilde{\tau}$ is denoted simply by τ.

For a faithful normal semifinite trace τ on $\mathcal{M}$, we denote

$$P_{\text{fin}}(\mathcal{M}, \tau) = \{p \in P(\mathcal{M}) : \tau(p) < \infty\}.$$

If $p \in P_{\text{fin}}(\mathcal{M}, \tau)$, then we say that p is a *τ-finite* projection. Note that any τ-finite projection is necessarily finite and $P_{\text{fin}}(\mathcal{M}, \tau)$ is a sublattice in $P_{\text{fin}}(\mathcal{M})$ (and hence in $P(\mathcal{M})$).

In the following proposition, we collect some properties of faithful normal semifinite traces on a von Neumann algebra and finite projections in this algebra.

Proposition 1.10.15. *Let $\mathcal{M}$ be a semifinite von Neumann algebra acting on a separable Hilbert space equipped with a faithful normal semifinite trace τ. For any finite projection $p \in P(\mathcal{M})$, we have that:*
(i) *there exists $z \in P(\mathcal{Z}(\mathcal{M}))$ such that $0 \neq zp$ is τ-finite;*
(ii) *there exists a central partition of unity $\{z_n\}_{n\in\mathbb{N}}$ in $\mathcal{M}$, such that pz_n is τ-finite for any $n \in \mathbb{N}$;*
(iii) *if ψ is a faithful normal state on $\mathcal{Z}(\mathcal{M})$ then for any $\varepsilon > 0$ there exists $z \in P(\mathcal{Z}(\mathcal{M}))$, such that $\tau(pz) < \infty$ and $\psi(z^\perp) < \varepsilon$.*

Proof. (i) Since the trace τ is semifinite, there exists a nonzero τ-finite projection $q \in P(\mathcal{M})$, such that $q \leq p$.

By Zorn's lemma, there exists a maximal family $\Gamma = \{q_i\}_{i\in I}$ of pairwise orthogonal projections in $\mathcal{M}$, which are equivalent to q and are majorized by p. We claim that I is a finite set. Let

$$e = \sup_{i\in I} q_i.$$

It is clear that $e \leq p$. Suppose that I is infinite. If $i_0 \in I$, then there exists a bijection $\Psi : I \to I \setminus \{i_0\}$. Then $e = \sup_{i\in I} q_i \sim \sup_{i\in\Psi(I)} q_i = e - q_{i_0} < e$. This implies that e is an infinite projection, too. Since $e \leq p$ and p is finite, this is a contradiction. Thus, I is a finite set. In particular, since each q_i is τ-finite, it follows that e is τ-finite, too.

By Theorem 1.9.5, there exists a projection $z \in P(\mathcal{Z}(\mathcal{M}))$, such that $z(p-e) \precsim zq$ and $z^\perp q \precsim z^\perp(p-e)$.

Suppose that $zp = 0$ and $z(p-e) = 0$. Then $zq = zqp = qzp = 0$, and so $q = z^\perp q \precsim z^\perp(p-e) = p - e$. This means that there exists $q_0 \sim q$, $q_0 \leq p - e$, which contradicts maximality of Γ. Thus, zp and $z(p-e)$ cannot be simultaneously zero.

If $z(p - e) \neq 0$, then the majorization $z(p - e) \precsim zq$ implies that

$$0 < \tau(zp) \leq \tau(ze) + \tau(zq) < \infty.$$

If $z(p-e) = 0$ and $zp \neq 0$, then $0 < \tau(zp) = \tau(ze) < \infty$. Thus, in both cases z is the desired central projection.

(ii) By Zorn's lemma and part (i), there exists a maximal family $\Gamma = \{z_i\}_{i\in I}$ of pairwise orthogonal central projections from $P(\mathcal{Z}(\mathcal{M}))$, such that $0 < \tau(pz_i) < \infty$. If $z := \sup_{i\in I} z_i$, then maximality of Γ and part (i) implies that $zp = p$. Therefore, $\Gamma \cup \{z^\perp\}$ is the required central partition of unity. By assumption, $\mathcal{M}$ acts on a separable Hilbert space. In particular, the required central partition of unity is at most countable.

(iii) Let $\Gamma = \{z_n\}_{n\in\mathbb{N}}$ be a central partition as in part (ii) and let $\varepsilon > 0$ be arbitrary. Since ψ is a normal state on $\mathcal{Z}(\mathcal{M})$, there exists N, such that $\sum_{i=N+1}^{\infty} \psi(z_i) < \varepsilon$. Taking $z = \sum_{i=1}^{N} z_i$, we have that $\tau(pz) < \infty$ and $\psi(z^\perp) < \varepsilon$, as required. □

For convenience, we formulate a corollary of Proposition 1.10.15 for a finite von Neumann algebra $\mathcal{M}$ (applied to the finite projection **1**).

Corollary 1.10.16. *Let $\mathcal{M}$ be a finite von Neumann algebra equipped with a faithful normal semifinite trace τ. There exists a central partition of unity $\{z_n\}_{n\in\mathbb{N}}$, such that $\tau(z_n) < +\infty$ for every $n \in \mathbb{N}$.*

Another corollary of Proposition 1.10.15 now follows.

Corollary 1.10.17. *Let $\mathcal{M}$ be a semifinite von Neumann algebra acting on a separable Hilbert space and let $p \in P(\mathcal{M})$ be a finite projection in $\mathcal{M}$. Then there exists a normal faithful semifinite trace τ on $\mathcal{M}$, such that $\tau(p) < \infty$.*

Proof. Since $\mathcal{M}$ is a semifinite von Neumann algebra, there exists a faithful normal semifinite trace τ_1 on $\mathcal{M}$.

Let p be a finite projection in $\mathcal{M}$. Proposition 1.10.15(ii) implies that there exists a central partition of unity $\{z_n\}_{n\in\mathbb{N}}$, such that $\tau_1(pz_n) < \infty$ for any $n \in \mathbb{N}$. Setting

$$\tau(x) = \sum_{\substack{n\in\mathbb{N},\\ pz_n\neq 0}} \frac{\tau_1(z_n x)}{2^n \tau_1(pz_n)} + \tau_1\Big(x \cdot \sum_{\substack{n\in\mathbb{N},\\ pz_n=0}} z_n\Big), \quad x \in \mathcal{M},$$

one can easily check that τ is a faithful normal semifinite trace on $\mathcal{M}$, such that $\tau(p) < \infty$. □

1.11 Dimension function on a von Neumann algebra

As follows from Theorem 1.10.6, in a semifinite von Neumann algebra $\mathcal{M}$, one can measure the "size" of a projection $p \in \mathcal{M}$ by evaluating the extended center-valued trace on p. In this section, we introduce the so-called dimension function, which serves this purpose for any (not necessarily) semifinite von Neumann algebra $\mathcal{M}$. On the semifinite summand of an arbitrary von Neumann algebra $\mathcal{M}$, the dimension function $\mathcal{D}$ coincides with the reduction of the faithful normal semifinite extended center-valued trace $\mathcal{T}$ to $P(\mathcal{M})$. On the type *III* summand of $\mathcal{M}$, the dimension function of any projection takes either infinite or zero value. The dimension function $\mathcal{D}$ is key for the introduction of the local measure topology $t(\mathcal{M})$ on the algebra of all locally measurable operators affiliated with $\mathcal{M}$ (see Section 4.3 below).

Let $\mathcal{M}$ be an arbitrary von Neumann algebra acting on a separable Hilbert space. Let φ be a $*$-isomorphism from $\mathcal{Z}(\mathcal{M})$ onto the $*$-algebra $L_\infty(\Omega,\Sigma,\mu)$, where (Ω,Σ,μ) is a probability space (see Remark 1.8.11). As before, denote by $L^+(\Omega,\Sigma,\mu)$ the set of all measurable functions defined on (Ω,Σ,μ) and taking values in the extended half-line $[0,\infty]$ (functions that are equal almost everywhere are identified). For $f \in L^+(\Omega)$, the support projection $s(f)$ is defined as (the equivalence class) χ_A, where $A = \{\omega \in \Omega : f(\omega) \neq 0\} \in \Sigma$.

Definition 1.11.1. Let $\mathcal{M}$ be a von Neumann algebra acting on a separable Hilbert space and let (Ω,Σ,μ) be a probability space, such that $\mathcal{Z}(\mathcal{M})$ is $*$-isomorphic to $L_\infty(\Omega,\Sigma,\mu)$ with φ denoting this $*$-isomorphism. A mapping $\mathcal{D}: P(\mathcal{M}) \to L^+(\Omega,\Sigma,\mu)$ is said to be a *dimension function* (on $\mathcal{M}$) if the following hold:
(i) If $p \in P(\mathcal{M})$ and $p \neq 0$, then $\mathcal{D}(p) \neq 0$;
(ii) The function $\mathcal{D}(p)$ is almost everywhere finite if and only if p is finite projection;
(iii) $\mathcal{D}(p \vee q) = \mathcal{D}(p) + \mathcal{D}(q)$ if $pq = 0$;
(iv) $\mathcal{D}(p) = \mathcal{D}(q)$ if $p \sim q$, $p, q \in P(\mathcal{M})$;
(v) $\mathcal{D}(zp) = \varphi(z)\mathcal{D}(p)$ for any $z \in P(\mathcal{Z}(\mathcal{M}))$ and $p \in P(\mathcal{M})$;
(vi) If $p_i, p \in P(\mathcal{M})$, $i \in I$, and $p_i \uparrow p$, then $\mathcal{D}(p) = \sup_{i\in I} \mathcal{D}(p_i)$.

Before showing that a dimension function exists for any von Neumann algebra $\mathcal{M}$, we fix some notation. If z is a central projection in $\mathcal{M}$, then $\varphi(z)$ is a projection in $L_\infty(\Omega,\Sigma,\mu)$, i. e., $\varphi(z) = \chi_A$ for some measurable set $A \in \Sigma$. We shall use the notation

$$\infty \cdot \varphi(z),$$

to denote the function in $L^+(\Omega,\Sigma,\mu)$ such that

$$(\infty \cdot \varphi(z))(\omega) = \begin{cases} \infty, & \omega \in A, \\ 0, & \text{otherwise}, \end{cases} \tag{1.6}$$

where $A \in \Sigma$ is such that $\chi_A = \varphi(z)$.

We also recall here that $s_c(p) \in P(\mathcal{Z}(\mathcal{M}))$ is the central support of a projection $p \in P(\mathcal{M})$, in particular, $s_c(p) = 0$ if and only if $p = 0$. Therefore, since φ is an $*$-isomorphism between $\mathcal{Z}(\mathcal{M})$ and $L_\infty(\Omega,\Sigma,\mu)$, it follows that $\varphi(s_c(p)) = 0$ if and only if $p = 0$.

As the following theorem shows, a dimension function exists on any von Neumann algebra and it extends the faithful normal semifinite center-valued trace defined on its semifinite summand when the latter is reduced to $P(\mathcal{M})$.

Theorem 1.11.2. *Let $\mathcal{M}$ be a von Neumann algebra and let $z \in P(\mathcal{Z}(\mathcal{M}))$ be such that $\mathcal{M}z$ is a semifinite von Neumann algebra and $\mathcal{M}z^{\perp}$ is a type III von Neumann algebra. Denoting by $\mathcal{T}$ a faithful normal semifinite extended center-valued trace $\mathcal{T}$ on the algebra $\mathcal{M}z$, the function $\mathcal{D} : P(\mathcal{M}) \to L^{+}(\Omega, \Sigma, \mu)$, defined by the formula*

$$\mathcal{D}(p) = \mathcal{T}(zp) + \infty \cdot \varphi(s_c(z^{\perp}p)), \quad p \in P(\mathcal{M}),$$

is a dimension function on $\mathcal{M}$.

Proof. Since the extended center-valued trace $\mathcal{T}$ is faithful, it follows that property (i) in Definition 1.11.1 is satisfied.

To prove (ii) in Definition 1.11.1, assume that $p \in P_{\mathrm{fin}}(\mathcal{M})$. We have that $pz^{\perp} = 0$ and, therefore, $\mathcal{D}(p) = \mathcal{T}(p)$. Referring to Proposition 1.10.7, we conclude that $\mathcal{D}(p) = \mathcal{T}(p)$ has finite values a. e. Conversely, assume that $\mathcal{D}(p)$ takes finite values a. e. In this case, we have $z^{\perp}p = 0$, i. e., $zp = p$. Therefore, $\mathcal{T}(p) = \mathcal{D}(p)$ is finite a. e. Referring to Proposition 1.10.7 once again, we conclude that p is finite.

Next, assume that $p, q \in P(\mathcal{M})$ are such that $pq = 0$. Since $pq = 0$, it follows that $p \vee q = p + q$. Using Proposition 1.9.2(v), we obtain

$$s_c(z^{\perp}(p+q)) = s_c(z^{\perp}p + z^{\perp}q) = s_c(z^{\perp}p) + s_c(z^{\perp}q).$$

Thus, we have that

$$\begin{aligned} \mathcal{D}(p \vee q) = \mathcal{D}(p+q) &= \mathcal{T}(zp + zq) + \infty \cdot s_c(z^{\perp}(p+q)) \\ &= \mathcal{T}(zp) + \mathcal{T}(zq) + \infty \cdot s_c(z^{\perp}p) + \infty \cdot s_c(z^{\perp}q) \\ &= \mathcal{D}(p) + \mathcal{D}(q), \end{aligned}$$

proving (iii) in Definition 1.11.1

Next, if $p \sim q$, $p, q \in P(\mathcal{M})$, then $zp \sim zq$ and $z^{\perp}p \sim z^{\perp}q$. On one hand, the definition of the extended center-valued trace $\mathcal{T}$ implies that $\mathcal{T}(zp) = \mathcal{T}(zq)$. On the other hand, the equivalence $z^{\perp}p \sim z^{\perp}q$ together with Theorem 1.9.4(ii) implies that $s_c(z^{\perp}p) = s_c(z^{\perp}q)$. Therefore,

$$\mathcal{D}(p) = \mathcal{T}(zp) + \infty \cdot \varphi(s_c(z^{\perp}p)) = \mathcal{T}(zq) + \infty \cdot \varphi(s_c(z^{\perp}q)) = \mathcal{D}(q).$$

This verifies (iv) in Definition 1.11.1.

To check (v) in Definition 1.11.1, we note that for any $z' \in P(\mathcal{Z}(\mathcal{M}))$ and $p \in P(\mathcal{M})$, we have that

$$\mathcal{D}(z'p) = \mathcal{T}(z \cdot z'p) + \infty \cdot \varphi(s_c(z^{\perp} \cdot z'p)) = \mathcal{T}(z' \cdot zp) + \infty \cdot \varphi(z' \cdot s_c(z^{\perp}p))$$

$$= \varphi(z')\mathcal{T}(zp) + \varphi(z')(\infty \cdot \varphi(s_c(p))) = \varphi(z')\mathcal{D}(p),$$

as required.

Finally, to verify (vi) in Definition 1.11.1, assume that $p_i, p \in P(\mathcal{M})$ are such that $p_i \uparrow p$. Then $zp_i \uparrow zp$ and $z^\perp p_i \uparrow z^\perp p$. Since the extended center-valued trace $\mathcal{T}$ is normal, it follows that $\mathcal{T}(zp) = \sup_{i\in I} \mathcal{T}(zp_i)$. By Proposition 1.9.2(v), we have that $s_c(z^\perp p_i) \uparrow s_c(z^\perp p)$ and, therefore, $\varphi(s_c(z^\perp p_i)) \uparrow \varphi(s_c(z^\perp p))$. Thus,

$$\mathcal{D}(p_i) = \mathcal{T}(zp_i) + \infty \cdot \varphi(s_c(z^\perp p_i)) \uparrow \mathcal{T}(zp) + \infty \cdot \varphi(s_c(z^\perp p)),$$

as required.

Thus, $\mathcal{D}$ is a dimension function on $\mathcal{M}$. □

Suppose that $\mathcal{M}$ is a semifinite von Neumann algebra. Then the central projection $z \in P(\mathcal{Z}(\mathcal{M}))$ in Theorem 1.11.2 is **1**. In particular, $\mathcal{D}(p) = \mathcal{T}(p)$ for any $p \in P(\mathcal{M})$, showing that for a semifinite von Neumann algebra the dimension function coincides with the reduction of the extended center-valued trace $\mathcal{T}$ to $P(\mathcal{M})$.

Assume now that $\mathcal{M}$ is a type *III* von Neumann algebra. Then the central projection $z \in P(\mathcal{Z}(\mathcal{M}))$ in Theorem 1.11.2 is 0. Therefore, the function $\mathcal{D} : P(\mathcal{M}) \to L^+(\Omega, \Sigma, \mu)$, defined by $\mathcal{D}(p) = \infty \cdot \varphi(s_c(p))$, is a dimension function on $\mathcal{M}$. As we will show next, for a type *III* von Neumann algebra any dimension function has this form. We start with such a formula for any properly infinite projection in an arbitrary von Neumann algebra. We recall that the function $\infty \cdot \varphi(z)$, $z \in P(\mathcal{Z}(\mathcal{M}))$ is defined in (1.6).

Proposition 1.11.3. *Let $\mathcal{M}$ be a von Neumann algebra and let $\mathcal{D}$ be a dimension function on $\mathcal{M}$. For any properly infinite projection $p \in P(\mathcal{M})$, we have that*

$$\mathcal{D}(p) = \infty \cdot \varphi(s_c(p)).$$

Proof. Let $p \in P(\mathcal{M})$ be a fixed properly infinite projection. Let $s(\mathcal{D}(p))$ be the suuport projection of $\mathcal{D}(p)$ and let $A \in \Omega$ be such that $\chi_A = s(\mathcal{D}(p))$. Suppose that there exists $X \in \Sigma$, such that $X \subset A$, $\mu(X) > 0$ and $\mathcal{D}(p)|_X < \infty$. Setting $z = \varphi^{-1}(\chi_X) \neq 0$, we have that $\mathcal{D}(zp) = \varphi(z)\mathcal{D}(p) = \chi_X\mathcal{D}(p)$. Hence, $\mathcal{D}(zp)$ takes finite values a. e., and so $zp \in P_{\text{fin}}(\mathcal{M})$ is a finite projection. Since p is properly infinite, it follows that $zp = 0$. This means that $\mathcal{D}(p) = 0$ on X. The assertion follows immediately. □

Corollary 1.11.4. *If $\mathcal{M}$ is a type III von Neumann algebra and $\mathcal{D}_1$, $\mathcal{D}_2$ are two dimension functions on $\mathcal{M}$, then $\mathcal{D}_1 = \mathcal{D}_2$.*

Proof. Since $\mathcal{M}$ is a type *III* von Neumann algebra, it follows that any nonzero projection $p \in P(\mathcal{M})$ is properly infinite. Therefore, by Proposition 1.11.3 we have that

$$\mathcal{D}_1(p) = \infty \cdot \varphi(s_c(p)) = \mathcal{D}_2(p).$$

Thus, $\mathcal{D}_1 = \mathcal{D}_2$. □

Remark 1.11.5. *If $\mathcal{M}$ is a properly infinite von Neumann algebra, it follows that the function $\mathcal{D}(\mathbf{1})$ takes infinite value almost everywhere. Indeed, since $\mathbf{1}$ is a properly infinite projection, it follows from Proposition* 1.11.3 *that $\mathcal{D}(\mathbf{1}) = \infty \cdot \varphi(s_c(\mathbf{1}))$. Clearly, $s_c(\mathbf{1}) = \mathbf{1}$, and hence, the assertion follows.*

We note several useful properties of a dimension function $\mathcal{D}$ on $\mathcal{M}$, which shall be used repeatedly.

Proposition 1.11.6.
(i) *If $p_n \in P(\mathcal{M})$, then $\mathcal{D}(\sup_{n\in\mathbb{N}} p_n) \le \sum_{n=1}^{\infty} \mathcal{D}(p_n)$. In addition, the equality holds if $p_n p_m = 0$;*
(ii) *If p_n are finite projections from $P(\mathcal{M})$ and $p_n \downarrow 0$, then $\mathcal{D}(p_n) \downarrow 0$ almost everywhere;*
(iii) *$\mathcal{D}(p) = 0$ if and only if $p = 0$;*
(iv) *If $p \precsim q$, then $\mathcal{D}(p) \le \mathcal{D}(q)$;*
(v) *If $p, q \in P(\mathcal{M})$, $\mathcal{D}(q) \le \mathcal{D}(p)$, and p is a finite projection, then $q \precsim p$;*
(vi) *$s(\mathcal{D}(p)) = \varphi(s_c(p))$ for any $p \in P(\mathcal{M})$.*

Proof. (i) Let $p, q \in P(\mathcal{M})$. If $p \le q$, then we can write $q = (q-p) + p$ with $(q-p)p = 0$. Therefore, referring to (iii) in Definition 1.11.1, we obtain $\mathcal{D}(q) = \mathcal{D}(q-p) + \mathcal{D}(p)$. We infer from the Kaplansky's identity (see Theorem 1.9.4(v), that $p \vee q - p \sim q - p \wedge q$. Using now (iv) in Definition 1.11.1, we infer that

$$\mathcal{D}(p \vee q) = \mathcal{D}(p) + \mathcal{D}(q) - \mathcal{D}(p \wedge q) \le \mathcal{D}(p) + \mathcal{D}(q).$$

By induction, we obtain that

$$\mathcal{D}\left(\bigvee_{k=1}^{n} p_k\right) \le \sum_{k=1}^{n} \mathcal{D}(p_k)$$

for any $p_1, \dots, p_n \in P(\mathcal{M})$.

To obtain the inequality for an infinite sequence $\{p_n\}_{n\in\mathbb{N}} \subset P(\mathcal{M})$, let $q_n = \bigvee_{k=1}^{n} p_k$. As $q_n \uparrow \sup_{n\in\mathbb{N}} p_n$, part (vi) in Definition 1.11.1 implies that

$$\mathcal{D}\left(\sup_{n\in\mathbb{N}} p_n\right) = \sup_{n\in\mathbb{N}}(\mathcal{D}(q_n)) \le \sup_{n\in\mathbb{N}} \sum_{k=1}^{n} \mathcal{D}(p_k) = \sum_{n=1}^{\infty} \mathcal{D}(p_n).$$

If $p_n p_m = 0$, $n \ne m$, then the equality follows directly from (iii) and (vi) in Definition 1.11.1.

(ii) Suppose that $\{p_n\}_{n\in\mathbb{N}} \subset P_{\text{fin}}(\mathcal{M})$ is such that $p_n \downarrow 0$ as $n \to \infty$. Let $q_n = p_1 - p_n$, $n > 1$. Then $\{q_n\}_{n\in\mathbb{N}} \subset P_{\text{fin}}(\mathcal{M})$ and $q_n \uparrow p_1$. By (vi) in Definition 1.11.1, we have that

$$\mathcal{D}(p_1) - \mathcal{D}(p_n) = \mathcal{D}(q_n) \uparrow \mathcal{D}(p_1), \quad n \to \infty.$$

Therefore, $-\mathcal{D}(p_n) \uparrow 0$, and so $\mathcal{D}(p_n) \downarrow 0$ almost everywhere.

(iii) If $\mathcal{D}(p) = 0$ a. e., then by (ii) in Definition 1.11.1 we have that $p = 0$. It remains to show that $\mathcal{D}(0) = 0$ a. e. Writing $0 = 0 \vee 0$ and referring to part (i), we have that

$$\mathcal{D}(0) = \mathcal{D}(0 \vee 0) = \mathcal{D}(0) + \mathcal{D}(0).$$

Note that $\mathcal{D}(0)$ is finite a. e. since 0 is a finite projection. Thus, $\mathcal{D}(0) = 0$ a. e., as required.

(iv) If $p \precsim q$ with $p_0 \in P(\mathcal{M})$, such that $p \sim p_0 \leq q$, then $\mathcal{D}(p) = \mathcal{D}(p_0) \leq \mathcal{D}(q)$, proving the required estimate.

(v) Suppose that $p \in P_{\mathrm{fin}}(\mathcal{M})$, $q \in P(\mathcal{M})$ are such $\mathcal{D}(q) \leq \mathcal{D}(p)$. Suppose on the contrary that q is not majorized by p. By Theorem 1.9.5, there exists $z \in P(\mathcal{Z}(\mathcal{M}))$, such that $pz \prec qz \neq 0$, i. e., $qz = r + v^*v$, $vv^* = pz$, $vr = 0$, $r \neq 0$. We now have

$$\varphi(z)\mathcal{D}(q) = \mathcal{D}(qz) = \mathcal{D}(r) + \mathcal{D}(pz) > \varphi(z)\mathcal{D}(p).$$

However, by assumption we have that

$$\varphi(z)\mathcal{D}(q) \leq \varphi(z)\mathcal{D}(p).$$

Since the functions $\varphi(z)\mathcal{D}(q)$ and $\varphi(z)\mathcal{D}(p)$ are finite-valued a. e., it follows that $\mathcal{D}(r) = 0$. Referring to (iii), we conclude that $r = 0$, which is a contradiction. Therefore, $q \precsim p$.

(vi) Since $p \leq s_c(p)$, it follows that $s(\mathcal{D}(p)) \leq \varphi(s_c(p))$. Suppose that $\varphi(s_c(p)) - s(\mathcal{D}(p)) > 0$. Hence, there exists a projection $0 \neq z \in P(\mathcal{Z}(\mathcal{M}))$, such that $\varphi(z) = \varphi(s_c(p)) - s(\mathcal{D}(p))$. In this case, $z \leq s_c(p)$, and so by Proposition 1.9.2(vi) we have that $zupu^* \neq 0$ for some $u \in U(\mathcal{M})$. Consequently,

$$0 \neq \mathcal{D}(zupu^*) = \varphi(z)\mathcal{D}(p) = (\varphi(s_c(p)) - s(\mathcal{D}(p)))\mathcal{D}(p) = 0,$$

which is a contradiction. Thus, $s(\mathcal{D}(p)) = \varphi(s_c(p))$, as required. □

We note that a faithful normal semifinite extended center-valued trace $\mathcal{T}$ on the semifinite summand is unique only up to multiplication by an a. e. positive and finite function (see Theorem 1.10.6). Therefore, the dimension function defined in Theorem 1.11.2 is not unique. As we will show in Theorem 1.11.13, the dimension function is also unique up to multiplication by an a. e. positive and finite function. We start with some technical lemmas.

Lemma 1.11.7. *Let $\mathcal{M}$ be a type I_n von Neumann algebra. Let $\mathcal{D}_1$ and $\mathcal{D}_2$ be dimension functions on $\mathcal{M}$. If $\mathcal{D}_1(\mathbf{1}) = \mathcal{D}_2(\mathbf{1})$, then $\mathcal{D}_1 = \mathcal{D}_2$.*

Proof. Let $q \in P(\mathcal{M})$. Referring to [143, Theorem 2.4.3], we have that $q = \sum_{i=1}^{n} z_i p_i$, where $z_i \in P(\mathcal{Z}(\mathcal{M}))$, $p_i \in P(\mathcal{M})$, $p_i p_j = 0$, $p_i \sim p_j$, $i, j = 1, \ldots, n$, $i \neq j$, $\sum_{i=1}^{n} p_i = \mathbf{1}$. Since $p_i \sim p_j$, it follows that $\mathcal{D}_1(p_i) = \mathcal{D}_1(p_j)$ for all $1 \leq i, j \leq n$. Therefore,

$$\mathcal{D}_1(p_i) = \frac{1}{n}\sum_{i=1}^{n} \mathcal{D}_1(p_j) = \frac{1}{n}\mathcal{D}_1(\mathbf{1})$$

and

$$\mathcal{D}_1(q) = \sum_{i=1}^{n} \mathcal{D}(z_i p_i) = \sum_{i=1}^{n} \varphi(z_i)\mathcal{D}_1(p_i) = \left(\frac{1}{n}\sum_{i=1}^{n} \varphi(z_i)\right)\mathcal{D}_1(\mathbf{1}).$$

Similarly,

$$\mathcal{D}_2(q) = \left(\frac{1}{n}\sum_{i=1}^{n} \varphi(z_i)\right)\mathcal{D}_2(\mathbf{1}).$$

Since $\mathcal{D}_1(\mathbf{1}) = \mathcal{D}_2(\mathbf{1})$, it follows that $\mathcal{D}_1(q) = \mathcal{D}_2(q)$. Since $q \in P(\mathcal{M})$ is arbitrary, the assertion follows. □

As demonstrated in Theorem 1.10.2 for a type II_1 von Neumann algebra $\mathcal{M}$, there exists a canonical center-valued trace $\mathcal{T}$ on $\mathcal{M}$. In the following lemma, we show that in this case, any dimension function $\mathcal{D}$ on $\mathcal{M}$ is proportional to the reduction of $\mathcal{T}$ to $P(\mathcal{M})$.

Lemma 1.11.8. *Let $\mathcal{M}$ be a type II_1 von Neumann algebra. Let $\mathcal{D}$ be a dimension function on $\mathcal{M}$ and let $\mathcal{T}$ be the canonical center-valued trace on $\mathcal{M}$. We have*

$$\mathcal{D}(q) = \varphi(\mathcal{T}(q))\mathcal{D}(\mathbf{1}), \quad q \in P(\mathcal{M}).$$

Proof. Take $q \in P(\mathcal{M})$ and note that $0 \le \mathcal{T}(q) \le \mathbf{1}$. Fix $n \in \mathbb{N}$ and let $z_k = \chi_{[\frac{k}{n},\frac{k+1}{n})}(\mathcal{T}(q))$ for $0 \le k \le n-1$. By [93, Lemma 6.5.6], we can write $\mathbf{1} = \sum_{l=0}^{n-1} e_l$, where $\{e_l\}_{l=0}^{n-1}$ are pairwise orthogonal and pairwise equivalent projections. We set

$$q_{\text{lower}} = \sum_{k=1}^{n-1} z_k\left(\sum_{l=0}^{k-l} e_l\right), \quad q_{\text{upper}} = \sum_{k=0}^{n-1} z_k\left(\sum_{l=0}^{k} e_l\right).$$

It is clear that

$$\mathcal{T}(q_{\text{lower}}) \le \mathcal{T}(q) \le \mathcal{T}(q_{\text{upper}})$$

and, therefore, Proposition 1.10.3 implies that

$$q_{\text{lower}} \preceq q \preceq q_{\text{upper}}.$$

Hence, by Proposition 1.11.6(iv) we have that

$$\mathcal{D}(q_{\text{lower}}) \le \mathcal{D}(q) \le \mathcal{D}(q_{\text{upper}}).$$

Clearly,

$$\mathcal{D}(q_{\text{lower}}) = \sum_{k=1}^{n-1} \frac{k}{n} \varphi(z_k)\mathcal{D}(\mathbf{1}) = \varphi(\mathcal{T}(q_{\text{lower}}))\mathcal{D}(1),$$

$$\mathcal{D}(q_{\text{upper}}) = \sum_{k=0}^{n-1} \frac{k+1}{n} \varphi(z_k)\mathcal{D}(\mathbf{1}) = \varphi(\mathcal{T}(q_{\text{upper}}))\mathcal{D}(1).$$

Hence,

$$\begin{aligned}|\mathcal{D}(q) - \varphi(\mathcal{T}(q))\mathcal{D}(\mathbf{1})| &\leq \mathcal{D}(q_{\text{upper}}) - \mathcal{D}(q_{\text{lower}})\\ &= (\varphi(\mathcal{T}(q_{\text{upper}})) - \varphi(\mathcal{T}(q_{\text{lower}})))\mathcal{D}(\mathbf{1}) \leq 1/n.\end{aligned}$$

Since $n \in \mathbb{N}$ is arbitrary, the assertion follows. □

A combination of Lemma 1.11.7 and Lemma 1.11.8 implies the following result, which shows that for a finite von Neumann algebra $\mathcal{M}$, the value of a dimension function on the finite projection $\mathbf{1}$ completely defines its value on all projections from $\mathcal{M}$.

Lemma 1.11.9. *Let $\mathcal{M}$ be a finite von Neumann algebra. Let $\mathcal{D}_1$ and $\mathcal{D}_2$ be dimension functions on $\mathcal{M}$. If $\mathcal{D}_1(\mathbf{1}) = \mathcal{D}_2(\mathbf{1})$, then $\mathcal{D}_1 = \mathcal{D}_2$.*

Proof. By Theorem 1.9.14 and Theorem 1.9.17, there exists a central partition of unity $\{z_n\}_{n\in\mathbb{N}_0}$, such that $z_0\mathcal{M}$ is type II_1 and $z_n\mathcal{M}$ is type I_n for all $n \in \mathbb{N}$ (some z_n may vanish). Note that $\mathcal{D}_1(z_n) = \varphi(z_n)\mathcal{D}_1(\mathbf{1}) = \varphi(z_n)\mathcal{D}_2(\mathbf{1}) = \mathcal{D}_2(z_n)$ for all $n \in \mathbb{N}_0$. Hence, the assertion follows from a combination of Lemma 1.11.7 and Lemma 1.11.8. □

Before we prove an analogue of Lemma 1.11.9 for a semifinite (but not necessarily finite) von Neumann algebra, we prove an auxiliary lemma.

Lemma 1.11.10. *Let $p, q \in P_{\text{fin}}(\mathcal{M})$ be such that $s_c(p) = \mathbf{1}$. There exists a sequence $\{q_n\}_{n\in\mathbb{N}} \subset P_{\text{fin}}(\mathcal{M})$ of pairwise orthogonal projections such that*

$$q = \sum_{n=1}^{\infty} q_n$$

and $q_n \preceq p$, $n \in \mathbb{N}$.

Proof. By Zorn's lemma, there exists a maximal family of pairwise orthogonal nonzero projections $\{q_i\}$, such that

$$q_i \preceq p, q' := \sum_i q_i \leq q.$$

Suppose that $q_0 = q - q' > 0$. By Theorem 1.9.5, there exists a central projection z, such that $zp \preceq zq_0$, $(1-z)q_0 \preceq (1-z)p$. If $z = 0$, then $q_0 \preceq p$, which contradicts the maximality of $\{q_i\}$. If $z \neq 0$, then $zp \neq 0$ and there exists $r \neq 0$, such that $r \sim zp \leq p$, and $r \leq zq_0 \leq q_0$. This again contradicts maximality of $\{q_i\}$.

Thus, $q_0 = 0$, which proves that $q = \sum_i q_i$. Since q is finite and $\mathcal{M}$ acts on a separable Hilbert space, Corollary 1.10.17 implies that there exists a faithful normal semifinite trace τ on $\mathcal{M}$, such that $\sum_i \tau(q_i) = \tau(q) < \infty$. Therefore, the family $\{q_i\}$ is at most countable. This completes the proof. □

The following result is an analogue of Lemma 1.11.9 for a semifinite (but not necessarily finite) von Neumann algebra $\mathcal{M}$. In contrast to Lemma 1.11.9, the value of a dimension function on any finite projection with central support **1** completely defines its value on all projections from $\mathcal{M}$.

Lemma 1.11.11. *Let $\mathcal{M}$ be a semifinite von Neumann algebra and let $p \in P_{\text{fin}}(\mathcal{M})$ be such that $s_c(p) = \mathbf{1}$. If $\mathcal{D}_1$ and $\mathcal{D}_2$ are dimension functions on $\mathcal{M}$, such that $\mathcal{D}_1(p) = \mathcal{D}_2(p)$, then $\mathcal{D}_1 = \mathcal{D}_2$.*

Proof. Let $q \in P_{\text{fin}}(\mathcal{M})$ and let $\{q_n\}_{n\in\mathbb{N}}$ be a sequence constructed in Lemma 1.11.10. Let $\{r_n\}_{n\in\mathbb{N}} \subset P(\mathcal{M})$ be such that $r_n \sim q_n$ and $r_n \leq p$ for every $n \in \mathbb{N}$. By Definition 1.11.1(vi) and (iv), we have

$$\mathcal{D}_j(q) = \sum_{n=1}^{\infty} \mathcal{D}_j(q_n) = \sum_{n=1}^{\infty} \mathcal{D}_j(r_n), \quad j = 1, 2.$$

Clearly, $\mathcal{D}_1$ and $\mathcal{D}_2$ are dimension functions on $p\mathcal{M}p$. By Lemma 1.11.9, $\mathcal{D}_1 = \mathcal{D}_2$ on $p\mathcal{M}p$. In particular, $\mathcal{D}_1(r_n) = \mathcal{D}_2(r_n)$ for every $n \in \mathbb{N}$. Thus, $\mathcal{D}_1(q) = \mathcal{D}_2(q)$ for every $q \in P_{\text{fin}}(\mathcal{M})$. Taking into account that any projection in $\mathcal{M}$ is supremum of a net of finite projections and appealing to Definition 1.11.1(vi), we complete the proof. □

With Lemma 1.11.9 and Lemma 1.11.11 at hand, we are now ready to prove that the dimension function on a semifinite von Neumann algebra $\mathcal{M}$ is essentially unique.

Lemma 1.11.12. *Let $\mathcal{M}$ be a semifinite von Neumann algebra. A dimension function $\mathcal{D}$ on $\mathcal{M}$ is unique up to a multiplication by $L^+(\Omega, \Sigma, \mu)$ in the sense that if $\mathcal{D}_1$ and $\mathcal{D}_2$ are two-dimension functions on $\mathcal{M}$, then there exists $c \in L^+(\Omega, \Sigma, \mu)$, such that*

$$0 < c(w) < \infty, \quad a.e.\ \omega \in \Omega$$

and

$$\mathcal{D}_1(p) = c\mathcal{D}_2(p) \quad a.e., p \in P(\mathcal{M}).$$

Proof. By Lemma 1.9.11, we can find $p \in P_{\text{fin}}(\mathcal{M})$, such that $s_c(p) = \mathbf{1}$. Since p is a finite projection, it follows that $\mathcal{D}_1(p)$ and $\mathcal{D}_2(p)$ are finite almost everywhere. By Proposition 1.11.6(vi), we have that $s(\mathcal{D}_j(p)) = \varphi(s_c(p)) = \varphi(\mathbf{1}) = \mathbf{1}$ for $j = 1, 2$. Therefore, there exists a function $c \in L^+(\Omega, \Sigma, \mu)$, such that $0 < c < \infty$ a. e. and

$$\mathcal{D}_1(p) = c\mathcal{D}_2(p).$$

We define

$$\mathcal{D}_3(q) = c\mathcal{D}_2(q), \quad q \in P(\mathcal{M}).$$

Clearly, $\mathcal{D}_3$ is a dimension function on $\mathcal{M}$.

We have $\mathcal{D}_1(p) = \mathcal{D}_3(p)$ for a given $p \in P_{\text{fin}}(\mathcal{M})$ such that $s_c(p) = \mathbf{1}$. By Lemma 1.11.11, $\mathcal{D}_1 = \mathcal{D}_3 = c\mathcal{D}_2$. This completes the proof. □

Theorem 1.11.13. *Let $\mathcal{M}$ be a von Neumann algebra acting on a separable Hilbert space and let (Ω, Σ, μ) be a probability space, such that $\mathcal{Z}(\mathcal{M})$ is $*$-isomorphic to $L_\infty(\Omega, \Sigma, \mu)$. A dimension function $\mathcal{D}$ on $\mathcal{M}$ is unique up to a multiplication by $L^+(\Omega, \Sigma, \mu)$ in the sense that if $\mathcal{D}_1$ and $\mathcal{D}_2$ are two dimension functions on $\mathcal{M}$, then there exists $c \in L^+(\Omega, \Sigma, \mu)$, such that*

$$0 < c(w) < \infty \quad a.e.\ \omega \in \Omega$$

and

$$\mathcal{D}_1(p) = c\mathcal{D}_2(p) \quad a.e. \tag{1.7}$$

for any $p \in P(\mathcal{M})$.

Proof. By Theorem 1.9.14, there exists $z \in P(\mathcal{Z}(\mathcal{M}))$, such that $z\mathcal{M}$ is semifinite and $z^\perp\mathcal{M}$ is type III. As $\varphi(z) \in L_\infty(\Omega, \Sigma, \mu)$ is a projection, there exists a measurable set $A \in \Sigma$, such that $\varphi(z) = \chi_A$. Setting $B = \Sigma \setminus A$, we write $\varphi(\mathbf{1} - z) = \chi_{\Omega \setminus A}$.

Suppose that $\mathcal{D}_1$ and $\mathcal{D}_2$ are two-dimension functions on $\mathcal{M}$. We can define dimension functions $\mathcal{D}_j^z : P(z\mathcal{M}) \to L^+(A, \Sigma|_A, \mu|_A), j = 1, 2$, on $z\mathcal{M}$, by setting

$$\mathcal{D}_j^z(p) = \mathcal{D}_j(p)|_A, \quad p \in P(\mathcal{M}).$$

As $z\mathcal{M}$ is a semifinite von Neumann algebra, Lemma 1.11.12 implies that there exists $c_z \in L^+(A, \Sigma|_A, \mu|_A)$, such that $\mathcal{D}_1^z = c_z\mathcal{D}_2^z$.

Similarly, for $j = 1, 2$, we can define dimension functions

$$\mathcal{D}_j^{z^\perp} : P(z^\perp\mathcal{M}) \to L^+(B, \Sigma|_B, \mu|_B),$$

by setting

$$\mathcal{D}_j^{z^\perp}(z^\perp p) = \mathcal{D}_j(z^\perp p)|_B, \quad p \in P(\mathcal{M}).$$

As $z^\perp\mathcal{M}$ is a type III von Neumann algebra, Corollary 1.11.4 implies that $\mathcal{D}_1^{z^\perp} = \mathcal{D}_2^{z^\perp}$.

We now define $c \in L^+(\Omega, \Sigma, \mu)$, by setting

$$c(\omega) = \begin{cases} c_z(\omega), & \omega \in A, \\ 1, & \text{otherwise.} \end{cases}$$

Since $0 < c_z < \infty$ a. e., it follows that $0 < c < \infty$ a. e. Furthermore, by Definition 1.11.1(iii), for any $p \in P(\mathcal{M})$ we have that

$$\begin{aligned}\mathcal{D}_1(p) &= \mathcal{D}_1(zp) + \mathcal{D}_1(z^\perp p) = \varphi(z)\mathcal{D}_1(zp) + \varphi(z^\perp)\mathcal{D}_1(z^\perp p)\\ &= \mathcal{D}_1(zp)|_A + \mathcal{D}_1(z^\perp p)|_B = \mathcal{D}_1^z(zp) + \mathcal{D}_1^{z^\perp}(z^\perp p)\\ &= c_z\mathcal{D}_2^z(zp) + \mathcal{D}_2^{z^\perp}(z^\perp p) = c_z\mathcal{D}_2(zp)|_A + \mathcal{D}_2(z^\perp p)|_B = c\mathcal{D}_2(p).\end{aligned}$$

This proves the assertion. □

In the particular case of a factor $\mathcal{M}$, we obtain the following uniqueness result of the dimension function. See [149, Chapter V, Corollary 2.32] for a similar result for traces on semifinite factors.

Corollary 1.11.14. *Suppose that $\mathcal{M}$ is a factor and let $\mathcal{D}_1$ and $\mathcal{D}_2$ be dimension functions on $\mathcal{M}$. There exists a constant $c > 0$, such that $\mathcal{D}_1 = c\mathcal{D}_2$.*

1.12 Direct integrals of von Neumann algebras

In this section, we recall elements of the reduction theory for von Neumann algebras, including direct integrals of Hilbert spaces, bounded operators, von Neumann algebras, and traces. We will recall basic properties of these constructs as well as some auxiliary results for the future chapters.

Throughout this section, we assume that w is a fixed σ-finite positive measure on a standard Borel space Z, namely a Polish space endowed with the Borel σ-algebra.

Definition 1.12.1. A *measurable field of Hilbert spaces* is a function $\zeta \to H_\zeta$, $\zeta \in Z$, where each H_ζ is a separable Hilbert space, together with a set S of vector fields $\zeta \to \xi_\zeta \in H_\zeta$ that satisfy:
(i) the function $\zeta \to \langle \xi_\zeta, \gamma_\zeta \rangle$ is measurable for all $\xi, \gamma \in S$;
(ii) if $\zeta \to \gamma_\zeta$ is a vector field and the function $\zeta \to \langle \xi_\zeta, \gamma_\zeta \rangle$ is measurable for each $\xi \in S$, then $\{\zeta \to \gamma_\zeta\} \in S$.

The vector fields from the set S are said to be *measurable vector fields*.

A measurable field of Hilbert spaces gives rise to another Hilbert space as follows. Let $\zeta \to H_\zeta$, $\zeta \in Z$, be a measurable field of Hilbert spaces. The set K of all measurable vector fields ζ, such that the function $\zeta \mapsto \|\xi_\zeta\|^2$ is w-integrable, is a complex vector space. This space can be endowed with a pre-Hilbert space structure with pre-inner product given by the formula

$$\langle \xi, \gamma \rangle = \int_Z^\oplus \langle \xi_\zeta, \gamma_\zeta \rangle \, dw(\zeta), \quad \xi, \gamma \in K.$$

If $\langle \xi, \xi \rangle = 0$ for some $\xi \in K$, then $\xi_\zeta = 0$ for almost every $\zeta \in Z$. Identifying two elements of K, which are equal almost everywhere, we arrive at the definition of the direct integral of Hilbert spaces.

Definition 1.12.2. Let $\zeta \to H_\zeta$, $\zeta \in Z$, be a measurable field of Hilbert spaces. The *direct integral Hilbert space*

$$H = \int_Z^{\oplus} H_\zeta \, dw(\zeta)$$

consists of all classes (equivalence relation is the equality almost everywhere) measurable vector fields $\xi \in S$ for which the function $\zeta \to \|\xi_\zeta\|^2$ is integrable with respect to w.

By [61, Section II,1.5, Proposition 5], the space $H = \int_Z^{\oplus} H_\zeta \, dw(\zeta)$ is a separable Hilbert space. In what follows, we shall use the notation $\xi = \int_Z^{\oplus} \xi_\zeta \, dw(\zeta) \in \int_Z^{\oplus} H_\zeta \, dw(\zeta)$ to denote a class of measurable vector fields.

If (Z, w) is discrete, then Z is necessarily countable. In this case, the direct integral Hilbert space $H = \int_Z^{\oplus} H_\zeta \, dw(\zeta)$ coincides with the direct sum of Hilbert spaces $\bigoplus_{\zeta \in Z} H_\zeta$ (see Section 1.8).

Fix a measurable field of Hilbert space $\zeta \to H_\zeta$ and let $H = \int_Z^{\oplus} H_\zeta \, dw(\zeta)$ be the direct integral Hilbert space.

Next, we recall the definition of measurable fields of bounded operators. For every $\zeta \in Z$, let x_ζ be a bounded linear operator in H_ζ. The mapping $\zeta \to x_\zeta$ will be called a *field of bounded operators* on Z or simply a *field of operators* on Z.

Definition 1.12.3.
(i) A field $\zeta \to x_\zeta \in \mathcal{B}(H_\zeta)$ of bounded operators is said to be *measurable* if for every measurable vector field ξ the field $\zeta \to x_\zeta \xi_\zeta$ is measurable. In this case, the map $\zeta \to \|x_\zeta\|_{\mathcal{B}(H_\zeta)}$ is measurable;
(ii) If $\zeta \to x_\zeta$ is a measurable field of bounded operators and the function

$$\zeta \mapsto \|x_\zeta\|_{\mathcal{B}(H_\zeta)} \tag{1.8}$$

is essentially bounded (with respect to w), then the field $\zeta \to x_\zeta$ is said to be an *(essentially) bounded measurable field of bounded operators.*

Every bounded measurable field $\zeta \to x_\zeta$ of bounded operators describes a bounded linear operator x on the direct integral Hilbert space H, defined by $(x\xi)_\zeta = x_\zeta \xi_\zeta$. The norm of x equals the essential supremum of the map (1.8). Furthermore, two bounded measurable fields of bounded operators define the same operator in H if and only if they are equal almost everywhere [61, Section II.2.5]. In what follows, for a bounded measurable field of bounded operators $\zeta \to x_\zeta$, we denote the corresponding class of equivalence also by $\zeta \to x_\zeta$.

Operators on the direct integral of Hilbert spaces $H = \int_Z^\oplus H_\zeta \, dw(\zeta)$ defined in this way form a special class of operators in H.

Definition 1.12.4.

(i) A bounded linear operator x on $H = \int_Z^\oplus H_\zeta \, dw(\zeta)$ is said to be *decomposable* if it is defined by a bounded measurable field $\zeta \to x_\zeta$. In this case, we write

$$x = \int_Z^\oplus x_\zeta \, dw(\zeta). \tag{1.9}$$

The set of decomposable operators in H is denoted by $\mathcal{R}$;

(ii) The *diagonal operators* are the decomposable operators $x = \int_Z^\oplus x_\zeta \, dw(\zeta)$ for which each x_ζ is a scalar multiple of the identity operator $\mathbf{1}_{H_\zeta}$ on H_ζ. The set of all diagonal operators is denoted by $\mathcal{C}$.

For a decomposable operator $x = \int_Z^\oplus x_\zeta \, dw(\zeta)$, the operators x_ζ are defined up to sets of measure zero. In particular, for a set $A \subset Z$ of measure zero, the operators x_ζ, $\zeta \in A$ can be chosen arbitrarily.

The set of all decomposable operators can be endowed with obvious pointwise operations of addition, multiplication, and taking adjoints. For these operations, we have the following result.

Proposition 1.12.5. *For decomposable operators $x = \int_Z^\oplus x_\zeta \, dw(\zeta)$ and $y = \int_Z^\oplus y_\zeta \, dw(\zeta)$, we have:*

(i) $\alpha x + \beta y = \int_Z^\oplus (\alpha x_\zeta + \beta y_\zeta) \, dw(\zeta)$, $\alpha, \beta \in \mathbb{C}$;

(ii) $xy = \int_Z^\oplus x_\zeta y_\zeta \, dw(\zeta)$;

(iii) $x^* = \int_Z^\oplus x_\zeta^* \, dw(\zeta)$.

Using Proposition 1.12.5, one can easily see that both the set $\mathcal{R}$ of decomposable operators and the set $\mathcal{C}$ of diagonal operators on H are $*$-algebras with respect to almost everywhere pointwise operations. Furthermore, the algebra $\mathcal{C}$ of all diagonal operators is a von Neumann algebra isomorphic to $L^\infty(Z, w)$ (see, e. g., [61, Section II, Section 2.4]).

Theorem 1.12.6 ([61, Section II, Section 2.5]). *A bounded operator x on H is decomposable if and only if x commutes with every bounded diagonalizable operator in H. In particular, the commutant of the algebra $\mathcal{C}$ of all diagonal operators is the algebra $\mathcal{R}$ of decomposable operators.*

Theorem 1.12.6 implies, in particular, that the algebra $\mathcal{R}$ of all decomposable operators is a von Neumann algebra acting on the Hilbert space H.

Next, we move to measurable fields of von Neumann algebras and decomposable von Neumann algebras.

We say that a mapping $\zeta \to \mathcal{M}_\zeta$, $\zeta \in Z$ is a field of von Neumann algebras if $\mathcal{M}_\zeta$ is a von Neumann algebra acting on the Hilbert space H_ζ for every $\zeta \in Z$. Note that since every H_ζ is a separable Hilbert space, $\mathcal{M}_\zeta$ is necessarily countably generated.

Definition 1.12.7. A field of von Neumann algebras $\zeta \to \mathcal{M}_\zeta$ over Z is said to be *measurable* if there exists a sequence $\zeta \to x_\zeta^{(1)}, \zeta \to x_\zeta^{(2)}, \dots$ of measurable fields of operators, such that $\mathcal{M}_\zeta$ is the von Neumann algebra generated by $\{x_\zeta^{(n)}\}_{n\in\mathbb{N}}$.

Without loss of generality, the measurable fields $\zeta \to x_\zeta^{(1)}, \zeta \to x_\zeta^{(2)}, \dots$ may be assumed to be essentially bounded.

Every measurable field $\zeta \to \mathcal{M}_\zeta$ of von Neumann algebras defines a von Neumann algebra acting on the Hilbert space $H = \int_Z^\oplus H_\zeta \, dw(\zeta)$ in the following way. Let $\mathcal{M}$ be the algebra of all decomposable operators $x = \int_Z^\oplus x_\zeta \, dw(\zeta)$ in H, such that $x_\zeta \in \mathcal{M}_\zeta$ for almost all $\zeta \in Z$. By [61, Section II, Section 3.2, Proposition 1], $\mathcal{M}$ is a von Neumann algebra such that $\mathcal{C} \subset \mathcal{M} \subset \mathcal{R}$, generated by $\mathcal{C}$ and a countable family of elements. Furthermore, if $\zeta \to \mathcal{N}_\zeta$ is another measurable field of von Neumann algebras, which defines the same von Neumann algebra $\mathcal{M}$, then $\mathcal{M}_\zeta = \mathcal{N}_\zeta$ almost everywhere [61, Section II, Section 3.2, Proposition 1(ii)].

Definition 1.12.8. A von Neumann algebra in the Hilbert space $H = \int_Z^\oplus H_\zeta \, dw(\zeta)$ is said to be *decomposable* if it is defined by a measurable field $\zeta \to \mathcal{M}_\zeta$ of von Neumann algebras. In this case, we write

$$\mathcal{M} = \int_Z^\oplus \mathcal{M}_\zeta \, dw(\zeta).$$

A von Neumann algebra $\mathcal{M}$ in H is decomposable if and only if it is a von Neumann algebra generated by $\mathcal{C}$ and a countable family of decomposable operators [61, Section II, Section 3.2, Theorem 2].

For a decomposable algebra $\mathcal{M} = \int_Z^\oplus \mathcal{M}_\zeta \, dw(\zeta)$, the von Neumann algebras $\mathcal{M}_\zeta$ are defined up to a set of measure zero. Furthermore (see [61, Section II, Section 3.3]), the field $\zeta \to \mathcal{M}_\zeta'$ is also a measurable field of von Neumann algebras and

$$\mathcal{M}' = \int_Z^\oplus \mathcal{M}_\zeta' \, dw(\zeta).$$

If $\mathcal{C}$ is the center of $\mathcal{M} = \int_Z^\oplus \mathcal{M}_\zeta \, dw(\zeta)$ (and hence of $\mathcal{M}'$), then almost everywhere $\mathcal{M}_\zeta$ and $\mathcal{M}_\zeta'$ are factors [61, Section II, Section 3.3. Theorem 3].

The fundamental importance of the theory of decomposable von Neumann algebras is that any von Neumann algebra acting on a separable Hilbert space can be written down as a decomposable von Neumann algebra defined by a measurable field of factors. Let H be a separable (complex) Hilbert space and let $\mathcal{A}$ be a commutative von Neumann algebra in H. By [61, Section II, Section 6.1, Theorem 2], there exists a compact metrizable space Z, a positive measure w on Z with support Z, a w-measurable field $\zeta \to H_\zeta$ of (nonzero complex) Hilbert spaces over Z, and an isomorphism of H onto $\int_Z^\oplus H_\zeta \, dw(\zeta)$, which transforms $\mathcal{A}$ into the algebra of all diagonalizable operators. In particular, given a von Neumann algebra $\mathcal{M}$ acting on H, we can decompose H with respect to the commutative algebra $\mathcal{Z}(\mathcal{M})$ and write $\mathcal{Z}(\mathcal{M})$ as the algebra of all diagonalizable operators. This then leads to the following reduction theorem.

Theorem 1.12.9 ([61, Section II, Section 6.1, Corollary]). *Let H be a separable (complex) Hilbert space and let $\mathcal{M}$ be a von Neumann algebra acting on H. There exists a compact metrizable space Z, a positive measure w on Z with support Z, a w-measurable field $\zeta \to H_\zeta$ of (nonzero complex) Hilbert spaces over Z, a measurable field $\zeta \to \mathcal{M}_\zeta$ of factors in H_ζ, and an isomorphism H onto $\int_Z^\oplus H_\zeta \, dw(\zeta)$, which transforms the algebra $\mathcal{M}$ into the algebra $\int_Z^\oplus \mathcal{M}_\zeta \, dw(\zeta)$, and the algebra $\mathcal{Z}(\mathcal{M})$ into the algebra of all diagonalizable operators. In this case, $\mathcal{Z}(\mathcal{M}) \cong L_\infty(Z, w)$.*

With Theorem 1.12.9 in mind, we shall speak of the (direct integral) decomposition of a (separable) Hilbert space H and a von Neumann algebra $\mathcal{M}$ acting on H relative to the center $\mathcal{Z}(\mathcal{M})$ of $\mathcal{M}$.

For future purposes, we also note the following result, which shows that a decomposition of a von Neumann algebra relative to its centre preserves the types of von Neumann algebras $\mathcal{M}_\zeta$.

Theorem 1.12.10 ([93, Theorem 14.1.21]). *Let Z be a Polish space, such that $\mathcal{Z}(\mathcal{M}) \cong L_\infty(Z, w)$ and let $\mathcal{M} = \int_Z^\oplus \mathcal{M}_\zeta \, dw(\zeta)$ be the decomposition of a von Neumann algebra $\mathcal{M}$ relative to its centre. The algebra $\mathcal{M}$ is of type $I_n, n \in \mathbb{N} \cup \{\infty\}, II_1, II_\infty$, or III if and only if $\mathcal{M}_\zeta$ is a factor of type $I_n, n \in \mathbb{N} \cup \{\infty\}, II_1, II_\infty$, or III almost everywhere.*

Next, we recall the notion of a measurable field of traces.

Suppose $\mathcal{M} = \int_Z^\oplus \mathcal{M}_\zeta \, dw(\zeta)$ is a decomposable von Neumann algebra and $\zeta \mapsto \tau_\zeta$ is a field of traces, each τ_ζ being a trace on $(\mathcal{M}_\zeta)_+$ taking values in $[0, +\infty]$. A field of traces $\zeta \to \tau_\zeta$ is said to be *measurable* if for every $x = \int_Z^\oplus x_\zeta \, dw(\zeta) \in \mathcal{M}$, the function $\zeta \mapsto \tau_\zeta(x_\zeta)$ is measurable. In this case (see [61, Section II.5.1]), setting

$$\tau(x) = \int_Z \tau_\zeta(x_\zeta) \, dw(\zeta), \quad x = \int_Z^\oplus x_\zeta \, dw(\zeta) \in \mathcal{M}_+,$$

defines a trace on $\mathcal{M}_+$, which is denoted by

$$\tau = \int_Z^\oplus \tau_\zeta \, dw(\zeta).$$

Theorem 1.12.11 ([61, Section II.5.2]). *Let $\mathcal{M} = \int_Z^{\oplus} \mathcal{M}_\zeta \, dw(\zeta)$ be a decomposable von Neumann algebra and let τ be a faithful normal semifinite trace on $\mathcal{M}$. There exists a measurable field $\zeta \to \tau_\zeta$ of faithful normal semifinite traces τ_ζ on $\mathcal{M}_\zeta$, such that*

$$\tau = \int_Z^{\oplus} \tau_\zeta \, dw(\zeta).$$

Note that for a faithful normal *finite* trace $\tau = \int_Z^{\oplus} \tau_\zeta \, dw(\zeta)$, after redefining the measure, if necessary, we may without loss of generality assume each τ_ζ is a tracial state (see [61, Section II.5.2]).

We now fix some additional notation. For a measurable subset $A \subset Z$ and a measurable vector field $\zeta \to \xi_\zeta$, the vector field $\zeta \to \eta_\zeta$, defined by

$$\eta_\zeta = \begin{cases} x_\zeta, & \zeta \in A; \\ 0, & \text{otherwise}, \end{cases}$$

is also a measurable vector field. In particular, we can define the vector $\int_Z^{\oplus} \eta_\zeta \, dw(\zeta) \in H$. In what follows, we shall use the notation

$$\int_A^{\oplus} \xi_\zeta \, dw(\zeta) := \int_Z^{\oplus} \eta_\zeta \, dw(\zeta).$$

For a measurable field of operators $\zeta \to x_\zeta$, von Neumann algebras $\zeta \to \mathcal{M}_\zeta$ and traces $\zeta \to \mathcal{M}_\zeta$, we similarly define $\int_A^{\oplus} x_\zeta \, dw(\zeta)$.

In the following proposition, we collect several properties of projections in a decomposable von Neumann algebra. Note that Proposition 1.12.5 implies that $p = \int_Z^{\oplus} p_\zeta \, dw(\zeta)$ is a projection in a decomposable von Neumann algebra $\mathcal{M} = \int_Z^{\oplus} \mathcal{M}_\zeta \, dw(\zeta)$ if and only if p_ζ is a projection in $\mathcal{M}_\zeta$ for almost every $\zeta \in Z$.

Proposition 1.12.12. *Let $\mathcal{M} = \int_Z^{\oplus} \mathcal{M}_\zeta \, dw(\zeta)$ be a decomposable von Neumann algebra and let $p = \int_Z^{\oplus} p_\zeta \, dw(\zeta)$ be a projection in $\mathcal{M}$. We have*
(i) *p is finite if and only if p_ζ is finite for almost every ζ;*
(ii) *p is properly infinite if and only if p_ζ is either properly infinite or zero for almost every ζ;*
(iii) *$s_c(p) = \int_Z^{\oplus} s_c(p_\zeta) \, dw(\zeta)$ a. e.*

Proof. (i) Suppose that p is a nonzero finite projection in $\mathcal{M}$. Clearly, p belongs to the semifinite direct summand of $\mathcal{M}$, and so without loss of generality we can assume that $\mathcal{M}$ is a semifinite von Neumann algebra. Corollary 1.10.17 implies that there exists a faithful normal semifinite trace τ on $\mathcal{M}$, such that $\tau(p) < \infty$.

By Theorem 1.12.11, there exists a measurable field $\zeta \to \tau_\zeta$ of faithful normal semifinite traces τ_ζ on $\mathcal{M}_\zeta$, such that

$$\tau = \int_Z^{\oplus} \tau_\zeta \, dw(\zeta).$$

Since

$$\int_Z \tau_\zeta(p_\zeta) \, dw(\zeta) = \tau(p) < \infty,$$

it follows that $\tau_\zeta(p_\zeta) < \infty$ a. e. In particular, we infer that p_ζ is finite for almost every ζ.

Conversely, assume that p_ζ is finite for almost every $\zeta \in Z$. Suppose, on the contrary, that p is infinite. There exists $q \in P(\mathcal{M})$ such that $q < p$ and $q \sim p$. Hence, there exists a partial isometry $v \in \mathcal{M}$, such that $vv^* = q < p = v^*v$. Writing

$$v = \int_Z^{\oplus} v_\zeta \, dw(\zeta)$$

we infer that

$$\int_Z^{\oplus} v_\zeta v_\zeta^* \, dw(\zeta) = vv^* < v^*v = p = \int_Z^{\oplus} p_\zeta \, dw(\zeta).$$

By [93, Proposition 14.1.9], we have that $v_\zeta v_\zeta^* < v_\zeta^* v_\zeta = p_\zeta$ on a set Z_0 of nonzero measure. Hence, for all $\zeta \in Z_0$ the projections p_ζ are infinite. The obtained contradiction implies that p is a finite projection.

(ii) If p is properly infinite, then $p\mathcal{M}p$ is a properly infinite von Neumann algebra. Referring to [61, Chapter 5, Section 4, Theorem 5], we obtain that $p_\zeta \mathcal{M}_\zeta p_\zeta$ is a properly infinite von Neumann algebra a. e. and, therefore, p_ζ is a properly infinite or zero projection a. e.

Conversely, assume that p_ζ is a properly infinite or zero projection a. e. Let $z = \int_Z^{\oplus} z_\zeta \, dw(\zeta) \in P(\mathcal{Z}(\mathcal{M}))$ be such that $pz \in P_{\text{fin}}(\mathcal{M})$. By Proposition 1.12.5, we have that $pz = \int_Z^{\oplus} p_\zeta z_\zeta \, dw(\zeta)$. Since pz is finite, it follow from part (ii) that $p_\zeta z_\zeta \in P_{\text{fin}}(\mathcal{M}_\zeta)$ a. e. Since p_ζ is properly infinite, it follows that $p_\zeta z_\zeta = 0$ a. e., and so $pz = 0$. Thus, p is a properly infinite projection.

Part (iii) is proved in [93, Lemma 14.1.20(v)]. □

As we also note next, the equivalence of two projections in a decomposable von Neumann algebra is also almost everywhere pointwise.

Proposition 1.12.13. *Let* $\mathcal{M} = \int_Z^{\oplus} \mathcal{M}_\zeta \, dw(\zeta)$ *be a decomposable von Neumann algebra. Let*

$$p = \int_Z^{\oplus} p_\zeta \, dw(\zeta), \quad q = \int_Z^{\oplus} q_\zeta \, dw(\zeta)$$

be projections in $\mathcal{M}$. Projections p and q are equivalent if and only if $p_\zeta \sim q_\zeta$ for almost every ζ.

Proof. Assume that $p \sim q$. By [93, Lemma 14.1.20], we have that $p_\zeta \sim q_\zeta$ for almost all $\zeta \in Z$.

Conversely, assume that $p_\zeta \sim q_\zeta$ for almost all $\zeta \in Z$. By Theorem 1.9.9(vi), there exists a unique central projection, such that pz is finite and $p(\mathbf{1} - z)$ is properly infinite. Since $z \in P(\mathcal{Z}(\mathcal{M}))$, it follows that there exists a measurable field $\zeta \to z_\zeta$ of central projections in $\mathcal{M}_\zeta$, such that $z = \int_Z^{\oplus} z_\zeta \, dw(\zeta)$.

Since $p_\zeta \sim q_\zeta$ a. e., it follows that $p_\zeta z_\zeta \sim q_\zeta z_\zeta$ and $p_\zeta(\mathbf{1}_{H_\zeta} - z_\zeta) \sim q_\zeta(\mathbf{1}_{H_\zeta} - z_\zeta)$ a. e. By Proposition 1.12.12(ii), the projection $p_\zeta(\mathbf{1}_{H_\zeta} - z_\zeta)$ (and hence $q_\zeta(\mathbf{1}_{H_\zeta} - z_\zeta)$) is either properly infinite or zero for almost all ζ. Hence, referring to Proposition 1.12.12(ii) once again, we infer that $q(\mathbf{1} - z)$ is a properly infinite projection. Furthermore, since $s_c(p_\zeta(\mathbf{1}_{H_\zeta} - z_\zeta)) = s_c(q_\zeta(\mathbf{1}_{H_\zeta} - z_\zeta))$ for a. e. $\zeta \in Z$, Proposition 1.12.12(iii) implies that

$$\begin{aligned} s_c(p(\mathbf{1} - z)) &= \int_Z^{\oplus} s_c(p_\zeta(\mathbf{1}_{H_\zeta} - z_\zeta)) \, dw(\zeta) = \int_Z^{\oplus} s_c(p_\zeta(\mathbf{1}_{H_\zeta} - z_\zeta)) \, dw(\zeta) \\ &= s_c(q(\mathbf{1} - z)), \end{aligned}$$

showing that $p(\mathbf{1} - z)$ and $q(\mathbf{1} - z)$ are properly infinite projections with equal central supports. Since H is a separable Hilbert, Theorem 1.9.9(viii) implies that $p(\mathbf{1}-z) \sim q(\mathbf{1}-z)$.

By Theorem 1.9.4(vi), it is sufficient to show that $pz \sim qz$. Without loss of generality, we assume that $z = \mathbf{1}$, so that p, q are finite projections, such that $p_\zeta \sim q_\zeta$ for almost every $\zeta \in Z$. Without loss of generality, we can further assume that $\mathcal{M}$ is a semifinite von Neumann algebra.

By Corollary 1.10.17, there exists a faithful normal semifinite trace τ on $\mathcal{M}$ such that $\tau(p \vee q) < \infty$. In particular, $\tau(p), \tau(q) < \infty$. By Theorem 1.12.11, we can write

$$\tau = \int_Z^{\oplus} \tau_\zeta \, dw(\zeta)$$

for some measurable field $\zeta \to \tau_\zeta$ of faithful normal semifinite traces τ_ζ on $\mathcal{M}_\zeta$.

By Theorem 1.9.5, there exists a central projection $z_1 \leq \mathbf{1}$ such that $pz_1 \prec qz_1$ and $q(\mathbf{1} - z_1) \prec p(\mathbf{1} - z_1)$. Let $r \in P(\mathcal{M}z_1)$ be such that $pz_1 \sim r \leq qz_1$. We have

$$\tau(pz_1) = \tau(r) \leq \tau(qz_1).$$

Writing

$$z_1 = \int_Z^{\oplus} z_\zeta^{(1)}\, dw(\zeta),$$

for some measurable field $\zeta \to z_\zeta^{(1)}$ of central projections, we obtain

$$pz_1 = \int_Z^{\oplus} p_\zeta z_\zeta^{(1)}\, dw(\zeta),$$

$$r = rz_1 = \int_Z^{\oplus} r_\zeta z_\zeta^{(1)}\, dw(\zeta),$$

$$qz_1 = \int_Z^{\oplus} q_\zeta z_\zeta^{(1)}\, dw(\zeta).$$

Hence,

$$\tau(pz_1) = \int_Z \tau_\zeta(p_\zeta z_\zeta^{(1)})\, dw(\zeta), \quad \tau(qz) = \int_Z \tau_\zeta(q_\zeta z_\zeta^{(1)})\, dw(\zeta).$$

Since $p_\zeta \sim q_\zeta$ almost everywhere, it follows that $\tau_\zeta(p_\zeta z_\zeta^{(1)}) = \tau_\zeta(q_\zeta z_\zeta^{(1)})$ for almost all $\zeta \in Z$. Therefore, $\tau(r) = \tau(pz_1) = \tau(qz_1)$, and so $\tau(qz_1 - r) = 0$. Since τ is faithful and since $r \le qz_1$, it follows that $r = qz_1$. Thus, $pz_1 \sim qz_1$. Similarly, $q(\mathbf{1} - z_1) \sim p(\mathbf{1} - z_1)$. Referring to Theorem 1.9.4(vi), we infer that $p \sim q$, as required. □

We now turn to the spectral theory and functional calculus of (self-adjoint) decomposable operators.

Proposition 1.12.5 immediately implies the following.

Lemma 1.12.14. *Let $x = \int_Z^{\oplus} x_\zeta\, dw(\zeta)$ be a bounded decomposable operator. The operator x is a normal operator if and only if x_ζ is normal for almost all ζ.*

We start with a preliminary lemma on the spectrum of decomposable operators.

Lemma 1.12.15. *Suppose*

$$x = \int_Z^{\oplus} x_\zeta\, dw(\zeta)$$

is a bounded decomposable operator. For almost every $\zeta \in Z$, we have $\sigma(x_\zeta) \subseteq \sigma(x)$.

Proof. Let $\lambda \in \mathbb{C} \setminus \sigma(x)$ be fixed. Since all bounded decomposable operators form a von Neumann algebra, it follows that $(x - \lambda\mathbf{1})^{-1} \in \mathcal{R}$ is a bounded decomposable operator, i. e., $(x - \lambda\mathbf{1})^{-1} = \int_Z^{\oplus} y_\zeta\, dw(\zeta)$ for some bounded measurable field $\zeta \to y_\zeta$ of bounded operators. Using Proposition 1.12.5, we may write

$$\int_Z^{\oplus} \mathbf{1}_{H_\zeta}\, dw(\zeta) = \mathbf{1} = (x - \lambda\mathbf{1})(x - \lambda\mathbf{1})^{-1} = \int_Z^{\oplus} (x_\zeta - \lambda\mathbf{1}_{H_\zeta})y_\zeta\, dw(\zeta),$$

$$\int_Z^{\oplus} \mathbf{1}_{H_\zeta}\, dw(\zeta) = \mathbf{1} = (x - \lambda\mathbf{1})^{-1}(x - \lambda\mathbf{1}) = \int_Z^{\oplus} y_\zeta(x_\zeta - \lambda\mathbf{1}_{H_\zeta})\, dw(\zeta).$$

Thus, for almost every $\zeta \in Z$, we have that $x_\zeta - \lambda\mathbf{1}_{H_\zeta}$ is invertible and $(x_\zeta - \lambda\mathbf{1}_{H_\zeta})^{-1} = y_\zeta$. Therefore, $\lambda \in \mathbb{C} \setminus \sigma(x_\zeta)$, proving that $\sigma(x_\zeta) \subset \sigma(x)$, for almost every $\zeta \in Z$. □

Lemma 1.12.16. *Let $x = \int_Z^{\oplus} x_\zeta\, dw(\zeta)$ be a bounded normal decomposable operator. Using Lemma 1.12.15 and Lemma 1.12.14, by redefining x_ζ for ζ in a null set, if necessary, we may suppose x_ζ is normal and has a spectrum contained in $\sigma(x)$ for all ζ. Suppose $f : \sigma(x) \to \mathbb{C}$ is a continuous function. In the continuous functional calculus, we have*

$$f(x) = \int_Z^{\oplus} f(x_\zeta)\, dw(\zeta).$$

Proof. Take a sequence $\{g_k\}_{k\in\mathbb{N}}$ of polynomials in z and $\overline{z}$, such that $g_k(z,\overline{z})$ converges uniformly to $f(z)$ for all $z \in \sigma(x)$. Letting $\varepsilon_k = \max_{z\in\sigma(x)} |f(z) - g_k(z,\overline{z})|$, we have $\lim_{k\to\infty} \varepsilon_k = 0$. However, $\|f(x) - g_k(x,x^*)\| = \varepsilon_k$ and for each ζ, since $\sigma(x_\zeta) \subseteq \sigma(x)$, we have $\|f(x_\zeta) - g_k(x_\zeta, x_\zeta^*)\| \leq \varepsilon_k$. Thus,

$$\left\| \int_Z^{\oplus} f(x_\zeta)\, dw(\zeta) - \int_Z^{\oplus} g_k(x_\zeta, x_\zeta^*)\, dw(\zeta) \right\| \leq \varepsilon_k.$$

By Proposition 1.12.5, we have

$$g_k(x, x^*) = \int_Z^{\oplus} g_k(x_\zeta, x_\zeta^*)\, dw(\zeta).$$

Taking $k \to \infty$, we complete the proof. □

We next consider spectral projections of decomposable operators.

Proposition 1.12.17. *Suppose $x = \int_Z^{\oplus} x_\zeta\, dw(\zeta)$ is a bounded normal decomposable operator, and as above, assume without loss of generality x_ζ is normal and has spectrum contained in $\sigma(x)$ for all ζ. If B is a Borel subset of $\mathbb{C}$, then*

$$E_x(B) = \int_Z^{\oplus} E_{x_\zeta}(B)\, dw(\zeta). \tag{1.10}$$

Proof. First, suppose that B is a nonempty open, bounded, and rectangular subset in $\mathbb{C}$. Let $\{f_n\}_{n\in\mathbb{N}}$ be an increasing sequence of continuous functions on $\mathbb{C}$, each taking values in $[0,1]$ and vanishing outside B and such that f_n converges pointwise to χ_B as $n \to \infty$. By Lemma 1.12.16, we have

$$f_n(x) = \int_Z^{\oplus} f_n(x_\zeta)\, dw(\zeta).$$

Since f_n is increasing to χ_B, by the spectral theorem, $f_n(x)$ converges in the strong operator topology to $E_x(B)$. Similarly, for every ζ, $f_n(x_\zeta)$ converges in the strong operator topology to $E_{x_\zeta}(B)$, for all ζ. Thus, by [61, Section II.2.3, Proposition 4], $f_n(x)$ converges strongly to $\int_Z^{\oplus} E_{x_\zeta}(B)\, dw(\zeta)$. This proves the equality (1.10) when B is an open rectangle.

We now show that the set β of Borel sets B with the property (1.10) is a σ-algebra. First, if $B \in \beta$, then

$$\begin{aligned} E_x(B^c) &= \mathbf{1} - E_x(B) = \mathbf{1} - \int_Z^{\oplus} E_{x_\zeta}(B)\, dw(\zeta) \\ &= \int_Z^{\oplus} (\mathbf{1} - E_{x_\zeta}(B))\, dw(\zeta) = \int_Z^{\oplus} E_{x_\zeta}(B^c)\, dw(\zeta), \end{aligned}$$

so $B^c \in \beta$. Now let $\{B_n\}_{n\in\mathbb{N}}$ be a sequence of sets from β. For any $i, j \in \mathbb{N}$, we have

$$\begin{aligned} E_x(B_i \cup B_j) &= E_x(B_i) + E_x(B_j) - E_x(B_i)E_x(B_j) \\ &= \int_Z^{\oplus} (E_{x_\zeta}(B_i) + E_{x_\zeta}(B_j) - E_{x_\zeta}(B_i)E_{x_\zeta}(B_j))\, dw(\zeta) \\ &= \int_Z^{\oplus} E_{x_\zeta}(B_i \cup B_j)\, dw(\zeta), \end{aligned}$$

so $B_i \cup B_j \in \beta$. Hence, β is closed under finite unions. Thus, for every n, we have

$$E_x\left(\bigcup_{i=1}^{n} B_i\right) = \int_Z^{\oplus} E_{x_\zeta}\left(\bigcup_{i=1}^{n} B_i\right) dw(\zeta).$$

Since $E_x(\bigcup_{i=1}^{n} B_i)$ converges in the strong operator topology to $E_x(\bigcup_{i=1}^{\infty} B_i)$, and for each ζ, $E_{x_\zeta}(\bigcup_{i=1}^{n} B_i)$ converges in strong operator topology to $E_{x_\zeta}(\bigcup_{i=1}^{\infty} B_i)$, applying again [61, Section II.2.3, Proposition 4], we infer that

$$E_x\left(\bigcup_{i=1}^{\infty} B_i\right) = \int_Z^{\oplus} E_{x_\zeta}\left(\bigcup_{i=1}^{\infty} B_i\right) dw(\zeta).$$

Thus, β is a σ-algebra.

As β is the σ-algebra containing all bounded open rectangles, it follows that β contains the Borel σ-algebra of $\mathbb{C}$. □

From the above result, it is easy to show that an analogue of Lemma 1.12.16 holds for the Borel functional calculus.

Proposition 1.12.18. *Assume that $x = \int_Z^{\oplus} x_\zeta \, dw(\zeta)$ is a bounded normal decomposable operator. Using Lemma 1.12.15 and Lemma 1.12.14, by redefining x_ζ for ζ in a null set, if necessary, we may suppose x_ζ is normal and has a spectrum contained in $\sigma(x)$ for all ζ. For any bounded Borel function $f : \sigma(x) \to \mathbb{C}$, we have*

$$f(x) = \int_Z^{\oplus} f(x_\zeta) \, dw(\zeta).$$

Proof. Let $\varepsilon > 0$ and let $g = \sum_{j=1}^{n} a_j \chi_{B_j}$ be a Borel measurable simple function such that $\sup_{z \in \sigma(x)} |f(z) - g(z)| < \varepsilon$.

Referring to Proposition 1.12.17, we have

$$g(x) = \int_Z^{\oplus} g(x_\zeta) \, dw(\zeta).$$

We have that $\|g(x) - f(x)\| < \varepsilon$. Moreover, for all ζ we have $\|g(x_\zeta) - f(x_\zeta)\|_{\mathcal{B}(H_\zeta)} < \varepsilon$, and so we get

$$\left\| \int_Z^{\oplus} g(x_\zeta) \, dw(\zeta) - \int_Z^{\oplus} f(x_\zeta) \, dw(\zeta) \right\|_{\mathcal{B}(H)} \leq \varepsilon.$$

This yields

$$\left\| f(x) - \int_Z^{\oplus} f(x_\zeta) \, dw(\zeta) \right\|_{\mathcal{B}(H)} < 2\varepsilon.$$

Letting $\varepsilon \to 0$, we complete the proof. □

1.13 Direct integrals of unbounded operators

In this section, we introduce measurable fields of closed (not necessarily bounded) operators and unbounded decomposable operators. We study the spectral theory of unbounded decomposable operators and their (Borel) functional calculus.

As in Section 1.12, we assume that w is a fixed σ-finite positive measure on a standard Borel space Z, namely a Polish space endowed with the Borel σ-algebra. Let $H = \int_Z^{\oplus} H_\zeta dw(\zeta)$ be a direct integral Hilbert space.

Recall that the characteristic matrix $(p_{ij})_{i,j=1,2}$ of a closed operator x in a Hilbert space K is the matrix representation of the projection onto the graph of x in $K \times K$ (see Definition 1.3.10). We suggest the reader review properties of the characteristic matrix $(p_{ij})_{i,j=1,2}$ in Section 1.3 as they will be used extensively in this section.

We say that $\zeta \to x_\zeta$ is a *field of closed operators* if for every $\zeta \in Z$, x_ζ is a closed linear operator (not necessarily bounded) in H_ζ.

Definition 1.13.1. A field of closed operators $\zeta \to x_\zeta$ is said to be *measurable* if for every $i, j = 1, 2$, the field $\zeta \mapsto (p_\zeta)_{ij}$, $\zeta \in Z$, is a measurable field of bounded operators.

Remark 1.13.2. *Since* $\|p_{ij}\|_{\mathcal{M}} \le 1$, $i, j = 1, 2$ *for the characteristic matrix* $(p_{ij})_{i,j=1}^2$ *of any closed operator, it follows that for a measurable field* $\zeta \to x_\zeta$ *the field* $\zeta \to (p_\zeta)_{ij}$, $\zeta \in Z$, *is an essentially bounded measurable field of (bounded) operators (as defined in Definition* 1.12.3).

We first show that in the case of fields of bounded operators this definition agrees with the classical definition of measurability of fields of bounded operators (see Definition 1.12.3). We start with an auxiliary result.

Lemma 1.13.3. *Let* $\zeta \to x_\zeta$ *be a bounded measurable field of self-adjoint bounded operators such that* x_ζ *is injective for all* ζ. *The field* $\zeta \to x_\zeta^{-1}\xi_\zeta$ *is measurable for every measurable vector field* $\zeta \to \xi_\zeta$, *such that* $\xi_\zeta \in \operatorname{dom}(x_\zeta^{-1})$ *for all* $\zeta \in Z$.

Proof. As $\zeta \to x_\zeta$ is a bounded measurable field of operators, we can define a self-adjoint decomposable operator $x = \int_Z^\oplus x_\zeta \, dw(\zeta)$. If $\xi = \int_Z^\oplus \xi_\zeta \, dw(\zeta)$ is such that $x\xi = 0$, then $\int_Z \|x_\zeta \xi_\zeta\|^2 \, dw(\zeta) = 0$ and, therefore, $x_\zeta \xi_\zeta = 0$ almost everywhere. Since x_ζ is injective, it follows that $\xi_\zeta = 0$ almost everywhere, proving that $\xi = 0$. Hence, x is injective, too.

Now let $\zeta \to \xi_\zeta$ be a measurable vector field such that $\xi_\zeta \in \operatorname{dom}(x_\zeta^{-1})$ for all ζ. Let $y_\zeta = x_\zeta^{-1}\xi_\zeta$. It is sufficient to show that $\zeta \to \langle y_\zeta, \psi_\zeta \rangle$ is a measurable function for all measurable vector fields $\zeta \to \psi_\zeta$. To prove this, it is clearly sufficient to show this for all $\psi \in M$, where M is a dense subset of $\int_Z^\oplus H_\zeta \, dw(\zeta)$. Let M be the range of x. Since x is injective, it follows that $M^\perp = \ker(x^*) = \ker(x) = \{0\}$, and so M is dense in H. For $\psi \in M$, we have $\psi = x\delta$, for some $\delta = \int_Z^\oplus \delta_\zeta \, dw(\zeta) \in H$. Therefore, we have

$$\zeta \to \langle x_\zeta^{-1}\xi_\zeta, \psi_\zeta \rangle = \langle x_\zeta^{-1}\xi_\zeta, x_\zeta \delta_\zeta \rangle = \langle \xi_\zeta, \delta_\zeta \rangle.$$

Since both $\zeta \to \xi_\zeta$ and $\zeta \to \delta_\zeta$ are measurable fields, it follows that $\zeta \to \langle x_\zeta^{-1}\xi_\zeta, \psi_\zeta \rangle$ is measurable, too. This proves the assertion. □

Proposition 1.13.4. *Let* $\zeta \to x_\zeta$ *be an essentially bounded field of bounded operators and* $((p_\zeta)_{ij})$ *the characteristic matrix of* x_ζ. *The field* $\zeta \to x_\zeta$ *of bounded operators is measurable (in the sense of Definition* 1.12.3) *if and only if the fields of characteristic matrices* $\zeta \to (p_\zeta)_{ij}$ *are measurable for every* $i, j = 1, 2$.

Proof. Suppose first that $\zeta \to x_\zeta$ is a bounded measurable field of bounded operators. By Proposition 1.12.5, the fields $\zeta \to \mathbf{1}_{H_\zeta} + x_\zeta^* x_\zeta$ and $\zeta \to \mathbf{1}_{H_\zeta} + x_\zeta x_\zeta^*$ are bounded measurable fields. Lemma 1.13.3 implies that the fields $\zeta \to (\mathbf{1}_{H_\zeta} + x_\zeta^* x_\zeta)^{-1}$, $\zeta \to (\mathbf{1}_{H_\zeta} + x_\zeta x_\zeta^*)^{-1}$ are measurable. By Proposition 1.3.15, we have that

$$(p_\zeta)_{11} = (x_\zeta^* x_\zeta + \mathbf{1}_{H_\zeta})^{-1}, \qquad (p_\zeta)_{12} = x_\zeta^* (x_\zeta x_\zeta^* + \mathbf{1}_{H_\zeta})^{-1},$$
$$(p_\zeta)_{21} = x_\zeta (x_\zeta^* x_\zeta + \mathbf{1}_{H_\zeta})^{-1}, \qquad (p_\zeta)_{22} = \mathbf{1}_{H_\zeta} - (x_\zeta x_\zeta^* + \mathbf{1}_{H_\zeta})^{-1}.$$

In particular, for every $i, j = 1, 2$, the field $\zeta \to (p_\zeta)_{ij}$ is measurable, as required.

Conversely, suppose that for every $i, j = 1, 2$ the field $\zeta \to (p_\zeta)_{ij}$, $i, j = 1, 2$ is measurable. Let $\zeta \to \xi_\zeta$ be a measurable vector field. Using again Proposition 1.3.15 and the equality $x_\zeta (x_\zeta^* x_\zeta + \mathbf{1}_{H_\zeta})^{-1} = (x_\zeta x_\zeta^* + \mathbf{1}_{H_\zeta})^{-1} x_\zeta$, we have that

$$(\mathbf{1}_{H_\zeta} - (p_\zeta)_{22}) x_\zeta \xi_\zeta = (p_\zeta)_{21} \xi_\zeta.$$

Since the field $\zeta \to (p_\zeta)_{21} \xi_\zeta$ is a (bounded) measurable vector field and since $\mathbf{1}_{H_\zeta} - (p_\zeta)_{22} = (x_\zeta x_\zeta^* + \mathbf{1}_{H_\zeta})^{-1}$ has an inverse for all ζ, it follows from Lemma 1.13.3 that

$$\zeta \to x_\zeta \xi_\zeta = (\mathbf{1}_{H_\zeta} - (p_\zeta)_{22})^{-1} (p_\zeta)_{21} \xi_\zeta$$

is a measurable field. Thus, the field $\zeta \to x_\zeta$ is measurable (as defined in Definition 1.12.3). □

Thus, in the case of bounded operators the definition of measurability in terms of the characteristic matrix agrees with the classical definition of measurability.

We also introduce a weaker notion of measurability of fields of closed operators.

Definition 1.13.5. A field $\zeta \to x_\zeta$ of closed operators is said to be *weakly measurable* if for every measurable field $\zeta \to \xi_\zeta$, such that $\xi_\zeta \in \mathrm{dom}(x_\zeta)$ for all $\zeta \in Z$ the field $\zeta \to x_\zeta \xi_\zeta$ is measurable.

As we show in the following proposition, measurability of a field of closed operators implies weak measurability of this field.

Proposition 1.13.6. *Every measurable field $\zeta \to x_\zeta$ of closed operators is weakly measurable.*

Proof. Suppose that $\zeta \to \xi_\zeta$ is such that $\xi_\zeta \in \mathrm{dom}(x_\zeta)$ for all $\zeta \in Z$. By Proposition 1.3.11, we have $x_\zeta \xi_\zeta = (p_\zeta)_{21} \xi_\zeta + (p_\zeta)_{22} x_\zeta \xi_\zeta$. Therefore,

$$(\mathbf{1}_{H_\zeta} - (p_\zeta)_{22}) x_\zeta \xi_\zeta = (p_\zeta)_{21} \xi_\zeta.$$

Referring to Proposition 1.3.11, we infer that $(\mathbf{1}_{H_\zeta} - (p_\zeta)_{22})$ has an inverse for all ζ. Hence, Lemma 1.13.3 implies that

$$\zeta \to x_\zeta \xi_\zeta = (\mathbf{1}_{H_\zeta} - (p_\zeta)_{22})^{-1} (p_\zeta)_{21} \xi_\zeta$$

is a measurable field, as required. □

Proposition 1.13.4 implies, in particular, that for bounded operators weak measurability is equivalent to measurability. However, as the following example shows, for unbounded operators this is no longer the case.

Example 1.13.7. *Let $(Z, w) = (\mathbb{R}, m)$, where m is the Lebesgue measure on $\mathbb{R}$ and let $H_\zeta = K$ for a fixed separable Hilbert space K. Suppose that y_0 and y_1 are densely defined closed symmetric (unbounded) operators in K, such that $y_0 \subsetneq y_1$. Let A be a nonmeasurable set of $\mathbb{R}$ (with respect to the Lebesgue measure). The field $\zeta \to x_\zeta$, $\zeta \in \mathbb{R}$, defined by*

$$x_\zeta = \begin{cases} y_0, & \zeta \in A, \\ y_1, & \zeta \in \mathbb{R} \setminus A \end{cases}$$

of closed densely defined operators is weakly measurable, but not measurable.

Proof. We first show that the field $\zeta \to x_\zeta$ is weakly measurable.

Let $\zeta \to \xi_\zeta$ be a measurable field, such that $\xi_\zeta \in \operatorname{dom}(x_\zeta)$ for all $\zeta \in \mathbb{R}$. Since $y_0 \subset y_1$ and y_1 is symmetric, it follows that

$$y_0 \subset y_1 \subset y_1^* \subset y_0^*.$$

In particular, $\xi_\zeta \in \operatorname{dom}(y_0^*)$ for all $\zeta \in \mathbb{R}$. Therefore, for any $\eta \in \operatorname{dom}(y_0)$ we have

$$\langle \eta, x_\zeta \xi_\zeta \rangle = \langle y_0 \eta, \xi_\zeta \rangle.$$

This implies that $\zeta \to \langle \eta, x_\zeta \xi_\zeta \rangle$ is measurable. Since $\operatorname{dom}(y_0)$ is dense, it follows that $\zeta \to x_\zeta \xi_\zeta$ is measurable, too (see, e. g., [61, Part II, Chapter 1, Section 3]). Thus, $\zeta \to x_\zeta$ is a weakly measurable field.

Next, we show that $\zeta \to x_\zeta$ is not measurable. We claim that the field $\zeta \to (p_\zeta)_{11}$ is not measurable. By Proposition 1.3.15, we have that

$$(p_\zeta)_{11} = (\mathbf{1}_K + x_\zeta^* x_\zeta)^{-1} = \begin{cases} (\mathbf{1}_K + y_0^* y_0)^{-1}, & \zeta \in A, \\ (\mathbf{1}_K + y_1^* y_1)^{-1}, & \zeta \in \mathbb{R} \setminus A. \end{cases}$$

For any $\xi, \eta \in K$ setting,

$$a_j = \langle \xi, (\mathbf{1}_K + y_j^* y_j)^{-1} \eta \rangle, \quad j = 1, 2,$$

and we have that

$$\langle \xi, (\mathbf{1}_K + x_\zeta^* x_\zeta)^{-1} \eta \rangle = \langle \xi, (\mathbf{1}_K + y_0^* y_0)^{-1} \eta \rangle \chi_A + \langle \xi, (\mathbf{1}_K + y_1^* y_1)^{-1} \eta \rangle \chi_{\mathbb{R} \setminus A}$$

$$= a_0\chi_A + a_1\chi_{\mathbb{R}\setminus A}.$$

Since A is not measurable, it is sufficient to find $\xi,\eta \in K$, such that $a_0 \neq a_1$.

Suppose the contrary, that for any $\xi,\eta \in K$, we have $a_0 = a_1$. By definition of a_0 and a_1, we have

$$\langle \xi, (\mathbf{1}_K + y_0^*y_0)^{-1}\eta\rangle = \langle \xi, (\mathbf{1}_K + y_1^*y_1)^{-1}\eta\rangle, \quad \xi,\eta \in K.$$

This implies that $(\mathbf{1}_K + y_0^*y_0)^{-1} = (\mathbf{1}_K + y_1^*y_1)^{-1}$, and so $y_0^*y_0 = y_1^*y_1$. Since

$$\mathrm{dom}(|y_0|^2) = \mathrm{dom}(y_0^*y_0) = \mathrm{dom}(y_1^*y_1) = \mathrm{dom}(|y_1|^2),$$

it follows that $\mathrm{dom}(|y_0|) = \mathrm{dom}(|y_1|)$, and so

$$\mathrm{dom}(y_0) = \mathrm{dom}(|y_0|) = \mathrm{dom}(|y_1|) = \mathrm{dom}(y_1).$$

The latter equality contradicts the assumption that $y_0 \subsetneq y_1$. This contradiction implies that $\zeta \to (p_\zeta)_{11}$ is not measurable, as required. □

Definition 1.13.8. Suppose that $\zeta \to x_\zeta$ is a field of closed operators (not necessarily measurable or weakly measurable). In $H = \int_Z H_\zeta\, dw(\zeta)$, define the operator x maximally defined by the field $\zeta \to x_\zeta$ as follows:

$$(x\xi)(\zeta) = x_\zeta\xi_\zeta,$$

$$\xi \in \mathrm{dom}(x) = \Big\{\xi \in \int_Z^{\oplus} H_\zeta\, dw(\zeta) : \xi_\zeta \in \mathrm{dom}(x_\zeta) \text{ a. e.},$$

$$\zeta \to x_\zeta\xi_\zeta \text{ is measurable}, \int_Z \|x_\zeta\xi_\zeta\|_{H_\zeta}^2\, dw(\zeta) < \infty\Big\}.$$

We say that x is *associated* with the field $\zeta \to x_\zeta$.

We note that for a weakly measurable (in particular, measurable) field $\zeta \to x_\zeta$ the domain of the associated operator x can be written as

$$\mathrm{dom}(x) = \Big\{\{\xi_\zeta\} : \xi_\zeta \in \mathrm{dom}(x_\zeta) \text{ a. e.}, \int_Z \|x_\zeta\xi_\zeta\|_{H_\zeta}^2\, dw(\zeta) < \infty\Big\}.$$

As the following lemma shows, the operator x associated with a field $\zeta \to x_\zeta$ of closed operators is also closed.

Lemma 1.13.9. *Assume that $\zeta \to x_\zeta$ is a (not necessarily measurable or weakly measurable) field of closed operators. The operator x associated with the field $\zeta \to x_\zeta$ (as defined in Definition* 1.13.8*) is closed.*

Proof. Suppose that $\{\xi_n\}_{n\in\mathbb{N}} \in \operatorname{dom} x$ is such that $\xi_n \to \xi$ and $x\xi_n \to \eta$ in H. Writing $\xi_n = \int_Z^\oplus \xi_\zeta^{(n)}\, dw(\zeta)$, $\xi = \int_Z^\oplus \xi_\zeta\, dw(\zeta)$, and $\eta = \int_Z^\oplus \eta_\zeta\, dw(\zeta)$, we have

$$\int_Z (\|x_\zeta \xi_\zeta^{(n)} - \eta_\zeta\|_{H_\zeta}^2 + \|\xi_\zeta^{(n)} - \xi_\zeta\|_{H_\zeta}^2)\, dw(\zeta) \to 0.$$

Therefore, there exists a subsequence $\{\xi_{n_k}\}_{k\in\mathbb{N}}$, such that

$$(\|x_\zeta \xi_\zeta^{(n_k)} - \eta_\zeta\|_{H_\zeta}^2 + \|\xi_\zeta^{(n_k)} - \xi_\zeta\|_{H_\zeta}^2) \to 0$$

almost everywhere as $n_k \to \infty$. Since x_ζ is closed, it implies that for almost all ζ we have $\xi_\zeta \in \operatorname{dom} x_\zeta$ and $\eta_\zeta = x_\zeta \xi_\zeta$. Hence, $\xi \in \operatorname{dom}(x)$ and $x\xi = \eta$, proving that x is closed. □

We now introduce a class of decomposable operators as those associated with measurable fields of closed operators.

Definition 1.13.10. An operator x in the Hilbert space $H = \int_Z^\oplus H_\zeta\, dw(\zeta)$ is called *decomposable* if there exists a measurable field $\zeta \to x_\zeta$ of closed operators, such that x is the operator associated with $\zeta \to x_\zeta$, i. e.,

$$(x\xi)(\zeta) = x_\zeta \xi_\zeta,$$

$$\xi \in \operatorname{dom}(x) = \left\{\xi \in \int_Z^\oplus H_\zeta\, dw(\zeta) : \xi_\zeta \in \operatorname{dom}(x_\zeta) \text{ a. e.}, \int_Z \|x_\zeta \xi_\zeta\|_{H_\zeta}^2\, dw(\zeta) < \infty\right\}.$$

In this case, we write

$$x = \int_Z^\oplus x_\zeta\, dw(\zeta).$$

By Lemma 1.13.9, any decomposable operator is necessarily closed.

Note that for bounded measurable fields this definition of decomposable operators agrees with Definition 1.12.4.

Remark 1.13.11. *Suppose that x is the operator associated with a field $\zeta \to x_\zeta$ of closed linear operators and let $y = \int_Z^\oplus f(\zeta)\mathbf{1}_{H_\zeta}\, dw(\zeta)$ be a bounded diagonalizable operator (i. e., f is a bounded w-measurable function on Z). In this case, we have $yx \subset xy$. Indeed, for any $\xi = \int_Z^\oplus \xi_\zeta\, dw(\zeta) \in \operatorname{dom}(x)$, we have that $f(\zeta)\xi_\zeta \in \operatorname{dom}(x_\zeta)$ a. e., $\zeta \to x_\zeta f(\zeta)\xi_\zeta = f(\zeta)x_\zeta\xi_\zeta$ is measurable, and*

$$\int_Z \|x_\zeta f(\zeta)\xi_\zeta\|_{H_\zeta}^2\, dw(\zeta) \le \|f\|_\infty \int_Z \|x_\zeta \xi_\zeta\|_{H_\zeta}^2\, dw(\zeta) < \infty.$$

Thus, $\operatorname{dom}(yx) = \operatorname{dom}(x) \subset \operatorname{dom}(xy)$. *Furthermore,*

$$((yx)\xi)(\zeta) = y(x_\zeta \xi_\zeta) = f(\zeta)(x_\zeta \xi_\zeta) = x_\zeta f(\zeta)\xi_\zeta = ((xy)\xi)(\zeta),$$

which implies that $yx \subset xy$.

Recall that a bounded operator on $H = \int_Z^\oplus H_\zeta \, dw(\zeta)$ is decomposable if and only if it commutes with all bounded diagonalizable operators (see Theorem 1.12.6). A similar criterion holds for any closed linear operators on H. To prove this, we first establish an auxiliary result showing that the characteristic matrix of a decomposable operator is also a (bounded) decomposable operator.

Proposition 1.13.12. *If* $x = \int_Z^\oplus x_\zeta \, dw(\zeta)$ *is a closed decomposable operator, then the characteristic matrix* (p_{ij}) *of* x *is also decomposable and*

$$p_{ij} = \int_Z^\oplus (p_\zeta)_{ij} \, dw(\zeta),$$

where $((p_\zeta)_{ij})$ *denotes the characteristic matrix of* x_ζ.

Proof. By assumption for each $i, j = 1, 2$, the fields $\zeta \to (p_\zeta)_{ij}$ are bounded measurable fields of bounded operators. Hence, there exist bounded operators $q_{ij} \in \mathcal{M}$, $i, j = 1, 2$, such that $q_{ij} = \int_Z^\oplus (p_\zeta)_{ij} \, dw(\zeta)$. We claim that (q_{ij}) is the characteristic matrix of a closed operator in H. By Proposition 1.3.11, it is sufficient to show that $(q_{ij})_{i,j=1}^2$ is a projection and $\ker(\mathbf{1} - q_{22}) = \{0\}$. Since $((p_\zeta)_{ij})$ is the characteristic matrix for x_ζ, it follows from Proposition 1.12.5 that

$$q_{ij}^* = \int_Z^\oplus (p_\zeta)_{ij}^* \, dw(\zeta) = \int_Z^\oplus (p_\zeta)_{ji} \, dw(\zeta) = q_{ji},$$

$$\sum_{k=1}^{2} (q_{ik} q_{kj}) = \int_Z^\oplus \sum_{k=1}^{2} (p_\zeta)_{ik} (p_\zeta)_{kj} \, dw(\zeta) = \int_Z^\oplus (p_\zeta)_{ij} \, dw(\zeta) = q_{ij}.$$

Thus, $(q_{ij})_{i,j=1}^2$ is a projection.

Moreover, since $((p_\zeta)_{ij})$ is the characteristic matrix for x_ζ, it follows that $(\mathbf{1}_{H_\zeta} - (p_\zeta)_{22})$ is invertible for every ζ (see Proposition 1.3.11). Lemma 1.13.3 implies that $\mathbf{1} - q_{22}$ is invertible, too. Thus, by Proposition 1.3.11, (q_{ij}) is the characteristic matrix of a closed operator y.

To complete the proof, it remains to show that $y = x$. Since both x and y are closed, it is sufficient to show that $\Gamma(y) = \Gamma(x)$. Note that for $\xi = \int_Z^\oplus \xi_\zeta \, dw(\zeta), \gamma = \int_Z^\oplus \gamma_\zeta \, dw(\zeta) \in H$, we have the inclusion $(\xi, \gamma) \in \Gamma(y)$ if and only if $\xi = q_{11}\xi + q_{12}\gamma$ and $\gamma = q_{21}\xi + q_{22}\gamma$. In this case, $\xi_\zeta = (p_{11})_\zeta \xi_\zeta + (p_{12})_\zeta \gamma_\zeta$ and $\gamma_\zeta = (p_{21})_\zeta \xi_\zeta + (p_{22})_\zeta \gamma_\zeta$ for almost all ζ. The last two relations are equivalent to the inclusion $(\xi_\zeta, y_\zeta) \in \Gamma(x_\zeta)$ for almost all ζ, which is in turn equivalent to $(\xi, y) \in \Gamma(x)$. Thus $\Gamma(y) = \Gamma(x)$, as required. □

Lemma 1.13.13. *If $x = \int_Z^\oplus x_\zeta \, dw(\zeta) \in \mathcal{M}$ is a bounded decomposable operator, then x has an inverse if and only if almost all x_ζ have an inverse.*

Proof. If almost all x_ζ have an inverse, then x has an inverse (cf. proof of Lemma 1.13.3).

Conversely, suppose that x has an inverse, or equivalently, $s_l(x^*) = \mathbf{1}$.

Let $\{\xi_n\}_{n\in\mathbb{N}} = \{\int_Z^\oplus \xi_\zeta^{(n)} \, dw(\zeta)\}_{n\in\mathbb{N}}$ be a dense sequence of measurable vector fields such that the functions $\zeta \to \|\xi_\zeta^{(n)}\|^2$ are integrable and $L(Z)$ the set of all continuous functions on Z with compact support. Let M be the set of all elements of the form $\zeta \to f(\zeta)\xi_\zeta^{(n)}$, $n \in \mathbb{N}$, and $f \in L(Z)$. Clearly, M is dense in H. Since x has inverse, it follows that $x^*(M)$ is dense in H.

Let C be any compact subset of Z and f an element of $L(Z)$, which has the value 1 on C. Since $x^*(M)$ is dense in H, it follows that, for every $n \in \mathbb{N}$, there exists a sequence of elements $\{y_{n,k}\}_{k\geq1}$ in M such that $x^* y_{n,k} \to f\xi_n$ as $k \to \infty$ in the metric of H. Hence, there exists a subsequence $\{\delta_{n,k}\}_{k\geq1}$ of $\{y_{n,k}\}_{k\geq1}$, such that $(x^*)_\zeta(\delta_\zeta^{(n,k)}) \to f(\zeta)\xi_\zeta^{(n)}$ for all $\zeta \in N_n$, where N_n is a w-null set. Hence, $s_l(x_\zeta^*)\xi_\zeta^{(n)} = \xi_\zeta^{(n)}$ for all $\zeta \in C \setminus N_n$. Let $N = \bigcup_{n\in\mathbb{N}} N_n$. Clearly, $s_l(x_\zeta^*)\xi_\zeta^{(n)} = \xi_\zeta^{(n)}$ for all n and $\zeta \in C \setminus N$, i. e., $s_l(x_\zeta^*) = \mathbf{1}_{H_\zeta}$ for all $\zeta \in C \setminus N$. Thus, x_ζ has an inverse for all $\zeta \in C \setminus N$. □

As in Section 1.12, let us denote by $\mathcal{R}$ the von Neumann algebra of all bounded decomposable operators in H and by $\mathcal{C}$ the von Neumann algebra of all bounded diagonalizable operators.

Theorem 1.13.14. *A closed linear operator in $H = \int_Z^\oplus H_\zeta \, dw(\zeta)$ is decomposable if and only if x commutes with any bounded diagonalizable operator.*

Proof. Assume first that x is a closed linear operator, which commutes with any operator in $\mathcal{C}$. By Proposition 1.3.17, the characteristic matrix $(p_{ij})_{i,j=1}^2$ of x commutes with every bounded diagonalizable operator in H. Theorem 1.12.6 implies that for every $i, j = 1, 2$ the bounded operator p_{ij} is decomposable for each $i, j = 1, 2$. Denote by $(p_{ij})_\zeta$ a bounded measurable field of bounded operators, such that $p_{ij} = \int_Z^\oplus (p_{ij})_\zeta \, dw(\zeta)$. Since $p_{ij}^* = p_{ji}$, it follows that $(p_{ij})_\zeta^* = (p_{ji})_\zeta$ for almost all ζ. Similarly, since $\sum_{k=1}^2 p_{ik}p_{kj} = p_{ij}$, it follows that $\sum_{k=1}^2 (p_{ik})_\zeta (p_{kj})_\zeta = (p_{ij})_\zeta$ for almost all ζ. Furthermore, since $(p_{ij})_{i,j=1}^2$ is the characteristic matrix of a closed operator, it follows that $\mathbf{1} - p_{22}$ is invertible (see Proposition 1.3.11). Hence, by Lemma 1.13.13, the operator $\mathbf{1} - (p_{22})_\zeta$ has inverse for almost all ζ. In other words, there exists a measure zero set $N \subset Z$, such that for all $\zeta \in Z \setminus N$ the matrix $((p_{ij})_\zeta)_{i,j=1,2}$ is the characteristic matrix of a closed linear operator (see Proposition 1.3.11). For all $\zeta \in Z \setminus N$, we let y_ζ be a closed operator in H_ζ with the characteristic matrix $((p_{ij})_\zeta)_{i,j=1,2}$. Define a field of closed linear operators $\zeta \to x_\zeta$, such that $x_\zeta = y_\zeta$ for $\zeta \in Z \setminus N$ and $x_\zeta = 0$ for $\zeta \in N$. By definition, $\zeta \to x_\zeta$ is a measurable field of closed linear operators.

Let $y = \int_Z^{\oplus} x_\zeta \, dw(\zeta)$ be the decomposable operator associated with the measurable field $\zeta \to x_\zeta$ (see Definition 1.13.10). By Lemma 1.13.9, the operator y is closed. Denote by $\{q_{ij}\}$ the characteristic matrix of y. By Proposition 1.13.12, we have that

$$q_{ij} = \int_Z^{\oplus} (p_{ij})_\zeta \, dw(\zeta) = p_{ij}.$$

Thus, the two closed operators x and y have the same characteristic matrix, and so $x = y$, which proves that x is decomposable.

Conversely, assume that x is decomposable. By Remark 1.13.11, we have that $yx \subset xy$ for any bounded diagonalizable operator y, as required. □

Remark 1.13.15. *We note that if x is a self-adjoint decomposable operator, then $f(x)$ is a decomposable operator for any Borel function f on $\sigma(x)$. Indeed, by Theorem* 1.13.14, *x is decomposable if and only if x commutes with all bounded diagonalizable operators. By Theorem* 1.4.8, *this implies that spectral projections of x (and hence of $f(x)$) commute with all bounded diagonalizable operators. Therefore, using again Theorem* 1.4.8, *we obtain that the self-adjoint operator $f(x)$ commutes with all bounded diagonalizable operators, and is therefore decomposable. In Theorem* 1.13.24 *below, we will show that as expected the Borel functional calculus acts fiberwise.*

We now proceed to the study of properties of decomposable operators. We first show that for a decomposable operator $x = \int_Z^{\oplus} x_\zeta \, dw(\zeta)$, the measurable field $\zeta \to x_\zeta$ is defined uniquely up to a set of measure zero.

Proposition 1.13.16. *Assume that x and y are two decomposable operators with $x = \int_Z^{\oplus} x_\zeta \, dw(\zeta)$ and $y = \int_Z^{\oplus} y_\zeta \, dw(\zeta)$. We have $x \subset y$ if and only if $x_\zeta \subset y_\zeta$ almost everywhere. In particular, $x = y$ if and only if $x_\zeta = y_\zeta$ almost everywhere.*

Proof. Let $p_\zeta = ((p_\zeta)_{ij})_{i,j=1}^2$ and $q_\zeta = ((q_\zeta)_{ij})_{i,j=1}^2$ be the characteristic matrices of x_ζ and y_ζ, respectively, and let $p = (p_{ij})_{i,j=1}^2$ and $q = (q_{ij})_{i,j=1}^2$ be the characteristic matrices for x and y, respectively. By Proposition 1.13.12, we have

$$p_{ij} = \int_Z^{\oplus} (p_\zeta)_{ij} \, dw(\zeta), \quad q_{ij} = \int_Z^{\oplus} (q_\zeta)_{ij} \, dw(\zeta). \tag{1.11}$$

Since $x \subset y$ if and only if $\Gamma(x) \subset \Gamma(y)$ and x is closed, it follows that $x \subset y$ if and only if $pq = p$. The equalities (1.11) imply that $x \subset y$ if and only if $p_\zeta q_\zeta = p_\zeta$ almost everywhere. Therefore, $x \subset y$ if and only if $x_\zeta \subset y_\zeta$ almost everywhere. □

Proposition 1.13.16 guarantees that for a decomposable operator $x = \int_Z^{\oplus} x_\zeta \, dw(\zeta)$ the measurable field $\zeta \to x_\zeta$ is defined up to a set of measure zero. In particular, given a measure zero set $N \subset Z$, the closed operators $x_\zeta, \zeta \in N$ can be chosen arbitrarily.

Next we establish a necessary and sufficient condition for a decomposable operator to be densely defined.

Proposition 1.13.17. *A decomposable operator* $x = \int_Z^{\oplus} x_\zeta \, dw(\zeta)$ *is densely defined if and only if* x_ζ *is densely defined for almost all* ζ.

Proof. Denote by $(p_{ij})_{i,j=1}^2$ and $((p_\zeta)_{ij})_{i,j=1}^2$ the characteristic matrices of x and x_ζ, respectively. By Proposition 1.3.12, x (resp., x_ζ) is densely defined if and only if p_{11} (resp., $(p_\zeta)_{11}$) has inverse. By Proposition 1.13.12, we have that

$$p_{ij} = \int_Z^{\oplus} (p_\zeta)_{ij} \, dw(\zeta).$$

By Lemma 1.13.13, the bounded decomposable operator p_{11} has inverse if and only if almost all $(p_\zeta)_{ij}$ has inverse. Thus, x is densely defined if and only if almost every x_ζ is densely defined. □

The above proposition together with Lemma 1.13.9 guarantees that the operator associated with a measurable field of closed densely defined operators is necessarily closed and densely defined. In the following example, we will show that the operator associated with a weakly measurable field of closed densely defined operators is, generally speaking, not densely defined. In particular, this example shows the necessity of defining measurability in terms of characteristic matrices rather than a seemingly natural definition of weak measurability.

Example 1.13.18. *Let* $Z = [0,1]$ *with the Lebesgue measure on* $[0,1]$ *and let* $H_\zeta = K$ *for all* $\zeta \in Z$ *and some separable Hilbert space* K. *Suppose that* a_0 *is a densely defined closed symmetric operator in a separable Hilbert space* K *with deficiency indices* $(1,1)$ *(see, e. g., [126]). Let* a_1 *be a self-adjoint extension of* a_0. *Since the deficiency indices of* a_0 *are equal to 1, it follows that there exists a unit vector* $\eta_0 \in \ker(a_0^* - i\mathbf{1}_K)$. *Setting* $b = \mathbf{1} + p_{\eta_0}$, *where* p_{η_0} *is the projection onto the space spanned by* η_0, *we define*

$$y_0 = a_0 + i\mathbf{1}_K, \quad y_1 = a_1 + i\mathbf{1}_K, \quad y_2 = by_1.$$

For a nonmeasurable set $A \subset [0,1]$, *define the field*

$$x_\zeta = \begin{cases} y_1^*, & \zeta \in A, \\ y_2^*, & \zeta \in [0,1] \setminus A. \end{cases}$$

We now have:

(i) *The field* $\zeta \to x_\zeta$ *is weakly measurable;*

(ii) *The field* $\zeta \to x_\zeta^*$ *is not weakly measurable;*

(iii) *The operator* x *associated with* $\zeta \to x_\zeta$ *is not densely defined.*

Proof. We first show that $\zeta \to x_\zeta$ is weakly measurable. Let $\zeta \to \xi_\zeta$ be a measurable field, such that $\xi_\zeta \in \operatorname{dom}(x_\zeta)$ for all $\zeta \in [0,1]$. As $\operatorname{dom}(y_0) = \operatorname{dom}(a_0)$ is dense in K, it is sufficient to show that $\zeta \to \langle \xi, x_\zeta \xi_\zeta \rangle$ is measurable for all $\xi \in \operatorname{dom}(a_0)$. Note that

$$x_\zeta^* = \begin{cases} y_1, & \zeta \in A, \\ y_2, & \zeta \in [0,1] \setminus A, \end{cases}$$

and so $\operatorname{dom}(a_0) = \operatorname{dom}(y_0) \subset \operatorname{dom}(y_i)$, $i = 1, 2$. Therefore, for any $\xi \in \operatorname{dom}(a_0)$, we have

$$\langle \xi, x_\zeta \xi_\zeta \rangle = \langle \xi, x_\zeta^{**} \xi_\zeta \rangle = \langle x_\zeta^* \xi, \xi_\zeta \rangle. \tag{1.12}$$

Since a_1 is an extension of a_0, it follows that $y_1 \xi = y_0 \xi$ for all $\xi \in \operatorname{dom}(a_0)$. Since $\eta_0 \in \ker(a_0^* - i\mathbf{1}_K) = \ker(y_0^*)$, it follows that

$$\begin{aligned} y_2 \xi &= y_1 \xi + \langle y_1 \xi, \eta_0 \rangle \eta_0 = y_0 \xi + \langle y_0 \xi, \eta_0 \rangle \eta_0 \\ &= y_0 \xi + \langle \xi, y_0^* \eta_0 \rangle \eta_0 = y_0 \xi \end{aligned}$$

for all $\xi \in \operatorname{dom}(a_0)$. Combining the latter with equality (1.12), we obtain that

$$\langle \xi, x_\zeta \xi_\zeta \rangle = \langle x_\zeta^* \xi, \xi_\zeta \rangle = \begin{cases} \langle y_1 \xi, \xi_\zeta \rangle, & \zeta \in A \\ \langle y_2 \xi, \xi_\zeta \rangle, & \zeta \in [0,1] \setminus A \end{cases} = \langle y_0 \xi, \xi_\zeta \rangle.$$

This proves that the field $\zeta \to x_\zeta$ is weakly measurable.

Next, we show that the field $\zeta \to x_\zeta^*$ is not weakly measurable. Since a_1 is self-adjoint, it follows that $\ker(y_1^*) = \ker(a_1 - i\mathbf{1}_K) = \{0\}$, and so $\operatorname{ran}(y_1)$ is dense in K. Therefore, there exists $\xi_0 \in \operatorname{dom}(y_1)$, such that $\langle y_1 \xi_0, \eta_0 \rangle \neq 0$. Since $\operatorname{dom}(y_2) = \operatorname{dom}(y_1)$, it follows that $\xi_0 \in \operatorname{dom}(y_2)$ and

$$\langle y_2 \xi_0, \eta_0 \rangle = \langle (\mathbf{1}_K + p_{\eta_0}) y_1 \xi_0, \eta_0 \rangle = \langle y_1 \xi_0, \eta_0 \rangle + \langle y_1 \xi_0, \eta_0 \rangle \langle \eta_0, \eta_0 \rangle = 2 \langle y_1 \xi_0, \eta_0 \rangle.$$

Using the fact that $\operatorname{dom}(x_\zeta^*) = \operatorname{dom}(y_1)$ for all ζ, we infer that

$$\langle x_\zeta^* \xi_0, \eta_0 \rangle = \begin{cases} \langle y_1 \xi_0, \eta_0 \rangle, & \zeta \in A, \\ \langle y_2 \xi_0, \eta_0 \rangle, & \zeta \in [0,1] \setminus A \end{cases} = \begin{cases} \langle y_1 \xi_0, \eta_0 \rangle, & \zeta \in A, \\ 2 \langle y_1 \xi_0, \eta_0 \rangle, & \zeta \in [0,1] \setminus A. \end{cases}$$

Since $\langle y_1 \xi_0, \eta_0 \rangle \neq 0$ and A is not measurable, it follows that $\zeta \to \langle x_\zeta^* \xi_0, \eta_0 \rangle$ is not measurable, proving that the field $\zeta \to x_\zeta^*$ is not weakly measurable.

Finally, we show that the operator x associated with the weakly measurable field $\zeta \to x_\zeta$ is not densely defined. Assume the contrary that $\operatorname{dom}(x)$ is dense. We claim that this implies that the field $\zeta \to x_\zeta^*$ is weakly measurable.

Let $\zeta \to \eta_\zeta$ be a measurable field, such that $\eta_\zeta \in \mathrm{dom}(x_\zeta^*)$. Take an arbitrary element $g \in K$, consider the constant field $\zeta \to g$ and let $\xi = \int_Z^\oplus g\, dw(\zeta)$. Since $\mathrm{dom}(x)$ is dense, it follows that there exists a sequence $\{\xi_n\}_{n\in\mathbb{N}} = \{\int_Z^\oplus \xi_\zeta^{(n)}\, dw(\zeta)\}_{n\in\mathbb{N}} \in \mathrm{dom}(x)$, such that

$$\|\xi - \xi_n\|^2 = \int\limits_{[0,1]} \|g - \xi_\zeta^{(n)}\|^2\, dw(\zeta) \to 0, \quad n \to \infty.$$

Consequently, there exists a subsequence $\{\xi_{n_k}\}$ of $\{\xi_n\}$, such that $\|g - \xi_\zeta^{(n_k)}\| \to 0$ almost everywhere as $k \to \infty$. Since $\xi_\zeta^{(n_k)} \in \mathrm{dom}(x_\zeta)$ and $\eta_\zeta \in \mathrm{dom}(x_\zeta^*)$, it follows that

$$\langle x_\zeta \xi_\zeta^{(n_k)}, \eta_\zeta\rangle = \langle \xi_\zeta^{(n_k)}, x_\zeta^* \eta_\zeta\rangle \to \langle g, x_\zeta^* \eta_\zeta\rangle, \quad k \to \infty.$$

By part (i), the field $\zeta \to x_\zeta$ is weakly measurable, and so the function $\zeta \to \langle x_\zeta \xi_\zeta^{(n_k)}, \eta_\zeta\rangle$ is measurable. Hence, $\zeta \to \langle g, x_\zeta^* \eta_\zeta\rangle$ is measurable for all $g \in K$. Since g is arbitrary, it follows that $\zeta \to x_\zeta^*$ is measurable. This is a contradiction with part (ii). Thus, $\mathrm{dom}(x)$ is not dense. □

Next, we show that any operator associated with an arbitrary (not necessarily weakly measurable) field of closed operators is necessarily decomposable.

Proposition 1.13.19. *Suppose $\zeta \to y_\zeta$ is a field of closed operators. The operator x associated with $\zeta \to y_\zeta$ (see Definition 1.13.8) is decomposable. In this case, writing $x = \int_Z^\oplus x_\zeta\, dw(\zeta)$ for some measurable field $\zeta \to x_\zeta$, we have that $x_\zeta \subset y_\zeta$ almost everywhere.*

Proof. By Remark 1.13.11, x commutes with every bounded diagonalizable operator, and hence, by Theorem 1.13.14, x is decomposable.

Let $\zeta \to x_\zeta$ be a measurable field of closed operators, such that $x = \int_Z^\oplus x_\zeta\, dw(\zeta)$. Let (p_{ij}) and $((p_\zeta)_{ij})$ be the characteristic matrices of x and x_ζ, respectively. By Proposition 1.13.12, we have that $p_{ij} = \int_Z^\oplus (p_\zeta)_{ij}\, dw(\zeta)$.

Since $p_{11}(H) \subset \mathrm{dom}(x)$, it follows that

$$p_{11}(H) \subset \left\{\zeta \to \xi_\zeta : \xi_\zeta \in \mathrm{dom}(y_\zeta), \zeta \to y_\zeta x_\zeta \text{ is measurable}, \int\limits_Z \|y_\zeta \xi_\zeta\|^2\, dw(\zeta) < \infty\right\}.$$

Therefore, since $p_{11}(H) = \int_Z^\oplus (p_\zeta)_{11}(H_\zeta)\, dw(\zeta)$, we conclude that $(p_\zeta)_{11}(H) \subset \mathrm{dom}(y_\zeta)$ for almost all ζ. Hence, the operator $y_\zeta (p_\zeta)_{11}$ is everywhere defined and closed for almost all ζ.

It is clear that the operator xp_{11} is the operator associated with the field $\{y_\zeta (p_\zeta)_{11}\}$. Similarly, the operator xp_{12} is the operator associated with the field $\{y_\zeta (p_\zeta)_{12}\}$.

Let (ξ_n) be a dense sequence of measurable vector fields such that the functions $\zeta \to \|(\xi_n)_\zeta\|^2$ are w-integrable. Let C be any compact subset of Z and χ_C its characteristic function. Let $y_n = \chi_C \xi_n$. Since xp_{11} and xp_{12} are everywhere defined, we have that $y_n \in \mathrm{dom}(xp_{11})$ and $y_n \in \mathrm{dom}(xp_{12})$ for all $n \in \mathbb{N}$. Therefore, $(y_n)_\zeta \in \mathrm{dom}(y_\zeta(p_{11})_\zeta)$ and $(y_n)_\zeta \in \mathrm{dom}(y_\zeta(p_{12})_\zeta)$ for all $\zeta \in C \setminus N_n$ where N_n is a w-null set. Let $N = \bigcup_{n\in\mathbb{N}} N_n$. Then $(y_n)_\zeta = (\xi_n)_\zeta \in \mathrm{dom}(y_\zeta(p_{11})_\zeta)$ and $(y_n)_\zeta = (\xi_n)_\zeta \in \mathrm{dom}(y_\zeta(p_{12})_\zeta)$ for all $\zeta \in C \setminus N$ and all $n \in \mathbb{N}$. On the other hand, by Proposition 1.3.11, we have $xp_{11} = p_{21}$, $xp_{12} = p_{22}$ and, therefore, $(xp_{11}y_n)_\zeta = (p_{21})_\zeta(y_n)_\zeta$ and $(xp_{12}y_n)_\zeta = (p_{22})_\zeta(y_n)_\zeta$ for all $n \in \mathbb{N}$ and almost all ζ. Hence, $y_\zeta(p_{11})_\zeta(\xi_n)_\zeta = (p_{21})_\zeta(\xi_n)_\zeta$ and $y_\zeta(p_{12})_\zeta(\xi_n)_\zeta = (p_{22})_\zeta(\xi_n)_\zeta$ for all $n \in \mathbb{N}$ and all $\zeta \in C \setminus N'$, where N' is a w-null set. Since $y_\zeta(p_{11})_\zeta$ and $y_\zeta(p_{12})_\zeta$ are closed and $(\xi_n)_\zeta$ is dense in H_ζ, it follows that $y_\zeta(p_{11})_\zeta = (p_{21})_\zeta$ and $y_\zeta(p_{12})_\zeta = (p_{22})_\zeta$ for all $\zeta \in C \setminus N'$. Proposition 1.3.11 implies that $x_\zeta \subset y_\zeta$ for all $\zeta \in C \setminus N'$. □

We note that for an operator x associated with a (not necessarily measurable) field of closed and densely defined operators $\zeta \to y_\zeta$, the measurable field $\zeta \to x_\zeta$, which defines x, may consist of closed operators, which are not densely defined. Indeed, otherwise, Proposition 1.13.17 would guarantee that the operator x is densely defined, too. However, as shown in Example 1.13.18, there exists a (weakly measurable) field of closed operators, such that the associated operator x is not densely defined.

We now turn to the basics of spectral theory of decomposable operators. We recall that for a decomposable operator $x = \int_Z^\oplus x_\zeta \, dw(\zeta)$, the measurable field $\zeta \to x_\zeta$ is defined up to measure zero, and on a measure zero set the operators x_ζ may be chosen arbitrarily. In particular, in what follows, for any operation $y \mapsto f(y)$ specified below, we choose x_ζ to be such that $f(x_\zeta)$ is well-defined.

Theorem 1.13.20. *Suppose $x = \int_Z^\oplus x_\zeta \, dw(\zeta)$ is a decomposable operator in $H = \int_Z^\oplus H_\zeta \, dw(\zeta)$. We have:*

(i) *x^* exists if and only if x_ζ^* exists almost everywhere. In this case, the field $\zeta \to x_\zeta^*$ (defined almost everywhere) is measurable and $x^* = \int_Z^\oplus x_\zeta^* \, dw(\zeta)$;*

(ii) *x is self-adjoint (resp., symmetric or normal), if and only if almost all x_ζ are self-adjoint (resp., symmetric or normal);*

(iii) *x^{-1} exists if and only if x_ζ^{-1} exists almost everywhere. In this case, the field $\zeta \to x_\zeta^{-1}$ (defined almost everywhere) is measurable and $x^{-1} = \int_Z^\oplus x_\zeta^{-1} \, dw(\zeta)$.*

Proof. Let (p_{ij}) and $((p_\zeta)_{ij})$ be the characteristic matrices of x and x_ζ, respectively. By Proposition 1.13.12, we have that

$$p_{ij} = \int_Z^\oplus (p_\zeta)_{ij} \, dw(\zeta). \tag{1.13}$$

(i) Since for a closed operator y the adjoint y^* exists if and only if y is densely defined, the first part of the statement follows from Proposition 1.13.17. Assume now that x^* and x_ζ^* exists for almost all ζ. If x commutes with all bounded diagonalizable operators, then x^* also commutes with all bounded diagonalizable operators. Therefore, by

Theorem 1.13.14, the operator x^* is decomposable. Let $\zeta \to y_\zeta$ be a measurable field such that $x^* = \int_Z^\oplus y_\zeta \, dw(\zeta)$. Denoting by (q_{ij}) and $((q_\zeta)_{ij})$, the characteristic matrices of x^* and y_ζ, respectively, Proposition 1.13.12 implies that

$$q_{ij} = \int_Z^\oplus (q_\zeta)_{ij} \, dw(\zeta). \tag{1.14}$$

By Proposition 1.3.14, we have that

$$q_{11} = \mathbf{1} - p_{22}, \quad q_{12} = p_{21}, \quad q_{21} = p_{12}, \quad q_{22} = \mathbf{1} - p_{11}.$$

Hence, equalities (1.13) and (1.14) imply that

$$(q_\zeta)_{11} = \mathbf{1}_{H_\zeta} - (p_\zeta)_{22}, \quad (q_\zeta)_{12} = (p_\zeta)_{21}, \quad (q_\zeta)_{21} = (p_\zeta)_{12}, \quad (q_\zeta)_{22} = \mathbf{1}_{H_\zeta} - (p_\zeta)_{11}$$

for almost all ζ. Appealing once again to Proposition 1.3.14, we infer that $y_\zeta = x_\zeta^*$ a. e.

(ii) This part follows from part (i) and Proposition 1.13.16.

(iii) By Proposition 1.3.16, x^{-1} exists if and only if $\mathbf{1} - p_{11}$ has an inverse. By Lemma 1.13.13 and equality (1.13), the operator $\mathbf{1} - p_{11}$ has an inverse if and only if $\mathbf{1}_{H_\zeta} - (p_\zeta)_{11}$ has an inverse for almost all ζ. Referring to Proposition 1.3.16 once again, we infer that x^{-1} exists if and only if x_ζ^{-1} exists for almost all ζ.

Assume now that x^{-1} exists. By Proposition 1.13.16, without loss of generality, we assume that x_ζ^{-1} exists for all ζ. By Theorem 1.13.14, the operator x^{-1} is decomposable, i. e., we can write $x^{-1} = \int_Z^\oplus y_\zeta \, dw(\zeta)$ for some measurable field $\zeta \to y_\zeta$. Denote by (q_{ij}) and $((q_\zeta)_{ij})$ the characteristic matrices of x^{-1} and y_ζ, respectively. By Proposition 1.3.16, we have that

$$q_{11} = p_{22}, q_{12} = p_{21}, q_{21} = p_{12}, q_{22} = p_{21}.$$

On the other hand, Proposition 1.13.12 implies that $q_{ij} = \int_Z^\oplus (q_\zeta)_{ij} \, dw(\zeta)$. Therefore, equality (1.13) implies that

$$(q_\zeta)_{11} = (p_\zeta)_{22}, (q_\zeta)_{12} = (p_\zeta)_{21}, (q_\zeta)_{21} = (p_\zeta)_{12}, (q_\zeta)_{22} = (p_\zeta)_{21}.$$

Referring once again to Proposition 1.3.16, we infer that $y_\zeta = x_\zeta^{-1}$, as required. □

Proposition 1.13.21. *If $x = \int_Z^\oplus x_\zeta \, dw(\zeta)$ is a normal decomposable operator, then $x^n = \int_Z^\oplus x_\zeta^n \, dw(\zeta)$ for all $n \in \mathbb{N}$.*

Proof. By Theorem 1.13.20(ii), the operators x_ζ are normal for almost all ζ. By Proposition 1.13.16, without loss of generality, we may assume that x_ζ is normal for all $\zeta \in Z$.

Since x is decomposable, it commutes with all bounded diagonalizable operators, and so x^n also commutes with all bounded diagonalizable operators. Hence, by Theorem 1.13.14 x^n is decomposable. Let $\{y_\zeta\}$ be a measurable field such that $x^n = \int_Z^\oplus y_\zeta \, dw(\zeta)$.

By Theorem 1.13.20 and Proposition 1.13.16, without loss of generality, we may assume that all y_ζ are normal.

On the other hand, denote by a the operator associated with the field x_ζ^n (see Definition 1.13.8). By Proposition 1.13.19, there exists a measurable field $\zeta \to a_\zeta$, such that $a = \int_Z^\oplus a_\zeta\, dw(\zeta)$ and $a_\zeta \subset x_\zeta^n$ for almost all ζ.

By definition of a, we have that $x^n \subset a$. Therefore, by Proposition 1.13.16, we have than $y_\zeta \subset a_\zeta$. Since $a_\zeta \subset x_\zeta^n$ for almost all ζ, it follows that $y_\zeta \subset x_\zeta^n$ for almost all ζ. Since both y_ζ and x_ζ^n are normal and any normal operator is maximal (see, e. g., [140, Proposition 3.26(iv)]), the latter inclusion implies that $y_\zeta = x_\zeta^n$ for almost all ζ. Thus, $x^n = \int_Z^\oplus x_\zeta^n\, dw(\zeta)$. □

Proposition 1.13.21 implies, in particular, the following.

Corollary 1.13.22. *Suppose that a decomposable operator $x = \int_Z^\oplus x_\zeta\, dw(\zeta)$ is self-adjoint. We have $x \geq 0$ if and only if $x_\zeta \geq 0$ a. e.*

Proof. If $x_\zeta \geq 0$ a. e., then clearly $x \geq 0$. Suppose now that $x \geq 0$. Since x is self-adjoint, it follows that $x = (x^{1/2})^2$ for a self-adjoint positive operator $x^{1/2}$. Furthermore, by Remark 1.13.15, the operator $x^{1/2}$ is decomposable.

Let $x^{1/2} = \int_Z^\oplus y_\zeta\, dw(\zeta)$ for a measurable field $\zeta \to y_\zeta$. Since $x^{1/2}$ is self-adjoint, it follows that almost all y_ζ are self-adjoint, too (see Theorem 1.13.20(ii)). By Proposition 1.13.21, we have that

$$\int_Z^\oplus x_\zeta\, dw(\zeta) = x = (x^{1/2})^2 = \int_Z^\oplus y_\zeta^2\, dw(\zeta).$$

The uniqueness of representation of decomposable operators (see Proposition 1.13.16) implies that $x_\zeta = y_\zeta^2$ for almost all ζ. Since almost all y_ζ are self-adjoint, it follows that $x_\zeta \geq 0$ for almost all ζ. □

Next, we consider spectral projections of decomposable operators.

Proposition 1.13.23. *Let $x = \int_Z^\oplus x_\zeta\, dw(\zeta)$ be a self-adjoint decomposable operator. For every Borel subset $B \subset \mathbb{R}$, the field $\zeta \to E_{x_\zeta}(B)$ is measurable and*

$$E_x(B) = \int_Z^\oplus E_{x_\zeta}(B)\, dw(\zeta).$$

Proof. By Theorem 1.13.20, without loss of generality, we can assume that all x_ζ are self-adjoint.

Using the Cayley transform to go from unbounded self-adjoint operators to unitary operators, we easily obtain the result from the functional calculus of bounded decomposable operators (see Proposition 1.12.18).

Namely, let γ be the Cayley transform. For any Borel set $B \subset \mathbb{R}$, we have that

$$E_x(B) = E_{\gamma(x)}(\gamma(B)), \quad E_{x_\zeta}(B) = E_{\gamma(x_\zeta)}(\gamma(B)).$$

Hence, by Proposition 1.12.18, it is sufficient to show that

$$\gamma(x) = \int_Z^{\oplus} \gamma(x_\zeta)\, dw(\zeta).$$

Since $\gamma(x_\zeta)$ is unitary for all ζ, it follows from Proposition 1.12.5 that the operator $\int_Z^{\oplus} \gamma(x_\zeta)\, dw(\zeta)$ is unitary.

Let $\xi = \int_Z^{\oplus} \xi_\zeta\, dw(\zeta) \in \operatorname{dom}(x)$. We have

$$(x - i\mathbf{1})\xi = \int_Z^{\oplus} (x_\zeta - i\mathbf{1}_{H_\zeta})\xi_\zeta\, dw(\zeta)$$

and, therefore,

$$\begin{aligned}
&\left(\left(\int_Z^{\oplus} (x_\zeta + i\mathbf{1}_{H_\zeta})(x_\zeta - i\mathbf{1}_{H_\zeta})^{-1} dw(\zeta)\right)(x - i\mathbf{1})\right)\xi_\eta \\
&= \left(\int_Z^{\oplus} (x_\zeta + i\mathbf{1}_{H_\zeta})(x_\zeta - i\mathbf{1}_{H_\zeta})^{-1}\, dw(\zeta)\right)((x_\eta - i\mathbf{1}_{H_\zeta})\xi_\eta) \\
&= ((x_\eta + i\mathbf{1}_{H_\zeta})(x_\eta - i\mathbf{1}_{H_\zeta})^{-1})((x_\eta - i\mathbf{1}_{H_\zeta})\xi_\eta) = (x_\eta + i\mathbf{1}_{H_\zeta})\xi_\eta \\
&= ((x + i\mathbf{1})\xi)_\eta, \quad \eta \in Z.
\end{aligned}$$

Thus, the two unitary operators $(x + i\mathbf{1})(x - i\mathbf{1})^{-1}$ and

$$\int_Z^{\oplus} (x_\zeta + i\mathbf{1}_{H_\zeta})(x_\zeta - i\mathbf{1}_{H_\zeta})^{-1}\, dw(\zeta)$$

agree on a dense subset of H, so they must be equal, as required. □

Via a standard argument, Proposition 1.13.23 implies the Borel function calculus formula for self-adjoint decomposable operators.

Theorem 1.13.24. *Assume that a decomposable operator $x = \int_Z^{\oplus} x_\zeta\, dw(\zeta)$ is self-adjoint. For every (possibly unbounded) Borel measurable function $f : \mathbb{R} \to \mathbb{R}$ the field $\zeta \to f(x_\zeta)$ is measurable and*

$$f(x) = \int_Z^{\oplus} f(x_\zeta)\, dw(\zeta).$$

With the Borel functional calculus of decomposable operators established, we can prove the polar decomposition for decomposable operators.

Theorem 1.13.25. *Suppose that a decomposable operator $x = \int_Z^\oplus x_\zeta \, dw(\zeta)$ is densely defined and let $x = v|x|$ be its polar decomposition. We have*

$$|x| = \int_Z^\oplus |x_\zeta| \, dw(\zeta), \quad v = \int_Z^\oplus v_\zeta \, dw(\zeta),$$

where $x_\zeta = v_\zeta |x_\zeta|$ is the polar decomposition of x_ζ for a. e. $\zeta \in Z$.

Proof. Theorem 1.13.24 immediately implies the equality $|x| = \int_Z^\oplus |x_\zeta| \, dw(\zeta)$. We now show that v is a decomposable operator. Since x is decomposable, Theorem 1.13.14 implies that x commutes with any bounded diagonalizable operator. Let u be a diagonalizable operator, which is a unitary. Since $u^{-1}xu = x$, it follows that

$$x = u^{-1}xu = u^{-1}v|x|u = (u^{-1}vu)(u^{-1}|x|u).$$

Uniqueness of the polar decomposition for the closed linear operator x (see Theorem 1.5.4) guarantees that $u^{-1}vu = v$ and, therefore, v commutes with any unitary diagonalizable operator. Hence, v commutes with any bounded diagonalizable operator. Theorem 1.12.6 guarantees that v is decomposable.

Therefore, there exists a bounded measurable field $\zeta \to v_\zeta$ of bounded operators, such that $v = \int_Z^\oplus v_\zeta \, dw(\zeta)$. By Proposition 1.12.5, we have that

$$E_{|x|}(0, +\infty) = v^* v = \int_Z^\oplus v_\zeta^* v_\zeta \, dw(\zeta),$$

$$E_{|x^*|}(0, +\infty) = vv^* = \int_Z^\oplus v_\zeta v_\zeta^* \, dw(\zeta).$$

By Proposition 1.13.23, it follows that $v_\zeta^* v_\zeta = E_{|x_\zeta|}(0, +\infty)$ and $v_\zeta v_\zeta^* = E_{|x_\zeta^*|}(0, +\infty)$ for almost all ζ, i. e., for almost all ζ the operator v_ζ is a partial isometry with initial space $E_{|x_\zeta|}(0, +\infty)(H_\zeta)$ and final space $E_{|x_\zeta^*|}(0, +\infty)(H_\zeta)$. It remains to show that v_ζ is a partial isometry, such that $x_\zeta = v_\zeta |x_\zeta|$ for a. e. $\zeta \in Z$.

Let a be the operator associated with the field $\zeta \to v_\zeta |x_\zeta|$ (see Definition 1.13.8). We first show that $vv^* a = a$. Since vv^* is bounded operator, it follows that $\mathrm{dom}(vv^* a) = \mathrm{dom}(a)$. For any $\xi = \int_Z^\oplus \xi_\zeta \, dw(\zeta) \in \mathrm{dom}(a)$, we have that

$$(vv^* a)\xi_\zeta = vv^*(v_\zeta |x_\zeta| \xi_\zeta) = v_\zeta v_\zeta^* v_\zeta |x_\zeta| \xi_\zeta = v_\zeta |x_\zeta| \xi_\zeta = a\xi_\zeta,$$

for a. e. $\zeta \in Z$. Thus, $vv^* a = a$.

By Proposition 1.13.19, there exists a measurable field $\zeta \to a_\zeta$, such that $a = \int_Z^\oplus a_\zeta\, dw(\zeta)$ and $a_\zeta \subset v_\zeta |x_\zeta|$ for a. e. ζ. Then $v_\zeta^* a_\zeta \subset |x_\zeta|$ a. e., and so by Proposition 1.13.16, we have that $v^* a \subset |x|$, implying that $a = vv^* a \subset v|x| = x$. On the other hand, $\operatorname{dom}(x_\zeta) = \operatorname{dom}(|x_\zeta|) = \operatorname{dom}(v_\zeta |x_\zeta|)$ a. e., and so $\operatorname{dom}(x) \subset \operatorname{dom}(a)$. Thus, $a = x$. Referring to Proposition 1.13.16 once again, we infer that $x_\zeta = v_\zeta |x_\zeta|$, which concludes the proof. □

Bibliographical notes

An excellent exposition of the theory of von Neumann algebras and extended bibliographical comments can be found in many monographs, e. g., [49, 60, 61, 92, 93, 134, 135, 147, 149–151].

The theory of unbounded self-adjoint operators and their functional calculi can be also found in numerous monographs. We refer the reader to, e. g., [43, 64, 66, 114, 127, 129, 140, 147].

Here, we give bibliographical notes only to less known results, which are not typically covered in the above cited sources.

The characteristic matrix of an operator was first implicitly used in [153]. Later in his fundamental paper [146], M. Stone presented a very detailed study of the characteristic matrix and its properties.

The dimension function was introduced in the fundamental papers by Murray and von Neumann and was further generalized by I. Segal [141]. The uniqueness result in Theorem 1.11.13 is formally new (although certainly known to experts).

The exposition of the theory of direct integrals of von Neumann algebras can be found in [61] and [93]. The spectral theory and functional calculus of bounded decomposable operators presented in Section 1.12 is basic. Proposition 1.12.17 is a special case of [106, Proposition 1.4]. The proof of this result is taken from [68, Proposition 3.4]. For more advanced results in this direction, we refer the reader to [17, 54, 55, 68, 106].

The notion of measurability of fields of closed operators was introduced by A. Nussbaum in [117]. This type of measurability was later termed Nussbaum measurability [73, 74]. The fundamental result of Theorem 1.13.14 that a closed operator in $H = \int_Z^\oplus H_\zeta\, dw(\zeta)$ is decomposable if and only if it commutes with all bounded diagonalizable operators in H was proved by Nussbaum [117, Corollary 4]. Section 1.13 mainly follows [117]. The counterexamples in Section 1.13 (see Example 1.13.7, Example 1.13.18) are taken from [73].

Theorem 1.13.20 was proved by Nussbaum in [117, Theorem 3]. For spectral theory and functional calculus of closed decomposable operators, we refer to [54, 68, 106, 107]. The exposition of these results in Section 1.13 follows [68].

2 Classes of unbounded operators

In this chapter, we introduce and discuss in depth three different classes of unbounded operators affiliated with a von Neumann algebra $\mathcal{M}$. The first two classes are measurable, and locally measurable operators affiliated with $\mathcal{M}$. The third class of τ-measurable operators can be introduced when $\mathcal{M}$ is a semifinite von Neumann algebra equipped with a faithful normal semifinite trace τ. It is shown that each class of these operators form a $*$-algebra containing $\mathcal{M}$. We establish some necessary and sufficient conditions for these three algebras to coincide. In the final section, we introduce the classes of compact and τ-compact operators affiliated with a von Neumann algebra $\mathcal{M}$ (in the latter case the algebra $\mathcal{M}$ is again assumed to be equipped with a faithful normal semifinite trace τ). These extend the classical notion of a compact operator in a Hilbert space.

2.1 Unbounded operators affiliated with a von Neumann algebra

To properly introduce the classes $S(\mathcal{M})$ and $LS(\mathcal{M})$ of measurable and locally measurable operators, we need first to discuss unbounded operators, which are affiliated with $\mathcal{M}$. In this section, we present a thorough exposition of the class of such operators and describe their essential properties.

Let H be a Hilbert space, and let $\mathcal{B}(H)$ be the $*$-algebra of all bounded linear operators on H. Let $\mathcal{M}$ be a von Neumann algebra acting on H. As before, we denote by $P(\mathcal{M})$ the complete lattice of all projections from $\mathcal{M}$, by $\mathbf{1}$ the unit of $\mathcal{M}$, and by $U(\mathcal{M}')$ the unitary group of the commutant algebra $\mathcal{M}'$.

Definition 2.1.1. A linear operator x with the domain $\mathrm{dom}(x)$ acting in the Hilbert space H is said to be *affiliated* with the von Neumann algebra $\mathcal{M}$ (notation: $x\eta\mathcal{M}$), if $ux \subseteq xu$ for every $u \in U(\mathcal{M}')$, i. e., if $u(\mathrm{dom}(x)) \subset \mathrm{dom}(x)$ and $ux(\xi) = xu(\xi)$ for every $\xi \in \mathrm{dom}(x)$.

Since $u^{-1} = u^*$ for any unitary, it follows that $u(\mathrm{dom}(x)) \subset \mathrm{dom}(x)$ for all $u \in U(\mathcal{M}')$ if and only if $u(\mathrm{dom}(x)) = \mathrm{dom}(x)$. Therefore, a linear operator x is affiliated with $\mathcal{M}$ if and only if $ux = xu$ for all $u \in U(\mathcal{M}')$, or equivalently, if $x = uxu^*$ for all $u \in U(\mathcal{M}')$. Furthermore, since any operator in $\mathcal{M}'$ is a linear combination of at most four unitaries, an operator x is affiliated with $\mathcal{M}$ if and only if $yx \subset xy$ for all $y \in \mathcal{M}'$. In particular, if x is bounded, then x is affiliated with $\mathcal{M}$ if and only if $x \in \mathcal{M}$.

In the next proposition, we list some basic properties of operators affiliated with a von Neumann algebra $\mathcal{M}$.

Proposition 2.1.2. *Let x and y be linear operators in the Hilbert space H.*
(i) *If $x, y\eta\mathcal{M}$, then $(x+y)\eta\mathcal{M}$, $xy\eta\mathcal{M}$, and $(\lambda x)\eta\mathcal{M}$ for all $\lambda \in \mathbb{C}$;*
(ii) *If x is closable and $x\eta\mathcal{M}$, then $\overline{x}\eta\mathcal{M}$;*
(iii) *If x is densely defined and $x\eta\mathcal{M}$, then $x^*\eta\mathcal{M}$;*

https://doi.org/10.1515/9783111599687-003

(iv) *If x is self-adjoint, then $x\eta\mathcal{M}$ if and only if $E_x(B) \in \mathcal{M}$ for all Borel sets $B \subset \mathbb{R}$;*

(v) *If x is self-adjoint and $x\eta\mathcal{M}$, then $f(x)\eta\mathcal{M}$ for any Borel function $f : \mathbb{R} \to \mathbb{C}$;*

(vi) *If x is closed and densely defined with the polar decomposition $x = v|x|$, then $x\eta\mathcal{M}$ if and only if $v \in \mathcal{M}$ and $|x|\eta\mathcal{M}$. In this case, both the left support $s_l(x)$ and the right support $s_r(x)$ belong to $\mathcal{M}$ and $s_r(x) \sim s_l(x)$ (with respect to $\mathcal{M}$).*

Proof. Assertions (i) and (ii) follow immediately from definitions of sum, product, and closure of linear operators.

(iii) As discussed above, $x\eta\mathcal{M}$ if and only if $ux = xu$ for any $u \in U(\mathcal{M}')$. Since u is a bounded operator, it follows from Theorem 1.3.3(vi) that $x^*u^* = (ux)^* = (xu)^* = u^*x^*$ for any $u \in U(\mathcal{M}')$. Since $u \in U(\mathcal{M}')$ if and only if $u^* \in U(\mathcal{M}')$, it follows that $ux^* = x^*u$ for any $u \in U(\mathcal{M}')$, proving that x^* is affiliated with $\mathcal{M}$.

(iv) Suppose that x is a self-adjoint operator and let $u \in U(\mathcal{M}')$. Theorem 1.4.8 guarantees that $ux \subset xu$ if and only if $uE_x(B) = E_x(B)u$ for any Borel set $B \subset \mathbb{R}$. Hence, $x\eta\mathcal{M}$ if and only if $E_x(B)$ commutes with all unitaries in the commutant $\mathcal{M}'$ for any Borel set $B \subset \mathbb{R}$. Since the latter condition holds if and only if $E_x(B) \in \mathcal{M}$, the assertion follows.

(v) By Proposition 1.4.7, the spectral measure of $f(x)$ is given by

$$E_{f(x)}(B) = E_x(f^{-1}(B \cap f(\mathbb{R})))$$

for all Borel set $B \subset \mathbb{R}$. Therefore, (v) is an immediate consequence of (iv).

(vi) Let x be a closed and densely defined operator with the polar decomposition $x = v|x|$. If $v \in \mathcal{M}$ (and so, $v\eta\mathcal{M}$) and $|x|\eta\mathcal{M}$, then part (i) implies that $x\eta\mathcal{M}$. Conversely, assume that $x\eta\mathcal{M}$ and let $u \in U(\mathcal{M}')$ be arbitrary. Since $u^{-1}xu = x$, it follows that

$$x = u^{-1}xu = u^{-1}v|x|u = (u^{-1}vu)(u^{-1}|x|u).$$

Uniqueness of the polar decomposition for the closed linear operator x (see Proposition 1.5.5) guarantees that

$$u^{-1}vu = v, \quad u^{-1}|x|u = |x|.$$

Hence, $v \in \mathcal{M}'' = \mathcal{M}$ and $|x|\eta\mathcal{M}$, as required.

Now, assume that $x\eta\mathcal{M}$. By Theorem 1.5.4, we have $s_r(x) = v^*v \in \mathcal{M}$, $s_l(x) = vv^* \in \mathcal{M}$. Since $v \in \mathcal{M}$, this shows that $s_l(x) \sim s_r(x)$ with respect to $\mathcal{M}$. □

We now demonstrate the notion of affiliated operators by two examples. We start with a very simple example of $\mathcal{M} = \mathcal{B}(H)$.

Example 2.1.3. *Let H be a Hilbert space and consider the von Neumann algebra $\mathcal{M} = \mathcal{B}(H)$. Since $\mathcal{B}(H)' = \mathbb{C}\mathbf{1}$, any linear (possibly unbounded) operator in H is affiliated with $\mathcal{B}(H)$.*

The second example we consider here is when $\mathcal{M}$ is the algebra of all (classes of a. e. equal) essentially bounded functions on a measure space (Ω, Σ, μ) acting by multiplication on $L_2(\Omega, \Sigma, \mu)$. Let $L_0(\Omega, \Sigma, \mu)$ be the $*$-algebra of all (classes of a. e. equal) complex-valued μ-a. e. finite μ-measurable functions on Ω. When the underlying measure space (Ω, Σ, μ) is clear from the context, we shall use notation $L_0(\Omega)$ instead of $L_0(\Omega, \Sigma, \mu)$.

Example 2.1.4. *Let (Ω, Σ, μ) be a σ-finite measure space and let*

$$\mathcal{M} = \{M_f : f \in L_\infty(\Omega, \Sigma, \mu)\},$$

which is $*$-isomorphic to $L_\infty(\Omega, \Sigma, \mu)$ (see Theorem 1.8.10). Then any closed densely defined operator affiliated with $\mathcal{M}$ coincides with the operators of multiplication M_f by a uniquely determined measurable function $f \in L_0(\Omega, \Sigma, \mu)$ with the domain

$$\mathrm{dom}(M_f) = \{g \in L_2(\Omega, \Sigma, \mu) : gf \in L_2(\Omega, \Sigma, \mu)\}.$$

Proof. We first show that the operator M_f of multiplication by $f \in L_0(\Omega, \Sigma, \mu)$ is affiliated with $\mathcal{M}$. Clearly, the domain $\mathrm{dom}(M_f)$ is a dense subspace in $L_2(\Omega, \Sigma, \mu)$ and M_f is closed. Let $u \in U(\mathcal{M}') = U(\mathcal{M})$. Then $u = M_g$ for some $g \in L_\infty(\Omega, \Sigma, \mu)$ with $|g| = \chi_\Omega$. If $\xi \in \mathrm{dom}(M_f)$, $fg\xi \in L_2(\Omega, \Sigma, \mu)$, and so, $u(\mathrm{dom}(M_f)) \subset \mathrm{dom}(M_f)$. Moreover,

$$M_f u\xi = fg\xi = gf\xi = uM_f\xi, \quad \xi \in \mathrm{dom}(M_f),$$

showing that $uM_f \subseteq M_f u$ for all $u \in U(\mathcal{M}')$. Hence, $M_f \eta \mathcal{M}$.

To prove the converse, assume first that x is a positive self-adjoint operator on $L_2(\Omega, \Sigma, \mu)$, affiliated with $\mathcal{M}$. By Proposition 2.1.2(iv), we have that the spectral projections $E_x(B)$ belong to $\mathcal{M}$ for all Borel sets $B \subseteq \mathbb{R}$. Define

$$p_n = E_x[n-1, n), \quad x_n = \int\limits_{[n-1,n)} \lambda dE_x(\lambda), \quad n \in \mathbb{N}.$$

By the spectral theorem, x_n is bounded and positive. Therefore, there exists $0 \leq g_n \in L_\infty(\Omega, \Sigma, \mu)$, such that $x_n = M_{g_n}$, $n \in \mathbb{N}$. Furthermore, since $p_n \in P(\mathcal{M})$ there exists $A_n \in \Sigma$, such that $p_n = M_{\chi_{A_n}}$, $n \in \mathbb{N}$. Appealing to the spectral theorem again, we have that $p_n p_m = 0$ whenever $n \neq m$ and $x_n = x_n p_n$ for all $n \in \mathbb{N}$. This implies that $\chi_{A_n}\chi_{A_m} = 0$, $n \neq m$, and $g_n = g_n \chi_{A_n}$ for all $n \in \mathbb{N}$. Therefore, it may be assumed that the sets $\{A_n\}_{n \in \mathbb{N}}$ are pairwise disjoint and that the function g_n is supported on A_n. Define $g \subset L_0(\Omega, \Sigma, \mu)$ by setting $g = \sum_{n=1}^\infty g_n$, with the sum taken pointwise on Ω. If $\xi \in \mathrm{dom}(x)$, then $x\xi = \sum_{n=1}^\infty x_n \xi$ as a norm convergent series in $L_2(\Omega, \Sigma, \mu)$, and so

$$x\xi = \sum_{n=1}^\infty M_{g_n}\xi = \sum_{n=1}^\infty g_n \xi = \left(\sum_{n=1}^\infty g_n\right)\xi.$$

This shows that $\xi \in \operatorname{dom}(M_g)$ and $x\xi = M_g\xi$. Hence, $x \subseteq M_g$. For the proof of the converse inclusion, suppose that $\xi \in \operatorname{dom}(M_g)$, i. e., $\sum_{n=1}^{\infty} g_n\xi \in L_2(\Omega, \Sigma, \mu)$. The dominated convergence theorem guarantees that the series $\sum_{n=1}^{\infty} g_n\xi$ is convergent in $L_2(\Omega, \Sigma, \mu)$, and so

$$\sum_{n=1}^{\infty} x_n\xi = \sum_{n=1}^{\infty} g_n\xi = M_g\xi$$

as a norm convergent series. This implies that $\xi \in \operatorname{dom}(x)$ and $x\xi = M_g\xi$. Thus, $M_g \subseteq x$, and so $x = M_g$.

Now, let x be any closed densely defined operator, which is affiliated with $\mathcal{M}$ and let $x = v|x|$ be its polar decomposition. By Proposition 2.1.2(vi), we have that $v \in \mathcal{M}$ and $|x|\eta\mathcal{M}$. So, $v = M_h$ for some $h \in L_\infty(\Omega, \Sigma, \mu)$ and $|x| = M_g$ for some $0 \le g \in L_0(\Omega, \Sigma, \mu)$, by the special case treated above. Therefore, $x = M_f$ with $f = hg$.

Finally, if $f_1, f_2 \in L_0(\Omega, \Sigma, \mu)$ are such that $M_{f_1} = M_{f_2}$, then $f_1\chi_F = f_2\chi_F$ for all sets F of finite measure, which implies that $f_1 = f_2$ in $L_0(\Omega, \Sigma, \mu)$ (since in a σ-finite measure space, local μ-null sets in X are μ-null sets). □

To conclude this section, we prove a useful property of spectral projections of affiliated operators, which we will use repeatedly in the future. We first prove an auxiliary lemma.

Lemma 2.1.5. *If x is a closable operator affiliated with $\mathcal{M}$ and $p \in P(\mathcal{M})$ is such that $p(H) \subset \operatorname{dom}(x)$, then xp is closed and $xp \in \mathcal{M}$.*

Proof. Since x and p are affiliated with $\mathcal{M}$, Proposition 2.1.2(i) implies that xp is affiliated with $\mathcal{M}$. Since the operator x is closable and $p(H) \subset \operatorname{dom}(x)$, it follows that xp is a closed operator with $\operatorname{dom}(xp) = H$. By the closed graph theorem (see e. g. [129]), we have that $xp \in \mathcal{B}(H)$. Thus, $xp \in \mathcal{M}$. □

Proposition 2.1.6. *Let x be a closed densely defined operator affiliated with $\mathcal{M}$. If $p \in P(\mathcal{M})$ is such that $p(H) \subseteq \operatorname{dom}(x)$ and $\lambda \in [0, \infty)$ is such that $\|xp\|_{\mathcal{M}} \le \lambda$, then $E_{|x|}(\lambda, \infty) \precsim (E_{|x|}(\lambda, \infty) \vee p) - p$. In particular, $E_{|x|}(\lambda, \infty) \precsim p^\perp$.*

Proof. Let $x = v|x|$ be the polar decomposition of x. Referring to Proposition 2.1.2(vi), we obtain that $v \in \mathcal{M}$ and $|x|\eta\mathcal{M}$. By Lemma 2.1.5, the operator xp is bounded.

Fix $\lambda \ge \|xp\|_{\mathcal{M}}$. Then

$$\||x|p\|_{\mathcal{M}} = \|v^*xp\|_{\mathcal{M}} \le \|v^*\|_{\mathcal{M}}\|xp\|_{\mathcal{M}} \le \lambda.$$

We define

$$q = p \wedge E_{|x|}(\lambda, \infty).$$

Suppose that $q \neq 0$.

By the spectral theorem, we have that

$$E_{|x|}(\lambda,\infty)|x|E_{|x|}(\lambda,\infty) \geq \lambda E_{|x|}(\lambda,\infty)$$

and, therefore,

$$q|x|q \geq \lambda q.$$

Fix $0 \neq \xi \in q(H)$. We have

$$\langle |x|\xi, \xi\rangle = \langle q|x|q\xi, \xi\rangle \geq \langle \lambda q\xi, \xi\rangle = \lambda\|\xi\|^2.$$

On the other hand,

$$\langle |x|\xi, \xi\rangle = \langle |x|p\xi, p\xi\rangle \leq \||x|p\xi\|\|p\xi\| \leq \||x|p\|_{\mathcal{M}}\|\xi\|^2 \leq \lambda\|\xi\|^2.$$

Therefore,

$$\langle |x|\xi, \xi\rangle = \lambda\|\xi\|^2.$$

Since, in addition,

$$\langle |x|\xi, \xi\rangle \leq \||x|\xi\|\|\xi\| = \||x|p\xi\|\|\xi\| \leq \lambda\|\xi\|^2 = \langle |x|\xi, \xi\rangle,$$

we infer that

$$\langle |x|\xi, \xi\rangle = \lambda\|\xi\|^2 = \||x|\xi\|\|\xi\|.$$

Therefore, the vectors $|x|\xi$ and ξ are collinear. Since $\langle |x|\xi, \xi\rangle = \lambda\|\xi\|^2$, it follows that $|x|\xi = \lambda\xi$.

Let e be the projection onto the closure of the subspace $\mathcal{M}'(\xi) = \{y\xi : y \in \mathcal{M}'\}$. A direct verification implies that $e\eta\mathcal{M}$, and so $e \in \mathcal{M}$. Furthermore, since

$$|x|u\xi = u|x|\xi = u\lambda\xi = \lambda u\xi, \quad u \in U(\mathcal{M}'),$$

it follows that $|x|e = \lambda e$. Therefore, $0 \neq e \leq E_{|x|}(\{\lambda\})$, and so $\xi \in E_{|x|}(\{\lambda\})(H)$. On the other hand, by the definition of q, we have that $\xi \in E_{|x|}(\lambda,\infty)(H)$. Since the projections $E_{|x|}(\{\lambda\})$ and $E_{|x|}(\lambda,\infty)$ are orthogonal, it follows that $\xi = 0$. This contradicts with the fact that $\xi \neq 0$.

Therefore,

$$p \wedge E_{|x|}(\lambda,\infty) = q = 0.$$

Hence, Kaplansky's identity (see Theorem 1.9.4(v)) implies that

$$E_{|x|}(\lambda,\infty) = E_{|x|}(\lambda,\infty) - (E_{|x|}(\lambda,\infty) \wedge p) \sim (E_{|x|}(\lambda,\infty) \vee p) - p \leq p^{\perp},$$

as required. □

As we show next, Proposition 2.1.6 implies in particular, that for any operator x affiliated with $\mathcal{M}$, the spectral projections for $|x|$ and $|x^*|$ are equivalent (with respect to $\mathcal{M}$).

Proposition 2.1.7. *Suppose that x is a closed densely defined operator affiliated with $\mathcal{M}$. Then $E_{|x|}(\lambda,\infty) \sim E_{|x^*|}(\lambda,\infty)$ for every $\lambda > 0$.*

Proof. Let $x = v|x|$ be the polar decomposition of x. Then $|x^*| = v|x|v^*$ (see (1.4)).

By Proposition 2.1.2(iii) and (vi), we have that $v \in \mathcal{M}$ and $x^*, |x^*|, |x|\eta\mathcal{M}$. Referring to Proposition 2.1.2(iv), we obtain that $E_{|x|}(B), E_{|x^*|}(B) \in \mathcal{M}$ for any Borel set $B \subset \mathbb{R}$.

Fix $\lambda > 0$ and let $q = E_{|x|}(0,\lambda]$. By spectral theory, we have that $q(H) \subset \operatorname{dom}(|x|)$. Therefore, referring to Lemma 2.1.5 we obtain that $|x|q \in \mathcal{M}$. The spectral theorem also implies that $|x|q \leq \lambda q$, and so $v(|x|q)v^* \leq \lambda vqv^*$.

Note that

$$q = E_{|x|}(0,\lambda] \leq E_{|x|}(0,\infty) = s(|x|) = v^*v.$$

Therefore,

$$|x^*|vqv^* = v|x|v^* \cdot vqv^* = v(|x|q)v^* \leq \lambda vqv^*,$$

i. e.,

$$\||x^*|vqv^*\|_{\mathcal{M}} \leq \lambda.$$

Employing Proposition 2.1.6, we infer that

$$\begin{aligned} E_{|x^*|}(\lambda,\infty) &\precsim E_{|x^*|}(\lambda,\infty) \vee vqv^* - vqv^* \leq vv^* - vqv^* = \\ &= v(v^*v)v^* - vqv^* = vE_{|x|}(0,\infty)v^* - vE_{|x|}(0,\lambda]v^* = \\ &= vE_{|x|}(\lambda,\infty)v^* = vE_{|x|}(\lambda,\infty) \cdot E_{|x|}(\lambda,\infty)v^* \\ &\sim E_{|x|}(\lambda,\infty)v^*vE_{|x|}(\lambda,\infty) = E_{|x|}(\lambda,\infty). \end{aligned}$$

Thus, $E_{|x^*|}(\lambda,\infty) \precsim E_{|x|}(\lambda,\infty)$.

Replacing x with x^*, we infer that $E_{|x|}(\lambda,\infty) \precsim E_{|x^*|}(\lambda,\infty)$ and, therefore, $E_{|x|}(\lambda,\infty) \sim E_{|x^*|}(\lambda,\infty)$. □

2.2 Measurable operators affiliated with a von Neumann algebra

In general, the set of all closed densely defined operators affiliated with a von Neumann algebra $\mathcal{M}$, neither has the structure of an algebra nor of a vector space. Indeed, if $\mathcal{M} = \mathcal{B}(H)$, then this set is equal to the collection of all closed densely defined operators in H (see Example 2.1.3), which does not have a natural structure of a vector space if H is infinite-dimensional. In this section, we introduce a class of measurable operators with respect to $\mathcal{M}$ and prove that they form a $*$-algebra with respect to the strong sum, strong product, multiplication by scalars, and taking adjoints.

Let $\mathcal{M}$ be a von Neumann algebra acting on a Hilbert space H.

Definition 2.2.1. A closed linear operator x affiliated with $\mathcal{M}$ is said to be *measurable* with respect to the von Neumann algebra $\mathcal{M}$, if there exists a sequence of projections $\{p_n\}_{n\in\mathbb{N}} \subset P(\mathcal{M})$ such that $p_n \uparrow \mathbf{1}$, $p_n(H) \subset \operatorname{dom}(x)$ and the projection $p_n^\perp$ is finite for every $n \in \mathbb{N}$.

The collection of all operators which are measurable with respect to $\mathcal{M}$ is denoted by $S(\mathcal{M})$.

If the von Neumann algebra $\mathcal{M}$ is fixed and no confusion may arise, we shall simply say that x is a measurable operator.

A sequence $\{p_n\}_{n\in\mathbb{N}}$ satisfying Definition 2.2.1 is called a *determining sequence* for the operator x (in this case operator x is said to be *strongly defined* by the sequence $\{p_n\}_{n\in\mathbb{N}}$). Since $p_n \uparrow \mathbf{1}$ and $p_n(H) \subset \operatorname{dom}(x)$, it follows that any measurable operator is necessarily densely defined. Furthermore, if x is a measurable operator and $\{p_n\}_{n\in\mathbb{N}}$ is a determining sequence for x, then Lemma 2.1.5 guarantess that $xp_n \in \mathcal{M}$ for any $n \in \mathbb{N}$.

Evidently, a bounded operator $x \in \mathcal{B}(H)$ is measurable if and only if $x \in \mathcal{M}$.

Proposition 2.2.2. *If x is a closed densely defined operator in H and $x = v|x|$ is the polar decomposition of the operator x, then the operator x is measurable with respect to the von Neumann algebra $\mathcal{M}$ if and only if $v \in \mathcal{M}$ and the operator $|x|$ is measurable with respect to $\mathcal{M}$.*

Proof. Proposition 2.1.2(vi) guarantees that $x\eta\mathcal{M}$ if and only if $|x|\eta\mathcal{M}$ and $v \in \mathcal{M}$. Since $\operatorname{dom}(x) = \operatorname{dom}(|x|)$, it follows that x is strongly defined by a sequence $\{p_n\}_{n\in\mathbb{N}}$ if and only if $|x|$ is strongly defined by the same sequence $\{p_n\}_{n\in\mathbb{N}}$. Thus, x is measurable if and only if $|x|$ is measurable. □

In the following proposition, we provide a criterion for measurability of a closed linear operator x affiliated with the von Neumann algebra $\mathcal{M}$.

Proposition 2.2.3. *Let x be a closed and densely defined operator affiliated with a von Neumann algebra $\mathcal{M}$. The following conditions are equivalent:*
(i) *the operator x is measurable with respect to $\mathcal{M}$;*
(ii) *there exists a projection $p \in P(\mathcal{M})$, such that $p(H) \subseteq \operatorname{dom}(x)$ and $p^\perp$ is finite;*
(iii) *for some $\lambda > 0$, the spectral projection $E_{|x|}(\lambda, \infty)$ is finite.*

Proof. It is evident that (i) implies (ii).

Suppose that condition (ii) holds. Since $p(H) \subset \operatorname{dom}(x)$, Lemma 2.1.5 implies that $xp \in \mathcal{M}$. Let $\lambda > 0$ be such that $\lambda \geq \|xp\|_{\mathcal{M}}$. Proposition 2.1.6 guarantees that $E_{|x|}(\lambda, \infty) \precsim p^\perp$. Therefore, by Theorem 1.9.9(i), the spectral projection $E_{|x|}(\lambda, \infty)$ is finite, proving (iii).

Now assume that (iii) holds for some $\lambda > 0$. Let $\{\lambda_n\}_{n\in\mathbb{N}}$ be any sequence in $\mathbb{R}$ such that $\lambda \leq \lambda_n \uparrow \infty$ and define $p_n = E_{|x|}[0, \lambda_n]$ for $n = 1, 2, \ldots$. It follows from Proposition 2.1.2(iv) and (vi) that $|x|\eta\mathcal{M}$ and $p_n \in P(\mathcal{M})$ for all $n \in \mathbb{N}$. Furthermore, spectral theory guarantees that $p_n \uparrow \mathbf{1}$ and $p_n(H) \subseteq \operatorname{dom}(|x|) = \operatorname{dom}(x)$ for all $n \in \mathbb{N}$. Since, in addition, $p_n^\perp = E_{|x|}(\lambda_n, \infty) \le E_{|x|}(\lambda, \infty)$, it follows that $p_n^\perp$ is finite for all $n \in \mathbb{N}$. Hence, $\{p_n\}_{n\in\mathbb{N}}$ is a determining sequence for x, and so x is measurable with respect to $\mathcal{M}$. □

An immediate corollary of Proposition 2.2.3(iii) and Proposition 2.1.7 is that $S(\mathcal{M})$ is closed under taking adjoints.

Corollary 2.2.4. *Suppose that x is a measurable operator. Then x^* is also a measurable operator.*

Proof. By Proposition 2.1.2(iii) we have that $x^*\eta\mathcal{M}$. By Proposition 2.2.3(iii) there exists $\lambda > 0$, such that $E_{|x|}(\lambda, \infty)$ is finite. Referring to Proposition 2.1.7, we obtain that the projection $E_{|x^*|}(\lambda, \infty) \sim E_{|x|}(\lambda, \infty)$ is finite too. Thus, using Proposition 2.2.3(iii) once again we conclude that x^* is measurable. □

If the von Neumann algebra $\mathcal{M}$ is finite (in particular, if $\mathcal{M}$ is commutative), then condition (iii) of Proposition 2.2.3 is fulfilled for every closed densely defined operator x affiliated with $\mathcal{M}$. Therefore, we have the following result.

Corollary 2.2.5. *If $\mathcal{M}$ is a finite von Neumann algebra, then every closed densely defined operator x acting in H and affiliated with $\mathcal{M}$ is measurable with respect to $\mathcal{M}$.*

In general, a symmetric densely defined operator x affiliated with a von Neumann algebra is not necessarily essentially self-adjoint and there may exist distinct self-adjoint extensions of x (see, e. g., [126, Chapter X]). However, for symmetric measurable operators this is no longer the case as any symmetric measurable operator is necessarily self-adjoint. To show this, we will need the following lemma. For later purposes, this result will be formulated in a slightly more general form than is needed at present.

Lemma 2.2.6. *Suppose that $\{e_n\}_{n\in\mathbb{N}}$ and $\{f_n\}_{n\in\mathbb{N}}$ are sequences of finite projections in $P(\mathcal{M})$, such that $e_n \downarrow 0$ and $f_n \downarrow 0$. If $\{p_n\}_{n\in\mathbb{N}}$ and $\{q_n\}_{n\in\mathbb{N}}$ are sequences in $P(\mathcal{M})$, such that $p_n \precsim e_n$ and $q_n \precsim f_n$ for all $n \in \mathbb{N}$, then $\bigwedge_n (p_n \vee q_n) = 0$.*

Proof. Denote by $\mathcal{D}$ the dimension function on $P(\mathcal{M})$ (see Section 1.11). By Proposition 1.11.6(i) and (iv), we have that

$$\mathcal{D}\left(\bigwedge_{k=1}^{n}(p_k \vee q_k)\right) \le \mathcal{D}(p_n \vee q_n) \le \mathcal{D}(p_n) + \mathcal{D}(q_n) \le \mathcal{D}(e_n) + \mathcal{D}(f_n)$$

for all $n \in \mathbb{N}$. Since e_n and f_n are finite projections, functions $\mathcal{D}(e_n)$ and $\mathcal{D}(f_n)$ are finite almost everywhere for all $n \in \mathbb{N}$ (see Definition 1.11.1(ii)). Since $e_n \downarrow 0$ and $f_n \downarrow 0$ as $n \to \infty$, Proposition 1.11.6(ii) implies that $\mathcal{D}(e_n) \downarrow 0$ and $\mathcal{D}(f_n) \downarrow 0$ almost everywhere. Consequently, $\mathcal{D}(\bigwedge_{k=1}^n (p_k \vee q_k)) \downarrow 0$ for $n \to \infty$ and, therefore, $\mathcal{D}(\bigwedge_{n\in\mathbb{N}}(p_k \vee q_k)) = 0$. Referring to Proposition 1.11.6(iii), we conclude that $\bigwedge_{n\in\mathbb{N}}(p_k \vee q_k) = 0$. □

Proposition 2.2.7. *If x is a symmetric measurable operator, then x is self-adjoint.*

Proof. By Theorem 1.3.5, it is sufficient to prove that $\mathrm{ran}(x\pm i\mathbf{1})^{\perp} = \{0\}$. Let $L = \mathrm{ran}(x-i\mathbf{1})^{\perp}$ and p be the projection onto L. Note that $p = \mathbf{1} - s_l(x - i\mathbf{1})$, and hence, by Proposition 2.1.2(vi) we have that $p \in P(\mathcal{M})$.

We claim that $\mathrm{dom}(x) \cap L = \{0\}$. Indeed, if $\xi \in \mathrm{dom}(x) \cap L$, then $\langle (x - i\mathbf{1})\xi, \xi\rangle = 0$, and so $\langle x\xi, \xi\rangle = i\langle \xi, \xi\rangle$. Since x is symmetric, we have that $\langle x\xi, \xi\rangle \in \mathbb{R}$ and, therefore, $\xi = 0$. This proves the claim.

Let $\{p_n\}_{n\in\mathbb{N}}$ be a determining sequence for x. Since $\mathrm{dom}(x) \cap L = \{0\}$, it follows that $p \wedge p_n = 0$. Therefore, by Kaplansky's identity (Theorem 1.9.4(v)), we have that $p \sim p \vee p_n - p_n \le p_n^{\perp}$, i. e., $p \precsim p_n^{\perp}$. Since the projections $p_n^{\perp}$ are finite and $p_n^{\perp} \downarrow 0$, it follows from Lemma 2.2.6 that $p = 0$, which implies that $\mathrm{ran}(x - i\mathbf{1})^{\perp} = \{0\}$. The proof of the equality $\mathrm{ran}(x + i\mathbf{1})^{\perp} = \{0\}$ is similar and is therefore omitted. □

Our next aim is to show that set $S(\mathcal{M})$ has a natural structure of a $*$-algebra. To show this, we need to prove a few technical results.

Definition 2.2.8. A subspace D of H is said to be *strongly dense* in H (with respect to $\mathcal{M}$), if there exists a sequence $\{p_n\}_{n\in\mathbb{N}}$ in $P(\mathcal{M})$ such that $p_n \uparrow \mathbf{1}$, $p_n(H) \subseteq D$ and $p_n^{\perp}$ is a finite projection for every $n \in \mathbb{N}$. The sequence $\{p_n\}_{n\in\mathbb{N}}$ is called a *determining sequence* for D.

It is clear from the definitions that a closed operator x affiliated with $\mathcal{M}$ is measurable if and only if the domain $\mathrm{dom}(x)$ is strongly dense in H. In this case, the sequence $\{p_n\}$ is determining sequence for both x and $\mathrm{dom}(x)$. Since $p_n \uparrow \mathbf{1}$ as $n \to \infty$ for any determining sequence $\{p_n\}_{n\in\mathbb{N}}$, it follows that any strongly dense subspace is dense in H.

The results of the following lemma plays an important role in establishing algebraic structure in the set $S(\mathcal{M})$ of all measurable operators.

Lemma 2.2.9.
(i) *If linear subspaces D_1 and D_2 are strongly dense in H with determining sequences $\{p_n\}_{n\in\mathbb{N}}$ and $\{q_n\}_{n\in\mathbb{N}}$, respectively, then the linear subspace $D_1 \cap D_2$ is also strongly dense in H with determining sequence $\{p_n \wedge q_n\}_{n\in\mathbb{N}}$;*
(ii) *If x is a measurable operator and D is a strongly dense subspace in H, then the linear space*

$$x^{-1}(D) = \{\xi \in \mathrm{dom}(x) : x\xi \in D\}$$

is also a strongly dense subspace in H.

Proof. (i) Let $\{p_n\}_{n\in\mathbb{N}}$ and $\{q_n\}_{n\in\mathbb{N}}$ be determining sequences in $P(\mathcal{M})$ for D_1 and D_2, respectively. Defining $e_n = p_n \wedge q_n$ in $P(\mathcal{M})$ for $n = 1, 2, \ldots$, it is clear that $e_n(H) \subseteq D_1 \cap D_2$. Since $p_n^{\perp}$ and $q_n^{\perp}$ are finite projections, the projection $e_n^{\perp} = p_n^{\perp} \vee q_n^{\perp}$ is also finite for all $n \in \mathbb{N}$. Since, in addition, $p_n^{\perp} \downarrow 0$ and $q_n^{\perp} \downarrow 0$, Lemma 2.2.6 implies that $e_n^{\perp} \downarrow 0$, equivalently, $e_n \uparrow \mathbf{1}$. Thus, $D_1 \cap D_2$ is strongly dense in H with determining sequence $\{e_n\}_{n\in\mathbb{N}}$.

(ii) Let $\{p_n\}_{n\in\mathbb{N}}$ and $\{q_n\}_{n\in\mathbb{N}}$ be determining sequences in $P(\mathcal{M})$ for $\operatorname{dom}(x)$ and D, respectively. For each $n \in \mathbb{N}$, the operator $x_n = xp_n$ belongs to $\mathcal{M}$, and so $q_n^\perp x_n \in \mathcal{M}$. Therefore, the projection $f_n = \mathbf{1} - s_r(q_n^\perp x_n)$ belongs to $P(\mathcal{M})$. Note that

$$x_n f_n = q_n x_n f_n + q_n^\perp x_n f_n = q_n x_n f_n, \quad n \in \mathbb{N}. \tag{2.1}$$

Furthermore,

$$f_n^\perp = s_r(q_n^\perp x_n) \sim s_l(q_n^\perp x_n) \leq q_n^\perp,$$

and so $f_n^\perp \precsim q_n^\perp$ for all $n \in \mathbb{N}$. Since $q_n^\perp$ is finite, it follows that $f_n^\perp$ is a finite projection, $n \in \mathbb{N}$.

We claim that the sequence $\{e_n\}_{n\in\mathbb{N}}$ in $P(\mathcal{M})$, defined by $e_n = f_n \wedge p_n$ for all n, is a determining sequence for $x^{-1}(D)$.

To show that $\{e_n\}_{n\in\mathbb{N}}$ is increasing, we note that $e_n = f_n e_n = p_n e_n$ and $q_{n+1}^\perp q_n = 0$ and, therefore, equality (2.1) implies that

$$\begin{aligned} q_{n+1}^\perp x_{n+1} e_n &= q_{n+1}^\perp x_{n+1} p_n e_n = q_{n+1}^\perp x_n e_n \\ &= q_{n+1}^\perp x_n f_n e_n = q_{n+1}^\perp q_n x_n f_n e_n = 0. \end{aligned}$$

Hence, $e_n \leq \mathbf{1} - s_r(q_{n+1}^\perp x_{n+1}) = f_{n+1}$. Moreover, $e_n \leq p_n \leq p_{n+1}$, and so $e_n \leq f_{n+1} \wedge p_{n+1} = e_{n+1}$, proving that $\{e_n\}_{n\in\mathbb{N}}$ is increasing.

Next, since $e_n^\perp = f_n^\perp \vee p_n^\perp$ and the projections $f_n^\perp$ and $p_n^\perp$ are finite, it follows that $e_n^\perp$ is also finite. It is clear that $e_n(H) \subseteq p_n(H) \subseteq \operatorname{dom}(x)$. Then (2.1) implies that

$$xe_n\xi = xp_n e_n \xi = x_n e_n \xi = x_n f_n e_n \xi = q_n x_n f_n e_n \xi \in q_n(H) \subseteq D,$$

for any $\xi \in H$ and this shows that $e_n(H) \subseteq x^{-1}(D)$ for all $n \in \mathbb{N}$.

It remains to be shown that $e_n \uparrow \mathbf{1}$. Since $e_n^\perp = f_n^\perp \vee p_n^\perp, f_n^\perp \precsim q_n^\perp$ for all n and $\{p_n^\perp\}_{n\in\mathbb{N}}$, $\{q_n^\perp\}_{n\in\mathbb{N}}$ are two sequences of finite projections satisfying $p_n^\perp \downarrow 0$, $q_n^\perp \downarrow 0$. It follows from Lemma 2.2.6 that $e_n^\perp = f_n^\perp \vee p_n^\perp \downarrow 0$ and, therefore, $e_n \uparrow \mathbf{1}$, as required. □

Proposition 2.2.10. *Let x and y be measurable operators with respect to a von Neumann algebra $\mathcal{M}$.*
(i) *If $x \subseteq y$, then $x = y$;*
(ii) *If D is a strongly dense subspace of $\operatorname{dom}(x)$, then D is a core of x;*
(iii) *If D is a strongly dense subspace of $\operatorname{dom}(x) \cap \operatorname{dom}(y)$ and $x_{|D} = y_{|D}$, then $x = y$.*

Proof. (i) Since $x \subset y$, it follows that $y^* \subseteq x^*$. By Corollary 2.2.4, the operators x^* and y^* are measurable. In particular, $\operatorname{dom}(y^*)$ is strongly dense. Setting

$$D_0 = \mathrm{dom}(x) \cap y^{-1}(\mathrm{dom}(y^*))$$

and employing Lemma 2.2.9(i) and (ii) we infer that D_0 is strongly dense in H.

If $\xi \in D_0$, then $x\xi = y\xi \in \mathrm{dom}(y^*)$ and $y^*y\xi = x^*y\xi = x^*x\xi$. Denote by z the restriction of the self-adjoint operator x^*x to D_0. Then z is symmetric, affiliated with $\mathcal{M}$, and $\mathrm{dom}(z) = D_0$ is strongly dense in H. In particular, the closure $\overline{z}$ is a symmetric measurable operator. By Proposition 2.2.7, the closure $\overline{z}$ is self-adjoint. Since $z \subseteq x^*x$, it is also clear that $\overline{z} = x^*x$, as self-adjoint operators are maximal symmetric (see, e. g., [140]). Similarly, it follows that $\overline{z} = y^*y$, and hence, $x^*x = y^*y$. This implies that $|x| = |y|$. Thus, $\mathrm{dom}(x) = \mathrm{dom}(|x|) = \mathrm{dom}(|y|) = \mathrm{dom}(y)$. Therefore, $x = y$.

(ii) Suppose that x is measurable and that $D \subseteq \mathrm{dom}(x)$ is strongly dense and let $y = x_{|D}$ be the restriction of x onto D. Since D is strongly dense in H, there exists a determining sequence $\{p_n\}_{n\in\mathbb{N}}$ for D. We define

$$D_0 = \bigcup_{n\in\mathbb{N}} p_n(H), \quad y_0 = x|_{D_0}.$$

It is clear that y_0 is affiliated with $\mathcal{M}$ and has strongly dense domain. By construction of y and y_0, we have inclusions $y_0 \subseteq y \subseteq x$. In particular, y_0 is closable, $\overline{y_0}$ is measurable, and $\overline{y_0} \subset \overline{y} \subset x$. Therefore, part (i) implies that $\overline{y_0} = x$ and so, $x = \overline{y}$. This shows that D is a core for $\overline{x}$.

(iii) By part (ii), D is a core for x and y. The assertion now follows since $x = \overline{x|_D} = \overline{y|_D} = y$ by the definition of a core of a closed linear operator. □

In the following proposition, we show that for two measurable operators the strong sum and strong product are well-defined and measurable, too. This result allows us to define the addition and multiplication operations on the set of all measurable operators, which turn $S(\mathcal{M})$ into an algebra.

Proposition 2.2.11. *If operators x and y are measurable with respect to a von Neumann algebra $\mathcal{M}$, then the strong sum $x \dotplus y$ and strong product $x \cdot y$ are well-defined and measurable with respect to $\mathcal{M}$.*

Proof. By Proposition 2.1.2(i), the operators $x + y$ and xy are affiliated with $\mathcal{M}$.

It follows from Lemma 2.2.9(i) that $\mathrm{dom}(x + y) = \mathrm{dom}(x) \cap \mathrm{dom}(y)$ is strongly dense. We claim that $x + y$ is closable. Since $\mathrm{dom}(x + y)$ is strongly dense, it follows that $x + y$ densely defined, and so $x^* + y^* \subseteq (x + y)^*$. By Corollary 2.2.4, x^* and y^* are measurable and, therefore, $\mathrm{dom}(x^* + y^*) = \mathrm{dom}(x^*) \cap \mathrm{dom}(y^*)$ is strongly dense in H. This implies that $(x + y)^*$ is densely defined and so $x + y$ is closable since $\overline{x + y} = (x + y)^{**}$. Thus, the strong sum $x \dotplus y = \overline{x + y}$ is well-defined. By Proposition 2.1.2(i) and (ii), we have that $(x \dotplus y)\eta\mathcal{M}$. Since $\mathrm{dom}(x + y) \subset \mathrm{dom}(x \dotplus y)$ and $\mathrm{dom}(x + y)$ is strongly dense, it follows that $x \dotplus y$ is measurable.

To show that $x \cdot y$ is well-defined and measurable, we first note that Lemma 2.2.9(i) and (ii) implies that $\mathrm{dom}(xy) = \mathrm{dom}(y) \cap y^{-1}(\mathrm{dom}(x))$ is strongly dense in H. Furthermore, $y^*x^* \subseteq (xy)^*$ and x^* and y^* are measurable. Hence, referring to Lemma 2.2.9 once again, we infer that

$$\mathrm{dom}(y^*x^*) = \mathrm{dom}(x^*) \cap (x^*)^{-1}(\mathrm{dom}(y^*))$$

is strongly dense. This implies, in particular, that $(xy)^*$ is densely defined and hence, xy is closable. Thus, the strong product $x \cdot y = \overline{xy}$ is well-defined. By Proposition 2.1.2(i) and (ii), we have that $(x \cdot y)\eta\mathcal{M}$. Since $\mathrm{dom}(xy) \subset \mathrm{dom}(x \cdot y)$ and $\mathrm{dom}(xy)$ is strongly dense, it follows that $x \cdot y$ is measurable. □

It is now possible to introduce the algebraic structure in the set $S(\mathcal{M})$ of all operators, which are measurable with respect to the von Neumann algebra $\mathcal{M}$.

If $x \in S(\mathcal{M})$ and $\lambda \in \mathbb{C}, \lambda \neq 0$, then $\lambda x \in S(\mathcal{M})$. If $\lambda = 0$, then $0x$ need not be closed, but it is closable with closure the zero operator 0 on H. Therefore, we set $0x = 0$ for all $x \in S(\mathcal{M})$. With these operations of scalar multiplication, strong addition and strong multiplication, the set $S(\mathcal{M})$ has the structure of an algebra.

Theorem 2.2.12. *The set $S(\mathcal{M})$ is a $*$-algebra over $\mathbb{C}$ with unit element* **1**, *with respect to the operations of scalar multiplication, strong sum and strong product and the $*$-operation of taking adjoints. The von Neumann algebra $\mathcal{M}$ is a $*$-subalgebra of $S(\mathcal{M})$.*

Proof. By Proposition 2.2.10, in order to check the algebraic identities, it suffices to verify these equalities on some strongly dense subspaces. The operators coincide on strongly dense subspaces and, therefore, they are identical. To illustrate this, we show the associative law for addition and a distributive law.

Let $x, y, z \in S(\mathcal{M})$. By Lemma 2.2.9(i), the subspace

$$D = \mathrm{dom}(x) \cap \mathrm{dom}(y) \cap \mathrm{dom}(z)$$

is strongly dense and on D, the operators $(x \dotplus y) \dotplus z$ and $x \dotplus (y \dotplus z)$ coincide with the algebraic sum $x + y + z$. Proposition 2.2.10(iii) implies that $(x \dotplus y) \dotplus z$ and $x \dotplus (y \dotplus z)$ are equal.

To show the distributive law, we note that by Lemma 2.2.9(i) and (ii) the space

$$D = x^{-1}(\mathrm{dom}(z)) \cap y^{-1}(\mathrm{dom}(z))$$

is a strongly dense subspace of H. Since the operators $z \cdot (x \dotplus y)$ and $z \cdot x \dotplus z \cdot y$ coincide on D with the operator $zx + zy$, the distributive law follows from Proposition 2.2.10(iii).

To show that $S(\mathcal{M})$ is a $*$-algebra, we note that Corollary 2.2.4 implies that $S(\mathcal{M})$ is closed under taking adjoints. The mapping $x \longmapsto x^*$ is an involution in $S(\mathcal{M})$. Indeed, it is clear that $(x^*)^* = x$ and $(\lambda x)^* = \bar{\lambda} x^*$ for all $x \in S(\mathcal{M})$ and $\lambda \in \mathbb{C}$. Given $x, y \in S(\mathcal{M})$, the operator $x + y$ is densely defined, and so $x^* + y^* \subseteq (x + y)^* = (x \dotplus y)^*$. Hence, the two measurable operators $x^* \dotplus y^*$ and $(x \dotplus y)^*$ coincide on the strongly dense subspace

$\mathrm{dom}(x^*) \cap \mathrm{dom}(y^*)$, and so $(x \dotplus y)^* = x^* \dotplus y^*$. Via a similar argument, one can show that $(x \cdot y)^* = y^* \cdot x^*$ for all $x, y \in S(\mathcal{M})$.

Evidently, $\mathcal{M} = S(\mathcal{M}) \cap \mathcal{B}(H)$. Therefore, the assertion that $\mathcal{M}$ is a $*$-subaglebra in $S(\mathcal{M})$ is trivial. □

We note that if $x \in S(\mathcal{M})$ and $y \in \mathcal{M} \subset S(\mathcal{M})$, then the operators $x + y$ and xy are closed and, therefore, $x \dotplus y = x + y$ and $x \cdot y = xy$.

From now on, the strong sum $x \dotplus y$ and the strong product $x \cdot y$ of two elements $x, y \in S(\mathcal{M})$ will be denoted simply by $x + y$ and xy, respectively, unless stated otherwise.

We now illustrate the notion of measurable operators by two simple examples.

Example 2.2.13. *Let $\mathcal{M} = \mathcal{B}(H)$, where H is any Hilbert space. As follows from Example 2.1.3, all closed and densely defined operators in H are affiliated with $\mathcal{M}$. Note that a projection $p \in P(\mathcal{M})$ is finite if and only if its range* $\mathrm{ran}(p)$ *is finite-dimensional. Suppose that the linear subspace D of H is strongly dense. By definition, there exists a sequence $\{p_n\}_{n \in \mathbb{N}}$ in $P(\mathcal{M})$ such that $p_n \uparrow \mathbf{1}$, $p_n(H) \subseteq D$ and $p_n^\perp$ is finite for all n. Since $p_n^\perp \downarrow 0$ and each $p_n^\perp$ has finite-dimensional range, there must be an n_0 such that $p_{n_0}^\perp = 0$ and hence, $p_{n_0} = \mathbf{1}$. This shows that $D = H$. Consequently, $S(\mathcal{M}) = \mathcal{M} = \mathcal{B}(H)$.*

Example 2.2.14. *Let (Ω, Σ, μ) be a σ-finite measure space and let $\mathcal{M}$ be the von Neumann algebra $L_\infty(\Omega, \Sigma, \mu)$ acting by multiplication on $H = L_2(\Omega, \Sigma, \mu)$, i. e., $\mathcal{M} = \{M_f : f \in L_\infty(\Omega, \Sigma, \mu)\}$. As seen in Example 2.1.4, the collection of all closed densely defined linear operators affiliated with $\mathcal{M}$ is given by $\{M_f : f \in L_0(\Omega, \Sigma, \mu)\}$. Since $\mathcal{M}$ is commutative (and hence, finite), it follows from Corollary 2.2.5, that every closed densely defined operator affiliated with $\mathcal{M}$ is measurable. Consequently,*

$$S(\mathcal{M}) = \{M_f : f \in L_0(\Omega, \Sigma, \mu)\}.$$

In the next theorem, we show that for $\mathcal{M} = L_\infty(\Omega, \Sigma, \mu)$ the algebras $S(\mathcal{M})$ and $L_0(\Omega, \Sigma, \mu)$ are, in fact, $*$-isomoprhic.

Theorem 2.2.15. *Let (Ω, Σ, μ) be a σ-finite measure space, let $\mathcal{M} = \{M_f : f \in L_\infty(\Omega, \Sigma, \mu)\}$ be a commutative von Neumann algebra acting on the Hilbert space $H = L_2(\Omega, \Sigma, \mu)$ by the rule $M_f(g)(\omega) = f(\omega)g(\omega)$, where $g \in H$, $\omega \in \Omega$. Then the mapping $\Phi : L_0(\Omega, \Sigma, \mu) \to S(\mathcal{M})$, defined by*

$$\Phi(f) = M_f, \quad f \in L_0(\Omega, \Sigma, \mu),$$

is a $*$-isomorphism.

Proof. Since $\mathcal{M}$ is a commutative von Neumann algebra (and so, it is finite), Corollary 2.2.5 implies that $S(\mathcal{M})$ coincides with the set of all closed densely defined operator affiliated with $\mathcal{M}$. As shown in Example 2.1.4, the map Φ is bijective.

If $f, h \in L_0(\Omega, \Sigma, \mu)$, and $g \in \mathrm{dom}(M_f) \cap \mathrm{dom}(M_h)$, then for almost all $\omega \in \Omega$ the equalities

$$M_{(f+h)}(g)(\omega) = (f(\omega) + h(\omega))g(\omega) = M_f(g)(\omega) + M_h(g)(\omega)$$

hold. By Lemma 2.2.9(i), the subspace $\mathrm{dom}(M_f) \cap \mathrm{dom}(M_h)$ is strongly dense, and so Proposition 2.2.10(iii) implies that

$$\Phi(f + h) = M_{(f+h)} = M_f \dotplus M_h = \Phi(f) \dotplus \Phi(h).$$

Similarly,

$$\Phi(fh) = \Phi(f) \cdot \Phi(h) \quad \text{and} \quad \Phi(\lambda f) = \lambda \Phi(f)$$

for all $\lambda \in \mathbb{C}$.

Moreover, $\mathrm{dom}(\Phi(f)) = \mathrm{dom}(\Phi(\bar{f}))$ and for all $g_1, g_2 \in \mathrm{dom}(\Phi(f))$ we have

$$\begin{aligned}\langle \Phi(f)(g_1), g_2 \rangle_{L_2(\Omega,\Sigma,\mu)} &= \int_\Omega f g_1 \overline{g_2} d\mu = \int_\Omega g_1 \overline{\bar{f} g_2} d\mu \\ &= \langle g_1, \bar{f} g_2 \rangle_{L_2(\Omega,\Sigma,\mu)} = \langle g_1, \Phi(\bar{f}) g_2 \rangle_{L_2(\Omega,\Sigma,\mu)}.\end{aligned}$$

Using again Proposition 2.2.10(iii), we obtain that $\Phi(\bar{f}) = \Phi(f)^*$.

Hence, Φ is a $*$-isomorphism from the $*$-algebra $S(\Omega, \Sigma, m)$ onto the $*$-algebra $S(\mathcal{M})$. □

We also give a simple description of the algebra $S(\mathcal{M})$ for a type I_n von Neumann algebra $\mathcal{M}$. Recall (see Proposition 1.9.16) that any type I_n von Neumann algebra is $*$-isomorphic to the algebra $M_{n\times n}(\mathcal{Z}(\mathcal{M}))$.

Proposition 2.2.16. *Let $\mathcal{M}$ be a von Neumann algebra of type I_n. Then the algebra $S(\mathcal{M})$ is $*$-isomorphic to the algebra $M_{n\times n}(S(\mathcal{Z}(\mathcal{M})))$.*

Proof. By Proposition 1.9.16, there exists a family of matrix units $\{e_{ij}\}_{i,j=1}^n \subset \mathcal{M}$, such that

$$x = \sum_{i,j=1}^n x_{ij} e_{ij}, \quad x_{ij} \in \mathcal{Z}(\mathcal{M}), \forall x \in \mathcal{M}. \tag{2.2}$$

Furthermore,

$$x_{ij} = \sum_{r=1}^n e_{ri} x e_{jr}, \quad i, j = 1, \dots, n.$$

Now let $a \in S(\mathcal{M})$. For $i, j = 1, \dots, n$ consider the element $a_{ij} = \sum_{r=1}^{n_k} e_{ri} a e_{j,r}$. By (2.2), we have that any such element commutes with $\mathcal{M}$. By the spectral theorem, the spectral projections of the real and imaginary part of a_{ij} commute with all elements from $\mathcal{M}$ and, therefore, they belong to $\mathcal{Z}(\mathcal{M})$. By Proposition 2.1.2(iv), we have that a_{ij} is affiliated with

$\mathcal{Z}(\mathcal{M})$. Since $\mathcal{Z}(\mathcal{M})$ is finite von Neumann algebra, it follows that $a_{ij} \in S(\mathcal{Z}(\mathcal{M}))$. Since $a = \sum_{i,j=1}^{n} a_{ij}e_{ij}$, it follows that

$$\sum_{i,j=1}^{n} S(\mathcal{Z}(\mathcal{M}))e_{ij} = S(\mathcal{M}).$$

It remains to define an isomorphism $\Phi : S(\mathcal{M}) \to M_{n\times n}(S(\mathcal{Z}(\mathcal{M})))$, by setting $\Phi(a) = (a_{ij})$, where $a_{ij} = \sum_{r=1}^{n} e_{ri}ae_{jr}$. □

In conclusion of this section, we note the following useful property of the $*$-algebra $S(\mathcal{M})$.

Proposition 2.2.17. *If z is a central projection from von Neumann algebra $\mathcal{M}$, then $zS(\mathcal{M}) = S(z\mathcal{M})$.*

Proof. Suppose first that $x \in S(\mathcal{M})$ and let $\{p_n\}_{n\in\mathbb{N}}$ be a determining sequence for x. Then $(zx)(p_n z) \in \mathcal{M}z$, $p_n^{\perp}z \in P_{\text{fin}}(\mathcal{M}z)$, and $p_n z \uparrow z$. Hence, $zx \in S(z\mathcal{M})$ showing that $zS(\mathcal{M}) \subset S(z\mathcal{M})$.

Conversely, let $x \in S(z\mathcal{M})$ and let $\{p_n\}_{n\in\mathbb{N}} \subset P(z\mathcal{M})$ be a determining sequence for x. The operators x, p_n are acting on a dense subspace of $z(H)$. We define their extensions acting in H by setting $\widetilde{p_n} = p_n + z^{\perp}$ and $\tilde{x} = x + 0|_{z^{\perp}(H)}$. Then $\tilde{x}$ is a densely defined closed operator affiliated with $\mathcal{M}$. Furthermore, $\widetilde{p_n} \uparrow z + z^{\perp} = \mathbf{1}$, $\widetilde{p_n}^{\perp} = p_n^{\perp}z \in P_{\text{fin}}(\mathcal{M})$, and $\tilde{x}\widetilde{p_n} = xp_n \in z\mathcal{M} \subset \mathcal{M}$. Therefore, $\tilde{x} \in S(\mathcal{M})$ and $x = \tilde{x}z$, showing that $x \in zS(\mathcal{M})$. Hence, $S(z\mathcal{M}) \subset zS(\mathcal{M})$, which completes the proof. □

2.3 Locally measurable operators affiliated with a von Neumann algebra

In this section, we introduce locally measurable operators and show that the set of all locally measurable operators form a $*$-algebra with respect to the strong sum and strong product. In contrast to the $*$-algebra of all measurable operators, the $*$-algebra of locally measurable operators possesses several additional topological and algebraic properties, which we shall describe in Chapter 3 and Chapter 4.

Let $\mathcal{M}$ be a von Neumann algebra acting on a (separable) Hilbert space H and let $\mathcal{Z}(\mathcal{M})$ be the center of $\mathcal{M}$. We denote by (Ω, Σ, μ) a probability space, such that $\mathcal{Z}(\mathcal{M})$ is $*$-isomorphic to $L_\infty(\Omega, \Sigma, \mu)$. By φ we denote a $*$-isomorpism from $\mathcal{Z}(\mathcal{M})$ onto $L_\infty(\Omega, \Sigma, \mu)$ (see Remark 1.8.11). We let $\mathcal{D} : P(\mathcal{M}) \to L^{+}(\Omega, \Sigma, \mu)$ denote the dimension function on $\mathcal{M}$.

Definition 2.3.1. A closed linear operator x acting in the Hilbert space H is said to be *locally measurable* with respect to the von Neumann algebra $\mathcal{M}$ if $x\eta\mathcal{M}$ and there exists a sequence of central projections $\{z_n\}_{n\in\mathbb{N}} \subset P(\mathcal{Z}(\mathcal{M}))$ such that $z_n \uparrow \mathbf{1}$ and $xz_n \in S(\mathcal{M})$ for all $n \in \mathbb{N}$.

The set of all operators locally measurable with respect to the von Neumann algebra $\mathcal{M}$ is denoted by $LS(\mathcal{M})$.

If the von Neumann algebra $\mathcal{M}$ is fixed and no confusion may arise, we shall simply say that x is a locally measurable operator.

We note that any locally measurable operator is necessarily densely defined in H. Indeed, since $z_n \uparrow \mathbf{1}$, it follows that $H_0 = \bigcup_n z_n(H)$ is a dense subspace in H. For every $n \in \mathbb{N}$, we have that $\operatorname{dom}(z_n x) \cap z_n(H) = \operatorname{dom}(x) \cap z_n(H) \subset \operatorname{dom}(x)$. Since $xz_n \in S(\mathcal{M}z_n)$, it follows that $\operatorname{dom}(xz_n) = z_n \operatorname{dom}(x)$ is dense in $z_n(H)$ for all $n \in \mathbb{N}$. Therefore, $\operatorname{dom}(x)$ is dense in H.

Remark 2.3.2. *Since $S(\mathcal{M})$ is an algebra (see Theorem 2.2.12), it follows that a closed operator x affiliated with $\mathcal{M}$ is locally measurable with respect to $\mathcal{M}$ if and only if there exists a central partition of unity $\{z_n\}_{n\in\mathbb{N}} \subset P(\mathcal{Z}(\mathcal{M}))$, such that $z_n x$ is measurable. Indeed, if $\{z_n\}_{n\in\mathbb{N}} \subset P(\mathcal{Z}(\mathcal{M}))$ is such that $z_n \uparrow \mathbf{1}$, then setting*

$$z_1' = z_1, \quad z_n' = z_n - z_{n-1}, \quad n = 2, \dots,$$

we have that $\{z_n'\}_{n\in\mathbb{N}}$ is a central partition of unity and $z_n' x \in S(\mathcal{M})$. Conversely, if $\{z_n\}_{n\in\mathbb{N}}$ is a central partition of unity such that $z_n x \in S(\mathcal{M})$, $n \in \mathbb{N}$, then $\{z_n'\}_{n\in\mathbb{N}}$ defined by $z_n' = \sum_{k=1}^{n} z_n$, is increasing sequence of central projections satisfying $\sup_{n\in\mathbb{N}} z_n' = \mathbf{1}$ and $z_n' x \in S(\mathcal{M})$.

It is clear that $\mathcal{M} \subset S(\mathcal{M}) \subset LS(\mathcal{M})$ for any von Neumann algebra $\mathcal{M}$. For necessary and sufficient conditions for any of the inclusions $\mathcal{M} \subset S(\mathcal{M}) \subset LS(\mathcal{M})$ to be equalities, see Section 2.5 below. In particular, see Proposition 2.5.4 (and its proof) for an example of locally measurable operator, which is not measurable.

An elementary sufficient condition for the equality $S(\mathcal{M}) = LS(\mathcal{M})$ is presented in the following proposition.

Proposition 2.3.3. *If $\mathcal{M}$ is a factor or a finite von Neumann algebra, then $S(\mathcal{M}) = LS(\mathcal{M})$.*

Proof. If $\mathcal{M}$ is a finite von Neumann algebra, then as shown in Corollary 2.2.5 any closed operator affiliated with $\mathcal{M}$ is measurable and, therefore, locally measurable.

If $\mathcal{M}$ is a factor, then the only nonzero central projection is $\mathbf{1}$, and so a closed operator is measurable if and only if it is locally measurable. □

We now present a version of Proposition 2.2.2 for locally measurable operators.

Proposition 2.3.4. *If x is a closed operator and $x = v|x|$ is the polar decomposition of the operator x, then the operator x is locally measurable with respect to $\mathcal{M}$ if and only if $v \in \mathcal{M}$ and $|x|$ is locally measurable with respect to $\mathcal{M}$.*

Proof. Let $x \in LS(\mathcal{M})$ and let $\{z_n\}_{n\in\mathbb{N}}$ be a sequence of central projections from $P(\mathcal{Z}(\mathcal{M}))$ such that $z_n \uparrow \mathbf{1}$ and $xz_n \in S(\mathcal{M})$. Since $x\eta\mathcal{M}$, by Proposition 2.1.2(vi) we have that $v \in \mathcal{M}$ and $|x|\eta\mathcal{M}$, in particular, $|x|z_n = z_n|x|$. Since $xz_n = (vz_n)|x|z_n$, and

$$(vz_n)^*(vz_n) = z_n v^* v z_n = z_n s_r(x) z_n = s_r(xz_n),$$

the uniqueness of the polar decomposition of the operator xz_n (see Theorem 1.5.4) implies that $|x|z_n = |xz_n| \in S(\mathcal{M})$ for all $n \in \mathbb{N}$. Hence, $|x| \in LS(\mathcal{M})$.

Conversely, let $|x| \in LS(\mathcal{M})$, $v \in \mathcal{M}$. By Proposition 2.1.2(vi), $x\eta\mathcal{M}$. Let $\{e_n\}_{n\in\mathbb{N}}$ be a sequence of central projections from $P(\mathcal{Z}(\mathcal{M}))$ such that $e_n \uparrow \mathbf{1}$ and $|x|e_n \in S(\mathcal{M})$. Again using the equality $xe_n = (ve_n)|x|e_n$ and uniqueness of polar decomposition of the operator xe_n, we obtain that $|xe_n| = |x|e_n \in S(\mathcal{M})$, in addition, $ve_n \in \mathcal{M}$. By Proposition 2.1.2(vi), we have that $xe_n \in S(\mathcal{M})$ for all $n \in \mathbb{N}$, which implies the inclusion $x \in LS(\mathcal{M})$. □

In the following proposition, we present various necessary and sufficient conditions for local measurability of a closed linear operator.

Proposition 2.3.5. *Let x be a closed densely defined operator affiliated with a von Neumann algebra $\mathcal{M}$. The following conditions are equivalent:*
(i) *The operator x is locally measurable with respect to $\mathcal{M}$;*
(ii) *There exists an increasing sequence $\{z_n\}_{n\in\mathbb{N}}$ of central projections from $\mathcal{M}$, such that $\sup_{n\in\mathbb{N}} z_n = \mathbf{1}$ and $z_n E_{|x|}[n,\infty)$ are finite for all $n \in \mathbb{N}$;*
(iii) *$\mathcal{D}(E_{|x|}(\lambda,\infty)) \downarrow 0$ almost everywhere as $\lambda \to \infty$;*
(iv) *There exists a sequence $\{p_n\}_{n\in\mathbb{N}} \subset P(\mathcal{M})$, such that $p_n \uparrow \mathbf{1}$, $p_n(H) \subset \mathrm{dom}(x)$ and $\mathcal{D}(p_n^\perp) \downarrow 0$ as $n \to \infty$ almost everywhere;*
(v) *There exist a sequence of projections $\{q_n\}_{n\in\mathbb{N}} \subset P(\mathcal{M})$ and a sequence of central projections $\{z_n\}_{n\in\mathbb{N}} \subset P(\mathcal{Z}(\mathcal{M}))$, such that $q_n \uparrow \mathbf{1}$, $z_n \uparrow \mathbf{1}$, $q_n(H) \subseteq \mathrm{dom}(x)$ and $z_n q_n^\perp$ are finite projections for all $n \in \mathbb{N}$.*

Proof. (i)⇒(ii). Let $\{z_k\}_{k\in\mathbb{N}} \subset P(\mathcal{Z}(\mathcal{M}))$ be such that $z_k \uparrow \mathbf{1}$ and $z_k x \in S(\mathcal{M})$ for all $k \in \mathbb{N}$. For every $n \in \mathbb{N}$, we define

$$w_n = \sup\{z_k : z_k E_{|x|}[n,\infty) \in P_{\mathrm{fin}}(\mathcal{M})\}.$$

By Theorem 1.9.9(vii), we have that $w_n E_{|x|}[n,\infty) \in P_{\mathrm{fin}}(\mathcal{M})$. If $n < m$, then $w_n E_{|x|}[m,\infty) \le w_n E_{|x|}[n,\infty) \in P_{\mathrm{fin}}(\mathcal{M})$, and so $w_n \le w_m$.

Since $z_k x$ is measurable for any $k \in \mathbb{N}$, Proposition 2.2.3(iii) implies that for every $k \in \mathbb{N}$ there exists $n \in \mathbb{N}$, such that $z_k E_{|x|}[n,\infty) = E_{|z_k x|}[n,\infty) \in P_{\mathrm{fin}}(\mathcal{M})$. Therefore, $z_k \le w_n$. Since $z_k \uparrow \mathbf{1}$, it follows that $w_n \uparrow \mathbf{1}$, as required.

(ii)⇒(iii). Let $\{z_n\}_{n\in\mathbb{N}}$ be an increasing sequence of central projections from $\mathcal{M}$, such that $\sup_{n\in\mathbb{N}} z_n = \mathbf{1}$ and $z_n E_{|x|}[n,\infty)$ are finite for all $n \in \mathbb{N}$. To show that $\mathcal{D}(E_{|x|}(\lambda,\infty)) \downarrow 0$ almost everywhere as $\lambda \to \infty$, it is sufficient to show that $\mathcal{D}(E_{|x|}[n,\infty)) \downarrow 0$ almost everywhere as $n \to \infty$.

Since $E_{|x|}[n,\infty) \downarrow 0$ as $n \to \infty$ and $z_n E_{|x|}[n,\infty)$ is finite, Proposition 1.11.6(ii) implies that for any $k \in \mathbb{N}$, $\mathcal{D}(z_k E_{|x|}[n,\infty)) \downarrow 0$ as $n \to \infty$ a. e. For any $k \in \mathbb{N}$ and $n > k$, we have that $\chi_{A_k}\mathcal{D}(E_{|x|}[n,\infty)) = \varphi(z_k)\mathcal{D}(z_nE_{|x|}[n,\infty))$ where $\varphi(z_k) = \chi_{A_k}$ for some measurable $A_k \subset \Omega$. Therefore, $\mathcal{D}(E_{|x|}[n,\infty)) \downarrow 0$ as $n \to \infty$ almost everywhere on A_k. Since $\sup_{k\in\mathbb{N}} z_k = \mathbf{1}$, it follows that $\sup_{k\in\mathbb{N}} \varphi(z_k) = \mathbf{1}$ almost everywhere, and so $\Omega \setminus \bigcup_{n\in\mathbb{N}} A_n$ has measure zero. Thus, $\mathcal{D}(E_{|x|}[n,\infty)) \downarrow 0$ as $n \to \infty$ almost everywhere on Ω.

Implication (iii)⇒(iv) follows from the spectral theorem by taking $p_n = E_{|x|}[0,n]$, $n \in \mathbb{N}$.

(iv)⇒(v). For every $n \in \mathbb{N}$, we set $\Omega_n = \{\omega \in \Omega : \mathcal{D}(p_n^\perp)(\omega) > n\}$. By assumption, we have that $\mu(\bigcap_n \Omega_n) = 0$. Let $z_n \in P(\mathcal{Z}(\mathcal{M}))$ be such that $\varphi(\mathbf{1} - z_n) = \chi_{\Omega_n}$. Then $z_n \uparrow \mathbf{1}$ and $\mathcal{D}(z_np_n^\perp) \le n$, implying that $z_np_n^\perp \in P_{\mathrm{fin}}(\mathcal{M})$.

(v)⇒(i). For every $n_0 \in \mathbb{N}$, we have that $z_{n_0}q_n \uparrow z_{n_0}$ as $n \to \infty$, $z_{n_0}q_n(H) \subset z_{n_0}(\mathrm{dom}(x)) = \mathrm{dom}(z_{n_0}x)$ and $z_{n_0}q_n^\perp \in P_{\mathrm{fin}}(\mathcal{M}z_{n_0})$ for $n \ge n_0$. By Proposition 2.2.3(ii), we have that $z_{n_0}x \in S(z_{n_0}\mathcal{M})$. Referring to Proposition 2.2.17, we conclude that $z_{n_0}x \in z_{n_0}S(\mathcal{M})$, proving that $x \in LS(\mathcal{M})$. □

Proposition 2.3.5(iii) and Proposition 2.1.7 imply that $LS(\mathcal{M})$ is closed under taking adjoints.

Corollary 2.3.6. *If x is a locally measurable operator, then x^* is also a locally measurable operator.*

Proof. By Proposition 2.1.2(iii), we have that $x^*\eta\mathcal{M}$. Proposition 2.3.5(iii) implies that $\mathcal{D}(E_{|x|}(\lambda,\infty)) \downarrow 0$ almost everywhere as $\lambda \to \infty$. By Proposition 2.1.7, we have that $E_{|x^*|}(\lambda,\infty) \sim E_{|x|}(\lambda,\infty)$. Since the dimension function $\mathcal{D}$ coincides on equivalent projections (see Proposition 1.11.6(iv)), it follows that $\mathcal{D}(E_{|x^*|}(\lambda,\infty)) \downarrow 0$ almost everywhere as $\lambda \to \infty$. Thus, referring to Proposition 2.3.5(iii) once again we obtain that x^* is locally measurable. □

In the following proposition, we prove an analogue of Proposition 2.2.7 and Proposition 2.2.10(i) for locally measurable operators. We first state an auxiliary lemma.

Lemma 2.3.7. *Suppose that y is a locally measurable operator and x is a closed operator affiliated with $\mathcal{M}$, such that $xz_n = yz_n$ for some central partition of unity $\{z_n\}_{n\in\mathbb{N}} \subset P(\mathcal{Z}(\mathcal{M}))$. Then x is locally measurable and $x = y$.*

Proof. Let $\{z_n\}_{n\in\mathbb{N}}$ be a central partition of unity, such that $z_nx = z_ny$. Since y is locally measurable, there exists a central partition $\{z_n'\}_{n\in\mathbb{N}} \subset P(\mathcal{Z}(\mathcal{M}))$, such that $z_n'y \in S(\mathcal{M})$ for all $n \in \mathbb{N}$ (see Remark 2.3.2).

Define

$$w_{n,m} = z_nz_m', \quad n, m \in \mathbb{N}.$$

Then $\{w_{n,m}\}_{n,m=1}^\infty$ is a central partition of unity, such that $w_{n,m}x = w_{n,m}y \in S(\mathcal{M})$. In particular, x is locally measurable (see Remark 2.3.2).

Let $\xi \in \mathrm{dom}(y)$, then $\xi = \lim_{n\to\infty} \xi_n$ where $\xi_n = \sum_{k=1}^n z_k\xi$, $z_k\xi \in \mathrm{dom}(z_ny)$. Since $z_nx = z_ny$, it follows that $z_n\xi, \xi_n \in \mathrm{dom}(x)$ and $y\xi_n = x\xi_n$. Since y is closed, it follows that

$$y\xi = \lim_{n\to\infty} y\xi_n = \lim_{n\to\infty} x\xi_n.$$

Therefore, the sequences $\{\xi_n\} \subset \mathrm{dom}(x)$ and $\{x\xi_n\}$ converge in H. Since x is closed, it follows that $\xi = \lim_{n\to\infty} \xi_n \in \mathrm{dom}(x)$ and

$$x\xi = \lim_{n\to\infty} x\xi_n = \lim_{n\to\infty} y\xi_n = y\xi,$$

showing that $y \subset x$. The converse inclusion $x \subset y$ is proved similarly and, therefore, $x = y$. □

Proposition 2.3.8.
(i) *If x, y are locally measurable and $x \subset y$, then $x = y$;*
(ii) *If x is a symmetric locally measurable operator, then x is self-adjoint.*

Proof. (i) If x and y are locally measurable operators, then there exists a central partition of unity $\{z_n\}_{n\in\mathbb{N}}$, such that $z_n x, z_n y \in S(\mathcal{M})$. Since $x \subset y$, it follows that $z_n x \subset z_n y$. Since $z_n x$, $z_n y$ are measurable, Proposition 2.2.10(i) implies that $z_n x = z_n y$. Appealing to Lemma 2.3.7, we conclude that $x = y$.

(ii) If x is symmetric and locally measurable, then by Proposition 2.1.2(iii) we have that $x^* \eta \mathcal{M}$. Since $x \subset x^*$ and x is locally measurable, Proposition 2.3.5(iv) implies that x^* is locally measurable, too. Hence, the assertion follows from part (i). □

As noted in Remark 2.3.2 for a locally measurable operator x, there exists a central partition of unity $\{z_n\}_{n\in\mathbb{N}}$, such that $z_n x \in S(\mathcal{M})$. In the following proposition, we show that the reverse is also true. This result is crucial in the proof that $LS(\mathcal{M})$ forms a $*$-algebra with respect to the strong sum, strong product, and taking adjoints.

Proposition 2.3.9. *Let $\{z_n\}_{n\in\mathbb{N}}$ be a central partition of unity in $\mathcal{M}$ and let $\{x_n\}_{n\in\mathbb{N}}$ be a sequence of measurable operators, such that $s_c(x_n) \le z_n$, $n \in \mathbb{N}$. Then the operator defined as*

$$x\xi = \sum_{n=1}^{\infty} x_n\xi, \quad \xi \in \mathrm{dom}(x),$$

$$\mathrm{dom}(x) = \left\{\xi \in H : z_n\xi \in \mathrm{dom}(x_n),\ n \in \mathbb{N},\ \sum_{n=1}^{\infty} \|x_n\xi\|^2 < \infty\right\}$$

is a unique locally measurable operator, such that $z_n x = x_n$ for any $n \in \mathbb{N}$.

Proof. Since $\mathrm{dom}(x_n)$ is dense in $z_n(H)$ for any $n \in \mathbb{N}$, it follows that $\mathrm{dom}(x)$ is dense in H. It is clear that $z_n x = x_n$ for any $n \in \mathbb{N}$.

We claim that x is locally measurable. To show that x is affiliated with $\mathcal{M}$, we note that if $u \in \mathcal{M}'$, $\xi \in \mathrm{dom}(x)$, then

$$z_n u\xi = uz_n\xi \in uz_n\,\mathrm{dom}(x_n) = z_n u\,\mathrm{dom}(x_n) \subset z_n\,\mathrm{dom}(x_n)$$

for any $n \in \mathbb{N}$, i. e., $u\xi \in \mathrm{dom}(x)$. Therefore, $u(\mathrm{dom}(x)) \subset \mathrm{dom}(x)$. Since, in addition,

$$z_n ux\xi = uz_n x\xi = ux_n\xi = x_n u\xi = z_n xu\xi, \quad n \in \mathbb{N},$$

we conclude that $ux\xi = xu\xi$, and so $x\eta\mathcal{M}$.

To show that x is closed, assume that $\{\xi_m\}_{m\in\mathbb{N}} \subset \mathrm{dom}(x)$, $\xi_m \to \xi$, and $x\xi_m \to \eta$. Then, for every $n \in \mathbb{N}$, we have that $z_n\xi_m \to z_n\xi$ and $x_n z_n \xi_m = z_n x\xi_m \to z_n\eta$ as $m \to \infty$. Since x_n is closed for every $n \in \mathbb{N}$, it follows that $z_n\xi \in \mathrm{dom}(x_n)$ and $z_n\eta = x_n\xi$ for any $n \in \mathbb{N}$. Then

$$\sum_{n=1}^{\infty} \|x_n\xi\|^2 = \sum_{n=1}^{\infty} \|z_n\eta\|^2 = \|\eta\|^2 < \infty.$$

Hence, $\xi \in \mathrm{dom}(x)$ and $\eta = x\xi$, proving that x is a closed operator.

By construction, $xz_n = x_n$ for every $n \in \mathbb{N}$. Since $\{z_n\}_{n\in\mathbb{N}}$ is a central partition of unity and x_n is measurable for every $n \in \mathbb{N}$, it follows that x is locally measurable (see Remark 2.3.2). Uniqueness of $x \in LS(\mathcal{M})$ follows from Lemma 2.3.7. □

Proposition 2.3.9 implies, in particular, the following result.

Proposition 2.3.10. *Given any central partition $\{z_n\}_{n\in\mathbb{N}}$ of unity and any family $\{x_n\}_{n\in\mathbb{N}} \subset LS(\mathcal{M})$, there exists a unique element $x \in LS(\mathcal{M})$, such that $z_n x = z_n x_n$ for all $n \in \mathbb{N}$.*

Proof. For every $n \in \mathbb{N}$, we define $y_n = z_n x_n$. Then $y_n \in LS(\mathcal{M})$ and $s_c(y_n) \le z_n$ for every $n \in \mathbb{N}$. Since every y_n is locally measurable, there exists central partition $\{z_{n,k}\}_{n,k\in\mathbb{N}}$ of unity, such that $z_{n,k}y_n \in S(\mathcal{M})$ for all $n, k \in \mathbb{N}$, and $\sum_{k=1}^{\infty} z_{n,k} = z_n$. By Proposition 2.3.9, there exists unique $x \in LS(\mathcal{M})$, such that $z_{n,k}x = z_{n,k}y_n$ for all $n, k \in \mathbb{N}$. Since $\sum_{k=1}^{\infty} z_{n,k} = z_n$, it follows that $x \in LS(\mathcal{M})$ is a unique locally measurable operator such that $z_n x = z_n y_n = x_n$. □

Next, we show that for $x, y \in LS(\mathcal{M})$, their strong sum and strong product are well-defined and are locally measurable operators.

Proposition 2.3.11. *Let $x, y \in LS(\mathcal{M})$. Then the strong sum $x \dotplus y$ and strong product $x \cdot y$ are well-defined and are locally measurable operators. Furthermore, for a central partition of unity $\{z_n\}_{n\in\mathbb{N}}$, such that $z_n y, z_n y \in S(\mathcal{M})$, the operators $x \dotplus y$, and $x \cdot y$ are unique locally measurable operators, such that*

$$z_n(x \dotplus y) = (z_n x) \dotplus (z_n y), \quad z_n(x \cdot y) = (z_n x) \cdot (z_n y)$$

for all $n \in \mathbb{N}$.

Proof. Let $\{z_n\}_{n\in\mathbb{N}}$ be a central partition of unity, such that $z_n x, z_n y \in S(\mathcal{M})$ for all $n \in \mathbb{N}$. By Proposition 2.2.11, the strong sum $z_n x \dotplus z_n y$ is measurable.

To prove that $x + y$ is closable, let $\{\xi_k\}_{k\in\mathbb{N}} \subset \mathrm{dom}(x + y) = \mathrm{dom}(x) \cap \mathrm{dom}(y)$ be such that $\xi_k \to 0$ and $(x + y)\xi_k \to \eta$ for some $\eta \in H$ as $k \to \infty$. Then for every $n \in \mathbb{N}$, we have that $z_n\xi_k \to 0$ and $z_n(x+y)\xi_k \to z_n\eta$ as $k \to \infty$. Since $z_n(x+y) = z_nx + z_ny$ is closable, it follows that $z_n\eta = 0$ for every $n \in \mathbb{N}$. Therefore, $\eta = 0$, proving that $x+y$ is a closable operator.

By Proposition 2.1.2(i) and (ii), we have that $x+y, \overline{x+y}\eta\mathcal{M}$.

By Proposition 2.3.9, there exists a locally measurable operator w with

$$\operatorname{dom}(w) = \left\{\xi \in H : z_n\xi \in \operatorname{dom}(z_nx \dotplus z_ny),\ n \in \mathbb{N},\ \sum_{n=1}^{\infty} \|(z_nx \dotplus z_ny)\xi\|^2 < \infty\right\},$$

such that $z_nw = z_nx \dotplus z_ny$.

We claim that $x+y \subseteq w$. Let $\xi \in \operatorname{dom}(x+y) = \operatorname{dom}(x) \cap \operatorname{dom}(y)$. Then $z_n\xi \in z_n \operatorname{dom}(x) \cap \operatorname{dom}(y) \subset \operatorname{dom}(z_nx \dotplus z_ny)$,

$$\begin{aligned}\left(\sum_{n=1}^{\infty} \|(z_nx + z_ny)\xi\|^2\right)^{1/2} &\leq \left(\sum_{n=1}^{\infty} \|z_nx\xi\|^2\right)^{1/2} + \left(\sum_{n=1}^{\infty} \|z_ny\xi\|^2\right)^{1/2} \\ &= \|x\xi\| + \|y\xi\| < \infty,\end{aligned}$$

and

$$(x+y)\xi = \sum_{n=1}^{\infty} z_n(x+y)\xi = \sum_{n=1}^{\infty} (z_nx + z_ny)\xi = \sum_{n=1}^{\infty} z_nw\xi = w\xi,$$

as required.

Therefore, $\overline{x+y} \subset w$. In particular, $z_n\overline{(x+y)} \subset z_nw$. Since $z_nw = z_nx \dotplus z_ny$, we have that

$$z_nw = \overline{z_n(x+y)} \subset z_n\overline{(x+y)} \subset z_nw,$$

i. e., $z_n\overline{(x+y)} = z_nw$ for all $n \in \mathbb{N}$. By Lemma 2.3.7, we obtain that $\overline{x+y}$ is a locally measurable operator and $\overline{x+y} = w$.

Next, we show that the strong product is well-defined and is a locally measurable operator. Arguing as above, one can show that xy is a closable operator and $xy, \overline{xy}\eta\mathcal{M}$.

By Proposition 2.3.9, there exists a locally measurable operator w' with

$$\operatorname{dom}(w') = \left\{\xi \in H : z_n\xi \in \operatorname{dom}(z_nxy),\ n \in \mathbb{N},\ \sum_{n=1}^{\infty} \|(z_nxy)\xi\|^2 < \infty\right\},$$

such that $z_nw' = z_nxy$.

We claim that $xy \subseteq w'$. Let $\xi \in \mathrm{dom}(xy)$. Then $z_n\xi \in \mathrm{dom}(z_nxy)$ and

$$\|w'\xi\|^2 = \sum_{n=1}^{\infty}\|z_nw'\xi\|^2 = \sum_{n=1}^{\infty}\|(z_nxy)\xi\|^2 \le \sum_{n=1}^{\infty}\|xy(z_n\xi)\|^2 = \|xy\xi\|^2 < \infty,$$

and

$$(xy)\xi = \sum_{n=1}^{\infty} z_n(xy)\xi = \sum_{n=1}^{\infty} xy(z_n\xi) = w'\xi,$$

as required.

Therefore, $\overline{xy} \subset w'$. In particular, $z_n\overline{xy} \subset z_nw'$. Since w' is locally measurable and $z_nw' = z_n(xy)$, we have that

$$z_nw' = \overline{z_nxy} \subset z_n\overline{xy} \subset z_nw',$$

i. e., $z_n\overline{xy} = z_nw'$ for all $n \in \mathbb{N}$. By Lemma 2.3.7, we obtain that $\overline{xy}$ is a locally measurable operator and $\overline{xy} = w'$.

The uniqueness assertion follows by Proposition 2.3.9. □

Proposition 2.3.11 and Corollary 2.3.6 imply that for two locally measurable operators x and y, their strong sum $x \dotplus y = \overline{x+y}$, strong product $x \cdot y = \overline{xy}$, and adjoint x^* are locally measurable operators. Now using the fact that $S(\mathcal{M})$ is a $*$-algebra with respect to these operations, we show that $LS(\mathcal{M})$ is also a $*$-algebra.

Theorem 2.3.12. *Let $\mathcal{M}$ be an arbitrary von Neumann algebra acting on a Hilbert space H. The set $LS(\mathcal{M})$ of all locally measurable operators with respect to $\mathcal{M}$ is a $*$-algebra over the field $\mathbb{C}$ of complex numbers with unit* **1** *with respect to the operations of strong sum, strong product, scalar multiplication, and taking adjoint.*

Proof. By Proposition 2.3.11 and Corollary 2.3.6, $LS(\mathcal{M})$ is closed under strong sum, strong product, scalar multiplication, and taking adjoints. The axioms of all algebraic operations follow from Theorem 2.2.12 and Lemma 2.3.7. □

It immediately follows from the definitions of the $*$-algebras $LS(\mathcal{M})$ and $S(\mathcal{M})$ that $S(\mathcal{M})$ is a $*$-subalgebra of $LS(\mathcal{M})$.

From now on, the strong sum $x \dotplus y$ and the strong product $x \cdot y$ of two elements $x, y \in LS(\mathcal{M})$ will be denoted simply by $x+y$ and xy, respectively, unless stated otherwise.

Since $LS(\mathcal{M})$ is a $*$-algebra, for any $x \in LS(\mathcal{M})$, we have that $x = \mathrm{Re}(x) + i\,\mathrm{Im}(x)$, where

$$\mathrm{Re}(x) = \frac{x + x^*}{2}, \quad \mathrm{Im}(x) = \frac{x - x^*}{2i}.$$

2.4 τ-measurable operators affiliated with a von Neumann algebra

In the case when we deal with a semifinite von Neumann algebra $\mathcal{M}$ equipped with a faithful normal semifinite trace τ, there is another kind of measurability one can consider. This is τ-measurability. In the present section, we describe algebraic properties of the $*$-algebra $S(\mathcal{M},\tau)$ of all τ-measurable operators and show that in general the latter algebra is a proper subalgebra in the $*$-algebra $S(\mathcal{M})$ (and hence in $LS(\mathcal{M})$).

Throughout this section, we assume that $\mathcal{M}$ is a semifinite von Neumann algebra acting on a Hilbert space H equipped with a faithful normal semifinite trace τ. The definition of a τ-measurable operator is a verbatim repetition of Definition 2.2.1 replacing the requirement of finite projections by τ-finite projections.

Definition 2.4.1. A closed linear operator x affiliated with $\mathcal{M}$ is said to be *τ-measurable* with respect to $(\mathcal{M},\tau)$ if there exists a sequence of projections $\{p_n\}_{n\in\mathbb{N}} \subset P(\mathcal{M})$, such that $p_n \uparrow \mathbf{1}$, $p_n(H) \subset \operatorname{dom}(x)$ and $\tau(p_n^\perp) < \infty$ for every $n \in \mathbb{N}$.

The set of all τ-measurable operators is denoted by $S(\mathcal{M},\tau)$.

Since any τ-finite projection is finite, it follows that $S(\mathcal{M},\tau) \subset S(\mathcal{M})$. Furthermore, if $x \in \mathcal{M}$, then $\operatorname{dom}(x) = H$ and so x is τ-measurable (by taking $p_n = \mathbf{1}$ for all $n \in \mathbb{N}$). Thus,

$$\mathcal{M} \subset S(\mathcal{M},\tau) \subset S(\mathcal{M}).$$

For necessary and sufficient conditions for these inclusions to be equalities, see Section 2.5.

The following proposition is similar to Proposition 2.2.2 and Proposition 2.2.3. It provides necessary and sufficient conditions for τ-measurability.

Proposition 2.4.2. *Let x be a closed and densely defined operator affiliated with a von Neumann algebra $\mathcal{M}$. The following conditions are equivalent:*

(i) *the operator x is τ-measurable with respect to $\mathcal{M}$;*

(ii) *in the polar decomposition $x = v|x|$, we have that $v \in \mathcal{M}$ and $|x|$ is τ-measurable;*

(iii) *there exists a projection $p \in P(\mathcal{M})$, such that $p(H) \subseteq \operatorname{dom}(x)$ and $\tau(p^\perp)$ is finite;*

(iv) $\tau(E_{|x|}(\lambda,\infty)) < \infty$ *for some $\lambda > 0$;*

(v) $\tau(E_{|x|}(\lambda,\infty)) \to 0$ *as $\lambda \to \infty$;*

(vi) *For every $\varepsilon > 0$, there exists a projection $p \in P(\mathcal{M})$, such that $p(H) \subset \operatorname{dom}(x)$ and $\tau(p^\perp) < \varepsilon$.*

Proof. (i)⇒(ii). Since $x\eta\mathcal{M}$, Proposition 2.1.2(vi) guarantees that $|x|\eta\mathcal{M}$ and $v \in \mathcal{M}$. Since $\operatorname{dom}(x) = \operatorname{dom}(|x|)$, the assertion follows.

(ii)⇒(iii). Let $\{p_n\}_{n\in\mathbb{N}}$ be such that $p_n \uparrow \mathbf{1}$, $p_n(H) \subset \operatorname{dom}(|x|)$, and $\tau(p_n^\perp) < \infty$ for all $n \in \mathbb{N}$. Since $\operatorname{dom}(x) = \operatorname{dom}(|x|)$, the assertion follows.

(iii)⇒(iv). Let $p \in P(\mathcal{M})$ be such that $p(H) \subseteq \mathrm{dom}(x)$ and $\tau(p^\perp) < \infty$. By Lemma 2.1.5, we have that $xp \in \mathcal{M}$. If $\lambda \in \mathbb{R}$ is such that $\lambda > \|xp\|_{\mathcal{M}}$, then it follows from Proposition 2.1.6 that $E_{|x|}(\lambda, \infty) \precsim p^\perp$, and so $\tau(E_{|x|}(\lambda, \infty)) \le \tau(p^\perp) < \infty$, as required.

(iv)⇒(v). Let $\lambda_0 > 0$ be such that $\tau(E_{|x|}(\lambda_0, \infty)) < \infty$. The convergence $E_{|x|}(\lambda, \infty) \downarrow 0$ as $\lambda \to \infty$ and the normality of the trace implies the convergence $\tau(E_{|x|}(\lambda, \infty)) \to 0$ as $\lambda \to \infty$, proving that (v) holds.

(v)⇒(vi). Let $\varepsilon > 0$ be arbitrary. By assumption, there exists $\lambda > 0$ such that $\tau(E_{|x|}(\lambda, \infty)) \le \varepsilon$. Setting $p = E_{|x|}[0, \lambda]$, it is clear that $p(H) \subseteq \mathrm{dom}(|x|) = \mathrm{dom}(x)$ and $\tau(p^\perp) \le \varepsilon$, which proves (vi).

Finally, to show that (vi) implies (i), let $q_n \in P(\mathcal{M})$ be such that $q_n(H) \subset \mathrm{dom}(x)$ and $\tau(q_n^\perp) \le 2^{-n}$. We define

$$p_n = \bigwedge_{k=n}^{\infty} q_k, \quad n \in \mathbb{N}.$$

It is clear that $p_n \le p_{n+1}$ and $p_n(H) \subseteq q_n(H) \subseteq D$ for all n. Furthermore,

$$\tau(p_n^\perp) = \tau\left(\bigvee_{k=n}^{\infty} q_n^\perp\right) \le \sum_{k=n}^{\infty} \tau(q_n^\perp) \le \sum_{k=n}^{\infty} 2^{-k} = 2^{-n+1},$$

for all $n \in \mathbb{N}$. Therefore, $\tau(\bigwedge_{n\in\mathbb{N}} p_n^\perp) = 0$ and so $\bigwedge_{n\in\mathbb{N}} p_n^\perp = 0$. Hence, $p_n^\perp \downarrow 0$, i. e., $p_n \uparrow \mathbf{1}$. Thus, x is τ-measurable, as required. □

We now show that $S(\mathcal{M}, \tau)$ is a $*$-subalgebra of $S(\mathcal{M})$. We first state a simple result. We leave its verification to the reader.

Lemma 2.4.3. *If $x, y \in LS(\mathcal{M})$, then*

$$\begin{aligned} xy = {} & 2^{-1}((2i)^{-1}((y^* + ix)(y^* + ix)^* - (y^* - ix)(y^* - ix)^*) \\ & + 2^{-1}((y^* + x)(y^* + x)^* - (y^* - x)(y^* - x)^*)). \end{aligned}$$

Theorem 2.4.4. *The set $S(\mathcal{M}, \tau)$ is a $*$-subalgebra in $S(\mathcal{M})$ (and hence, in $LS(\mathcal{M})$).*

Proof. Let $x \in S(\mathcal{M}, \tau)$, $\alpha \in \mathbb{C}$, then by Proposition 2.4.2(iv) there exists $\lambda > 0$, such that $\tau(E_{|x|}(\lambda, \infty)) < \infty$. We have

$$\tau(E_{|\alpha x|}(\lambda|\alpha|, \infty)) = \tau(E_{|x|}(\lambda, \infty)) < \infty$$

and, by Proposition 2.1.7, we have that

$$\tau(E_{|x|}(\lambda, \infty)) = \tau(E_{|x^*|}(\lambda, \infty)).$$

Referring to Proposition 2.4.2(iv) again, we obtain that $\alpha x, x^* \in S(\mathcal{M}, \tau)$.

Let $x, y \in S(\mathcal{M}, \tau)$. By Proposition 2.4.2(iii), there exist projections $p, q \in P(\mathcal{M})$, such that $p(H) \subseteq \operatorname{dom}(x)$, $q(H) \subseteq \operatorname{dom}(y)$ and $\tau(p^\perp), \tau(q^\perp) < \infty$. Since $(p \wedge q)(H) \subseteq \operatorname{dom}(x+y)$ and

$$\tau((p \wedge q)^\perp) = \tau(p^\perp \vee q^\perp) \le \tau(p^\perp) + \tau(q^\perp) < \infty,$$

Proposition 2.4.2(iii) implies that $x + y \in S(\mathcal{M}, \tau)$.

To show that $xy \in S(\mathcal{M}, \tau)$ for $x, y \in S(\mathcal{M}, \tau)$, we first note that if $x \in S(\mathcal{M}, \tau)$, then $x^*x \in S(\mathcal{M}, \tau)$. Indeed, by Proposition 2.4.2(iv), $x \in S(\mathcal{M}, \tau)$ if and only if $\tau(E_{|x|}(\lambda, \infty)) < \infty$ for some $\lambda > 0$. Since $E_{x^*x}(\lambda^2, \infty) = E_{|x|^2}(\lambda^2, \infty) = E_{|x|}(\lambda, \infty)$, it follows that $\tau(E_{|x|^2}(\lambda^2, \infty)) < \infty$. Hence, another application of Proposition 2.4.2(iv) implies that $x^*x \in S(\mathcal{M}, \tau)$.

Now, let $x, y \in S(\mathcal{M}, \tau)$. By the proven above, we have that $y^* + x, y^* - x, y^* + ix, y^* - ix \in S(\mathcal{M}, \tau)$. Hence, the inclusion $xy \in S(\mathcal{M}, \tau)$ follows from Lemma 2.4.3. □

If $\mathcal{M}$ is a factor with normal faithful semifinite trace τ and $p \in P(\mathcal{M})$, then $\tau(p) < \infty$ if and only if the projection p is finite in $\mathcal{M}$. Hence, Proposition 2.4.2(iv) and Proposition 2.2.3(iii) imply the following corollary.

Corollary 2.4.5. *Let $\mathcal{M}$ be a factor and τ be a faithful normal semifinite trace on $\mathcal{M}$. Then $S(\mathcal{M}, \tau) = S(\mathcal{M})$.*

We now present some simple examples of τ-measurable operators.

Example 2.4.6. *Let H be a Hilbert space and let $\mathcal{M} = \mathcal{B}(H)$, equipped with the standard trace $\operatorname{tr} : \mathcal{M}_+ \to [0, \infty]$. Since $\mathcal{B}(H)$ is a factor, Corollary 2.4.5 and Example 2.2.13 imply that $S(\mathcal{B}(H), \operatorname{tr}) = S(\mathcal{B}(H)) = \mathcal{B}(H)$.*

Example 2.4.7. *Let (Ω, Σ, μ) be a σ-finite measure space, $H = L_2(\Omega, \Sigma, \mu)$ and $\mathcal{M} = \{M_f : f \in L_\infty(\Omega, \Sigma, \mu)\}$ (see Example 2.1.4). By Theorem 1.8.10, the mapping $f \longmapsto M_f$ is a $*$-isomorphism from $L_\infty(\Omega, \Sigma, \mu)$ onto $\mathcal{M}$. As seen in Example 1.10.11, the functional $\tau : \mathcal{M}_+ \to [0, \infty]$, defined by*

$$\tau(M_f) = \int_\Omega f d\mu, \quad 0 \le f \in L_\infty(\Omega, \Sigma, \mu),$$

is a faithful normal semifinite trace on $\mathcal{M}$. By Theorem 2.2.15, the map $f \longmapsto M_f$, $f \in L_0(\Omega, \Sigma, \mu)$ is a $*$-isomorphism from $L_0(\Omega, \Sigma, \mu)$ onto $S(\mathcal{M})$. If $M_f \in S(\mathcal{M})$, then it follows from Proposition 2.4.2(iii) that $M_f \in S(\mathcal{M}, \tau)$ if and only if there exists a projection $p \in P(\mathcal{M})$, such that $p(H) \subseteq \operatorname{dom}(M_f)$ (equivalently, $M_f p \in \mathcal{M}$) and $\tau(p^\perp) < \infty$. Any projection $p \in P(\mathcal{M})$ is of the form $p = M_{\chi_A}$ for some $A \in \Sigma$ and $\tau(p) = \mu(A)$. Therefore, if $f \in L_0(\Omega, \Sigma, \mu)$, then $M_f \in S(\mathcal{M}, \tau)$ if and only if there exists $A \in \Sigma$, such that $\mu(X \setminus A) < \infty$ and $f\chi_A \in L_\infty(\Omega, \Sigma, \mu)$. Thus, if the $*$-subalgebra $S(\Omega, \Sigma, \mu)$ of $L_0(\Omega, \Sigma, \mu)$ is defined by

$$S(\Omega,\Sigma,\mu) = \{f \in L_0(\Omega,\Sigma,\mu) : \exists\, A \in \Sigma, \mu(\Omega \setminus A) < \infty, \\ f\chi_A \in L_\infty(\Omega,\Sigma,\mu)\}, \tag{2.3}$$

then the map $f \longmapsto M_f$ is a $*$-isomorphism from $S(\Omega,\Sigma,\mu)$ onto $S(\mathcal{M},\tau)$.

If the measure μ is finite, then it is clear that $S(\Omega,\Sigma,\mu) = L_0(\Omega,\Sigma,\mu)$, and so in this case, $S(\mathcal{M},\tau) = S(\mathcal{M})$. However, if μ is an infinite measure, then $S(\Omega,\Sigma,\mu) \subsetneqq L_0(\Omega,\Sigma,\mu)$. Indeed, in this case there exists a pairwise disjoint sequence $\{A_n\}_{n\in\mathbb{N}}$ in Σ with $\mu(A_n) \geq 1$ for all n. The function $f = \sum_{n=1}^{\infty} n\chi_{A_n}$ belongs to $L_0(\Omega,\Sigma,\mu)$ but does not belong to $S(\Omega,\Sigma,\mu)$. Hence, in this case, the inclusion $S(\mathcal{M},\tau) \subseteq S(\mathcal{M})$ is proper.

2.5 Relations between the $*$-algebras $\mathcal{M}$, $S(\mathcal{M})$, $S(\mathcal{M},\tau)$, and $LS(\mathcal{M})$

As already shown, we have that $\mathcal{M} \subset S(\mathcal{M},\tau) \subset S(\mathcal{M}) \subset LS(\mathcal{M})$. In this section, we provide necessary and sufficient conditions for any of these inclusions to be equalities.

As before, we assume that $\mathcal{M}$ is a von Neumann algebra acting on a Hilbert space H.

We first establish necessary and sufficient conditions for the algebras $S(\mathcal{M})$ and $\mathcal{M}$ to coincide. The following proposition shows that, in general, the $*$-algebra $S(\mathcal{M})$ is substantially wider than the von Neumann algebra $\mathcal{M}$ itself.

Proposition 2.5.1. *If a von Neumann algebra $\mathcal{M}$ admits a finite projection e, such that $e = \sup_{n\in\mathbb{N}} e_n$ for some increasing sequence $\{e_n\}_{n\in\mathbb{N}} \subset P(\mathcal{M})$ with $e \neq e_n$, $n \in \mathbb{N}$, then $S(\mathcal{M}) \neq \mathcal{M}$.*

Proof. Let

$$\{e_n\}_{n\in\mathbb{N}} \subset P(\mathcal{M}), \quad e_n \uparrow e, \quad e_n \neq e$$

for all $n \in \mathbb{N}$. For the projection,

$$p_n := e^{\perp} + e_n, \quad n \in \mathbb{N},$$

we have that $p_n \uparrow \mathbf{1}$ and $p_n^{\perp} = e - e_n \leq e$, in particular, the projection $p_n^{\perp}$ is finite in $\mathcal{M}$.

We consider in H everywhere dense linear subspace $\operatorname{dom}(x) = \bigcup_{n\in\mathbb{N}} p_n(H)$, and define a linear operator x on $\operatorname{dom}(x)$ by setting $x\xi = n\xi$ for all $\xi \in (p_n - p_{n-1})(H)$, $n \in \mathbb{N}$, where $p_0 = 0$.

We claim that $\overline{x}$ is measurable with respect to $\mathcal{M}$. To show that x is closable, let $\{\xi_k\}_{k\in\mathbb{N}} \subset \operatorname{dom}(x)$, $\eta \in H$ be such that $\xi_k \to 0$ and $x\xi_k \to \eta$ as $k \to \infty$. Then for every fixed n, we have that $p_n\xi_k \to 0$ and $p_n x\xi_k - p_n\eta \to 0$ as $k \to \infty$,

$$\|p_n(\eta)\|_H \leq \|p_n x\xi_k - p_n\eta\|_H + \|xp_n\xi_k\|_H \to 0,$$

as $k \to \infty$, it follows that $p_n\eta = 0$ for all n. Therefore, the convergence $p_n \uparrow \mathbf{1}$ implies that $\eta = 0$. Hence, the operator x admits closure $\overline{x}$, and by the definition of x, the operator $\overline{x}$ is a positive operator affiliated with $\mathcal{M}$.

Since

$$\|\overline{x}p_n\|_{\mathcal{M}} = \|xp_n\|_{\mathcal{M}} \leq n < n+1,$$

By Proposition 2.1.6 we have that $E_{\overline{x}}(n+1,\infty) \precsim p_n^{\perp}$. Since $p_n^{\perp}$ is finite, Theorem 1.9.9(i) implies that the projection $E_{\overline{x}}(n+1,\infty)$ is finite too. Hence, Proposition 2.2.3(iii) guarantees that $\overline{x} \in S(\mathcal{M})$.

Since $e_n \neq e$ for all $n \in \mathbb{N}$, there exist integers $n_1 < n_2 < \cdots$, such that $p_{n_{k+1}} - p_{n_k} \neq 0$, in particular, $\|\,|\overline{x}(\xi_k)\|\,\| = n_k$ for some nonzero $\xi_k \in (p_{n_{k+1}} - p_{n_k})(H)$. It means that the operator $\overline{x}$ is unbounded and, therefore, $S(\mathcal{M}) \neq \mathcal{M}$. □

As a corollary of Proposition 2.5.1, we obtain that for a type II von Neumann algebra $\mathcal{M}$, the $*$-algebra $S(\mathcal{M})$ is strictly larger than $\mathcal{M}$.

Proposition 2.5.2. *If $\mathcal{M}$ is a nonatomic algebra containing a nonzero finite projection, then $S(\mathcal{M}) \neq \mathcal{M}$. In particular, if $\mathcal{M}$ is of type II, then $S(\mathcal{M}) \neq \mathcal{M}$.*

Proof. Let $e \in P(\mathcal{M})$ be a nonzero finite projections. Since $\mathcal{M}$ is nonatomic, there exists a sequence of nonzero projections $\{q_n\}_{n\in\mathbb{N}} \subset P(\mathcal{M})$, such that $q_n \leq e$, $q_nq_m = 0$ for $n \neq m$ and $e = \sup_{n\in\mathbb{N}} q_n$. For the sequence of projections $e_n = \sum_{m=1}^{n} q_m$, we have that $e_n \uparrow e$ and $e_n \neq e$ for all $n \in \mathbb{N}$. Since e is a finite projection, Proposition 2.5.1 immediately implies that $S(\mathcal{M}) \neq \mathcal{M}$.

If $\mathcal{M}$ is of type II, the lattice $P(\mathcal{M})$ has no atoms and there exists a non-zero finite projection. Therefore, the claim follows from the proven above. □

If a von Neumann algebra $\mathcal{M}$ is of type III, then any nonzero projection is infinite and, therefore, Proposition 2.2.3(iii) implies that only bounded operators can be measurable with respect to $\mathcal{M}$, so that $S(\mathcal{M}) = \mathcal{M}$.

The following theorem presents a necessary and sufficient condition guaranteeing that the $*$-algebras $S(\mathcal{M})$ and $\mathcal{M}$ coincide.

Theorem 2.5.3. *For a von Neumann algebra $\mathcal{M}$, the following conditions are equivalent:*
(i) *$S(\mathcal{M}) = \mathcal{M}$;*
(ii) *$\mathcal{M}$ is representable in the form of direct sum $\mathcal{M} = \bigoplus_{n=0}^{m} \mathcal{M}_n$, where $\mathcal{M}_0$ is a von Neumann algebra of type III, and $\mathcal{M}_n$ are factors of type I, $n = 1, 2, \ldots, m$ and m is an integer (some of summands may vanish).*

Proof. Suppose first that $S(\mathcal{M}) = \mathcal{M}$. By Theorem 1.9.14 and Proposition 1.9.6, the von Neumann algebra $\mathcal{M}$ can be written as direct sum of atomic algebra $\mathcal{N}$ of type I, atomless algebra $\mathcal{N}_1$ of type I, type II algebra $\mathcal{N}_2$, and type III algebra $\mathcal{M}_0$. Combining Proposition 2.5.2 and Proposition 2.2.17, we infer that the summands $\mathcal{N}_1$ and $\mathcal{N}_2$ must vanish. Thus,

$$\mathcal{M} = \mathcal{M}_0 \oplus \mathcal{N},$$

where $\mathcal{M}_0$ is a von Neumann algebra of type *III* and $\mathcal{N}$ is an atomic von Neumann algebra of type *I*.

It remains to show that $\mathcal{N}$ is direct sum of finitely many type *I* factors. Let $\{q_i\}_{i\in J}$ be the set of all atoms in $P(\mathcal{Z}(\mathcal{N}))$, where J is some index set. Since $\mathcal{Z}(q_i\mathcal{N}) = q_i\mathcal{Z}(\mathcal{N}) = q_i\mathbb{C}$, it follows that $q_i\mathcal{N}$ is a factor of type *I* for every $i \in J$.

Suppose that the set J is infinite. We may choose a nonzero finite projection $e_i \in \mathcal{M}_i$, $i \in I$, and set $e = \sup_{i\in J} e_i$. Since $e_i = e_i q_i$, $q_i q_j = 0$ for $i \neq j$ and $q_i \in P(\mathcal{Z}(\mathcal{N}))$, the projection e is a finite projection in the von Neumann algebra $\mathcal{M}$ (see Theorem 1.9.9(vii)). Therefore, by Proposition 2.5.1 we have that $S(\mathcal{M}) \neq \mathcal{M}$, which is not the case. Consequently, the set J is finite, as required.

Conversely, suppose that $\mathcal{M}$ is of the form of a direct sum $\bigoplus_{n=0}^{m} \mathcal{M}_n$, where $\mathcal{M}_0$ is a von Neumann algebra of type *III*, and $\mathcal{M}_n$ are factors of the type *I*, $n = 1, 2, \ldots, m$. If $x \in S(\mathcal{M})$, then there exists a sequence $\{p_n\}_{n\in\mathbb{N}} \subset P(\mathcal{M})$, such that $p_n \uparrow \mathbf{1}$, $p_n(H) \subseteq \operatorname{dom}(x)$ and the projection $p_n^\perp$ is finite in $\mathcal{M}$ for all $n \in \mathbb{N}$. Let z_i be the unit of the algebra $\mathcal{M}_i$, $i = 1, \ldots, m$. Then $z_i p_n^\perp \in \mathcal{M}_i$ and $\dim(z_i p_n^\perp(H)) < \infty$ for any $i = 1, \ldots, m$. Since $p_n^\perp = p_n^\perp z_1 + \cdots + p_n^\perp z_m$, it follows that

$$\dim(p_n^\perp(H)) = \dim(z_1 p_n^\perp(H)) + \cdots + \dim(z_m p_n^\perp(H)) < \infty.$$

Since $p_n^\perp \downarrow 0$, it follows that $p_n^\perp = 0$ for all sufficiently large n. It means that $\operatorname{dom}(x) = H$ and $x \in \mathcal{M}$, i. e., $S(\mathcal{M}) = \mathcal{M}$. □

Next, we consider when the algebras $S(\mathcal{M})$ and $LS(\mathcal{M})$ coincide. We first establish a sufficient condition for the algebra $LS(\mathcal{M})$ to be strictly larger than $S(\mathcal{M})$. In particular, we construct a locally measurable operator which is not measurable.

Proposition 2.5.4. *Assume that $\mathcal{M}$ is a von Neumann algebra and there exists a sequence of central projections $\{z_n\}_{n\in\mathbb{N}} \subset P(\mathcal{Z}(\mathcal{M}))$, such that $z_n \uparrow \mathbf{1}$ and $(\mathbf{1} - z_n)$ is not a finite projection in $\mathcal{M}$ for all $n \in \mathbb{N}$, then $LS(\mathcal{M}) \neq S(\mathcal{M})$.*

Proof. Consider a central partition of unity $\{z_n'\}_{n\in\mathbb{N}}$ such that

$$z_1' = z_1, \quad z_n' = z_n - z_{n-1}, \quad n = 2, 3, \ldots.$$

By Proposition 2.3.9, there exists $x \in LS(\mathcal{M})$, such that $xz_n' = nz_n'$. It is clear that x is positive.

Since $xz_n = \sum_{k=1}^{n} xz_n' = \sum_{k=1}^{n} kz_n'$, it follows that $E_x[n, +\infty) \geq \mathbf{1} - z_n$ for any $n \in \mathbb{N}$. Since $\mathbf{1} - z_n$ is not finite for any $n \in \mathbb{N}$, Proposition 2.2.3(iii) implies that $x \notin S(\mathcal{M})$. □

Proposition 2.5.4 allows us to distinguish the class of von Neumann algebras $\mathcal{M}$, for which the equality $LS(\mathcal{M}) = S(\mathcal{M})$ fails.

Corollary 2.5.5. *If a von Neumann algebra $\mathcal{M} = \bigoplus_{n\in\mathbb{N}} \mathcal{M}_n$ is the direct sum of an infinite number of infinite von Neumann algebras $\mathcal{M}_n$, $n \in \mathbb{N}$, then*

$$LS(\mathcal{M}) \neq S(\mathcal{M}).$$

Proof. Denote by z_n the central projections from $\mathcal{M}$, such that $z_n\mathcal{M} = \mathcal{M}_n$, $n \in \mathbb{N}$. The family $\{z_n\}_{n\in\mathbb{N}}$ is a family of nonzero pairwise orthogonal central projections from $\mathcal{M}$, such that $\sup_{n\in\mathbb{N}} z_n = \mathbf{1}$. Taking $q_m = \sum_{k=1}^m z_k$, we have an increasing sequence of central projections from $\mathcal{M}$, so that the projection $(\mathbf{1} - q_m)$ is not finite in $\mathcal{M}$ for all $m \in \mathbb{N}$. If $q_m \uparrow \mathbf{1}$, then Proposition 2.5.4 implies that $LS(\mathcal{M}) \neq S(\mathcal{M})$. If the convergence $q_m \uparrow \mathbf{1}$ does not hold, we set $q := \sup_{m\in\mathbb{N}} q_m \neq \mathbf{1}$. Then, setting $g_m = q_m + q^\perp$, we again obtain a sequence of central projections from $\mathcal{M}$, such that $g_m \uparrow \mathbf{1}$ and the projection $(\mathbf{1} - g_m)$ is not finite in $\mathcal{M}$, for all $m \in \mathbb{N}$. Thus, we again have $LS(\mathcal{M}) \neq S(\mathcal{M})$. □

In the following theorem, we characterize von Neumann algebras such that the $*$-algebras $LS(\mathcal{M})$ and $S(\mathcal{M})$ coincide.

Theorem 2.5.6. *Let $\mathcal{M}$ be an arbitrary von Neumann algebra. The following conditions are equivalent:*
(i) $LS(\mathcal{M}) = S(\mathcal{M})$;
(ii)

$$\mathcal{M} = \bigoplus_{n=0}^{m} \mathcal{M}_n,$$

where $\mathcal{M}_0$ is a finite von Neumann algebra and $\mathcal{M}_n$ are factors of type I_∞, II_∞, III, $n = 1, 2, \dots, m$, and $m \in \mathbb{N}$ (some of summands may vanish).

Proof. Assume first that $S(\mathcal{M}) = LS(\mathcal{M})$. Choose a central projection $z_0 \in P(\mathcal{Z}(\mathcal{M}))$, such that $\mathcal{M} = z_0\mathcal{M} + (\mathbf{1} - z_0)\mathcal{M}$, where $\mathcal{M}_0 := z_0\mathcal{M}$ is a finite von Neumann algebra and the algebra $\mathcal{N} := (\mathbf{1} - z_0)\mathcal{M}$ has no nonzero finite central projections, i. e., the von Neumann algebra $\mathcal{N}$ is a properly infinite algebra. If the Boolean algebra $P(\mathcal{Z}(\mathcal{N}))$ contains infinitely many elements, then there exists a sequence $\{z_n\}_{n\in\mathbb{N}}$ of nonzero projections from $P(\mathcal{Z}(\mathcal{N}))$, such that $z_n z_m = 0$ for $n \neq m$ and $\sup_{n\in\mathbb{N}} z_n = \mathbf{1} - z_0$. For the central projections $p_n = \sum_{m=0}^n z_m$ from $\mathcal{M}$, we have that $p_n \uparrow \mathbf{1}$ and $(\mathbf{1} - p_n)$ is a nonzero projection from $\mathcal{N}$, so that $(\mathbf{1} - p_n)$ is not finite projection in $\mathcal{M}$, $n \in \mathbb{N}$. Consequently, by Proposition 2.5.4, we have that $LS(\mathcal{M}) \neq S(\mathcal{M})$, which contradicts the assumption.

Thus, the Boolean algebra $P(\mathcal{Z}(\mathcal{N}))$ contains only finite numbers of elements. Let $\{q_n\}_{n=1}^m$ be the set of all atoms in the Boolean algebra $P(\mathcal{Z}(\mathcal{N}))$ and $\mathcal{M}_n = q_n\mathcal{N} = q_n\mathcal{M}$. It is clear that $\mathcal{M} = \bigoplus_{n=0}^m \mathcal{M}_n$, where for every $n = 1, 2, \dots, m$ the algebra $\mathcal{M}_n$ is not a finite factor, in particular, $\mathcal{M}_n$ is one of the type I_∞, II_∞, or III.

Conversely, assume that $\mathcal{M} = \bigoplus_{n=0}^m \mathcal{M}_n$, where $\mathcal{M}_0$ is a finite von Neumann algebra and $\mathcal{M}_n$ are factors one of the types I_∞, II_∞, or III, $n = 1, 2, \dots, m$. Denote by q_n the unit in the algebra $\mathcal{M}_n$, $n = 0, 1, \dots, m$, and suppose that $x \in LS(\mathcal{M})$. Choose a sequence $\{z_k\}_{k\in\mathbb{N}}$ of central projections in $\mathcal{M}$, such that $z_k \uparrow \mathbf{1}$ and $xz_k \in S(\mathcal{M})$, $k \in \mathbb{N}$. Since $\mathcal{M}_n$ are factors, $n = 1, 2, \dots, m$, there exists an integer k_0 such that $q_n z_k = q_n$ for all $k \geq k_0$, $n = 1, 2, \dots, m$. In particular,

$$x(\mathbf{1} - q_0) = \sum_{n=1}^{m} xq_n = \sum_{n=1}^{m} xz_{k_0} q_n \in S(\mathcal{M}).$$

Since q_0 is a finite central projection, it follows from Corollary 2.2.5 that $LS(q_0\mathcal{M}) = S(q_0\mathcal{M})$ and, therefore, $xq_0 \in S(\mathcal{M})$. Consequently, $x = xq_0 + x(\mathbf{1} - q_0) \in S(\mathcal{M})$, proving that $LS(\mathcal{M}) = S(\mathcal{M})$. □

Combining Theorem 2.5.6 and Theorem 2.5.3, we obtain a necessary and sufficient condition for the $*$-algebras $LS(\mathcal{M})$ and $\mathcal{M}$ to coincide.

Theorem 2.5.7. *Let $\mathcal{M}$ be an arbitrary von Neumann algebra. The following conditions are equivalent:*

(i) $LS(\mathcal{M}) = \mathcal{M}$;

(ii) *The algebra $\mathcal{M}$ is presented in the form of direct sum $\mathcal{M} = \bigoplus_{n=1}^{m} \mathcal{M}_n$, where $\mathcal{M}_n$ are factors of type I or type III, $n = 1, 2, \dots, m$, $m \in \mathbb{N}$ (some summands may vanish).*

Proof. Note that since $\mathcal{M} \subset S(\mathcal{M}) \subset LS(\mathcal{M})$, the equality $\mathcal{M} = LS(\mathcal{M})$ holds if and only if $\mathcal{M} = S(\mathcal{M}) = LS(\mathcal{M})$. Hence, the assertion follows from Theorem 2.5.6 and Theorem 2.5.3. □

The algebra $S(\mathcal{M}, \tau)$ of all τ-measurable operators depends, in general, on the choice of the trace τ. This may be illustrated by the following example. Suppose that (Ω, Σ, μ) is a measure space, where the measure μ is σ-finite but not finite and let $\mathcal{M} = \{M_f : f \in L_\infty(\Omega, \Sigma, \mu)\}$, as in Example 2.4.7. There exists a finite measure ν on Σ, which is equivalent with μ (see, e. g., Remark 1.8.11). If $\tau_1 : \mathcal{M}^+ \to [0, \infty]$ is defined by setting $\tau_1(M_f) = \int_\Omega f d\nu$, $0 \leq f \in L_\infty(\Omega, \Sigma, \mu)$, then τ_1 is a finite normal faithful trace on $\mathcal{M}$. From Proposition 2.4.2, it is clear that $S(\mathcal{M}, \tau_1) = S(\mathcal{M})$, whereas $S(\mathcal{M}, \tau) \subsetneq S(\mathcal{M})$, since μ is not a finite measure (see Example 2.4.7).

We now establish a necessary and sufficient condition for the algebras $S(\mathcal{M}, \tau_1)$ and $S(\mathcal{M}, \tau_2)$ to coincide. We recall that for every faithful normal semifinite trace τ on $\mathcal{M}$ we denote

$$P_{\text{fin}}(\mathcal{M}, \tau) = \{p \in P(\mathcal{M}) : \tau(p) < \infty\}.$$

Proposition 2.5.8. *Let $\tau_j, j = 1, 2$ be faithful normal semifinite traces on $\mathcal{M}$. The following conditions are equivalent:*

(i) $S(\mathcal{M}, \tau_1) = S(\mathcal{M}, \tau_2)$;

(ii) $P_{\text{fin}}(\mathcal{M}, \tau_1) = P_{\text{fin}}(\mathcal{M}, \tau_2)$.

Proof. The implication (ii)⇒(i) follows from Definition 2.4.1. To prove the converse implication, it is sufficient to show that the inclusion $S(\mathcal{M}, \tau_1) \subset S(\mathcal{M}, \tau_2)$ implies that $P_{\text{fin}}(\mathcal{M}, \tau_1) \subset P_{\text{fin}}(\mathcal{M}, \tau_2)$.

Suppose that $S(\mathcal{M}, \tau_1) \subset S(\mathcal{M}, \tau_2)$ and suppose the contrary that the inclusion $P_{\text{fin}}(\mathcal{M}, \tau_1) \subset P_{\text{fin}}(\mathcal{M}, \tau_2)$ does not hold. Then there exists a projection $p \in P(\mathcal{M})$, such that $\tau_2(p) = \infty$ and $\tau_1(p) < \infty$. Since the trace τ_2 is semifinite, there exists an increasing sequence of projections $\{e_n\}_{n\in\mathbb{N}}$, such that

$$\tau_2(e_n) < \infty, \quad \sup_{n\in\mathbb{N}} e_n = e \leq p, \quad \tau_2(e) = \infty,$$

in particular, $e_n \neq e$ for all $n \in \mathbb{N}$. As in the proof of Proposition 2.5.1, we define a linear operator x on the dense linear subspace $\operatorname{dom}(x) = \bigcup_{n\in\mathbb{N}} p_n(H)$, by setting $x\xi = n\xi$ for all $\xi \in (p_n - p_{n-1})(H)$, where $p_n = e^{\perp} + e_n$, $n \in \mathbb{N}$, $p_0 = 0$. The closure $\overline{x}$ of the operator x is a positively defined operator and $\overline{x} \in S(\mathcal{M})$. Furthermore, $E_{\overline{x}}(-\infty, n+1) = p_n$. Since

$$\tau_2(p_n^{\perp}) = \tau_2(e - e_n) = \infty, \quad \tau_1(p_n^{\perp}) \leq \tau_1(p) < \infty, \quad n = 1, 2, \ldots$$

it follows that

$$\overline{x} \in S(\mathcal{M}, \tau_1) \setminus S(\mathcal{M}, \tau_2),$$

that contradicts to the inclusion $S(\mathcal{M}, \tau_1) \subset S(\mathcal{M}, \tau_2)$. Hence, we conclude that $P(\mathcal{M}, \tau_1) \subset P(\mathcal{M}, \tau_2)$.

□

The following result is proved similar to Proposition 2.5.8. We leave the details to the reader.

Proposition 2.5.9. *For a faithful normal semifinite trace τ, the following conditions are equivalent:*
(i) $S(\mathcal{M}) = S(\mathcal{M}, \tau)$;
(ii) $P_{\text{fin}}(\mathcal{M}, \tau) = P_{\text{fin}}(\mathcal{M}) = \{p \in P(\mathcal{M}) : p \text{ is finite}\}$.

2.6 Compact and τ-compact operators affiliated with a von Neumann algebra

As observed in Proposition 2.2.3(iii), a closed densely defined operator x, affiliated with a von Neumann algebra $\mathcal{M}$, belongs to $S(\mathcal{M})$ if and only if there exists $\lambda > 0$ such that $E_{|x|}(\lambda, \infty)$ is a finite projection. In general, if $x \in S(\mathcal{M})$, then the projection $E_{|x|}(\lambda, \infty)$ may be not finite for some λ. Indeed, if $\mathcal{M}$ is an infinite von Neumann algebra, then the projection $E_{\mathbf{1}}(\lambda, \infty) = \mathbf{1}$, $0 < \lambda < 1$, is not finite.

Similarly, if $\mathcal{M}$ is equipped with a faithful normal semifinite trace τ, then for $x \in S(\mathcal{M}, \tau)$ there exists $\lambda > 0$ such that $E_{|x|}(\lambda, \infty)$ is τ-finite (see Proposition 2.4.2) and if the trace τ is not finite, then for some $x \in S(\mathcal{M}, \tau)$ there may exist λ, such that $E_{|x|}(\lambda, \infty)$ is not τ-finite.

In this section, we distinguish two special $*$-subalgebras in $S(\mathcal{M})$, and respectively in $S(\mathcal{M}, \tau)$, with the property that the spectral projections $E_{|x|}(\lambda, \infty)$ are finite (resp., τ-finite) for all $\lambda > 0$.

Definition 2.6.1. Let $\mathcal{M}$ be a von Neumann algebra.
(i) A closed densely defined operator x affiliated with $\mathcal{M}$ is said to be *compact* (relatively to $\mathcal{M}$) if the spectral projections $E_{|x|}(\lambda, \infty)$ are finite for all $\lambda > 0$. The collection of all compact (relative to $\mathcal{M}$) operators is denoted by $S_0(\mathcal{M})$;
(ii) Assume that $\mathcal{M}$ is a semifinite von Neumann algebra equipped with a faithful normal semifinite trace τ. A closed densely defined operator x affiliated with $\mathcal{M}$ is said to be *τ-compact* if the spectral projections $E_{|x|}(\lambda, \infty)$ are τ-finite for all $\lambda > 0$. The collection of all τ-compact operators is denoted by $S_0(\mathcal{M}, \tau)$.

As noted above, $S_0(\mathcal{M}) \subset S(\mathcal{M})$ and $S_0(\mathcal{M}, \tau) \subset S(\mathcal{M}, \tau)$. Since every τ-finite projection is finite, it follows that $S_0(\mathcal{M}, \tau) \subset S_0(\mathcal{M})$. It is also evident that if $\mathbf{1} \in P_{\text{fin}}(\mathcal{M})$, then $S_0(\mathcal{M})$ is equal to the set of all closed densely defined operators affiliated with $\mathcal{M}$. So, in this case, $S_0(\mathcal{M}) = S(\mathcal{M})$. However, if $\mathbf{1} \notin P_{\text{fin}}(\mathcal{M})$, then $S_0(\mathcal{M}) \neq S(\mathcal{M})$, as is clear from the discussion preceding Definition 2.6.1. Similarly, if $\tau(\mathbf{1}) < \infty$, then $S_0(\mathcal{M}, \tau) = S(\mathcal{M}, \tau) = S(\mathcal{M})$ and if $\tau(\mathbf{1}) = \infty$, then $S_0(\mathcal{M}, \tau) \neq S(\mathcal{M}, \tau)$.

From the definition, it is also clear that, if x is a closed densely defined linear operator in H and $x\eta\mathcal{M}$, then $x \in S_0(\mathcal{M})$ (resp., $x \in S_0(\mathcal{M}, \tau)$) if and only if $|x| \in S_0(\mathcal{M})$ (resp., $|x| \in S_0(\mathcal{M}, \tau)$).

The proposition that follows presents another characterization of compact and τ-compact operators.

Proposition 2.6.2. *If x is a closed densely defined operator in H affiliated with the von Neumann algebra $\mathcal{M}$, then $x \in S_0(\mathcal{M})$ (resp., $x \in S_0(\mathcal{M}, \tau)$) if and only if for every $\varepsilon > 0$ there exists a projection $p \in P(\mathcal{M})$ such that $p^\perp \in P_{\text{fin}}(\mathcal{M})$ (resp., $p^\perp \in P_{\text{fin}}(\mathcal{M}, \tau)$), $p(H) \subseteq \text{dom}(x)$ and $\|xp\|_{\mathcal{M}} \leq \varepsilon$.*

Proof. We prove the assertion for compact operators only, as the proof for τ-compact operators is a verbatim repetition.

Assume that $x \in S_0(\mathcal{M})$ and let $\varepsilon > 0$ be given. Since x is compact, it follows that for the projection $p := E_{|x|}[0, \varepsilon]$ we have that $p^\perp = E_{|x|}(\varepsilon, \infty)$ is finite. By spectral theory, we have that $p \in P(\mathcal{M})$, $p(H) \subseteq \text{dom}(|x|) = \text{dom}(x)$ and $\||x|p\|_{\mathcal{M}} \leq \varepsilon$. If $x = v|x|$ is the polar decomposition of x, then $\|xp\|_{\mathcal{M}} = \|v|x|p\|_{\mathcal{M}} \leq \||x|p\|_{\mathcal{M}} \leq \varepsilon$. Thus, p is the required projection.

For the proof of the converse implication, let $\lambda > 0$ be given. By assumption, there exists $p \in P(\mathcal{M})$, such that $p^\perp \in P_{\text{fin}}(\mathcal{M})$, $p(H) \subseteq \text{dom}(x)$, and $\|xp\|_{\mathcal{M}} \leq \lambda$. It follows from Proposition 2.1.6 that $E_{|x|}(\lambda, \infty) \precsim p^\perp$, and so $E_{|x|}(\lambda, \infty) \in P_{\text{fin}}(\mathcal{M})$. Since λ is arbitrary, it proves that $x \in S_0(\mathcal{M})$. $\square$

Remark 2.6.3. *If x is a closed densely defined operator affiliated with $\mathcal{M}$ satisfying $s_r(x) \in P_{\text{fin}}(\mathcal{M})$ or $s_l(x) \in P_{\text{fin}}(\mathcal{M})$, then $x \in S_0(\mathcal{M})$. Indeed, since $s_r(x) \sim s_l(x)$ it follows that $s_r(x) \in P_{\text{fin}}(\mathcal{M})$ if and only if $s_l(x) \in P_{\text{fin}}(\mathcal{M})$. Assume that $s_r(x) \in P_{\text{fin}}(\mathcal{M})$. Letting $p = \mathbf{1} - s_r(x)$, we have that $xp = 0$ and $p^\perp \in P_{\text{fin}}(\mathcal{M})$. Hence, Proposition 2.6.2 implies that $x \in S_0(\mathcal{M})$.*

Similarly, if $s_r(x) \sim s_l(x) \in P_{\text{fin}}(\mathcal{M}, \tau)$, then $x \in S_0(\mathcal{M}, \tau)$.

In the following proposition we show that $S_0(\mathcal{M})$ and $S_0(\mathcal{M}, \tau)$ are $*$-algebras. In fact, the algebra $S_0(\mathcal{M})$ is a two-sided $*$-ideal in $S(\mathcal{M})$, while $S_0(\mathcal{M}, \tau)$ is a two-sided $*$-ideal in $S(\mathcal{M}, \tau)$.

Proposition 2.6.4. *Let $\mathcal{M}$ be a von Neumann algebra.*
(i) *$S_0(\mathcal{M})$ is a two-sided $*$-ideal in $S(\mathcal{M})$ and $S(\mathcal{M}) = S_0(\mathcal{M}) + \mathcal{M}$;*
(ii) *If $\mathcal{M}$ is equipped with a faithful normal semifinite trace τ, then $S_0(\mathcal{M}, \tau)$ is a two-sided $*$-ideal in $S(\mathcal{M}, \tau)$ and $S(\mathcal{M}, \tau) = S_0(\mathcal{M}, \tau) + \mathcal{M}$.*

Proof. We prove the assertion for $S_0(\mathcal{M})$ only, as the proof for $S_0(\mathcal{M}, \tau)$ is similar.

We first show that $S_0(\mathcal{M})$ is a linear subspace in $S(\mathcal{M})$. If $x \in S_0(\mathcal{M})$ and $0 \neq \alpha \in \mathbb{C}$, then $E_{|\alpha x|}(\lambda, \infty) = E_{|x|}(\frac{\lambda}{|\alpha|}, \infty) \in P_{\text{fin}}(\mathcal{M})$ for any $\lambda > 0$. Therefore, $\alpha x \in S_0(\mathcal{M})$.

Now let $x, y \in S_0(\mathcal{M})$ and fix $\varepsilon > 0$. By Proposition 2.6.2, there exist projections $p, q \in P(\mathcal{M})$, such that

$$p(H) \subset \text{dom}(x), \quad xp \in \mathcal{B}(H), \quad \|xp\|_{\mathcal{M}} < \varepsilon/2, \quad p^\perp \in P_{\text{fin}}(\mathcal{M}),$$

and

$$q(H) \subset \text{dom}(y), \quad yq \in \mathcal{B}(H), \quad \|yq\|_{\mathcal{M}} < \varepsilon/2, \quad q^\perp \in P_{\text{fin}}(\mathcal{M}).$$

For the projection $e = p \wedge q$ from $P(\mathcal{M})$, we have (see Theorem 1.9.9(ii))

$$e^\perp = p^\perp \vee q^\perp \in P_{\text{fin}}(\mathcal{M}).$$

Since $x, y \in S_0(\mathcal{M}) \subset S(\mathcal{M})$, it follows that $x+y \in S(\mathcal{M})$, in addition, $(x+y)e = xe+ye \in \mathcal{M}$ and

$$\|(x+y)e\|_{\mathcal{M}} \le \|xpe\|_{\mathcal{M}} + \|yqe\|_{\mathcal{M}} < \varepsilon.$$

Referring to Proposition 2.6.2 once again, we obtain that $x + y \in S_0(\mathcal{M})$, proving that $S_0(\mathcal{M})$ is a linear subspace in $LS(\mathcal{M})$.

By Proposition 2.1.7, we have that $E_{|x|}(\lambda, \infty) \sim E_{|x^*|}(\lambda, \infty)$ for any $\lambda > 0$ and any closed densely defined operator x affiliated with $\mathcal{M}$. Hence, by definition of compact operators, we have that $x \in S_0(\mathcal{M})$ if and only if $x^* \in S_0(\mathcal{M})$.

Next, we show that $S_0(\mathcal{M})$ is a subalgebra in $S(\mathcal{M})$. Note that if $x \in S_0(\mathcal{M})$, then $|x| \in S_0(\mathcal{M})$. Since $E_{|x|^2}(\lambda^2, \infty) = E_{|x|}(\lambda, \infty)$ for any $\lambda > 0$, it follows that $E_{|x|^2}(\lambda^2, \infty) \in P_{\text{fin}}(\mathcal{M})$

for all $\lambda > 0$ and, therefore, $x^*x = |x|^2 \in S_0(\mathcal{M})$. Employing Lemma 2.4.3, we conclude that $xy \in S_0(\mathcal{M})$ for any $x, y \in S_0(\mathcal{M})$. Thus, $S_0(\mathcal{M})$ is a $*$-subalgebra in $LS(\mathcal{M})$.

Next, we show that any $x \in S(\mathcal{M})$ can be written as $x = x_1 + x_2$ with $x_1 \in S_0(\mathcal{M})$ and $x_2 \in \mathcal{M}$. By Proposition 2.2.3(iii), there exists $\lambda > 0$, such that $E_{|x|}(\lambda, \infty) \in P_{\text{fin}}(\mathcal{M})$. We set $x_1 := xE_{|x|}(\lambda, \infty)$, $x_2 := xE_{|x|}[0, \lambda]$. By spectral theory, we have that $x_2 \in \mathcal{M}$. Since $s_r(x_1) \leq E_{|x|}(\lambda, \infty) \in P_{\text{fin}}(\mathcal{M})$, Remark 2.6.3 implies that $x_1 \in S_0(\mathcal{M})$. Since $x = xE_{|x|}(\lambda, \infty) + xE_{|x|}[0, \lambda) = x_1 + x_2$, the claim follows.

Finally, we show that $S_0(\mathcal{M})$ is a two-sided ideal in $S(\mathcal{M})$. Let $x \in S_0(\mathcal{M})$ and $y \in S(\mathcal{M})$. Since $S(\mathcal{M}) = S_0(\mathcal{M}) + \mathcal{M}$, we may write $y = y_1 + y_2$ with $y_1 \in S_0(\mathcal{M})$ and $y_2 \in \mathcal{M}$. Since $S_0(\mathcal{M})$ is a subalgebra, it follows that $xy_1, y_1x \in S_0(\mathcal{M})$. To show that $y_2x \in S_0(\mathcal{M})$, let $\varepsilon > 0$ be arbitrary. By Proposition 2.6.2, there exists $p \in P(\mathcal{M})$, such that $p^\perp \in P_{\text{fin}}(\mathcal{M})$ and $\|xp\|_{\mathcal{M}} \leq \frac{\varepsilon}{\|y_2\|_{\mathcal{M}}}$. Then $\|y_2xp\|_{\mathcal{M}} \leq \|y_2\|_{\mathcal{M}}\|xp\|_{\mathcal{M}} \leq \varepsilon$ and, therefore, referring to Proposition 2.6.2 once again we have that $y_2x \in S_0(\mathcal{M})$. Repeating the argument, we infer that $xy_2 = (y_2^*x^*)^* \in S_0(\mathcal{M})$. Thus, $xy, yx \in S_0(\mathcal{M})$, proving that $S_0(\mathcal{M})$ is a (two-sided) ideal in $S(\mathcal{M})$. □

Remark 2.6.5. *As we will show later in Chapter 4 (see Theorem 4.3.23), the algebra $S_0(\mathcal{M})$ is a two-sided ideal in the algebra $LS(\mathcal{M})$, too. We note however that, in general, $S_0(\mathcal{M}, \tau)$ is not an ideal in $LS(\mathcal{M})$. Indeed, consider the example $\mathcal{M} = \{M_f : f \in L_\infty(\Omega, \Sigma, \mu)\}$, where $(\Omega, \Sigma, \mu) = ([1, \infty), \Sigma, m)$ with m denoting the classical Lebesgue measure and let trace τ be given by $\tau(M_f) = \int_{[1,\infty)} f\, dm$. By Example 2.2.14 and Proposition 2.3.3, we have that $LS(\mathcal{M}) = S(\mathcal{M}) = \{M_f : f \in L_0(\Omega, \Sigma, \mu)\}$. Let $f(t) = \frac{1}{t}$, $t \in [1, \infty)$. We have that $E_f(\lambda, \infty) = \chi_{[1,\lambda]}$ and, therefore, $M_f \in S_0(\mathcal{M}, \tau)$. Then for $g(t) = t$, we have that $M_g \in S(\mathcal{M})$ and*

$$M_f M_g = M_{fg} = \mathbf{1} \notin S_0(\mathcal{M}, \tau).$$

We now give an example of compact and τ-compact operators.

Example 2.6.6. *Let H be a Hilbert space and let $\mathcal{M} = \mathcal{B}(H)$, equipped with the standard trace* tr. *Observe that $S(\mathcal{B}(H), \mathrm{tr}) = S(\mathcal{B}(H)) = \mathcal{B}(H)$, as already seen in Example 2.4.6. Then $S_0(\mathcal{B}(H), \mathrm{tr}) = S_0(\mathcal{B}(H)) = K(H)$, where $K(H)$ denotes the ideal of all compact operators on H.*

Proof. First, let $x \in K(H)$ be given with polar decomposition $x = v|x|$. Since $|x| = v^*x$, it is clear that also $|x| \in K(H)$. If x is a finite rank operator, then the right and left supports of x have finite trace. As noted in Remark 2.6.3, it follows that $x \in S_0(\mathcal{B}(H), \mathrm{tr})$. Therefore, it may be assumed that $\mathrm{ran}(x)$ (equivalently, $\mathrm{ran}(|x|)$) is infinite-dimensional. By the spectral theorem for compact self-adjoint operators, there exists an orthonormal system $\{\xi_n\}_{n\in\mathbb{N}}$ in H, such that $|x| = \sum_{n=1}^{\infty} \lambda_n \langle \cdot, \xi_n \rangle \xi_n$ where the series converges in the operator norm and $\{\lambda_n\}_{n\in\mathbb{N}}$ is the sequence of nonzero eigenvalues of $|x|$ (repeated according to multiplicity) satisfying $\lambda_n \downarrow 0$. Let $\varepsilon > 0$ be given and $N \in \mathbb{N}$ be such that

$\lambda_N \geq \varepsilon$. We have that $E_{|x|}(\varepsilon, \infty) = \sum_{n=1}^{N} \langle \cdot, \xi_n \rangle \xi_n$, which implies that $E_{|x|}(\varepsilon, \infty)$ is a finite projection. This shows that $x \in S_0(\mathcal{B}(H), \mathrm{tr})$.

Now assume that $x \in S_0(\mathcal{B}(H), \mathrm{tr})$. For every $\varepsilon > 0$, there exists a projection p such that $\mathrm{tr}(p^{\perp}) < \infty$, i. e., $p^{\perp}$ is a finite rank projection, and $\|xp\|_{\mathcal{B}(H)} \leq \varepsilon$, i. e., $\|x - xp^{\perp}\|_{\mathcal{B}(H)} \leq \varepsilon$. Hence, x can be approximated in the operator norm by finite rank operators, and so $x \in K(H)$. □

Proposition 2.6.7. *Assume that $\mathcal{M}$ is a semifinite von Neumann algebra. The strong operator closures of the ideals $S_0(\mathcal{M}, \tau) \cap \mathcal{M}$ and $S_0(\mathcal{M}) \cap \mathcal{M}$ coincide with $\mathcal{M}$.*

Proof. We first show the claim for $S_0(\mathcal{M}, \tau) \cap \mathcal{M}$. Suppose first that $p \in P(\mathcal{M})$. Since the trace τ is semifinite, there exists a family $\{p_i\}_{i \in I} \subset P_{\mathrm{fin}}(\mathcal{M}, \tau)$ of pairwise orthogonal projections, such that

$$p = \bigvee_{i \in I} p_i.$$

By Proposition 1.1.2 (and discussion after), we have that $p = \sum_{i \in I} p_i$, where the series converges with respect to the strong operator topology. Since $\{p_i\}_{i \in I} \subset P_{\mathrm{fin}}(\mathcal{M}, \tau) \subset S_0(\mathcal{M}, \tau) \cap \mathcal{M}$, it follows that $p \in \overline{(S_0(\mathcal{M}, \tau) \cap \mathcal{M})}^{so}$. Hence,

$$P(\mathcal{M}) \subset \overline{(S_0(\mathcal{M}, \tau) \cap \mathcal{M})}^{so}.$$

By spectral theory, any self-adjoint operator $x \in \mathcal{M}$ can be approximated in the uniform norm (and so in the strong operator topology) by a linear combination of projections in $\mathcal{M}$. Therefore, $\mathcal{M}_h \subset \overline{(S_0(\mathcal{M}, \tau) \cap \mathcal{M})}^{so}$. Writing now an arbitrary $x \in \mathcal{M}$ as a linear combination of two self-adjoint operators from $\mathcal{M}$, we infer that

$$\overline{(S_0(\mathcal{M}, \tau) \cap \mathcal{M})}^{so} = \mathcal{M}.$$

Next, to prove the assertion for $S_0(\mathcal{M}) \cap \mathcal{M}$, we note that $S_0(\mathcal{M}, \tau) \cap \mathcal{M} \subset S_0(\mathcal{M}) \cap \mathcal{M}$. Therefore, the assertion follows from the one proven for the algebra $S_0(\mathcal{M}, \tau)$. □

Bibliographical notes

The notion of affiliated operators appeared in 1936 in the foundational paper by Murray and von Neumann [111]. In [111, Chapter XVI], they introduced (densely defined) operators affiliated with a type II_1 factor $\mathcal{M}$ and showed that such a class of operators forms a $*$-algebra (denoted in [111] by $U(\mathcal{M})$) with respect to the strong sum, strong product, and passing to the adjoint operations. Precisely, these $*$-algebras were later termed *Murray–von Neumann algebras*.

Note that the same $*$-algebra featured in the posthumously published paper [159, p. 491]. As stated in the introduction to [159]:

This manuscript was written in the period 1935–1937 when von Neumann's interest in continuous geometry was most intense, and was found after his death. It seems likely that he intended to include it in the Colloquium volume, which he never completed.

In 1952, Dye [67, p. 249], while working on the Radon–Nikodym type theorem for states on finite W^*-algebras, introduced the $*$-algebra (denoted by $C(\mathcal{M})$) of all closed densely defined operators affiliated with a finite von Neumann algebra.

In 1953 [141], Segal introduced the notion of measurable operators and strongly dense subspaces with respect to a von Neumann algebra $\mathcal{M}$. In this paper, he showed that the class $S(\mathcal{M})$ of all measurable operators with respect to a von Neumann algebra $\mathcal{M}$ is a $*$-algebra with respect to the strong sum, strong product, and passing to the adjoint operations. In the special case, when $\mathcal{M}$ is a type II_1 algebra, any operator affiliated with $\mathcal{M}$ is automatically measurable. Thus, Segal recovered the result of Murray and von Neumann [111].

The detailed description of properties of measurable operators can be found in [64] and [152]. A beautiful application of algebras of measurable operators to the representation theory of unimodular groups is given in [145].

The notion of locally measurable operators was first introduced by Sankaran in 1959 in [137, p. 338, Definition 2.2] for von Neumann algebras with a countably decomposable center. This notion was later extended to the case on an arbitrary von Neumann algebra by Yeadon in 1973 in [162]. In [63], Dixon presented an alternative approach for definition of the $*$-algebras $LS(\mathcal{M})$ by means of quasimeasurable operators. In [162], Yeadon proved that the class of locally measurable operators and that of quasimeasurable operators with respect to a von Neumann algebra $\mathcal{M}$ coincide.

In 1974 [116], Nelson introduced the class of τ-measurable operators affiliated with a semifinite von Neumann algebra $\mathcal{M}$ equipped with a faithful normal semifinite trace τ. In this paper, he showed that this class also forms a $*$-algebra (which he denoted by $\widetilde{M}$). Nelson introduced the algebra of τ-measurable operators in terms of completion of the von Neumann algebra $\mathcal{M}$ with a certain topology (namely, the measure topology introduced later in Chapter 4).

Numerous properties of τ-measurable operators are given in [70, 108, 152] and the recent monograph [64]. In fact, the main features of the theory (at least in the setting of finite von Neumann algebras) were surmised by A. Grothendieck yet in 1955 (see the reprint of his penetrating work in [75]).

In 1968–1971 [51, 121, 144], Breuer, Ovčinnikov, and Sonis independently of each other introduced the $*$-algebras of compact and τ-compact operators affiliated with a von Neumann algebra (in the later case, $\mathcal{M}$ was assumed to be equipped with a faithful normal semifinite trace τ). In 1975, the class of τ-compact operators was reintroduced by Yeadon [164].

In 1993, A. Ströh, G. P. West in [148] introduced an algebra (denoted by $\mathcal{K}(\widetilde{M}, \tau)$) of τ-compact operators. They did not reference earlier Breuer, Ovčinnikov, and Sonis papers mentioned above; instead they credited Fack and Kosaki's paper [70].

The theory of measurable and locally measurable operators is also developed for AW^*-algebras. The definition of these AW^*-algebras does not use the fact that such algebras act on a Hilbert space but only via the inner algebraic properties. For a detailed discussion of AW^*-algebras and (locally) measurable operators associated with these algebras see [39, 130, 131]. See also [139] for some other generalizations of unbounded operator algebras.

Finally, we note that the results of the present chapter hold without the assumption that the von Neumann algebra $\mathcal{M}$ is acting on a separable Hilbert space. This would require passing from a sequence of central projections $\{z_n\}_{n\in\mathbb{N}}$ to a net $\{z_i\}_{i\in I}$ throughout.

3 Properties of locally measurable operators

In this chapter, we establish the fundamental properties of locally measurable operators, which will be used throughout the remainder of the book.

We start with one of the fundamental properties of the algebra $LS(\mathcal{M})$ of all locally measurable operators, which shows that $LS(\mathcal{M})$ is the maximal possible bimodule over a given von Neumann algebra $\mathcal{M}$. In particular, any algebra (with respect to the strong sum and strong product) of unbounded operators affiliated with $\mathcal{M}$, which contains $\mathcal{M}$ itself, must be contained in $LS(\mathcal{M})$.

In Section 3.2, we establish the functional calculus of self-adjoint locally measurable operators. In Section 3.3, we show that the natural partial order on a von Neumann algebra $\mathcal{M}$ can be extended to the algebra $LS(\mathcal{M})$ and the subspace of all self-adjoint locally measurable operators equipped with this partial order is a conditionally monotone complete lattice. In this section, we also introduce the class of absolutely solid $*$-subalgebras of $LS(\mathcal{M})$ as well as bimodules over $\mathcal{M}$. The subalgebras $S(\mathcal{M})$, $S(\mathcal{M},\tau)$, $S_0(\mathcal{M})$ as well as $S_0(\mathcal{M},\tau)$, introduced in the previous chapter, are all examples of absolutely solid $*$-subalgebras of $LS(\mathcal{M})$ and $\mathcal{M}$-bimodules.

In Section 3.4, we show that the $*$-algebra of locally measurable operators is closed with respect to the operation of direct integration, i. e., if $\mathcal{M} = \int_Z^{\oplus} \mathcal{M}_\zeta \, dw(\zeta)$ is a decomposable von Neumann algebra, then any locally measurable operator affiliated with $\mathcal{M}$ can be written as a decomposable operator of a measurable field of locally measurable operators affiliated with $\mathcal{M}_\zeta$.

In the final two sections, the discussion is specialized to the commutator estimates for locally measurable operators. These commutator estimates will play an important role in the description of derivations in the future chapters.

3.1 Maximality of $LS(\mathcal{M})$ as a bimodule over $\mathcal{M}$

In this section, we prove that $LS(\mathcal{M})$ is the largest possible bimodule over a given von Neumann algebra $\mathcal{M}$.

We start with yet another sufficient condition of local measurability of a closed densely defined operator affiliated with $\mathcal{M}$, established by Yeadon. Recall that if x is a closed operator acting in a Hilbert space and $y \in \mathcal{B}(H)$, then the operator xy is necessarily closed. However, the operator yx may be not even closable. It turns out that closability of yx for all $y \in \mathcal{M}$ is a sufficient condition for the operator x to be locally measurable with respect to $\mathcal{M}$.

As before, we assume that $\mathcal{M}$ is a von Neumann algebra acting on a separable Hilbert space. Let φ be a $*$-isomorphism from $\mathcal{Z}(\mathcal{M})$ onto the $*$-algebra $L_\infty(\Omega,\Sigma,\mu)$, where (Ω,Σ,μ) is a probability space (see Remark 1.8.11). Recall that $\mathcal{D} : P(\mathcal{M}) \to L^+(\Omega,\Sigma,\mu)$ is the dimension function on $\mathcal{M}$.

We first prove an auxiliary lemma.

https://doi.org/10.1515/9783111599687-004

Lemma 3.1.1. *Suppose that x is a closed densely defined operator affiliated with a von Neumann algebra $\mathcal{M}$. If x is not locally measurable, then there exists an increasing sequence $\{n_k\}_{k\in\mathbb{N}} \subset \mathbb{N}$ and a sequence of nonzero, mutually equivalent projections $\{r_k\}_{k\in\mathbb{N}}$ in $\mathcal{M}$, such that $r_k \le E_{|x|}[n_k, n_{k+1})$ for all $k \in \mathbb{N}$.*

Proof. Since x is not locally measurable, it is not measurable. Therefore, by Proposition 2.2.3(iii), the projections $E_{|x|}[n,\infty)$ are infinite for all $n \in \mathbb{N}$. By Theorem 1.9.9(vi) for every $n \in \mathbb{N}$, there exists the largest projection $q_n \in P(\mathcal{Z}(\mathcal{M}))$, such that $q_n E_{|x|}[n,\infty)$ is finite. We define

$$e = \mathbf{1} - \bigvee_{n\in\mathbb{N}} q_n.$$

Since x is not locally measurable, Proposition 2.3.5(ii) implies that $e \neq 0$. Since $\mathcal{M}$ is acting on a separable Hilbert space, we have that e is countably decomposable. Note that $e \in P(\mathcal{Z}(\mathcal{M}))$ is a nonzero projection, such that $fE_{|x|}[n,\infty)$ is infinite for any nonzero $f \in P(\mathcal{Z}(\mathcal{M}))$ with $f \le e$ and any $n \in \mathbb{N}$. Replacing e by a suitable nonzero subprojection in $\mathcal{Z}(\mathcal{M})$, we may assume that either:

(i) the von Neumann algebra $\mathcal{M}e$ is semifinite

or

(ii) the von Neumann algebra $\mathcal{M}e$ is of type III.

We consider the cases (i) and (ii) separately and construct $\{n_k\}$ and $\{r_k\}$ by induction.

Assume first that the algebra $\mathcal{M}e$ is semifinite. We let $n_1 = 0$ and assume that the integers n_k are already constructed. By Lemma 1.9.11, there exists a finite projection $g \in P(\mathcal{M})$, such that $s_c(g) = e$. By Theorem 1.9.5 for all $n > n_k$, there exists a largest projection $p_{k,n}$ in $P(\mathcal{Z}(\mathcal{M}))$, such that

$$p_{k,n} \le e, \quad p_{k,n} g \preceq p_{k,n} E_{|x|}[n_k, n).$$

We let

$$p_k = \sup_{n>n_k} p_{k,n}$$

and prove that $p_k = e$.

We first claim that

$$(e - p_k)E_{|x|}[n_k, n) \preceq (e - p_k)g, \quad \text{for all } n > n_k. \tag{3.1}$$

Indeed, if that is not the case then by Theorem 1.9.5 there exists $z \in P(\mathcal{Z}(\mathcal{M}))$, such that

$$z(e - p_k)g \prec z(e - p_k)E_{|x|}[n_k, n), \quad z(e - p_k) \neq 0.$$

Then $p_{k,n} < p_{k,n} + z(e - p_k) \le e$ and

$$(p_{k,n} + z(e - p_k))g \preceq (p_{k,n} + z(e - p_k))E_{|x|}[n_k, n),$$

which contradicts the choice of $p_{k,n}$. Thus, (3.1) holds.

Since $E_{|x|}[n_k, n) \uparrow E_{|x|}[n_k, \infty)$ as $n \to \infty$, Definition 1.11.1(vi) and majorization in (3.1) implies that

$$\mathcal{D}((e - p_k)E_{|x|}[n_k, \infty)) = \sup_{n > n_k} \mathcal{D}((e - p_k)E_{|x|}[n_k, n)) \leq \mathcal{D}((e - p_k)g).$$

Since the projection $(e - p_k)g$ is finite, it follows that $\mathcal{D}((e - p_k)g)$ is finite almost everywhere and, therefore, $\mathcal{D}((e - p_k)E_{|x|}[n_k, \infty))$ is finite almost everywhere, too, proving that the projection $(e - p_k)E_{|x|}[n_k, \infty)$ is finite (see Definition 1.11.1(ii)). Hence, by the choice of e, we conclude that $e - p_k = 0$, i. e., $e = p_k$ as required.

Let ψ be a normal state on $\mathcal{Z}(\mathcal{M})e$ (see Example 1.10.12). Since $p_{n,k} \uparrow p_k = e$, it follows that $\psi(p_{k,n}) \uparrow \psi(e)$. We now choose n_{k+1}, such that

$$\psi(p_{k,n_{k+1}}) > (1 - 2^{-(k+2)})\psi(e),$$
$$\psi(e - p_{k,n_{k+1}}) < 2^{-(k+2)}\psi(e).$$

Let $p = \inf_{k \in \mathbb{N}} p_{k,n_{k+1}} \leq e$. We have

$$e - p = \sup_{k \in \mathbb{N}}(e - p_{k,n_{k+1}}),$$
$$\psi(e - p) \leq \sum_{k=1}^{\infty} \psi(e - p_{k,n_{k+1}}) < \frac{1}{2}\psi(e),$$

and

$$\psi(p) > \frac{1}{2}\psi(e) > 0.$$

Thus, p is a nonzero projection in $\mathcal{Z}(\mathcal{M})$, such that $pg \preceq pE_{|x|}[n_k, n_{k+1})$ for all $k \in \mathbb{N}$. To conclude the proof for the case of semifinite von Neumann algebra $\mathcal{M}e$, it remains to choose mutually equivalent nonzero projections $r_k \in \mathcal{M}$, such that $r_k \leq pE_{|x|}[n_k, n_{k+1})$ and $r_k \sim pg$ for all $k \in \mathbb{N}$.

Now, assume that $\mathcal{M}e$ is of type *III*. Let $n_1 = 0$ and assume that n_k is known for some fixed k. We first note that since $\mathcal{M}e$ is type *III* algebra, and by the choice of $q_n \in P(\mathcal{Z}(\mathcal{M}))$, the projection $q_n E_{|x|}[n, \infty)$ is finite, it follows that $q_n E_{|x|}[n, \infty) = 0$ for any $n \in \mathbb{N}$. Therefore, $s_c(E_{|x|}[n, \infty)) = \mathbf{1} - q_n$, $n \in \mathbb{N}$.

We define $p_{k,n} = e \cdot s_c(E_{|x|}[n_k, n))$ for all $n > n_k$ and set

$$p_k = \sup_{n > n_k} p_{k,n}.$$

Since $E_{|x|}[n_k, n) \uparrow E_{|x|}[n_k, \infty)$ as $n \to \infty$, it follows that

$$s_c(E_{|x|}[n_k, n)) \uparrow s_c(E_{|x|}[n_k, \infty)) = \mathbf{1} - q_{n_k} \geq e$$

and, therefore, $p_{k,n} = e \cdot s_c(E_{|x|}[n_k, n)) \uparrow e \cdot (\mathbf{1} - q_{n_k}) = e$. Thus, $p_k = e$.

Let ψ be a normal state on $\mathcal{Z}(\mathcal{M})e$. Then $\psi(p_{k,n}) \uparrow \psi(e)$. We now choose n_{k+1} so that

$$\psi(p_{k,n_{k+1}}) > (1 - 2^{-(k+2)})\psi(e),$$
$$\psi(e - p_{k,n_{k+1}}) < 2^{-(k+2)}\psi(e).$$

Let $p = \inf_{k \in \mathbb{N}} p_{k,n_{k+1}} \leq e$. Then

$$e - p = \sup_{k \in \mathbb{N}} (e - p_{k,n_{k+1}}),$$
$$\psi(e - p) \leq \sum_{k=1}^{\infty} \psi(e - p_{k,n_{k+1}}) < \frac{1}{2}\psi(e),$$

and

$$\psi(p) > \frac{1}{2}\psi(e) > 0.$$

It follows that $p \in \mathcal{Z}(\mathcal{M})$ is a nonzero and satisfies $s_c(pE_{|x|}[n_k, n_{k+1})) = p$.

For every $k \in \mathbb{N}$, we set $r_k = pE_{|x|}[n_k, n_{k+1})$. Since p is a central projection, we have that $r_k \leq E_{|x|}[n_k, n_{k+1})$ for all $k \in \mathbb{N}$. Since $s_c(r_k) = s_c(pE_{|x|}[n_k, n_{k+1})) = p$, it follows from Theorem 1.9.9(viii) that r_k are pairwise equivalent, as required. □

Theorem 3.1.2. *Let $\mathcal{M}$ be a von Neumann algebra and let x be a closed densely defined operator affiliated with $\mathcal{M}$. If the algebraic product yx is closable for all $y \in \mathcal{M}$, then $x \in LS(\mathcal{M})$.*

Proof. Suppose that x is a closed densely defined operator affiliated with $\mathcal{M}$, such that yx is closable for all $y \in \mathcal{M}$. Assume on the contrary that x is not locally measurable. By Lemma 3.1.1, there exist an increasing sequence of nonnegative integers $\{n_k\}_{k \in \mathbb{N}}$, and a sequence of nonzero, mutually equivalent projections $\{r_k\}_{k \in \mathbb{N}}$ in $\mathcal{M}$, such that $r_k \leq E_{|x|}[n_k, n_{k+1})$ for all $k \in \mathbb{N}$.

Let $\{v_k\}_{k \in \mathbb{N}} \subset \mathcal{M}$ be a sequence of partial isometries, such that $v_k v_k^* = r_1$, $v_k^* v_k = r_k$ for each $k \in \mathbb{N}$.

Since $v_k r_k = v_k$ for all $k \in \mathbb{N}$, for any $\xi \in H$, and $n, m \in \mathbb{N}$, we have

$$\left\| \sum_{k=n}^{n+m} k^{-3/4} v_k \xi \right\| = \left\| \sum_{k=n}^{n+m} k^{-3/4} v_k r_k \xi \right\| \leq \sum_{k=n}^{n+m} k^{-3/4} \|r_k \xi\|$$
$$\leq \left(\left(\sum_{k=n}^{n+m} k^{-3/2} \right) \left(\sum_{k=n}^{n+m} \|r_k \xi\|^2 \right) \right)^{1/2} \leq \left(\sum_{k=n}^{n+m} k^{-3/2} \right)^{1/2} \|\xi\|.$$

Since the series $\sum_{k=1}^{\infty} k^{-3/2}$ converges, we may define $v \in \mathcal{M}$, by setting

$$v := \sum_{k=1}^{\infty} k^{-3/4} v_k,$$

where the series converges in the uniform norm.

Let $0 \neq \xi_0 \in r_1(H)$. For every $k \geq 2$, there exists a vector $\xi_k \in r_k(H)$, such that $n_k v \xi_k = \xi_0$. Since $k^{3/4} v|_{r_k(H)} = v_k|_{r_k(H)}$ is isometry, it follows that $\|\xi_k\| = n_k^{-1} k^{3/4} \|\xi_0\|$. Since $k \leq n_k$, it follows that $\xi_k \to 0$ as $k \to \infty$.

Let $f = \sum_{k=1}^{\infty} n_k \chi_{[n_k, n_{k+1})}$. For the operator $y := f(|x|)$, the functional calculus and Proposition 2.1.2(v) imply that y is a closable positive densely defined operator affiliated with $\mathcal{M}$ and $\||x|^{-1}y\|_{\mathcal{M}} \leq 1$. Therefore, $\eta_k := |x|^{-1} y \xi_k \to 0$.

Since $\xi_k \in r_k(H)$ and $r_k \leq E_{|x|}[n_k, n_{k+1})$ for every $k \in \mathbb{N}$, it follows that $\xi_k = E_{|x|}[n_k, n_{k+1})\xi_k$ and, therefore,

$$y\xi_k = \sum_{k=1}^{\infty} n_k E_{|x|}[n_k, n_{k+1})\xi_k = n_k E_{|x|}[n_k, n_{k+1})\xi_k = n_k \xi_k,$$

for every $k \in \mathbb{N}$. Since

$$v|x|\eta_k = v|x||x|^{-1} y\xi_k = vy\xi_k = n_k v\xi_k = \xi_0$$

and $\xi_0 \neq 0$, it follows that the operator $v|x|$ is not closable.

Finally, if $x = w|x|$ is the polar decomposition of x, and $y_0 = vw^*$, then $y_0 \in \mathcal{M}$ and $y_0 x = (vw^*)x = v(w^* x) = v|x|$. Since $v|x|$ is not closable, it follows that the operator $y_0 x$ is not closable, too, which is a contradiction. Thus, x is a locally measurable operator. □

We now introduce the notion of bimodules over a given von Neumann algebra.

Definition 3.1.3. Let $\mathcal{E}$ be a set of closed densely defined operators affiliated with a von Neumann algebra $\mathcal{M}$. We say that $\mathcal{E}$ is an $\mathcal{M}$-*bimodule* (or bimodule over $\mathcal{M}$) if:
(i) $\mathcal{E}$ is closed with respect to strong addition and scalar multiplication;
(ii) $\mathcal{E}$ is closed with respect to strong multiplication by elements of $\mathcal{M}$.

The following theorem immediately results from Theorem 3.1.2 and shows that $LS(\mathcal{M})$ is the largest possible bimodule over $\mathcal{M}$.

Theorem 3.1.4. *Let $\mathcal{M}$ be a von Neumann algebra and let $\mathcal{E}$ be an $\mathcal{M}$-bimodule. Then $\mathcal{E} \subset LS(\mathcal{M})$.*

Proof. Let $x \in \mathcal{E}$. Since $\mathcal{E}$ is closed with respect to strong multiplication by elements of $\mathcal{M}$, Theorem 3.1.2 implies that $x \in LS(\mathcal{M})$. Thus, $\mathcal{E} \subset LS(\mathcal{M})$. □

Remark 3.1.5. *In view of Theorem* 3.1.4, *the definition of an $\mathcal{M}$-bimodule can be reformulated as follows: a linear subspace $\mathcal{E}$ of $LS(\mathcal{M})$ is an $\mathcal{M}$-bimodule if $uxv \in \mathcal{E}$ whenever $x \in \mathcal{E}$ and $u, v \in \mathcal{M}$.*

In the following proposition, some of the elementary properties of $\mathcal{M}$-bimodules are collected together.

Proposition 3.1.6. *If $\mathcal{E} \subseteq LS(\mathcal{M})$ is an $\mathcal{M}$-bimodule, then the following assertions hold:*
(i) *If $x \in LS(\mathcal{M})$, then $x \in \mathcal{E}$ if and only if $|x| \in \mathcal{E}$;*
(ii) *If $x \in LS(\mathcal{M})$, then $x \in \mathcal{E}$ if and only if $x^* \in \mathcal{E}$.*

Proof. To prove (i), let x be a locally measurable operator with the polar decomposition $x = v|x|$. Clearly, if $|x| \in \mathcal{E}$, then $x \in \mathcal{E}$. Conversely, if $x \in \mathcal{E}$, then the equality $|x| = v^*x$ implies that $|x| \in \mathcal{E}$. Thus, $x \in \mathcal{E}$ if and only if $|x| \in \mathcal{E}$.

(ii) Since $x^{**} = x$ for any $x \in LS(\mathcal{M})$ (see Theorem 1.3.4), it is sufficient to show that $x^* \in \mathcal{E}$ if $x \in \mathcal{E}$. Let $x \in \mathcal{E}$ with polar decomposition $x = v|x|$. By part (i), we have that $|x| \in \mathcal{E}$. Therefore, the equality $x^* = |x|v^*$ implies that $x^* \in \mathcal{E}$, as required. □

Next, we shall study properties of $\mathcal{M}$-bimodules that are equipped with some topology. We first recall the definitions of F-norms and quasi-norms (see, e. g., [95]) and some of their properties.

Definition 3.1.7. An F-norm $\|\cdot\|_E$ on a vector space E over the field $\mathbb{C}$ is a function $\|\cdot\|_E : E \to [0,\infty)$, such that for all $x, y \in E$ the following properties hold:
(i) $\|x\| = 0 \Leftrightarrow x = 0$;
(ii) $\|\alpha x\| \leqslant \|x\|, \forall\, \alpha \in \mathbb{C}, |\alpha| \leq 1$;
(iii) $\lim_{\alpha\to 0} \|\alpha x\| = 0$;
(iv) $\|x + y\| \leq \|x\| + \|y\|$.

The couple $(E, \|\cdot\|)$ is called an *F-normed* space. If $(E, \|\cdot\|_E)$ is an F-normed space, and E is complete with respect to the metric generated by $\|\cdot\|_E$, we say that $(E, \|\cdot\|_E)$ is an *F-space*.

We note that every F-normed space $(E, \|\cdot\|_E)$ is metrizable and conversely every metrizable space can be equipped with an F-norm (see, e. g., [95]). In general, the topology generated by an F-norm is neither locally bounded nor locally convex.

In the case when the topological vector space is locally bounded, the topology can be defined via a quasi-norm.

Definition 3.1.8. A function $\|\cdot\| : E \to [0,\infty)$ on a vector space E over the field $\mathbb{C}$ is called a *quasi-norm* on E, if there exists a constant $C \geq 1$, such that for all $x, y \in E$, $\alpha \in \mathbb{C}$ the following properties hold:
(i) $\|x\| = 0 \Leftrightarrow x = 0$;
(ii) $\|\alpha x\| = |\alpha|\|x\|$;
(iii) $\|x + y\| \leq C(\|x\| + \|y\|)$.

The couple $(E, \|\cdot\|)$ is called a quasi-normed space and the least of all constants C satisfying (iii) above is called the *modulus of concavity* of the quasi-norm $\|\cdot\|$ and denoted by C_E.

If $(E, \|\cdot\|)$ is a quasi-normed space, then by the Aoki–Rolewicz theorem (see [95]) there exists an equivalent quasi-norm $\|\cdot\|_1$ on E, such that $\|x+y\|_1^p \leq \|x\|_1^p + \|y\|_1^p$, $x, y \in E$, where $2^{1/p} = 2C_E$. In particular, $\|\cdot\|_1^p$ is an F-norm on E, which is equivalent to the quasi-norm $\|\cdot\|$, and so E is a metrizable topological vector space with locally bounded topology. Conversely, any locally bounded metrizable topological vector space can be equipped with a quasi-norm, which generates this topology.

Next we introduce the notion of an F-normed (resp., quasi-normed) bimodule over a given von Neumann algebra that combines the properties of a module and an F-normed (resp., quasi-normed) space.

Definition 3.1.9. Let $\mathcal{E}$ be an $\mathcal{M}$-bimodule. If $\mathcal{E}$ is equipped with an F-norm $\|\cdot\|_{\mathcal{E}}$, satisfying

$$\|uxv\|_{\mathcal{E}} \leq \big\|\|u\|_{\mathcal{M}}\|v\|_{\mathcal{M}}x\big\|_{\mathcal{E}}, \quad x \in \mathcal{E},\ u, v \in \mathcal{M}, \tag{3.2}$$

then $\mathcal{E}$ is called an *F-normed $\mathcal{M}$-bimodule* of locally measurable operators. If $\|\cdot\|_{\mathcal{E}}$ is a (quasi-)norm, then $(\mathcal{E}, \|\cdot\|_{\mathcal{E}})$ is called a (quasi-)normed $\mathcal{M}$-bimodule.

The following proposition complements Proposition 3.1.6.

Proposition 3.1.10. *Let $(\mathcal{E}, \|\cdot\|_{\mathcal{E}})$ be an F-normed bimodule. Then for all $x \in \mathcal{E}$, we have*

$$\|x\|_{\mathcal{E}} = \|x^*\|_{\mathcal{E}} = \||x|\|_{\mathcal{E}}.$$

Proof. Let $x = v|x|$ be the polar decomposition of x. Then $|x| = v^*x$ and, therefore,

$$\||x|\|_{\mathcal{E}} = \|v^*x\| \leq \big\|\|v^*\|_{\mathcal{M}}x\big\|_{\mathcal{E}} = \|x\|_{\mathcal{E}} = \|v|x|\|_{\mathcal{E}} \leq \big\|\|v\|_{\mathcal{M}}|x|\big\|_{\mathcal{E}} = \||x|\|_{\mathcal{E}},$$

showing that $\|x\|_{\mathcal{E}} = \||x|\|_{\mathcal{E}}$. Now using the equality $x^* = |x|v^*$ we infer that

$$\|x^*\|_{\mathcal{E}} = \||x|v^*\|_{\mathcal{E}} \leq \big\|\|v^*\|_{\mathcal{M}}|x|\big\|_{\mathcal{E}} = \|x\|_{\mathcal{E}}.$$

Applying the same argument with x replaced by x^*, we conclude the proof. □

3.2 Functional calculus for locally measurable operators

One of the important properties of the cone $\mathcal{M}_h$ consisting of all self-adjoint operators from the given von Neumann algebra $\mathcal{M}$ is its invariance with respect to functional calculus defined by real-valued bounded Borel functions on $\mathbb{R}$. It is natural to expect that a similar property holds for algebra of locally measurable operators. In this section, we extend the functional calculus from Section 1.4 to the $*$-algebras of locally measurable, measurable, and τ-measurable operators, and show that the respective algebras remain invariant with respect to suitable versions of the functional calculi involving unbounded measurable functions.

Let $\mathcal{M}$ be an arbitrary von Neumann algebra acting on a Hilbert space H. As in Section 1.4, we denote by $\mathcal{B}(\sigma(x))$ the $*$-algebra of all Borel complex-valued functions, defined on the spectrum $\sigma(x)$ of a self-adjoint operator x. The $*$-subalgebra consisting of all bounded functions in $\mathcal{B}(\sigma(x))$ is denoted by $\mathcal{B}_b(\sigma(x))$. Furthermore, $\mathcal{B}_{bc}(\sigma(x))$ is defined to be the $*$-subalgebra of $\mathcal{B}(\sigma(x))$ consisting of all Borel functions, which are bounded on compact subsets of $\sigma(x)$.

Theorem 3.2.1. *Suppose that x is a self-adjoint operator affiliated with $\mathcal{M}$ and let $f \in \mathcal{B}_{bc}(\sigma(x))$. Then:*
(i) *If x is locally measurable, then $f(x)$ is locally measurable;*
(ii) *If x is measurable, then $f(x)$ is measurable;*
(iii) *If $\mathcal{M}$ is equipped with a faithful normal semifinite trace τ and x is τ-measurable, then $f(x)$ is τ-measurable.*

Furthermore, the mapping $f \mapsto f(x)$ is a $*$-homomorphism from $\mathcal{B}_{bc}(\sigma(x))$ into $LS(\mathcal{M})$ (resp., $S(\mathcal{M})$ and $S(\mathcal{M},\tau)$).

Proof. Given $f \in \mathcal{B}_{bc}(\sigma(x))$, Proposition 2.1.2(v) implies that $f(x)\eta\mathcal{M}$. Furthermore, $f(x)$ is closed and densely defined.

Assume first that x is locally measurable. By Proposition 2.3.5(ii), there exists a sequence $\{z_n\}_{n\in\mathbb{N}} \subset P(\mathcal{Z}(\mathcal{M}))$, such that $z_n \uparrow \mathbf{1}$ and $z_n E_{|x|}[n,\infty)$ is finite. Define $q_n = E_{|x|}[0,n]$. By the spectral theorem, we have that $q_n \uparrow \mathbf{1}$ as $n \to \infty$. Furthermore, since f is bounded on the compact interval $[-n,n]$, $n \in \mathbb{N}$, we have

$$\int\limits_{\sigma(x)} |f(\lambda)|^2 d\langle E_x(\lambda)q_n\xi, q_n\xi\rangle = \int\limits_{\sigma(x)} |\chi_{[-n,n]}(\lambda)f(\lambda)|^2 \langle E_x(\lambda)\xi,\xi\rangle < \infty,$$

for any $\xi \in H$ and, therefore, $q_n(H) \subset \operatorname{dom}(f(x))$. Appealing to Proposition 2.3.5(v), we conclude that $f(x) \in LS(\mathcal{M})$.

Next, let x be measurable. By Proposition 2.2.3(iii), we have that there exists λ, such that the projection $E_{|x|}(\lambda,\infty)$ is finite. Since f is bounded on $\sigma(x) \cap [-\lambda,\lambda]$, there exists $0 < \rho \in \mathbb{R}$, such that $|\mu| > \lambda$ whenever $\mu \in \sigma(x)$ and $|f(\mu)| > \rho$. Hence,

$$\begin{aligned} E_{|f(x)|}(\rho,\infty) &= E_{|f|(x)}(\rho,\infty) = E_x\{\mu \in \sigma(x) : |f(\mu)| > \rho\} \\ &\leq E_x\{\mu \in \sigma(x) : |\mu| > \lambda\} = E_{|x|}(\lambda,\infty). \end{aligned} \tag{3.3}$$

Thus, there exists $\rho > 0$, such that $E_{|f|}(\rho,\infty)$ is finite. Appealing to Proposition 2.2.3(iii) again, we conclude that $f(x)$ is measurable.

If $\mathcal{M}$ is equipped with a faithful normal semifinite trace τ and $x \in S(\mathcal{M},\tau)$, then the proof is verbatim repetition of the case of measurable operators with reference to Proposition 2.4.2(iv) instead.

Theorem 1.4.4 implies that the mapping $f \longmapsto f(a)$ is a $*$-homomorphism from $\mathcal{B}_{bc}(\sigma(a))$ into the respective $*$-algebra $LS(\mathcal{M})$ (resp., $S(\mathcal{M})$ and $S(\mathcal{M},\tau)$). □

Remark 3.2.2. *Suppose that $\mathcal{M}$ is a finite von Neumann algebra. It follows from Corollary 2.2.5 and Proposition 2.3.3 that $LS(\mathcal{M}) = S(\mathcal{M})$ and $LS(\mathcal{M})$ coincides with the collection of all closed densely defined operators affiliated with $\mathcal{M}$. Therefore, it follows from Proposition 2.1.2(v), that $f(x) \in S(\mathcal{M})$ for any self-adjoint $x \in S(\mathcal{M})$ and all $f \in \mathcal{B}(\sigma(x))$. If the von Neumann algebra $\mathcal{M}$ is not finite, then the latter statement does not hold in general, as can be easily seen by considering the von Neumann algebra $\mathcal{B}(H)$, whenever H is infinite-dimensional.*

Given a self-adjoint operator $x \in LS(\mathcal{M})$, the above theorem implies, in particular, that $f(x)$ and $g(x)$ are commuting elements of $LS(\mathcal{M})$, for any two functions $f,g \in \mathcal{B}_{bc}(\sigma(a))$. In the next proposition, we establish a necessary and sufficient condition for two self-adjoint operators $x,y \in LS(\mathcal{M})$ to commute.

Proposition 3.2.3. *For self-adjoint operators $x,y \in LS(\mathcal{M})$, the following are equivalent:*
(i) *x and y commute in $LS(\mathcal{M})$;*
(ii) *$E_x(B_1)E_y(B_2) = E_y(B_2)E_x(B_1)$ for all Borel sets B_1, B_2 in $\mathbb{R}$;*
(iii) *$f(x)g(y) = g(y)f(x)$ for all $f,g \in \mathcal{B}_{bc}(\mathbb{R})$.*

Proof. Let $x,y \in LS(\mathcal{M})$ be self-adjoint. We note that Theorem 3.2.1(i) guarantees that the Cayley transform $\gamma(x) = (x - i\mathbf{1})(x + i\mathbf{1})^{-1}$ is a locally measurable operator. Since $\gamma(x)$ is also bounded operator, it follows that $\gamma(x) \in \mathcal{M}$. Since $xy = yx$, it follows that $(x \pm i\mathbf{1})y = y(x \pm i\mathbf{1})$ and $(x + i\mathbf{1})^{-1}y = y(x + i\mathbf{1})^{-1}$. Hence, we have that $\gamma(x)y = y\gamma(x)$. Conversely, suppose that $\gamma(x)y = y\gamma(x)$. Since $x = i(\mathbf{1} + \gamma(x))(\mathbf{1} - \gamma(x))^{-1}$, it follows that $xy = yx$ in $LS(\mathcal{M})$. Thus, we proved that

$$xy = yx \Leftrightarrow \gamma(x)y = y\gamma(x). \tag{3.4}$$

(i)⇔(ii). By (3.4), the equality $xy = yx$ holds if and only if $\gamma(x)y = y\gamma(x)$. By Theorem 1.4.8, the latter is equivalent to the equality $\gamma(x)E_y(B_2) = E_y(B_2)\gamma(x)$ for any Borel set $B_2 \subset \mathbb{R}$. Since $E_y(B) \in \mathcal{M} \subset LS(\mathcal{M})$, employing (3.4) again we obtain that $\gamma(x)E_y(B_2) = E_y(B_2)\gamma(x)$ if and only if $xE_y(B_2) = E_y(B_2)x$. Referring once again to Theorem 1.4.8, we obtain that $xE_y(B_2) = E_y(B_2)x$ if and only if $E_x(B_1)E_y(B_2) = E_y(B_2)E_x(B_1)$ for all Borel sets B_1, B_2 in $\mathbb{R}$, proving that (i) and (ii) are equivalent.

(ii)⇒(iii). Let $f,g \in \mathcal{B}_{bc}(\mathbb{R})$ be given. Considering the real and imaginary parts of f and g separately, we may assume, without loss of generality, that both f and g are real-valued. By Proposition 1.4.7, we have that the spectral measure $E_{f(x)}$ of $f(x)$ is given by $E_{f(x)}(B) = E_x(f^{-1}(B \cap f(\mathbb{R})))$, $B \subseteq \mathbb{R}$, and similarly for $E_{g(y)}$. Thus, assumption of (ii) is satisfied for the operators $f(x)$ and $g(y)$. Since (ii) implies (i), it follows that $f(x)$ and $g(y)$ commute in $LS(\mathcal{M})$.

The implication (iii)⇒(i) is evident by taking $f(\lambda) = g(\lambda) = \lambda$, $\lambda \in \mathbb{R}$. □

Note that it follows in particular from the above proposition that the self-adjoint operators $x, y \in LS(\mathcal{M})$ commute in $LS(\mathcal{M})$ if and only if $xE_y(B) = E_y(B)x$ in $LS(\mathcal{M})$ for all Borel sets $B \subseteq \mathbb{R}$. We also obtain the following description of the center $\mathcal{Z}(LS(\mathcal{M}))$ of the algebra $LS(\mathcal{M})$.

Corollary 3.2.4. *For a von Neumann algebra $\mathcal{M}$, we have that*

$$\mathcal{Z}(LS(\mathcal{M})) := \{x \in LS(\mathcal{M}) : xy = yx \ \forall y \in LS(\mathcal{M})\} = S(\mathcal{Z}(\mathcal{M})).$$

Proof. Suppose that $x \in LS(\mathcal{M})$ is such that $xy = yx$ for all $y \in LS(\mathcal{M})$. Without loss of generality, we may assume that x is self-adjoint. Therefore, by Proposition 3.2.3, we have that the spectral projections E_x of x commute with all elements of $\mathcal{M}$. Therefore, $E_x(B) \in \mathcal{M}' \cap \mathcal{M} = \mathcal{Z}(\mathcal{M})$ for all Borel sets $B \subset \sigma(x)$. By Proposition 2.1.2(iv), we have that x is affiliated with $\mathcal{Z}(\mathcal{M})$. Since $\mathcal{Z}(\mathcal{M})$ is a commutative (and hence finite) algebra, it follows that x is measurable with respect to $\mathcal{Z}(\mathcal{M})$ (see Corollary 2.2.5).

Conversely, let $x \in S(\mathcal{Z}(\mathcal{M}))$. Without loss of generality, x is self-adjoint. Since x is affiliated with $\mathcal{Z}(\mathcal{M}) = \mathcal{M}' \cap \mathcal{M}$, Proposition 2.1.2(iv) implies that $E_x(B) \in \mathcal{M}'$ for any Borel set $B \subset \sigma(x)$. Therefore, by Proposition 3.2.3, we have that x commutes with any self-adjoint element of $LS(\mathcal{M})$, and so x commutes with any element of $LS(\mathcal{M})$. □

For a commutative von Neumann algebra $\mathcal{M}$, Corollary 3.2.4 combined with Corollary 2.2.5 implies the following.

Corollary 3.2.5. *If $\mathcal{M}$ is a commutative von Neumann algebra, then the algebra $LS(\mathcal{M})$ is also commutative (and $LS(\mathcal{M})$ consists of all closed densely defined operators affiliated with $\mathcal{M}$).*

Proof. If $\mathcal{M}$ is a commutative von Neumann algebra, then $\mathcal{Z}(\mathcal{M}) = \mathcal{M}$ and, therefore, by Corollary 3.2.4 and Proposition 2.3.3 we have that

$$LS(\mathcal{M}) = S(\mathcal{M}) = S(\mathcal{Z}(\mathcal{M})) = \mathcal{Z}(LS(\mathcal{M})).$$

Thus, $LS(\mathcal{M})$ is a commutative algebra. The second assertion follows from Corollary 2.2.5. □

The next result is concerned with the functional calculus of self-adjoint compact and τ-compact operators. For this purpose, we define

$$\mathcal{B}_0(\sigma(a)) = \Big\{f \in \mathcal{B}(\sigma(a)) : \lim_{\lambda \to 0} f(\lambda) = 0\Big\}.$$

Note that $\mathcal{B}_0(\sigma(a))$ is a $*$-subalgebra in $\mathcal{B}(\sigma(a))$.

Proposition 3.2.6. *Suppose that $x \in S(\mathcal{M})$ is self-adjoint and let $f \in \mathcal{B}_0(\sigma(x))$.*
(i) *If $x \in S_0(\mathcal{M})$, then $f(x) \in S_0(\mathcal{M})$;*
(ii) *If $\mathcal{M}$ is equipped with a faithful normal semifinite trace τ and $x \in S_0(\mathcal{M},\tau)$, then $f(x) \in S_0(\mathcal{M},\tau)$.*

Furthermore, the map $f \longmapsto f(x)$ is a $*$-homomorphism from $\mathcal{B}_0(\sigma(x))$ into $S_0(\mathcal{M})$ (resp., into $S_0(\mathcal{M},\tau)$).

Proof. We prove the statement for a compact operator $x \in S_0(\mathcal{M})$, as the proof for a τ-compact operator is similar.

Let $f \in \mathcal{B}_0(\sigma(x))$ be given. Proposition 2.1.2(v) implies that $f(x)\eta\mathcal{M}$. Moreover, $f(x)$ is closed and densely defined. Therefore, to show that $f(x) \in S_0(\mathcal{M})$, it is sufficient to prove that the spectral projection $E_{|f(x)|}(\lambda,\infty)$ is finite for all $\lambda > 0$. Given $\lambda > 0$, observe that

$$E_{|f(x)|}(\lambda,\infty) = E_{|f|(x)}(\lambda,\infty) = E_x(\{\mu \in \sigma(x) : |f(\mu)| > \lambda\}).$$

Since $\lim_{\mu\to 0} f(\mu) = 0$, there exists $\alpha > 0$, such that $|f(\mu)| \leq \lambda$ for all $\mu \in \sigma(a)$ with $|\mu| \leq \alpha$. Consequently,

$$\begin{aligned} E_x(\{\mu \in \sigma(x) : |f(\mu)| > \lambda\}) &\leq E_x(\{\mu \in \sigma(x) : |\mu| > \alpha\}) \\ &= E_{|x|}(\alpha,\infty). \end{aligned}$$

Since the projection $E_{|x|}(\alpha,\infty)$ is finite, it follows that the projection $E_{|f(x)|}(\lambda,\infty)$ is finite, too. The properties of the functional calculus immediately imply that the map $f \longmapsto f(x)$ is a $*$-homomorphism from $\mathcal{B}_0(\sigma(x))$ into $S_0(\mathcal{M})$. □

Remark 3.2.7. *In general, for a self-adjoint $x \in S_0(\mathcal{M})$ (resp., $x \in S_0(\mathcal{M},\tau)$) and $f \in \mathcal{B}_{bc}(\sigma(x))$ the operator $f(x)$ may not belong to $S_0(\mathcal{M})$ (resp., $S_0(\mathcal{M},\tau)$). This can be easily seen on the example when $\mathcal{M} = B(H)$ and τ is the standard trace* tr. *As shown in Example* 2.6.6, *in this case $S_0(\mathcal{M}) = S_0(\mathcal{M},\tau) = K(H)$, where $K(H)$ is the ideal of all compact operators in H. In this case, for a compact self-adjoint operator x and the function $f(\lambda) = 1$, $\lambda \in \mathbb{R}$, $f(x)$ is not compact (whenever, H is infinite-dimensional).*

3.3 Order properties of locally measurable operators

In this section, we introduce a natural partial order on the algebra of self-adjoint locally measurable operators $LS(\mathcal{M})$ extending the classical ordering on $\mathcal{M}$. As in the classical (commutative) integration theory, this ordering plays an indispensable role in the theory of algebras of unbounded operators. We shall select a cone of positive locally measurable operators, which is similar to the cone of positive measurable functions on a measure space. This cone contains the subcone of all positive elements from the von

Neumann algebra $\mathcal{M}$. We show that an arbitrary locally measurable operator is a linear combination of four positive ones.

Let $\mathcal{M}$ be an arbitrary von Neumann algebra acting on a Hilbert space H and let $LS(\mathcal{M})$ (resp., $S(\mathcal{M})$) be the $*$-algebra of all locally measurable (resp., measurable) operators with respect to $\mathcal{M}$. If $\mathcal{M}$ is equipped with a faithful normal semifinite trace τ, we let $S(\mathcal{M},\tau)$ to be the algebra of all τ-measurable operators.

As before, for the operations of strong sum $x \dot{+} y$ and strong product $x \cdot y$ of operators x, y from $LS(\mathcal{M})$, we will use notation $x + y$, xy.

We denote by $LS_h(\mathcal{M})$ (resp., $S_h(\mathcal{M})$ and $S_h(\mathcal{M},\tau)$) the collection of all self-adjoint operators from $LS(\mathcal{M})$ (resp., $S(\mathcal{M})$ and $S(\mathcal{M},\tau)$).

Recall that an operator x acting in the Hilbert space H is called positive, whenever $\langle x\xi, \xi\rangle \geq 0$ for all $\xi \in \operatorname{dom}(x)$. The set of positive operators from $LS(\mathcal{M})$ (resp., from $S(\mathcal{M})$ or $S(\mathcal{M},\tau)$) we denote by $LS_+(\mathcal{M})$ (resp., $S_+(\mathcal{M})$ or $S_+(\mathcal{M},\tau)$).

If $x \in LS_h(\mathcal{M})$ satisfies $x \geq 0$ as well as $-x \geq 0$, then $\langle x\xi, \xi\rangle = 0$ for all $\xi \in \operatorname{dom}(x)$, and so via polarization, it follows that $\langle x\xi, \eta\rangle = 0$ for all $\xi, \eta \in \operatorname{dom}(x)$. Hence, $x = 0$. This shows that $LS_+(\mathcal{M}) \cap (-LS_+(\mathcal{M})) = \{0\}$. Furthermore, $x + y \in LS_+(\mathcal{M})$ and $\lambda x \in LS_+(\mathcal{M})$ whenever $x, y \in LS_+(\mathcal{M})$ and $0 \leq \lambda \in \mathbb{R}$. Consequently, $LS_+(\mathcal{M})$ is a proper cone in $LS_h(\mathcal{M})$.

We define a partial order on $LS_h(\mathcal{M})$ by setting $x \leq y$, if $(y - x) \geq 0$. For the inequality $x \leq y$, we will also use notation $y \geq x$. With respect to this ordering, $LS_h(\mathcal{M})$ is a partially ordered vector space. Clearly, this partial ordering extends the partial order in the space $\mathcal{M}_h$.

As we show in the following proposition, similar to operators from $\mathcal{M}$, an operator $x \in LS(\mathcal{M})$ is positive if and only if $x = y^*y$ for some $y \in LS(\mathcal{M})$. The algebras $S(\mathcal{M})$ and $S(\mathcal{M},\tau)$ also satisfy this property.

Proposition 3.3.1. *Let $\mathcal{M}$ be an arbitrary von Neumann algebra. Then*

$$LS_+(\mathcal{M}) = \{x^*x : x \in LS(\mathcal{M})\} = \{x^2 : x \in LS_h(\mathcal{M})\};$$
$$S_+(\mathcal{M}) = \{x^*x : x \in S(\mathcal{M})\} = \{x^2 : x \in S_h(\mathcal{M})\}.$$

If $\mathcal{M}$ is equipped with a faithful normal semifinite trace τ, then

$$S_+(\mathcal{M},\tau) = \{x^*x : x \in S(\mathcal{M},\tau)\} = \{x^2 : x \in S_h(\mathcal{M},\tau)\}.$$

Proof. We will prove the statement for the algebra $LS(\mathcal{M})$ only, as the proof for $S(\mathcal{M})$ and $S(\mathcal{M},\tau)$ is similar.

Since $x^*x \geq 0$ for any closed densely defined operator x, we have that

$$\{x^2 : x \in LS_h(\mathcal{M})\} \subset \{x^*x : x \in LS(\mathcal{M})\} \subset LS_+(\mathcal{M}).$$

On the other hand, if $y \in LS_+(\mathcal{M})$, then by Theorem 3.2.1(i) the operator $y^{1/2}$ also belongs to $LS(\mathcal{M})$, and by functional calculus, it is also positive. Consequently, $y = y^{1/2}y^{1/2} \in \{x^2 : x \in LS_h(\mathcal{M})\}$, which implies the inclusion

$$LS_+(\mathcal{M}) \subset \{x^2 : x \in LS_h(\mathcal{M})\},$$

which suffices to conclude the proof. □

Recall that for a self-adjoint operator x acting in the Hilbert space H, the positive self-adjoint operators x_+ and x_- are defined by

$$x_+ = \int_{\mathbb{R}} \lambda_+ dE_x(\lambda) = xE_x[0,\infty), \quad x_- = \int_{\mathbb{R}} \lambda_- dE_x(\lambda) = -xE_x(-\infty,0],$$

where $\lambda_+ = \max(\lambda, 0)$ and $\lambda_- = \max(-\lambda, 0)$ (see Section 1.4). Theorem 3.2.1(i) implies that for $x \in LS_h(\mathcal{M})$ (resp., $S_h(\mathcal{M})$ or $S_h(\mathcal{M},\tau)$) the positive and negative part x_+, x_- belong to $LS_+(\mathcal{M})$ (resp., to $S_+(\mathcal{M})$ and $S_+(\mathcal{M},\tau)$). Note, furthermore, that the equalities

$$x_+ = \frac{1}{2}(|x| + x), \quad x_- = \frac{1}{2}(|x| - x), \quad x_+x_- = 0$$

hold in $LS(\mathcal{M})$. As in the case of bounded operators, the decomposition $x = x_+ - x_-$ of an element $x \in LS_h(\mathcal{M})$ as difference of two disjoint positive elements is unique in the following sense.

Proposition 3.3.2. *If $x \in LS_h(\mathcal{M})$ and $y, z \in LS(\mathcal{M})$ are positive operators, such that $x = y - z$ and $yz = 0$, then $y = x_+$ and $z = x_-$.*

Proof. The equality $yz = 0$ implies that also $zy = (yz)^* = 0$ and, therefore,

$$x^2 = (y - z)^2 = y^2 + z^2 = (y + z)^2.$$

By the uniqueness of the positive square root, this implies that $|x| = (x^2)^{1/2} = y + z$, and so $y = \frac{1}{2}(|x| + x) = x_+$ and $z = \frac{1}{2}(|x| - x) = x_-$. □

We note the following simple corollary of Theorem 3.2.1.

Corollary 3.3.3. *If $x \in LS_h(\mathcal{M})$ is such that $x \geq \varepsilon\mathbf{1}$ for some $\varepsilon > 0$, then x is invertible in $LS(\mathcal{M})$ and $0 \leq x^{-1} \leq \frac{1}{\varepsilon}\mathbf{1}$.*

Proof. Since $x \geq \varepsilon\mathbf{1}$, it follows that $\sigma(x) \subset [\varepsilon, \infty)$ and, therefore, the function $f(t) = t^{-1}$, $t \in \sigma(x)$, belongs to $\mathcal{B}_b(\sigma(x)) \subset \mathcal{B}_{bc}(\sigma(x))$. Hence, by Theorem 3.2.1(i), we have that $f(x) = x^{-1}$ exists in $LS(\mathcal{M})$. Furthermore, by Theorem 1.4.2(iv), the operator x^{-1} is a bounded operator on H with $\|x^{-1}\|_{\mathcal{M}} \leq \frac{1}{\varepsilon}$. Thus, $x^{-1} \in \mathcal{M}$ and $0 \leq x^{-1} \leq \frac{1}{\varepsilon}\mathbf{1}$. □

We note also that under the assumption of the above corollary the operator x^α, $\alpha > 0$ is also invertible with the inverse

$$(x^{\alpha})^{-1} = (x^{-1})^{\alpha} = x^{-\alpha} \in \mathcal{M}. \tag{3.5}$$

The following proposition collects some basic properties of the partially ordered vector space $(LS_h(\mathcal{M}), \leq)$.

Proposition 3.3.4.
(i) *If $x, y \in LS_+(\mathcal{M})$ and $xy = yx$, then $xy \geq 0$;*
(ii) *If $x, y \in LS_h(\mathcal{M})$, $a \in LS(\mathcal{M})$ and $x \leq y$, then $a^*xa \leq a^*ya$;*
(iii) *If $x \in LS_+(\mathcal{M})$ and x is invertible in the $*$-algebra $LS(\mathcal{M})$, then $x^{-1} \in LS_+(\mathcal{M})$;*
(iv) *If $x, y \in LS_+(\mathcal{M})$, $0 \leq x \leq y$ and x, y are invertible in the $*$-algebra $LS(\mathcal{M})$, then $y^{-1} \leq x^{-1}$;*
(v) *If $x, y \in LS_h(\mathcal{M})$ and $x \leq y$, then $s(x_+) \preceq s(y_+)$ and $E_x(\lambda, +\infty) \preceq E_y(\lambda, +\infty)$ for any $\lambda \in \mathbb{R}$.*

Proof. (i) Let $x, y \in LS_+(\mathcal{M})$ be such that $xy = yx$. By Proposition 3.2.3, we have that $x^{1/2}y^{1/2} = y^{1/2}x^{1/2}$ and, therefore,

$$xy = (x^{1/2}x^{1/2})(y^{1/2}y^{1/2}) = x^{1/2}y^{1/2}y^{1/2}x^{1/2} = (y^{1/2}x^{1/2})^*(y^{1/2}x^{1/2}) \geq 0.$$

(ii) Since $y - x \in LS_+(\mathcal{M})$, Theorem 3.2.1(i) implies that $(y-x)^{1/2} \in LS_+(\mathcal{M})$. Therefore, Proposition 3.3.1 implies that

$$a^*(y-x)a = ((y-x)^{1/2}a)^*((y-x)^{1/2}a) \geq 0,$$

that is $a^*xa \leq a^*ya$.

(iii) Let $x \in LS_+(\mathcal{M})$ be invertible in $LS(\mathcal{M})$. By Theorem 3.2.1(i), we have that $x^{1/2} \in LS_+(\mathcal{M})$, in addition, $xx^{1/2} = x^{1/2}x$ (see Proposition 3.2.3), which implies that $x^{1/2}x^{-1} = x^{-1}x^{1/2}$. Consequently,

$$(x^{-1}x^{1/2})x^{1/2} = \mathbf{1} = x^{1/2}(x^{-1}x^{1/2}).$$

Thus, $x^{1/2}$ is invertible in $LS(\mathcal{M})$ and $(x^{1/2})^{-1} = x^{-1}x^{1/2}$. Hence,

$$x^{-1} = (x^{1/2}x^{1/2})^{-1} = ((x^{1/2})^{-1})^2 \geq 0.$$

(iv) Let $x, y \in LS_+(\mathcal{M})$ be invertible in $LS(\mathcal{M})$ and let $0 \leq x \leq y$. By Theorem 3.2.1(i), we have that $(y^{-1})^{1/2}$ is a locally measurable operator. Therefore, $y^{1/2}$ is also invertible with $(y^{1/2})^{-1} = (y^{-1})^{1/2} = y^{-1/2}$. Similarly, $x^{-1/2} \subset LS(\mathcal{M})$.

Multiplying the inequality $0 \leq x \leq y$ by $y^{-1/2}$ on both sides and using (iii), we have that $0 \leq y^{-1/2}xy^{-1/2} \leq \mathbf{1}$. Writing

$$y^{-1/2}xy^{-1/2} = (x^{1/2}y^{-1/2})^*(x^{1/2}y^{-1/2})$$

implies that $x^{1/2}y^{-1/2} \in \mathcal{M}$ and $\|(x^{1/2}y^{-1/2})^*\|_{\mathcal{M}} = \|x^{1/2}y^{-1/2}\|_{\mathcal{M}} \leq 1$. Hence,

$$0 \leq x^{1/2}y^{-1}x^{1/2} = (x^{1/2}y^{-1/2})(x^{1/2}y^{-1/2})^* \leq \mathbf{1}.$$

Multiplying the latter inequality by $x^{-1/2}$ on both sides and using (iii) once again, we obtain that $0 \leq y^{-1} \leq x^{-1}$.

(v) Let $x, y \in LS_h(\mathcal{M})$ be such that $x \leq y$. Writing $y = y_+ - y_-$ with $y_\pm \in LS_+(\mathcal{M})$, we have that $x \leq y_+ - y_- \leq y_+$.

We set $p = s(x_+) \wedge (s(y_+)^\perp)$. Then by (iii), we have

$$(x_+^{1/2}p)^*(x_+^{1/2}p) = px_+p = pxp \leq py_+p = 0.$$

Therefore, $x_+^{1/2}p = 0$ and so $px_+ = 0$, proving that $p \leq s(x_+)^\perp$. Since $p \leq s(x_+)$ by definition, it follows that $p = 0$. Therefore, referring to Kaplansky's identity (see Theorem 1.9.4(v)), we conclude that

$$\begin{aligned} s(x_+) &= s(x_+) - (s(x_+) \wedge (s(y_+)^\perp)) \sim (s(x_+) \vee (s(y_+)^\perp)) - s(y_+)^\perp \\ &\leq \mathbf{1} - s(y_+)^\perp = s(y_+), \end{aligned}$$

proving the first part of the statement.

Next, let $\lambda \in \mathbb{R}$ be arbitrary. Since $E_x(\lambda, +\infty) = s((x - \lambda\mathbf{1})_+)$ for any $x \in LS_h(\mathcal{M})$ and $x - \lambda\mathbf{1} \leq y - \lambda\mathbf{1}$, it follows that

$$E_x(\lambda, +\infty) = s((x - \lambda\mathbf{1})_+) \preceq s((y - \lambda\mathbf{1})_+) = E_y(\lambda, +\infty),$$

as required. □

In relation to Proposition 3.3.4(v), we also prove the following result.

Lemma 3.3.5. *If $x \in LS_h(\mathcal{M})$ and $p \in P(\mathcal{M})$ are such that $pxp \geq \lambda p$ for some $\lambda \in \mathbb{R}$, then*

$$p \preceq E_x[\lambda, \infty), \quad E_x(-\infty, \lambda) \preceq \mathbf{1} - p.$$

Proof. It is sufficient to show that inequality $pxp \geq 0$ implies that

$$p \preceq E_x[0, \infty), \quad E_x(-\infty, 0) \preceq \mathbf{1} - p.$$

For simplicity, we denote $e = E_x(-\infty, 0]$ and define $q = p \wedge e$. Since $pxp \geq 0$ and $q \leq p$, it follows from Proposition 3.3.4(ii) that $qxq = qpxpq \geq 0$. Since $exe = -x_- \leq 0$ and $q \leq e$, appealing again to Proposition 3.3.4(ii), we conclude that $qxq \leq 0$. Thus, $qxq = 0$.

We have $x_- = -xe$ and, therefore, $x_-q = -xeq = -xq$. Hence, $qx_-q = -qxq = 0$. In other words, $|x_-^{\frac{1}{2}}q| = 0$. This implies that $x_-^{\frac{1}{2}}q = 0$, and so $x_-q = 0$. Therefore,

$$xq = xeq = -x_-q = 0,$$

i. e., $q \leq (\mathbf{1} - s(x)) = n(x)$.

On the other hand, by Kaplansky's identity, we have

$$p - q = p - p \wedge e \sim p \vee e - e \leq \mathbf{1} - e,$$

i. e., $p - q \preceq \mathbf{1} - e$. Since the projections q and $p - q$ are orthogonal, and the projections $\mathbf{1} - e = E_x(0, \infty)$ and $n(x)$ are orthogonal, Theorem 1.9.4(vi) implies that

$$p = (p - q) + q \preceq (\mathbf{1} - e) + n(x) = E_x[0, \infty),$$

which proves the first assertion.

To prove the second assertion, we denote $f = E_x(-\infty, 0)$ and let $r = p \wedge f$. Since $f = E_x(-\infty, 0) \leq E_x(-\infty, 0] = e$, it follows that $r = p \wedge f \leq p \wedge e = q$. Using the inequality $q \leq n(x)$, we obtain that $r \leq n(x)$. However, $r \leq f$ and, therefore, $r \leq f \wedge n(x) = 0$. Using Kaplansky's identity once again, we conclude that

$$f = f - p \wedge f \sim p \vee f - p \leq \mathbf{1} - p,$$

as required. □

Next, we show that $LS_h(\mathcal{M})$ equipped with the partial order $\leq$ is a conditionally monotone complete.

Theorem 3.3.6. *If $\{x_i\}_{i \in I}$ is an increasing net of operators from $LS_h(\mathcal{M})$ and there exists $x \in LS_h(\mathcal{M})$, such that $x_i \leq x$ for every $i \in I$, then $\sup_{i \in I} x_i$ exists in $LS_h(\mathcal{M})$. Similarly, if $\{y_i\}_{i \in I}$ is a decreasing net of operators from $LS_h(\mathcal{M})$ and there exists $y \in LS_h(\mathcal{M})$, such that $y \leq y_i$ for every $i \in I$, then $\inf_{i \in I} y_i$ exists in $LS_h(\mathcal{M})$.*

Proof. We prove the statement only for the existence of the least upper bound, since the statement for the greatest lower bound can be proven similarly.

Without loss of generality, we may assume that $\mathbf{1} \leq x_i \leq x$ for all $i \in I$. By Corollary 3.3.3, we have that x is invertible in $LS(\mathcal{M})$ and $x^{-1} \in \mathcal{M}$. Furthermore, by (3.5), we have that

$$x^{-1/2} = (x^{1/2})^{-1} = (x^{-1})^{1/2}.$$

By Proposition 3.3.4(ii), the mapping $\Phi : LS_h(\mathcal{M}) \to LS_h(\mathcal{M})$, defined by setting

$$\Phi(y) = x^{-1/2} y x^{-1/2}, \quad y \in LS_h(\mathcal{M}),$$

preserves the partial order. Note that Φ is bijective with inverse defined by

$$\Phi^{-1}(y) = x^{1/2} y x^{1/2}, \quad y \in LS_h(\mathcal{M}).$$

Since the net $\{x_i\}_{i \in I}$ increases and $x_i \leq x$, it follows that the net $\{\Phi(x_i)\}_{i \in I}$ increases and $0 \leq \Phi(x_i) \leq \Phi(x) = \mathbf{1}$, in particular, $\Phi(x_i) \in \mathcal{M}$ for every $i \in I$. Since the von

Neumann algebra $\mathcal{M}$ is conditionally monotone complete (see Theorem 1.8.3), it follows that there exists $a = \sup_{i\in I} \Phi(x_i)$ in $\mathcal{M}_h$.

Since the mapping $\Phi : LS_h(\mathcal{M}) \to LS_h(\mathcal{M})$ is a bijection, we may define $b = \Phi^{-1}(a) \in LS_h(\mathcal{M})$, which is the least upper bound in $LS_h(\mathcal{M})$ of the increasing net $\{\Phi^{-1}(\Phi(x_i))\} = \{x_i\}$, i. e., $b = \sup_{i\in I} x_i$. □

In the next proposition, we collect several properties of the partially ordered set $(LS_h(\mathcal{M}), \leq)$ related to its conditional monotone completeness.

Proposition 3.3.7.
(i) *If $x \in LS_+(\mathcal{M})$, $\{\varepsilon_n\}_{n\in\mathbb{N}} \subset \mathbb{R}$, $\varepsilon_n \downarrow 0$, then $\varepsilon_n x \downarrow 0$;*
(ii) *If $\{x_i\}_{i\in I} \subset LS_h(\mathcal{M})$, $x_i \uparrow x \in LS_h(\mathcal{M})$, $a \in LS(\mathcal{M})$, then $(a^* x_i a) \uparrow (a^* x a)$;*
(iii) *If $x \in LS_h(\mathcal{M})$, $\{p_i\}_{i\in I} \subset P(\mathcal{M})$, $p_i \uparrow p \in P(\mathcal{M})$ and $p_i x p_i = 0$ for all $i \in I$, then $pxp = 0$. In addition, if $x \geq 0$, then $px = xp = 0$.*

Proof. (i) It is sufficient to show that for every $x \in LS_+(\mathcal{M})$ the convergence $n^{-1}x \downarrow 0$ holds as $n \to \infty$. Since the sequence $\{n^{-1}x\}_{n\in\mathbb{N}}$ decreases and $n^{-1}x \geq 0$, Theorem 3.3.6 implies that there exists $y \in LS_+(\mathcal{M})$, such that $n^{-1}x \downarrow y$ for all $n \in \mathbb{N}$. The sequence of positive operators $\{ny\}$ increases and $ny \leq x$. Using again Theorem 3.3.6, we obtain that $\{ny\} \uparrow a$ for some $a \in LS_+(\mathcal{M})$. Consequently,

$$a + y = \sup_{n\in\mathbb{N}}\{ny + y\} = \sup_{n\in\mathbb{N}}\{(n+1)y\} = a,$$

which implies the equality $y = 0$, as required.

(ii) Without loss of generality, we may assume that $0 \leq x_i \leq x$. Since $x_i \leq x$, Proposition 3.3.4(ii) implies that $a^* x_i a \leq a^* x a$ for all $i \in I$. By Theorem 3.3.6, we have that $y := \sup_i (a^* x_i a) \in LS(\mathcal{M})$ and referring to Proposition 3.3.4(ii) again we note the inequality

$$y := \sup_i (a^* x_i a) \leq a^* x a.$$

Assume first that a is an invertible operator in the $*$-algebra $LS(\mathcal{M})$. Then the inequality $a^* x_i a \leq y$ implies that

$$x_i = (a^{-1})^* (a^* x_i a) a^{-1} \leq (a^{-1})^* y a^{-1}.$$

Hence, $x = \sup_i x_i \leq (a^{-1})^* y a^{-1}$ for all $i \in I$, which implies the inequality

$$a^* x a \leq a^* (a^{-1})^* y a^{-1} a = y.$$

Thus, we obtain the desired equality $\sup_i (a^* x_i a) = y = a^* x a$.

Next, assume that $a \in LS(\mathcal{M})$ is positive. Fix a natural number n. By Corollary 3.3.3, the positive locally measurable operator $b_n = a + n^{-1}\mathbf{1}$ is invertible. By the proven above, the equality $b_n x b_n = \sup_i b_n x_i b_n$ holds for every $n \in \mathbb{N}$. Since $(\mathbf{1} - a)x_i(\mathbf{1} - a) \geq 0$, we have $x_i a + a x_i \leq (a x_i a + x_i)$. Therefore,

$$\begin{aligned} b_n x_i b_n &= (a + n^{-1}\mathbf{1})x_i(a + n^{-1}\mathbf{1}) = a x_i a + n^{-1}(x_i a + a x_i) + n^{-2} x_i \\ &\leq a x_i a + n^{-1}(a x_i a + x_i) + n^{-2} x_i \end{aligned}$$

for every $n \in \mathbb{N}$. Thus,

$$\begin{aligned} b_n x b_n &= \sup_i b_n x_i b_n \leq \sup_i a x_i a + \sup_i (n^{-1}(a x_i a + x) + n^{-2} x_i) \\ &\leq \sup_i a x_i a + n^{-1}(axa + x) + n^{-2} x. \end{aligned} \tag{3.6}$$

On the other hand, the inequality $(a + \mathbf{1})x(a + \mathbf{1}) \geq 0$ implies that $ax + xa \geq -axa - x$ and, therefore,

$$\begin{aligned} b_n x b_n &= (a + n^{-1}\mathbf{1})x(a + n^{-1}\mathbf{1}) = axa + n^{-1}(ax + xa) + n^{-2} x \\ &\geq axa - n^{-1}(axa + x) + n^{-2} x. \end{aligned} \tag{3.7}$$

Combining (3.6) and (3.7), we conclude that

$$axa \leq \sup_i a x_i a + 2n^{-1}(axa + x), \quad n \in \mathbb{N}.$$

By part (i), we have that $n^{-1}(axa + x) \downarrow 0$ and, therefore, $axa \leq \sup_i a x_i a \leq axa$, i. e., the equality $axa = \sup_i a x_i a$ holds.

Finally, let a be an arbitrary operator from $LS(\mathcal{M})$. For the positive operator $h = (a^* x a - y) \in LS_+(\mathcal{M})$, we have

$$a^* x_i a + h \leq y + h = a^* x a.$$

Multiplying the last equality on the left by a and on the right by a^*, we obtain

$$a a^* x_i a a^* + a h a^* \leq a a^* x a a^*.$$

Since the operator aa^* is positive, by the proved above we have that $\sup_i a a^* x_i a a^* = a a^* x a a^*$ and, therefore,

$$a a^* x a a^* + a h a^* \leq a a^* x a a^*,$$

which implies the equality

$$a h^{1/2} (a h^{1/2})^* = a h a^* = 0.$$

Hence, $a h^{1/2} = 0$. Since $y \in LS_+(\mathcal{M})$, the inequality $h = (a^* x a - y) \leq a^* x a$ holds, which in turn, implies that

$$0 \le h^2 = h^{1/2} \cdot h \cdot h^{1/2} \le h^{1/2} \cdot a^* x a \cdot h^{1/2} = 0,$$

i. e., $h = 0$. It means that $\sup_i a^* x_i a = a^* x a$, as required.

(iii) Since $p_i x p_i = 0$ for all $i \in I$ and $p_i \le p_j$, $i \le j, i, j \in I$, it follows that $p_i x p_j = p_i p_j x p_j = 0$ for all $j \ge i$, i. e., $\mathbf{1} - s_r(p_i x) \ge p_j$ for all j. Therefore,

$$\mathbf{1} - s_r(p_i x) \ge \sup_j p_j = p.$$

The latter inequality implies that $p_i x p = p_i x(\mathbf{1} - s_r(p_i(x)))p = 0$.

Since x is self-adjoint, it follows that $p x p_i = 0$ for all $i \in I$, i. e., $\mathbf{1} - s_r(px) \ge p_i$ for all $i \in I$. Thus,

$$\mathbf{1} - s_r(px) \ge \sup_i p_i = p,$$

implying that $pxp = 0$.

If $x \ge 0$, then $(x^{1/2}p)^*(x^{1/2}p) = pxp = 0$ and, therefore, $x^{1/2}p = 0$, which implies that $xp = 0$. □

The celebrated Heinz inequality (see, e. g., [140, Proposition 10.14]) states that the function $t \mapsto t^\alpha$ is operator-monotone for $0 \le \alpha \le 1$, i. e., if x, y are positive self-adjoint operators acting in H and $x \le y$, then $x^\alpha \le y^\alpha$. The following proposition establishes the connection between locally measurable operators $x^{1/2}$ and $y^{1/2}$ provided that $0 \le x \le y$, $x, y \in LS(\mathcal{M})$.

Proposition 3.3.8. *Suppose that $x, y \in LS_h(\mathcal{M})$ and $0 \le x \le y$. Then $x^{1/2} = ay^{1/2}$ for some $a \in \mathcal{M}$ with $\|a\|_{\mathcal{M}} \le 1$.*

Proof. By Proposition 3.3.4(v), we have that $s(x) \le s(y)$. Therefore, passing, if necessary, to the reduced algebra $s(y)\mathcal{M}s(y)$ we may assume that $s(y) = \mathbf{1}$.

For every $n \in \mathbb{N}$, we denote $p_n = E_y[1/n, n]$. By the spectral theorem, we have that $p_n \uparrow s(y) = \mathbf{1}$ and, therefore, the linear subspace $H_0 = \bigcup_{n \in \mathbb{N}} p_n(H)$ is dense in H and $H_0 \subset \operatorname{dom}(y) \cap \operatorname{dom}(y^{1/2}) \subset \operatorname{dom}(y^{1/2})$. Furthermore, Proposition 3.3.4(ii) implies that $0 \le p_n x p_n \le p_n y p_n \le n p_n$. Therefore,

$$p_n x p_n = (x^{1/2} p_n)^* \cdot x^{1/2} p_n \in \mathcal{M}$$

and

$$\|(x^{1/2} p_n)^* \cdot x^{1/2} p_n\|_{\mathcal{M}} = \|p_n x p_n\|_{\mathcal{M}} \le n.$$

Hence, $x^{1/2} p_n \in \mathcal{M}$ and $\|x^{1/2} p_n\|_{\mathcal{M}} \le \sqrt{n}$ for all $n \in \mathbb{N}$. In particular, $H_0 \subset \operatorname{dom}(x^{1/2})$.

Since $y^{1/2} p_n \le \sqrt{n} p_n$ and

$$y^{1/2}(p_n(H)) = p_n y^{1/2}(p_n(H)) \subset p_n(H)$$

for all $n \in \mathbb{N}$, we have $y^{1/2}(H_0) \subset H_0$. Thus, we may define a linear mapping $b : y^{1/2}(H_0) \to H$ by setting

$$b(y^{1/2}(\xi)) = x^{1/2}(\xi), \quad \xi \in H_0.$$

If $\xi \in H_0$ is such that $y^{1/2}(\xi) = 0$, then

$$\|x^{1/2}(\xi)\|_H^2 = \langle x^{1/2}(\xi), x^{1/2}(\xi)\rangle = \langle x(\xi), \xi\rangle \leq \langle y(\xi), \xi\rangle = \|y^{1/2}(\xi)\|_H^2$$

implying that $x^{1/2}(\xi) = 0$. Therefore, the operator b is well-defined. In addition, for every $\xi \in H_0$, we have

$$\|b(y^{1/2}(\xi))\|_H^2 = \|x^{1/2}(\xi)\|_H^2 \leq \|y^{1/2}(\xi)\|_H^2,$$

i. e., b is a continuous linear operator on $y^{1/2}(H_0)$ and $\|b\|_{y^{1/2}(H_0)\to H} \leq 1$.

Since H_0 is dense in H, the operator b extends uniquely to a linear operator $a : H \to H$, such that $\|a\|_{\mathcal{M}} \leq 1$. Furthermore, $ay^{1/2}(\xi) = x^{1/2}(\xi)$ for all $\xi \in H_0$, by construction.

If u is a unitary operator from the commutant $\mathcal{M}'$, then $u(p_n(H)) = p_n(H)$ for all $n \in \mathbb{N}$ and, therefore, $u(H_0) = H_0$. If $\eta \in H_0$, then $\eta = y^{1/2}(\xi)$ for some $\xi \in H_0$ and

$$\begin{aligned} u^{-1}au(\eta) &= u^{-1}auy^{1/2}(\xi) = u^{-1}ay^{1/2}u(\xi) \\ &= u^{-1}x^{1/2}u(\xi) = u^{-1}ux^{1/2}(\xi) = x^{1/2}(\xi) = ay^{1/2}(\xi) = a(\eta). \end{aligned}$$

Therefore, $u^{-1}au = a$, i. e., $a\eta\mathcal{M}$. Since a is bounded, it follows that $a \in \mathcal{M}$.

Since $p_n ay^{1/2}p_n = p_n x^{1/2}p_n$ for all $n \in \mathbb{N}$ and $p_n \uparrow \mathbf{1}$, by Proposition 3.3.7(iii), we have $ay^{1/2} = x^{1/2}$. □

We note a useful corollary of Proposition 3.3.8.

Corollary 3.3.9. *Suppose that $x, y \in LS_h(\mathcal{M})$ and $0 \leq x \leq y$. Then $x = aya^*$ for some $a \in \mathcal{M}$ with $\|a\|_{\mathcal{M}} \leq 1$.*

Proof. By Proposition 3.3.8, there exists $a \in \mathcal{M}$ with $\|a\|_{\mathcal{M}} \leq 1$, such that $x^{1/2} = ay^{1/2}$. Therefore,

$$x = x^{1/2}(x^{1/2})^* = ay^{1/2} \cdot y^{1/2}a^* = aya^*,$$

as required. □

Next, we present some properties of $\mathcal{M}$-bimodules related to the partial order $\leq$.

Proposition 3.3.10. *Let $\mathcal{E} \subseteq LS(\mathcal{M})$ be an F-normed $\mathcal{M}$-bimodule and let $x \in LS(\mathcal{M})$ and $y \in \mathcal{E}$ be such that $|x| \leq |y|$. Then $x \in \mathcal{E}$ and $\|x\|_{\mathcal{E}} \leq \|y\|_{\mathcal{E}}$.*

Proof. By Corollary 3.3.9, there exists $a \in \mathcal{M}$, such that $|x| = a|y|a^*$ and $\|a\|_{\mathcal{M}} \leq 1$. Referring to Proposition 3.1.6(i), we have that $x \in \mathcal{E}$. Furthermore, by Proposition 3.1.10, we obtain that

$$\|x\|_{\mathcal{E}} = \||x|\|_{\mathcal{E}} = \|a|y|a^*\|_{\mathcal{E}} \leq \|\|a\|_{\mathcal{M}}\|a^*\|_{\mathcal{M}}|y|\|_{\mathcal{E}} \leq \||y|\|_{\mathcal{E}} = \|y\|_{\mathcal{E}},$$

as required. □

In the following propositions, we collect properties of F-normed bimodules in relation to projections in $\mathcal{M}$. For an $\mathcal{M}$-bimodule $\mathcal{E}$, we denote

$$P(\mathcal{E}) = \{p \in P(\mathcal{M}) : p \in \mathcal{E}\}.$$

Proposition 3.3.11. *Let $(\mathcal{E}, \|\cdot\|_{\mathcal{E}})$ be an F-normed $\mathcal{M}$-bimodule. Then:*
(i) *if $p \in P(\mathcal{E})$ and $q \in P(\mathcal{M})$, such that $q \precsim p$, then $q \in P(\mathcal{E})$ and $\|q\|_{\mathcal{E}} \leq \|p\|_{\mathcal{E}}$;*
(ii) *if $p_1, \ldots, p_n \in P(\mathcal{E})$, then $\sup_{i=1,\ldots,n} p_i \in P(\mathcal{E})$ and $\|\sup_{i=1,\ldots,n} p_i\|_{\mathcal{E}} \leq \sum_{i=1}^n \|p_i\|_{\mathcal{E}}$. If $\|\cdot\|_{\mathcal{E}}$ is a quasi-norm on $\mathcal{E}$, then $\|\sup_{i=1,\ldots,n} p_i\|_{\mathcal{E}} \leq \sum_{i=1}^n C_{\mathcal{E}}^i \|p_i\|_{\mathcal{E}}$, where $C_{\mathcal{E}}$ is modulus of concavity of the quasi-norm $\|\cdot\|_{\mathcal{E}}$.*

Proof. (i) If $q \precsim p$, then there exists $q_1 \in P(\mathcal{M})$, such that $q_1 \leq p$ and $q \sim q_1$. It follows from $q_1 \leq p$ that $q_1 = pq_1 \in P(\mathcal{E})$. In particular, by Proposition 3.3.10, we have that $\|q_1\|_{\mathcal{E}} \leq \|p\|_{\mathcal{E}}$. It remains to show that $q \in \mathcal{E}$ and $\|q\|_{\mathcal{E}} = \|q_1\|_{\mathcal{E}}$.

Let $v \in \mathcal{M}$ be a partial isometry, such that $q_1 = v^*v$ and $q = vv^*$. This implies that $q_1 = v^*qv$ and $q = vq_1v^*$. Hence, $q \in \mathcal{E}$ and

$$\|q\|_{\mathcal{E}} = \|vq_1v^*\|_{\mathcal{E}} \leq \|\|v\|_{\mathcal{M}}\|v^*\|_{\mathcal{M}}q\|_{\mathcal{E}} \leq \|q\|_{\mathcal{E}}.$$

Similarly, $q_1 = v^*qv$ implies that $\|q_1\|_{\mathcal{E}} \leq \|q\|_{\mathcal{E}}$ and so, $\|q\|_{\mathcal{E}} = \|q_1\|_{\mathcal{E}}$, as required.

(ii) We prove both the cases of an F-norm and a quasi-norm simultaneously. For this purpose, for an F-norm $\|\cdot\|_{\mathcal{E}}$, we denote $C_{\mathcal{E}} = 1$.

We first show that if $p, q \in P(\mathcal{E})$, then $p \vee q \in P(\mathcal{E})$ and $\|p \vee q\|_{\mathcal{E}} \leq C_{\mathcal{E}}(\|p\|_{\mathcal{E}} + \|q\|_{\mathcal{E}})$. Since $p - p \wedge q \leq p$, it is clear that $p - p \wedge q \in P(\mathcal{E})$ and $\|p - p \wedge q\|_{\mathcal{E}} \leq \|p\|_{\mathcal{E}}$. Kaplansky's identity (see Theorem 1.9.4(v)) and part (i) imply that $p \vee q - q \sim p - p \wedge q \in P(\mathcal{E})$ and $\|p \vee q - q\|_{\mathcal{E}} = \|p - p \wedge q\|_{\mathcal{E}}$. Since $p \vee q = (p \vee q - q) + q$, it is now clear that $p \vee q \in P(\mathcal{E})$ and

$$\|p \vee q\|_{\mathcal{E}} \leq C_{\mathcal{E}}(\|p \vee q - q\|_{\mathcal{E}} + \|q\|_{\mathcal{E}}) \leq C_{\mathcal{E}}(\|p\|_{\mathcal{E}} + \|q\|_{\mathcal{E}}),$$

as required.

Assume now that $p_1, \ldots, p_n \in P(\mathcal{E})$. Proceeding by induction, we have that $\sup_{i=1,\ldots,n} p_i \in P(\mathcal{E})$ and

$$\Big\|\sup_{1\le i\le n} p_i\Big\|_{\mathcal{E}} = \Big\|p_1 \vee \Big(\sup_{2\le i\le n} p_i\Big)\Big\|_{\mathcal{E}} \le C_{\mathcal{E}}\|p_1\|_{\mathcal{E}} + C_{\mathcal{E}}\Big\|\sup_{1\le i\le n-1} p_{i+1}\Big\|_{\mathcal{E}}$$
$$\le C_{\mathcal{E}}\|p_1\|_{\mathcal{E}} + C_{\mathcal{E}}\sum_{i=1}^{n-1} C_{\mathcal{E}}^i\|p_{i+1}\|_{\mathcal{E}} = \sum_{i=1}^{n} C_{\mathcal{E}}^i\|p_i\|_{\mathcal{E}}.$$

This completes the proof. □

By means of the partial order in $LS_h(\mathcal{M})$, we distinguish fundamental class of absolutely solid $*$-subalgebras of the $*$-algebra $LS(\mathcal{M})$. As we will show in Corollary 3.3.15, all other algebras of unbounded operators introduced so far, are examples of absolutely solid $*$-subalgebras of $LS(\mathcal{M})$.

Definition 3.3.12. A $*$-subalgebra $\mathcal{A}$ in $LS(\mathcal{M})$ is said to be an *absolutely solid $*$-subalgebra* if $|x| \le |y|$, $x \in LS(\mathcal{M})$, $y \in \mathcal{A}$ implies that $x \in \mathcal{A}$.

One can easily see that the von Neumann algebra $\mathcal{M}$ is an absolutely solid $*$-subalgebra of $LS(\mathcal{M})$.

It is clear that for an absolutely solid $*$-subalgebra $\mathcal{A}$ we have that $x \in \mathcal{A}$ if and only if $|x| \in \mathcal{A}$. In fact, as the following proposition shows, this is one of the necessary and sufficient conditions for a $*$-subalgebra to be absolutely solid.

Proposition 3.3.13. *For a $*$-subalgebra $\mathcal{A}$ of $LS(\mathcal{M})$, the following conditions are equivalent:*
(i) *$\mathcal{A}$ is an absolutely solid $*$-subalgebra of $LS(\mathcal{M})$;*
(ii) *For $x \in LS(\mathcal{M})$, the operator $|x|$ is in $\mathcal{A}$ if and only if $x \in \mathcal{A}$;*
(iii) *$axb \in \mathcal{A}$ for any $x \in \mathcal{A}$ and $a, b \in \mathcal{M}$.*

Proof. The implication (i)⇒(ii) is evident.

(ii)⇒(iii). If $x \in \mathcal{A}$ and u is a unitary operator from $\mathcal{M}$, then $|ux| = |x| \in \mathcal{A}$, which implies $ux \in \mathcal{A}$. Since $\mathcal{A}$ is $*$-subalgebra, we have that $x^*u^* = (ux)^* \in \mathcal{A}$. Since $\mathcal{A}$ is a subspace of $LS(\mathcal{M})$ and every element from $\mathcal{M}$ is linear combination of four unitary operators from $\mathcal{M}$, it follows that ax, $xa \in \mathcal{A}$ for all $a \in \mathcal{M}$, $x \in \mathcal{A}$, as required.

(iii)⇒(i). Let $|x| \le |y|$, $x \in LS(\mathcal{M})$ and $y \in \mathcal{A}$. If $y = u|y|$ is the polar decomposition of y, then $|y| = u^*y$. Therefore, since $u \in \mathcal{M}$ and $y \in \mathcal{A}$, the assumption implies that $y \in \mathcal{A}$. By Corollary 3.3.9, there exists an operator $a \in \mathcal{M}$, such that $|x| = a|y|a^*$. By assumption, $a|y|a^* \in \mathcal{A}$ and, therefore, $|x| \in \mathcal{A}$. Using the polar decomposition for $x = u|x|$ and referring to the assumption once again, we obtain that $x \in \mathcal{A}$, as required. □

Remark 3.3.14. *Proposition* 3.3.13(iii) *implies that any absolutely solid $*$-subalgebra of $LS(\mathcal{M})$ is an $\mathcal{M}$-bimodule. In particular, one can introduce absolutely solid subalgebras as $\mathcal{M}$-bimodules, which are simultaneously $*$-subalgebras of $LS(\mathcal{M})$.*

Using Proposition 3.3.13, it is easy to show that the $*$-algebras $S(\mathcal{M})$, and $S(\mathcal{M}, \tau)$, $S_0(\mathcal{M})$, $S_0(\mathcal{M}, \tau)$ are absolutely solid $*$-subalgebras of $LS(\mathcal{M})$.

Corollary 3.3.15. *The $*$-algebras $S(\mathcal{M})$, $S_0(\mathcal{M})$, $S(\mathcal{M},\tau)$, and $S_0(\mathcal{M},\tau)$ are absolutely solid $*$-subalgebras of $LS(\mathcal{M})$.*

Proof. Since $\mathcal{M}$ is a $*$-subalgebra of $S(\mathcal{M})$, it follows that $axb \in S(\mathcal{M})$ (resp., $axb \in S(\mathcal{M},\tau)$) for any $x \in S(\mathcal{M})$ (resp., in $S(\mathcal{M},\tau)$) and $a, b \in \mathcal{M}$. Therefore, the assertions for $S(\mathcal{M})$ and $S(\mathcal{M},\tau)$ follow from Proposition 3.3.13(iii).

It immediately follows from the definition that an operator x is compact (resp., τ-compact) if and only if $|x|$ is compact (resp., τ-compact). Therefore, the assertion for $S_0(\mathcal{M})$ and $S_0(\mathcal{M},\tau)$ follows from Proposition 3.3.13(ii). □

Remark 3.3.16. *We note that if $\mathcal{A}$ is an absolutely solid $*$-subalgebra of $LS(\mathcal{M})$ and $x \in \mathcal{A}$ is self-adjoint, then $E_x(\lambda,\infty), E_x(-\infty,-\lambda) \in \mathcal{A}$ for any $\lambda > 0$. Indeed, since $\lambda > 0$, the function $f(t) = \frac{1}{t}\chi_{(\lambda,\infty)}(t)$ is bounded, and so $f(x) \in \mathcal{M}$. Writing $E_x(\lambda,\infty) = x \cdot f(x)$ and referring to Proposition* 3.3.13(iii), *we infer that $E_x(\lambda,\infty) \in \mathcal{A}$. A similar argument ensures that $E_x(-\infty,-\lambda) \in \mathcal{A}$.*

In conclusion of this section, we prove a convenient necessary condition for the inclusion $x \in LS_h(\mathcal{M})$ via partial order on $LS_h(\mathcal{M})$.

Proposition 3.3.17. *Let $x \in LS_h(\mathcal{M})$. If $\{p_n\}_{n\in\mathbb{N}} \subset P(\mathcal{M})$ is a sequence of pairwise equivalent projections, such that*

$$p_n x p_n \geq n p_n, \quad n \in \mathbb{N},$$

then $p_n = 0$ for all $n \in \mathbb{N}$.

Proof. For every $n \in \mathbb{N}$, we define

$$q_n = p_n \wedge E_x(-\infty, n-1).$$

By assumption, $p_n x p_n \geq n p_n$ and, therefore, Proposition 3.3.4(ii) implies that

$$q_n x q_n = q_n p_n x p_n q_n \geq n q_n p_n q_n = n q_n.$$

The spectral theory guarantees that

$$xE_x(-\infty, n-1) \leq (n-1)E_x(-\infty, n-1), \quad n \in \mathbb{N},$$

and so referring to Proposition 3.3.4(ii) again, we obtain that $q_n x q_n \leq (n-1)q_n$ for all $n \in \mathbb{N}$. Thus, $nq_n \leq (n-1)q_n$, $n \in \mathbb{N}$, implying that $q_n = p_n \wedge E_x(-\infty, n-1) = 0$, $n \in \mathbb{N}$.

By Kaplansky's identity, we have

$$\begin{aligned} p_n = p_n - (p_n \wedge E_x(-\infty, n-1)) &\sim (p_n \vee E_x(-\infty, n-1)) - E_x(-\infty, n-1) \\ &\leq \mathbf{1} - E_x(-\infty, n-1) = E_x[n-1, \infty), \end{aligned}$$

i. e., $p_n \precsim E_x[n-1,\infty)$ for all $n \in \mathbb{N}$. Since the projections $p_n, n \in \mathbb{N}$ are pairwise equivalent, it follows that

$$p_1 \precsim E_x[n-1,\infty), \quad n \in \mathbb{N}.$$

By Proposition 1.11.6(iv), we have that

$$\mathcal{D}(p_1) \leq \mathcal{D}(E_x[n-1,\infty)), \quad n \in \mathbb{N}.$$

Since $x \in LS(\mathcal{M})$, Proposition 2.3.5(iii) implies that $\mathcal{D}(E_x[n-1,\infty)) \downarrow 0$ almost everywhere as $n \to \infty$. Therefore, taking limit $n \to \infty$, we obtain that $\mathcal{D}(p_1) = 0$, proving that $p_1 = 0$. Since the projections $p_n, n \in \mathbb{N}$ are pairwise equivalent, it follows that $p_n = 0$ for all $n \in \mathbb{N}$. □

3.4 Direct integrals of locally measurable operators

In this section, we show that for a decomposable von Neumann algebra $\mathcal{M} = \int_Z^\oplus \mathcal{M}_\zeta \, dw(\zeta)$, acting on a direct integral Hilbert space $\int_Z^\oplus H_\zeta \, dw(\zeta)$, any locally measurable operators affiliated with $\mathcal{M}$ is a decomposable operator associated with a measurable field $\zeta \to x_\zeta$ of locally measurable operators affiliated with $\mathcal{M}_\zeta$, i. e.,

$$LS(\mathcal{M}) = \int_Z^\oplus LS(\mathcal{M}_\zeta) \, dw(\zeta).$$

Decomposing $\mathcal{M}$ over its center (so that a. e. $\mathcal{M}_\zeta$ is a factor) allows us then to write any $x \in LS(\mathcal{M})$ as decomposable operator associated with a measurable field $\zeta \to x_\zeta$ of measurable operators $x_\zeta \in S(\mathcal{M}_\zeta)$.

We refer the reader to Section 1.12 and Section 1.13 for the background material on direct integrals of von Neumann algebras and unbounded operators.

Throughout this section, we let w to be a fixed σ-finite positive measure on a standard Borel space Z and assume that $H = \int_Z^\oplus H_\zeta \, dw(\zeta)$ is a direct integral Hilbert space (see Definition 1.12.2).

As in Section 1.12, let us denote by $\mathcal{R}$ the von Neumann algebra of all bounded decomposable operators in H and by $\mathcal{C}$ the von Neumann algebra of all bounded diagonalizable operators. By Theorem 1.12.6, the algebra $\mathcal{C}$ is the commutant of $\mathcal{R}$ and so, using the notion of affiliated operators, Theorem 1.13.14 can be reformulated as follows.

Theorem 3.4.1. *A closed linear operator in H is decomposable if and only if it is affiliated to the von Neumann algebra of all bounded decomposable operators.*

Next, we describe operators affiliated with a decomposable von Neumann algebra.

Proposition 3.4.2. *Suppose*

$$\mathcal{M} = \int_Z^{\oplus} \mathcal{M}_\zeta \, dw(\zeta)$$

is a decomposable von Neumann algebra and let x be a closed densely defined operator in H. Then x is affiliated to $\mathcal{M}$ if and only if:
(i) *x is decomposable;*
(ii) *writing out the decomposition as $x = \int_Z^{\oplus} x_\zeta \, dw(\zeta)$ we have that x_ζ is affiliated to $\mathcal{M}_\zeta$ for almost every ζ.*

Proof. Suppose first that x is a decomposable operator with decomposition $x = \int_Z^{\oplus} x_\zeta \, dw(\zeta)$ given by a measurable field $\zeta \to x_\zeta$ of closed operators, such that x_ζ is affiliated with $\mathcal{M}_\zeta$ for almost all ζ. Since x is densely defined, Proposition 1.13.17 guarantees that x_ζ is densely defined for almost all ζ.

Let $x = v|x|$ and $x_\zeta = v_\zeta |x_\zeta|$ be the polar decompositions of x and x_ζ, respectively. Theorem 1.13.25 implies that $v = \int_Z^{\oplus} v_\zeta \, dw(\zeta)$. Since x_ζ is affiliated with $\mathcal{M}_\zeta$ for almost all ζ, it follows that $v_\zeta \in \mathcal{M}_\zeta$ for almost all ζ, and so $v \in \mathcal{M}$.

By Proposition 1.13.23 for any Borel subset $B \subset \mathbb{R}$, we have that $E_{|x|}(B) = \int_Z^{\oplus} E_{|x_\zeta|}(B) \, dw(\zeta)$. Since $E_{|x_\zeta|}(B) \in \mathcal{M}_\zeta$ for almost every ζ, it follows that $E_{|x|}(B) \in \mathcal{M}$ for any Borel subset $B \subset \mathbb{R}$. Proposition 2.1.2(iv) guarantees that x is affiliated with $\mathcal{M}$.

Conversely, suppose that x is affiliated with $\mathcal{M}$. Let $x = v|x|$ be the polar decomposition of x. Since $v \in \mathcal{M}$ and all spectral projections $E_{|x|}(-\infty, \lambda), \lambda \in \mathbb{R}$ are in $\mathcal{M}$, they all commute with all bounded diagonalizable operators. By Theorem 1.4.8, it follows that $|x|$ commutes with all bounded diagonalizable operators and, therefore, $x = v|x|$ also commutes with all bounded diagonalizable operators. Theorem 1.13.14 implies that x is decomposable. Let $\zeta \to x_\zeta$ be a measurable field of closed operators, such that $x = \int_Z^{\oplus} x_\zeta \, dw(\zeta)$. By Proposition 1.13.17, almost every x_ζ is densely defined.

By Theorem 1.13.25, we have that $v = \int_Z^{\oplus} v_\zeta \, dw(\zeta)$, where $x_\zeta = v_\zeta |x_\zeta|$ is the polar decomposition of x_ζ for almost all ζ. Since $v \in \mathcal{M}$, it follows that $v_\zeta \in \mathcal{M}_\zeta$.

Using Proposition 1.13.23 for every Borel set B, we have that $E_{|x|}(B) = \int_Z^{\oplus} E_{|x_\zeta|}(B) \, dw(\zeta)$. Since $E_{|x|}(B) \in \mathcal{M}$, for every Borel subset B there is a null set N_B, such that $E_{|x_\zeta|}(B) \in \mathcal{M}_\zeta$ for all $\zeta \notin N_B$. Let N be the union of the sets N_B as B ranges over the open intervals with rational endpoints in $\mathbb{R}$. Then N is a null set and for all $\zeta \notin N$ we have $E_{|x_\zeta|}(a, b) \in \mathcal{M}_\zeta$ for all rational numbers $a < b$. Therefore, $E_{|x_\zeta|}(B) \in \mathcal{M}_\zeta$ for all Borel subsets $B \subset \mathbb{R}$ and all $\zeta \notin N$. Thus, we have that x_ζ is affiliated to $\mathcal{M}_\zeta$ for almost every ζ, which completes the proof. □

The following theorem is the main result of this section and gives a description of locally measurable operators affiliated with a decomposable von Neumann algebra.

Theorem 3.4.3. *Let*

$$\mathcal{M} = \int_Z^{\oplus} \mathcal{M}_\zeta \, dw(\zeta)$$

be a decomposable von Neumann algebra. Then $x \in LS(\mathcal{M})$ if and only if:
(i) *x is decomposable;*
(ii) *writing out the decomposition as $x = \int_Z^{\oplus} x_\zeta \, dw(\zeta)$ we have that $x_\zeta \in LS(\mathcal{M}_\zeta)$ for almost every ζ.*

Proof. Assume first that $x \in LS(\mathcal{M})$. Since x is affiliated with $\mathcal{M}$, it follows from Proposition 3.4.2 that

$$x = \int_Z^{\oplus} x_\zeta \, dw(\zeta),$$

where almost every x_ζ is affiliated with $\mathcal{M}_\zeta$. We claim that $x_\zeta \in LS(\mathcal{M}_\zeta)$ for almost every $\zeta \in Z$.

Since $x \in LS(\mathcal{M})$, Proposition 2.3.5(ii) guarantees that there exists an increasing sequence of central projections $\{z_n\}_{n\in\mathbb{N}} \in P(\mathcal{Z}(\mathcal{M}))$, such that $\sup_{n\in\mathbb{N}} z_n = \mathbf{1}$ and the projections $E_{|x|}(n,\infty)z_n$ are finite for $n \in \mathbb{N}$. By Theorem 1.13.25 and Proposition 1.13.23, we have that

$$E_{|x|}(n,\infty) = \int_Z^{\oplus} E_{|x_\zeta|}(n,\infty) \, dw(\zeta).$$

Writing $z_n = \int_Z^{\oplus} z_\zeta^{(n)} \, dw(\zeta)$ and referring to Proposition 1.12.5, we infer that

$$E_{|x|}(n,\infty)z_n = \int_Z^{\oplus} E_{|x_\zeta|}(n,\infty)z_\zeta^{(n)} \, dw(\zeta).$$

Proposition 1.12.12(i) implies that $E_{|x_\zeta|}(n,\infty)z_\zeta^{(n)}$ is finite for almost all ζ. Since, in addition, for almost all ζ we have

$$z_\zeta^{(n)} \in P(\mathcal{Z}(\mathcal{M}_\zeta)), \quad z_\zeta^{(n)} \uparrow \mathbf{1}_{H_\zeta}, n \to \infty,$$

referring to Proposition 2.3.5(ii) once again we conclude that for almost all ζ the operator x_ζ is locally measurable.

Conversely, suppose that $x = \int_Z^{\oplus} x_\zeta \, dw(\zeta)$ for some measurable field $\zeta \to x_\zeta$ with $x_\zeta \in LS(\mathcal{M}_\zeta)$ for almost all ζ. For every $n \in \mathbb{N}$, we set

$$e_n = E_{|x|}(n,\infty), \quad e_\zeta^{(n)} = E_{|x_\zeta|}(n,\infty).$$

By Theorem 1.13.25 and Proposition 1.13.23, we have that

$$e_n = \int_Z^{\oplus} e_\zeta^{(n)}\, dw(\zeta).$$

By Theorem 1.9.9(vi), there exists the unique central projection $z_n \in P(\mathcal{Z}(\mathcal{M}))$, such that $z_n e_n$ is properly infinite and $(\mathbf{1} - z_n)e_n$ is finite. For each $n \in \mathbb{N}$, we may write $z_n = \int_Z^{\oplus} z_\zeta^{(n)}\, dw(\zeta)$, for some measurable field $\zeta \to z_\zeta^{(n)}$ of operators. Note that by Proposition 1.12.5, $z_\zeta^{(n)}$ is almost everywhere a central projection. By Proposition 1.12.12(i) and (ii), we have that for each $n \in \mathbb{N}$, the projection $z_\zeta^{(n)} e_\zeta^{(n)}$ is properly infinite or zero and the projection $\mathbf{1} - z_\zeta^{(n)} e_\zeta^{(n)}$ is finite almost everywhere.

Setting

$$q_n = \mathbf{1} - s_c(z_n e_n), \quad q_\zeta^{(n)} = \mathbf{1} - s_c(z_\zeta^{(n)} e_\zeta^{(n)}), \quad n \in \mathbb{N},$$

and referring to Proposition 1.12.12(iii), we have

$$q_n = \int_Z^{\oplus} q_\zeta^{(n)}\, dw(\zeta). \tag{3.8}$$

Note that

$$e_n q_n = e_n - s_c(z_n e_n) e_n = (\mathbf{1} - z_n) e_n,$$

and so, by the choice of z_n, we have that $e_n q_n \in P_{\text{fin}}(\mathcal{M})$. Since $e_n \downarrow 0$, it follows that $z_n e_n \downarrow 0$ and, therefore, the sequences $\{q_n\}_{n\in\mathbb{N}}$ is increasing. It remains to show that $q_n \uparrow \mathbf{1}$.

Since $x_\zeta \in LS(\mathcal{M}_\zeta)$ for a. e. ζ, there exists $\{w_\zeta^{(m)}\}_{m\in\mathbb{N}} \subset P(\mathcal{Z}(\mathcal{M}_\zeta))$, such that $w_\zeta^{(m)} \uparrow \mathbf{1}$ and $e_\zeta^{(n_m)} w_\zeta^{(m)} \in P_{\text{fin}}(\mathcal{M}_\zeta)$ a. e. We have

$$e_\zeta^{(n_m)} w_\zeta^{(m)} = e_\zeta^{(n_m)} (\mathbf{1} - z_\zeta^{(n_m)}) w_\zeta^{(m)} + e_\zeta^{(n_m)} z_\zeta^{(n_m)} w_\zeta^{(m)},$$

and since both $e_\zeta^{(n_m)} w_\zeta^{(m)}$ and $\mathbf{1} - z_\zeta^{(n_m)} e_\zeta^{(n_m)}$ are finite, it follows that $z_\zeta^{(n_m)} e_\zeta^{(n_m)} w_\zeta^{(m)}$ is finite, too, for a. e. $\zeta \in Z$. Since $z_\zeta^{(n_m)} e_\zeta^{(n_m)}$ is properly infinite, it follows that $z_\zeta^{(n_m)} w_\zeta^{(m)} = 0$, and so

$$w_\zeta^{(m)} q_\zeta^{(n_m)} = w_\zeta^{(m)} - w_\zeta^{(m)} z_\zeta^{(n_m)} s_c(e_\zeta^{(n_m)}) = w_\zeta^{(m)}.$$

Thus, $w_\zeta^{(m)} \le q_\zeta^{(n_m)}$. Since $w_\zeta^{(m)} \uparrow \mathbf{1}$ and $q_\zeta^{(m)} \uparrow$, it follows that $q_\zeta^{(m)} \uparrow \mathbf{1}$, and so by (3.8) we conclude that $q_m \uparrow \mathbf{1}$, proving that $x \in LS(\mathcal{M})$. □

Remark 3.4.4. *To simplify notation, for a decomposable von Neumann algebra $\mathcal{M} = \int_Z^{\oplus} \mathcal{M}_\zeta \, dw(\zeta)$, we shall write*

$$LS(\mathcal{M}) = \int_Z^{\oplus} LS(\mathcal{M}_\zeta) \, dw(\zeta),$$

in the sense that any $x \in LS(\mathcal{M})$ can be written as in Theorem 3.4.3.

When $\mathcal{M}$ is decomposed over its center (i. e., (Z, w) such that $\mathcal{Z}(\mathcal{M}) \cong L_\infty(Z, w)$), then almost every $\mathcal{M}_\zeta$ is a factor (see Theorem 1.12.9) and, therefore, $LS(\mathcal{M}_\zeta) = S(\mathcal{M}_\zeta)$. Thus, Theorem 3.4.3 immediately implies the following.

Corollary 3.4.5. *Suppose that (Z, w), is such that $\mathcal{Z}(\mathcal{M}) \cong L_\infty(Z, w)$, so that $\mathcal{M} = \int_Z^{\oplus} \mathcal{M}_\zeta \, dw(\zeta)$ is decomposed over its center. Then*

$$LS(\mathcal{M}) = \int_Z^{\oplus} S(\mathcal{M}_\zeta) \, dw(\zeta),$$

i. e., $x \in LS(\mathcal{M})$ if and only if x is a decomposable operator associated with a measurable field of measurable operators affiliated with $\mathcal{M}_\zeta$.

Corollary 3.4.5 allows a constructive description of the algebra of locally measurable operators in some cases.

Example 3.4.6. *Let $Z = [0, 1]$ with the Lebesgue measure on $[0, 1]$ and let $\mathcal{M}_\zeta = \mathcal{B}(H)$, $\zeta \in Z$, for a fixed separable (infinite-dimensional) Hilbert space H. By Example 2.2.13, we have that $S(\mathcal{M}_\zeta) = \mathcal{B}(H)$ for all $\zeta \in Z$. Therefore, by Corollary 3.4.5, an operator x is locally measurable with respect to $\mathcal{M} = \int_{[0,1]}^{\oplus} \mathcal{B}(H) \, dw(\zeta)$ if and only if $x = \int_Z^{\oplus} x_\zeta \, dw(\zeta)$ for a (not necessarily bounded) measurable field $\zeta \to x_\zeta$ with $x_\zeta \in \mathcal{B}(H_\zeta)$.*

Now, we consider a particular case, when Z is discrete. Note that since (Z, w) is Polish, it follows that Z is countable.

Corollary 3.4.7. *Suppose that Z is discrete and let $\mathcal{M}_\zeta, \zeta \in Z$, be a collection of von Neumann algebras. We have*

$$LS\Big(\bigoplus_{\zeta \in Z} \mathcal{M}_\zeta\Big) = \prod_{\zeta \in Z} LS(\mathcal{M}_\zeta),$$

where the direct product $\prod_{\zeta \in Z} LS(\mathcal{M}_\zeta)$ is defined as

$$\prod_{\zeta \in Z} LS(\mathcal{M}_\zeta) = \{\{x_\zeta\}_{\zeta \in Z} : x_\zeta \in LS(\mathcal{M}_\zeta), \zeta \in Z\}.$$

Proof. If Z is discrete, then every field $\zeta \to x_\zeta$ of closed operators is measurable. In particular, the von Neumann algebra $\mathcal{M} = \int_Z^{\oplus} \mathcal{M}_\zeta \, dw(\zeta)$ coincides with the direct sum

$\bigoplus_{\zeta\in Z}\mathcal{M}_\zeta$, and by Theorem 3.4.3, an operator x is locally measurable with respect to $\mathcal{M} = \int_Z^{\oplus}\mathcal{M}_\zeta\, dw(\zeta)$ if and only if $x = \int_Z^{\oplus} x_\zeta d\omega(\zeta)$ for a field $\zeta \to x_\zeta$ with $x_\zeta \in LS(\mathcal{M}_\zeta)$. Thus, $x = \{x_\zeta\}$, as required. □

We note that the assertion of Corollary 3.4.7 does not hold for the algebras of all measurable operators. Indeed, if $\mathcal{M}_j = \mathcal{M}$ is a von Neumann algebra of type *III* for every $j \in \mathbb{N}$, then the direct sum $\mathcal{M} = \bigoplus_{j\in\mathbb{N}}\mathcal{M}_j$ of von Neumann algebras $\mathcal{M}_j$ is also a von Neumann algebra of type *III* and, therefore, by Theorem 2.5.3 we have that $S(\mathcal{M}) = \mathcal{M}$ and $S(\mathcal{M}_j) = \mathcal{M}_j$ for all $j \in \mathbb{N}$. For every nonzero $x \in \mathcal{M}$, we have

$$\{jx\}_{j\in\mathbb{N}} \in \prod_{j\in\mathbb{N}} S(\mathcal{M}_j), \quad \{jx\}_{j\in\mathbb{N}} \notin S\Big(\prod_{j\in\mathbb{N}}\mathcal{M}_j\Big).$$

In conclusion of this section, we prove that the strong sum and strong product for decomposable locally measurable operators are defined pointwise (see Proposition 1.12.5 for bounded decomposable operators).

Theorem 3.4.8. *Let $\mathcal{M} = \int_Z^{\oplus}\mathcal{M}_\zeta\, dw(\zeta)$ be a decomposable von Neumann algebra. If $x, y \in LS(\mathcal{M})$,*

$$x = \int_Z^{\oplus} x_\zeta\, dw(\zeta), \quad y = \int_Z^{\oplus} y_\zeta\, dw(\zeta),$$

then

$$x \dotplus y = \int_Z^{\oplus}(x_\zeta \dotplus y_\zeta)\, dw(\zeta), \quad x \cdot y = \int_Z^{\oplus}(x_\zeta \cdot y_\zeta)\, dw(\zeta).$$

Proof. Let $x = \int_Z^{\oplus} x_\zeta\, dw(\zeta), y = \int_Z^{\oplus} y_\zeta\, dw(\zeta)$ where $\zeta \to x_\zeta$ and $\zeta \to y_\zeta$ are measurable fields of locally measurable operators affiliated with $\mathcal{M}_\zeta$. By Proposition 2.3.11, the strong sum $x_\zeta \dotplus y_\zeta$ is well-defined. Hence, by Lemma 1.13.9, the operator a maximally defined by the field $\{x_\zeta \dotplus y_\zeta\}$ is closed. By Proposition 1.13.19, the operator a is decomposable, i. e., there exists a measurable field $\zeta \to a_\zeta$, such that $a_\zeta \subset x_\zeta \dotplus y_\zeta$ and $a = \int_Z^{\oplus} a_\zeta\, dw(\zeta)$.

Since the fields $\zeta \to x_\zeta$, $\zeta \to y_\zeta$ are measurable, it follows that they are weakly measurable (see Proposition 1.13.6). In particular, for any $\{\xi_\zeta\} \in \operatorname{dom}(x + y)$, we have that $\xi_\zeta \in \operatorname{dom}(x_\zeta) \cap \operatorname{dom}(y_\zeta)$, $\zeta \to x_\zeta\xi_\zeta$, and $\zeta \to y_\zeta\xi_\zeta$ are measurable and $\int_Z \|x_\zeta\xi_\zeta\|^2_{H_\zeta} dw(\zeta), \int_Z \|y_\zeta\xi_\zeta\|^2_{H_\zeta} dw(\zeta) < \infty$. Therefore, $\xi_\zeta \in \operatorname{dom}(x_\zeta \dotplus y_\zeta), \zeta \to (x_\zeta \dotplus y_\zeta)\xi_\zeta = x_\zeta\xi_\zeta + y_\zeta\xi_\zeta$ is measurable and $\int_Z \|(x_\zeta \dotplus y_\zeta)\xi_\zeta\|^2_{H_\zeta} dw(\zeta) < \infty$, i. e., $\{\xi_\zeta\} \in \operatorname{dom}(a)$. Furthermore, for any $\{\xi_\zeta\} \in \operatorname{dom}(x + y)$, we have that

$$((x + y)\xi)(\zeta) = (x_\zeta + y_\zeta)\xi_\zeta = (x_\zeta \dotplus y_\zeta)\xi_\zeta = (a\xi)(\zeta),$$

that is $x \dotplus y \subset a$. Hence, for the strong sum $x \dotplus y = \overline{x + y}$, we have $x \dotplus y \subset a$. Since $x \dotplus y$ is locally measurable, Theorem 3.4.3 implies that $x \dotplus y = \int_Z^{\oplus} b_\zeta \, dw(\zeta)$ for some measurable field $\zeta \to b_\zeta$ of locally measurable operators. Since

$$\int_Z^{\oplus} b_\zeta \, dw(\zeta) = x \dotplus y \subset a = \int_Z^{\oplus} a_\zeta dw(\zeta),$$

Proposition 1.13.16 implies that $b_\zeta \subset a_\zeta \subset x_\zeta + y_\zeta$ for a. e. ζ. Since b_ζ and $x_\zeta \dotplus y_\zeta$ are locally measurable, Proposition 2.3.8(i) implies that $b_\zeta = x_\zeta \dotplus y_\zeta$ a. e., proving that $x \dotplus y = \int_Z^{\oplus} x_\zeta \dotplus y_\zeta \, dw(\zeta)$.

The proof of equality $x \cdot y = \int_Z^{\oplus} (x_\zeta \cdot y_\zeta) \, dw(\zeta)$ is similar and is, therefore, omitted. □

3.5 Commutator equalities for locally measurable operators

In this section, we show that if $\mathcal{M}$ has no semifinite properly infinite direct summand, then for any $x = x^* \in LS(\mathcal{M})$ there exists a self-adjoint central element $c \in \mathcal{Z}(LS(\mathcal{M}))$ and a unitary $u \in \mathcal{M}$, such that

$$\big|[u, x]\big| = |x - c| + u^* |x - c| u. \tag{3.9}$$

Throughout this section, we assume that $\mathcal{M}$ is a von Neumann algebra acting on a separable Hilbert space H. The direct integral decomposition of $\mathcal{M}$ relative to its centre $\mathcal{Z}(\mathcal{M})$ constitutes a fundamental part of our proof of the commutator equality (3.9).

We first present an example which demonstrates the main idea behind the proof of (3.9) for the case when $\mathcal{M}$ is identified with the algebra $M_{n \times n}(\mathbb{C})$ of all $n \times n$ complex matrices (here, $\mathcal{M}$ acts on $\mathbb{C}^n$ by multiplication on a $n \times n$ matrix). In this case, $LS(\mathcal{M}) = M_{n \times n}(\mathbb{C})$ and $\mathcal{Z}(\mathcal{M}) = \mathbb{C}\mathbf{1}$.

Example 3.5.1. *Let $x = x^* \in M_{n \times n}(\mathbb{C})$. There exists a unitary matrix u and $\lambda_0 \in \mathbb{C}$, such that*

$$\big|[u, x]\big| = |x - \lambda_0 \mathbf{1}| + u^* |x - \lambda_0 \mathbf{1}| u.$$

Proof. Fix $x = x^* \in M_{n \times n}(\mathbb{C})$ and select a unitary matrix $v \in M_{n \times n}(\mathbb{C})$, such that

$$v^* x v = \begin{pmatrix} \lambda_1 & 0 & \dots & 0 \\ 0 & \lambda_2 & \dots & 0 \\ \vdots & \vdots & \ddots & \vdots \\ 0 & 0 & \dots & \lambda_n \end{pmatrix} \in M_{n \times n}(\mathbb{C}),$$

where $\lambda_n \le \lambda_{n-1} \le \dots \le \lambda_1$.

Let the unitary matrix $u \in M_{n\times n}(\mathbb{C})$ be counterdiagonal, i. e.,

$$u = \begin{pmatrix} 0 & \dots & 0 & 1 \\ 0 & \dots & 1 & 0 \\ \vdots & \ddots & \vdots & \vdots \\ 1 & \dots & 0 & 0 \end{pmatrix}.$$

Observe that

$$u^*v^*xvu = \begin{pmatrix} \lambda_n & 0 & \dots & 0 \\ 0 & \lambda_{n-1} & \dots & 0 \\ \vdots & \vdots & \ddots & \vdots \\ 0 & 0 & \dots & \lambda_1 \end{pmatrix}.$$

Therefore,

$$\big|[x, vu]\big| = \big|u^*v^*xvu - x\big| = \begin{pmatrix} |\lambda_n - \lambda_1| & 0 & \dots & 0 \\ 0 & |\lambda_{n-1} - \lambda_2| & \dots & 0 \\ \vdots & \vdots & \ddots & \vdots \\ 0 & 0 & \dots & |\lambda_1 - \lambda_n| \end{pmatrix}.$$

If n is odd, then for all $1 \le k \le n$, we have

$$|\lambda_k - \lambda_{n+1-k}| = |\lambda_k - \lambda_0| + |\lambda_{n+1-k} - \lambda_0|$$

for $\lambda_0 = \lambda_{(n+1)/2}$.

If n is even, then for all $1 \le k \le n$, we have

$$|\lambda_k - \lambda_{n+1-k}| = |\lambda_k - \lambda_0| + |\lambda_{n+1-k} - \lambda_0|$$

for every $\lambda_0 \in [\lambda_{n/2}, \lambda_{n/2+1}]$.

Therefore, for every $n \in \mathbb{N}$, we have

$$\big|[v^*xv, u]\big| = u^*v^*|x - \lambda_0\mathbf{1}|vu + v^*|x - \lambda_0\mathbf{1}|v.$$

It is clear, that vuv^* is a unitary and

$$\begin{aligned} \big|[x, vuv^*]\big| &= \big|(vuv^*)^*x(vuv^*) - x\big| = v\big|u^*v^*xvu - v^*xv\big|v^* = v\big|[v^*xv, u]\big|v^* \\ &= (vuv^*)^*|x - \lambda_0\mathbf{1}|(vuv^*) + |x - \lambda_0\mathbf{1}|, \end{aligned}$$

as required. □

First, we establish an auxiliary lemma, which proves the commutator equality (3.9) under an additional assumption of symmetry of the supports of negative and positive parts of the operator x. We note that this result holds for an arbitrary von Neumann algebra $\mathcal{M}$. We shall use this lemma repeatedly.

Lemma 3.5.2. *Let $\mathcal{M}$ be a von Neumann algebra and let $x = x^* \in LS(\mathcal{M})$ be such $s(x_+)$ and $s(x_-)$ are equivalent. Then there exists a unitary $u \in \mathcal{M}$, such that*

$$|[x,u]| = |x| + u^{-1}|x|u.$$

The same assertion holds if $s(x_+)$ and $s(x_-)+n(x)$ (or $s(x_+)+n(x)$ and $s(x_-)$) are equivalent.

Proof. Assume that $x = x^* \in LS(\mathcal{M})$ is such that $s(x_+)$ and $s(x_-)$ are equivalent. In this case, there exists a partial isometry $v \in \mathcal{M}$, such that $v^*v = s(x_+)$ and $vv^* = s(x_-)$. We claim that

$$u = v + v^* + n(x)$$

is the required unitary.

Note that since $v = vs(x_+) = vE_X(0,\infty)$ and $v = s(x_-)v = E_X(-\infty,0)v$, one can easily show that u is a unitary operator.

To show that the commutator equality is satisfied, we first note that

$$s(x_-)u = s(x_-)v + s(x_-)v^* + s(x_-)n(x) = v + s(x_-)s(x_+)v^* = v.$$

Similarly, $us(x_+) = v$, and $s(x_+)u = v^* = us(x_-)$ and, therefore,

$$s(x_-)u = us(x_+), \quad s(x_+)u = us(x_-).$$

Hence, we have that

$$\begin{aligned} x_+ + u^{-1}x_-u &= x_+ + u^{-1}s(x_-)x_-s(x_-)u \\ &= s(x_+)x_+s(x_+) + s(x_+)u^{-1}x_-us(x_+) \\ &= s(x_+)(x_+ + u^{-1}x_-u)s(x_+), \end{aligned}$$

and similarly

$$x_- + u^{-1}x_+u = s(x_-)(x_- + u^{-1}x_+u)s(x_-).$$

Thus, $x_+ + u^{-1}x_-u$, $x_- + u^{-1}x_+u$ are disjoint and positive. Therefore, referring to Proposition 3.3.2, we conclude that

$$\begin{aligned} |[x,u]| = |x - u^{-1}xu| &= (x_+ + u^{-1}x_-u) + (x_- + u^{-1}x_+u) \\ &= (x_+ + x_-) + u^{-1}(x_- + x_+)u = |x| + u^{-1}|x|u, \end{aligned}$$

as required. □

Having established the auxiliary Lemma 3.5.2, we move onto proving the commutator equality (3.9). We first establish this result for type I_{fin} algebras.

Proposition 3.5.3. *Let $\mathcal{M}$ be a type I_{fin} von Neumann algebra. For any $x = x^* \in LS(\mathcal{M})$ there exist a self-adjoint central element $c \in \mathcal{Z}(LS(\mathcal{M}))$ and a unitary $u \in \mathcal{M}$, such that*

$$\big|[u, x]\big| = |x - c| + u^*|x - c|u.$$

Proof. Since $\mathcal{M}$ is a type I_{fin} algebra, there exists a central partition of unity $\{z_n\}_{n\in\mathbb{N}}$, such that $z_n\mathcal{M}$ has type I_n for all $n \in \mathbb{N}$ (see Theorem 1.9.17). Since z_n are central and pairwise orthogonal, it is sufficient to prove the assertion under the assumption that $z_n = \mathbf{1}$ for some $n \in \mathbb{N}$. In this case, $\mathcal{M}$ is a type I_n von Neumann algebra.

Let Z be a Polish space, such that $\mathcal{Z}(\mathcal{M}) \cong L_\infty(Z, w)$ and

$$\mathcal{M} = \int_Z^{\oplus} \mathcal{M}_\zeta \, dw(\zeta),$$

where almost every $\mathcal{M}_\zeta$ is factor (see Theorem 1.12.9). By Theorem 1.12.10, we have that $\mathcal{M}_\zeta$ is a factor of type I_n for almost all ζ, i. e., $\mathcal{M}_\zeta = M_{n\times n}(\mathbb{C})$ for almost every $\zeta \in Z$. Assume that $x = x^* \in LS(\mathcal{M})$. Then, by Theorem 3.4.3 and Theorem 1.13.20(ii) we have that

$$x = \int_Z^{\oplus} x_\zeta \, dw(\zeta),$$

with $x_\zeta = x_\zeta^* \in M_{n\times n}(\mathbb{C}) = LS(M_{n\times n}(\mathbb{C}))$ for almost every $\zeta \in Z$.

For $\zeta \in Z$, let $\{\lambda_k(\zeta)\}_{k=1}^n$ be the sequence of eigenvalues of x_ζ taken in the decreasing order of absolute values counting multiplicities. We claim that the fields $\zeta \to \lambda_k(\zeta)$ and $\zeta \to E_{x_\zeta}(\{\lambda_k(\zeta)\})$ are measurable for every $k = 1, \ldots, n$. Indeed, let A be a countable dense set in the unit sphere of $\mathbb{C}^n = H_\zeta$. We have $\lambda_1(\zeta) = \sup_{\xi\in A}\langle x_\zeta\xi, \xi\rangle$ and, therefore, the field $\zeta \to \lambda_1(\zeta)$ is measurable. Therefore, the field $\zeta \to x_\zeta - \lambda_1(\zeta)\mathbf{1}_{H_\zeta}$ is measurable too. By Proposition 1.13.23, the field

$$\zeta \to E_{x_\zeta - \lambda_1(\zeta)\mathbf{1}_{H_\zeta}}(\{0\}) = E_{x_\zeta}(\{\lambda_1(\zeta)\})$$

is measurable. Hence, the field $\zeta \to x_\zeta - \lambda_1(\zeta)E_{x_\zeta}(\{\lambda_1(\zeta)\})$ is measurable. Applying the above argument to the field $\zeta \to x_\zeta - \lambda_1(\zeta)E_{x_\zeta}(\{\lambda_1(\zeta)\})$ and repeating finitely many times, we infer the claim.

Let $d_\zeta \in M_{n\times n}(\mathbb{C})$ be the diagonal matrix with $\{\lambda_k(\zeta)\}_{k=1}^n$ on the diagonal. By the preceding paragraph, the mapping $\zeta \to d_\zeta$ is measurable and, therefore, we may define

$$d = \int_Z^{\oplus} d_\zeta \, dw(\zeta) \in LS(\mathcal{M}).$$

Since x_ζ and d_ζ are unitarily equivalent for almost all $\zeta \in Z$, it follows that there exists unitary $v \in \mathcal{M}$, such that $vdv^{-1} = x$. Indeed, for every partition (we denote the partition by π) $n = k_1 + (k_2 - k_1) + \cdots + (k_l - k_{l-1})$, set

$$Z_\pi = \{\zeta \in Z : \lambda_{k_{m-1}+1}(\zeta) = \lambda_{k_m}(\zeta),\ \lambda_{k_{m-1}}(\zeta) > \lambda_{k_m}(\zeta),\ 1 \le m \le l\}.$$

Set

$$p_{\pi,m} = \int_{Z_\pi}^{\oplus} E_{x_\zeta}(\{\lambda_{k_m}(\zeta)\})\, dw(\zeta), \quad q_{\pi,m} = \int_{Z_\pi}^{\oplus} \left(\sum_{k=k_{m-1}+1}^{k_m} E_{k,k} \right) dw(\zeta).$$

Since the center-valued traces of $p_{\pi,m}$ and $q_{\pi,m}$ coincide, it follows that $p_{\pi,m} \sim q_{\pi,m}$. Hence, there exists a partial isometry $v_{\pi,m}$ such that $v_{\pi,m}^* v_{\pi,m} = q_{\pi,m}$ and $v_{\pi,m} v_{\pi,m}^* = p_{\pi,m}$. It is now immediate that $v = \sum_\pi \sum_{m=1}^{l_\pi} v_{m,\pi}$ is unitary and $vq_{\pi,m}v^{-1} = p_{\pi,m}$. Hence, $vdv^{-1} = x$.

Since for any unitary $u \in \mathcal{M}$ and $\lambda \in \mathbb{C}$, we have that

$$\begin{aligned} |[x, vuv^{-1}]| &= |(vuv^{-1})^{-1} x (vuv^{-1}) - x| = v|u^{-1}v^{-1}xvu - v^{-1}xv|v^{-1} \\ &= v|[v^{-1}xv, u]|v^{-1} = v|[d, u]|v^{-1} \end{aligned}$$

and

$$|x - \lambda \mathbf{1}| + (vuv^{-1})^{-1}|x - \lambda \mathbf{1}|(vuv^{-1}) = v(|d - \lambda \mathbf{1}| + u^{-1}|d - \lambda \mathbf{1}|u)v^{-1},$$

it is sufficient to prove the assertion for $x = d$.

Assume now that $x = d = \int_Z^{\oplus} d_\zeta\, dw(\zeta)$. Let

$$u_\zeta = \begin{pmatrix} 0 & 0 & \dots & 0 & 1 \\ 0 & 0 & \dots & 1 & 0 \\ \vdots & \vdots & \ddots & \vdots & \vdots \\ 0 & 1 & \dots & 0 & 0 \\ 1 & 0 & \dots & 0 & 0 \end{pmatrix}, \quad \zeta \in Z.$$

Clearly, $\zeta \to u_\zeta$ is a measurable field, since it is a constant field. We define unitary element $u \in \mathcal{M}$ by setting

$$u = \int_Z^{\oplus} u_\zeta\, dw(\zeta).$$

We claim that the unitary u satisfies the desired commutator identity, i. e., for some $c \in \mathcal{Z}(LS(\mathcal{M}))$, we have

$$\big|[d,u]\big| = |d-c| + u^*|d-c|u.$$

Observe that

$$u_\zeta^{-1} d_\zeta u_\zeta = \begin{pmatrix} \lambda_n(\zeta) & 0 & \dots & 0 \\ 0 & \lambda_{n-1}(\zeta) & \dots & 0 \\ \vdots & \vdots & \ddots & \vdots \\ 0 & 0 & \dots & \lambda_1(\zeta) \end{pmatrix}$$

and, therefore,

$$\begin{aligned} \big|[d_\zeta, u_\zeta]\big| &= \big|u_\zeta^{-1} d_\zeta u_\zeta - d_\zeta\big| \\ &= \begin{pmatrix} |\lambda_n(\zeta) - \lambda_1(\zeta)| & 0 & \dots & 0 \\ 0 & |\lambda_{n-1}(\zeta) - \lambda_2(\zeta)| & \dots & 0 \\ \vdots & \vdots & \ddots & \vdots \\ 0 & 0 & \dots & |\lambda_1(\zeta) - \lambda_n(\zeta)| \end{pmatrix}. \end{aligned}$$

For every $\zeta \in Z$, we define

$$c_\zeta = \begin{cases} \lambda_{\frac{n+1}{2}}(\zeta), & n \text{ is odd}, \\ \frac{1}{2}(\lambda_{\frac{n}{2}}(\zeta) + \lambda_{\frac{n}{2}+1}(\zeta)), & n \text{ is even}. \end{cases}$$

A simple direct computation yields

$$\big|\lambda_k(\zeta) - \lambda_{n-1-k}(\zeta)\big| = \big|\lambda_k(\zeta) - c_\zeta\big| + \big|\lambda_{n-1-k}(\zeta) - c_\zeta\big|, \quad 1 \le k \le n,\ \zeta \in Z.$$

Therefore,

$$\big|d_\zeta - u_\zeta^{-1} d_\zeta u_\zeta\big| = |d_\zeta - c_\zeta \mathbf{1}_{H_\zeta}| + u_\zeta^{-1}|d_\zeta - c_\zeta \mathbf{1}_{H_\zeta}| u_\zeta. \tag{3.10}$$

By the proven above, the field $\zeta \to \lambda_k(\zeta)$ is measurable for every $k = 1, \dots, n$, and, therefore, $\zeta \to c_\zeta$ is measurable, too. By Theorem 3.4.3, we may define $c \in LS(\mathcal{M})$ by setting

$$c = \int_Z^{\oplus} c_\zeta \mathbf{1}_{H_\zeta}\, dw(\zeta).$$

By Theorem 3.4.8 and Theorem 1.13.25, we have that

$$\big|[u,d]\big| = \big|d - u^{-1}du\big| = \int_Z^{\oplus} \big|d_\zeta - u_\zeta^{-1} d_\zeta u_\zeta\big|\, dw(\zeta),$$

$$|d-c| = \int_Z^{\oplus} |d_\zeta - c_\zeta \mathbf{1}_{H_\zeta}| \, dw(\zeta),$$

$$u^{-1}|d-c|u = \int_Z^{\oplus} u_\zeta^{-1}|d_\zeta - c_\zeta \mathbf{1}_{H_\zeta}|u_\zeta \, dw(\zeta).$$

Using equality (3.10) and Theorem 3.4.8, we conclude that

$$|[u,d]| = |d-c| + u^{-1}|d-c|u,$$

as required. □

Next, we prove the commutator equality (3.9) for type II_1 algebras.

Proposition 3.5.4. *Let $\mathcal{M}$ be a type II_1 von Neumann algebra and let $x = x^* \in LS(\mathcal{M})$. There exists a self-adjoint central element $c \in \mathcal{Z}(LS(\mathcal{M}))$ and a unitary $u \in \mathcal{M}$, such that*

$$|[u,x]| = |x-c| + u^*|x-c|u.$$

Proof. Let Z be a Polish space, such that $\mathcal{Z}(\mathcal{M}) \cong L_\infty(Z, w)$ and

$$\mathcal{M} = \int_Z^{\oplus} \mathcal{M}_\zeta \, dw(\zeta),$$

where almost every $\mathcal{M}_\zeta$ is factor (see Theorem 1.12.9). By Theorem 1.12.10, $\mathcal{M}_\zeta$ is a II_1-factor for almost every $\zeta \in Z$. Since $\mathcal{M}$ is a finite von Neumann algebra, there exists a faithful normal finite trace τ on $\mathcal{M}$ (see Corollary 1.10.17). Using Theorem 1.12.11, we write

$$\tau = \int_Z^{\oplus} \tau_\zeta dw(\zeta),$$

where $\zeta \to \tau_\zeta$ is a measurable field of faithful normal finite traces on $\mathcal{M}_\zeta$. Without loss of generality, we may assume that $\tau_\zeta(\mathbf{1}_{H_\zeta}) = 1$ for almost every $\zeta \in Z$. Without loss of generality, we further assume that $s(x) = \mathbf{1}$. By Corollary 3.4.5 and Theorem 1.13.20(ii), we may write

$$x = \int_Z^{\oplus} x_\zeta \, dw(\zeta),$$

where $x_\zeta = x_\zeta^* \in S(\mathcal{M}_\zeta)$ for almost every $\zeta \in Z$.

We define

$$c_\zeta = \inf\left\{\lambda : \tau_\zeta(E_{x_\zeta}(\lambda, \infty)) \le \frac{1}{2}\right\}.$$

We claim that $\zeta \to c_\zeta \mathbf{1}_{H_\zeta}$ is a measurable field. To this end, it is sufficient to show that the function $\zeta \to c_\zeta$ is measurable.

By Proposition 1.13.23, the field $\zeta \to E_{x_\zeta}(\lambda, \infty)$ is measurable, and so since $\zeta \to \tau_\zeta$ is a measurable field of traces, it follows that $\zeta \to \tau_\zeta(E_{x_\zeta}(\lambda, \infty))$ is measurable. For any $\mu \in \mathbb{R}$, we have that

$$\begin{aligned}\{\zeta : c_\zeta \geq \mu\} &= \left\{\zeta : \tau_\zeta(E_{x_\zeta}(\lambda, \infty)) \leq \frac{1}{2} \ \forall \lambda > \mu\right\} \\ &= \left\{\zeta : \tau_\zeta\left(E_{x_\zeta}\left(\mu + \frac{1}{m}, \infty\right)\right) \leq \frac{1}{2} \ \forall m \in \mathbb{N}\right\} \\ &= \bigcap_{m \in \mathbb{N}} \left\{\zeta : \tau_\zeta\left(E_{x_\zeta}\left(\mu + \frac{1}{m}, \infty\right)\right) \leq \frac{1}{2}\right\},\end{aligned}$$

proving that the function $\zeta \to c_\zeta$ is measurable. Thus, the field $\zeta \to c_\zeta \mathbf{1}_{H_\zeta}$ is a measurable field of bounded operators.

By Theorem 3.4.3, we may define $c \in LS(\mathcal{M})$ by setting

$$c = \int_Z^{\oplus} c_\zeta \mathbf{1}_{H_\zeta} \, dw(\zeta).$$

Theorem 3.4.8 guarantess that c commutes with any element of $LS(\mathcal{M})$, that is $c \in \mathcal{Z}(LS(\mathcal{M}))$.

Refering to Theorem 3.4.8 once again, we have that

$$x - c = \int_Z^{\oplus} (x_\zeta - c_\zeta \mathbf{1}_{H_\zeta}) \, dw(\zeta).$$

In particular, by Proposition 1.13.23, we have that the fields

$$\begin{aligned}&\zeta \to E_{x_\zeta}(c_\zeta, \infty) = E_{x_\zeta - c_\zeta \mathbf{1}_{H_\zeta}}(0, \infty), \\ &\zeta \to E_{x_\zeta}(-\infty, c_\zeta) = E_{x_\zeta - c_\zeta \mathbf{1}_{H_\zeta}}(-\infty, 0), \\ &\zeta \to E_{x_\zeta}\{c_\zeta\} = E_{x_\zeta - c_\zeta \mathbf{1}_{H_\zeta}}\{0\}\end{aligned}$$

are all measurable. Thus, we may define projections p, q, and $r \in \mathcal{M}$ by setting

$$p = \int_Z^{\oplus} E_{x_\zeta}(c_\zeta, \infty) \, dw(\zeta), \quad q = \int_Z^{\oplus} E_{x_\zeta}(-\infty, c_\zeta) \, dw(\zeta),$$

$$r = \int_Z^{\oplus} E_{x_\zeta}\{c_\zeta\} \, dw(\zeta).$$

If $\mathcal{T}$ is the canonical center-valued trace on $\mathcal{M}$ (see Theorem 1.10.2), then taking into account the definition of c_ζ, p, and q, we obtain

$$\mathcal{T}(p) \le \frac{1}{2}\mathbf{1}, \quad \mathcal{T}(q) \le \frac{1}{2}\mathbf{1}.$$

Since $\mathcal{M}$ is type II_1 algebra, we may write $r = r_1 + r_2$, where $r_1, r_2 \in P(\mathcal{M})$ are such that $r_1 r_2 = 0$ and

$$\mathcal{T}(r_1) = \frac{1}{2}\mathbf{1} - \mathcal{T}(p), \quad \mathcal{T}(r_2) = \frac{1}{2}\mathbf{1} - \mathcal{T}(q).$$

Set

$$p_0 = p + r_1, \quad q_0 = q + r_2.$$

Since $p + q + r = \mathbf{1}$, we have that

$$q_0 = \mathbf{1} - (s(x) - q - r_2) = \mathbf{1} - (p + q + r - q - r_2) = \mathbf{1} - (p + r_1) = \mathbf{1} - p_0.$$

Furthermore,

$$\mathcal{T}(p_0) = \mathcal{T}(p + r_1) = \frac{1}{2}\mathbf{1}, \quad \mathcal{T}(q_0) = \mathcal{T}(q + r_2) = \frac{1}{2}\mathbf{1},$$

so that $\mathcal{T}(p_0) = \mathcal{T}(q_0)$. By Proposition 1.10.3, we have that $p_0 \sim q_0$. Let v be a partial isometry, such that $v^* v = p_0$ and $vv^* = q_0$. Define a self-adjoint operator

$$u = v + v^*.$$

Since p_0 and q_0 are orthogonal, it follows that

$$u^2 = v^2 + v^* v + vv^* + (v^*)^2 = p_0 + q_0 = p + q + r = s(x) = \mathbf{1},$$

i. e., u is unitary. We claim that

$$u^{-1}|x - c|u + |x - c| = \big|[x, u]\big|.$$

Set $y = x - c = \int_Z^{\oplus} (x_\zeta - c_\zeta \mathbf{1}_{H_\zeta})\, dw(\zeta)$. By Proposition 1.13.23, we have that

$$p = E_y(0, \infty), \quad q = E_y(-\infty, 0), \quad r = E_y\{0\}.$$

Note that every subprojection of $n(y) = E_y\{0\}$ commutes with x. In particular, so do r_1 and r_2. Thus, y commutes with $p_0 = p + r_1$ and $q_0 = q + r_2$. Since, in addition, $p_0 u = u q_0$ and $q_0 u = u p_0$, we have that

$$u^{-1} yu \cdot p_0 = u^{-1} y \cdot q_0 u = u^{-1} \cdot q_0 y \cdot u = p_0 \cdot u^{-1} yu.$$

Similarly, one can show that q_0 commutes with both y and $u^{-1} xu$.

Thus, we obtain that

$$(y - u^{-1}yu)p_0 = p_0(y - u^{-1}yu), \quad (y - u^{-1}yu)q_0 = q_0(y - u^{-1}yu)$$

and, therefore, since p_0 and q_0 are orthogonal and $p_0 + q_0 = \mathbf{1}$, we may write

$$\begin{aligned} |y - u^{-1}yu| &= |y - u^{-1}yu|p_0 + |y - u^{-1}yu|q_0 \\ &= |(y - u^{-1}yu)p_0| + |(y - u^{-1}yu)q_0|. \end{aligned} \tag{3.11}$$

By construction, we have that

$$yp_0 \geq 0, \quad yq_0 \leq 0.$$

Since $u^{-1}yup_0 = u^{-1}yq_0u$ and $u^{-1}yuq_0 = u^{-1}yp_0u$, we also have that

$$u^{-1}yu \cdot p_0 \leq 0, \quad u^{-1}yu \cdot p_0 \geq 0.$$

Hence, equality (3.11) implies that

$$|y - u^{-1}yu| = (y - u^{-1}yu)p_0 - (y - u^{-1}yu)q_0 = |y| + u^{-1}|y|u.$$

In other words, we have

$$|[x, u]| = |[y, u]| = |y - u^{-1}yu| = |y| + u^{-1}|y|u = |x - c| + u^{-1}|x - c|u,$$

as required. □

Finally, we prove the commutator equality (3.9) for the type *III* von Neumann algebras.

Proposition 3.5.5. *Let $\mathcal{M}$ be a type III von Neumann algebra and let $x = x^* \in LS(\mathcal{M})$. There exist a self-adjoint central element $c \in \mathcal{Z}(LS(\mathcal{M}))$ and a unitary $u \in \mathcal{M}$, such that*

$$|[u, x]| = |x - c| + u^*|x - c|u.$$

Proof. Let $\Gamma = \{z_i\}_{i \in I}$ be a maximal family of pairwise orthogonal central projections in $\mathcal{M}$, such that either $xz_i \in \mathcal{Z}(LS(\mathcal{M}z_i))$ or $s_c(E_x(-\infty, \lambda_i])z_i = s_c(E_x(\lambda_i, +\infty))z_i = z_i$ for some $\lambda_i \in \mathbb{R}$. Since $\mathcal{M}$ is acting on a separable Hilbert space, it follows that the family Γ is at most countable.

We claim that $\sum_{i \in I} z_i = \mathbf{1}$. Suppose, by contradiction, that $z := \mathbf{1} - \sum_{i \in I} z_i \neq 0$. Fix $t \in \mathbb{R}$. Assume that $z_0 := s_c(E_x(-\infty, t])s_c(E_x(t, +\infty))z \neq 0$. Then $s_c(E_x(-\infty, t])z_0 = s_c(E_x(t, +\infty))z_0 = z_0$, which contradicts with the maximality of Γ. Hence, $z_0 = s_c(E_x(-\infty, t])s_c(E_x(t, +\infty))z = 0$.

By Proposition 1.9.2(v), we have that

$$\begin{aligned} s_c(E_x(-\infty,t])z + s_c(E_x(t,+\infty))z &= (s_c(E_x(-\infty,t]) \vee s_c(E_x(t,+\infty)))z \\ &= s_c(E_x(-\infty,t] \vee E_x(t,\infty))z = s_c(\mathbf{1})z = z. \end{aligned}$$

Since

$$E_x(-\infty,t]z \leq s_c(E_x(-\infty,t])z, \quad E_x(t,+\infty)z \leq s_c(E_x(t,+\infty))z$$

and $E_x(-\infty,t]z + E_x(t,+\infty)z = z$, it follows that

$$E_x(-\infty,t]z = s_c(E_x(-\infty,t])z, \quad E_x(t,+\infty)z = s_c(E_x(t,+\infty))z.$$

Thus, for any $t \in \mathbb{R}$, we have that $E_x(-\infty,t]z = s_c(E_x(-\infty,t])z \in \mathcal{Z}(\mathcal{M}z)$. Referring to Proposition 3.2.3, we conclude that $xz \in \mathcal{Z}(LS(\mathcal{M}z))$, which again contradicts with the maximality of Γ.

Thus, $\sum_{i\in I} z_i = \mathbf{1}$, as required.

Let $i \in I$ be fixed. If $s_c(E_x(-\infty,\lambda_i])z_i = s_c(E_x(\lambda_i,+\infty))z_i$ for some $\lambda_i \in \mathbb{R}$, then Theorem 1.9.9(viii) implies that $E_x(-\infty,\lambda_i]z_i \sim E_x(\lambda_i,+\infty)z_i$ (since $\mathcal{M}$ acts on a separable Hilbert space). By Lemma 3.5.2, there exists $u_i \in U(\mathcal{M}z_i)$ where $c_i = \lambda_i z_i$, such that

$$|[xz_i, u_i]| = |xz_i - c_i| + u_i^* |xz_i - c_i| u_i.$$

If $xz_i \in \mathcal{Z}(LS(\mathcal{M}z_i))$, then we define $c_i = xz_i$, $u_i = z_i$.

By Proposition 2.3.10, we may define $c \in \mathcal{Z}(LS(\mathcal{M}))$, by setting

$$c = \sum_{i\in I} c_i \in \mathcal{Z}(LS(\mathcal{M})).$$

Define also

$$u = \sum_{i\in I} u_i \in U(\mathcal{M}).$$

Since $\sum_{i\in I} z_i = \mathbf{1}$, using Proposition 2.3.10 once again, we conclude that

$$|[u,x]| = |x - c| + u^* |x - c| u,$$

as required. □

Gathering together the results of Proposition 3.5.3, Proposition 3.5.4, and Proposition 3.5.5 we establish the commutator equality (3.9) for any von Neumann algebra without a semifinite properly infinite summand.

Theorem 3.5.6. *Let $\mathcal{M}$ be an arbitrary von Neumann algebra, which does not contain a semifinite properly infinite summand. For any $x = x^* \in LS(\mathcal{M})$, there exist $c \in \mathcal{Z}(LS(\mathcal{M}))$ and a unitary $u \in \mathcal{M}$, such that*

$$|[x,u]| = u^{-1}|x-c|u + |x-c|. \tag{3.12}$$

Proof. By the decomposition theorem (see Theorem 1.9.14), we may find central partition of unity z_k, $k = 1,2,3$, such that $z_1\mathcal{M}$ is type I_{fin} algebra, $z_2\mathcal{M}$ is type II_1 algebra, and $z_3\mathcal{M}$ is type III algebra.

Combining Proposition 3.5.3, Proposition 3.5.4 and Proposition 3.5.5, we have that there exist $u_i \in z_i\mathcal{M}$ and $c_i \in \mathcal{Z}(LS(z_i\mathcal{M}))$, $i = 1,2,3$, such that

$$|[z_i x, u_i]| = u_i^{-1}|z_i x - c_i|u_i + |z_i x - c_i|, \quad i = 1,2,3.$$

We define $u = u_1 + u_2 + u_3$ and $c = c_1 + c_2 + c_3 \in \mathcal{Z}(LS(\mathcal{M}))$. Since $\{z_i\}$ is a central partition of unity in $\mathcal{M}$, it follows that u is a unitary and

$$\begin{aligned}|[x,u]| &= \left|\sum_{i=1}^{3}[z_i x, u_i]\right| = \sum_i |[z_i x, u_i]| = \sum_{i=1}^{3}\left(u_i^{-1}|z_i x - c_i|u_i + |z_i x - c_i|\right) \\ &= u^{-1}|x-c|u + |x-c|,\end{aligned}$$

as required. □

3.6 Commutator estimates for locally measurable operators

As proved in Theorem 3.5.6, if a von Neumann algebra has no semifinite properly infinite summand, then for any self-adjoint $x \in LS(\mathcal{M})$, there exist $c \in \mathcal{Z}(LS(\mathcal{M}))$ and a unitary $u \in \mathcal{M}$, such that

$$|[x,u]| = u^{-1}|x-c|u + |x-c|.$$

In this section, we show that such an equality does not hold if $\mathcal{M}$ has a semifinite properly infinite summand and prove that in the general case we instead have commutator estimates

$$|[x,u]| \geq (1-\varepsilon)|x-c|, \tag{3.13}$$

for any $\varepsilon > 0$, an appropriate unitary u, and a central element $c \in \mathcal{Z}(LS(\mathcal{M}))$.

We start this section with showing that the commutator equality established in Theorem 3.5.6 does not hold for a general von Neumann algebra.

Example 3.6.1. *Let $\mathcal{M}$ be a semifinite infinite factor. Fix a sequence $\{p_n\}_{n\in\mathbb{N}}$ of pairwise orthogonal finite projections $p_1, p_2, \ldots, p_n, \ldots$, such that $\sup_{n\in\mathbb{N}} p_n = \mathbf{1}$ and set*

$$x := \sum_{n\in\mathbb{N}} n^{-1}p_n.$$

It is clear that x is a bounded, self-adjoint, and compact (with respect to $\mathcal{M}$) operator. Moreover, the support of x, $s(x)$ is equal to $\mathbf{1}$. Suppose that

$$\big|[x,u]\big| \geq |x - \lambda\mathbf{1}| \tag{3.14}$$

for some $\lambda \in \mathbb{C}$ and some unitary $u \in \mathcal{M}$.

Since $[x,u]$ is also compact and $S_0(\mathcal{M})$ is an absolutely solid $*$-subalgebra of $LS(\mathcal{M})$, inequality (3.14) implies that $x - \lambda\mathbf{1}$ is compact, too. If we assume that λ is nonzero, then $\mathbf{1} = \frac{1}{\lambda}x - \frac{1}{\lambda}(x-\lambda\mathbf{1}) \in S_0(\mathcal{M})$, which is not possible since $\mathcal{M}$ is infinite algebra. Hence, $\lambda = 0$, i. e.,

$$\big|u^{-1}xu - x\big| = \big|[x,u]\big| \geq |x| = x. \tag{3.15}$$

Let $y = u^{-1}xu - x$ and define

$$e_+ := s(y_+), \quad e_- := \mathbf{1} - e_+.$$

We claim that $e_+ = \mathbf{1}$. By inequality (3.15) and the definition of e_-, we have that $e_-(x - u^{-1}xu)e_- \geq e_-xe_-$, or equivalently, $e_-u^{-1}xue_- \leq 0$. On the other hand, since $u^{-1}xu \geq 0$, it follows that $e_-u^{-1}xue_- \geq 0$. Hence, $e_-u^{-1}xue_- = 0$, or equivalently, $x^{1/2}ue_- = 0$, which in turn implies $xue_-u^{-1} = 0$. Thus,

$$x = xu(e_+ + e_-)u^{-1} = xue_+u^{-1}.$$

The latter equality and definition of the support projection imply that $ue_+u^{-1} \geq s(x) = \mathbf{1}$. Hence, $e_+ = \mathbf{1}$, as required.

Next, by the definition of e_+, the functional calculus and inequality (3.15), we have that

$$e_+(u^{-1}xu - x)e_+ = e_+\big|u^{-1}xu - x\big|e_+ \geq e_+xe_+.$$

Since $e_+ = \mathbf{1}$, it follows that $u^{-1}xu \geq 2x$. Hence,

$$1 = \|x\|_{\mathcal{M}} \geq 2\|x\|_{\mathcal{M}} = 2,$$

which is a contradiction. Thus, inequality (3.14) is false.

Throughout this section, unless stated otherwise, we assume that $\mathcal{M}$ is a semifinite properly infinite von Neumann algebra acting on a separable Hilbert space H. We first prove an auxiliary result for maximal abelian subalgebras of $\mathcal{M}$.

Theorem 3.6.2. *Let $\mathcal{M}$ be a finite von Neumann algebra. If $\mathcal{A}$ is a maximal abelian subalgebra in $\mathcal{M}$ and if $p \in P(\mathcal{M})$, then there exists $q \in \mathcal{A}$, such that $p \sim q$.*

Proof. Let $\mathcal{T}$ be the canonical center-valued trace on $\mathcal{M}$ (see Theorem 1.10.2). Since $\mathcal{M}$ is a finite von Neumann algebra, Theorem 1.9.14 implies that there exists a sequence $\{z_k\}_{k\in\mathbb{N}} \subset P(\mathcal{Z}(\mathcal{M}))$ of pairwise orthogonal central projections, such that:
(1) $\mathcal{M}z_1$ has type II_1;
(2) $\mathcal{M}z_k$ has type I_{n_k} for all $k \geq 2$ and distinct family of integers $n_k \in \mathbb{N}$;
(3) $\sum_{k\in\mathbb{N}} z_k = \mathbf{1}$.

Here, some z_k may vanish. It suffices to prove the assertion separately for each algebra $\mathcal{M}z_k$, $k \in \mathbb{N}$. We may assume without loss of generality that $z_k = \mathbf{1}$ for some $k \in \mathbb{N}$.

If $z_1 = \mathbf{1}$, then $\mathcal{M}$ has type II_1. By [143, Theorem 5.6.2], there exists $q \in \mathcal{A}$, such that $\mathcal{T}(q) = \mathcal{T}(p)$. Proposition 1.10.3 implies that $p \sim q$, as required.

Assume now that $z_k = \mathbf{1}$ for some $k \geq 2$, so that $\mathcal{M}$ is a type I_{n_k} algebra. By Proposition 1.9.16, we have that $\mathcal{M} = M_{n_k\times n_k}(L_\infty(\Omega,\Sigma,\mu))$ for some σ-finite measure space (Ω,Σ,μ). Let $D_{n_k\times n_k}(L_\infty(\Omega,\Sigma,\mu))$ be the subalgebra of all diagonal matrices with entries from $L_\infty(\Omega,\Sigma,\mu)$. By [143, Theorem 2.4.3], without loss of generality, $\mathcal{A} = D_{n_k\times n_k}(L_\infty(\Omega,\Sigma,\mu))$.

If $p \in \mathcal{M}$ is a projection, then $p(\omega) \in M_{n_k\times n_k}(\mathbb{C})$ is a projection for almost every $\omega \in \Omega$. Hence, $\mathrm{tr}(p(\omega)) \in \{0,1,\ldots,n_k\}$ for almost every $\omega \in \Omega$. Now, let

$$A_l = \{\omega \in \Omega : \mathrm{tr}(p(\omega)) = l\}, \quad 1 \leq l \leq n_k.$$

Each A_l is a measurable set. Let e_{ij}, $1 \leq i,j \leq n_k$, be matrix units in $M_{n_k\times n_k}(\mathbb{C})$. Define a projection $q \in \mathcal{A}$ by setting

$$q(\omega) = \sum_{m=1}^{l} e_{mm}, \quad \omega \in A_l.$$

Clearly, we have $\mathcal{T}(p) = \mathcal{T}(q)$. Referring to Proposition 1.10.3 once again, we conclude that $p \sim q$. □

We now move onto proving the commutator estimates (3.13). We first prove (3.13) under additional assumption that x is a compact operator relative to $\mathcal{M}$. In the following lemma, we first show that for a self-adjoint compact operator x associated with $\mathcal{M}$ and any given finite projection p we may find an equivalent projection q which commutes with x.

Lemma 3.6.3. *Let $\mathcal{M}$ be a semifinite properly infinite von Neumann algebra, let $x = x^* \in S_0(\mathcal{M})$ with $s(x) = \mathbf{1}$ and let $r \in P(\mathcal{M})$ be a finite projection. Then there exists a projection r_0, such that $r_0 \sim r$ and $r_0x = xr_0$.*

Proof. For every $n \in \mathbb{Z}$, we set $p_n = E_{|x|}(2^{-n},\infty)$. By assumption, x is compact and, therefore, p_n is finite for every $n \in \mathbb{Z}$. By Theorem 1.9.5, for every $n \in \mathbb{Z}$, there exists a central projection z_n, such that

$$rz_n \preceq p_n z_n, \quad p_n(\mathbf{1} - z_n) \preceq r(\mathbf{1} - z_n). \tag{3.16}$$

Since p_n is finite, we have that rz_n, $n \in \mathbb{Z}$, is finite.

Define the central projection $z \in P(\mathcal{Z}(\mathcal{M}))$ by setting

$$z = \inf_{n \in \mathbb{Z}} z_n.$$

Since $rz_n \preceq p_n z_n$ and $z \leq z_n$, Theorem 1.9.4(iii) implies that

$$rz \preceq p_n z, \quad n \in \mathbb{Z}. \tag{3.17}$$

By assumption, the von Neumann algebra $\mathcal{M}$ is semifinite and, therefore, since the projection p_0 is finite, Corollary 1.10.17 guarantees that there exists a faithful normal semifinite trace τ on $\mathcal{M}$, such that $\tau(p_0) < \infty$. It follows from the definition of the projections p_n that $\tau(p_n z) \to 0$ as $n \to -\infty$. The majorization (3.17) guarantees that $\tau(rz) \leq \tau(p_n z) \to 0$ as $n \to -\infty$. Hence,

$$rz = 0. \tag{3.18}$$

Define $w \in P(\mathcal{Z}(\mathcal{M}))$ by setting

$$w = \inf_{n \in \mathbb{Z}} (\mathbf{1} - z_n).$$

Since $p_n(\mathbf{1} - z_n) \preceq r(\mathbf{1} - z_n)$ and $w \leq \mathbf{1} - z_n$, it follows that $p_n w \preceq rw$ for all $n \in \mathbb{Z}$. Since $\sup_{n \in \mathbb{Z}} p_n = \mathbf{1}$, it follows that $\mathcal{D}(w) = \mathcal{D}(\sup_{n \in \mathbb{Z}} p_n w) \leq \mathcal{D}(rw)$ where $\mathcal{D}$ be a dimension function. By assumption, r is finite, and so rw is finite too. Since dimension functions take a.e. finite values only on finite projections (see Definition 1.11.1), we infer that the projection w is finite, too. Since $\mathcal{M}$ is properly infinite algebra and w is a central projection, it follows that $w = 0$. By the definition of w, the latter equality implies that

$$\sup_{n \in \mathbb{Z}} z_n = \mathbf{1}.$$

Let us define a sequence $\{w_n\}_{n \in \mathbb{Z}} \subset P(\mathcal{Z}(\mathcal{M}))$ by setting

$$w_n = z_n \wedge \Big(\bigwedge_{k<n} (\mathbf{1} - z_k) \Big), \quad n \in \mathbb{Z}.$$

It is clear that $\{w_n\}_{n \in \mathbb{Z}}$ are pairwise orthogonal and $w_n \leq z_n$ for every $n \in \mathbb{Z}$. An exercise in Boolean algebra operation yields

$$\mathbf{1} = z \vee \Big(\bigvee_{k \in \mathbb{Z}} w_k \Big).$$

It is sufficient to prove the assertion separately in each algebra $\mathcal{M}w_k, k \in \mathbb{Z}$. Indeed, if $r_{0,k}$ is the projection in the algebra $w_k\mathcal{M}$ satisfying $r_{0,k} \sim rw_k$ and $r_{0,k}x = xr_{0,k}$, then setting $r_0 = \sum_{k\in\mathbb{Z}} r_{0,k} \in \sum_{k\in\mathbb{Z}} w_k\mathcal{M}$, we have that $r_0 \sum_{k\in\mathbb{Z}} w_k \sim r \sum_{k\in\mathbb{Z}} w_k$ and $r_0x = xr_0$. In addition, in the algebra $z\mathcal{M}$, we have that $zr = 0$. Hence, for the direct summand $(\mathbf{1} - \sum_{k\in\mathbb{Z}} w_k)\mathcal{M} \subset z\mathcal{M}$ we may take zero projection, which clearly is equivalent to the projection $rz = 0$ and commutes with x.

Thus, in what follows, without loss of generality we assume $w_k = \mathbf{1}$ for some fixed $k \in \mathbb{Z}$. Under this assumption, we have that $z_k = \mathbf{1}$ and, therefore, (3.16) implies that $r \preceq p_k$. In particular, there exists r_1, such that $r_1 \leq p_k$ and $r_1 \sim r$. Replacing r with r_1, we may assume without loss of generality that $r_1 \leq p_k$. Since p_k is finite, the von Neumann algebra $\mathcal{N} = p_k\mathcal{M}p_k$ is finite. Let $\mathcal{A}$ be a maximal Abelian subalgebra in $\mathcal{N}$ containing all spectral projections of xp_k. By Theorem 3.6.2 (applied to the finite von Neumann algebra $\mathcal{N}$ and to the maximal Abelian subalgebra $\mathcal{A}$), there exists a projection $r_0 \in \mathcal{A}$, such that $r_0 \sim r$. By the definition of the algebra $\mathcal{A}$, it follows that $r_0x = xr_0$, which suffices to conclude the proof. □

In the following lemma we develop the idea of Lemma 3.6.3 even further.

Lemma 3.6.4. *Let $\mathcal{M}$ be a semifinite properly infinite von Neumann algebra and let $x = x^* \in S_0(\mathcal{M})$. Let $r_1, r_2 \in \mathcal{M}$ be projections, such that $r_1 \preceq r_2$. If $r_2x = xr_2$, then there exists $r_3 \leq r_2$, such that $r_1 \sim r_3$ and $r_3x = xr_3$.*

Proof. By the assumption, the projection r_2 commutes with x and, therefore, r_2 commutes with $n(x)$ and $s(x)$. Let $z_1 \in P(\mathcal{Z}(\mathcal{M}))$ be the maximal central projection (possibly zero), such that r_1z_1 is properly infinite and the central support of r_1z_1 is z_1. Let z_2 be the maximal central projection (possibly zero), such that r_2z_2 is finite. Since $z_2r_1 \preceq z_2r_2$, it follows that z_2r_1 is a finite projection. Therefore, the choice of z_1 implies that $z_1z_2 = 0$. Let z_3 be the maximal central projection, such that $z_3 \leq \mathbf{1} - z_1 - z_2$, $r_2n(x)z_3$ is properly infinite and central support of $r_2n(x)z_3$ is z_3. Set $z_4 = \mathbf{1} - z_1 - z_2 - z_3$.

By construction, we have that $\sum_{k=1}^4 z_k = \mathbf{1}$ and

(i) r_1z_1 is properly infinite and the central support of r_1z_1 is z_1;
(ii) r_1z_2 and r_2z_2 are finite;
(iii) r_1z_3 is finite, $r_2 \cdot n(x) \cdot z_3$ is properly infinite and central support of $r_2 \cdot n(x) \cdot z_3$ is z_3;
(iv) r_1z_4 is finite, $r_2 \cdot s(x) \cdot z_4$ is properly infinite and central support of $r_2 \cdot s(x) \cdot z_4$ is z_4.

Here, the projections r_1z_3 and r_1z_4 are finite due to the maximality of z_1.

It suffices to prove the assertion separately in each algebra $z_k\mathcal{M}, 1 \leq k \leq 4$. That is, we may assume without loss of generality that $z_k = \mathbf{1}$ for some $1 \leq k \leq 4$.

Assume that $z_1 = \mathbf{1}$. Then r_1 is properly infinite and the central support of r_1 is $\mathbf{1}$. Since $r_1 \preceq r_2$, then r_2 is also properly infinite and the central support of r_2 is $\mathbf{1}$. Hence, by Theorem 1.9.9(viii), we have that $r_1 \sim r_2$. In this case, we may take $r_3 = r_2$, to obtain the projection r_3 satisfying $r_3 \sim r_1$ and $r_3x = xr_3$.

Assume now that $z_2 = \mathbf{1}$. In this case, r_2 is finite. Since $r_1 \preceq r_2$, there exists $r_4 \leq r_2$, such that $r_4 \sim r_1$. By replacing r_1 with r_4, we may assume without loss of generality that $r_1 \leq r_2$. Let $\mathcal{A}$ be a maximal Abelian subalgebra in the finite von Neumann algebra $\mathcal{N} = r_2\mathcal{M}r_2$ containing all spectral projections of xr_2. Using Theorem 3.6.2 with the finite von Neumann algebra $\mathcal{N}$ and the maximal Abelian subalgebra $\mathcal{A}$, we may find a projection $r_3 \in \mathcal{A}$, such that $r_3 \sim r_1$. By the definition of $\mathcal{A}$, we have that $r_3x = xr_3$.

Next, assume that $z_3 = \mathbf{1}$. In this case r_1 is finite, while the projection $r_2 \cdot n(x)$ is properly infinite with central support equal to the identity. Hence, by Lemma 1.9.12, we have that $r_1 \preceq r_2 \cdot n(x)$. Let $r_3 \leq r_2 \cdot n(x)$ be such that $r_3 \sim r_1$. Since $r_3 \leq n(x)$, it follows that r_3 commutes with x, as required.

Finally, let $z_4 = \mathbf{1}$. Similar to the previous case, we have that $r_1 \preceq r_2 \cdot s(x)$. Let $r_4 \leq r_2 \cdot s(x)$ be such that $r_1 \sim r_4$. By replacing r_1 with r_4, we may assume without loss of generality that $r_1 \leq r_2 \cdot s(x)$. By replacing $\mathcal{M}$ with $(r_2 \cdot s(x))\mathcal{M}(r_2 \cdot s(x))$, we may assume without loss of generality that $r_2 = \mathbf{1}$ and $s(x) = \mathbf{1}$. Appealing to Lemma 3.6.3, we complete the proof for the case $z_4 = \mathbf{1}$. □

Next, we introduce an auxiliary notion of so-called (x, ε)-good partition for a positive compact operator $x \in S_0(\mathcal{M})$. For a fixed $\varepsilon \in (0, 1)$ and a positive compact operator $x \in S_0(\mathcal{M})$, an (x, ε)-good partition $\{p_n\}$ allows us to construct a unitary u satisfying inequality (3.13).

Definition 3.6.5. Let $\mathcal{M}$ be a semifinite properly infinite von Neumann algebra. Let $x \in S_0(\mathcal{M})$ be positive with $s(x) = \mathbf{1}$. A partition of unity $\{p_n\}_{n\in\mathbb{N}_0} \subset \mathcal{M}$ is said to be (x, ε)-*good partition* for some $\varepsilon > 0$ if:
(i) $xp_n = p_nx$ for all $n \in \mathbb{N}_0$;
(ii) $p_{2n} \sim p_{2n+1}$ for all $n \in \mathbb{N}_0$;
(iii) for all $n \in \mathbb{N}_0$, we have

$$p_{2n} \leq E_x(\varepsilon^{2n}, \infty), \quad p_{2n+1} \leq E_x[0, \varepsilon^{2n+1}].$$

Remark 3.6.6. *We note that if $x \in S_0(\mathcal{M})$ is positive and $\{p_n\}$ is any family of projections which satisfies the assumptions of Definition* 3.6.5(ii) *and* (iii) *for some $\varepsilon > 0$, then every projection p_n is necessarily finite. Indeed, since x is compact, it follows that the projection $E_x(\varepsilon^{2n}, \infty)$ is finite for any $\varepsilon > 0$ and $n \in \mathbb{N}_0$. Since $p_{2n} \leq E_x(\varepsilon^{2n}, \infty)$, it follows that p_{2n} is finite too. The equivalence $p_{2n} \sim p_{2n+1}$ implies that the projections p_{2n+1} are finite too.*

We first show that such for any positive $x \in S_0(\mathcal{M})$ and any ε, an (x, ε)-good partition always exists.

Proposition 3.6.7. *Let $\mathcal{M}$ be a semifinite properly infinite von Neumann algebra. Let $x \in S_0(\mathcal{M})$ be positive with $s(x) = \mathbf{1}$. For every $\varepsilon \in (0, 1)$, there exists an (x, ε)-good partition of unity.*

Proof. Note that since $E_x(0,\infty) = s(x) = \mathbf{1}$, it follows that $E_x(\{0\}) = 0$, and so $E_x(0,\lambda) = E_x[0,\lambda)$ for any $\lambda > 0$.

Fix $\varepsilon > 0$. Since x is compact, it follows that the projection $E_x(\varepsilon,\infty)$ is finite. Hence, by Proposition 1.9.10(i), the projection $E_x(0,\varepsilon] = \mathbf{1} - E_x(\varepsilon,\infty)$ is a properly infinite projection with central support equal to $\mathbf{1}$.

Set $p_0 = E_x(1,\infty)$. Using again compactness of x, we have that p_0 is a finite projection. Hence, by Lemma 1.9.12, we obtain that $p_0 \preceq E_x(0,\varepsilon]$. Since $E_x(0,\varepsilon]$ commutes with x, we may employ Lemma 3.6.4 (with $r_1 = p_0$ and $r_2 = E_x(0,\varepsilon]$) to find $p_1 \leq E_x(0,\varepsilon]$, such that $p_1p_0 = 0$, $p_1 \sim p_0$ and $xp_1 = p_1x$.

We now construct an (x,ε)-good sequence $\{p_n\}_{n\in\mathbb{N}_0}$ by induction on n. For $n = 0$ and $n = 1$, the projections p_0 and p_1 are already constructed in the preceding paragraph.

Fix $n \in \mathbb{N}_0$ and assume that $\{p_k\}_{k=0}^{2n-1}$ are constructed. Set

$$r_n = \inf_{k=0,\dots,2n-1} (\mathbf{1} - p_k).$$

Since $\{p_k\}_{k=0}^{2n-1}$ are pairwise orthogonal, it follows that $r_n = \prod_{k=0}^{2n-1}(\mathbf{1} - p_k)$. We define

$$p_{2n} = E_x(\varepsilon^{2n},\infty)\cdot r_n, \quad P_{2n} = E_x(0,\varepsilon^{2n+1}]\cdot r_n.$$

By assumption of induction, every projection p_k, $0 \leq k < 2n-1$ commutes with x, and hence, so does r_n. Since r_n commutes with spectral projections of x, it follows that p_{2n} and P_{2n} are also projections.

By Remark 3.6.6, the projections $\{p_k\}_{k=0}^{2n-1}$, $0 \leq k < 2n$ are finite. Since they are also pairwise orthogonal, it follows that

$$\mathbf{1} - r_n = \sum_{k=0}^{2n-1} p_k \tag{3.19}$$

is also finite. In particular, using the fact that $s(x) = E_x(0,\infty) = \mathbf{1}$, we may write

$$\begin{aligned} P_{2n} &= E_x(0,\varepsilon^{2n+1}] - E_x(0,\varepsilon^{2n+1}](\mathbf{1} - r_n) \\ &= \mathbf{1} - (E_x(\varepsilon^{2n+1},\infty) + E_x(0,\varepsilon^{2n+1}]\cdot(\mathbf{1} - r_n)). \end{aligned} \tag{3.20}$$

Since both projections $E_x(\varepsilon^{2n+1},\infty)$ and $E_x(0,\varepsilon^{2n+1}]\cdot(\mathbf{1} - r_n)$ are finite, equation (3.20) combined with Proposition 1.9.10(i) implies that P_{2n} is a properly infinite projection with the central support equal to $\mathbf{1}$. Using Lemma 1.9.12 for the finite projection p_{2n} and the properly infinite projection P_{2n} with $s_c(P_{2n}) = \mathbf{1}$, we obtain that $p_{2n} \preceq P_{2n}$. By Lemma 3.6.4 (with $r_1 = p_{2n}$ and $r_2 = P_{2n}$), there exists $p_{2n+1} \leq P_{2n}$, such that $p_{2n+1} \sim p_{2n}$ and $p_{2n+1}x = xp_{2n+1}$.

We claim that the obtained sequence $\{p_k\}_{k=0}^{2n+1}$ satisfies all three conditions of Definition 3.6.5. By construction, $p_{2n} \sim p_{2n+1}$ and

$$p_{2n} \le E_x(\varepsilon^{2n}, \infty), \quad p_{2n+1} \le P_{2n} \le E_x(0, \varepsilon^{2n+1}] \le E_x(0, \varepsilon^{2n}].$$

Hence, $p_{2n}p_{2n+1} = 0$. Since $r_n p_k = 0$ for $0 \le k < 2n$, it follows that

$$p_{2n}p_k = p_{2n+1}p_k = 0, \quad 0 \le k < 2n,$$

as required.

Thus, by induction, there exists a sequence $\{p_n\}_{n\in\mathbb{N}_0}$ of finite pairwise orthogonal projections satisfying all three conditions of Definition 3.6.5. To conclude the proof, we only need to show that $\{p_n\}_{n\in\mathbb{N}_0}$ is a partition of unity.

Note that the definition of p_{2n} implies

$$\begin{aligned} E_x(\varepsilon^{2n}, \infty) &= (E_x(\varepsilon^{2n}, \infty) \cdot (\mathbf{1} - r_n)) \vee (E_x(\varepsilon^{2n}, \infty) \cdot r_n) \\ &= (E_x(\varepsilon^{2n}, \infty) \cdot (\mathbf{1} - r_n)) \vee p_{2n} \\ &\le (\mathbf{1} - r_n) \vee p_{2n} \overset{(3.19)}{=} \left(\sum_{k=0}^{2n-1} p_k\right) \vee p_{2n} \\ &= \sum_{k=0}^{2n} p_k. \end{aligned}$$

Since $\varepsilon \in (0,1)$, passing $n \to \infty$, we obtain

$$E_x(0, \infty) \le \sum_{k=0}^{\infty} p_k.$$

By assumption, $E_x(0, \infty) = s(x) = \mathbf{1}$ and, therefore,

$$\mathbf{1} \le \sum_{k=0}^{\infty} p_k, \quad \text{or equivalently,} \quad \sum_{k=0}^{\infty} p_k = \mathbf{1}.$$

This concludes the proof that the sequence $\{p_n\}_{n\in\mathbb{N}_0}$ is an (x, ε)-good partition of unity. □

Now using the existence of an (x, ε)-good partition of unity for a positive compact operator $x \in S_0(\mathcal{M})$, we prove the commutator estimates (3.13) for a positive compact operator.

Proposition 3.6.8. *Let $\mathcal{M}$ be a semifinite properly infinite von Neumann algebra and let $x \in S_0(\mathcal{M})$ be a positive operator. For every $\varepsilon > 0$, there exists a unitary element $u \in \mathcal{M}$, such that*

$$\big|[u, x]\big| \ge (1 - \varepsilon)x. \tag{3.21}$$

Proof. Fix $\varepsilon \in (0,1)$. Without loss of generality, we may assume that $s(x) = \mathbf{1}$. Then, by Proposition 3.6.7, there exists an (x, ε)-good partition of unity $\{p_n\}_{n\in\mathbb{N}_0}$. By the definition of (x,ε)-good partition of unity for every $n \in \mathbb{N}_0$, there exists a partial isometry $v_n \in \mathcal{M}$, such that $v_n^* v_n = p_{2n}$ and $v_n v_n^* = p_{2n+1}$. Since the projections $\{p_n\}_{n\in\mathbb{N}_0}$ are pairwise orthogonal, we have that

$$\begin{aligned} v_n v_m &= v_n p_{2n} \cdot p_{2m+1} v_m = v_n \cdot p_{2n} p_{2m+1} \cdot v_m = 0, \quad n,m \in \mathbb{N}_0, \\ v_n v_m^* &= v_n p_{2n} \cdot p_{2m} v_m^* = v_n \cdot p_{2n} p_{2m} \cdot v_m^* = 0, \quad n \neq m, \\ v_n^* v_m &= v_n^* p_{2n+1} \cdot p_{2m+1} v_m = v_n^* \cdot p_{2n+1} p_{2m+1} \cdot v_m = 0, \quad n \neq m, \\ v_n^* v_m^* &= v_n^* p_{2n+1} \cdot p_{2m} v_m^* = v_n^* \cdot p_{2n+1} p_{2m} \cdot v_m^* = 0, \quad n,m \in \mathbb{N}_0. \end{aligned} \tag{3.22}$$

We define

$$u = \sum_{n=0}^{\infty} (v_n + v_n^*) \in \mathcal{M},$$

where the series converges in the strong operator topology as a series of uniformly bounded family of pairwise disjoint operators. We claim that u is the unitary operator satisfying (3.21).

It follows from the definition of u that $u = u^*$. In addition, by (3.22) and the fact that $\{p_n\}_{n\in\mathbb{N}_0}$ is a partition of unity, we have that

$$\begin{aligned} u^2 &= \sum_{n,m=0}^{\infty} (v_n v_m + v_n v_m^* + v_n^* v_m + v_n^* v_m^*) \\ &= \sum_{n=0}^{\infty} (v_n^2 + v_n v_n^* + v_n^* v_n + (v_n^*)^2) \\ &= \sum_{n=0}^{\infty} (p_{2n} + p_{2n+1}) = \mathbf{1}. \end{aligned}$$

Thus, $u \in \mathcal{M}$ is indeed a unitary.

To show the inequality (3.21), we first note that by the definition of u and properties of $\{v_n\}_{n\in\mathbb{N}_0}$, we may write

$$u p_{2n} = p_{2n+1} u, \quad u p_{2n+1} = p_{2n} u, \quad n \in \mathbb{N}_0.$$

Hence, using the fact that $\{p_n\}_{n\in\mathbb{N}_0}$ is an (x,ε)-good partition of unity (and so every p_n commutes with x), we infer that the operators x and $u^{-1}xu$ commute with every p_k, $k \in \mathbb{N}_0$. Therefore, the operator $|x - u^{-1}xu|$ also commutes with every $\{p_k\}_{k\in\mathbb{N}_0}$.

By the definition of an (x,ε)-good partition of unity (see Definition 3.6.5(iii)) for every $n \in \mathbb{N}_0$, we have that

$$x p_{2n} \geq \varepsilon^{2n} p_{2n}, \quad x p_{2n+1} \leq \varepsilon^{2n+1} p_{2n+1}, \tag{3.23}$$

and so

$$\begin{aligned}(u^{-1}xu)p_{2n} &= u^{-1}\cdot xp_{2n+1}\cdot u \le \varepsilon^{2n+1}p_{2n},\\ (u^{-1}xu)p_{2n+1} &= u^{-1}\cdot xp_{2n}\cdot u \ge \varepsilon^{2n}p_{2n+1}.\end{aligned} \tag{3.24}$$

Therefore, combining these estimates with the fact that $|x - u^{-1}xu|$ commutes with every p_k, $k \in \mathbb{N}_0$, for every $n \in \mathbb{N}_0$ we obtain

$$\begin{aligned}|x - u^{-1}xu|p_{2n} \ge (x - u^{-1}xu)p_{2n} &\overset{(3.24)}{\ge} xp_{2n} - \varepsilon^{2n+1}p_{2n}\\ &\overset{(3.23)}{\ge} (1-\varepsilon)xp_{2n}\end{aligned}$$

and

$$\begin{aligned}|x - u^{-1}xu|p_{2n+1} \ge (u^{-1}xu - x)p_{2n+1} &\overset{(3.24)}{\ge} \varepsilon^{2n}p_{2n+1} - xp_{2n+1}\\ &\overset{(3.23)}{\ge} \left(\frac{1}{\varepsilon} - 1\right)xp_{2n+1} \ge (1-\varepsilon)xp_{2n+1}.\end{aligned}$$

It remains to show that $|[u,x]| \ge (1-\varepsilon)x$. Let $\xi \in \operatorname{dom} x$. Then $\xi \in \operatorname{dom}(|[u,x]|) = \operatorname{dom}(x)$. Since $\{p_n\}_{n\in\mathbb{N}_0}$ is a partition of unity, we have that $\xi = \sum_{k=1}^{\infty} p_k\xi$. Furthermore, since the operators x and $|x - u^{-1}xu|$ are closed, it follows that $x\xi = \sum_{k=1}^{\infty} xp_k\xi$ and $|x - u^{-1}xu|\xi = \sum_{k=1}^{\infty} |x - u^{-1}xu|p_k\xi$. Hence,

$$\begin{aligned}\langle |[u,x]|\xi,\xi\rangle = \langle |x - u^{-1}xu|\xi,\xi\rangle &= \left\langle \sum_{k=0}^{\infty} |x - u^{-1}xu|p_k\xi, p_k\xi\right\rangle\\ &= \sum_{k=0}^{\infty}\langle |x - u^{-1}xu|p_k\xi, p_k\xi\rangle \ge (1-\varepsilon)\sum_{k=0}^{\infty}\langle xp_k\xi,\xi\rangle\\ &= (1-\varepsilon)\langle x\xi,\xi\rangle.\end{aligned}$$

Since $\xi \in \operatorname{dom}(x)$ is arbitrary, it follows that $|[u,x]| \ge (1-\varepsilon)x$, as required. □

Next, we prove a version of Proposition 3.6.8 for general self-adjoint compact operator. Before we prove this result, we firstly show that there is a convenient central partition of unity $\{w_k\}_{k=1}^{6}$, such that in each of the reduced algebras $w_k\mathcal{M}$, $k = 1,\dots,6$, we may apply the previously established results.

Lemma 3.6.9. *Let $\mathcal{M}$ be a semifinite properly infinite von Neumann algebra. Let $x = x^* \in S_0(\mathcal{M})$ with $s(x) = \mathbf{1}$. Then there exists a central partition of unity $\{w_k\}_{k=1}^{6}$, such that:*

(i) *The projection $s(xw_1)$ is finite;*
(ii) *The projections $s(x_-w_2)$ and $s(x_+w_2)$ are properly infinite with central supports equal to w_2;*
(iii) *The projection $s(x_-w_3)$ is finite, and the projections $n(xw_3)$ and $s(x_+w_3)$ are properly infinite with central supports equal to w_3;*
(iv) *The projection $s(x_+w_4)$ is finite, and the projections $n(xw_4)$ and $s(x_-w_4)$ are properly infinite with central supports equal to w_4;*

(v) *The projections* $s(x_-w_5)$, $n(xw_5)$ *are finite, and the projection* $s(x_+w_5)$ *is properly infinite with central support equal to* w_5*;*
(vi) *The projections* $s(x_+w_6)$, $n(xw_6)$ *are finite, and the projection* $s(x_-w_6)$ *is properly infinite with central support equal to* w_6.

Proof. To find the required partition of unity $\{w_k\}_{k=1}^6$, we start with projections z_0, z_1, z_2 (possibly zero) defined in the following way. Let $z_0 \in P(\mathcal{Z}(\mathcal{M}))$ be the maximal central projection, such that $s(x_+)z_0$ is properly infinite and central support of x_+z_0 is z_0. Similarly, let z_1 be the maximal central projection, such that $s(x_-)z_1$ is properly infinite and central support of x_-z_1 is z_1. Also, let z_2 be the maximal central projection, such that $n(x)z_2$ is properly infinite and central support of $n(x)z_2$ is z_2. Since the von Neumann algebra $\mathcal{M}$ is properly infinite and $s(x) = \mathbf{1}$, it follows that $\sup_{k=0,1,2} z_k = \mathbf{1}$, i. e., projections z_0, z_1, z_2 cannot be simultaneously zero.

Now we define

$$w_1 = (\mathbf{1} - z_0) \cdot (\mathbf{1} - z_1), \qquad w_2 = z_0 \cdot z_1,$$
$$w_3 = (\mathbf{1} - z_1) \cdot z_0 \cdot z_2, \qquad w_4 = (\mathbf{1} - z_0) \cdot z_1 \cdot z_2,$$
$$w_5 = (\mathbf{1} - z_1) \cdot (\mathbf{1} - z_2) \cdot z_0, \qquad w_6 = (\mathbf{1} - z_0) \cdot (\mathbf{1} - z_2) \cdot z_1.$$

It is clear that $\{w_k\}_{k=1}^6$ are pairwise orthogonal central projections with $\sum_{k=1}^6 w_k = \mathbf{1}$.

By construction, we have that $s(xw_1)$ is finite and $s(x_-w_2)$, $s(x_+w_2)$ are properly infinite projections with central support equal to w_2. Furthermore, for the projection w_3, the inequality $w_3 \le z_0$ (resp., $w_3 \le z_2$) implies that $s(x_+w_3)$ (resp., $n(x)w_3$) is a properly infinite projection with central support equal to w_3, while the inequality $w_3 \le \mathbf{1} - z_1$ implies that $s(x_-)w_3$ is finite. Repeating the argument, we obtain that $\{w_k\}_{k=1}^6$ is a required central partition of unity. □

In the following proposition we show that the assumption of positivity of the operator x in Proposition 3.6.8 can be dropped.

Proposition 3.6.10. *Let* $\mathcal{M}$ *be a semifinite properly infinite von Neumann algebra and let* $x \in S_0(\mathcal{M})$ *be self-adjoint. For every* $\varepsilon > 0$*, there exists a unitary* $u \in \mathcal{M}$*, such that*

$$\big|[u, x]\big| \ge (1 - \varepsilon)|x|.$$

Proof. Without loss of generality, we may assume that $s(x) = \mathbf{1}$. Let $\{w_k\}_{k=1}^6$ be the central partition of unity given by Lemma 3.6.9. It suffices to prove the assertion separately for each xw_k, $k = 1, \dots, 6$ and, therefore, in what follows, we may assume that $w_k = \mathbf{1}$ for some $k = 1, \dots, 6$.

Suppose that $w_1 = \mathbf{1}$. In this case, $s(x)$ is finite. Hence, since $\mathcal{M}$ is a properly infinite von Neumann algebra, Proposition 1.9.10(i) implies that $n(x) = \mathbf{1} - s(x)$ is a properly infinite projection with central support equal to $\mathbf{1}$. Hence, by Lemma 1.9.12, we have that

$s(x) \preceq n(x)$. Let $p \in P(\mathcal{M})$ be such that $p \leq n(x)$ and $p \sim s(x)$. Choose a partial isometry v, such that $v^*v = s(x)$ and $vv^* = p$ and define a unitary element $u \in \mathcal{M}$ by setting

$$u = v + v^* + (n(x) - p).$$

One can check that the operators x and $u^{-1}xu$ are orthogonal and, therefore,

$$\big|[u, x]\big| = \big|x - u^{-1}xu\big| = |x| + u^{-1}|x|u \geq (1 - \varepsilon)|x|.$$

This proves the assertion for the case $w_1 = \mathbf{1}$.

In the case $w_2 = \mathbf{1}$, Theorem 1.9.9(viii) implies that $s(x_+) \sim s(x_-)$. Hence, the assertion follows from Lemma 3.5.2. Similarly, for $w_3 = \mathbf{1}$ (resp., $w_4 = \mathbf{1}$) we have that $s(x_+) \sim s(x_-) + n(x)$ (resp., $s(x_+) + n(x) \sim s(x_-)$). Hence, another application of Lemma 3.5.2 provides the required unitary operator for w_3 and w_4.

Suppose now that $w_5 = \mathbf{1}$. Since both projections $s(x_-), n(x)$ are finite (and so, $s(x_-) + n(x)$ is finite, too) and the projection $s(x_+)$ is properly infinite, Lemma 1.9.12 implies that $s(x_-) + n(x) \preceq s(x_+)$. By Lemma 3.6.4, there exists $p \leq s(x_+)$, such that $px = xp$ and $p \sim s(x_-) + n(x) =: q$. Applying Lemma 3.5.2 to the operator $x(p + q)$ in the algebra $(p + q)\mathcal{M}(p + q)$, we obtain a unitary element $u_1 \in (p + q)\mathcal{M}(p + q)$, such that

$$\big|[u_1, x(p + q)]\big| \geq |x|(p + q).$$

Since $p + q$ is finite, Proposition 1.9.10(i) implies that $(\mathbf{1} - p - q)$ is properly infinite. In addition, the operator $x(\mathbf{1} - p - q)$ is positive (and compact). Hence, by Proposition 3.6.8 applied to the operator $x(\mathbf{1} - p - q)$ and the algebra $(\mathbf{1} - p - q)\mathcal{M}(\mathbf{1} - p - q)$, there exists a unitary element $u_2 \in (\mathbf{1} - p - q)\mathcal{M}(\mathbf{1} - p - q)$, such that

$$\big|[u_2, x(\mathbf{1} - p - q)]\big| \geq (1 - \varepsilon)|x|(\mathbf{1} - p - q).$$

Setting $u = u_1 + u_2$, we complete the proof for the case $w_5 = \mathbf{1}$.

The case, $w_6 = \mathbf{1}$ can be completed similar to the case $w_5 = \mathbf{1}$. □

Finally, we prove the commutator estimates (3.13) for an arbitrary locally measurable operator affiliated with a semifinite properly infinite algebra.

Theorem 3.6.11. *Let $\mathcal{M}$ be an arbitrary von Neumann algebra and let $x = x^* \in LS(\mathcal{M})$. There exists a self-adjoint central element $c \in \mathcal{Z}(LS(\mathcal{M}))$, such that for every $\varepsilon > 0$, there exists a unitary $u \in \mathcal{M}$ with*

$$\big|[u, x]\big| \geq (1 - \varepsilon)|x - c|.$$

Proof. Assume first that $\mathcal{M}$ is a semifinite properly infinite algebra. By Theorem 1.10.13(ii), there exists a faithful normal semifinite trace τ on $\mathcal{M}$.

Let Z be a Polish space, such that $\mathcal{Z}(\mathcal{M}) \cong L_\infty(Z, w)$ and

$$\mathcal{M} = \int_Z^{\oplus} \mathcal{M}_\zeta \, dw(\zeta),$$

where almost every $\mathcal{M}_\zeta$ is factor (see Theorem 1.12.9). By Theorem 1.12.10 and Theorem 1.12.11, $\mathcal{M}_\zeta$ is a semifinite properly infinite factor for almost every $\zeta \in Z$ and

$$\tau = \int_Z^{\oplus} \tau_\zeta \, dw(\zeta),$$

for a measurable field $\zeta \to \tau_\zeta$ of faithful normal semifinite traces on $\mathcal{M}_\zeta$.

Let $x \in LS(\mathcal{M})$ be fixed. By Theorem 3.4.3, we may write

$$x = \int_Z^{\oplus} x_\zeta \, dw(\zeta),$$

where $x_\zeta \in S(\mathcal{M}_\zeta)$ for almost every $\zeta \in Z$.

For every $\zeta \in Z$, we define

$$c_\zeta^+ = \inf\{\lambda : E_{x_\zeta}(\lambda, \infty) \text{ is finite}\}, \quad c_\zeta^- = \sup\{\lambda : E_{x_\zeta}(-\infty, \lambda) \text{ is finite}\}.$$

We claim that both fields $\zeta \to c_\zeta^+ \mathbf{1}_{H_\zeta}$, $\zeta \to c_\zeta^- \mathbf{1}_{H_\zeta}$ are measurable. By Proposition 1.13.23, we have that $\zeta \to E_{x_\zeta}(\lambda, \infty)$ is measurable for any $\lambda \in \mathbb{R}$. Since $\zeta \to \tau_\zeta$ is measurable, we have that $\zeta \to \tau_\zeta(E_{x_\zeta}(\lambda, \infty))$ is measurable for any $\lambda \in \mathbb{R}$. For any μ, we may write

$$\begin{aligned}
\{\zeta : c_\zeta^+ \geq \mu\} &= \{\zeta : E_{x_\zeta}(\lambda, \infty) \text{ is finite for all } \lambda > \mu\} \\
&= \left\{\zeta : E_{x_\zeta}\left(\mu + \frac{1}{m}, \infty\right) \text{ is finite for all } m \in \mathbb{N}\right\} \\
&= \bigcap_{m \in \mathbb{N}} \left\{\zeta : E_{x_\zeta}\left(\mu + \frac{1}{m}, \infty\right) \text{ is finite}\right\} \\
&= \bigcap_{m \in \mathbb{N}} \bigcup_{n \in \mathbb{N}} \left\{\zeta : \tau_\zeta\left(E_{x_\zeta}\left(\mu + \frac{1}{m}, \infty\right)\right) < n\right\}.
\end{aligned}$$

Hence, the field $\zeta \to c_\zeta^+ \mathbf{1}_{H_\zeta}$ is measurable. Similarly, $\zeta \to c_\zeta^- \mathbf{1}_{H_\zeta}$ is measurable. In addition,

$$c_\zeta^- \leq c_\zeta^+, \quad \zeta \in Z. \tag{3.25}$$

Set

$$c = \int_Z^{\oplus} c_\zeta \mathbf{1}_{H_\zeta}\, dw(\zeta), \quad c_\zeta = \frac{1}{2}(c_\zeta^+ + c_\zeta^-).$$

By Corollary 3.4.5 and Theorem 3.4.8, we have that $c \in \mathcal{Z}(LS(\mathcal{M}))$.

We claim that c is the central element we are looking for. To this end, we define

$$A = \{\zeta \in Z : E_{x_\zeta}(c_\zeta, \infty), E_{x_\zeta}(-\infty, c_\zeta) \text{ are infinite}\}$$

and

$$B = Z \backslash A.$$

We define a central projection

$$z = \int_A^{\oplus} \mathbf{1}_{H_\zeta}\, dw(\zeta)$$

and prove the assertion separately for xz and $x(\mathbf{1} - z)$.

For the operator xz, we claim that the operator $y := (x - c)z$ satisfies the conditions of Lemma 3.5.2 in the algebra $z\mathcal{M}$. Note that by Theorem 3.4.8, we have that

$$y = \int_A^{\oplus} (x_\zeta - c_\zeta \mathbf{1}_{H_\zeta})\, dw(\zeta).$$

By the definition of A, we have that $E_{x_\zeta}(c_\zeta, \infty)$ and $E_{x_\zeta}(-\infty, c_\zeta)$ are both infinite in a factor $\mathcal{M}_\zeta$ and, therefore, they both are properly infinite. Hence, $E_{x_\zeta}(c_\zeta, \infty) \sim E_{x_\zeta}(-\infty, c_\zeta)$ for almost all $\zeta \in Z$ (see Theorem 1.9.9(viii)). Now using Proposition 1.12.13 and Proposition 1.13.23, we obtain that

$$\begin{aligned} s(y_+) &= \int_A^{\oplus} E_{x_\zeta - c_\zeta \mathbf{1}_{H_\zeta}}(0, \infty)\, dw(\zeta) \\ &= \int_A^{\oplus} E_{x_\zeta}(c_\zeta, \infty)\, dw(\zeta) \sim \int_A^{\oplus} E_{x_\zeta}(-\infty, c_\zeta)\, dw(\zeta) = s(y_-). \end{aligned}$$

Applying Lemma 3.5.2 to the element y and algebra $z\mathcal{M}$, we construct a unitary element $u_1 \in z\mathcal{M}$, such that

$$|(x - c)z| + u_1^{-1}|(x - c)z|u_1 = |[(x - c)z, u_1]| = |[xz, u_1]|.$$

In particular,

$$|(x-c)z| \le |[(x-c)z, u_1]| = |[xz, u_1]|. \tag{3.26}$$

Next, we turn to the algebra $(\mathbf{1}-z)\mathcal{M}$. We claim that the operator

$$y := (x-c)(\mathbf{1}-z)$$

is compact with respect to the algebra $(\mathbf{1}-z)\mathcal{M}$. We note that by Theorem 3.4.8 we have that

$$y = \int_B^{\oplus} (x_\zeta - c_\zeta \mathbf{1}_{H_\zeta})\, dw(\zeta).$$

In particular, by Proposition 1.13.23, we have that

$$E_y(\lambda, \infty) = \int_B^{\oplus} E_{x_\zeta - c_\zeta \mathbf{1}_{H_\zeta}}(\lambda, \infty)\, dw(\zeta). \tag{3.27}$$

First, if $\zeta \in B$, then at least one of the projections $E_{x_\zeta}(c_\zeta, \infty)$, $E_{x_\zeta}(-\infty, c_\zeta)$ is finite. Assume that $E_{x_\zeta}(c_\zeta, \infty)$ is finite (the case when $E_{x_\zeta}(-\infty, c_\zeta)$ is finite can be treated similarly). Then, by definition of c_ζ^+ and c_ζ, we have

$$c_\zeta^+ \le c_\zeta = \frac{1}{2}(c_\zeta^+ + c_\zeta^-),$$

or equivalently,

$$c_\zeta^+ \le c_\zeta^-.$$

Combining this with inequality (3.25), we conclude that $c_\zeta^+ = c_\zeta^- = c_\zeta$ for every $\zeta \in B$.

By the definition of c_ζ^- and c_ζ^+, we have

$$E_{x_\zeta}(c_\zeta + \varepsilon, \infty), \quad E_{x_\zeta}(-\infty, c_\zeta - \varepsilon) \quad \text{are finite for } \zeta \in B$$

for all $\varepsilon > 0$. In other words, for any $\varepsilon > 0$ and $\zeta \in B$, the projection $E_{|x_\zeta - c_\zeta \mathbf{1}_{H_\zeta}|}(\varepsilon, \infty)$ is finite. Referring to (3.27) and Proposition 1.12.12(i), we conclude that $E_{|y|}(\varepsilon, \infty)$ is finite for every $\varepsilon > 0$, proving that $y \in S_0(\mathcal{M}(\mathbf{1}-z))$.

Hence, we may apply Proposition 3.6.10 to the element y and algebra $\mathcal{M}(\mathbf{1}-z)$, and construct a unitary element $u_2 \in \mathcal{M}(\mathbf{1}-z)$, such that

$$(1-\varepsilon)|(x-c)(\mathbf{1}-z)| \le |[x(\mathbf{1}-z), u_2]|.$$

Setting $u = u_1 + u_2$, and combining the latter inequality with (3.26), we obtain that

$$\begin{aligned}|[x,u]| &= |[xz,u_1]| + |[x(\mathbf{1}-z),u_2]| \\ &\geq |(x-c)z| + (1-\varepsilon)|(x-c)(\mathbf{1}-z)| \geq (1-\varepsilon)|x-c|,\end{aligned}$$

which completes the proof for the case of a semifinite properly infinite algebra $\mathcal{M}$.

Assume now that $\mathcal{M}$ is an arbitrary von Neumann algebra. By the decomposition theorem (see Theorem 1.9.14), we may find a central projection $z \in P(\mathcal{Z}(\mathcal{M}))$, such that $z\mathcal{M}$ is a semifinite properly infinite von Neumann algebra and $(\mathbf{1}-z)\mathcal{M}$ has no semifinite properly infinite direct summand.

By Theorem 3.5.6, there exists a unitary $u_1 \in (\mathbf{1}-z)\mathcal{M}$ and a central element $c_1 \in \mathcal{Z}(LS((\mathbf{1}-z)\mathcal{M}))$, such that

$$|[(\mathbf{1}-z)x,u_1]| = u_1^{-1}|(\mathbf{1}-z)x - c_1|u + |(\mathbf{1}-z)x - c_1| \geq (1-\varepsilon)|(\mathbf{1}-z)x - c_1|.$$

By the result just proved, there exists a unitary $u_2 \in z\mathcal{M}$ and a central element $c_2 \in \mathcal{Z}(LS(z\mathcal{M}))$, such that

$$|[zx,u_2]| \geq (1-\varepsilon)|zx - c_2|.$$

We set $u = u_1 + u_2$ and $c = c_1 + c_2 \in \mathcal{Z}(LS(\mathcal{M}))$. Since z and $\mathbf{1}-z$ are pairwise orthogonal, it follows that u is unitary and

$$|[x,u]| \geq (1-\varepsilon)|x-c|,$$

as required. □

Bibliographical notes

In 1970–1971 [62, 63], Dixon introduced the so-called extended C^*-algerbas (or EC^*-algebras) and extended W^*-algebras (or EW^*-algebras) and showed that if $\mathcal{A}^e$ is an EW^*-algebra over the W^*-algebra $\mathcal{A}$ on a separable Hilbert space, then $\mathcal{A}^e$ is a subalgebra of $LS(\mathcal{A})$ [63, p. 63, Theorem 5.3]. For the case of a general Hilbert space, this result was proved in [165].

In 1974, Yeadon in [163] proved a far stronger result that $LS(\mathcal{M})$ is the largest possible bimodule over a given von Neumann algebra $\mathcal{M}$ (see Theorem 3.1.2). The proof of this result presented in this monograph follows the original proof by Yeadon. We note that the assumption that $\mathcal{M}$ acts on a separable Hilbert space is not necessary for this result.

Topological, order-theoretic, and analytical aspects of Murray–von Neumann algebras have been treated in great detail in a recent survey due to Nayak [115]. A detailed exposition of the algebra of τ-measurable operators and its properties may be found in [64].

Theorem 3.3.6 and Proposition 3.3.8 were proved in [162].

The direct integral decomposition of operators affiliated with a decomposable von Neumann algebra (see Proposition 3.4.2) was proved in passing by M. Lennon in [105, Theorem 3.1]. Using a different approach, this result was proved in [68]. The proof of Proposition 3.4.2 presented here follows [68]. The direct integral decomposition of locally measurable operators affiliated with a decomposable von Neumann algebra (see Theorem 3.4.3) was proved by M. Lennon in [105, Theorem 3.1]. We note Corollary 3.4.7, which asserts that the class of locally measurable operators is closed with respect to the direct sum operation, a property that holds even for a not necessarily countable family of von Neumann algebras [109, Proposition 4.3].

Theorem 3.4.8 is proved by M. Lennon in [105, Proposition 4.1]. This result is based on another paper [107], where Lennon established necessary and sufficient conditions for the existence of the strong sum and strong product of two closed decomposable operators. This result extends Nussbaum's earlier result for decomposable operators associated with (not necessarily essentially bounded) measurable fields of bounded operators [117, Theorem 8]. For earlier results of sums and products of decomposable operators, see also [54, Proposition 3.6], [62, Theorem 3.12].

The commutator equalities of Section 3.5 were first established in [37] for the case when the von Neumann algebra $\mathcal{M}$ is a factor. The general case of commutator equalities in Section 3.5 as well as commutator inequalities in Section 3.6 were established in [36].

We note that results of Section 3.5 and Section 3.6 hold for an arbitrary von Neumann algebra $\mathcal{M}$ [36]. Our exposition in these sections is limited to von Neumann algebras acting on separable Hilbert spaces due to availability of direct integral techniques, which the authors find appealing.

4 Topologies on algebras of unbounded operators

In this chapter, we introduce two natural topologies on algebras of unbounded operators affiliated with a von Neumann algebra $\mathcal{M}$. The first topology is the topology t_τ of convergence in the measure on the algebra $S(\mathcal{M}, \tau)$ (in the case when $\mathcal{M}$ is assumed to be equipped with a faithful normal semifinite trace τ) and the second one is the topology $t(\mathcal{M})$ of convergence locally in the measure on the algebra $LS(\mathcal{M})$. As we will show in this chapter, in the case of a commutative von Neumann algebra $L_\infty(\Omega, \Sigma, \mu)$ of all essentially bounded measurable functions on a σ-finite space (Ω, Σ, μ), with the trace τ given by integration with respect to the measure μ, the measure topology t_τ is precisely the classical topology of convergence in measure for measurable functions on (Ω, Σ, μ) and the local measure topology $t(L_\infty(\Omega, \Sigma, \mu))$ is the classical topology of convergence locally in measure for measurable functions.

The measure topology (resp., the local measure topology) turns the $*$-algebra $S(\mathcal{M}, \tau)$ (resp., $LS(\mathcal{M})$) into a complete metrizable topological $*$-algebra in which the von Neumann algebra $\mathcal{M}$ is dense. Thus, the algebras $S(\mathcal{M}, \tau)$ and $LS(\mathcal{M})$ can be viewed as completions of $\mathcal{M}$ with respect to t_τ and $t(\mathcal{M})$, respectively.

In the present chapter, we shall study the properties of these two topologies, relations between them, as well as order properties of these topologies.

4.1 The singular value function of τ-measurable operators

To introduce the measure topology t_τ, we shall use the so-called singular value function of a τ-measurable operator. In this section, we present the basic properties of the singular value function. The exposition follows along the lines of [64]. We refer the reader to [64] for more details on the singular value function, in particular, and noncommutative integration in general.

Suppose that $p : \mathbb{R} \to [-\infty, \infty]$ is a decreasing function (i. e., $p(t_1) \geq p(t_2)$ whenever $t_1 \leq t_2$). For $t \in \mathbb{R}$, set

$$p(t+) = \lim_{s \downarrow t} p(s) = \sup_{s > t} p(s),$$
$$p(t-) = \lim_{s \uparrow t} p(s) = \inf_{s < t} p(s).$$

Evidently, $p(t+) \leq p(t) \leq p(t-)$ for all $t \in \mathbb{R}$ and $p(t_2-) \leq p(t_1+)$ whenever $t_1 < t_2$ in $\mathbb{R}$.

Definition 4.1.1. Let $p : \mathbb{R} \to [-\infty, \infty]$ be a decreasing function. The *right-continuous inverse* q of p is defined by setting

$$q(t) = \inf\{s \in \mathbb{R} : p(s) \leq t\}, \quad t \in \mathbb{R}.$$

https://doi.org/10.1515/9783111599687-005

Note that $q(t) = -\infty$ if and only if $p(s) \leq t$ for all $s \in \mathbb{R}$ and that $q(t) = \infty$ if and only if $p(s) > t$ for all $s \in \mathbb{R}$. Since p is a decreasing function, the set $\{s \in \mathbb{R} : p(s) \leq t\}$ is an interval of the form (a, ∞) or $[a, \infty)$, where $a \in [-\infty, \infty]$. For each $t \in \mathbb{R}$, the end point a of this interval is given by $q(t)$.

Remark 4.1.2. *Two simple observations will be used frequently:*
(i) *if $q(t) > -\infty$ and $s \in \mathbb{R}$ is such that $s < q(t)$, then $p(s) > t$;*
(ii) *if $q(t) < \infty$ and $s \in \mathbb{R}$ is such that $q(t) < s$, then $p(s) \leq t$.*

In other words,

$$(-\infty, q(t)) \subseteq \{s \in \mathbb{R} : p(s) > t\} \subseteq (-\infty, q(t)].$$

This implies immediately that an alternative formula for $q(t)$ is given by

$$q(t) = \sup\{s \in \mathbb{R} : p(s) > t\}, \quad t \in \mathbb{R}.$$

Lemma 4.1.3. *If $p : \mathbb{R} \to [-\infty, \infty]$ is a decreasing function, then the right-continuous inverse $q : \mathbb{R} \to [-\infty, \infty]$ has the following properties: if $s \in \mathbb{R}$ and $p(s) \in \mathbb{R}$, then*

$$q(p(s)) \leq s \leq q(p(s)-).$$

Proof. Evidently, $q(p(s)) = \inf\{u \in \mathbb{R} : p(u) \leq p(s)\} \leq s$. If $t \in \mathbb{R}$ is such that $t < p(s)$, then it follows from Remark 4.1.2(b) that $s \leq q(t)$, and so

$$q(p(s)-) = \inf\{q(t) : t < p(s)\} \geq s. \qquad \square$$

Throughout this section, $\mathcal{M}$ is assumed to be a semifinite von Neumann algebra on a Hilbert space H with a given faithful normal semifinite trace τ. Recall that $S(\mathcal{M}, \tau)$ is the $*$-algebra of all τ-measurable operators. If $x \in S(\mathcal{M}, \tau)$, then the *spectral distribution function* $d(|x|) = d(\cdot; |x|)$ of $|x|$ is defined by setting

$$d(s; |x|) = \tau(E_{|x|}(s, \infty)), \quad s \geq 0.$$

Evidently, the function $d(|x|) : [0, \infty) \to [0, \infty]$ is decreasing. If $s \in [0, \infty)$ and $s_n \downarrow s$, then $E_{|x|}(s_n, \infty) \uparrow E_{|x|}(s, \infty)$ in $\mathcal{M}_+$, and so $\tau(E_{|x|}(s_n, \infty)) \uparrow \tau(E_{|x|}(s, \infty))$. Hence, $d(|x|)$ is right continuous on $[0, \infty)$. Since $x \in S(\mathcal{M}, \tau)$, it follows from Proposition 2.4.2(iv) and (v) that there exists $\lambda > 0$ such that $d(\lambda; |x|) < \infty$ and that $d(s; |x|) \to 0$ as $s \to \infty$.

Definition 4.1.4. For $x \in S(\mathcal{M}, \tau)$, *the singular value function* $\mu(x)$ is defined to be the right-continuous inverse of the spectral distribution function $d(|x|)$, i. e.,

$$\mu(t; x) = \inf\{s \geq 0 : d(s; |x|) \leq t\}, \quad t \geq 0.$$

It follows from the definition that the function $\mu(x) : [0,\infty) \to [0,\infty]$ is decreasing and right continuous. Note that, by definition, $\mu(x) = \mu(|x|)$ for all $x \in S(\mathcal{M},\tau)$. Since $d(s;|x|) \to 0$ as $s \to \infty$, it is clear that $\mu(t;x) < \infty$ for all $t > 0$ (note that the equality $\mu(0;x) = \infty$ may occur; see Lemma 4.1.7(i) below). From the properties of the right-continuous inverse, it also follows that

$$d(s;|x|) = m\{t \geq 0 : \mu(t;x) > s\}, \quad s \geq 0, \tag{4.1}$$

where m is the Lebesgue measure on $\mathbb{R}$.

Furthermore, if $\tau(\mathbf{1}) < \infty$, then $d(s;|x|) \leq \tau(\mathbf{1})$ for all $s \geq 0$, and so $\mu(t;x) = 0$ for all $t \geq \tau(\mathbf{1})$. Therefore, in the case that $\tau(\mathbf{1}) < \infty$, one could consider $\mu(x)$ as a function on the interval $[0,\tau(\mathbf{1}))$, but in the present exposition, $\mu(x)$ will always be considered as a function on $[0,\infty)$.

We now give examples of the singular value function for specific operators.

Example 4.1.5. *Let $(\mathcal{M},\tau)$ be a semifinite von Neumann algebra and suppose that $x = \sum_{j=1}^m \alpha_j p_j$, where $p_1,\dots,p_m \in P(\mathcal{M})$ with $p_j p_k = 0$ whenever $j \neq k$, and $0 < \alpha_j \in \mathbb{R}$ $(j = 1,\dots,m)$ are such that $\alpha_j \neq \alpha_k$ whenever $j \neq k$. For the computation of $\mu(x)$, it may be assumed that $\alpha_1 > \alpha_2 > \cdots > \alpha_m > 0$. Setting $p_{m+1} = \mathbf{1} - \sum_{j=1}^m p_j$ and $\alpha_{m+1} = 0$, the spectral measure of x is then given by*

$$E_x = \sum_{j=1}^{m+1} p_j \delta_{\alpha_j},$$

where δ_{α_j} denotes the Dirac measure at the point α_j. Since

$$E_x(\lambda,\infty) = \sum_{\alpha_j > \lambda} p_j, \quad \lambda \geq 0,$$

the spectral distribution function of x is given by

$$d(\lambda;x) = \tau(E_x(\lambda,\infty)) = \sum_{\alpha_j > \lambda} \tau(p_j), \quad \lambda \geq 0.$$

Defining $k = \min\{1 \leq j \leq m : \tau(p_j) = \infty\}$ (if $\tau(p_j) < \infty$ for all $1 \leq j \leq m$, set $k = m+1$, in which case $\alpha_k = 0$), it follows that

$$d(\lambda;x) = \begin{cases} 0 & \text{if } \lambda \geq \alpha_1, \\ \sum_{i=1}^j \tau(p_i) & \text{if } \alpha_{j+1} \leq \lambda < \alpha_j \quad (j = 1,\dots,k-1), \\ \infty & \text{if } 0 \leq \lambda < \alpha_k. \end{cases}$$

Define $\rho_j = \sum_{i=1}^j \tau(p_i)$ for $j = 1,\dots,m$ and $\rho_0 = 0$. It is now easily verified that $\mu(x)$ is given by

$$\mu(x) = \sum_{j=1}^{k-1} \alpha_j \chi_{[\rho_{j-1},\rho_j)} + \alpha_k \chi_{[\rho_{k-1},\infty)} \tag{4.2}$$

(note that if $\tau(p_j) < \infty$ for all $1 \le j \le m$, then $k-1 = m$ and the last term is equal to zero in the above formula). Note, in particular, if $x = \mathbf{1}$ and $\tau(\mathbf{1}) = \infty$, then $d(\lambda; \mathbf{1}) = 0$ for all $\lambda \ge 1$ and $d(\lambda; \mathbf{1}) = \infty$ whenever $0 \le \lambda < 1$, and so $\mu(\mathbf{1}) = \chi_{[0,\infty)}$.

Example 4.1.6. *Now suppose that $\mathcal{M} = \mathcal{B}(H)$ for a Hilbert space H, equipped with the standard trace* tr. *Suppose that $x \in \mathcal{B}(H)$ is a positive compact operator. From the spectral theorem, it follows that x can be written as*

$$x = \sum_j \alpha_j p_j,$$

where the series converges in the uniform norm, and $\alpha_1 > \alpha_2 > \cdots > 0$ is the (finite or infinite) sequence of distinct nonzero eigenvalues of x and each p_j is the orthogonal projection onto the eigenspace corresponding to α_j. Note that $\operatorname{tr}(p_j) < \infty$ is the dimension of the eigenspace corresponding to α_j. Since the spectral measure of x is given by

$$E_x = \sum_j p_j \delta_{\alpha_j} + \Big(\mathbf{1} - \sum_j p_j\Big)\delta_0,$$

a calculation analogous to that in Example 4.1.5 above shows that

$$\mu(x) = \sum_j \alpha_j \chi_{[\rho_{j-1},\rho_j)},$$

where $\rho_j = \sum_{i=1}^{j} \operatorname{tr}(p_i)$ for all j and $\rho_0 = 0$. Since the length of the interval $[\rho_{j-1}, \rho_j)$ is equal to the multiplicity of the eigenvalue α_j, this may also be written as

$$\mu(x) = \sum_j \lambda_j \chi_{[j-1,j)},$$

where $\lambda_1 \ge \lambda_2 \ge \cdots > 0$ is the sequence of nonzero eigenvalues of x, repeated according to multiplicity. If $x \in \mathcal{B}(H)$ is an arbitrary compact operator, then $|x|$ is also compact and the eigenvalues of $|x|$ are called the singular values *of x. Accordingly,*

$$\mu(x) = \sum_j \mu_j \chi_{[j-1,j)},$$

where $\mu_1 \ge \mu_2 \ge \cdots > 0$ is the sequence of nonzero singular values of x, repeated according to multiplicity.

For later reference, two simple properties of $\mu(x)$ are formulated in the next lemma.

Lemma 4.1.7. *Suppose that $x \in S(\mathcal{M}, \tau)$.*

(i) $\mu(0;x) < \infty$ *if and only if* $x \in \mathcal{M}$, *in which case* $\mu(0;x) = \|x\|_{\mathcal{M}}$;
(ii) *If* $\lambda_0 = \inf\{s \geq 0 : d(s;|x|) < \infty\}$, *then* $\lim_{t\to\infty} \mu(t;x) = \lambda_0$.

Proof. (i) If $x \in S(\mathcal{M},\tau)$ and $s \geq 0$, then it is clear that $E_{|x|}(s,\infty) = 0$ if and only if $x \in \mathcal{M}$ and $\|x\|_{\mathcal{M}} \leq s$. Since $\mu(0;x) = \inf\{s \geq 0 : E_{|x|}(s,\infty) = 0\}$, the assertion follows immediately.

(ii) If $t \geq 0$, then $\{s \geq 0 : d(s;|x|) \leq t\} \subseteq [\lambda_0,\infty)$, and so $\mu(t;x) \geq \lambda_0$ for all $t \geq 0$. It follows from $s > \lambda_0$ that $d(s;|x|) < \infty$, and hence, by Lemma 4.1.3, $\mu(d(s;|x|);x) \leq s$. Since $\mu(x)$ is decreasing, this suffices to show that $\lim_{t\to\infty} \mu(t;x) = \lambda_0$. □

It follows, in particular, from (ii) in the lemma above that if $x \in S(\mathcal{M},\tau)$, then $d(s;|x|) < \infty$ for all $s > 0$ if and only if $\lim_{t\to\infty} \mu(t;x) = 0$. Recall from Definition 2.6.1 that the collection of all $x \in S(\mathcal{M},\tau)$ with the property that $d(s;|x|) < \infty$ for all $s > 0$ is denoted by $S_0(\mathcal{M},\tau)$ (which is a $*$-closed two-sided ideal in $S(\mathcal{M},\tau)$; see Proposition 2.6.4). These observations immediately yield the following result.

Proposition 4.1.8. *The subalgebra* $S_0(\mathcal{M},\tau)$ *of* $S(\mathcal{M},\tau)$ *satisfies*

$$S_0(\mathcal{M},\tau) = \Big\{x \in S(\mathcal{M},\tau) : \lim_{t\to\infty} \mu(t;x) = 0\Big\}.$$

In the next proposition, we collect the main properties of the singular value function.

Proposition 4.1.9. *If* $x \in S(\mathcal{M},\tau)$, *then:*
(i) *The function* $t \mapsto \mu(t;x)$, $t > 0$, *is decreasing and right continuous;*
(ii) $\mu(t;x) \to \|x\|_{\mathcal{M}}$ *when* $t \to 0$ *for bounded* x *and* $\mu(t;x) \to \infty$ *when* $t \to 0$ *for unbounded* x;
(iii) $\mu(t;x) = \mu(t;|x|)$ *for all* $t > 0$;
(iv) *If* $\alpha \in \mathbb{C}$, *then* $\mu(t;\alpha x) = |\alpha|\mu(t;x)$ *for all* $t > 0$;
(v) *If* $0 \leq y \leq x$, *then* $\mu(t;y) \leq \mu(t;x)$ *for all* $t > 0$;
(vi) *If* $\tau(\mathbf{1}) = 1$, *then* $\mu(t;x) = 0$ *for all* $t > 1$;
(vii) *If* $a, b \in \mathcal{M}$, *then*

$$\mu(axb) \leq \|a\|_{\mathcal{M}}\|b\|_{\mathcal{M}}\mu(t;x), \quad \mu(t;x^*) = \mu(t;x); \tag{4.3}$$

(viii) *Let* $y \in S(\mathcal{M},\tau)$ *and* $t, s > 0$. *Then*

$$\mu(t+s;x+y) \leq \mu(t;x) + \mu(s;y),$$

and

$$\mu(t+s;xy) \leq \mu(t;x)\mu(s;y);$$

(ix) *If* $x \geq 0$ *and if* $f : \mathbb{R}_+ \to \mathbb{R}_+$ *is a continuous and increasing function, such that* $f(0) = 0$, *then* $\mu(f(x)) = f(\mu(x))$;

(x) *The operation* $x \to \mu(t; x)$ *is continuous in the uniform norm on* $\mathcal{M}$. *More precisely,*

$$|\mu(s; x) - \mu(s; y)| \leq \|x - y\|_{\mathcal{M}}, \quad \forall x, y \in \mathcal{M}, \ \forall s > 0.$$

Proof. Part (iii) follows from Definition 4.1.4; parts (i) and (ii) follow from the reasoning after Definition 4.1.4.

(iv) We have

$$d(s; |ax|) = \tau(E_{|ax|}(s, \infty)) = \tau\left(E_{|x|}\left(\frac{s}{|a|}, \infty\right)\right) = d\left(\frac{s}{|a|}; |x|\right)$$

for $s \geq 0$. Therefore,

$$\begin{aligned}
\mu(t; ax) &= \inf\{s \geq 0 : d(s; |ax|) \leq t\} \\
&= \inf\left\{s \geq 0 : d\left(\frac{s}{|a|}; |x|\right) \leq t\right\} \\
&= |a| \inf\left\{\frac{s}{|a|} \geq 0 : d\left(\frac{s}{|a|}; |x|\right) \leq t\right\} \\
&= |a| \inf\{s \geq 0 : d(s; |x|) \leq t\} \\
&= |a| \mu(t; x).
\end{aligned}$$

(v) By Proposition 3.3.4(v), we have

$$d(s; y) = \tau(E_{|y|}(s, \infty)) \leq \tau(E_{|x|}(s, \infty)) = d(s; x)$$

for $s \geq 0$. Therefore,

$$\{s \geq 0 : d(s; x) \leq t\} \subset \{s \geq 0 : d(s; y) \leq t\}.$$

Hence,

$$\mu(t; y) = \inf\{s \geq 0 : d(s; y) \leq t\} \leq \inf\{s \geq 0 : d(s; x) \leq t\} = \mu(t; x).$$

(vi) If $\tau(\mathbf{1}) = 1$, then $d(s; |x|) \leq 1$ for $s \geq 0$. Therefore, $\mu(t; x) = \inf\{s \geq 0 : d(s; x) \leq t\} = 0$ for $t > 1$.

(vii) It follows from Proposition 2.1.7 that $d(s; x) = d(s; x^*)$ for $s \geq 0$. Therefore, $\mu(t; x^*) = \mu(t; x)$.

Further, $|ax|^2 = x^* a^* ax \leq \|a\|_{\mathcal{M}}^2 |x|^2$. Therefore, by Proposition 3.3.4(v), we have

$$E_{|ax|}(s, \infty) = E_{|ax|^2}(s^2, \infty) \preceq E_{\|a\|_{\mathcal{M}}^2 |x|^2}(s^2, \infty) = E_{\|a\|_{\mathcal{M}} |x|}(s, \infty)$$

for $s \geq 0$. Hence, it follows from part (iv) that

$$\mu(t; ax) \leq \mu(t; \|a\|_{\mathcal{M}} x) = \|a\|_{\mathcal{M}} \mu(t; x).$$

Then

$$\begin{aligned}\mu(t; axb) &= \mu(t; a(xb)) \leq \|a\|_{\mathcal{M}} \mu(t; xb) = \|a\|_{\mathcal{M}} \mu(t; (xb)^*) = \|a\|_{\mathcal{M}} \mu(t; b^* x^*) \\ &\leq \|a\|_{\mathcal{M}} \|b^*\|_{\mathcal{M}} \mu(t; x^*) = \|a\|_{\mathcal{M}} \|b\|_{\mathcal{M}} \mu(t; x),\end{aligned}$$

as required.

(viii) Let $\lambda_1, \lambda_2 > 0$, $p = E_{|x|}[0, \lambda_1]$, $q = E_{|y|}[0, \lambda_2]$, $\tau(p^\perp) \leq t$, $\tau(q^\perp) \leq s$, $e = p \wedge q$. Then

$$\tau(e^\perp) = \tau(p^\perp \vee q^\perp) \leq t + s, \quad \|(x+y)e\|_{\mathcal{M}} \leq \|xp\|_{\mathcal{M}} + \|yq\|_{\mathcal{M}} \leq \lambda_1 + \lambda_2.$$

It follows from Proposition 2.1.6 that

$$d(\lambda_1 + \lambda_2; |x+y|) = \tau(E_{|x+y|}(\lambda_1 + \lambda_2, \infty)) \leq \tau(e^\perp) \leq t + s.$$

Hence,

$$\begin{aligned}&\{\lambda_1 \geq 0 : d(\lambda_1, |x|) \leq t\} + \{\lambda_2 \geq 0 : d(\lambda_2, |y|) \leq s\} \\ &\quad \subset \{\lambda \geq 0 : d(\lambda, |x+y|) \leq t + s\}.\end{aligned}$$

Therefore,

$$\mu(t + s; x + y) \leq \mu(t; x) + \mu(s; y).$$

To prove the inequality for $\mu(xy)$, we let $f = s_r(p^\perp y)^\perp$. Then $p^\perp y f = 0$, and so $yf = pyf$. Furthermore,

$$f^\perp = s_r(p^\perp y) \sim s_l(p^\perp y) \leq p^\perp,$$

and hence, $\tau(f^\perp) \leq \tau(p^\perp)$. Setting $r = f \wedge q$, it follows that

$$\tau(r^\perp) = \tau(f^\perp \vee q^\perp) \leq \tau(f^\perp) + \tau(q^\perp) \leq t + s.$$

Moreover, since

$$xyr = xyfr = xpyfr = xpyqr \in \mathcal{M},$$

it follows that

$$\|xyr\|_{\mathcal{M}} \leq \|xp\|_{\mathcal{M}} \|yq\|_{\mathcal{M}} \leq \lambda_1 \lambda_2.$$

It follows from Proposition 2.1.6 that

$$d(\lambda_1 \lambda_2; |xy|) = \tau(E_{|xy|}(\lambda_1 \lambda_2, \infty)) \leq \tau(r^\perp) \leq t + s.$$

Hence,

$$\{\lambda_1 \geq 0 : d(\lambda_1; |x|) \leq t\} \cdot \{\lambda_2 \geq 0 : d(\lambda_2; |y|) \leq s\}$$
$$\subset \{\lambda \geq 0 : d(\lambda; |xy|) \leq t + s\}.$$

Therefore,

$$\mu(t + s; xy) \leq \mu(t; x)\mu(s; y),$$

as required.

(ix) By Proposition 1.4.7, we have that

$$E_{f(x)}(s, \infty) = E_x(f^{-1}(s), \infty),$$

and so

$$d(s; f(x)) = d(f^{-1}(s); x).$$

Hence,

$$d(s; f(x)) \leq t \Leftrightarrow d(f^{-1}(s); x) \leq t.$$

Therefore,

$$f(\mu(t; x)) = \inf\{f(s) : s \geq 0,\ d(s; |x|) \leq t\}$$
$$= \inf\{s \geq 0 : d(f^{-1}(s); |x|) \leq t\} = \mu(t; f(x)).$$

(x) The function $\mu(x - y)$ is right continuous. Hence, it follows from (viii) that

$$\mu(s; x) \leq \mu(0; x - y) + \mu(s, y), \quad \mu(s, y) \leq \mu(0; y - x) + \mu(s, x).$$

Since $\mu(0; x - y) = \mu(0; y - x)$, Lemma 4.1.7(i) implies that

$$|\mu(s; x) - \mu(s; y)| \leq \mu(0; x - y) = \|x - y\|_{\mathcal{M}}. \qquad \square$$

4.2 Topology of convergence in measure on the algebra of τ-measurable operators

A natural topology on the algebra $L_0(\Omega, \Sigma, \mu)$ of all measurable functions on a measure space (Ω, Σ, μ) is the topology t_μ of convergence in measure μ. When $\mathcal{M}$ is a semifinite von Neumann algebra equipped with faithful normal semifinite trace τ, a noncommutative version of the topology t_μ on the $*$-algebra $S(\mathcal{M}, \tau)$ is the topology t_τ of convergence in measure defined by the trace τ. In this section, we present a detailed description of the properties of the topology t_τ, in particular, we show that the pair $(S(\mathcal{M},\tau), t_\tau)$ is a complete metrizable topological $*$-algebra in which $\mathcal{M}$ is a dense subalgebra.

Throughout this section, we assume that $\mathcal{M}$ is a semifinite von Neumann algebra acting on a (separable) Hilbert space H, equipped with a faithful normal semifinite trace τ.

Recall that the generalized singular value function $\mu(x)$ of a τ-measurable operator x is defined in Definition 4.1.4 and is the right-continuous inverse of the distribution function $\lambda \mapsto \tau(E_{|x|}(\lambda,\infty))$. We shall introduce the measure topology t_τ via an F-norm on $S(\mathcal{M},\tau)$. See Definition 3.1.7 for the notion of F-norms.

Before we define the F-norm, which generates the measure topology on $S(\mathcal{M},\tau)$, we prove an auxiliary lemma.

Lemma 4.2.1. *For any $x \in S(\mathcal{M},\tau)$, we have*

$$\inf_{t>0}\{t+\mu(t;x)\} = \inf_{\lambda>0}\{\lambda + \tau(E_{|x|}(\lambda,\infty))\}.$$

Proof. Let $a = \inf_{t>0}\{t+\mu(t;x)\}$ and $\varepsilon > 0$. There exists $t > 0$ such that $t + \mu(t,x) < a + \frac{\varepsilon}{2}$. Since $\mu(t,x) < a - t + \frac{\varepsilon}{2}$ and $\mu(x)$ is the right-continuous inverse of the distribution function (see Definition 4.1.4), it follows that

$$\tau\left(E_{|x|}\left(a - t + \frac{\varepsilon}{2},\infty\right)\right) \le t.$$

Setting $\lambda = a - t + \frac{\varepsilon}{2}$, we have that

$$\lambda + \tau(E_{|x|}(\lambda,\infty)) \le a - t + \frac{\varepsilon}{2} + t < a + \varepsilon.$$

Since ε is arbitrary, it follows that

$$\inf_{\lambda>0}\{\lambda + \tau(E_{|x|}(\lambda,\infty))\} \le a = \inf_{t>0}\{t+\mu(t;x)\}.$$

Similarly, setting $b = \inf_{\lambda>0}\{\lambda + \tau(E_{|x|}(\lambda,\infty))\}$, for any $\varepsilon > 0$, we can find $\lambda > 0$, such that $\lambda + \tau(E_{|x|}(\lambda,\infty)) < b + \frac{\varepsilon}{2}$. Therefore, the definition of the generalized singular value function implies that

$$\mu\left(b - \lambda + \frac{\varepsilon}{2};x\right) = \inf\left\{s \in \mathbb{R} : \tau(E_{|x|}(s,\infty)) \le b - \lambda + \frac{\varepsilon}{2}\right\} \le \lambda,$$

and so for $t = b - \lambda + \frac{\varepsilon}{2}$ we have

$$t + \mu(t;x) \le b - \lambda + \frac{\varepsilon}{2} + \lambda < b + \varepsilon.$$

Thus,

$$\inf_{t>0}\{t+\mu(t;x)\} \le \inf_{\lambda>0}\{\lambda+\tau(E_{|x|}(\lambda,\infty))\},$$

which proves the assertion. □

Using Lemma 4.2.1, we now define the F-norm $\|\cdot\|_{S(\mathcal{M},\tau)}$ on the $*$-algebra $S(\mathcal{M},\tau)$.

Definition 4.2.2. For an operator $x \in S(\mathcal{M},\tau)$, we define

$$\|x\|_{S(\mathcal{M},\tau)} = \inf_{t>0}\{t+\mu(t;x)\} = \inf_{\lambda>0}\{\lambda+\tau(E_{|x|}(\lambda,\infty))\}.$$

As the following proposition shows, $\|\cdot\|_{S(\mathcal{M},\tau)}$ is indeed an F-norm on $S(\mathcal{M},\tau)$.

Proposition 4.2.3. *The mapping $\|\cdot\|_{S(\mathcal{M},\tau)} : S(\mathcal{M},\tau) \longrightarrow \mathbb{R}$ is an F-norm on $S(\mathcal{M},\tau)$.*

Proof. It is clear that $\|0\|_{S(\mathcal{M},\tau)} = 0$ and if $\|x\|_{S(\mathcal{M},\tau)} = 0$, then $\mu(0;x) = 0$, which guarantees that $\|x\|_{\mathcal{M}} = 0$ (see Lemma 4.1.7(i)), and so $x = 0$. Next, if $x \in S(\mathcal{M},\tau)$ and $\alpha \in \mathbb{C}$, $|\alpha| \le 1$, then Proposition 4.1.9(iv) implies that

$$\mu(t;\alpha x) = |\alpha|\mu(t;x) \le \mu(t;x), \quad t>0,$$

so that

$$\|\alpha x\|_{S(\mathcal{M},\tau)} = \inf_{t>0}\{t+\mu(t;\alpha x)\} \le \inf_{t>0}\{t+\mu(t;x)\} = \|x\|_{S(\mathcal{M},\tau)}.$$

Let $x \in S(\mathcal{M},\tau)$. For any $\varepsilon > 0$, there exists $\delta > 0$, such that $|\alpha|\mu(\frac{\varepsilon}{4};x) < \frac{\varepsilon}{2}$ for any $|\alpha| < \delta$. Therefore,

$$\mu(t;\alpha x) = |\alpha|\mu(t;x) \le |\alpha|\mu\left(\frac{\varepsilon}{4};x\right) < \frac{\varepsilon}{2}, \quad t \ge \frac{\varepsilon}{4},$$

so that $t + \mu(t;\alpha x) \le \varepsilon$ for any $t \in [\frac{\varepsilon}{4},\frac{\varepsilon}{2}]$. Hence, $\|\alpha x\|_{S(\mathcal{M},\tau)} \le \varepsilon$ for $|\alpha| < \delta$, i. e., $\lim_{\alpha\to 0}\|\alpha x\|_{S(\mathcal{M},\tau)} = 0$.

Finally, let $x,y \in S(\mathcal{M},\tau)$. For any $\varepsilon > 0$, there exist $t_1,t_2 > 0$, such that

$$t_1 + \mu(t_1;x) < \|x\|_{S(\mathcal{M},\tau)} + \frac{\varepsilon}{2}, \quad t_2 + \mu(t_2;y) < \|y\|_{S(\mathcal{M},\tau)} + \frac{\varepsilon}{2}.$$

Proposition 4.1.9(viii) implies that

$$\begin{aligned}\|x+y\|_{S(\mathcal{M},\tau)} &\le t_1+t_2+\mu(t_1+t_2;x+y) \le t_1+\mu(t_1;x)+t_2+\mu(t_2;y)\\ &< \|x\|_{S(\mathcal{M},\tau)} + \|y\|_{S(\mathcal{M},\tau)} + \varepsilon.\end{aligned}$$

Since ε is arbitrary, it follows that $\|x+y\|_{S(\mathcal{M},\tau)} \le \|x\|_{S(\mathcal{M},\tau)} + \|y\|_{S(\mathcal{M},\tau)}$. Thus, $\|\cdot\|_{S(\mathcal{M},\tau)}$ is an F-norm on $S(\mathcal{M},\tau)$. □

Since $\|\cdot\|_{S(\mathcal{M},\tau)}$ is an F-norm, we may define a metrizable vector topology on $S(\mathcal{M},\tau)$.

Definition 4.2.4. The metrizable vector topology t_τ on $S(\mathcal{M},\tau)$ generated by the F-norm $\|\cdot\|_{S(\mathcal{M},\tau)}$ is called the *measure topology*. If a sequence $\{x_n\}_{n\in\mathbb{N}} \subset S(\mathcal{M},\tau)$ converges to an operator $x \in S(\mathcal{M},\tau)$ with respect to the topology t_τ (notation: $x_n \xrightarrow{t_\tau} x$), then we say that the sequence $\{x_n\}_{n\in\mathbb{N}}$ converges to x *in measure*.

The measure topology can be equivalently defined via a base of neighborhoods of zero.

Proposition 4.2.5. *The sets*

$$\begin{aligned} U(\varepsilon,\delta) &= \{x \in S(\mathcal{M},\tau) : \exists p \in P(\mathcal{M}), \|xp\|_{\mathcal{M}} \le \varepsilon, \tau(p^\perp) \le \delta\} \\ &= \{x \in S(\mathcal{M},\tau) : \tau(E_{|x|}(\varepsilon,\infty)) \le \delta\}, \quad \varepsilon,\delta > 0, \end{aligned} \tag{4.4}$$

form a base of neighborhoods of zero for the measure topology t_τ and

$$U\left(\frac{\varepsilon}{2},\frac{\varepsilon}{2}\right) \subset \{x \in S(\mathcal{M},\tau) : \|x\|_{S(\mathcal{M},\tau)} \le \varepsilon\} \subset U(\varepsilon,\varepsilon)$$

for any $\varepsilon > 0$.

Proof. We first show that the sets $U(\varepsilon,\delta)$, $\varepsilon,\delta > 0$ have two equivalent definitions as stated in (4.4).

Fix $\varepsilon,\delta > 0$ and let $x \in S(\mathcal{M},\tau)$ be such that $\tau(E_{|x|}(\varepsilon,\infty)) \le \delta$. Setting $p = E_{|x|}[0,\varepsilon]$ and using the polar decomposition for x, we have that $\|xp\|_{\mathcal{M}} \le \||x|p\|_{\mathcal{M}} \le \varepsilon$ and $\tau(p^\perp) = \tau(E_{|x|}(\varepsilon,\infty)) \le \delta$. Conversely, assume that there exists $p \in P(\mathcal{M})$, such that $\|xp\|_{\mathcal{M}} \le \varepsilon$ and $\tau(p^\perp) \le \delta$. By Proposition 2.1.6, we have that $E_{|x|}(\varepsilon,\infty) \preceq p^\perp$ and, therefore, $\tau(E_{|x|}(\varepsilon,\infty)) \le \delta$. Thus, $U(\varepsilon,\delta)$ can be defined in the two stated ways.

To prove that the sets $U(\varepsilon,\delta)$, $\varepsilon,\delta > 0$, form the base of neighborhoods of zero for the measure topology t_τ, it is sufficient to show the inclusions of neighborhoods $U(\cdot,\cdot)$ and balls of radius ε, $\varepsilon > 0$, in $S(\mathcal{M},\tau)$ with respect to the F-norm $\|\cdot\|_{S(\mathcal{M},\tau)}$.

Let $\varepsilon > 0$ be fixed, and let $x \in U(\frac{\varepsilon}{2},\frac{\varepsilon}{2})$. Then $\tau(E_{|x|}(\frac{\varepsilon}{2},\infty)) \le \frac{\varepsilon}{2}$ and, therefore, by definition of the F-norm $\|\cdot\|_{S(\mathcal{M},\tau)}$ we have that $\|x\|_{S(\mathcal{M},\tau)} \le \frac{\varepsilon}{2} + \tau(E_{|x|}(\frac{\varepsilon}{2},\infty)) \le \varepsilon$. Hence, $x \in \{y \in S(\mathcal{M},\tau) : \|y\|_{S(\mathcal{M},\tau)} \le \varepsilon\}$. Next, if $x \in S(\mathcal{M},\tau)$ is such that $\|x\|_{S(\mathcal{M},\tau)} \le \varepsilon$, then for any $n \in \mathbb{N}$, there exists $\lambda_n > 0$, such that $\lambda_n + \tau(E_{|x|}(\lambda_n,\infty)) < \varepsilon + \frac{1}{n}$. In particular, it follows that $\lambda_n < \varepsilon + \frac{1}{n}$ and $\tau(E_{|x|}(\lambda_n,\infty)) < \varepsilon + \frac{1}{n}$. Since $E_{|x|}(\varepsilon + \frac{1}{n},\infty) \le E_{|x|}(\lambda_n,\infty)$, it follows that $\tau(E_{|x|}(\varepsilon + \frac{1}{n},\infty)) < \varepsilon + \frac{1}{n}$. Since $\varepsilon + n^{-1} \downarrow \varepsilon$ as $n \to \infty$, it follows that

$$\tau(E_{|x|}(\varepsilon,+\infty)) = \tau\Big(\sup_{n\in\mathbb{N}} E_{|x|}(\varepsilon + n^{-1},+\infty)\Big) = \sup_{n\in\mathbb{N}} \tau(E_{|x|}(\varepsilon + n^{-1},+\infty)) \le \varepsilon,$$

i. e., $x \in U(\varepsilon,\varepsilon)$, as required. □

We note a simple corollary of the above proposition.

Corollary 4.2.6. *If $\|x\|_{S(\mathcal{M},\tau)} = \alpha$, then $\tau(E_{|x|}(\alpha,\infty)) \le \alpha$.*

Proof. By Proposition 4.2.5, the assumption that $\|x\|_{S(\mathcal{M},\tau)} = \alpha$ implies that $x \in U(\alpha,\alpha)$. Therefore, the second equivalent description of the neighborhoods $U(\cdot,\cdot)$ (see (4.4)) implies the assertion. □

Next, we present some fundamental properties of the F-norm $\|\cdot\|_{S(\mathcal{M},\tau)}$.

Proposition 4.2.7.

(i) *If $\mu(x) \leq \mu(y)$ for some $x, y \in S(\mathcal{M},\tau)$, then $\|x\|_{S(\mathcal{M},\tau)} \leq \|y\|_{S(\mathcal{M},\tau)}$. In particular, if $0 \leq x \leq y$, then $\|x\|_{S(\mathcal{M},\tau)} \leq \|y\|_{S(\mathcal{M},\tau)}$;*

(ii) *$\|x\|_{S(\mathcal{M},\tau)} = \|x^*\|_{S(\mathcal{M},\tau)} = \||x|\|_{S(\mathcal{M},\tau)}$ for any $x \in S(\mathcal{M},\tau)$;*

(iii) *$(S(\mathcal{M},\tau), \|\cdot\|_{S(\mathcal{M},\tau)})$ is an F-normed $\mathcal{M}$-bimodule, i. e., for any $a \in S(\mathcal{M},\tau)$ and $x, y \in \mathcal{M}$, we have that*

$$\|xay\|_{S(\mathcal{M},\tau)} \leq \big\|\|x\|_{\mathcal{M}}\|y\|_{\mathcal{M}}a\big\|_{S(\mathcal{M},\tau)};$$

(iv) *If $p, q \in P(\mathcal{M})$ and $p \preceq q$, then $\|p\|_{S(\mathcal{M},\tau)} \leq \|q\|_{S(\mathcal{M},\tau)}$;*

(v) *For any $x \in S(\mathcal{M},\tau)$ and $\varepsilon > 0$, there exists $\delta > 0$ such that $\|xy\|_{S(\mathcal{M},\tau)} < \varepsilon$ for all $\|y\|_{S(\mathcal{M},\tau)} < \delta$;*

(vi) *For any $x, y \in S(\mathcal{M},\tau)$, we have that*

$$\|xy\|_{S(\mathcal{M},\tau)} \leq 2\max\{\|x\|_{S(\mathcal{M},\tau)}\|y\|_{S(\mathcal{M},\tau)}, \|x\|_{S(\mathcal{M},\tau)} + \|y\|_{S(\mathcal{M},\tau)}\}.$$

Proof. (i) If $\mu(x) \leq \mu(y)$, $x, y \in S(\mathcal{M},\tau)$, then

$$\|x\|_{S(\mathcal{M},\tau)} = \inf_{t>0}\{t + \mu(t;x)\} \leq \inf_{t>0}\{t + \mu(t;y)\} = \|y\|_{S(\mathcal{M},\tau)},$$

as required. If $0 \leq x \leq y$, then by Proposition 4.1.9(v), we have that $\mu(x) \leq \mu(y)$, and so $\|x\|_{S(\mathcal{M},\tau)} \leq \|y\|_{S(\mathcal{M},\tau)}$.

To prove (ii) and (iii), we note that Proposition 4.1.9(iii) and (vii) imply that $\mu(x) = \mu(x^*) = \mu(|x|)$ and $\mu(xay) \leq \|x\|_{\mathcal{M}}\|y\|_{\mathcal{M}}\mu(a)$. Hence, the assertions follow from part (i).

(iv) As shown in (iii), $(S(\mathcal{M},\tau), \|\cdot\|_{S(\mathcal{M},\tau)})$ is an F-normed bimodule over $\mathcal{M}$. Therefore, the assertion follows from Proposition 3.3.11(i).

(v) Since x is τ-measurable, by Proposition 2.4.2(v), we have that $\tau(E_{|x|}(\lambda,\infty)) \to 0$ as $\lambda \to \infty$. Hence, there exists Λ, such that for the projection $e := E_{|x|}(\Lambda,\infty)$ we have

$$\tau(e) < \frac{\varepsilon}{4}.$$

We set $\delta = \frac{\varepsilon}{2(\Lambda+2)}$ and assume that $y \in S(\mathcal{M},\tau)$ is such that $\|y\|_{S(\mathcal{M},\tau)} < \delta$. By Proposition 4.2.5, there exists a projection $q \in P(\mathcal{M})$, such that $\|yq\|_{\mathcal{M}} < \delta$ and $\tau(q^\perp) < \delta$.

For the projection $p := s_r(ey)^\perp$, we have that

$$\begin{aligned} yp &= y - yp^\perp = y - eyp^\perp - e^\perp yp^\perp \\ &= y - ey - e^\perp yp^\perp = e^\perp y - e^\perp yp^\perp \\ &= e^\perp yp. \end{aligned}$$

Therefore,

$$\begin{aligned}\|xy(p\wedge q)\|_{\mathcal{M}} &= \|xyp(p\wedge q)\|_{\mathcal{M}} = \|xe^{\perp}yp(p\wedge q)\|_{\mathcal{M}}\\ &\le \|xE_{|x|}[0,\Lambda]\|_{\mathcal{M}}\|yq\|_{\mathcal{M}} \le \delta\Lambda < \frac{\varepsilon}{2}.\end{aligned}$$

Since $s_r(w)\sim s_l(w)$ for any w affiliated with $\mathcal{M}$ (see Proposition 2.1.2(vi)), it follows that

$$\begin{aligned}\tau((p\wedge q)^{\perp}) &\le \tau(p^{\perp})+\tau(q^{\perp}) < \tau(s_r(ey))+\delta\\ &= \tau(s_l(ey))+\delta \le \tau(e)+\delta < \frac{\varepsilon}{2}.\end{aligned}$$

Appealing to Proposition 4.2.5, we conclude that $\|xy\|_{S(\mathcal{M},\tau)} < \varepsilon$, as required.

(vi) Let $\lambda_x = \|x\|_{S(\mathcal{M},\tau)}$ and $\lambda_y = \|y\|_{S(\mathcal{M},\tau)}$. We denote

$$e = E_{|x|}[0,\lambda_x],\quad f = E_{|y|}[0,\lambda_y],\quad p = (s_r(ey))^{\perp},\quad q = f\wedge p.$$

As in part (v), we have that $yp = e^{\perp}yp$, and so

$$\begin{aligned}\|xyq\|_{\mathcal{M}} &= \|xypq\|_{\mathcal{M}} = \|xe^{\perp}ypq\|_{\mathcal{M}}\\ &\le \|xe^{\perp}\|_{\mathcal{M}}\|yp\|_{\mathcal{M}} \le \lambda_x\lambda_y = \|x\|_{S(\mathcal{M},\tau)}\|y\|_{S(\mathcal{M},\tau)}.\end{aligned}$$

Furthermore, by Corollary 4.2.6 we have that

$$\begin{aligned}\tau(q^{\perp}) &\le \tau(f^{\perp})+\tau(p^{\perp}) \le \tau(E_{|x|}(\lambda_x,\infty))+\tau(E_{|y|}(\lambda_y,\infty))\\ &\le \|x\|_{S(\mathcal{M},\tau)}+\|y\|_{S(\mathcal{M},\tau)}.\end{aligned}$$

Thus,

$$xy\in U(\|x\|_{S(\mathcal{M},\tau)}\|y\|_{S(\mathcal{M},\tau)}, \|x\|_{S(\mathcal{M},\tau)}+\|y\|_{S(\mathcal{M},\tau)}),$$

where $U(\cdot,\cdot)$ is defined in (4.4). By Proposition 4.2.5, we obtain that

$$\|xy\|_{S(\mathcal{M},\tau)} \le 2\max\{\|x\|_{S(\mathcal{M},\tau)}\|y\|_{S(\mathcal{M},\tau)}, \|x\|_{S(\mathcal{M},\tau)}+\|y\|_{S(\mathcal{M},\tau)}\},$$

as required. □

The following theorem establishes a fundamental property of the vector space $(S(\mathcal{M},\tau), \|\cdot\|_{S(\mathcal{M},\tau)})$.

Theorem 4.2.8. *The pair $(S(\mathcal{M},\tau), \|\cdot\|_{S(\mathcal{M},\tau)})$ is a topological $*$-algebra with jointly continuous multiplication and an F-space.*

Proof. By Proposition 4.2.7(ii), we have that the involution $x\mapsto x^*$, $x\in S(\mathcal{M},\tau)$ is continuous in $\|\cdot\|_{S(\mathcal{M},\tau)}$. Let $x_n\to x$ and $y_n\to y$ in $\|\cdot\|_{S(\mathcal{M},\tau)}$ as $n\to\infty$, then

$$\begin{aligned}\|x_n y_n - xy\|_{S(\mathcal{M},\tau)} \le \big\|(x_n - x)y\big\|_{S(\mathcal{M},\tau)} + \big\|x(y_n - y)\big\|_{S(\mathcal{M},\tau)}\\ + \big\|(x_n - x)(y_n - y)\big\|_{S(\mathcal{M},\tau)},\end{aligned}$$

and so Proposition 4.2.7(v) and (vi) guarantees that $\|x_n y_n - xy\|_{S(\mathcal{M},\tau)} \to 0$ as $n \to \infty$. Thus, $(S(\mathcal{M},\tau), \|\cdot\|_{S(\mathcal{M},\tau)})$ is a topological $*$-algebra. It remains to show completeness of $(S(\mathcal{M},\tau), \|\cdot\|_{S(\mathcal{M},\tau)})$.

Let $\{x_n\}_{n\in\mathbb{N}}$ be a Cauchy sequence in $(S(\mathcal{M},\tau), \|\cdot\|_{S(\mathcal{M},\tau)})$. Since $(S(\mathcal{M},\tau), \|\cdot\|_{S(\mathcal{M},\tau)})$ is a topological $*$-algebra, without loss of generality, we may assume that $x_n = x_n^*$ for every $n \in \mathbb{N}$. Passing to a subsequence, if necessary, we may further assume that $\|x_{n+1} - x_n\|_{S(\mathcal{M},\tau)} < \frac{1}{2^{n-1}}$ for all $n \in \mathbb{N}$. By Proposition 4.2.5 for every $n \in \mathbb{N}$, there exists a projection $p_n \in P(\mathcal{M})$, such that

$$\tau(p_n^\perp) \le \frac{1}{2^n}, \quad \|(x_{n+1} - x_n)p_n\|_{\mathcal{M}} \le \frac{1}{2^n}. \tag{4.5}$$

Let

$$q_n = \bigwedge_{k=n+1}^{\infty} p_k, \quad n \in \mathbb{N}, \quad \mathfrak{D} = \bigcup_{n\in\mathbb{N}} q_n(H).$$

It is clear that the sequence $\{q_n\}_{n\in\mathbb{N}}$ is increasing and the set $\mathfrak{D}$ is a linear subspace in the Hilbert space H. In addition, $uq_n = q_n u$, for any unitary operator $u \in \mathcal{M}'$ and for each $n \in \mathbb{N}$. Consequently, $u(\mathfrak{D}) \subset \mathfrak{D}$ for all unitary operators $u \in \mathcal{M}'$.

By (1.5) and the first inequality in (4.5), we have that

$$\tau(q_n^\perp) = \tau\Big(\bigvee_{k>n} p_k^\perp\Big) \le \sum_{k=n+1}^{\infty} \tau(p_k^\perp) \le \sum_{k=n+1}^{\infty} \frac{1}{2^k} = \frac{1}{2^n} \to 0.$$

Therefore, $\{q_n\}_{n\in\mathbb{N}}$ is a sequence of τ-finite projections, such that $q_n(H) \subset \mathfrak{D}$ and $q_n \uparrow \mathbf{1}$ as $n \to \infty$.

If $\xi \in \mathfrak{D}$, then $\xi \in q_{n_0}(H)$ for some $n_0 \in \mathbb{N}$ and, therefore, for every $m \ge n_0$ and $l \in \mathbb{N}$, the second inequality in (4.5) implies that

$$\begin{aligned}\big\|(x_{m+l} - x_m)(\xi)\big\|_H &= \big\|(x_{m+l} - x_m)q_{n_0}(\xi)\big\|_H\\ &\le \sum_{k=1}^{l} \big\|(x_{m+k} - x_{m+k-1})q_{n_0}(\xi)\big\|_H\\ &\le \sum_{k=1}^{l} \frac{1}{2^{m+k-1}}\big\|q_{n_0}(\xi)\big\|_H \le \sum_{k=m}^{\infty} \frac{1}{2^m}\big\|q_{n_0}(\xi)\big\|_H\\ &= \frac{1}{2^{m-1}}\big\|q_{n_0}(\xi)\big\|_H.\end{aligned} \tag{4.6}$$

That is, $\{x_n(\xi)\}_{n\in\mathbb{N}}$ is a Cauchy sequence in H, and so for every $\xi \in \mathfrak{D} \subset H$, there exists $x(\xi) \in H$, such that $\|x(\xi) - x_n(\xi)\|_H \to 0$ as $n \to \infty$. It is clear that the mapping $x : \mathfrak{D} \to H$ is a linear operator.

We claim that x is closable and its closure $\overline{x}$ is a τ-measurable operator, such that $\|x_n - \overline{x}\|_{S(\mathcal{M},\tau)} \to 0$ as $n \to \infty$.

For any unitary $u \in \mathcal{M}'$ and $\xi \in \mathfrak{D}$, we have that

$$ux(\xi) = u\Big(\lim_{n\to\infty} x_n(\xi)\Big) = \lim_{n\to\infty} ux_n(\xi) = \lim_{n\to\infty} x_n(u(\xi)) = xu(\xi),$$

i. e., x is affiliated with $\mathcal{M}$. Furthermore, the equality $x_n = x_n^*$, $n \in \mathbb{N}$ implies that for any $\xi, \zeta \in \mathfrak{D}$, we have

$$\langle x(\xi), \zeta\rangle_H = \lim_{n\to\infty} \langle x_n(\xi), \zeta\rangle_H = \lim_{n\to\infty} \langle \xi, x_n(\zeta)\rangle_H = \langle \xi, x(\zeta)\rangle_H,$$

i. e., x is a symmetric operator. In particular, x is closable. Since $x\eta\mathcal{M}$, Proposition 2.1.2(ii) implies that $\overline{x}\eta\mathcal{M}$. Since for the sequence $\{q_n\}_{n\in\mathbb{N}}$ constructed above, we have that $q_n(H) \subset \mathfrak{D} \subset \operatorname{dom}(\overline{x})$, $\tau(q_n^\perp) < \infty$ for all $n \in \mathbb{N}$ and $q_n \uparrow \mathbf{1}$ as $n \to \infty$, it follows from the definition that $\overline{x}$ is τ-measurable (see Definition 2.4.1).

By (4.6) for every $\xi \in q_n(H)$ and all $m \geq n$, $l \in \mathbb{N}$, we have that $\|(x_{m+l} - x_m)(\xi)\|_H \leq \frac{1}{2^{m-1}}\|\xi\|_H$. Passing to the limit as $l \to \infty$, we have that $\|(x - x_m)(\xi)\|_H \leq \frac{1}{2^{m-2}}\|\xi\|_H$, so that $\|(x - x_m)q_n\|_{\mathcal{M}} \leq \frac{1}{2^{m-1}} \leq \frac{1}{2^n}$. Since $\tau(q_n^\perp) \leq \frac{1}{2^n}$ for all $n \in \mathbb{N}$, Proposition 4.2.5 implies that $\|x_n - \overline{x}\|_{S(\mathcal{M},\tau)} \leq \frac{1}{2^{n-1}} \to 0$ as $n \to \infty$. Thus, $(S(\mathcal{M},\tau), \|\cdot\|_{S(\mathcal{M},\tau)})$ is complete. □

Next, we present the following simple characterisation of the measure convergence.

Proposition 4.2.9. *Let $\{x_n\}_{n\in\mathbb{N}} \subset S(\mathcal{M},\tau)$, $x \in S(\mathcal{M},\tau)$. The following conditions are equivalent:*

(i) $\|x - x_n\|_{S(\mathcal{M},\tau)} \to 0$ *as* $n \to \infty$;

(ii) $x_n \xrightarrow{t_\tau} x$ *as* $n \to \infty$;

(iii) $|x_n - x| \xrightarrow{t_\tau} 0$ *as* $n \to \infty$;

(iv) $\tau(E_{|x_n - x|}(\lambda, \infty)) \longrightarrow 0$ *as* $n \to \infty$ *for all* $\lambda > 0$;

(v) $\mu(t; x_n - x) \to 0$ *as* $n \to \infty$ *for any* $t > 0$;

(vi) *for any* $\varepsilon > 0$, *there exists* $N \in \mathbb{N}$, *such that for any* $n \geq N$, *there exists a projection* $p_n \in P(\mathcal{M})$, *such that* $\tau(p_n^\perp) \leq \varepsilon$, $(x_n - x)\cdot p_n \in \mathcal{M}$, *and* $\|(x_n - x)\cdot p_n\|_{\mathcal{M}} \leq \varepsilon$.

Proof. The equivalence of the first three assertions trivially follows from the definition on t_τ and $\|\cdot\|_{S(\mathcal{M},\tau)}$.

The equivalence of (i) and (iv) follows from Corollary 4.2.6, while equivalence of (iv) and (v) follows from the definition of $\mu(x)$, $x \in S(\mathcal{M},\tau)$.

The equivalence of (i) and (vi) follows from Proposition 4.2.5. □

Since $E_p(\varepsilon, \infty) = p$ for any projection $p \in P(\mathcal{M})$ and $\varepsilon \in (0,1)$, Proposition 4.2.9(iv) immediately implies the following characterization of convergence in measure of a sequence of projections.

Corollary 4.2.10. *A sequence of projections* $\{p_n\}_{n\in\mathbb{N}} \subset P(\mathcal{M})$ *converges in measure to* 0 *if and only if* $\tau(p_n) \to 0$ *as* $n \to \infty$.

Proposition 4.2.9 also implies a sufficient condition for convergence in measure.

Corollary 4.2.11. *Let* $x_n \in S(\mathcal{M},\tau)$. *If* $\tau(s_r(x_n)) = \tau(s_l(x_n)) \to 0$ *as* $n \to \infty$, *then* $x_n \to 0$ *in measure.*

Proof. Since $s_r(x_n) = s(|x_n|)$ for every $n \in \mathbb{N}$ and $E_{|x_n|}(\lambda,\infty) \le s(|x_n|)$ for any $\lambda > 0$, it follows that

$$\tau(E_{|x_n|}(\lambda,\infty)) \le \tau(s_r(x_n)) \to 0.$$

Proposition 4.2.9(iv) guarantees that $x_n \to 0$ in measure. □

We now present some simple properties of the measure topology.

Proposition 4.2.12.
(i) *The set* $S_h(\mathcal{M},\tau)$ *is closed in* $(S(\mathcal{M},\tau), t_\tau)$;
(ii) *For any* $x \in \mathcal{M}$, *we have that* $\|x\|_{S(\mathcal{M},\tau)} \le \|x\|_{\mathcal{M}}$. *In particular, the embedding of* $(\mathcal{M}, \|\cdot\|_{\mathcal{M}})$ *into* $(S(\mathcal{M},\tau), t_\tau)$ *is continuous;*
(iii) *The unit ball of* $\mathcal{M}$ *is closed in* $S(\mathcal{M},\tau)$ *with respect to the topology* t_τ;
(iv) *Suppose that* $\{x_n\}_{n\in\mathbb{N}} \subset S(\mathcal{M},\tau)$ *and* $\{p_n\}_{n\in\mathbb{N}} \subset P(\mathcal{M})$ *is such that* $p_n \to 0$ *in measure as* $n \to \infty$. *Then* $x_n p_n \to 0$ *in measure as* $n \to \infty$;
(v) *Suppose that* $x_n, y_n \in S(\mathcal{M},\tau)$ *and* $x_n \to 0$ *in measure. Passing to a subsequence* $\{x_{n_k}\}_{k\in\mathbb{N}}$ *of the sequence* $\{x_n\}_{n\in\mathbb{N}}$, *we have* $\|y_k x_{n_k}\|_{S(\mathcal{M},\tau)} \le 2^{-k}$ *for all* $k \in \mathbb{N}$.

Proof. Part (i) follows immediately since the involution $x \mapsto x^*$ is continuous in $(S(\mathcal{M},\tau), t_\tau)$.

(ii) By Proposition 4.1.9, we have that $\mu(t;x) \le \|x\|_{\mathcal{M}}, t \ge 0$, for any $x \in \mathcal{M}$. Therefore, for any $x \in \mathcal{M}$, we infer that

$$\|x\|_{S(\mathcal{M},\tau)} = \inf_{t>0}\{t + \mu(t;x)\} \le \inf_{t>0}\{t + \|x\|_{\mathcal{M}}\} = \|x\|_{\mathcal{M}}.$$

(iii) Suppose that $x_n \in \mathcal{M}, x \in S(\mathcal{M},\tau)$ are such that $\|x_n\|_{\mathcal{M}} \le 1$ and $\|x_n - x\|_{S(\mathcal{M},\tau)} \to 0$ as $n \to \infty$. Let $\varepsilon > 0$. By Proposition 4.2.9(v), there exists N, such that $\mu(\frac{\varepsilon}{2}; x - x_n) < \frac{1}{n}$ for all $n \ge N$. Hence, by Proposition 4.1.9(viii) for any $\varepsilon > 0$, we have that

$$\mu(\varepsilon, x) \le \mu\left(\frac{\varepsilon}{2}; x - x_n\right) + \mu\left(\frac{\varepsilon}{2}; x_n\right) \le \mu\left(\frac{\varepsilon}{2}; x - x_n\right) + 1 < 1 + \frac{1}{n}, \quad n \ge N,$$

and so $\mu(\varepsilon;x) \le 1$ for any $\varepsilon > 0$. Hence, Lemma 4.1.7(i) implies that $\|x\|_{\mathcal{M}} = \lim_{\varepsilon\to 0}\mu(\varepsilon;x) \le 1$.

(iv) By Corollary 4.2.10, we have that $\tau(p_n) \to 0$ as $n \to \infty$. For any $n \in \mathbb{N}$ we have that $s_r(x_n p_n) \le p_n$, and so $\tau(s_r(x_n p_n)) \to 0$ as $n \to \infty$. Referring to Corollary 4.2.11 we conclude that $x_n p_n \to 0$ in measure as $n \to \infty$.

(v) Since $(S(\mathcal{M},\tau),t_\tau)$ is a topological $*$-algebra, it follows that the sequence $\{y_1x_n\}_{n\in\mathbb{N}}$ converges to 0 in measure, i. e., there exists n_1 such that $\|y_1x_n\|_{S(\mathcal{M},\tau)} \le 2^{-k}$, $n \ge n_1$. In particular, $\|y_1x_{n_1}\|_{S(\mathcal{M},\tau)} \le 2^{-1}$. Similarly, the sequence $\{y_2x_n\}_{n>n_1}$ converges to 0 in measure, and so there exists $n_2 > n_1$ such that $\|y_1x_n\|_{S(\mathcal{M},\tau)} \le 2^{-2}$, $n \ge n_2$. In particular, $\|y_2x_{n_2}\|_{S(\mathcal{M},\tau)} \le 2^{-2}$. Repeating the argument, we obtain a strictly increasing sequence $\{n_k\}_{k\in\mathbb{N}}$, such that

$$\|y_kx_{n_k}\|_{S(\mathcal{M},\tau)} \le 2^{-k}, \quad k \in \mathbb{N}.$$

□

The following corollary explains the importance of the algebra $S(\mathcal{M},\tau)$ in noncommutative integration.

Corollary 4.2.13. *The von Neumann algebra $\mathcal{M}$ is dense in $S(\mathcal{M},\tau)$ in the measure topology t_τ. For an operator $x \in S(\mathcal{M},\tau)$, a sequence of bounded operators approximating x can be chosen to be $\{xE_{|x|}[0,n)\}_{n\in\mathbb{N}}$.*

Proof. Let $x \in S(\mathcal{M},\tau)$ with the polar decomposition $x = u|x|$. By Proposition 2.4.2(v), we have that $\tau(E_{|x|}[n,+\infty)) \to 0$ as $n \to \infty$. Therefore, Proposition 4.2.9(iv) implies that

$$|x| - |x|E_{|x|}[0,n) = |x|E_{|x|}[n,+\infty) \to 0, \quad n \to \infty,$$

in measure. Hence,

$$x - xE_{|x|}[0,n) = u\big(|x| - |x|E_{|x|}[0,n)\big) \to 0, \quad n \to \infty,$$

in measure. Since $xE_{|x|}[0,n) = u|x|E_{|x|}[0,n) \in \mathcal{M}$, we conclude that $\mathcal{M}$ is dense in $S(\mathcal{M},\tau)$ in the measure topology t_τ. □

In the following proposition, we show that the two-sided $*$-ideal $S_0(\mathcal{M},\tau)$ of all τ-compact operators is closed in $(S(\mathcal{M},\tau),t_\tau)$.

Proposition 4.2.14. *The two-sided $*$-ideal $S_0(\mathcal{M},\tau)$ of τ-compact operators is closed in $(S(\mathcal{M},\tau),t_\tau)$, so that $(S_0(\mathcal{M},\tau),t_\tau)$ is a topological $*$-algebra and an F-space.*

Proof. Let $x_n \in S_0(\mathcal{M},\tau)$, $n \in \mathbb{N}$, $x \in S(\mathcal{M},\tau)$ be such that x_n converges to x in measure as $n \to \infty$. Let $\lambda > 0$ be arbitrary. There exists $n \in \mathbb{N}$, such that $\|x_n - x\|_{S(\mathcal{M},\tau)} < \frac{\lambda}{2}$. By Proposition 4.2.5, there exists a projection $p \in P(\mathcal{M})$, such that $\|(x - x_n)p\|_{\mathcal{M}} < \frac{\lambda}{2}$ and $\tau(p^\perp) < \frac{\lambda}{2}$. Denoting $e = E_{|x_n|}[0,\frac{\lambda}{2}]$ and setting $q = p \wedge e$, we have

$$\|xq\|_{\mathcal{M}} \le \big\|(x - x_n)q\big\|_{\mathcal{M}} + \|x_nq\|_{\mathcal{M}} \le \big\|(x - x_n)p\big\|_{\mathcal{M}} + \big\||x_n|e\big\|_{\mathcal{M}} < \frac{\lambda}{2} + \frac{\lambda}{2} = \lambda.$$

Proposition 2.1.6 implies that $E_{|x|}(\lambda,\infty) \preceq q^\perp$, and so

$$\tau(E_{|x|}(\lambda,\infty)) \le \tau(q^\perp) \le \tau(p^\perp) + \tau\Big(E_{|x_n|}\Big(\frac{\lambda}{2},\infty\Big)\Big) < \infty,$$

where the last inequality holds since $x_n \in S_0(\mathcal{M},\tau)$. Thus, for any $\lambda > 0$ we have that $\tau(E_{|x|}(\lambda,\infty)) < \infty$. Referring to Definition 2.6.1, we conclude that $x \in S_0(\mathcal{M},\tau)$. Thus, $S_0(\mathcal{M},\tau)$ is closed in $(S(\mathcal{M},\tau),t_\tau)$. Since $(S(\mathcal{M},\tau),t_\tau)$ is a topological $*$-algebra and an F-space, it follows that $(S_0(\mathcal{M},\tau),t_\tau)$ is also a topological $*$-algebra and an F-space. □

Since the embedding of $(\mathcal{M},\|\cdot\|_{\mathcal{M}})$ into $(S(\mathcal{M},\tau),t_\tau)$ is continuous (see Proposition 4.2.12(ii)), Proposition 4.2.14 implies the following corollary.

Corollary 4.2.15. *The ideal $\mathcal{M}\cap S_0(\mathcal{M},\tau)$ equipped with the norm $\|\cdot\|_{\mathcal{M}}$ is a Banach space.*

We now give examples of the topologies t_τ on the two simplest examples of von Neumann algebras.

Example 4.2.16. *Let $\mathcal{M} = \mathcal{B}(H)$ and $\tau(x) = \mathrm{tr}(x)$ be the canonical trace on $\mathcal{B}(H)$. As shown in Example 2.4.6, $S(\mathcal{B}(H),\mathrm{tr}) = \mathcal{B}(H)$. By Proposition 4.2.12, the uniform norm convergence implies the measure convergence. Conversely, if $x_n \in \mathcal{B}(H)$, is such that $\|x_n\|_{S(\mathcal{M},\tau)} \to 0$ as $n \to \infty$, then by Proposition 4.2.9 we have that $\mu(\frac{1}{2},x_n) \to 0$ as $n \to \infty$. By Example 4.1.6, the generalized singular function $\mu(x)$ for any $x \in \mathcal{B}(H)$ is a step function and $\mu(t,x) = \|x\|_{\mathcal{M}}$ for any $t \in [0,1)$. Therefore, $\|x_n\|_{\mathcal{M}} = \mu(\frac{1}{2},x_n) \to 0$ as $n \to \infty$. Thus, for $(\mathcal{M},\tau) = (\mathcal{B}(H),\mathrm{tr})$, the measure topology coincides with the uniform norm topology.*

Example 4.2.17. *Let $\mathcal{M}$ be a commutative von Neumann algebra. By Theorem 1.8.10, $\mathcal{M}$ is identified with the von Neumann algebra $L_\infty(\Omega,\Sigma,\mu)$ of all essentially bounded measurable functions on some σ-finite measure space (Ω,Σ,μ). As seen in Example 1.10.11, $\mathcal{M}$ is equipped with the faithful normal semifinite trace $\tau(f) = \int_\Omega f d\mu$. By Example 2.4.7, the $*$-algebra $S(\mathcal{M},\tau)$ can be identified with the $*$-algebra*

$$S(\Omega,\Sigma,\mu) = \{f \in L_0(\Omega,\Sigma,\mu) : \mu(\{|f| > \lambda\}) < \infty \text{ for some } \lambda > 0\}.$$

Note that $E_{|f|}(\lambda,\infty) = \{\omega \in \Omega : |f| > \lambda\}$ for any $\lambda > 0$ and $f \in S(\Omega,\Sigma,\mu)$. Hence, for a sequence $\{f_n\} \in S(\Omega,\Sigma,\mu)$, Proposition 4.2.9 implies that $f_n \to 0$ in t_τ if and only if

$$\mu(\{\omega \in \Omega : |f_n| > \lambda\}) = \tau(E_{|f_n|}(\lambda,\infty)) \to 0,$$

as $n \to \infty$. Thus, $f_n \to 0$ in t_τ if and only if $f_n \to 0$ in measure μ. Hence, for a commutative von Neumann algebra $\mathcal{M} = L_\infty(\Omega,\Sigma,\mu)$, the convergence in the topology t_τ is the classical convergence in measure μ.

Remark 4.2.18. *Since the $*$-algebra $S(\mathcal{M},\tau)$ depends on the choice of the faithful normal semifinite trace τ (see Proposition 2.5.8 and discussion before that) and since $S(\mathcal{M},\tau)$ is the completion of $\mathcal{M}$ with respect to the measure topology (see Corollary 4.2.13), it follows that the measure topology t_τ itself depends on the choice of the faithful normal semifinite trace τ. Furthermore, the measure topology t_τ restricted to the von Neumann algebra $\mathcal{M}$ depends on τ.*

We conclude this section by discussing the relation between the measure topology t_τ and the partial order on $S_h(\mathcal{M},\tau)$.

Proposition 4.2.19.
(i) *The set $S_+(\mathcal{M},\tau)$ is closed in $(S(\mathcal{M},\tau),t_\tau)$;*
(ii) *If $\{x_n\}_{n\in\mathbb{N}} \subset S_h(\mathcal{M},\tau), x, y, z \in S_h(\mathcal{M},\tau), y \le x_n \le z$ for all $n \in \mathbb{N}$, and $x_n \to x$ in measure as $n \to \infty$, then $y \le x \le z$;*
(iii) *If a sequence $\{x_n\}_{n\in\mathbb{N}} \subset S_h(\mathcal{M},\tau)$, is increasing and converges to $x \in S_h(\mathcal{M},\tau)$ in measure, then $x_n \uparrow x$;*
(iv) *If $\{x_n\}_{n\in\mathbb{N}}, \{y_n\}_{n\in\mathbb{N}} \subset S(\mathcal{M},\tau), y_n \xrightarrow{t_\tau} 0, |x_n| \le |y_n|, n \in \mathbb{N}$, then $x_n \xrightarrow{t_\tau} 0$.*

Proof. (i) By Proposition 4.2.12(i), $S_h(\mathcal{M},\tau)$ is closed in $(S(\mathcal{M},\tau),t_\tau)$, and therefore, it is sufficient to show that $S_+(\mathcal{M},\tau)$ is closed in $S_h(\mathcal{M},\tau)$. Suppose that $x \in S_h(\mathcal{M},\tau)$ and $\{x_n\}_{n\in\mathbb{N}} \subset S_+(\mathcal{M},\tau)$ are such that $x_n \to x$ in measure. Let $p = E_x(-\infty,0]$. Since $x_n \ge 0$, it follows that $px_np \ge 0$, and so

$$0 \le x_- = -pxp \le p(x_n - x)p, \quad n \in \mathbb{N}.$$

Since $\|p(x_n - x)p\|_{S(\mathcal{M},\tau)} \to 0$ as $n \to \infty$, Proposition 4.2.7(i) implies that $\|x_-\|_{S(\mathcal{M},\tau)} \le \|p(x_n - x)p\|_{S(\mathcal{M},\tau)} \to 0$ as $n \to \infty$ and, therefore, $x_- = 0$. Thus, $x \ge 0$, as required.

Part (ii) follows immediately from part (i) and elementary properties of the partial order.

(iii) Let $\{x_n\}_{n\in\mathbb{N}}$ be an increasing sequence, such that $x_n \to x$ in measure as $n \to \infty$. Since $x_n \le x_m$ for $n \le m$, part (ii) implies that have that $x_n \le x$ for all $n \in \mathbb{N}$. Suppose now that $y \in S_h(\mathcal{M},\tau)$ is such that $x_n \le y$ for all $n \in \mathbb{N}$. Using part (ii) again, we infer that $x \le y$, i. e., x is the least upper bound of $\{x_n\}_{n\in\mathbb{N}}$, as required.

(iv) If $y_n \to 0$ in measure, then $\|y_n\|_{S(\mathcal{M},\tau)} \to 0$ as $n \to \infty$. Since $|x_n| \le |y_n|$, it follows that $\mu(x_n) \le \mu(y_n)$ for all $n \in \mathbb{N}$. Hence, by Proposition 4.2.7, we have that $\|x_n\|_{S(\mathcal{M},\tau)} \le \|y_n\|_{S(\mathcal{M},\tau)} \to 0$ as $n \to \infty$. □

In the following theorem, we prove a noncommutative analogue of the dominated convergence theorem for the measure topology.

Theorem 4.2.20. *Let $\{x_i\}_{i\in I}$ be a net in $S_+(\mathcal{M},\tau)$, such that $x_i \downarrow 0$ and $x_i \le x$ for some $x \in S_0(\mathcal{M},\tau)$. Then $x_i \xrightarrow{t_\tau} 0$.*

Proof. Since $S_0(\mathcal{M},\tau)$ is an absolutely solid $*$-subalgebra of $S(\mathcal{M},\tau)$, it follows that $x_i \in S_0(\mathcal{M},\tau)$. For a fixed $\lambda > 0$, we have

$$E_{x_i}(\lambda,\infty) \le \frac{1}{\lambda}x_iE_{x_i}(\lambda,\infty) \le \frac{1}{\lambda}x_i.$$

Setting $p_i = \bigvee_{j\ge i} E_{x_j}(\lambda,\infty)$, we have that $p_i \le \frac{1}{\lambda}x_i$. Since $x_i \downarrow 0$, we infer that $p_i \downarrow 0$.

By Proposition 3.3.4(v), we have that $E_{x_j}(\lambda,\infty) \le E_{x_i}(\lambda,\infty)$ for $j > i$, and so

$$\tau(p_i) = \inf_{j\ge i} \tau(E_{x_j}(\lambda,\infty)) = \tau(E_{x_i}(\lambda,\infty)).$$

Since $x_i \in S_0(\mathcal{M},\tau)$, it follows that $\tau(p_i) < \infty$. Since $p_i \downarrow 0$, we infer that $\tau(p_i) \to 0$ and, therefore, $\tau(E_{x_i}(\lambda,\infty)) \to 0$. Since λ is arbitrary, Proposition 4.2.9(iv) implies that $x_i \to 0$ in measure. □

Theorem 4.2.20 immediately implies the following simple corollary.

Corollary 4.2.21. *Let $\{x_i\}_{i\in I}$ be an increasing net in $S_+(\mathcal{M},\tau)$, such that $x_i \uparrow x$ for all $i \in I$ and some $0 \le x \in S_0(\mathcal{M},\tau)$. Then $x_i \to x$ in measure.*

Proof. Consider a net $y_i = x - x_i$, $i \in I$. Then $\{y_i\}_{i\in I}$ is a decreasing net, such that $y_i \downarrow 0$. Moreover, since x_i are positive, it follows that $y_i = x - x_i \le x \in S_0(\mathcal{M},\tau)$ for all $i \in I$. By Theorem 4.2.20, we have that $x - x_i = y_i \to 0$ in measure. Thus, $x_i \to x$ in measure. □

Remark 4.2.22. *Note that the assumption that $x_i \le x$ for some $x \in S_0(\mathcal{M},\tau)$ cannot be omitted in Theorem 4.2.20 and Corollary 4.2.21. Indeed, let H be a separable infinite-dimensional Hilbert space, $\mathcal{M} = \mathcal{B}(H)$, $\tau(x) = \mathrm{tr}(x)$ be a canonical trace on $\mathcal{B}(H)$. In this case, $S(\mathcal{M},\tau) = \mathcal{B}(H)$ and as we showed in Example 4.2.16 the measure topology is the uniform norm topology. Let $\{p_n\}_{n\in\mathbb{N}} \subset P(\mathcal{B}(H))$, $p_n \uparrow \mathbf{1}$, $\dim p_n(H) < \infty$. Then $\|\mathbf{1} - p_n\|_{\mathcal{M}} = 1$ for all $n \in \mathbb{N}$, i. e., the sequence $\{p_n\}$ does not converge to the unit operator $\mathbf{1}$ with respect to the measure topology t_τ. In particular, the order convergence $x_n \uparrow x$ or $x_n \downarrow x$, $\{x_n\}_{n\in\mathbb{N}} \subset S_h(\mathcal{M},\tau)$, $x \in S_h(\mathcal{M},\tau)$, generally speaking, does not imply measure convergence $x_n \to x$.*

4.3 Topology of convergence locally in measure on the algebra of locally measurable operators

In this section, we introduce the topology of convergence locally in measure $t(\mathcal{M})$ on the algebra $LS(\mathcal{M})$ and study its properties. This topology generalises the classical convergence locally in measure on the algebra $L_0(\Omega,\Sigma,\mu)$ of all μ-a. e. finite measurable functions on a measure space (Ω,Σ,μ) [72].

The definition of the local measure topology relies on the existence of dimension function $\mathcal{D}$ on $\mathcal{M}$ and its properties. We refer the reader to Section 1.11 for details on the dimension function.

Let $\mathcal{M}$ be a von Neumann algebra acting on a separable Hilbert space. As in Remark 1.8.11, we let (Ω,Σ,μ) be a probability space, such that the center $\mathcal{Z}(\mathcal{M})$ of the algebra $\mathcal{M}$ is isomorphic to $L_\infty(\Omega,\Sigma,\mu)$. We denote by φ the isomorphism of $\mathcal{Z}(\mathcal{M})$ onto $L_\infty(\Omega,\Sigma,\mu)$. We denote by ψ a (tracial) state on $\mathcal{Z}(\mathcal{M})$, such that $\psi(a) = \int_\Omega \varphi(a)d\mu$, $a \in \mathcal{Z}(\mathcal{M})$ (see Example 1.10.12).

Recall that for the measure topology t_τ, the central role is played by the trace τ, which allows us to measure the "size" of projections (see Definition 4.2.2). Roughly speaking, for the local measure topology, we replace the trace by the dimension function $\mathcal{D}$. Since the dimension function of any projection is itself a function on (Ω, Σ, μ), there is an additional technical difficulty. We note that one (of many) differences between the measure topology t_τ and the local measure topology $t(\mathcal{M})$ is that the local measure topology $t(\mathcal{M})$ does not depend on the choice of dimension function (see Proposition 4.3.31 below), while the measure topology t_τ does depend on the choice of trace τ (see Remark 4.2.18).

Let $\mathcal{D} : P(\mathcal{M}) \to L^+(\Omega, \Sigma, \mu)$ be the dimension function (see Section 1.11). For a function $f \in L^+(\Omega, \Sigma, \mu)$ and $\lambda \in \mathbb{R}$, we use the notation $[f > \lambda]$ to denote the superlevel set $\{\omega \in \Omega : f(\omega) > \lambda\}$.

We start with an auxiliary lemma.

Lemma 4.3.1. *Let $x \in LS(\mathcal{M})$.*

(i) *If $\lambda \geq 0$ and $p \in P(\mathcal{M})$ is such that $\|xp\|_{\mathcal{M}} \leq \lambda$, then*

$$\mathcal{D}(E_{|x|}(\lambda, \infty)) \leq \mathcal{D}(p^\perp);$$

(ii) *If $\lambda_1 > \lambda_2 \geq 0$, then*

$$[\mathcal{D}(E_{|x|}(\lambda_1, \infty)) > \lambda_1] \subset [\mathcal{D}(E_{|x|}(\lambda_2, \infty)) > \lambda_2];$$

(iii) *If $\lambda \geq 0$, then*

$$[\mathcal{D}(E_{|x|}(\lambda, \infty)) > \lambda] = \bigcup_{t>0} [\mathcal{D}(E_{|x|}(\lambda + t, \infty)) > \lambda + t].$$

Proof. (i) Since $\|xp\|_{\mathcal{M}} \leq \lambda$, Proposition 2.1.6 implies that $E_{|x|}(\lambda, \infty) \preceq p^\perp$. Therefore, referring to Proposition 1.11.6(iv), we obtain

$$\mathcal{D}(E_{|x|}(\lambda, \infty)) \leq \mathcal{D}(p^\perp).$$

(ii) If $\lambda_1 > \lambda_2$, then $E_{|x|}(\lambda_1, \infty) \leq E_{|x|}(\lambda_2, \infty)$, and so

$$[\mathcal{D}(E_{|x|}(\lambda_1, \infty)) > \lambda_1] \subset [\mathcal{D}(E_{|x|}(\lambda_1, \infty)) > \lambda_2] \subset [\mathcal{D}(E_{|x|}(\lambda_2, \infty)) > \lambda_2].$$

(iii) To simplify notation, we define the function $f(\lambda) = \mathcal{D}(E_{|x|}(\lambda, \infty))$, $\lambda \geq 0$. It is clear that

$$[f(\lambda) > \lambda] = \bigcup_{n \in \mathbb{N}} \left[f(\lambda) > \lambda + \frac{2}{n} \right]. \tag{4.7}$$

Since $E_{|x|}(\lambda + \frac{1}{m}, \infty) \uparrow E_{|x|}(\lambda, \infty)$ as $m \to \infty$, $\lambda \geq 0$, Definition 1.11.1(vi) implies that

$$\sup_{m\in\mathbb{N}} f\left(\lambda+\frac{1}{m}\right)=\sup_{m\in\mathbb{N}} \mathcal{D}\left(E_{|x|}\left(\lambda+\frac{1}{m},\infty\right)\right)=\mathcal{D}(E_{|x|}(\lambda,\infty))=f(\lambda)$$

for all $\lambda\geq 0$. In particular, for every $n\in\mathbb{N}$, there exists $m\geq n$, such that $f(\lambda+\frac{1}{m})>f(\lambda)-\frac{1}{n}$. Therefore, if $\omega\in[f(\lambda)>\lambda+\frac{2}{n}]$, then

$$f\left(\lambda+\frac{1}{m}\right)(\omega)>f(\lambda)(\omega)-\frac{1}{n}>\lambda+\frac{1}{n}>\lambda+\frac{1}{m}.$$

Thus,

$$\left[f(\lambda)>\lambda+\frac{2}{n}\right]=\bigcup_{m\geq n}\left[f\left(\lambda+\frac{1}{m}\right)>\lambda+\frac{1}{m}\right].$$

Referring to (4.7), we conclude that

$$[f(\lambda)>\lambda]\subset\bigcup_{t>0}[f(\lambda+t)>\lambda+t].$$

The converse inclusion immediately follows from part (ii). □

Lemma 4.3.1 immediately implies the following property, which we will use repeatedly in the study of the local measure topology. We recall here that (Ω,Σ,μ) is a probability space, such that $\mathcal{Z}(\mathcal{M})$ is $*$-isomorphic to $L_\infty(\Omega,\Sigma,\mu)$.

Corollary 4.3.2. *For any $x\in LS(\mathcal{M})$, the function*

$$\lambda\mapsto\mu([\mathcal{D}(E_{|x|}(\lambda,\infty))>\lambda]),\quad \lambda\geq 0,$$

is nonincreasing and right continuous.

Proof. Lemma 4.3.1(ii) implies that the function

$$\lambda\mapsto\mu([\mathcal{D}(E_{|x|}(\lambda,\infty))>\lambda]),\quad \lambda\geq 0,$$

is nonincreasing. To prove the right continuity of this function, let $\lambda\geq 0$ be fixed and let $t>0$ be arbitrary. Combining Lemma 4.3.1(iii) with continuity (from below) of the measure μ we infer that

$$\begin{aligned}\mu([\mathcal{D}(E_{|x|}(\lambda,\infty))>\lambda])&=\mu\Big(\bigcup_{t>0}[\mathcal{D}(E_{|x|}(\lambda+t,\infty))>\lambda+t]\Big)\\&=\lim_{t\to 0}\mu([\mathcal{D}(E_{|x|}(\lambda+t,\infty))>\lambda+t]),\end{aligned}$$

as required. □

Now we introduce the sets $\{V(\lambda)\}_{\lambda>0}$, which will form the base of the local measure topology $t(\mathcal{M})$.

Definition 4.3.3. For $\lambda > 0$, we define

$$V(\lambda) = \{x \in LS(\mathcal{M}) : \mu([\mathcal{D}(E_{|x|}(\lambda, \infty)) > \lambda]) \leq \lambda\}.$$

By convention, μ is a probability measure on (Ω, Σ), and so any locally measurable operator $x \in LS(\mathcal{M})$ belongs to $V(\lambda)$ for $\lambda \geq 1$.

We have the following equivalent description of the sets $V(\cdot)$.

Proposition 4.3.4. *For every $\lambda > 0$, we have that*

(i)

$$V(\lambda) = \{x \in LS(\mathcal{M}) : \exists p \in P(\mathcal{M}), \|xp\|_{\mathcal{M}} \leq \lambda, \mu([\mathcal{D}(p^\perp) > \lambda]) \leq \lambda\};$$

(ii)

$$V(\lambda) = \{x \in LS(\mathcal{M}) : \exists p \in P(\mathcal{M}), z \in P(\mathcal{Z}(\mathcal{M}))\ \textit{such that}$$
$$\|x(\mathbf{1} - p)\|_{\mathcal{M}} \leq \lambda, pz \in P_{\text{fin}}(\mathcal{M}), \mathcal{D}(pz) \leq \lambda\varphi(z), \psi(\mathbf{1} - z) \leq \lambda\}.$$

Proof. (i) Let $x \in LS(\mathcal{M})$ be such that there exists a projection $p \in P(\mathcal{M})$ with $\|xp\|_{\mathcal{M}} \leq \lambda$ and $\mu([\mathcal{D}(p^\perp) > \lambda]) \leq \lambda$. By Lemma 4.3.1(i), we have that $\mathcal{D}(E_{|x|}(\lambda, \infty)) \leq \mathcal{D}(p^\perp)$ and, therefore,

$$[\mathcal{D}(E_{|x|}(\lambda, \infty)) > \lambda] \subset [\mathcal{D}(p^\perp) > \lambda].$$

Hence, by assumption,

$$\mu([\mathcal{D}(E_{|x|}(\lambda, \infty)) > \lambda]) \leq \mu([\mathcal{D}(p^\perp) > \lambda]) \leq \lambda,$$

which proves that

$$\{x \in LS(\mathcal{M}) : \exists p \in P(\mathcal{M}), \|xp\|_{\mathcal{M}} \leq \lambda, \mu([\mathcal{D}(p^\perp) > \lambda]) \leq \lambda\} \subset V(\lambda).$$

To show the converse inclusion, assume that $x \in LS(\mathcal{M})$ is such that $\mu([\mathcal{D}(E_{|x|}(\lambda, \infty)) > \lambda]) \leq \lambda$. Then for the projection $p = E_{|x|}[0, \lambda]$, we have that $\|xp\|_{\mathcal{M}} \leq \||x|p\|_{\mathcal{M}} \leq \lambda$ and

$$\mu([\mathcal{D}(p^\perp) > \lambda]) = \mu([\mathcal{D}(E_{|x|}(\lambda, \infty)) > \lambda]) \leq \lambda,$$

which proves the converse inclusion.

(ii) Assume that $x \in V(\lambda)$. By part (i), there exists $q \in P(\mathcal{M})$, such that $\|xq\|_{\mathcal{M}} \leq \lambda$, $\mu([\mathcal{D}(q^\perp) > \lambda]) \leq \lambda$. Setting $p = q^\perp$, $z = \varphi^{-1}(\chi_{[\mathcal{D}(q^\perp) \leq \lambda]})$, we have that $\|x(\mathbf{1} - p)\|_{\mathcal{M}} = \|xq\|_{\mathcal{M}} \leq \lambda$,

$$\mathcal{D}(pz) = \mathcal{D}(p)\varphi(z) = \mathcal{D}(p)\chi_{[\mathcal{D}(p) \leq \lambda]} \leq \lambda\chi_{[\mathcal{D}(p) \leq \lambda]} = \lambda\varphi(z)$$

and

$$\psi(\mathbf{1}-z)=\mu([\mathcal{D}(p)>\lambda])\leq\lambda.$$

Since $\mathcal{D}(pz)$ is an a. e. finite-valued function, it follows that pz is a finite projection (see Definition 1.11.1(ii)), which shows one of the inclusions.

To show the converse inclusion, assume that for $x\in LS(\mathcal{M})$ there exist projections $p\in\mathcal{M}$ and $z\in\mathcal{Z}(\mathcal{M})$, such that $\|x(\mathbf{1}-p)\|_{\mathcal{M}}\leq\lambda$, $pz\in P_{\mathrm{fin}}(\mathcal{M})$, $\mathcal{D}(pz)\leq\lambda\varphi(z)$, and $\psi(\mathbf{1}-z)\leq\lambda$. Setting $q=p^{\perp}z$, we have that $\|xq\|_{\mathcal{M}}\leq\lambda$ and

$$\begin{aligned}\mu([\mathcal{D}(q^{\perp})>\lambda])&=\mu([\mathcal{D}(pz)+\mathbf{1}-\varphi(z)>\lambda])\leq\mu([\lambda\varphi(z)+\mathbf{1}-\varphi(z)>\lambda])\\&=\mu([\varphi(z)(\lambda-1)>\lambda-1]).\end{aligned}$$

If $\lambda\geq1$, then $\mu([\mathcal{D}(q^{\perp})>\lambda])\leq1\leq\lambda$. If $\lambda<1$, then

$$\mu([\mathcal{D}(q^{\perp})>\lambda])\leq\mu([\varphi(z)<1])=\mu([\varphi(z)=0])=\psi(\mathbf{1}-z)\leq\lambda.$$

Therefore, by part (i), we conclude that $x\in V(\lambda)$. □

Next, we shall define an F-norm on $LS(\mathcal{M})$, which generates the local measure topology on $LS(\mathcal{M})$ (see Definition 3.1.7 for the notion of F-norms).

Definition 4.3.5. For any $x\in LS(\mathcal{M})$, we define

$$\|x\|_{LS(\mathcal{M})}=\inf\{\lambda: x\in V(\lambda)\}=\inf\{\lambda:\mu([\mathcal{D}(E_{|x|}(\lambda,\infty))>\lambda])\leq\lambda\}.$$

Since $V(1)=LS(\mathcal{M})$, it follows that $\|x\|_{LS(\mathcal{M})}\leq1$ for any $x\in LS(\mathcal{M})$. The mapping $\|\cdot\|_{LS(\mathcal{M})}:LS(\mathcal{M})\to[0,\infty)$ can be given a slightly different definition.

Proposition 4.3.6. *For any $x\in LS(\mathcal{M})$, we have*

$$\|x\|_{LS(\mathcal{M})}=\inf_{\lambda>0}\max\{\lambda,\mu([\mathcal{D}(E_{|x|}(\lambda,\infty))>\lambda])\}. \tag{4.8}$$

Proof. Let us denote $r=\inf_{\lambda>0}\max\{\lambda,\mu([\mathcal{D}(E_{|x|}(\lambda,\infty))>\lambda])\}$.

Let $t>\|x\|_{LS(\mathcal{M})}$. Then there exists $\lambda<t$ such that $\mu([\mathcal{D}(E_{|x|}(\lambda,\infty))>\lambda])\leq\lambda<t$. Hence, $\lambda\geq r$, and so $\|x\|_{LS(\mathcal{M})}\geq r$.

Conversely, for any $t>r$ there exists $\lambda<t$ such that $\mu([\mathcal{D}(E_{|x|}(\lambda,\infty))>\lambda])<t$. Hence, $\mu([\mathcal{D}(E_{|x|}(t,\infty))>t])<t$, and so $r\geq\|x\|_{LS(\mathcal{M})}$. □

We present a useful description of the base $\{V(\lambda)\}_{\lambda>0}$.

Lemma 4.3.7. *For any $\lambda>0$, we have*

$$V(\lambda)=\{x\in LS(\mathcal{M}):\|x\|_{LS(\mathcal{M})}\leq\lambda\}.$$

Proof. Let $x \in V(\lambda)$. Since $\|x\|_{LS(\mathcal{M})} = \inf\{\varepsilon : x \in V(\varepsilon)\}$, it is clear that $\|x\|_{LS(\mathcal{M})} \leq \lambda$.

Conversely, suppose that $\|x\|_{LS(\mathcal{M})} \leq \lambda$ for some $\lambda > 0$, i. e.,

$$\inf\{\varepsilon : \mu([\mathcal{D}(E_{|x|}(\varepsilon, \infty)) > \varepsilon]) \leq \varepsilon\} \leq \lambda.$$

Then for any $n \in \mathbb{N}$ there exists $t_n \leq \lambda + \frac{1}{n}$, such that

$$\mu([\mathcal{D}(E_{|x|}(t_n, \infty)) > t_n]) \leq \lambda + \frac{1}{n}.$$

Therefore, Lemma 4.3.1(ii) implies that

$$\mu\left(\left[\mathcal{D}\left(E_{|x|}\left(\lambda + \frac{1}{n}, \infty\right)\right) > \lambda + \frac{1}{n}\right]\right) \leq \mu([\mathcal{D}(E_{|x|}(t_n, \infty)) > t_n]) < \lambda + \frac{1}{n}.$$

By Corollary 4.3.2, we have that

$$\mu([\mathcal{D}(E_{|x|}(\lambda, \infty)) > \lambda]) = \lim_{n\to\infty} \mu\left(\left[\mathcal{D}\left(E_{|x|}\left(\lambda + \frac{1}{n}, \infty\right)\right) > \lambda + \frac{1}{n}\right]\right) \leq \lambda.$$

Thus, $x \in V(\lambda)$, as required. □

We are now in a position to prove that $\|\cdot\|_{LS(\mathcal{M})}$ is indeed an F-norm on $LS(\mathcal{M})$.

Proposition 4.3.8. *The mapping $\|\cdot\|_{LS(\mathcal{M})}$ is an F-norm on $LS(\mathcal{M})$.*

Proof. It is clear that $\|0\|_{LS(\mathcal{M})} = 0$. Conversely, assume that $\|x\|_{LS(\mathcal{M})} = 0$. By Corollary 4.3.2, the mapping

$$\lambda \mapsto \mu([\mathcal{D}(E_{|x|}(\lambda, \infty)) > \lambda]), \quad \lambda \geq 0,$$

is nonnegative and nonincreasing. Assuming that there exists $\varepsilon > 0$ such that $\mu([\mathcal{D}(E_{|x|}(\lambda, \infty)) > \lambda]) > \varepsilon$ for any $\lambda > 0$, we infer that $\|x\|_{LS(\mathcal{M})} \geq \varepsilon > 0$, which is a contradiction. Therefore,

$$\mu([\mathcal{D}(E_{|x|}(0, \infty)) > 0]) = \lim_{\lambda\downarrow 0} \mu([\mathcal{D}(E_{|x|}(\lambda, \infty)) > \lambda]) = 0,$$

i. e., $\mathcal{D}(E_{|x|}(0, \infty)) = 0$ a. e. Hence, referring to Proposition 1.11.6(iii), we obtain that $E_{|x|}(0, \infty) = 0$, which guarantees that $x = 0$.

To show that $\|\alpha x\|_{LS(\mathcal{M})} \leq \|x\|_{LS(\mathcal{M})}$ for any $x \in LS(\mathcal{M})$ and $\alpha \in \mathbb{C}$ with $|\alpha| \leq 1$, it is sufficient to show that $\alpha V(\lambda) \subset V(\lambda)$ for any $\lambda > 0$. Assume that $x \in V(\lambda)$ for some $\lambda > 0$ and let $\alpha \in \mathbb{C}$ be such that $|\alpha| \leq 1$. By Proposition 4.3.4, there exists $p \in P(\mathcal{M})$, such that $\|xp\|_{\mathcal{M}} \leq \lambda$ and $\mu([\mathcal{D}(p^{\perp}) > \lambda]) \leq \lambda$. Since $\|\alpha xp\|_{\mathcal{M}} \leq \|xp\|_{\mathcal{M}} \leq \lambda$, it follows that $\alpha x \in V(\lambda)$, as required.

Next, let $x \in LS(\mathcal{M})$, $\{\alpha_n\}_{n\in\mathbb{N}} \subset \mathbb{C}$ be such that $\alpha_n \to 0$ as $n \to \infty$. Note that $E_{|\alpha_n x|}(a, \infty) = E_{|x|}(\frac{a}{|\alpha_n|}, \infty)$ for any $a > 0$, and so Proposition 2.3.5(iii) implies that

$\mathcal{D}(E_{|a_n x|}(a,\infty)) \to 0$ in measure μ as $n \to \infty$. Hence, for any $\varepsilon > 0$ there exists $N \in \mathbb{N}$ such that $\mu([\mathcal{D}(E_{|a_n x|}(\varepsilon,\infty)) > \varepsilon]) < \varepsilon$ for all $n \geq N$. Since ε is arbitrary, it follows that $\|a_n x\|_{LS(\mathcal{M})} \to 0$ as $n \to \infty$.

Finally, to show the triangle inequality let $x, y \in LS(\mathcal{M})$ be arbitrary. We denote $\|x\|_{LS(\mathcal{M})} = \lambda_x$, $\|y\|_{LS(\mathcal{M})} = \lambda_y$ and define

$$p = E_{|x|}[0,\lambda_x] \wedge E_{|y|}[0,\lambda_y].$$

By the spectral theorem, we have that

$$\|xE_{|x|}[0,\lambda_x]\|_{\mathcal{M}} \leq \lambda_x, \quad \|yE_{|y|}[0,\lambda_y]\|_{\mathcal{M}} \leq \lambda_y,$$

and so

$$\|(x+y)p\|_{\mathcal{M}} \leq \|xE_{|x|}[0,\lambda_x]\|_{\mathcal{M}} + \|yE_{|y|}[0,\lambda_y]\|_{\mathcal{M}} \leq \lambda_x + \lambda_y.$$

By Lemma 4.3.7, we have that $x \in V(\lambda_x)$ and $y \in V(\lambda_y)$. Hence, the definition of $V(\cdot)$ implies that

$$\mu([\mathcal{D}(E_{|x|}(\lambda_x,\infty)) > \lambda_x]) \leq \lambda_x, \quad \mu([\mathcal{D}(E_{|y|}(\lambda_y,\infty)) > \lambda_y]) \leq \lambda_y.$$

Since

$$\mathcal{D}(p^\perp) = \mathcal{D}(E_{|x|}(\lambda_x,\infty) \vee E_{|y|}(\lambda_y,\infty)) \leq \mathcal{D}(E_{|x|}(\lambda_x,\infty)) + \mathcal{D}(E_{|y|}(\lambda_y,\infty)),$$

it follows that

$$\begin{aligned}\mu([\mathcal{D}(p^\perp) > \lambda_x + \lambda_y]) &\leq \mu([\mathcal{D}(E_{|x|}(\lambda_x,\infty)) > \lambda_x]) \\ &\quad + \mu([\mathcal{D}(E_{|y|}(\lambda_y,\infty)) > \lambda_y]) \leq \lambda_x + \lambda_y.\end{aligned}$$

Proposition 4.3.4(i) implies that $x + y \in V(\lambda_x + \lambda_y)$, and so

$$\|x+y\|_{LS(\mathcal{M})} \leq \lambda_x + \lambda_y = \|x\|_{LS(\mathcal{M})} + \|y\|_{LS(\mathcal{M})},$$

as required. □

Proposition 4.3.8 implies that $LS(\mathcal{M})$ can be equipped with a topology, generated by the F-norm $\|\cdot\|_{LS(\mathcal{M})}$.

Definition 4.3.9. The metrizable vector topology $t(\mathcal{M})$ on the $*$-algebra $LS(\mathcal{M})$ generated by the F-norm $\|\cdot\|_{LS(\mathcal{M})}$ is called the *local measure topology*. If a sequence $\{x_n\}_{n\in\mathbb{N}} \subset LS(\mathcal{M})$ converges to an operator $x \in LS(\mathcal{M})$ with respect to the topology $t(\mathcal{M})$ (notation: $x_n \xrightarrow{t(\mathcal{M})} x$), then we say that the sequence $\{x_n\}_{n\in\mathbb{N}}$ converges to x *locally in measure*.

As shown in Theorem 1.11.13, the dimension function $\mathcal{D}$ on $\mathcal{M}$ is unique up to multiplication on an a. e. nonzero and finite function. As we will show in Propisition 4.3.31 below, the local measure topology $t(\mathcal{M})$ does not depend on the choice of such a function. This is in sharp contrast to the measure topology t_τ, which does depend on the choice of trace τ (see Remark 4.2.18).

As we shall see in Section 4.4, in general, the local measure topology $t(\mathcal{M})$ and the measure topology t_τ (where τ is a faithful normal semifinite trace on $\mathcal{M}$) are two different topologies, with the measure topology being the stronger one. For the in depth discussion of the differences between the measure topology and local measure topology, we refer the reader to Section 4.4. At the same time, as we shall demonstrate in this section, these two topologies possess many of the same properties.

The local measure topology can be defined via a slightly different base of neighborhoods of zero, as shown in the following proposition.

Proposition 4.3.10. *The sets*

$$\begin{aligned} V(\lambda,\alpha,\beta) &= \{x \in LS(\mathcal{M}) : \exists p \in P(\mathcal{M}), z \in P(\mathcal{Z}(\mathcal{M})) \text{ such that} \\ &\qquad \|x(\mathbf{1}-p)\|_{\mathcal{M}} \le \lambda; pz \in P_{\text{fin}}(\mathcal{M}), \mathcal{D}(pz) \le \alpha\varphi(z); \psi(\mathbf{1}-z) \le \beta\} \\ &= \{x \in LS(\mathcal{M}) : \exists z \in P(\mathcal{Z}(\mathcal{M})), \mathcal{D}(zE_{|x|}(\lambda,\infty)) \le \alpha\varphi(z), \psi(\mathbf{1}-z) \le \beta\}, \\ &\qquad \lambda,\alpha,\beta > 0, \end{aligned} \tag{4.9}$$

form the base of neighborhoods of zero for the local measure topology $t(\mathcal{M})$.

Proof. By Proposition 4.3.4(ii), we have that $V(\lambda,\lambda,\lambda) = V(\lambda)$ for any $\lambda > 0$. Furthermore, by Lemma 4.3.7, we have that $V(\lambda) = \{x \in LS(\mathcal{M}) : \|x\|_{LS(\mathcal{M})} \le \lambda\}$. Hence, setting $\lambda_1 = \min\{\lambda,\alpha,\beta\}$, $\lambda_2 = \max\{\lambda,\alpha,\beta\}$, we have that

$$V(\lambda_1) \subset V(\lambda,\alpha,\beta) \subset V(\lambda_2),$$

proving that the system $\{V(\lambda,\alpha,\beta)\}_{\lambda,\alpha,\beta>0}$ forms the base of neighborhoods of zero for the local measure topology $t(\mathcal{M})$.

To prove the second equality in (4.9), let $\lambda,\alpha,\beta > 0$ be fixed. Assume that $x \in LS(\mathcal{M})$ is such that there exist $p \in P(\mathcal{M})$, $z \in P(\mathcal{Z}(\mathcal{M}))$, such that $\|x(\mathbf{1}-p)\|_{\mathcal{M}} \le \lambda$, $pz \in P_{\text{fin}}(\mathcal{M})$, $\mathcal{D}(pz) \le \alpha\varphi(z)$, and $\psi(\mathbf{1}-z) \le \beta$. By Proposition 2.1.6, we have that $E_{|x|}(\lambda,\infty) \le p$. By Theorem 1.9.4(iii) and Proposition 1.11.6(iv), we have that $\mathcal{D}(zE_{|x|}(\lambda,\infty)) \le \mathcal{D}(zp) \le \alpha\varphi(z)$, as required. Conversely, if there exists $z \in P(\mathcal{Z}(\mathcal{M}))$, such that $\mathcal{D}(zE_{|x|}(\lambda,\infty)) \le \alpha\varphi(z)$ and $\psi(\mathbf{1}-z) \le \beta$, then setting $p = E_{|x|}(\lambda,\infty)$ we have that $\|x(\mathbf{1}-p)\|_{LS(\mathcal{M})} = \||x|E_{|x|}[0,\lambda]\|_{\mathcal{M}} \le \lambda$, so that $x \in V(\lambda,\alpha,\beta)$. □

Before we establish further properties of the local measure topology, we note a basic formula of $\|p\|_{LS(\mathcal{M})}$ for projections $p \in P(\mathcal{M})$.

Proposition 4.3.11. *For a projection $p \in P(\mathcal{M})$, we have that*

$$\|p\|_{LS(\mathcal{M})} = \inf\{\lambda : \mu([\mathcal{D}(p) > \lambda]) \le \lambda\} = \inf_{\lambda>0} \max\{\lambda, \mu([\mathcal{D}(p) > \lambda])\}. \tag{4.10}$$

In particular, for a sequence of projections $\{p_n\}_{n\in\mathbb{N}} \subset P(\mathcal{M})$, *we have that* $\|p_n\|_{LS(\mathcal{M})} \to 0$ *if and only if* $\mathcal{D}(p_n) \to 0$ *in measure* μ.

Proof. We note that for a projection $p \in P(\mathcal{M})$ we have that $E_p(1,\infty) = 0$ and $E_p(\lambda,\infty) = p$ for $0 < \lambda < 1$, and so

$$\|p\|_{LS(\mathcal{M})} = \inf\{\lambda : \mu([\mathcal{D}(p) > \lambda]) \le \lambda\}.$$

Proposition 4.3.6 implies that

$$\|p\|_{LS(\mathcal{M})} = \inf_{\lambda>0} \max\{\lambda, \mu([\mathcal{D}(p) > \lambda])\},$$

as required.

To prove the assertion for convergence, let $\{p_n\}_{n\in\mathbb{N}} \subset P(\mathcal{M})$, be a sequence of projections, such that $\|p_n\|_{LS(\mathcal{M})} \to 0$ as $n \to \infty$. Fix $\varepsilon > 0$. Then there exists $N \in \mathbb{N}$, such that

$$\|p_n\|_{LS(\mathcal{M})} = \inf\{\lambda : \mu([\mathcal{D}(p_n) > \lambda]) \le \lambda\} < \varepsilon/2,$$

for all $n \ge N$. Choose $\varepsilon/2 < \lambda_0 < \varepsilon$, such that

$$\mu([\mathcal{D}(p_n) > \lambda_0]) \le \lambda_0, \quad n \ge N.$$

We have that

$$\mu([\mathcal{D}(p_n) \ge \varepsilon]) \le \mu([\mathcal{D}(p_n) > \lambda_0]) \le \lambda_0 < \varepsilon$$

for all $n \ge N$. Thus, $\mathcal{D}(p_n) \to 0$ in measure μ.

Conversely, assume that $\mathcal{D}(p_n) \to 0$ in measure μ. Then for any $\varepsilon > 0$ there exists $N \in \mathbb{N}$, such that $\mu([\mathcal{D}(p_n) \ge \varepsilon/2]) \le \varepsilon/2$ for all $n \ge N$. Then

$$\mu([\mathcal{D}(p_n) > \varepsilon]) \le \mu([\mathcal{D}(p_n) \ge \varepsilon/2]) \le \varepsilon/2 < \varepsilon, \quad n \ge N,$$

which proves that

$$\|p_n\|_{LS(\mathcal{M})} = \inf\{\lambda : \mu([\mathcal{D}(p_n) > \lambda]) \le \lambda\} \le \varepsilon$$

for all $n \ge N$, as required. □

Remark 4.3.12. *We note that, by Proposition* 2.3.5(iv), $\mathcal{D}(E_{|x|}(\lambda,\infty)) \downarrow 0$ *as* $\lambda \to \infty$, *almost everywhere, for any* $x \in LS(\mathcal{M})$. *Since* μ *is a probability measure, it follows that* $\mathcal{D}(E_{|x|}(\lambda,\infty)) \to 0$ *in measure* μ *as* $\lambda \to \infty$. *Hence, Proposition* 4.3.11 *implies that* $\|E_{|x|}(\lambda,\infty)\|_{LS(\mathcal{M})} \to 0$ *as* $\lambda \to \infty$ *for any* $x \in LS(\mathcal{M})$.

Next, we want to show that $(LS(\mathcal{M}), t(\mathcal{M}))$ is in fact a topological $*$-algebra, which is also complete. We first prove some properties of the F-norm $\|\cdot\|_{LS(\mathcal{M})}$. As seen in Proposition 4.2.7, the F-norm $\|\cdot\|_{S(\mathcal{M},\tau)}$ generating the measure topology, satisfies similar properties.

Proposition 4.3.13. *For the F-norm $\|\cdot\|_{LS(\mathcal{M})}$, we have:*

(i) $\|x\|_{LS(\mathcal{M})} = \|x^*\|_{LS(\mathcal{M})} = \||x|\|_{LS(\mathcal{M})}$ *for any $x \in LS(\mathcal{M})$;*

(ii) *For any $a \in LS(\mathcal{M})$ and $x, y \in \mathcal{M}$, we have*

$$\|xay\|_{LS(\mathcal{M})} \leq \big\|\|x\|_{\mathcal{M}}\|y\|_{\mathcal{M}}a\big\|_{LS(\mathcal{M})},$$

i. e., $(LS(\mathcal{M}), \|\cdot\|_{LS(\mathcal{M})})$ is an F-normed $\mathcal{M}$-bimodule;

(iii) *if $0 \leq x \leq y$, $x, y \in LS(\mathcal{M})$, then $\|x\|_{LS(\mathcal{M})} \leq \|y\|_{LS(\mathcal{M})}$;*

(iv) *if $p, q \in P(\mathcal{M})$ are such that $p \preceq q$, then $\|p\|_{LS(\mathcal{M})} \leq \|q\|_{LS(\mathcal{M})}$;*

(v) *if $\{p_n\}_{n\in\mathbb{N}} \subset P(\mathcal{M})$, then*

$$\Big\|\sup_{n\in\mathbb{N}} p_n\Big\|_{LS(\mathcal{M})} \leq \sum_{n=1}^{\infty} \|p_n\|_{LS(\mathcal{M})};$$

(vi) *if $x \in LS(\mathcal{M})$ and $\varepsilon > 0$, then there exists $\delta > 0$, such that $\|xy\|_{LS(\mathcal{M})} < \varepsilon$ whenever $\|y\|_{LS(\mathcal{M})} < \delta$;*

(vii) *for any $x, y \in LS(\mathcal{M})$, we have that*

$$\|xy\|_{LS(\mathcal{M})} \leq \max\big\{\|x\|_{LS(\mathcal{M})}\|y\|_{LS(\mathcal{M})}, \|x\|_{LS(\mathcal{M})} + \|y\|_{LS(\mathcal{M})}\big\}.$$

Proof. (i) By definition of $\|\cdot\|_{LS(\mathcal{M})}$, we have that $\|x\|_{LS(\mathcal{M})} = \||x|\|_{LS(\mathcal{M})}$ for any $x \in LS(\mathcal{M})$. By Proposition 2.1.7, we have that $E_{|x^*|}(t,\infty) \sim E_{|x|}(t,\infty)$ for any $t > 0$. Proposition 1.11.6(iv) implies that $\mathcal{D}(E_{|x^*|}(t,\infty)) = \mathcal{D}(E_{|x|}(t,\infty))$ for any $t > 0$ and, therefore,

$$\|x^*\|_{LS(\mathcal{M})} = \||x^*|\|_{LS(\mathcal{M})} = \||x|\|_{LS(\mathcal{M})} = \|x\|_{LS(\mathcal{M})},$$

as required.

(ii) Let $a \in LS(\mathcal{M})$, $x, y \in \mathcal{M}$, and let $\lambda := \|a\|_{LS(\mathcal{M})}$. Then, by Proposition 4.3.4, there exists a projection $p \in P(\mathcal{M})$, such that $\|ap\|_{\mathcal{M}} \leq \lambda$ and $\mu([\mathcal{D}(p^\perp) > \lambda]) \leq \lambda$. Since $\|(xa)p\|_{\mathcal{M}} \leq \|(\|x\|_{\mathcal{M}}a)p\|_{\mathcal{M}} \leq \lambda$, it follows that $\|xa\|_{LS(\mathcal{M})} \leq \|\|x\|_{\mathcal{M}}a\|_{LS(\mathcal{M})}$. Therefore, using part (i) repeatedly, we infer that

$$\begin{aligned}\|xay\|_{LS(\mathcal{M})} &\leq \|\|x\|_{\mathcal{M}}ay\|_{LS(\mathcal{M})} = \|\|x\|_{\mathcal{M}}y^*a^*\|_{LS(\mathcal{M})} \\ &\leq \big\|\|x\|_{\mathcal{M}}\|y\|_{\mathcal{M}}a^*\big\|_{LS(\mathcal{M})} = \big\|\|x\|_{\mathcal{M}}\|y\|_{\mathcal{M}}a\big\|_{LS(\mathcal{M})}.\end{aligned}$$

(iii) and (iv). By part (ii), $(LS(\mathcal{M}), \|\cdot\|_{LS(\mathcal{M})})$ is an F-normed bimodule over $\mathcal{M}$. Therefore, the assertions follow from Proposition 3.3.10 and Proposition 3.3.11(i), respectively.

(v) Let $\{p_n\}_{n\in\mathbb{N}} \subset P(\mathcal{M})$. Referring to Proposition 1.11.6(i), we have that $\mathcal{D}(\sup_{n\in\mathbb{N}} p_n) \leq \sum_{n=1}^{\infty} \mathcal{D}(p_n)$. Let $\lambda > 0$ be such that $\sum_{n=1}^{\infty} \|p_n\|_{LS(\mathcal{M})} < \lambda$. Choose a sequence $\{\lambda_n\}_{n\in\mathbb{N}} \subset (0,\infty)$, such that $\|p_n\|_{LS(\mathcal{M})} < \lambda_n$, $\sum_n \lambda_n = \lambda$. We have that $\mu([\mathcal{D}(p_n) > \lambda_n]) \le \lambda_n$ for any $n \in \mathbb{N}$.

If $\omega \in [\sum_{n=1}^{\infty} \mathcal{D}(p_n) > \lambda]$, then there exists such $n_0 \in \mathbb{N}$ that $\omega \in [\mathcal{D}(p_{n_0}) > \lambda_{n_0}]$. Hence,

$$\left[\sum_{n=1}^{\infty} \mathcal{D}(p_n) > \lambda\right] \subset \bigcup_{n\in\mathbb{N}} [\mathcal{D}(p_n) > \lambda_n].$$

Therefore,

$$\mu\left(\left[\sup_{n\in\mathbb{N}} \mathcal{D}(p_n) > \lambda\right]\right) \le \mu\left(\left[\sum_{n=1}^{\infty} \mathcal{D}(p_n) > \lambda\right]\right) \le \mu\left(\bigcup_{n\in\mathbb{N}} [\mathcal{D}(p_n) > \lambda_n]\right) \le \lambda.$$

Referring again to (4.10), we infer that

$$\left\|\sup_{n\in\mathbb{N}} p_n\right\|_{LS(\mathcal{M})} \le \lambda.$$

Since $\lambda > 0$ is an arbitrary number satisfying $\sum_{n=1}^{\infty} \|p_n\|_{LS(\mathcal{M})} < \lambda$, we conclude that

$$\left\|\sup_{n\in\mathbb{N}} p_n\right\|_{LS(\mathcal{M})} \le \sum_{n=1}^{\infty} \|p_n\|_{LS(\mathcal{M})},$$

as required.

(vi) Fix $x \in LS(\mathcal{M})$ and $\varepsilon > 0$. By Remark 4.3.12, there exists $\Lambda > 0$, such that $\|E_{|x|}(\Lambda,\infty)\|_{LS(\mathcal{M})} < \frac{\varepsilon}{2}$. Let $\delta = \frac{\varepsilon}{\Lambda+2}$ and let $y \in LS(\mathcal{M})$ be such that $\|y\|_{LS(\mathcal{M})} < \delta$. By Proposition 4.3.4, there exists $q \in P(\mathcal{M})$, such that $\|yq\|_{\mathcal{M}} < \delta$ and $\mu([\mathcal{D}(q^{\perp}) > \delta]) \le \delta$. By (4.10), we have $\|q^{\perp}\|_{LS(\mathcal{M})} \le \delta$.

We set $e = E_{|x|}(\Lambda,\infty)$, $p = (s_r(ey))^{\perp}$. We have

$$\begin{aligned} xyp &= xy - xyp^{\perp} = xy - xe^{\perp}yp^{\perp} - xeyp^{\perp} = xy - xe^{\perp}yp^{\perp} - xey \\ &= xe^{\perp}y - xe^{\perp}yp^{\perp} = xe^{\perp}yp, \end{aligned}$$

and, therefore,

$$\|xy(p\wedge q)\|_{\mathcal{M}} \le \|xypq\|_{\mathcal{M}} = \|xe^{\perp}ypq\|_{\mathcal{M}} \le \|xe^{\perp}\|_{\mathcal{M}}\|ypq\|_{\mathcal{M}} < \Lambda\delta < \varepsilon.$$

Furthermore, since $p^{\perp} = s_r(ey) \sim s_l(ey) \le e$, parts (iv) and (v) implies that

$$\begin{aligned} \|(p\wedge q)^{\perp}\|_{LS(\mathcal{M})} &= \|p^{\perp} \vee q^{\perp}\|_{LS(\mathcal{M})} \le \|p^{\perp}\|_{LS(\mathcal{M})} + \|q^{\perp}\|_{LS(\mathcal{M})} \\ &\le \|e\|_{LS(\mathcal{M})} + \|q^{\perp}\|_{LS(\mathcal{M})} < \frac{1}{2}\varepsilon + \delta < \varepsilon. \end{aligned}$$

Therefore, $\|xy\|_{LS(\mathcal{M})} < \varepsilon$ as required.

(vii) Let $\rho_x = \|x\|_{LS(\mathcal{M})}$, $\rho_y = \|y\|_{LS(\mathcal{M})}$. We denote

$$e = E_{|x|}[0,\rho_x], \quad f = E_{|y|}[0,\rho_y], \quad p = (s_r(ey))^{\perp}, \quad q = f \wedge p.$$

Repeating the argument in part (vi) verbatim, one obtains that

$$xyp = xe^{\perp}yp.$$

Therefore,

$$\begin{aligned} \|xyq\|_{\mathcal{M}} = \|xypq\|_{\mathcal{M}} &= \|xe^{\perp}ypq\|_{\mathcal{M}} = \|xe^{\perp}yfq\|_{\mathcal{M}} \\ &\leq \|xe^{\perp}\|_{\mathcal{M}}\|yf\|_{\mathcal{M}} \leq \rho_x\rho_y \end{aligned}$$

and

$$\|q^{\perp}\|_{LS(\mathcal{M})} \leq \|p^{\perp}\|_{LS(\mathcal{M})} + \|f^{\perp}\|_{LS(\mathcal{M})} \leq \|e^{\perp}\|_{LS(\mathcal{M})} + \|f^{\perp}\|_{LS(\mathcal{M})} \leq \rho_x + \rho_y,$$

where the second inequality follows from a combination of parts (iv) and (v) with the majorisation $p^{\perp} \sim s_l(ey) \leq e$. Thus, $\|xy\|_{LS(\mathcal{M})} \leq \max\{\rho_x\rho_y, \rho_x + \rho_y\}$, as required. □

Recall that when $\mathcal{M}$ is a semifinite von Neumann algebra equipped with a faithful normal semifinite trace, the $*$-algebra $S(\mathcal{M},\tau)$ is a complete topological $*$-algebra, and an F-space when equipped with the measure topology t_τ (see Theorem 4.2.8). As we show next, the same is true for the $*$-algebra $LS(\mathcal{M})$, equipped with the local measure topology (where $\mathcal{M}$ is arbitrary, not necessarily semifinite, von Neumann algebra).

Theorem 4.3.14. *Let $\mathcal{M}$ be an arbitrary von Neumann algebra. The pair $(LS(\mathcal{M}), t(\mathcal{M}))$ is a topological $*$-algebra with jointly continuous multiplication and an F-space.*

Proof. Proposition 4.3.13(i) immediately implies that the mapping $x \mapsto x^*$, $x \in LS(\mathcal{M})$ is continuous. To show that the product $(x,y) \mapsto xy$, $x,y \in LS(\mathcal{M})$ is jointly continuous, let $x_n \to x$ and $y_n \to y$ locally in measure. We have

$$\begin{aligned} \|x_ny_n - xy\|_{LS(\mathcal{M})} \leq & \|(x_n - x)y\|_{LS(\mathcal{M})} + \|x(y_n - y)\|_{LS(\mathcal{M})} \\ & + \|(x_n - x)(y_n - y)\|_{LS(\mathcal{M})}. \end{aligned}$$

Proposition 4.3.13(vi) guarantees that the first two terms on the right-hand side vanish, while Proposition 4.3.13(vii) guarantees that the third term vanishes, too. Hence, $x_ny_n \to xy$ locally in measure.

It remains to show completeness of the space $(LS(\mathcal{M}), \|\cdot\|_{LS(\mathcal{M})})$. Let $\{x_n\}_{n\in\mathbb{N}}$ be a Cauchy sequence in $(LS(\mathcal{M}), \|\cdot\|_{LS(\mathcal{M})})$. Since involution is $t(\mathcal{M})$-continuous on $LS(\mathcal{M})$, without loss of generality, we may assume that $x_n^* = x_n$. Passing to a subsequence, if necessary, we assume that $\|x_n - x_{n+1}\|_{LS(\mathcal{M})} < 2^{-n}$, $n \in \mathbb{N}$.

By Remark 4.3.12, we have that $\mu([\mathcal{D}(E_{|x_n|}(m,\infty)) > 0]) \downarrow 0$ for $m \to \infty$. Therefore, $\mu([\mathcal{D}(E_{|x_n|}(m(n),\infty)) > 0]) < 2^{-n}$ for some $m(n) \in \mathbb{N}$. We set

$$p_n = E_{|x_n|}[0, m(n)], \quad n \in \mathbb{N}.$$

By the functional calculus, we have that $x_n p_n \in \mathcal{M}$. Furthermore, since

$$\mu([\mathcal{D}(p_n^\perp) > 2^{-n}]) \le \mu([\mathcal{D}(p_n^\perp) > 0]) = \mu([\mathcal{D}(E_{|x_n|}(m(n),\infty)) > 0]) < 2^{-n},$$

equality (4.10) implies that $\|p_n^\perp\|_{LS(\mathcal{M})} \le 2^{-n}$.

Similarly, for every $n \in \mathbb{N}$ there exists a projection $e_n \in P(\mathcal{M})$, such that

$$\|(x_n - x_{n+1})e_n\|_{\mathcal{M}} < 2^{-n}, \quad \|e_n^\perp\|_{LS(\mathcal{M})} \le 2^{-n}, \quad n \in \mathbb{N}.$$

Define an increasing sequence of projections by setting

$$q_k = \inf_{n \ge k}(p_n \wedge e_n).$$

Referring to Proposition 4.3.13(v), we have that

$$\begin{aligned}\|q_k^\perp\|_{LS(\mathcal{M})} &= \Big\|\sup_{n\ge k} p_n^\perp \vee e_n^\perp\Big\|_{LS(\mathcal{M})} \le \sum_{n=k}^{\infty}(\|p_n^\perp\|_{LS(\mathcal{M})} + \|q_n^\perp\|_{LS(\mathcal{M})}) \\ &\le 2^{-k+2} \to 0, \quad k \to \infty.\end{aligned}$$

Hence, $q_k \uparrow \mathbf{1}$ as $k \to \infty$.

For every $k \in \mathbb{N}$, we have that $x_k q_k \in \mathcal{M}$ and $\|(x_{n+1} - x_n)q_k\|_{\mathcal{M}} < 2^{-n}$ for $n \ge k$. Hence, for every fixed $k \in \mathbb{N}$, the sequence $\{x_n q_k\}_{n\in\mathbb{N}}$ converges in $\mathcal{M}$ to some $y_k \in \mathcal{M}$. Since

$$y_k = \lim_{n\to\infty} x_n q_k = \lim_{n\to\infty} x_n q_m q_k = y_m q_k$$

for $m \ge k$, we may define an operator y on $\mathrm{dom}(y) = \bigcup_k q_k(H)$ by setting $yq_k = y_k$ for every $k \in \mathbb{N}$.

We claim that y is a symmetric operator affiliated with $\mathcal{M}$. For all $\xi, \eta \in \mathrm{dom}(y)$, we have that $\xi = q_{k_1}\xi$ and $\eta = q_{k_2}\eta$ for some $k_1, k_2 \in \mathbb{N}$. We have

$$\langle y\xi, \eta\rangle = \langle yq_{k_1}\xi, \eta\rangle = \langle y_{k_1}\xi, \eta\rangle = \lim_{n\to\infty}\langle x_n q_{k_1}\xi, \eta\rangle = \lim_{n\to\infty}\langle x_n\xi, \eta\rangle.$$

Since x_n is self-adjoint for every $n \in \mathbb{N}$, we infer that

$$\begin{aligned}\langle y\xi, \eta\rangle &= \lim_{n\to\infty}\langle x_n\xi, \eta\rangle = \lim_{n\to\infty}\langle \xi, x_n\eta\rangle = \lim_{n\to\infty}\langle \xi, x_n q_{k_2}\eta\rangle \\ &= \lim_{n\to\infty}\langle \xi, yq_{k_2}\eta\rangle = \langle \xi, y\eta\rangle.\end{aligned}$$

Thus, y is symmetric and, therefore, a closable operator. Next, let $u \in U(\mathcal{M}')$. For every $\xi \in \operatorname{dom}(y)$, we have that $uy\xi = uy_kq_k\xi = y_kq_ku\xi = y(u\xi)$, showing that y is affiliated with $\mathcal{M}$.

Let x denote the closure of y. By Proposition 2.1.2(ii), x is affiliated with $\mathcal{M}$. Since $q_k \uparrow \mathbf{1}$, $q_k(H) \subset \operatorname{dom}(y) \subset \operatorname{dom}(x)$ and $\mathcal{D}(q_k^\perp) \downarrow 0$ as $k \to \infty$, Proposition 2.3.5(iv) implies that x is locally measurable with respect to $\mathcal{M}$. Furthermore, we have

$$\begin{aligned}\|(x - x_k)q_k\|_{\mathcal{M}} &= \|(y - x_k)q_k\|_{\mathcal{M}} = \lim_{n\to\infty} \|(x_{k+n} - x_n)q_k\|_{\mathcal{M}} \\ &\le \sum_{n=0}^{\infty} \|(x_{k+p+1} - x_{k+p})q_k\|_{\mathcal{M}} < 2^{-k+1}.\end{aligned}$$

Since $\|q_k^\perp\|_{LS(\mathcal{M})} < 2^{-k+2}$, it follows that $\|x_k - x\|_{LS(\mathcal{M})} < 2^{-k+2}$, $k \in \mathbb{N}$, i. e., $x_k \to x$ locally in measure as $k \to \infty$, proving that $(LS(\mathcal{M}), \|\cdot\|_{LS(\mathcal{M})})$ is complete. □

The following corollary demonstrates the importance of the algebra $LS(\mathcal{M})$ from a topological point of view. For a similar result for the measure topology, we refer to Corollary 4.2.13.

Corollary 4.3.15. *The von Neumann algebra $\mathcal{M}$ is dense in $LS(\mathcal{M})$ with respect to the local measure topology. Namely, any $x \in LS(\mathcal{M})$ can be approximated in the local measure topology by the sequence $\{xE_{|x|}[0,n]\}_{n\in\mathbb{N}} \subset \mathcal{M}$.*

Proof. Let $x \in LS(\mathcal{M})$ and let $\varepsilon > 0$ be arbitrary. By Proposition 4.3.13(vi), there exists δ such that $\|xy\|_{LS(\mathcal{M})} < \varepsilon$ whenever $\|y\|_{LS(\mathcal{M})} < \delta$. By Remark 4.3.12, we have that $\mu([\mathcal{D}(E_{|x|}(n,\infty)) > 0]) \downarrow 0$ as $n \to \infty$. Therefore, there exists $n \in \mathbb{N}$, such that $\mu([\mathcal{D}(E_{|x|}(n,\infty)) > 0]) < \delta$. Since

$$\mu([\mathcal{D}(E_{|x|}(n,\infty)) > \delta]) \le \mu([\mathcal{D}(E_{|x|}(n,\infty)) > 0]) < \delta,$$

equality (4.10), and implies that $\|E_{|x|}(n,\infty)\|_{LS(\mathcal{M})} < \delta$. For each $n \in \mathbb{N}$, setting $x_n = xE_{|x|}[0,n] \in \mathcal{M}$, we have

$$\|x_n - x\|_{LS(\mathcal{M})} = \|xE_{|x|}(n,\infty)\|_{LS(\mathcal{M})} < \varepsilon,$$

and so $x_n \to x$ in the local measure topology as $n \to \infty$. □

In the following proposition, we collect some necessary and sufficient conditions for convergence locally in measure. For similar necessary and sufficient conditions for convergence in the measure topology, see Proposition 4.2.9.

Proposition 4.3.16. *Assume that $\{x_n\}_{n\in\mathbb{N}} \subset LS(\mathcal{M})$, $x \in LS(\mathcal{M})$. The following conditions are equivalent:*

(i) *x_n converges to x locally in measure as $n \to \infty$;*

(ii) *$E_{|x_n-x|}(\lambda,\infty) \to 0$ locally in measure as $n \to \infty$ for every $\lambda > 0$;*

(iii) *$\mathcal{D}(E_{|x_n-x|}(\lambda,\infty)) \to 0$ in measure μ as $n \to \infty$ for every $\lambda > 0$;*

(iv) *for every $\varepsilon > 0$, there exists $\{p_n\}_{n\in\mathbb{N}} \subset P(\mathcal{M})$, such that $p_n \to \mathbf{1}$ locally in measure and $\|(x_n - x)p_n\|_{\mathcal{M}} < \varepsilon$.*

Proof. Without loss of generality, we may assume that $x = 0$.

(i)$\Leftrightarrow$ (ii). Assume that $\|x_n\|_{LS(\mathcal{M})} \to 0$ as $n \to \infty$. For any $\lambda > 0$, we have that

$$E_{|x_n|}(\lambda, \infty) \leq \frac{1}{\lambda}|x_n|E_{|x_n|}(\lambda, \infty)$$

and, therefore, by Proposition 4.3.13(ii) and (iii), we have that

$$\|E_{|x_n|}(\lambda, \infty)\|_{LS(\mathcal{M})} \leq \left\|\frac{1}{\lambda}|x_n|\right\|_{LS(\mathcal{M})} \to 0, \quad n \to \infty.$$

Conversely, assume that $\|E_{|x_n|}(\lambda, \infty)\|_{LS(\mathcal{M})} \to 0$ as $n \to \infty$ for any $\lambda > 0$. For every fixed $\varepsilon > 0$, there exists $N \in \mathbb{N}$, such that $\|E_{|x_n|}(\varepsilon, \infty)\|_{LS(\mathcal{M})} < \varepsilon/2$ for all $n > N$. By (4.10), we have that

$$\|E_{|x_n|}(\varepsilon, \infty)\|_{LS(\mathcal{M})} = \inf\{t : \mu([\mathcal{D}(E_{|x_n|}(\varepsilon, \infty)) > t]) \leq t\},$$

and so there exists $t \in (\varepsilon/2, \varepsilon)$, such that $\mu([\mathcal{D}(E_{|x_n|}(\varepsilon, \infty)) > t]) \leq t$. Since $t < \varepsilon$, we have that

$$\mu([\mathcal{D}(E_{|x_n|}(\varepsilon, \infty)) > \varepsilon]) \leq \mu([\mathcal{D}(E_{|x_n|}(\varepsilon, \infty)) > t]) \leq t < \varepsilon.$$

Therefore, by definition of $\|\cdot\|_{LS(\mathcal{M})}$, we infer that

$$\|x_n\|_{LS(\mathcal{M})} = \inf\{\lambda : \mu([\mathcal{D}(E_{|x_n|}(\lambda, \infty)) > \lambda]) \leq \lambda\} \leq \varepsilon,$$

for all $n > N$. Since ε is arbitrary, it follows that $x_n \to 0$ locally in measure.

The equivalence (ii)$\Leftrightarrow$(iii) follows immediately from Proposition 4.3.11.

(ii)$\Leftrightarrow$(iv). Assume that for every $\varepsilon > 0$ we have that $E_{|x_n|}(\varepsilon, \infty) \to 0$ locally in measure as $n \to \infty$, or equivalently, $E_{|x_n|}[0, \varepsilon] \to \mathbf{1}$ locally in measure as $n \to \infty$. Since $\|x_n E_{|x_n|}[0, \varepsilon]\|_{\mathcal{M}} \leq \varepsilon$, the claim follows by taking $p_n = E_{|x_n|}[0, \varepsilon]$.

Conversely, assume that for every $\varepsilon > 0$ there exists $\{p_n\}_{n\in\mathbb{N}} \in P(\mathcal{M})$, such that $p_n \to \mathbf{1}$ locally in measure and $\|x_n p_n\|_{\mathcal{M}} < \varepsilon$. By Proposition 2.1.6, we have that $E_{|x_n|}(\varepsilon, \infty) \preceq p_n^{\perp}$. By Proposition 4.3.13(iv), we have that $\|E_{|x_n|}(\varepsilon, \infty)\|_{LS(\mathcal{M})} \leq \|p_n^{\perp}\|_{LS(\mathcal{M})} \to 0$ as $n \to \infty$, as required. □

Proposition 4.3.16 implies the following sufficient condition for local measure convergence and is similar to Corollary 4.2.11. We recall that $s_l(x)$ and $s_r(x)$ denote the left and right supports of $x \in LS(\mathcal{M})$, respectively.

Corollary 4.3.17. *If $\{x_n\}_{n\in\mathbb{N}} \subset LS(\mathcal{M})$ is such that $s_r(x_n) \xrightarrow{t(\mathcal{M})} 0$ (or equivalently, $s_l(x_n) \xrightarrow{t(\mathcal{M})} 0$) as $n \to \infty$, then $x_n \xrightarrow{t(\mathcal{M})} 0$.*

Proof. Since $s_r(x_n) \sim s_l(x_n)$ for any $n \in \mathbb{N}$, it follows that $\mathcal{D}(s_r(x_n)) = \mathcal{D}(s_l(x_n))$ and, therefore, by Proposition 4.3.11 we have that $s_r(x_n) \to 0$ locally in measure, if and only if $s_l(x_n) \to 0$ locally in measure.

Assume now that $s_r(x_n) \to 0$ locally in measure as $n \to \infty$. Note that for any $\lambda > 0$ we have that $E_{|x_n|}(\lambda, \infty) \le s(|x_n|) = s_r(x_n)$. Therefore, Proposition 4.3.13(iii) implies that $\|E_{|x_n|}(\lambda, \infty)\|_{LS(\mathcal{M})} \le \|s_r(x_n)\|_{LS(\mathcal{M})} \to 0$ as $n \to \infty$. Proposition 4.3.16(ii) immediately implies that $x_n \to 0$ locally in measure. □

In the following proposition, we collect some simple properties of the local measure topology. The reader is invited to compare the properties below with the corresponding properties of the topology of convergence in measure given in Proposition 4.2.12.

Proposition 4.3.18.

(i) *The sets $LS_h(\mathcal{M})$ and $LS_+(\mathcal{M}) = \{x \in LS(\mathcal{M}) : x \ge 0\}$ are closed in $(LS(\mathcal{M}), t(\mathcal{M}))$;*

(ii) *For any $x \in \mathcal{M}$, we have that $\|x\|_{LS(\mathcal{M})} \le \|x\|_{\mathcal{M}}$. In particular, the embedding of $(\mathcal{M}, \|\cdot\|_{\mathcal{M}})$ into $(LS(\mathcal{M}), t(\mathcal{M}))$ is continuous;*

(iii) *The span of $P(\mathcal{M})$ is dense in $(LS(\mathcal{M}), t(\mathcal{M}))$;*

(iv) *The unit ball of $\mathcal{M}$ is closed in $t(\mathcal{M})$;*

(v) *The topology $t(\mathcal{M})$ induces the topology $t(\mathcal{Z}(\mathcal{M}))$ on $S(\mathcal{Z}(\mathcal{M}))$;*

(vi) *Let $\{x_n\}_{n\in\mathbb{N}} \subset LS(\mathcal{M})$ and $\{p_n\}_{n\in\mathbb{N}} \subset P(\mathcal{M})$. If $p_n \to 0$ locally in measure as $n \to \infty$, then $x_n p_n, p_n x_n \to 0$ locally in measure as $n \to \infty$.*

Proof. (i) Since $(LS(\mathcal{M}), t(\mathcal{M}))$ is a topological $*$-algebra, it follows immediately that $LS_h(\mathcal{M})$ is $t(\mathcal{M})$-closed.

Let $x_n \in LS_+(\mathcal{M})$ be a sequence converging locally in measure to an operator $x \in LS(\mathcal{M})$. Setting $e = E_x(-\infty, 0)$, we have

$$0 \le x_- = -exe \le ex_n e - exe,$$

where the last inequality follows from the inequality $ex_n e \ge 0$. By Proposition 4.3.13(ii) and (iii), we have that

$$\|x_-\|_{LS(\mathcal{M})} \le \|e(x_n - x)e\|_{LS(\mathcal{M})} \le \|x_n - x\|_{LS(\mathcal{M})} \to 0, \quad n \to \infty.$$

Hence, $\|x_-\|_{LS(\mathcal{M})} = 0$, so that $x_- = 0$. Thus, $x = x_+ \ge 0$.

(ii) Note that for a bounded operator $x \in \mathcal{M}$, the spectral projection $E_{|x|}(\|x\|_{\mathcal{M}}, \infty)$ is 0 and, therefore, by the definition of the F-norm $\|\cdot\|_{LS(\mathcal{M})}$, we have that

$$\|x\|_{LS(\mathcal{M})} = \inf\{\lambda : \mu([\mathcal{D}(E_{|x|}(\lambda, \infty)) > \lambda]) \le \lambda\} \le \|x\|_{\mathcal{M}},$$

as required.

(iii) By Corollary 4.3.15, the algebra $\mathcal{M}$ is dense in $(LS(\mathcal{M}), t(\mathcal{M}))$ and, therefore, it is sufficient to show that the span of $P(\mathcal{M})$ is $t(\mathcal{M})$-dense in $\mathcal{M}$. Let $x \in \mathcal{M}$. Without loss of generality, $\|x\|_{\mathcal{M}} = 1$. Furthermore, since x can be written as a linear combination of positive elements from $\mathcal{M}$, we may further assume that x is positive.

For each $n \in \mathbb{N}$, we set $e_i = E_x(\frac{i-1}{n}, \frac{i}{n}]$, $i = 1, \ldots, n$, and define

$$y_n := \sum_{i=1}^{n} \frac{i}{n} e_i.$$

Then y_n belongs to the span of $P(\mathcal{M})$ for every $n \in \mathbb{N}$, and

$$\|x - y_n\|_{\mathcal{M}} = \left\| \sum_{i=1}^{n} \left(x - \frac{i}{n} \right) e_i \right\|_{\mathcal{M}} \leq \frac{\|x\|_{\mathcal{M}}}{n}$$

and, therefore, by (ii) we have that $\|x - y_n\|_{LS(\mathcal{M})} \leq \frac{\|x\|_{\mathcal{M}}}{n} \to 0$ as $n \to \infty$. Thus, the span of $P(\mathcal{M})$ is $t(\mathcal{M})$-dense in $\mathcal{M}_+$, as required.

(iv) Let $\{x_n\} \in \mathcal{M}$ and $x \in LS(\mathcal{M})$ be such that $x_n \to x$ in the local measure topology and $\|x_n\|_{\mathcal{M}} \leq 1$ for all $n \in \mathbb{N}$. Then $\mathbf{1} - x_n^* x_n \geq 0$, and so by part (i), we have that $\mathbf{1} - x^* x \geq 0$. Hence, $\|x\|_{\mathcal{M}} \leq 1$, as required.

(v) Let $x \in S(\mathcal{Z}(\mathcal{M})) = LS(\mathcal{Z}(\mathcal{M}))$. Since $E_{|x|}[\lambda, \infty) \in P(\mathcal{Z}(\mathcal{M}))$ for any $\lambda > 0$, it follows that $\|x\|_{LS(\mathcal{M})} = \|x\|_{LS(\mathcal{Z}(\mathcal{M}))}$.

(vi) It is clear that $s_r(x_n p_n) \leq p_n$. Therefore, by Proposition 4.3.13(iii), we have

$$\|s_r(x_n p_n)\|_{LS(\mathcal{M})} \leq \|p_n\|_{LS(\mathcal{M})} \to 0, \quad n \to \infty.$$

By Corollary 4.3.17, we have $x_n p_n \to 0$ locally in measure as $n \to \infty$.

Similarly, since $s_l(p_n x_n) \leq p_n$, we conclude that $p_n x_n \to 0$ in $LS(\mathcal{M})$ as $n \to \infty$. □

For a sequence of projections, convergence locally in measure can be also characterised as follows. This result complements Proposition 4.3.11.

Proposition 4.3.19. *Let $\{p_n\}_{n\in\mathbb{N}} \subset P(\mathcal{M})$. Then $p_n \to 0$ locally in measure as $n \to \infty$ if and only if there exists a sequence $\{z_n\}_{n\in\mathbb{N}} \subset P(\mathcal{Z}(\mathcal{M}))$, such that $z_n p_n \in P_{\mathrm{fin}}(\mathcal{M})$, $z_n^\perp \to 0$ locally in measure and $\mathcal{D}(z_n p_n) \to 0$ in measure μ as $n \to \infty$.*

Proof. Assume first that $p_n \to 0$ locally in measure as $n \to \infty$. By Theorem 1.9.9(vi) for every $n \in \mathbb{N}$, there exists a central projection z_n, such that $z_n p_n \in P_{\mathrm{fin}}(\mathcal{M})$, $z_n^\perp \leq s_c(p_n)$, and $z_n^\perp p_n$ is a properly infinite projection.

Proposition 1.11.6(vi) and Remark 1.11.5 imply that $s(\mathcal{D}(z_n^\perp p_n)) = s(\varphi(z_n^\perp))$ and $\mathcal{D}(z_n^\perp p_n) = \infty \cdot \varphi(z_n^\perp)$ for any $n \in \mathbb{N}$. Therefore, for any $\lambda > 0$ we have $[\mathcal{D}(z_n^\perp p_n) > \lambda] = [\mathcal{D}(z_n^\perp p_n) > 0]$, and so

$$[\mathcal{D}(z_n^\perp) > \lambda] = [\varphi(z_n^\perp) > \lambda] = s(\varphi(z_n^\perp)) = s(\mathcal{D}(z_n^\perp p_n)) = [\mathcal{D}(z_n^\perp p_n) > \lambda].$$

Thus, $\|z_n^\perp\|_{LS(\mathcal{M})} = \|z_n^\perp p_n\|_{LS(\mathcal{M})} \leq \|p_n\|_{LS(\mathcal{M})} \to 0$, showing that $z_n^\perp \to 0$ locally in measure.

Since $\|p_n\|_{LS(\mathcal{M})} \to 0$, it follows that $\|z_n p_n\|_{LS(\mathcal{M})} \le \|p_n\|_{LS(\mathcal{M})} \to 0$. By Proposition 4.3.11, we have that $\mathcal{D}(z_n p_n) \to 0$ in measure μ, as required.

Conversely, assume that there exists a sequence $\{z_n\}_{n\in\mathbb{N}} \in P(\mathcal{Z}(\mathcal{M}))$, such that $z_n p_n \in P_{\text{fin}}(\mathcal{M})$, $z_n^{\perp} \to 0$ locally in measure and $\mathcal{D}(z_n p_n) \to 0$ in measure. Proposition 4.3.18(vi) implies that $z_n^{\perp} p_n \to 0$ locally in measure. By (4.10), we have that $\|z_n p_n\|_{LS(\mathcal{M})} = \inf\{\lambda : \mu([\mathcal{D}(p_n z_n) > \lambda]) \le \lambda\} \to 0$. Hence, $\|p_n\|_{LS(\mathcal{M})} \le \|p_n z_n\|_{LS(\mathcal{M})} + \|p_n z_n^{\perp}\|_{LS(\mathcal{M})} \to 0$, proving that $p_n \to 0$ locally in measure as $n \to \infty$. □

We now describe the topology of convergence locally in measure on the two simplest examples.

Example 4.3.20. *Let $\mathcal{M}$ be a commutative von Neumann algebra. By Theorem* 1.8.10, *$\mathcal{M}$ is identified with the von Neumann algebra $L_\infty(\Omega, \Sigma, \mu)$ of all essentially bounded measurable functions on some σ-finite measure space (Ω, Σ, μ). In this case, the algebra $LS(\mathcal{M})$ is identified with the algebra $L_0(\Omega, \Sigma, \mu)$ of all μ-a. e. finite measurable functions on Ω (see Example* 2.1.4 *and Proposition* 2.3.3*). Then, the topology $t(L_\infty(\Omega, \Sigma, \mu))$ coincides with the topology of convergence locally in measure for measurable functions.*

Proof. Clearly, $\mathcal{Z}(\mathcal{M}) = L_\infty(\Omega, \Sigma, \mu)$. However, to define the dimension function $\mathcal{D}$, we replace the measure μ on the space (Ω, Σ) by an equivalent probability measure ν. For this reason, we make a distinction and assume that (Ω, Σ, μ) is a σ-finite measure space and denote by ν an equivalent probability measure on (Ω, Σ). We let $\mathcal{D} : P(\mathcal{M}) \to L_\infty(\Omega, \Sigma, \nu)$ be the dimension function. In this case, $\mathcal{D}(p) = p$, $p \in P(\mathcal{M})$.

By Proposition 4.3.16, it is sufficient to show that the convergence in the topology $t(\mathcal{M})$ coincides with the local measure convergence in $L_\infty(\Omega, \Sigma, \mu)$ on projections.

Let $\{p_n\}_{n\in\mathbb{N}} \subset P(\mathcal{M})$. Then $p_n = \chi_{A_n}$ for some $A_n \in \Sigma$. By Proposition 4.3.11, $\|p_n\|_{LS(\mathcal{M})} \to 0$ if and only if $\chi_{A_n} = \mathcal{D}(p_n) \to 0$ in measure ν. Since ν is a finite measure, it follows that $\|p_n\|_{LS(\mathcal{M})} \to 0$ if and only if every subsequence of $\{\chi_{A_n}\}$ has a subsequence, which converges to 0 ν-a. e. Since the measure ν and μ have the same null sets, it follows that ν-a. e. convergence is the same as μ-a. e. convergence. Therefore, $\|p_n\|_{LS(\mathcal{M})} \to 0$ if and only if every subsequence of $\{\chi_{A_n}\}$ has a subsequence, which converges to 0 ν-a. e., or, equivalently, if $\{\chi_{A_n}\}$ converges to 0 locally in measure μ. This proves the assertion. □

Example 4.3.21. *If $\mathcal{M}$ is a factor of type I or III, then the local measure topology $t(\mathcal{M})$ coincides with the topology of the uniform norm convergence.*

Proof. By Theorem 2.5.7 we have that $LS(\mathcal{M}) = \mathcal{M}$. Since $\mathcal{Z}(\mathcal{M})$ is trivial, the space (Ω, Σ, μ) consists of a single point $\Omega = \{\omega\}$. As before, we assume that $\mu(\{\omega\}) = 1$.

If $\mathcal{M}$ is type I algebra, then $\mathcal{D}(p)(\omega)$ is equal to the rank of a projection $p \in P(\mathcal{M})$. If $\mathcal{M}$ is type III, then $\mathcal{D}(p)(\omega) = +\infty$ for any nonzero projection $p \in P(\mathcal{M})$.

Since $\|x\|_{LS(\mathcal{M})} \le \|x\|_{\mathcal{M}}$ (see Proposition 4.3.18(ii)), it is sufficient to show that the convergence $\|x_n\|_{LS(\mathcal{M})} \to 0$ implies the convergence $\|x_n\|_{\mathcal{M}} \to 0$ for any sequence $\{x_n\}_{n\in\mathbb{N}} \subset \mathcal{M}$. Without loss of generality, we may assume that $x_n \ge 0$ and $\|x_n\|_{LS(\mathcal{M})} < \frac{1}{2}$

for any $n \in \mathbb{N}$. By definition of the F-norm $\|\cdot\|_{LS(\mathcal{M})}$ (see Definition 4.3.5), there exists $\lambda > 0$, such that $\mu([\mathcal{D}(E_{|x_n|}(\lambda,\infty)) > \lambda]) \le \lambda < 2\|x_n\|_{LS(\mathcal{M})} < 1$. We claim that $E_{|x_n|}(\lambda,\infty) = 0$. If $E_{|x_n|}(\lambda,\infty) \ne 0$, then $\mathcal{D}(E_{|x_n|}(\lambda,\infty))(\omega) \ge 1 > \lambda$ and, therefore, $\mu([\mathcal{D}(E_{|x|}(\lambda,\infty)) > \lambda]) = 1$, which is a contradiction. Hence, $E_{|x_n|}(\lambda,\infty) = 0$ for any $n \in \mathbb{N}$. By the spectral theorem, this implies that $\|x_n\|_{\mathcal{M}} \le \lambda < 2\|x_n\|_{LS(\mathcal{M})} \to 0$, as $n \to \infty$.

Thus, $x_n \to 0$ in the local measure topology if and only if $x_n \to 0$ in the uniform norm. □

Our next aim is to show that the $*$-algebra $S_0(\mathcal{M})$ of all compact operators affiliated with $\mathcal{M}$ is, in fact, a closed ideal in $(LS(\mathcal{M}), \|\cdot\|_{LS(\mathcal{M})})$. We start with an auxiliary lemma.

Lemma 4.3.22. *Let $x, y \in LS(\mathcal{M})$ and let $\alpha, \beta > 0$ be such that $2\alpha < \beta$. Then*

$$E_{|x+y|}(\beta,\infty) \preceq E_{|x|}(\alpha,\infty) \vee E_{|y|}(\alpha,\infty).$$

Proof. We define

$$q = E_{|x|}[0,\alpha] \wedge E_{|y|}[0,\alpha], \quad p = E_{|x+y|}(\beta,\infty) \wedge q.$$

By the spectral theorem, we have that $q(H) \subset \mathrm{dom}(x) \cap \mathrm{dom}(y) = \mathrm{dom}(x+y)$ and $p(H) \subset \mathrm{dom}(x+y)$.

Suppose that $p \ne 0$. Then there exists $\xi \in p(H)$ such that $\|\xi\| = 1$. We have

$$\big\||x+y|\xi\big\| \ge \beta, \quad \|x\xi\|, \|y\xi\| \le \alpha$$

and, therefore,

$$2\alpha \ge \|x\xi\| + \|y\xi\| \ge \big\|(x+y)\xi\big\| = \big\||x+y|\xi\big\| \ge \beta.$$

Since $2\alpha < \beta$, this is a contradiction. Hence, $p = 0$. Referring to Kaplansky's identity (see Theorem 1.9.4(v)), we conclude that

$$\begin{aligned} E_{|x+y|}(\beta,\infty) &= E_{|x+y|}(\beta,\infty) - p \sim E_{|x+y|}(\beta,\infty) \vee q - q \le q^{\perp} \\ &= E_{|x|}(\alpha,\infty) \vee E_{|y|}(\alpha,\infty), \end{aligned}$$

as required. □

Theorem 4.3.23. *The algebra $S_0(\mathcal{M})$ is a closed two-sided ideal in the topological algebra $(LS(\mathcal{M}), \|\cdot\|_{LS(\mathcal{M})})$. In particular, $(S_0(\mathcal{M}), \|\cdot\|_{LS(\mathcal{M})})$ is a topological $*$-algebra with jointly continuous multiplication and an F-space.*

Proof. Suppose that $\{x_n\}_{n\in\mathbb{N}} \subset S_0(\mathcal{M})$ is such that $x_n \to x$ in the local measure topology for some $x \in LS(\mathcal{M})$.

Let $\lambda > 0$ be fixed. By Proposition 4.3.16(ii), we have that the sequence $\{E_{|x-x_n|}(\frac{\lambda}{3},\infty)\}_{n\in\mathbb{N}}$ converges to 0 locally in measure as $n \to \infty$. By Proposition 4.3.19, there exists $z'_n \in P(\mathcal{Z}(\mathcal{M}))$, such that $z'_n \to \mathbf{1}$ locally in measure and $E_{|x-x_n|}(\frac{\lambda}{3},\infty)z'_n \in P_{\mathrm{fin}}(\mathcal{M})$. Then $\bigvee_{n\in\mathbb{N}} z'_n = \mathbf{1}$. We set

$$z_1 = z'_1, \quad z_{n+1} = \bigvee_{k=1}^{n+1} z'_k - \bigvee_{k=1}^{n} z'_k, \quad n \in \mathbb{N}.$$

It is clear that $z := \bigvee_{n\in\mathbb{N}} z_n = \bigvee_{n\in\mathbb{N}} z'_n$ and $z_n \leq z'_n$ for any $n \in \mathbb{N}$. Therefore, $z^\perp z'_n = 0$ for any $n \in \mathbb{N}$. Since $z'_n \to \mathbf{1}$ locally in measure, it follows that $z^\perp = z^\perp \cdot \mathbf{1} = \lim_{n\to\infty} z^\perp \cdot z'_n = 0$. Hence, $\bigvee_{n\in\mathbb{N}} z_n = \mathbf{1}$, and so $\{z_n\}_{n\in\mathbb{N}}$ is a central partition of unity.

Lemma 4.3.22 implies that

$$E_{|xz_n|}(\lambda,\infty) \prec E_{|x_nz_n|}\left(\frac{\lambda}{3},\infty\right) \vee E_{|z_nx-z_nx_n|}\left(\frac{\lambda}{3},\infty\right).$$

Since $E_{|x_nz_n|}(\frac{\lambda}{3},\infty) = E_{|x_n|}(\frac{\lambda}{3},\infty)z_n \in P_{\mathrm{fin}}(\mathcal{M})$ and $E_{|z_nx-z_nx_n|}(\frac{\lambda}{3},\infty) = E_{|x-x_n|}(\frac{\lambda}{3},\infty)z_n \in P_{\mathrm{fin}}(\mathcal{M})$ for any $n \in \mathbb{N}$, it follows that $E_{|x|}(\lambda,\infty)z_n = E_{|xz_n|}(\lambda,\infty) \in P_{\mathrm{fin}}(\mathcal{M})$ for any $n \in \mathbb{N}$. Since $\{z_n\}_{n\in\mathbb{N}}$ is a central partition of unity, it follows from Theorem 1.9.9(vii) that

$$E_{|x|}(\lambda,\infty) = \sum_{n=1}^{\infty} E_{|x|}(\lambda,\infty)z_n \in P_{\mathrm{fin}}(\mathcal{M}).$$

Since $\lambda > 0$ is arbitrary, it follows that $x \in S_0(\mathcal{M})$, proving that $S_0(\mathcal{M})$ is a closed subalgebra in $(LS(\mathcal{M}), \|\cdot\|_{LS(\mathcal{M})})$.

It remains to show that $S_0(\mathcal{M})$ is an ideal in $LS(\mathcal{M})$. For this purpose, let $x \in S_0(\mathcal{M})$, $y \in LS(\mathcal{M})$. By Corollary 4.3.15, the algebra $\mathcal{M}$ is dense in $LS(\mathcal{M})$ in the local measure topology. Therefore, there exists a sequence $\{y_n\}_{n\in\mathbb{N}} \subset \mathcal{M}$, such that $y_n \to y$ locally in measure as $n \to \infty$. Since $S_0(\mathcal{M})$ is an absolutely solid $*$-subalgebra in $LS(\mathcal{M})$ (see Corollary 3.3.15), Proposition 3.3.13(iii) implies that $xy_n \in S_0(\mathcal{M})$. Since $xy_n \to xy$ locally in measure as $n \to \infty$, by the proven above, we conclude that $xy \in S_0(\mathcal{M})$. Similarly, $yx \in S_0(\mathcal{M})$. Thus, $S_0(\mathcal{M})$ is a two-sided ideal in $LS(\mathcal{M})$. □

Since the embedding of $(\mathcal{M}, \|\cdot\|_{\mathcal{M}})$ into $(LS(\mathcal{M}), t(\mathcal{M}))$ is continuous (see Proposition 4.3.18(ii)), Theorem 4.3.23 implies the following corollary.

Corollary 4.3.24. *The ideal $\mathcal{M} \cap S_0(\mathcal{M})$ equipped with the norm $\|\cdot\|_{\mathcal{M}}$ is a closed ideal in $\mathcal{M}$.*

We now continue with the discussion of order properties of the local measure topology.

Proposition 4.3.25.
(i) *If $\{p_n\}_{n\in\mathbb{N}} \subset P_{\mathrm{fin}}(\mathcal{M})$, $p_n \downarrow 0$, then $p_n \to 0$ locally in measure;*
(ii) *If $\{z_n\}_{n\in\mathbb{N}} \subset P(\mathcal{Z}(\mathcal{M}))$ and $z_n \downarrow 0$, then $z_n \to 0$ locally in measure.*

Proof. (i) Since p_n are finite projections, Proposition 1.11.6(ii) implies that $\mathcal{D}(p_n) \downarrow 0$ almost everywhere as $n \to \infty$. Since the measure μ is finite, it follows that $\mathcal{D}(p_n) \to 0$ in measure μ. Proposition 4.3.11 implies that $\|p_n\|_{LS(\mathcal{M})} \to 0$.

(ii) Let $\{z_i\}_{i\in I} \subset P(\mathcal{Z}(\mathcal{M}))$ and $z_i \downarrow 0$. Then $\mu([\varphi(z_i) > 0]) \downarrow 0$. Referring again to (4.10), we conclude that $z_i \to 0$ locally in measure. □

Assume, for a moment, that $\mathcal{M}$ is a semifinite von Neumann algebra equipped with a faithful normal semifinite trace τ. As shown in Theorem 4.2.20, the order convergence $x_i \downarrow 0$ for a net of positive operators implies the measure convergence $x_i \to 0$, provided that every x_i is dominated by a τ-compact operator x. We also recall that $S_0(\mathcal{M}, \tau)$ is, in general, a proper subset of $S_0(\mathcal{M})$ (see Section 2.6). As we will demonstrate next, if the assumption $x \in S_0(\mathcal{M}, \tau)$ is replaced by a weaker assumption $x \in S_0(\mathcal{M})$, the order convergence $x_i \downarrow 0$ will imply only the local measure convergence $x_i \to 0$, which is weaker than the measure convergence (see Section 4.4).

Theorem 4.3.26. *Let $\{x_i\}_{i\in I} \subset LS(\mathcal{M})$ be a decreasing net, such that $x_i \downarrow 0$ and $x_i \le x$ for some $x \in S_0(\mathcal{M})$ and all $i \in I$. Then $x_i \to 0$ locally in measure.*

Proof. Since $x \in S_0(\mathcal{M})$, $x_i \le x$ and since $S_0(\mathcal{M})$ is an absolutely solid $*$-subalgebra (see Proposition 2.6.4), it follows that $x_i \in S_0(\mathcal{M})$.

Let $\lambda > 0$ be fixed. By spectral theory, we have

$$E_{x_i}(\lambda, +\infty) \le \frac{1}{\lambda} x_i E_{x_i}(\lambda, +\infty) \le \frac{1}{\lambda} x_i.$$

Setting $p_i = \bigvee_{j \ge i} E_{x_j}(\lambda, +\infty)$, we have that $p_i \le \frac{1}{\lambda} x_i$. Since $x_i \downarrow 0$, we infer that $p_i \downarrow 0$.

By Proposition 3.3.4(v), we have that $E_{x_j}(\lambda, +\infty) \preceq E_{x_i}(\lambda, +\infty)$ for $j > i$ and so, by Proposition 1.11.6(iv) and Definition 1.11.1(vi), we have

$$\mathcal{D}(p_i) = \bigvee_{j \ge i} \mathcal{D}(E_{x_j}(\lambda, +\infty)) = \mathcal{D}(E_{x_i}(\lambda, +\infty)).$$

Since $x_i \in S_0(\mathcal{M})$, it follows that the projection $E_{x_i}(\lambda, \infty)$ is finite for all $\lambda > 0$. By definition of the dimension function $\mathcal{D}$ (see Definition 1.11.1), we have that the projection p_i is also finite. Proposition 1.11.6(ii) guarantees that $\mathcal{D}(E_{x_i}(\lambda, +\infty)) = \mathcal{D}(p_i) \downarrow 0$ almost everywhere. Since the measure μ is finite, it follows $\mathcal{D}(E_{x_i}(\lambda, +\infty)) \downarrow 0$ in measure μ. Proposition 4.3.11 implies that $\|E_{x_i}(\lambda, \infty)\|_{LS(\mathcal{M})} \to 0$ as $i \to \infty$. Since λ is arbitrary, Proposition 4.3.16(ii) implies that $x_i \to 0$ locally in measure. □

Similar to the case of measure convergence, the dominated convergence result of Theorem 4.3.26 implies the following simple corollary. The proof of Corollary 4.3.27 is similar to the proof of Corollary 4.2.21 and is therefore omitted.

Corollary 4.3.27. *Let $\{x_i\}_{i\in I}$ be an increasing net in $LS_+(\mathcal{M})$, such that $x_i \uparrow x$ for all $i \in I$ and some $0 \le x \in S_0(\mathcal{M})$. Then $x_i \to x$ locally in measure.*

Remark 4.3.28. *As for the measure topology, the assumption that $x_i \le x$ for some $x \in S_0(\mathcal{M})$ cannot be omitted in Theorem 4.3.26. Indeed, if $\mathcal{M} = \mathcal{B}(H)$, then both the local measure topology and the measure topology coincide with the uniform norm topology (see Example 4.2.16 and Example 4.3.21) and so the counterexample of Remark 4.2.22 gives an example of a decreasing to zero net, which does not converge to zero locally in measure. Thus, as for the measure topology, for a general von Neumann algebra $\mathcal{M}$ the order convergence $x_i \downarrow x$, $x_i, x \in LS(\mathcal{M})$, $i \in I$, does not imply the convergence $x_i \to x$ in the local measure topology.*

Our next aim is to prove that the topology of convergence locally in measure agrees well with reduction by central projections, showing the stability of the local measure topology with respect to reduction by a central projection. We start with a preliminary lemma.

Lemma 4.3.29. *Assume that $0 \ne z \in P(\mathcal{Z}(\mathcal{M}))$ and let $\mathcal{D}$ be a dimension function on $P(\mathcal{M})$. We have that*
(i) *$\mathcal{D}_{\mathcal{M}z} := \mathcal{D}|_{P(\mathcal{M}z)}$ is a dimension function on $P(\mathcal{M}z)$;*
(ii) *the local measure topology $t(\mathcal{M}z)$ on $LS(\mathcal{M}z)$ defined by the dimension function $\mathcal{D}_{z\mathcal{M}}$ coincides with the restriction $t(\mathcal{M})|_{LS(\mathcal{M}z)}$ of the topology $t(\mathcal{M})$ to $LS(\mathcal{M}z)$.*

Proof. Let, as before, $\varphi : \mathcal{Z}(\mathcal{M}) \to L_\infty(\Omega, \Sigma, \mu)$ be a $*$-isomorhism for a finite measure space (Ω, Σ, μ). Let $A \in \Sigma$ be such that $\chi_A = \varphi(z)$. For any $w = zw \in \mathcal{Z}(\mathcal{M}z)$ we have that $\varphi(w) = \varphi(zw) = \varphi(z)\varphi(w) = \chi_A\varphi(w)$. Therefore, using natural identification of $\chi_A \cdot L_\infty(\Omega, \Sigma, \mu) \subset L_\infty(\Omega, \Sigma, \mu)$ with $L_\infty(A, \Sigma|_A, \mu|_A)$, we have that $\varphi(\mathcal{Z}(\mathcal{M}z))$ coincides with $L_\infty(A, \Sigma|_A, \mu|_A)$. One can easily verify that $\mathcal{D}_{\mathcal{M}z} := \varphi(z)\mathcal{D}$ is a dimension function on $P(\mathcal{M}z)$.

Furthermore, computing F-norms of elements $zx, x \in LS(\mathcal{M}z)$ in $LS(\mathcal{M})$ and in $LS(\mathcal{M}z)$ we have

$$\begin{aligned}
\|zx\|_{LS(\mathcal{M})} &= \inf\{\lambda : \mu([\mathcal{D}(E_{|zx|}(\lambda, \infty)) > \lambda]) \le \lambda\} \\
&= \inf\{\lambda : \mu([\mathcal{D}(zE_{|zx|}(\lambda, \infty)) > \lambda]) \le \lambda\} \\
&= \inf\{\lambda : \mu([\varphi(z)\mathcal{D}_{\mathcal{M}z}(E_{|zx|}(\lambda, \infty)) > \lambda]) \le \lambda\} \\
&= \inf\{\lambda : \mu|_A([\mathcal{D}_{\mathcal{M}z}(E_{|zx|}(\lambda, \infty)) > \lambda]) \le \lambda\} \\
&= \|zx\|_{LS(\mathcal{M}z)}.
\end{aligned}$$

Since the local measure topology $LS(\mathcal{M}z)$ is defined by the F-norm $\|\cdot\|_{LS(\mathcal{M}z)}$, the claim follows.

The equivalence of convergences $\|zx_n\|_{LS(\mathcal{M})} \to 0 \iff \|zx_n\|_{LS(\mathcal{M}z)} \to 0$, $\{x_n\}_{n\in\mathbb{N}} \subset LS(\mathcal{M})$, follows immediately. □

With this auxiliary lemma at hand, we prove the stability of the local measure topology with respect to reduction by a central partition of unity.

Theorem 4.3.30. *Let $\mathcal{M}$ be a von Neumann algebra and $\{z_i\}_{i\in\mathbb{N}}$ be a central partition of unity in $\mathcal{M}$. Then $x_n \overset{t(\mathcal{M})}{\to} x$ as $n \to \infty$ if and only if $z_i x_n \overset{t(z_i\mathcal{M})}{\to} z_i x$ for every $i \in \mathbb{N}$. Here, the local measure topology $t(\mathcal{M}z_i)$ on $LS(\mathcal{M}z_i)$ defined by the dimension function $\mathcal{D}_{z_i\mathcal{M}} = \mathcal{D}|_{P(\mathcal{M}z_i)}$.*

Proof. Without loss of generality, $x = 0$.

If $x_n \to 0$ in $t(\mathcal{M})$ as $n \to \infty$, then Lemma 4.3.29 implies that $z_i x_n \to 0$ in $t(z_i\mathcal{M})$ as $n \to \infty$ for every $i \in \mathbb{N}$.

Conversely, suppose that $z_i x_n \overset{t(z_i\mathcal{M})}{\to} 0$ as $n \to \infty$ for every $i \in \mathbb{N}$. Let $\lambda > 0$ be fixed.

For any $i \in \mathbb{N}$, let $\Omega_i \in \Sigma$, be such that $\chi_{\Omega_i} = \varphi(z_i)$. Since the measure μ is finite on Ω, it follows that there exists n_0, such that $\mu(\sum_{i=n_0+1}^{\infty}\Omega_i) < \frac{\lambda}{2}$. Since $z_i x_n \to 0$ in $t(z_i\mathcal{M})$ as $n \to \infty$, for every $i = 1, \dots n_0$, there exists $N \geq n_0$, such that $\|z_i x_n\|_{LS(z_i M)} \leq \frac{\lambda}{2N}$ for all $n \geq N$ and $i = 1, \dots N$. By Proposition 4.3.4 and Lemma 4.3.7 for every $i = 1, \dots, n_0$, there exists $p_i \in P(z_i\mathcal{M})$, such that $\|x_n p_i\|_{\mathcal{M}} \leq \frac{\lambda}{2N}$ and $\mu([\mathcal{D}(z_i - p_i) > \frac{\lambda}{2N}]) \leq \frac{\lambda}{2N}$.

Set $p = \sum_{i=1}^{N} p_i$. Note that

$$\begin{aligned}
[\mathcal{D}(p^{\perp}) > \lambda] &= \bigcup_{i\in\mathbb{N}} \Omega_i \cap [\mathcal{D}(p^{\perp}) > \lambda] = \bigcup_{i\in\mathbb{N}} [\mathcal{D}(z_i p^{\perp}) > \lambda] \\
&= \bigcup_{i\in\mathbb{N}} [\mathcal{D}(z_i - p_i) > \lambda] \subset \left(\bigcup_{i=1}^{N} [\mathcal{D}(z_i - p_i) > \lambda]\right) \cup \bigcup_{i=N+1}^{\infty} \Omega_i \\
&\subset \left(\bigcup_{i=1}^{N} \left[\mathcal{D}(z_i - p_i) > \frac{\lambda}{2N}\right]\right) \cup \bigcup_{i=N+1}^{\infty} \Omega_i.
\end{aligned}$$

Therefore,

$$\mu([\mathcal{D}(p^{\perp}) > \lambda]) \leq \sum_{i=1}^{N} \mu\left(\left[\mathcal{D}(z_i - p_i) > \frac{\lambda}{2N}\right]\right) + \mu\left(\bigcup_{i=N+1}^{\infty} \Omega_i\right) < \sum_{i=1}^{N} \frac{\lambda}{2N} + \frac{\lambda}{2} = \lambda.$$

Furthermore, since $p_i \in P(z_i\mathcal{M})$ and $\{z_i\}_{i\in\mathbb{N}}$ are pairwise orthogonal, it follows that

$$\|x_n p\|_{\mathcal{M}} = \sup_{i=1,\dots N} \|x_n p_i\|_{\mathcal{M}} \leq \frac{\lambda}{2N} \leq \lambda.$$

It follows that $x_n \in V(\lambda)$ for all $n \geq N$. Therefore, $\|x_n\|_{LS(\mathcal{M})} < \lambda$ for all $n \geq N$, and so $x_n \to 0$ in $t(\mathcal{M})$, as required. □

Given Theorem 4.3.30, we can now prove that the local measure topology $t(\mathcal{M})$ does not depend on the choice of such a dimension function $\mathcal{D}$ on $\mathcal{M}$. We reiterate, that this is in sharp contrast to the measure topology t_τ (see Remark 4.2.18).

Proposition 4.3.31. *The local measure topology does not depend on the choice of dimension function $\mathcal{D}$ on $P(\mathcal{M})$.*

Proof. Let $\mathcal{D}_1, \mathcal{D}_2$ be two dimension functions on $P(\mathcal{M})$. Denote by $V_{\mathcal{D}_j}(\cdot,\cdot,\cdot)$, the neighborhoods of zero as defined in Proposition 4.3.10 with $\mathcal{D}$ replaced by $\mathcal{D}_j, j = 1, 2$.

By Theorem 1.11.13, there exists an a. e. positive finite-valued measurable function c on (Ω, Σ, μ) with $s(c) = \mathbf{1}$, such that $\mathcal{D}_2 = c\mathcal{D}_1$. We may write $\Omega = \bigvee_n \Omega_n$ a. e., where $a_n < c|_{\Omega_n} \leq b_n$ are such that the intervals $(a_n, b_n]$ do not intersect and cover $\mathbb{R}_+$. Setting $z_n = \varphi^{-1}(\chi_{\Omega_n})$, it follows that $\{z_n\}_{n\in\mathbb{N}}$ is a central partition of unity. By Theorem 4.3.30, it is sufficient to prove the assertion for each of the algebras $\mathcal{M}z_n$, $n \in \mathbb{N}$. Since for every $n \in \mathbb{N}$, we have that $a_n\mathcal{D}_1 \leq \mathcal{D}_2 \leq b_n\mathcal{D}_1$. It follows that

$$V_{\mathcal{D}_1}(\lambda, \alpha, \beta) \subset V_{\mathcal{D}_2}(\lambda, b_n\alpha, \beta) \subset V_{\mathcal{D}_1}\left(\lambda, \frac{b_n}{a_n}\alpha, \beta\right),$$

for all $\lambda, \alpha, \beta > 0$. Therefore, the assertion follows from Proposition 4.3.10. □

Since the von Neumann algebra $\mathcal{M}$ is dense in $LS(\mathcal{M})$ in the local measure topology (see Corollary 4.3.15), Proposition 4.3.31 allows us to establish that the algebras $LS(\mathcal{M}_1)$ and $LS(\mathcal{M}_2)$ are $*$-isomorphic if the von Neumann algebras $\mathcal{M}_1$ and $\mathcal{M}_2$ are $*$-isomorphic.

Theorem 4.3.32. *Any $*$-isomorphism Φ between two von Neumann algebras $\mathcal{M}_1$ and $\mathcal{M}_2$ extends to a $*$-isomorphism between the $*$-algebras $LS(\mathcal{M}_1)$ and $LS(\mathcal{M}_2)$. This extension is continuous in the local measure topology.*

Proof. We denote by $(\Omega_j, \Sigma_j, \mu_j), j = 1, 2$, probability spaces, such that $\mathcal{Z}(\mathcal{M}_j)$ is $*$-isomorphic to $L_\infty(\Omega_j, \Sigma_j, \mu_j)$ via respective $*$-isomorphism. Let $\mathcal{D}_j : P(\mathcal{M}_j) \to L^+(\Omega_j, \Sigma_j, \mu_j)$ be the corresponding dimension functions.

Let $\Phi : \mathcal{M}_1 \to \mathcal{M}_2$ be a $*$-isomorphism. By [92, Theorem 4.1.8], Φ is an isometry. Since $\Phi : \mathcal{Z}(\mathcal{M}_1) \to \mathcal{Z}(\mathcal{M}_1)$ is a $*$-isomorphism of the commutative von Neumann algebras $\mathcal{Z}(\mathcal{M}_1)$ and $\mathcal{Z}(\mathcal{M}_2)$, Φ generates a $*$-isomorphism $\tilde{\Phi}$ between the algebras $L_\infty(\Omega_1, \Sigma_1, \mu_1)$, which is an isometry. We may naturally extend $\tilde{\Phi}$ to a $*$-isomorphism of the algebras $L^+(\Omega_1, \Sigma_1, \mu_1)$ and $L^+(\Omega_2, \Sigma_2, \mu_2)$. It is clear that the mapping $\tilde{\Phi} \circ \mathcal{D}_1 \circ \Phi^{-1} : P(\mathcal{M}_2) \to L^+(\Omega_2, \Sigma_2, \mu_2)$ is a dimension function on $\mathcal{M}_2$. Since the local measure topology does not depend on the choice of dimension function (see Proposition 4.3.31), it follows that the local measure topology $t(\mathcal{M}_2)$ can be equivalently defined via the dimension function $\tilde{\Phi} \circ \mathcal{D}_1 \circ \Phi^{-1}$. Hence, $\Phi : (\mathcal{M}_1, \|\cdot\|_{LS(\mathcal{M}_1)}) \to (\mathcal{M}_2, \|\cdot\|_{LS(\mathcal{M}_2)})$ is a homeomorphism. Since a von Neumann algebra $\mathcal{M}$ is dense in $LS(\mathcal{M})$ with respect to the local measure topology $t(\mathcal{M})$ (see Corollary 4.3.15), we may extend $\Phi : (\mathcal{M}_1, \|\cdot\|_{LS(\mathcal{M}_1)}) \to (\mathcal{M}, \|\cdot\|_{LS(\mathcal{M}_2)})$ by continuity up to a $*$-isomorphism $\Phi : LS(\mathcal{M}_1) \to LS(\mathcal{M}_2)$. Clearly, this extension is continuous in the respective local measure topologies. □

4.4 Relations between the measure topology and the local measure topology

In this section, we study how two topologies – the topology of convergence in measure and the topology of convergence locally in measure – are related.

First, we show that the topology of convergence in measure is stronger than the topology of convergence locally in measure (whenever both are defined).

Proposition 4.4.1. *Let $\mathcal{M}$ be a von Neumann algebra with faithful normal semifinite trace τ. If $\{x_n\}_{n\in\mathbb{N}} \subset S(\mathcal{M},\tau)$, $x \in S(\mathcal{M},\tau)$ are such that $x_n \to x$ in measure as $n \to \infty$, then $x_n \to x$ locally in measure.*

Proof. Without loss of generality, we assume that $x = 0$. Suppose the contrary that x_n does not converge to 0 locally in measure. Passing to a subsequence, if necessary, we assume that $\|x_n\|_{LS(\mathcal{M})} \geq \varepsilon > 0$ for all $n \in \mathbb{N}$ and some $\varepsilon > 0$.

Since $x_n \to 0$ in measure, Proposition 4.2.9(vi) guarantees that there exist $n_1 < n_2 < \cdots < n_k < \cdots$ and a projection $p_k \in P(\mathcal{M})$, such that $\|x_{n_k} p_k^\perp\|_{\mathcal{M}} \leq 2^{-k-1}$, $\tau(p_k) \leq 2^{-k-1}$. Setting $q_k = \bigvee_{i=k}^{\infty} p_i$, $k \in \mathbb{N}$, we have a decreasing sequence of projections. Since

$$\tau(q_k) \leq \sum_{i=k}^{\infty} \tau(p_i) \leq 2^{-k} \to 0, \quad k \to \infty,$$

it follows that $q_k \downarrow 0$ and q_k is τ-finite (in particular, is finite). Therefore, Proposition 4.3.25 implies the convergence $\|q_k\|_{LS(\mathcal{M})} \to 0$ as $k \to \infty$, or equivalently $q_k^\perp \to \mathbf{1}$ locally in measure. Furthermore, we have that

$$\|x_{n_k} q_k^\perp\|_{\mathcal{M}} \leq \|x_{n_k} p_k^\perp\|_{\mathcal{M}} \leq 2^{-k-1},$$

and so by Proposition 4.3.16(iv) we conclude that $x_{n_k} \to 0$ locally in measure, which is a contradiction. Thus, $x_n \to 0$ locally in measure. □

However, if the trace τ is finite, then the topological $*$-algebras $(LS(\mathcal{M}), t(\mathcal{M}))$ and $(S(\mathcal{M},\tau), t_\tau)$ coincide. To show this we firstly prove an auxiliary lemma.

Lemma 4.4.2. *Assume that the trace τ is finite and let $\{p_n\}_{n\in\mathbb{N}} \subset P(\mathcal{M})$ converge locally in measure to 0. Then $p_n \to 0$ in measure.*

Proof. By Corollary 4.2.10, it is sufficient to show that $\tau(p_n) \longrightarrow 0$ as $n \to \infty$. Passing to a subsequence, if necessary, we may assume that $\|p_n\|_{LS(\mathcal{M})} \leq 2^{-n}$ for all $n \in \mathbb{N}$.

We set

$$q_n = \bigvee_{k=n}^{\infty} p_k, \quad n \in \mathbb{N}.$$

It is clear that $\{q_n\}_{n\in\mathbb{N}}$ is a decreasing sequence of projections. Furthermore, since $\mathcal{D}(q_n) \leq \sum_{k=n}^{\infty} \mathcal{D}(p_k)$ (see Proposition 1.11.6(i)), Proposition 4.3.13(v) implies that

$$\|q_n\|_{LS(\mathcal{M})} \le \left\| \sum_{k=n}^{\infty} p_k \right\|_{LS(\mathcal{M})} \le \sum_{k=n}^{\infty} \|p_k\|_{LS(\mathcal{M})} \le 2^{-n+1}.$$

That is, $q_n \to 0$ locally in measure. Therefore, $q_n \downarrow 0$ (since otherwise Proposition 4.3.25(i) would give a contradiction). Hence, normality of the trace implies that $\tau(q_n) \downarrow 0$, and so $\tau(p_n) \to 0$, as required. □

Theorem 4.4.3. *Let $\mathcal{M}$ be a von Neumann algebra with a faithful normal finite trace τ. Then the topological $*$-algebra $(LS(\mathcal{M}), t(\mathcal{M}))$ coincides with the topological $*$-algebra $(S(\mathcal{M}, \tau), t_\tau)$.*

Proof. Since the trace τ is finite, it follows that the $*$-algebras $LS(\mathcal{M})$ and $S(\mathcal{M}, \tau)$ coincide (since they both consist of all closed densely defined operators affiliated with $\mathcal{M}$).

Since t_τ is stronger than $t(\mathcal{M})$ (see Proposition 4.4.1), it is sufficient to show that if $x_n \to 0$ in $t(\mathcal{M})$, then $x_n \to 0$ in t_τ for any sequence $\{x_n\}_{n\in\mathbb{N}} \subset S(\mathcal{M}, \tau)$.

Let $\{x_n\}_{n\in\mathbb{N}}$ be a sequence, such that $x_n \to 0$ in $t(\mathcal{M})$. By Proposition 4.3.16(ii), it follows that $E_{|x_n|}(\lambda, \infty) \to 0$ locally in measure for any $\lambda > 0$. By Lemma 4.4.2, we have that $E_{|x_n|}(\lambda, \infty) \to 0$ in measure for any $\lambda > 0$. Referring to Proposition 4.2.9(iv), we conclude that $x_n \to 0$ in measure, as required. □

Remark 4.4.4.

(i) *If τ is an infinite trace, then $t_\tau \neq t(\mathcal{M})$, in general. Indeed, let $\mathcal{M}$ be a commutative von Neumann algebra $L_\infty(\Omega, \Sigma, \mu)$ with an infinite measure μ. Let τ be a faithful normal semifinite trace on $\mathcal{M}$ given by integration with respect to μ (see Example 1.10.11). By Example 4.3.20, the topology $t(\mathcal{M})$ coincides with the topology of convergence locally in measure μ, while the topology t_τ coincides with the topology of convergence in measure μ (see Example 4.2.17). Since μ is an infinite measure, it follows that $t_\tau \neq t(\mathcal{M})$. Furthermore, for a general von Neumann algebra, such that $LS(\mathcal{M}) \neq S(\mathcal{M}, \tau)$, the reduction of $t(\mathcal{M})$ on $S(\mathcal{M}, \tau)$ does not coincide with t_τ. Indeed, if $t_\tau = t(\mathcal{M})|_{S(\mathcal{M},\tau)}$, then $t_\tau|_{\mathcal{M}} = t(\mathcal{M})|_{\mathcal{M}}$. However, Corollary 4.2.13 and Corollary 4.3.15 imply that*

$$LS(\mathcal{M}) = \overline{\mathcal{M}}^{t(\mathcal{M})} = \overline{\mathcal{M}}^{t_\tau} = S(\mathcal{M}, \tau),$$

which is a contradiction.

(ii) *Suppose that $\mathcal{M}$ is a finite von Neumann algebra equipped with a faithful normal semifinite trace τ. Then, by Corollary 1.10.16, there exists a central partition of unity $\{z_i\}_{i\in\mathbb{N}}$, such that the trace $\tau|_{z_i\mathcal{M}}$ is finite on $z_i\mathcal{M}$. By Theorem 4.3.30, $x_n \to x$, $x_n, x \in LS(\mathcal{M})$, in the local measure topology $t(\mathcal{M})$ if and only if $z_i x_n \to z_i x$ in the local measure topology $t(z_i\mathcal{M})$ for every $i \in \mathbb{N}$. By Theorem 4.4.3, this is equivalent to $z_i x_n \to z_i x$ in the measure topology $t_{\tau_{z_i}}$ for every $i \in \mathbb{N}$. Therefore, under these assumptions on the algebra $\mathcal{M}$, we may define the local measure topology on $LS(\mathcal{M})$ as follows: x_n converges to x locally in measure in $LS(\mathcal{M})$ if $zx_n \to zx$ in measure in $S(z\mathcal{M}, \tau)$ for every central projection $z \in \mathcal{Z}(\mathcal{M})$, such that $\tau(z_i) < \infty$.*

Next, our aim is to establish an Egorov-type theorem for the topology of convergence locally in measure on $LS(\mathcal{M})$ (see Theorem 4.4.7 and Theorem 4.4.9 below). The classical Egorov theorem for measurable functions establishes that the almost everywhere convergence is nearly uniform convergence for measurable functions (see, e. g., [78]). Depending on the type of the von Neumann algebra $\mathcal{M}$, we will show that local measure convergence $t(\mathcal{M})$ is nearly measure convergence t_τ (in the case when $\mathcal{M}$ is a semifinite algebra) or is nearly uniform convergence (when $\mathcal{M}$ is type III).

We first deal with the case of a semifinite von Neumann algebra. Let $\mathcal{M}$ be a semifinite von Neumann algebra (acting on a separable Hilbert space) equipped with a faithful normal semifinite trace τ. As before, we denote by φ an isomorphism between $\mathcal{Z}(\mathcal{M})$ and $L_\infty(\Omega, \Sigma, \mu)$, where (Ω, Σ, μ) is a probability space. We denote by ψ a faithful normal finite (tracial) state on $\mathcal{Z}(\mathcal{M})$, such that $\psi(x) = \int \varphi(x)d\mu$ for any $x \in \mathcal{Z}(\mathcal{M})$.

In the following lemma, we prove a Lusin-type theorem in the semifinite setting that a locally measurable operator is nearly τ-measurable.

Lemma 4.4.5. *Let $\mathcal{M}$ be a semifinite von Neumann algebra equipped with a faithful normal semifinite trace τ and let ψ be a faithful normal finite tracial state on $\mathcal{Z}(\mathcal{M})$. Assume that $\{x_n\}_{n\in\mathbb{N}} \subset LS(\mathcal{M})$. For any $\varepsilon > 0$, there exists a central projection $z \in \mathcal{M}$, such that $\psi(\mathbf{1} - z) < \varepsilon$ and $\{x_n z\}_{n\in\mathbb{N}} \subset S(z\mathcal{M}, \tau)$.*

Proof. We first establish the result in the special case when $\{x_n\}_{n\in\mathbb{N}}$ is a constant sequence $\{x\}_{n\in\mathbb{N}}$ for some $x \in LS(\mathcal{M})$. By definition, there exists a central partition of unity $\{z_i\}_{i\in\mathbb{N}}$, such that $xz_i \in S(\mathcal{M})$ for any $i \in \mathbb{N}$. By Proposition 2.2.3(ii), for any $i \in \mathbb{N}$, there exists $\lambda_i > 0$ and $p_i \in P(\mathcal{M})$, such that $\|p_i z_i x\|_\mathcal{M} \le \lambda_i$, $\mathbf{1} - p_i \in P_{\text{fin}}(\mathcal{M})$. By Proposition 1.10.15(ii), for each $i \in \mathbb{N}$ there exists central partition $\{z_{i,j}\}_{j\in\mathbb{N}}$ of the projection z_i, such that $\tau(z_{i,j}(\mathbf{1} - p_i)) < \infty$ for any $i, j \in \mathbb{N}$.

Since ψ is a normal state, there exists N, such that $\sum_{i=N+1}^\infty \psi(z_i) < \frac{\varepsilon}{2}$ and for every $i \in \mathbb{N}$, there exists N_i, such that $\sum_{j=N_i+1}^\infty \psi(z_{i,j}) < \frac{\varepsilon}{2^{i+1}}$.

Setting $z = \bigvee_{i=1}^N \bigvee_{j=1}^{N_i} z_{i,j}$, we have that $\psi(\mathbf{1} - z) < \varepsilon$. It remains to show that $xz \in S(z\mathcal{M}, \tau)$. To this end, we set $p = \mathbf{1} - z + \bigvee_{i=1}^N \bigvee_{j=1}^{N_i} z_{i,j} p_i$, $\lambda = \max_{i=1}^N \lambda_i$. Then

$$\tau(\mathbf{1} - p) = \sum_{i=1}^N \sum_{j=1}^{N_i} \tau(z_{i,j}(\mathbf{1} - p_i)) < \infty, \quad \|zxp\|_\mathcal{M} \le \lambda,$$

which guarantees that $xz \in S(z\mathcal{M}, \tau)$ (see Proposition 2.4.2(iii)).

Now let $\{x_n\}_{n\in\mathbb{N}}$ be an arbitrary sequence of locally measurable operators and let $\varepsilon > 0$ be fixed. By the proven above, there exists a sequence of central projections $\{z_n\}_{n\in\mathbb{N}}$, such that $\psi(\mathbf{1} - z_n) < \varepsilon \cdot 2^{-n-1}$ and $\{x_n z_n\}_{n\in\mathbb{N}} \subset S(\mathcal{M}, \tau)$.

Setting $z = \bigwedge_{n\in\mathbb{N}} z_n$, we have that $x_n z = x_n z_n z \in S(\mathcal{M}, \tau)$ and

$$\psi(\mathbf{1} - z) = \psi\Big(\bigvee_{n\in\mathbb{N}} (\mathbf{1} - z_n)\Big) \le \sum_{n=1}^\infty \psi(\mathbf{1} - z_n) < \sum_{n=1}^\infty \varepsilon \cdot 2^{-n-1} = \varepsilon,$$

as required. □

Lemma 4.4.6. *Let $\{p_n\}_{n\in\mathbb{N}} \subset P_{\text{fin}}(\mathcal{M},\tau)$ be such that $\|\mathcal{D}(p_n)\|_\infty \to 0$. Then there exists a subsequence $\{p_{n_k}\}_{k\in\mathbb{N}}$, such that $\tau(p_{n_k}) \to 0$ as $k \to \infty$.*

Proof. Choose a subsequence $\{p_{n_k}\}$ of $\{p_n\}$, such that

$$\|\mathcal{D}(p_{n_k})\|_\infty < 2^{-k}, \quad k \in \mathbb{N},$$

and set $q_k = \bigvee_{i=k}^{\infty} p_{n_k}$, $q = \inf_k q_k$, $k \in \mathbb{N}$.

By Proposition 1.11.6(i), we have that $\mathcal{D}(q_k) \le \sum_{i=k}^{\infty} \mathcal{D}(p_{n_k})$. Therefore, $\mathcal{D}(q_k) \in L_\infty(\Omega,\Sigma,\mu)$ and $\|\mathcal{D}(q_k)\|_\infty \le 2^{-k+1}$. In particular, $q_k \in P_{\text{fin}}(\mathcal{M})$ for any $k \in \mathbb{N}$.

Noting that $q_1 - q_k \uparrow q_1 - q$ as $k \to \infty$, the definition of the dimension function $\mathcal{D}$ (see Definition 1.11.1) implies that

$$\mathcal{D}(q_1) - \mathcal{D}(q_k) = \mathcal{D}(q_1 - q_k) \uparrow_k \mathcal{D}(q_1 - q) = \mathcal{D}(q_1) - \mathcal{D}(q)$$

and, therefore, $\mathcal{D}(q_k) \downarrow \mathcal{D}(q)$ as $k \to \infty$. Thus, $\mathcal{D}(q) = 0$, implying that $q = 0$.

Hence, we may infer that $q_1 - q_k \uparrow q_1$ as $k \to \infty$. Since $\tau(q_1) - \tau(q_k) = \tau(q_1 - q_k) \uparrow \tau(q_1)$ as $k \to \infty$, it follows that $\tau(q_k) \downarrow 0$, $k \to \infty$. Noting that $\tau(p_{n_k}) \le \tau(q_k)$, we complete the proof. □

We now prove an Egorov-type theorem in the semifinite setting.

Theorem 4.4.7. *Let $\mathcal{M}$ be a semifinite von Neumann algebra equipped with a faithful normal semifinite trace τ and let ψ be a faithful normal finite state on $\mathcal{Z}(\mathcal{M})$. Assume that $\{x_n\}_{n\in\mathbb{N}} \subset LS(\mathcal{M})$ is such that $x_n \to 0$ locally in measure. For any $\varepsilon > 0$, there exist a subsequence $\{x_{n_k}\}_{k\in\mathbb{N}}$ and a central projection $z \in P(\mathcal{Z}(\mathcal{M}))$, such that $\psi(\mathbf{1} - z) < \varepsilon$, $\{x_{n_k} z\}_{n\in\mathbb{N}} \subset S(z\mathcal{M},\tau)$ and $x_{n_k} z \to 0$ in measure in $S(z\mathcal{M},\tau)$.*

Proof. Let $\varepsilon > 0$ be arbitrary. Since $\|x_n\|_{LS(\mathcal{M})} \to 0$, there exists a subsequence $\{x_{n_k}\}_{k\in\mathbb{N}}$, such that $\|x_{n_k}\|_{LS(\mathcal{M})} \le \frac{\varepsilon}{2^{k+1}}$ for every $k \in \mathbb{N}$. Lemma 4.3.7 implies that $x_{n_k} \in V(\frac{\varepsilon}{2^{k+1}})$ for every $k \in \mathbb{N}$.

By Proposition 4.3.4(ii) for every $k \in \mathbb{N}$, there exist projections $p_k \in P(\mathcal{M})$ and $z_k' \in P(\mathcal{Z}(\mathcal{M}))$, such that

$$\|x_{n_k}(\mathbf{1} - p_k)\|_\infty \le \frac{\varepsilon}{2^{k+1}}, \quad \psi(\mathbf{1} - z_k') \le \varepsilon \cdot 2^{-k-1}, \quad \mathcal{D}(p_k z_k') \le \frac{\varepsilon}{2^{k+1}} \varphi(z_k').$$

Since $\mathcal{D}(p_k z_k')$ is almost everywhere finite, it follows that $p_k z_k'$ is a finite projection for every $k \in \mathbb{N}$ (see Definition 1.11.1(ii)). By Proposition 1.10.15(iii) for each $k \in \mathbb{N}$, there exists $z_k' \ge z_k \in P(\mathcal{Z}(\mathcal{M}))$, such that $\tau(p_k z_k) < \infty$ and $\psi(z_k' - z_k) < \varepsilon \cdot 2^{-k-1}$.

Note that

$$\psi(\mathbf{1} - z_k) < \varepsilon \cdot 2^{-k-1} + \varepsilon \cdot 2^{-k-1} = \varepsilon \cdot 2^{-k},$$

and

$$\mathcal{D}(p_k z_k) = \mathcal{D}(p_k z_k')\varphi(z_k) \le \varepsilon \cdot 2^{-k-1}\varphi(z_k' z_k) = \varepsilon \cdot 2^{-k-1}\varphi(z_k).$$

Therefore, setting

$$z = \bigwedge_{k\in\mathbb{N}} z_k$$

we have that $\psi(\mathbf{1} - z) < \varepsilon$. It is clear that

$$\|x_{n_k}(z - p_k z)\|_{\mathcal{M}} \le \|x_{n_k}(1 - p_k)\|_{\mathcal{M}} \le \varepsilon \cdot 2^{-k}, \quad \mathcal{D}(p_k z) \le \varepsilon \cdot 2^{-k}\varphi(z).$$

By Lemma 4.4.6, there is a subsequence $\{p_{k_l}\}_{l\in\mathbb{N}}$, such that $\tau(p_{k_l} z) \le \varepsilon \cdot 2^{-l}$. Therefore,

$$\|x_{n_{k_l}}(z - p_{k_l} z)\|_{\mathcal{M}} \le \varepsilon \cdot 2^{-k_l} \le \varepsilon \cdot 2^{-l}, \quad \tau(p_{k_l} z) \le \varepsilon \cdot 2^{-l}.$$

By Proposition 4.2.9(vi), we have that $x_{n_{k_l}} z \to 0$ in measure (in $S(z M, \tau|_{z\mathcal{M}})$) as $l \to \infty$. □

Now we move on to the Egorov-type theorem in the type *III* setting. We start with an analogue of Lemma 4.4.5 for a type *III* algebra $\mathcal{M}$. As before, we denote by ψ a faithful normal state on $\mathcal{Z}(\mathcal{M})$.

Lemma 4.4.8. *Let $\mathcal{M}$ be a type III von Neumann algebra and let ψ be a faithful normal finite state on $\mathcal{Z}(\mathcal{M})$. Assume that $\{x_n\}_{n\in\mathbb{N}} \subset LS(\mathcal{M})$. For any $\varepsilon > 0$, there exists a central projection $z \in P(\mathcal{Z}(\mathcal{M}))$, such that $\psi(\mathbf{1} - z) < \varepsilon$ and $\{x_n z\}_{n\in\mathbb{N}} \subset z\mathcal{M}$.*

Proof. Suppose first that the sequence $\{x_n\}_{n\in\mathbb{N}}$ consists of a single locally measurable operator x. By the definition of locally measurable operators, there exists a central partition of unity $\{z_n\}_{n\in\mathbb{N}} \subset P(\mathcal{Z}(\mathcal{M}))$, such that $x z_n \in S(\mathcal{M}) = \mathcal{M}$ (see Theorem 2.5.3). Since ψ is a normal state on $\mathcal{Z}(\mathcal{M})$, there exists N, such that $\sum_{n=N+1}^{\infty} \psi(z_n) < \varepsilon$. Setting $z = \sum_{n=1}^{N} z_n$, we have that $\psi(\mathbf{1} - z) < \varepsilon$ and $xz \in \mathcal{M}$, as required.

Now let $\{x_n\}_{n\in\mathbb{N}} \subset LS(\mathcal{M})$ be an arbitrary sequence. By the proven above, we may find central projections $\{z_n\}_{n\in\mathbb{N}}$, such that $\psi(\mathbf{1} - z_n) < \varepsilon \cdot 2^{-n}$ and $\{x_n z_n\}_{n\in\mathbb{N}} \subset \mathcal{M}$.

Setting $z = \bigwedge_{n\in\mathbb{N}} z_n$, we have that $x_n z = x_n z_n z \in \mathcal{M}$ and

$$\psi(\mathbf{1} - z) = \psi\Big(\bigvee_{n\in\mathbb{N}} (\mathbf{1} - z_n)\Big) \le \sum_{n=1}^{\infty} \psi(\mathbf{1} - z_n) < \sum_{n\in\mathbb{N}} \varepsilon \cdot 2^{-n} = \varepsilon,$$

as required. □

Now we present an Egorov-type theorem for the type *III* setting.

Theorem 4.4.9. *Assume that $\mathcal{M}$ is a type III von Neumann algebra with a faithful normal state ψ on $\mathcal{Z}(\mathcal{M})$. Let $\{x_n\}_{n\in\mathbb{N}} \subset LS(\mathcal{M})$ converge locally in measure to 0. Then for any $\varepsilon > 0$, there exists a subsequence $\{x_{n_k}\}_{k\in\mathbb{N}}$ and a central projection $z \in P(\mathcal{Z}(\mathcal{M}))$, such that $\psi(\mathbf{1} - z) < \varepsilon$, $\{x_{n_k}\}_{k\in\mathbb{N}} \subset z\mathcal{M}$ and $x_{n_k} z \to 0$ in the uniform norm as $k \to \infty$.*

Proof. As in the proof of Theorem 4.4.7, we may find a subsequence $\{x_{n_k}\}_{k\in\mathbb{N}}$, such that $x_{n_k} \in V(\varepsilon_k)$, $k \in \mathbb{N}$, where $0 < \varepsilon_k$ are such that

$$\sum_{k=1}^{\infty} \varepsilon_k \le \varepsilon.$$

By Proposition 4.3.4(ii) for every $k \in \mathbb{N}$, there exist projections $p_n \in P(\mathcal{M})$ and $z_n \in P(\mathcal{Z}(\mathcal{M}))$, such that

$$\|x_{n_k}(\mathbf{1} - p_k)\|_{\mathcal{M}} \le \varepsilon_n, \quad \psi(\mathbf{1} - z_k) \le \varepsilon_k, \quad \mathcal{D}(p_k z_k) \le \varepsilon_k \varphi(z_k).$$

The last inequality and Definition 1.11.1(ii) imply that $p_k z_k$ is a finite projection. Since the von Neumann algebra $\mathcal{M}$ is of type III, it follows that $p_k z_k = 0$.

Let $z = \bigwedge_{k\in\mathbb{N}} z_k$. We have that $\psi(\mathbf{1} - z) \le \sum_{k=1}^{\infty} \varepsilon_k$, $p_k z = 0$ for $k \in \mathbb{N}$ and, therefore,

$$\|x_{n_k} z\|_{\mathcal{M}} = \|x_{n_k}(\mathbf{1} - p_k) z\|_{\mathcal{M}} \le \varepsilon_k, \quad k \in \mathbb{N}.$$

Thus, $x_{n_k} z \to 0$ in the uniform norm as $k \to \infty$, as required. □

In conclusion of this chapter, we also note the following proposition.

Proposition 4.4.10. *If $\{x_n\}_{n\in\mathbb{N}} \subset LS(\mathcal{M})$ and $x_n \to 0$ locally in measure, then there exists a subsequence $\{x_{n_k}\}_{k\in\mathbb{N}}$ of $\{x_n\}_{n\in\mathbb{N}}$, such that $x_{n_k} = y_k + y_k'$, where $y_k \in \mathcal{M}$, $y_k' \in LS(\mathcal{M})$, $k \in \mathbb{N}$, $\|y_k\|_{\mathcal{M}} \to 0$, and $s(|y_k'|) \to 0$ locally in measure as $k \to \infty$.*

Proof. By Proposition 4.3.16(ii) for every $\lambda > 0$, we have that $E_{|x_n|}(\lambda, \infty) \to 0$ as $n \to \infty$ locally in measure. Therefore, there exists a subsequence $\{x_{n_k}\}_{k\in\mathbb{N}}$ of $\{x_n\}_{n\in\mathbb{N}}$, such that $\|E_{|x_{n_k}|}(\frac{1}{k}, \infty)\|_{LS(\mathcal{M})} \le \frac{1}{k}$ for all $k \in \mathbb{N}$.

We set

$$y_k = x_{n_k} E_{|x_{n_k}|}\left[0, \frac{1}{k}\right], \quad y_k' = x_{n_k} E_{|x_{n_k}|}\left(\frac{1}{k}, \infty\right).$$

By the spectral theorem, we have that $x_{n_k} = y_k + y_k'$, $y_k \in \mathcal{M}$, and $\|y_k\|_{\mathcal{M}} \le 1/k$.

It remains to show that $s(|y_k'|) \to 0$ locally in measure. By functional calculus, we have that

$$|y_k'| = |x_{n_k}| E_{|x_{n_k}|}\left(\frac{1}{k}, \infty\right)$$

and, therefore, $s(|y_k'|) \le E_{|x_{n_k}|}(\frac{1}{k}, \infty)$. Proposition 4.3.13(iii) implies that $\|s(|y_k'|)\|_{LS(\mathcal{M})} \le \|E_{|x_{n_k}|}(\frac{1}{k}, \infty)\|_{LS(\mathcal{M})} \le \frac{1}{k}$ for all $k \in \mathbb{N}$, proving that $s(|y_k'|) \to 0$ locally in measure. □

Bibliographical notes

In [75], Grothendieck introduced what he called “decreasing rearrangement” for bounded positive operators in a semifinite von Neumann algebra satisfying some additional assumptions. In that paper, Grothendieck presented (without proof) several fundamental properties of decreasing rearrangements, which underlie numerous properties of τ-measurable operators. The first systematic presentation of the generalized singular value function and its properties was given by V. Ovčinnikov [120]. Some of the results from [120] were proved earlier by M. Sonis in the special case of factors [144].

A detailed exposition on the generalized singular value function was later given by T. Fack and H. Kosaki in [70].

The measure topology t_τ was initially introduced by Segal [141] on the $*$-algebra $S(\mathcal{M})$ of all measurable operators associated with a von Neumann algebra $\mathcal{M}$ equipped with a faithful normal semifinite trace τ. In [141], Segal showed, in particular, that $(S(\mathcal{M}), t_\tau)$ is a Hausdorff metrizable topological ring with continuous involution and separately continuous multiplication. In [145], W. F. Stinespring proved joint continuity of multiplication with respect to the topology t_τ under some additional conditions. These additional conditions motivated E. Nelson [116] to introduce the algebra $S(\mathcal{M}, \tau)$ of τ-measurable operators as the completion of the von Neumann algebra $\mathcal{M}$ with respect to t_τ. In [116], it was proved that $(S(\mathcal{M}, \tau), t_\tau)$ is a complete metrizable topological $*$-algebra with jointly continuous multiplication. Properties of the topology t_τ on the $*$-algebra $S(\mathcal{M}, \tau)$ were also considered in [164]. For detailed exposition of other properties of this topology, we refer the reader to [64].

We note that the measure topology t_τ is usually introduced via a base of neighborhoods of zero $\{U(\cdot, \cdot)\}$ as defined in Proposition 4.2.5. It is then proved that this base defines a metrizable topology and, therefore, there exists an F-norm on $S(\mathcal{M}, \tau)$, which generates the topology t_τ. The F-norm $\|\cdot\|_{S(\mathcal{M},\tau)}$ we chose in Section 4.2 was introduced in [81].

The local measure topology $t(\mathcal{M})$ on the $*$-algebra $LS(\mathcal{M})$ was first introduced in 1964 by S. Sankaran in [138]. In this paper, Sankaran considered countably decomposable finite von Neumann algebras and used the dimension function to introduce a topology on $LS(\mathcal{M}) = S(\mathcal{M})$, which he termed “stochastic convergence.” For a general von Neumann algebra, the local measure topology was introduced by Yeadon in 1973 in [162]. We note that in Section 4.3 we assumed that a von Neumann algebra $\mathcal{M}$ acts on a separable Hilbert space, so that $\mathcal{M}$ (and $\mathcal{Z}(\mathcal{M})$) is countably generated. Yeadon's approach in [162] works for an arbitrary von Neumann algebra, in which case one has to introduce the topology $t(\mathcal{M})$ via a family of pseudometrics. The local measure topology may be introduced via a base of neighborhoods $\{V(\cdot, \cdot, \cdot)\}$ as defined in Proposition 4.3.10 (see, e. g., [23, 24]). We note that for a general von Neumann algebra, the local measure topology $t(\mathcal{M})$ is not necessarily metrizable.

For an equivalent description of the local measure topology in the case when $\mathcal{M}$ is the tensor product of $L_\infty(\Omega, \Sigma, \mu)$ with $\mathcal{B}(H)$, see [28].

For other possible topologies on algebras of unbounded operators, we refer the reader to [40–42, 53, 64, 65, 122, 162] among others.

5 Properties of derivations on algebras of locally measurable operators

This chapter presents the basics of the theory of derivations on algebras of locally measurable operators. The elementary notions of self-adjoint and reduced derivations $\delta^{(e)}$, $e \in P(\mathcal{M})$ are also introduced here. We discuss a particularly important reduction of derivations by central projections, $\delta^{(z)}, z \in P(\mathcal{Z}(\mathcal{M}))$, and we show that this procedure keeps track of whether a derivation is continuous in the local measure topology (or inner). Such a reduction of a derivation using a central partition of unity allows us to reduce the study of the Ayupov–Kadison–Liu problem to specific types of the von Neumann algebras separately.

In Section 5.2, we show that any derivation $\delta : \mathcal{M} \to LS(\mathcal{M})$ can be uniquely extended up to a derivation on the whole algebra $LS(\mathcal{M})$. In Section 5.3, we recall some of the classical results for derivations on algebras of bounded operators, which will be used in the subsequent results. In this section, we also present some results for derivations on ideals of von Neumann algebras.

The most important result of this chapter is established in Section 5.5. This result is a fundamental property of derivations showing that for a derivation δ on the algebra of locally measurable operators (resp., compact operators) continuity of δ in the local measure topology is a necessary and sufficient condition for the derivation δ to be inner (resp., spatial). A similar result also holds for the derivations on the algebra of τ-measurable operators and τ-compact operators if the local measure topology is replaced by the measure topology. This characterization of all inner/spatial derivations will be essential in the resolution of the Ayupov–Kadison–Liu problem in Chapter 7.

5.1 General properties of derivations

In this section, we introduce the main objective of this book, derivations on algebras of locally measurable operators, and study their basic properties.

Throughout this chapter, we assume that $\mathcal{M}$ is an arbitrary von Neumann algebra. Whenever the algebras $S(\mathcal{M}, \tau)$ and $S_0(\mathcal{M}, \tau)$ appear, it is assumed that the von Neumann algebra $\mathcal{M}$ is equipped with a faithful normal semifinite trace τ.

Definition 5.1.1. Let $\mathcal{A}$ be a subalgebra in $LS(\mathcal{M})$. A linear mapping $\delta : \mathcal{A} \to LS(\mathcal{M})$ is called a *derivation* on $\mathcal{A}$ with values in $LS(\mathcal{M})$ if δ satisfies the Leibniz rule

$$\delta(xy) = \delta(x)y + x\delta(y)$$

for any $x, y \in \mathcal{A}$.

https://doi.org/10.1515/9783111599687-006

Assume that $a \in \mathcal{A}$. The linear mapping $\delta_a(x) : \mathcal{A} \to \mathcal{A}$, defined by $\delta_a(x) := [a, x] = ax - xa$, $x \in \mathcal{A}$, is a derivation on $\mathcal{A}$ with values in $\mathcal{A}$. Indeed, for any $x, y \in \mathcal{A}$ we have that

$$\begin{aligned}\delta_a(xy) &= axy - xya = axy - xay + xay - xya = (ax - xa)y + x(ay - ya) \\ &= \delta_a(x)y + x\delta_a(y).\end{aligned}$$

Derivations defined in this way form a special class of derivations, as recognised in the following definition.

Definition 5.1.2. Let $\mathcal{A}$ be a subalgebra in $LS(\mathcal{M})$.

(i) A derivation $\delta : \mathcal{A} \to LS(\mathcal{M})$ is said to be an *inner* derivation on $\mathcal{A}$ if $\delta = \delta_a = [a, \cdot]$ for some $a \in \mathcal{A}$;

(ii) A derivation $\delta : \mathcal{A} \to LS(\mathcal{M})$ is said to be *spatial* if $\delta = \delta_a = [a, \cdot]$ for some $a \in LS(\mathcal{M})$, such that $a \notin \mathcal{A}$.

In either case, we say that the derivation δ_a is *implemented* by a.

Since the operation of multiplication is continuous with respect to the local measure topology, it immediately follows that any inner derivation of $\mathcal{A}$ is continuous with respect to the local measure topology. Similarly, if $\mathcal{M}$ is a semifinite von Neumann algebra equipped with a faithful normal semifinite trace τ, then any inner derivation on $S(\mathcal{M}, \tau)$ is continuous with respect to the measure topology. In Section 5.5 below, we will show that the converse also holds.

Let $\mathcal{A}$ be a $*$-subalgebra in $LS(\mathcal{M})$, let δ be a derivation on $\mathcal{A}$ with values in $LS(\mathcal{M})$. Let us define a mapping $\delta^* : \mathcal{A} \to LS(\mathcal{M})$, by setting

$$\delta^*(x) := (\delta(x^*))^*, \quad x \in \mathcal{A}.$$

A direct verification shows that δ^* is also a derivation on $\mathcal{A}$.

Definition 5.1.3. A derivation δ on $\mathcal{A}$ is said to be *self-adjoint*, if $\delta = \delta^*$.

Every derivation δ on $\mathcal{A}$ can be represented in the form

$$\delta = \mathrm{Re}(\delta) + i\,\mathrm{Im}(\delta),$$

where

$$\mathrm{Re}(\delta) = \frac{\delta + \delta^*}{2}, \quad \mathrm{Im}(\delta) = \frac{\delta - \delta^*}{2i}$$

are self-adjoint derivations on $\mathcal{A}$.

As $(LS(\mathcal{M}), t(\mathcal{M}))$ and $(S(\mathcal{M}, \tau), t_\tau)$ are topological $*$-algebras, the following result holds.

Proposition 5.1.4. *Let $\mathcal{A}$ be a $*$-subalgebra in $LS(\mathcal{M})$ and let $\delta : \mathcal{A} \to LS(\mathcal{M})$ be a derivation.*

(i) *The derivation δ is continuous with respect to the local measure topology $t(\mathcal{M})$ (resp., the measure topology t_τ) if and only if the self-adjoint derivations* $\mathrm{Re}(\delta)$ *and* $\mathrm{Im}(\delta)$ *are continuous with respect to $t(\mathcal{M})$ (resp., t_τ);*
(ii) *The derivation $\delta : \mathcal{A} \to \mathcal{A}$ is inner if and only if* $\mathrm{Re}(\delta)$ *and* $\mathrm{Im}(\delta)$ *are inner derivations.*

Proof. (i) By Theorem 4.3.14, the involution $* : x \to x^*$ is continuous in the local measure topology. By definition, $\delta^* = * \circ \delta \circ *$. If δ is continuous with respect to the local measure topology, then δ^* is also continuous with respect to the local measure topology. Therefore, $\mathrm{Re}(\delta)$ and $\mathrm{Im}(\delta)$ are continuous with respect to the local measure topology, too. Conversely, if $\mathrm{Re}(\delta)$ and $\mathrm{Im}(\delta)$ are continuous with respect to the local measure topology, then $\delta = \mathrm{Re}(\delta) + i\,\mathrm{Im}(\delta)$ is also continuous with respect to the local measure topology.

As $(S(\mathcal{M}, \tau), t_\tau)$ is a topological $*$-algebra (see Theorem 4.2.8), the proof for the measure topology t_τ is similar.

(ii) Suppose that $\delta = \delta_a$, $a \in \mathcal{A}$, is an inner derivation on $\mathcal{A}$. For every $x \in \mathcal{A}$, we have

$$\delta^*(x) = (\delta(x^*))^* = ([a, x^*])^* = (ax^* - x^*a)^* = xa^* - a^*x = [-a^*, x].$$

Since $\mathcal{A}$ is a $*$-algebra, we have that $-a^* \in \mathcal{A}$. Therefore, δ^* is an inner derivation implemented by $-a^*$. Writing

$$\mathrm{Re}(\delta)(x) = \frac{1}{2}(\delta(x) + \delta^*(x)) = \frac{1}{2}([a, x] + [-a^*, x]) = i[\mathrm{Im}(a^*), x],$$

and

$$\mathrm{Im}(\delta)(x) = \frac{1}{2i}(\delta(x) - \delta^*(x)) = \frac{1}{2i}([a, x] + [a^*, x]) = -i[\mathrm{Re}(a), x],$$

and noting that $\mathrm{Im}(a), \mathrm{Re}(a) \in \mathcal{A}$, we conclude that $\mathrm{Re}(\delta)$ and $\mathrm{Im}(\delta)$ are also inner derivations.

Conversely, if $\mathrm{Re}(\delta) = \delta_b$ and $\mathrm{Im}(\delta) = \delta_c$, $c, b \in \mathcal{A}$, are inner derivations, then

$$\delta(x) = \mathrm{Re}(\delta)(x) + i\,\mathrm{Im}(\delta)(x) = [b, x] + i[c, x] = [b + ic, x].$$

Thus, δ is also an inner derivation. □

We now show that any derivation $\delta : LS(\mathcal{M}) \to LS(\mathcal{M})$ reduces to a derivation on the center $\mathcal{Z}(LS(\mathcal{M}))$ of the algebra $LS(\mathcal{M})$.

Lemma 5.1.5. *Assume that $\delta : LS(\mathcal{M}) \to LS(\mathcal{M})$ is a derivation. Then $\delta(\mathcal{Z}(LS(\mathcal{M}))) \subset \mathcal{Z}(LS(\mathcal{M}))$. In particular, δ is a derivation on $\mathcal{Z}(LS(\mathcal{M}))$.*

Proof. Let $x \in LS(\mathcal{M})$ and $z \in \mathcal{Z}(LS(\mathcal{M}))$. By the Leibniz rule, we have

$$\delta(xz) = \delta(x)z + x\delta(z), \quad \delta(zx) = \delta(z)x + z\delta(x).$$

Therefore,

$$x\delta(z) = \delta(xz) - \delta(x)z, \quad \delta(z)x = \delta(zx) - z\delta(x) = \delta(xz) - \delta(x)z.$$

Consequently,

$$x\delta(z) = \delta(z)x, \quad x \in LS(\mathcal{M}),\ z \in \mathcal{Z}(LS(\mathcal{M})).$$

Hence,

$$\delta(z) \in \mathcal{Z}(LS(\mathcal{M})),$$

as required. □

Next, we study reductions of derivations by projections. Let $\mathcal{A}$ be a subalgebra in $LS(\mathcal{M})$, $0 \neq e \in P(\mathcal{M}) \cap \mathcal{A}$, let $\delta : \mathcal{A} \to \mathcal{A}$ be a derivation on $\mathcal{A}$, and let $\delta^{(e)}$ be a linear mapping from $e\mathcal{A}e := \{exe, x \in \mathcal{A}\}$ into $e\mathcal{A}e$ defined as follows:

$$\delta^{(e)}(x) := e\delta(x)e, \quad x \in e\mathcal{A}e \subset \mathcal{A}. \tag{5.1}$$

We now show that $\delta^{(e)}$ is a derivation.

Lemma 5.1.6. *Let $\mathcal{A}$ be a subalgebra in $LS(\mathcal{M})$, $0 \neq e \in P(\mathcal{M}) \cap \mathcal{A}$, and let δ be a derivation on $\mathcal{A}$. The mapping $\delta^{(e)} : e\mathcal{A}e \to e\mathcal{A}e$, defined by* (5.1)*, is a derivation on $e\mathcal{A}e$.*

Proof. Clearly, $\delta^{(e)}$ is a linear mapping on $e\mathcal{A}e$. If $x, y \in e\mathcal{A}e$, then $x, y \in \mathcal{A}$ and

$$\delta^{(e)}(xy) = e\delta(xy)e = e(\delta(x)y)e + e(x\delta(y))e = e\delta(x) \cdot ye + ex \cdot \delta(y)e.$$

Since $ye = ey$ and $ex = xe$, it follows that

$$\delta^{(e)}(xy) = e\delta(x) \cdot ey + xe \cdot \delta(y)e = e\delta(x)e \cdot y + x \cdot e\delta(y)e = \delta^{(e)}(x)y + x\delta^{(e)}(y),$$

as required. □

Definition 5.1.7. Let $\mathcal{A}$ be a subalgebra in $LS(\mathcal{M})$, $0 \neq e \in P(\mathcal{M}) \cap \mathcal{A}$, and let $\delta : \mathcal{A} \to \mathcal{A}$ be a derivation on $\mathcal{A}$. Then the derivation $\delta^{(e)} : e\mathcal{A}e \to e\mathcal{A}e$, defined by (5.1), is said to be a *reduced* derivation on $e\mathcal{A}e$.

When a projection e is a central projection in the algebra $\mathcal{M}$, the reduced derivation $\delta^{(e)}$ allows a different description.

Proposition 5.1.8. *Suppose that $\mathcal{A}$ is a subalgebra of $LS(\mathcal{M})$ and let $\delta : \mathcal{A} \to LS(\mathcal{M})$ be a derivation. If $z \in P(\mathcal{Z}(\mathcal{M})) \cap \mathcal{A}$, then $\delta(z) = 0$ and $\delta(zx) = z\delta(x)$ for all $x \in \mathcal{A}$. In particular, the reduced derivation $\delta^{(z)}$ coincides with the restriction $\delta|_{z\mathcal{A}}$ of the derivation δ onto $z\mathcal{A}$.*

Proof. By the Leibniz rule, we have that

$$\delta(z) = \delta(z^2) = \delta(z)z + z\delta(z) = 2z\delta(z),$$

where the last equality holds since z is a central projection. Multiplying by z on the left, we obtain

$$z\delta(z) = 2z^2\delta(z) = 2z\delta(z).$$

Hence, $z\delta(z) = 0$ and, therefore, $\delta(z) = 0$. Employing the Leibniz rule once again, we obtain that $\delta(zx) = \delta(z)x + z\delta(x) = z\delta(x)$. □

For the algebras $S_0(\mathcal{M})$ (resp., $S_0(\mathcal{M}, \tau)$) of all compact (resp., τ-compact), the assertion of Proposition 5.1.8 holds without the assumption that the central projection z belongs to the algebra.

Proposition 5.1.9. *Let $\mathcal{M}$ be a von Neumann algebra and let $\mathcal{A}$ be either $S_0(\mathcal{M})$ or $S_0(\mathcal{M}, \tau)$ (in the latter case $\mathcal{M}$ is assumed to be equipped with a faithful normal semifinite trace τ). If $\delta : \mathcal{A} \to \mathcal{A}$ is a derivation and $z \in P(\mathcal{Z}(\mathcal{M}))$, then $\delta(zx) = z\delta(x)$ for any $x \in \mathcal{A}$. In particular, the reduced derivation $\delta^{(z)}$ coincides with the restriction $\delta|_{zS_0(\mathcal{M})}$ (resp., $\delta|_{zS_0(\mathcal{M},\tau)}$).*

Proof. If $\mathcal{A} = S_0(\mathcal{M}, \tau)$, then $\mathcal{M}$ is necessarily a semifinite von Neumann algebra. If $\mathcal{A} = S_0(\mathcal{M})$, then without loss of generality one can assume that $\mathcal{M}$ is semifinite. Therefore, in what follows for either case, we assume that $\mathcal{M}$ is a semifinite von Neumann algebra.

Let $z \in P(\mathcal{Z}(\mathcal{M}))$ be arbitrary. Since $\mathcal{M}$ is semifinite, any central projection from $\mathcal{M}$ is the supremum of finite or τ-finite projections. Let $p \in P(\mathcal{M}) \cap \mathcal{A}$ be such that $p \leq \mathbf{1} - z$. Since $p \in \mathcal{A}$, using the Leibniz rule and the inclusion $z \in P(\mathcal{Z}(\mathcal{M}))$, we have

$$z\delta(p) = z\delta(p^2) = zp\delta(p) + z\delta(p)p = zp\delta(p) + \delta(p)zp = 0,$$

where the last equality holds since $p \leq \mathbf{1} - z$. For any $x \in \mathcal{A}$, another appeal to the Leibniz rule implies that

$$\delta(xz)p = \delta(xzp) - xz\delta(p) = 0.$$

Since $\mathbf{1} - z = \sup\{p \in P(\mathcal{M}) \cap \mathcal{A} : p \leq \mathbf{1} - z\}$, Proposition 3.3.7(iii) implies that

$$\delta(xz)(\mathbf{1} - z) = 0,$$

or equivalently,

$$\delta(xz) = \delta(xz)z, \quad z \in P(\mathcal{Z}(\mathcal{M})),\ x \in \mathcal{A}. \tag{5.2}$$

Now let $x \in \mathcal{A}$ be arbitrary. Writing $xz = x - x(\mathbf{1} - z)$ and using equality (5.2) for the central projection $\mathbf{1} - z$, we conclude that

$$\begin{aligned}\delta(xz) &= \delta(xz)z = \delta(x)z - \delta(x(\mathbf{1} - z))z \\ &= \delta(x)z - \delta(x(\mathbf{1} - z))(\mathbf{1} - z)z = \delta(x)z,\end{aligned}$$

as required. □

Now we show that if δ is a derivation on a subalgebra $\mathcal{A} \subset LS(\mathcal{M})$, which is continuous locally in measure (resp., inner), then the reduced derivation $\delta^{(z)}$ by a central projection $z \in P(\mathcal{Z}(\mathcal{M})) \cap \mathcal{A}$ is also continuous locally in measure (resp., inner).

Proposition 5.1.10. *Let $\delta : \mathcal{A} \to \mathcal{A}$ be a derivation on $\mathcal{A}$ and let $z \in P(\mathcal{Z}(\mathcal{M})) \cap \mathcal{A}$. Then:*
(i) *if δ is $t(\mathcal{M})$-continuous, then $\delta^{(z)}$ is $t(z\mathcal{M})$-continuous;*
(ii) *if δ is inner, then $\delta^{(z)}$ is inner.*

Proof. (i) If $x_n, x \in z\mathcal{A}$, and x_n converges to x with respect to $t(z\mathcal{M})$, then by Lemma 4.3.29, we have $x_n \xrightarrow{t(\mathcal{M})} x$. Therefore, since $(LS(\mathcal{M}), t(\mathcal{M}))$ is a topological algebra, it follows that

$$\delta^{(z)}(x_n) = z\delta(x_n) \xrightarrow{t(\mathcal{M})} z\delta(x) = \delta^{(z)}(x).$$

Referring to Lemma 4.3.29 once again, we conclude that $\delta^{(z)}(x_n)$ converges to $\delta^{(z)}(x)$ with respect to $t(z\mathcal{M})$. Thus, $\delta^{(z)}$ is $t(z\mathcal{M})$-continuous.

(ii) Let $\delta = \delta_a = [a, \cdot]$ for some $a \in \mathcal{A}$. Then

$$\delta^{(z)}(x) = z\delta(x) = z[a, x] = [za, x], \quad x \in z\mathcal{A}.$$

Since $za \in z\mathcal{A}$, it follows that $\delta^{(z)}$ is an inner derivation on $z\mathcal{A}$ implemented by za. □

In the following proposition, we show that if reduced derivations $\delta^{(z_n)}$ are continuous with respect to the local measure topology (resp., inner) on $z_n\mathcal{A}$ for all $n \in \mathbb{N}$ and if $\bigvee_{n\in\mathbb{N}} z_n = \mathbf{1}_{\mathcal{M}}$, then δ is necessarily continuous locally in measure (resp., is a spatial derivation). In the particular case of a finite partition of unity $\{z_n\}_{n=1}^k$, the derivation δ is inner. This is a crucial proposition which will be used repeatedly in the future. It allows us to use the decomposition of von Neumann algebras in the study of derivations on the (subalgebras of) $LS(\mathcal{M})$ by considering cases of a von Neumann algebra $\mathcal{M}$ of specific types separately.

Proposition 5.1.11. *Let $\mathcal{A}$ be a subalgebra of $LS(\mathcal{M})$ containing $P(\mathcal{Z}(\mathcal{M}))$ and let δ be a derivation $\delta : \mathcal{A} \to \mathcal{A}$. For a central partition $\{z_n\}_{n\in\mathbb{N}}$ of unity in $\mathcal{M}$, we have:*

(i) *If* $\delta^{(z_n)} : z_n\mathcal{A} \to z_n\mathcal{A}$ *is continuous in the topology* $t(z_n\mathcal{M})$ *for any* $n \in \mathbb{N}$, *then* $\delta : \mathcal{A} \to \mathcal{A}$ *is continuous in the topology* $t(\mathcal{M})$;

(ii) *If* $\delta^{(z_n)} = \delta_{d_n}$, $d_n \in z_n\mathcal{A}$, *is an inner derivation on* $z_n\mathcal{A}$ *for every* $n \in \mathbb{N}$, *then there exists an operator* $d \in LS(\mathcal{M})$, *such that* $\delta(x) = [d,x]$ *for all* $x \in \mathcal{A}$ *and* $z_nd = d_n$ *for every* $n \in \mathbb{N}$. *Furthermore, if the central partition of unity* $\{z_n\}_{n=1}^k$, $k \in \mathbb{N}$, *is finite, then* $d \in \mathcal{A}$ *and* δ *is inner.*

Proof. The assertion of part (i) follows directly from Theorem 4.3.30.

(ii) Since $\{z_n\}_{n\in\mathbb{N}}$ is a central partition of the unity $\mathbf{1}$ and $d_n \in z_n\mathcal{A} \subset z_nLS(\mathcal{M})$, Proposition 2.3.10 implies that there exists $d \in LS(\mathcal{M})$, such that $z_nd = d_n$ for every $n \in \mathbb{N}$. For any $y \in \mathcal{A}$ and $n \in \mathbb{N}$, we have that

$$z_n\delta(y) = \delta^{(z_n)}(z_ny) = [d_n, z_ny] = [z_nd, z_ny] = z_n[d,y].$$

Since $\sup_{n\in\mathbb{N}} z_n = \mathbf{1}$, Lemma 2.3.7 implies that $\delta(y) = [d,y]$.

If the partition $\{z_n\}_{n=1}^k$ is finite, then $d = \sum_{n=1}^k d_n \in \mathcal{A}$, and so the derivation δ is inner. □

In the following proposition, we show that in the case of a properly infinite von Neumann algebra $\mathcal{M}$ an analogue of Proposition 5.1.11(ii) holds for a derivation $\delta : LS(\mathcal{M}) \to LS(\mathcal{M})$ reduced by noncentral projections p and $\mathbf{1} - p$.

Proposition 5.1.12. *Let* $\mathcal{M}$ *be a properly infinite von Neumann algebra and let* $\delta : LS(\mathcal{M}) \to LS(\mathcal{M})$ *be a derivation. If for some* $p \in P(\mathcal{M})$, *the reduced derivations* $\delta^{(p)}$ *and* $\delta^{(1-p)}$ *are inner, then* δ *is also inner.*

Proof. By Proposition 1.9.10(ii), one can choose a central projection z such that $pz \sim z$ and $(\mathbf{1}-p)(\mathbf{1}-z) \sim \mathbf{1}-z$. By Proposition 5.1.11, it suffices to prove the assertion separately for the algebras $\mathcal{M}z$ and $\mathcal{M}(\mathbf{1}-z)$. In what follows, we assume without loss of generality that $z = \mathbf{1}$ (so that $p \sim \mathbf{1}$). Let u be a partial isometry such that $u^*u = \mathbf{1}$ and $uu^* = p$.

By the Leibniz rule, for any $x \in LS(p\mathcal{M}p)$ we have

$$\delta(x) = \delta(pxp) = \delta(p)xp + p\delta(x)p + px\delta(p) = \delta(p)x + x\delta(p) + \delta^{(p)}(x).$$

Since $\delta^{(p)}$ is inner, it follows that there exists $c \in LS(p\mathcal{M}p)$, such that

$$\delta(x) = \delta(p)x + x\delta(p) + [c,x], \quad x \in LS(p\mathcal{M}p). \tag{5.3}$$

For an arbitrary $x \in LS(\mathcal{M})$, we have

$$x = u^* \cdot uxu^* \cdot u, \quad uxu^* \in LS(p\mathcal{M}p).$$

Using the Leibniz rule and (5.3), we can write

$$\begin{aligned}\delta(x) &= \delta(u^*) \cdot uxu^* \cdot u + u^* \cdot \delta(uxu^*) \cdot u + u^* \cdot uxu^* \cdot \delta(u) \\ &= \delta(u^*)ux + u^* \cdot ([c, uxu^*] + \delta(p) \cdot uxu^* + uxu^* \cdot \delta(p)) \cdot u + xu^*\delta(u) \\ &= (\delta(u^*)u + u^*cu + u^*\delta(p)u)x + x(u^*\delta(u) - u^*cu + u^*\delta(p)u).\end{aligned} \tag{5.4}$$

Note that the Leibniz rule implies

$$p\delta(p)p = p \cdot p\delta(p) \cdot p = p \cdot \delta(p \cdot p) \cdot p - p \cdot \delta(p)p \cdot p = p\delta(p)p - p\delta(p)p = 0,$$

and, therefore,

$$u^*\delta(p)u = u^*p \cdot \delta(p) \cdot pu = u^* \cdot p\delta(p)p \cdot u = u^* \cdot 0 \cdot u = 0.$$

Thus, equality (5.4) can be written as

$$\delta(x) = (\delta(u^*)u + u^*cu)x + x(u^*\delta(u) - u^*cu).$$

Using the Leibniz rule once again and referring to Proposition 5.1.8, we conclude that

$$\begin{aligned}\delta(x) &= (\delta(\mathbf{1}) - u^*\delta(u) + u^*cu)x + x(u^*\delta(u) - u^*cu) \\ &= (u^*cu - u^*\delta(u))x - x(u^*cu - u^*\delta(u)) = [u^*cu - u^*\delta(u), x],\end{aligned}$$

proving that δ is an inner derivation. □

In conclusion of this section, we list several properties of derivations related to continuity (with respect to the measure topology and the local measure topology).

Lemma 5.1.13.
(i) *If $\delta : \mathcal{M} \to LS(\mathcal{M})$ is a derivation and $\{p_n\}_{n\in\mathbb{N}} \subset P(\mathcal{M})$ is such that $p_n \to 0$ locally in measure, then $\delta(p_n) \to 0$ locally in measure;*
(ii) *If $\delta : \mathcal{M} \to S(\mathcal{M}, \tau)$ is a derivation and $\{p_n\}_{n\in\mathbb{N}} \subset P(\mathcal{M})$ is such that $p_n \to 0$ in measure, then $\delta(p_n) \to 0$ in measure.*

Proof. By the Leibniz rule, we have that

$$\delta(p_n) = \delta(p_n^2) = \delta(p_n)p_n + p_n\delta(p_n).$$

If $p_n \to 0$ locally in measure, then Proposition 4.3.18(vi) guarantees that $\delta(p_n)p_n$ and $p_n\delta(p_n)$ converge to 0 in the local measure topology. Therefore, $\delta(p_n) \to 0$ locally in measure.

Similarly, if $p_n \to 0$ in measure, then Proposition 4.2.12(iv) implies that $\delta(p_n) \to 0$ in measure. □

The next two propositions show that a derivation δ on $LS(\mathcal{M})$ (resp., on $S(\mathcal{M}, \tau)$) is continuous locally in measure (resp., in measure) if and only if it is continuous from $(\mathcal{M}, \|\cdot\|_{\mathcal{M}})$ into $(LS(\mathcal{M}), t(\mathcal{M}))$ (resp., from $(\mathcal{M}, \|\cdot\|_{\mathcal{M}})$ into $(S(\mathcal{M}, \tau), t_\tau)$). The proofs of these results are verbatim repetitions of one another requiring only different references to Chapter 4.

Proposition 5.1.14. *Let $\delta : LS(\mathcal{M}) \to LS(\mathcal{M})$ be a derivation. Then the following assertions are equivalent:*
(i) *δ is $t(\mathcal{M})$-continuous;*
(ii) *$\delta|_{\mathcal{M}}$ is a continuous mapping from $(\mathcal{M}, \|\cdot\|_{\mathcal{M}})$ to $(LS(\mathcal{M}), t(\mathcal{M}))$.*

Proof. Suppose that $\delta|_{\mathcal{M}}$ is a continuous mapping from $(\mathcal{M}, \|\cdot\|_{\mathcal{M}})$ to $(LS(\mathcal{M}), t(\mathcal{M}))$. Let $\{x_n\}_{n\in\mathbb{N}} \in LS(\mathcal{M})$ be such that $x_n \to 0$ locally in measure. By Proposition 4.3.16(iv), there exists a sequence $\{p_n\}_{n\in\mathbb{N}} \in P(\mathcal{M})$ such that $p_n \to \mathbf{1}$ in measure and $\|x_n p_n\|_{\mathcal{M}} \to 0$ as $n \to \infty$. We can write

$$\delta(x_n) = \delta(x_n)p_n + \delta(x_n)(\mathbf{1} - p_n).$$

Since $p_n \to \mathbf{1}$ locally in measure, Proposition 4.3.18(vi) implies that $\delta(x_n)(\mathbf{1} - p_n) \to 0$ locally in measure as $n \to \infty$. Therefore, it is sufficient to show that $\delta(x_n)p_n \to 0$ locally in measure as $n \to \infty$.

By the Leibniz rule, we have that

$$\delta(x_n)p_n = \delta(x_n p_n) - x_n\delta(p_n).$$

The first term converges to zero locally in measure since $\|x_n p_n\|_{\mathcal{M}} \to 0$ and $\delta : (\mathcal{M}, \|\cdot\|_{\mathcal{M}}) \to (LS(\mathcal{M}), t(\mathcal{M}))$ is continuous. The second term converges to zero by Proposition 4.3.13(vi). Thus, $\delta : LS(\mathcal{M}) \to LS(\mathcal{M})$ is continuous in the local measure topology.

Conversely, suppose that $\delta : LS(\mathcal{M}) \to LS(\mathcal{M})$ is continuous in the local measure topology and let $\{x_n\}_{n\in\mathbb{N}} \subset \mathcal{M}$ be such that $\|x_n\|_{\mathcal{M}} \to 0$ as $n \to \infty$. By Proposition 4.3.18(ii), we have that $x_n \to 0$ locally in measure. Since δ is continuous locally in measure, it follows that $\delta(x_n) \to 0$ locally in measure, as required. □

Proposition 5.1.15. *Let $\delta : S(\mathcal{M}, \tau) \to S(\mathcal{M}, \tau)$ be a derivation. Then the following assertions are equivalent:*
(i) *δ is t_τ-continuous;*
(ii) *$\delta|_{\mathcal{M}}$ is a continuous mapping from $(\mathcal{M}, \|\cdot\|_{\mathcal{M}})$ to $(S(\mathcal{M}, \tau), t_\tau)$.*

Proof. Suppose that $\delta|_{\mathcal{M}}$ is a continuous mapping from $(\mathcal{M}, \|\cdot\|_{\mathcal{M}})$ to $(S(\mathcal{M}, \tau), t_\tau)$. Let $\{x_n\}_{n\in\mathbb{N}} \in S(\mathcal{M}, \tau)$ be such that $x_n \to 0$ in measure. By Proposition 4.2.9(vi) and Corollary 4.2.10, there exists a sequence $\{p_n\}_{n\in\mathbb{N}} \in P(\mathcal{M})$, such that $p_n \to \mathbf{1}$ in measure and $\|x_n p_n\|_{\mathcal{M}} \to 0$ as $n \to \infty$. We can write

$$\delta(x_n) = \delta(x_n)p_n + \delta(x_n)(\mathbf{1} - p_n).$$

Since $p_n \to \mathbf{1}$ in measure, Proposition 4.2.12(iv) implies that $\delta(x_n)(\mathbf{1}-p_n) \to 0$ in measure as $n \to \infty$. Therefore, it is sufficient to show that $\delta(x_n)p_n \to 0$ in measure as $n \to \infty$.

By the Leibniz rule, we have that

$$\delta(x_n)p_n = \delta(x_np_n) - x_n\delta(p_n).$$

The first term converges to zero in measure since $\|x_np_n\|_{\mathcal{M}} \to 0$ and $\delta : (\mathcal{M}, \|\cdot\|_{\mathcal{M}}) \to (S(\mathcal{M},\tau), t_\tau)$ is continuous. The second term converges to zero by Lemma 5.1.13 and Proposition 4.2.7(v). Thus, $\delta : LS(\mathcal{M}) \to LS(\mathcal{M})$ is continuous in the measure topology.

Conversely, suppose that $\delta : S(\mathcal{M},\tau) \to S(\mathcal{M},\tau)$ is continuous in the measure topology t_τ and let $\{x_n\}_{n\in\mathbb{N}} \subset \mathcal{M}$ be such that $\|x_n\|_{\mathcal{M}} \to 0$ as $n \to \infty$. By Proposition 4.2.12(ii), we have that $x_n \to 0$ in measure. Since δ is continuous in measure, it follows that $\delta(x_n) \to 0$ in measure, as required. □

5.2 Extension of derivations

In this section, we extend a derivation acting on a von Neumann algebra $\mathcal{M}$ with values in $LS(\mathcal{M})$ to a derivation from $LS(\mathcal{M})$ into $LS(\mathcal{M})$.

Definition 5.2.1. Let $\mathcal{A}$ and $\mathcal{B}$ be subalgebras of $LS(\mathcal{M})$, such that $\mathcal{A} \subset \mathcal{B}$ and let $\delta : \mathcal{A} \to LS(\mathcal{M})$ be a derivation. A derivation $\widetilde{\delta} : \mathcal{B} \to LS(\mathcal{M})$ is said to be an *extension of* δ if $\delta(x) = \widetilde{\delta}(x)$ for all $x \in \mathcal{A}$.

Let $\delta : \mathcal{M} \to LS(\mathcal{M})$ be a derivation. As shown in Corollary 4.3.15, any $x \in LS(\mathcal{M})$ can be approximated in the local measure topology by the sequence $\{xE_{|x|}[0,n]\}_{n\in\mathbb{N}} \subset \mathcal{M}$. It is natural to expect that extension of δ up to a derivation $\widetilde{\delta}$ on $LS(\mathcal{M})$ can be defined by

$$\widetilde{\delta}(x) = \lim_{n\to\infty} \delta(xE_{|x|}[0,n]),$$

if the limit on the right-hand side exists in the local measure topology and gives a well-defined mapping $\widetilde{\delta}$ on $LS(\mathcal{M})$. In the next two lemmas, we establish precisely that. In Proposition 5.2.4 below, we will show that $\widetilde{\delta}$ defined as above is a derivation on $LS(\mathcal{M})$.

Lemma 5.2.2. *Let $\mathcal{M}$ be a von Neumann algebra and let $\delta : \mathcal{M} \to LS(\mathcal{M})$ be a derivation. Let $x \in LS(\mathcal{M})$ and let $\{p_n\}_{n\in\mathbb{N}} \subset P(\mathcal{M})$ be an increasing sequence of projections, such that $\{xp_n\}_{n\in\mathbb{N}} \subset \mathcal{M}$ and $p_n \to \mathbf{1}$ locally in measure. Then the sequence $\{\delta(xp_n)\}_{n\in\mathbb{N}}$ is Cauchy in $(LS(\mathcal{M}), t(\mathcal{M}))$ and $\|\delta(p_nxp_n) - \delta(xp_n)\|_{LS(\mathcal{M})}$, $\|\delta(p_nx) - \delta(xp_n)\|_{LS(\mathcal{M})} \to 0$ as $n \to \infty$.*

Proof. We first prove that $\{\delta(xp_n)\}_{n\in\mathbb{N}}$ is a Cauchy sequence in the space $(LS(\mathcal{M}), t(\mathcal{M}))$. Let $n, m \in \mathbb{N}$ be arbitrary. Since the sequence $\{p_n\}_{n\in\mathbb{N}}$ is increasing, we can write

$$xp_n - xp_m = (xp_n - xp_m) \cdot (\mathbf{1} - p_{n\wedge m}).$$

By the Leibniz rule, we have

$$\delta(xp_n - xp_m) = \delta(xp_n - xp_m) \cdot (\mathbf{1} - p_{n\wedge m}) + (xp_n - xp_m) \cdot \delta(\mathbf{1} - p_{n\wedge m}). \tag{5.5}$$

Let $i : \mathbb{N} \to \mathbb{N}\times\mathbb{N}$ be a bijection. Denote $i(k) = (i_1(k), i_2(k))$, $k \in \mathbb{N}$. Define sequences $\{y_k\}_{k\in\mathbb{N}} \subset LS(\mathcal{M})$ and $\{q_k\}_{k\in\mathbb{N}} \subset P(\mathcal{M})$ by setting

$$y_k = \delta(xp_{i_1(k)} - xp_{i_2(k)}), \quad q_k = \mathbf{1} - p_{i_1(k)\wedge i_2(k)}, \quad k \in \mathbb{N}.$$

Since $p_n \to \mathbf{1}$ locally in measure as $n \to \infty$, it follows that $q_k \to 0$ locally in measure as $k \to \infty$. Proposition 4.3.18(vi) implies that $\|y_k q_k\|_{LS(\mathcal{M})} \to 0$ as $k \to \infty$. In other words,

$$\|\delta(xp_n - xp_m) \cdot (\mathbf{1} - p_{n\wedge m})\|_{LS(\mathcal{M})} = \|y_{i^{-1}(n,m)} \cdot q_{i^{-1}(n,m)}\|_{LS(\mathcal{M})} \to 0, \tag{5.6}$$

as $n, m \to \infty$.

Similarly, we have

$$\|(p_n - p_m) \cdot \delta(\mathbf{1} - p_{n\wedge m})\|_{LS(\mathcal{M})} \to 0, \quad n, m \to \infty.$$

Referring to Proposition 4.3.13(vi), we infer that

$$\|(xp_n - xp_m) \cdot \delta(\mathbf{1} - p_{n\wedge m})\|_{LS(\mathcal{M})} \to 0, \quad n, m \to \infty. \tag{5.7}$$

Combining (5.5), (5.6), and (5.7), we conclude that

$$\begin{aligned}\|\delta(xp_n - xp_m)\|_{LS(\mathcal{M})} &\le \|\delta(xp_n - xp_m) \cdot (\mathbf{1} - p_{n\wedge m})\|_{LS(\mathcal{M})} \\ &\quad + \|(xp_n - xp_m) \cdot \delta(\mathbf{1} - p_{n\wedge m})\|_{LS(\mathcal{M})} \to 0,\end{aligned}$$

as $n, m \to \infty$, proving that $\{\delta(xp_n)\}_{n\in\mathbb{N}}$ is a Cauchy sequence in the local measure topology.

Next, by the Leibniz rule we have that

$$\delta(xp_n - p_n xp_n) = \delta((\mathbf{1} - p_n)xp_n) = \delta(\mathbf{1} - p_n)xp_n + (\mathbf{1} - p_n)\delta(xp_n)$$

and, therefore,

$$\|\delta(xp_n - p_n xp_n)\|_{LS(\mathcal{M})} \le \|\delta(\mathbf{1} - p_n)xp_n\|_{LS(\mathcal{M})} + \|(\mathbf{1} - p_n)\delta(xp_n)\|_{LS(\mathcal{M})}.$$

By Lemma 5.1.13, we have that $\delta(\mathbf{1} - p_n) \to 0$ locally in measure. Referring to Proposition 4.3.13(ii) and (vi), we infer that

$$\|\delta(xp_n - p_n xp_n)\|_{LS(\mathcal{M})} \to 0, \quad n \to \infty.$$

Similarly, one can show that

$$\|\delta(p_nx - p_nxp_n)\|_{LS(\mathcal{M})} \to 0, \quad n \to \infty.$$

Finally, combining the latter two convergences, we conclude that

$$\begin{aligned}&\|\delta(p_nx) - \delta(xp_n)\|_{LS(\mathcal{M})}\\ &\quad \le \|\delta(p_nx - p_nxp_n)\|_{LS(\mathcal{M})} + \|\delta(p_nxp_n - xp_n)\|_{LS(\mathcal{M})} \to 0, \quad n \to \infty.\end{aligned}$$ □

Lemma 5.2.3. *Let $\mathcal{M}$ be a von Neumann algebra and let $\delta : \mathcal{M} \to LS(\mathcal{M})$ be a derivation. Let $x \in LS(\mathcal{M})$ and let $\{p_n\}_{n\in\mathbb{N}}, \{q_n\}_{n\in\mathbb{N}} \subset P(\mathcal{M})$ be increasing sequences of projections, such that $\{xp_n\}_{n\in\mathbb{N}}, \{xq_n\}_{n\in\mathbb{N}} \subset \mathcal{M}$ and $p_n, q_n \to \mathbf{1}$ with respect to the local measure topology. Then*

$$\lim_{n\to\infty} \delta(xp_n) = \lim_{n\to\infty} \delta(xq_n),$$

where the limit is taken with respect to the local measure topology $t(\mathcal{M})$.

Proof. Set

$$r_n = \mathbf{1} - p_n \wedge q_n, \quad n \in \mathbb{N}.$$

Since $p_n, q_n \to \mathbf{1}$ with respect to the local measure topology, Proposition 4.3.13(v) implies that

$$\begin{aligned}\|r_n\|_{LS(\mathcal{M})} &= \|\mathbf{1} - p_n \wedge q_n\|_{LS(\mathcal{M})} = \|(\mathbf{1} - p_n) \vee (\mathbf{1} - q_n)\|_{LS(\mathcal{M})}\\ &\le \|\mathbf{1} - p_n\|_{LS(\mathcal{M})} + \|\mathbf{1} - q_n\|_{LS(\mathcal{M})} \to 0, \quad n \to \infty.\end{aligned}$$

It is clear that

$$(p_n - q_n) = (p_n - q_n)r_n, \quad n \in \mathbb{N}.$$

Therefore, by the Leibniz rule, we have

$$\delta(xp_n - xq_n) = \delta(xp_n - xq_n) \cdot r_n + (xp_n - xq_n) \cdot \delta(r_n), \quad n \in \mathbb{N}. \tag{5.8}$$

Since $r_n \to 0$ with respect to the local measure topology, Proposition 4.3.18(vi) implies that

$$\|\delta(xp_n - xq_n) \cdot r_n\|_{LS(\mathcal{M})} \to 0. \tag{5.9}$$

By Lemma 5.1.13, we have that $\delta(r_n) \to 0$ locally in measure. By Proposition 4.3.13(ii), we have that $\|(p_n - q_n)\delta(r_n)\|_{LS(\mathcal{M})} \le 2\|\delta(r_n)\|_{LS(\mathcal{M})}$ for any $n \in \mathbb{N}$. Since $\|\delta(r_n)\|_{LS(\mathcal{M})} \to 0$ as $n \to \infty$, Proposition 4.3.13(vi) implies that

$$\|(xp_n - xq_n) \cdot \delta(r_n)\|_{LS(\mathcal{M})} = \|x \cdot (p_n - q_n)\delta(r_n)\|_{LS(\mathcal{M})} \to 0, \quad n \to \infty.$$

Combining the latter convergence with (5.8) and (5.9), we conclude that

$$\begin{aligned}\|\delta(xp_n - xq_n)\|_{LS(\mathcal{M})} &\leq \|\delta(xp_n - xq_n)\cdot r_n\|_{LS(\mathcal{M})}\\ &\quad + \|(xp_n - xq_n)\cdot\delta(r_n)\|_{LS(\mathcal{M})} \to 0, \quad n\to\infty,\end{aligned}$$

as required. □

Now, equipped with Lemma 5.2.3, we may extend any derivation $\delta : \mathcal{M} \to LS(\mathcal{M})$ up to a derivation $\widetilde{\delta}$ from $LS(\mathcal{M})$ into $LS(\mathcal{M})$.

Proposition 5.2.4. *Let $\mathcal{M}$ be a von Neumann algebra and let $\delta : \mathcal{M} \to LS(\mathcal{M})$ be a derivation. There exists a derivation $\widetilde{\delta} : LS(\mathcal{M}) \to LS(\mathcal{M})$ such that $\widetilde{\delta}|_{\mathcal{M}} = \delta$. Namely, $\widetilde{\delta}$ is defined as*

$$\widetilde{\delta}(x) = \lim_{n\to\infty} \delta(xE_{|x|}[0,n]),$$

where the limit is taken in the local measure topology.

Proof. Let $x \in LS(\mathcal{M})$. By the spectral theorem, we have that $E_{|x|}[0,n]x \in \mathcal{M}$ for any $n \in \mathbb{N}$. Furthermore, since x is locally measurable, Remark 4.3.12 guarantees that

$$\|\mathbf{1} - E_{|x|}[0,n]\|_{LS(\mathcal{M})} = \|E_{|x|}(n,\infty)\|_{LS(\mathcal{M})} \to 0, \quad n\to\infty.$$

By Lemma 5.2.2, the sequence $\{\delta(xE_{|x|}[0,n])\}_{n\in\mathbb{N}}$ is a Cauchy sequence in $(LS(\mathcal{M}), t(\mathcal{M}))$. Since $(LS(\mathcal{M}), t(\mathcal{M}))$ is complete, we can define

$$\widetilde{\delta}(x) = \lim_{n\to\infty} \delta(xE_{|x|}[0,n]).$$

We claim that $\widetilde{\delta} : LS(\mathcal{M}) \to LS(\mathcal{M})$ is a derivation, which extends δ.

We first show that $\widetilde{\delta}$ is a linear mapping. Let $x, y \in LS(\mathcal{M})$ be arbitrary. We define

$$e_n = E_{|x|}[0,n] \wedge E_{|y|}[0,n].$$

Clearly, $\{e_n\}_{n\in\mathbb{N}}$ is increasing. Since $E_{|x|}[0,n], E_{|y|}[0,n] \to \mathbf{1}$ locally in measure as $n\to\infty$, Proposition 4.3.13(v) implies that

$$\begin{aligned}\|\mathbf{1} - e_n\|_{LS(\mathcal{M})} &= \|\mathbf{1} - E_{|x|}[0,n] \wedge E_{|y|}[0,n]\|_{LS(\mathcal{M})}\\ &= \|(\mathbf{1} - E_{|x|}[0,n]) \vee (\mathbf{1} - E_{|y|}[0,n])\|_{LS(\mathcal{M})}\\ &\leq \|\mathbf{1} - E_{|x|}[0,n]\|_{LS(\mathcal{M})} + \|\mathbf{1} - E_{|y|}[0,n]\|_{LS(\mathcal{M})} \to 0,\end{aligned}$$

as $n\to\infty$.

By the spectral theorem, we have that $xe_n, ye_n, (x+y)e_n \in \mathcal{M}$ for any $n \in \mathbb{N}$. By Lemma 5.2.3, we have that

$$\widetilde{\delta}(x+y)=\lim_{n\to\infty}\delta((x+y)e_n)=\lim_{n\to\infty}(\delta(xe_n)+\delta(ye_n))=\widetilde{\delta}(x)+\widetilde{\delta}(y).$$

Similarly,

$$\widetilde{\delta}(\lambda x)=\lambda\widetilde{\delta}(x).$$

This proves that the mapping $\widetilde{\delta}: LS(\mathcal{M})\to LS(\mathcal{M})$ is linear.

Next, we show that $\widetilde{\delta}: LS(\mathcal{M})\to LS(\mathcal{M})$ is a derivation. Let $x,y\in LS(\mathcal{M})$ be arbitrary. We define

$$f_n=E_{|xy|}[0,n]\wedge E_{|x^*|}[0,n]\wedge E_{|y|}[0,n].$$

Clearly, $\{f_n\}_{n\in\mathbb{N}}\subset P(\mathcal{M})$ is an increasing sequence. Since xy, x^* and y are locally measurable operators, Remark 4.3.12 implies that $E_{|xy|}[0,n]$, $E_{|x^*|}(n,\infty)$, $E_{|y|}(n,\infty)\to 0$ locally in measure as $n\to\infty$. Therefore, by Proposition 4.3.13(v), we have that

$$\begin{aligned}\|\mathbf{1}-f_n\|_{LS(\mathcal{M})}&=\left\|E_{|xy|}(n,\infty)\vee E_{|x^*|}(n,\infty)\vee E_{|y|}(n,\infty)\right\|_{LS(\mathcal{M})}\\&\le\left\|E_{|xy|}(n,\infty)\right\|_{LS(\mathcal{M})}+\left\|E_{|x^*|}(n,\infty)\right\|_{LS(\mathcal{M})}\\&\quad+\left\|E_{|y|}(n,\infty)\right\|_{LS(\mathcal{M})}\to 0,\quad n\to\infty.\end{aligned}$$

By the spectral theorem, we have that $f_nx, yf_n, xyf_n\in\mathcal{M}$ for every $n\in\mathbb{N}$. Therefore, by Lemma 5.2.2 and Lemma 5.2.3, we have that

$$\widetilde{\delta}(xy)=\lim_{n\to\infty}\delta(xyf_n),\tag{5.10}$$

and

$$\widetilde{\delta}(x)=\lim_{n\to\infty}\delta(f_nx),\quad\widetilde{\delta}(y)=\lim_{n\to\infty}\delta(yf_n),\tag{5.11}$$

where the limits are understood in $LS(\mathcal{M})$.

By Lemma 5.2.2, we have that

$$\widetilde{\delta}(xy)=\lim_{n\to\infty}\delta(xyf_n)=\lim_{n\to\infty}\delta(f_nxyf_n),\tag{5.12}$$

with respect to the local measure topology.

Since $xf_n, yf_n, xyf_n\in\mathcal{M}$ and $\delta:\mathcal{M}\to LS(\mathcal{M})$ is a derivation, it follows that

$$\begin{aligned}\delta(f_nxyf_n)&=\delta(f_nx)\cdot yf_n+f_nx\cdot\delta(yf_n)\\&=\delta(f_nx)\cdot y+x\cdot\delta(yf_n)-\delta(f_nx)y(\mathbf{1}-f_n)-(\mathbf{1}-f_n)x\delta(yf_n).\end{aligned}$$

By Proposition 4.3.18(vi), the last two summands tend to 0 locally in measure as $n\to\infty$. Therefore, by (5.12) and (5.11), we conclude that

$$\begin{aligned}\widetilde{\delta}(xy) &= \lim_{n\to\infty} \delta(f_n xy f_n)\\ &= \lim_{n\to\infty} (\delta(f_n x)\cdot y + x\cdot \delta(y f_n))\\ &= \widetilde{\delta}(x)y + x\widetilde{\delta}(y),\end{aligned}$$

proving that $\widetilde{\delta} : LS(\mathcal{M}) \to LS(\mathcal{M})$ is a derivation.

Finally, to show that $\widetilde{\delta}|_{\mathcal{M}} = \delta$ we note that for $x \in \mathcal{M}$, there exists $n \in \mathbb{N}$, such that $xE_{|x|}[0,k] = x$ for all $k \geq n$, implying that

$$\widetilde{\delta}(x) = \lim_{n\to\infty} \delta(xE_{|x|}[0,n]) = \delta(x),$$

proving that $\widetilde{\delta}$ is an extension of δ. □

Lemma 5.2.5. *Let $\mathcal{M}$ be a von Neumann algebra and let $\mathcal{A}$ be a subalgebra in $LS(\mathcal{M})$, such that $\mathcal{M} \subset \mathcal{A}$. Assume that $\delta_1, \delta_2 : \mathcal{A} \to LS(\mathcal{M})$ are derivations, such that $\delta_1|_{\mathcal{M}} = \delta_2|_{\mathcal{M}}$. Then $\delta_1 = \delta_2$.*

Proof. Let $x \in \mathcal{A}$ and set $p_n = E_{|x|}(n,\infty)$, $n \in \mathbb{N}$. By Remark 4.3.12, we have that $p_n \to 0$ locally in measure as $n \to \infty$. By Lemma 5.1.13, we have that $\delta_j(p_n) \to 0$ locally in measure, $j = 1, 2$.

By the Leibniz rule, for any $n \in \mathbb{N}$, we have

$$\delta_j(x(\mathbf{1} - p_n)) = \delta_j(x)(\mathbf{1} - p_n) - x\delta_j(p_n), \quad j = 1, 2.$$

Since $p_n, \delta_j(p_n) \to 0$ locally in measure, it follows that $\delta_j(x(\mathbf{1} - p_n)) \to \delta_j(x)$ locally in measure as $n \to \infty$. Since $x(\mathbf{1} - p_n) \in \mathcal{M}$ and $\delta_1|_{\mathcal{M}} = \delta_2|_{\mathcal{M}}$, it follows that

$$\delta_1(x) = \lim_{n\to\infty} \delta_1(x(\mathbf{1} - p_n)) = \lim_{n\to\infty} \delta_2(x(\mathbf{1} - p_n)) = \delta_2(x).$$

Since $x \in \mathcal{A}$ is arbitrary, the assertion follows. □

Using now Proposition 5.2.4, we prove the main result of this section about extension of derivations.

Theorem 5.2.6. *Let $\mathcal{A}$ be a subalgebra of $LS(\mathcal{M})$, such that $\mathcal{M} \subset \mathcal{A}$ and let $\delta : \mathcal{A} \to LS(\mathcal{M})$ be a derivation. Then there exists a unique derivation $\widetilde{\delta} : LS(\mathcal{M}) \to LS(\mathcal{M})$, such that $\widetilde{\delta}(x) = \delta(x)$ for all $x \in \mathcal{A}$.*

Proof. Since $\mathcal{M} \subset \mathcal{A}$, the restriction $\delta|_{\mathcal{M}}$ of the derivation δ to $\mathcal{M}$ is a well-defined derivation from $\mathcal{M}$ into $LS(\mathcal{M})$. Hence, by Proposition 5.2.4, there exists a derivation $\widetilde{\delta} : LS(\mathcal{M}) \to LS(\mathcal{M})$, such that $\widetilde{\delta}|_{\mathcal{M}} = \delta|_{\mathcal{M}}$. By Lemma 5.2.5, we have $\widetilde{\delta}|_{\mathcal{A}} = \delta$. The uniqueness of extension $\widetilde{\delta} : LS(\mathcal{M}) \to LS(\mathcal{M})$ follows again from Lemma 5.2.5 (this time with $\mathcal{A} = LS(\mathcal{M})$). □

We conclude this section with an application of the commutator estimates established in Theorem 3.6.11 to derivations on algebras of locally measurable operators.

Theorem 5.2.7. *Let $\mathcal{A}$ be an absolutely solid $*$-subalgebra of $LS(\mathcal{M})$, such that $\mathcal{M} \subset \mathcal{A}$.*
(i) *If δ is a spatial derivation on $\mathcal{A}$, then δ is inner;*
(ii) *If all derivations on $LS(\mathcal{M})$ are inner, then all derivations on $\mathcal{A}$ are inner.*

Proof. (i) By Proposition 5.1.4, without loss of generality, we can assume that δ is a self-adjoint derivation. By assumption, there exists $a = a^* \in LS(\mathcal{M})$, such that $\delta(x) = [ia, x]$ for any $x \in \mathcal{A}$. By Theorem 3.6.11, there exists $c = c^* \in \mathcal{Z}(LS(\mathcal{M}))$ and unitary $u \in \mathcal{M}$, such that

$$\big|[a,u]\big| \geq \frac{1}{2}|a - c|. \tag{5.13}$$

Since $\mathcal{M} \subset \mathcal{A}$, it follows that $[a, u] = -i\delta(u) \subset \mathcal{A}$, and so, by Proposition 3.3.13(ii), we have that $|[a, u]| \in \mathcal{A}$. Since $\mathcal{A}$ is an absolutely solid algebra, inequality (5.13) implies that $b = a - c \in \mathcal{A}$. Since $c \in \mathcal{Z}(LS(\mathcal{M}))$, it follows that

$$\delta(x) = [ia, x] = \big[i(a - c), x\big] = [ib, x].$$

Thus, δ is an inner derivation.

(ii) Assume now that all derivations on $LS(\mathcal{M})$ are inner, and let $\delta : \mathcal{A} \to \mathcal{A}$ be a derivation. Since $\mathcal{M} \subset \mathcal{A}$, Theorem 5.2.6 implies that there exists a unique derivation $\delta_1 : LS(\mathcal{M}) \to LS(\mathcal{M})$, such that $\delta_1(x) = \delta(x)$ for all $x \in \mathcal{A}$. By assumption, there exists $a \in LS(\mathcal{M})$, such that $\delta(x) = \delta_1(x) = [a, x]$, i. e., δ is a spatial derivation on $\mathcal{A}$. By part (i), the derivation δ is inner. □

5.3 Derivations on algebras and ideals of bounded operators

In this section, we recall some classical results for derivations $\delta : \mathcal{A} \to \mathcal{A}$ for a C^*-algebra (or more precisely, von Neumann algebra) $\mathcal{A}$ and study derivations with values in ideals of a von Neumann algebra $\mathcal{M}$.

The following theorem establishes norm continuity of every derivation on a C^*-algebra $\mathcal{A}$.

Theorem 5.3.1 ([134, Chapter IV, Section 1, 4.1.3]). *Every derivation on a C^*-algebra is norm continuous.*

If $\mathcal{A}$ is a commutative algebra, then every inner derivation on $\mathcal{A}$ is identically zero. In fact, there are no other derivations on a commutative C^*-algebra.

Theorem 5.3.2 ([134, Chapter IV, Section 1, 4.1.2]). *If $\mathcal{A}$ is a commutative C^*-algebra and δ is a derivation on $\mathcal{A}$, then $\delta(x) = 0$ for all $x \in \mathcal{A}$.*

In the case when a C^*-algebra $\mathcal{A}$ is noncommutative, there may exist non-inner derivations. Let us give the following example from [134, Chapter IV, Section 1]. Let

H be an infinite-dimensional separable Hilbert space and let $K(H)$ be the closed two-sided ideal of all compact operators in the algebra $\mathcal{B}(H)$. Consider the C^*-subalgebra $\mathcal{A} = K(H) + \mathbb{C}\mathbf{1}$ in $\mathcal{B}(H)$ and set $\delta_a(x) = [a, x] = ax - xa$ for all $x \in \mathcal{A}$ and some $a \in \mathcal{B}(H)$. It is clear that δ_a is a derivation on the C^*-algebra $\mathcal{A}$, in addition, the derivation δ_a is inner on $\mathcal{A}$ if and only if $a \in \mathcal{A}$ [134, Chapter IV, Section 1, 4.1.8].

The following theorem permits extension of derivations defined on a C^*-subalgebra of $\mathcal{A}$ in $\mathcal{B}(H)$ up to a derivation, defined on the (wo)-closure $\overline{\mathcal{A}}^{wo}$ of the C^*-subalgebra $\mathcal{A}$.

Theorem 5.3.3 ([134, Chapter IV, Section 1, 4.1.4]). *Let $\mathcal{A}$ be a C^*-subalgebra in $\mathcal{B}(H)$ and let δ be a derivation on $\mathcal{A}$. Then δ can be extended up to a derivation δ_1 on the (wo)-closure $\overline{\mathcal{A}}^{wo}$ and δ_1 is (wo)-continuous on $\overline{\mathcal{A}}^{wo}$.*

Theorem 5.3.3 immediately implies that every derivation on a von Neumann subalgebra $\mathcal{M} \subseteq \mathcal{B}(H)$ is (wo)-continuous. This also follows from the following description of such derivations, which follows from a combination of [134, Chapter IV, Section 1, 4.1.6] and [166, Theorem 2].

Theorem 5.3.4. *Let δ be a derivation on a von Neumann algebra $\mathcal{M}$. Then δ is inner, i. e., there exists an element $a \in \mathcal{M}$, such that $\delta(x) = \delta_a(x) = [a, x]$ for all $x \in \mathcal{M}$. Moreover, an element $a \in \mathcal{M}$ can be chosen to satisfy the equality $\|a\|_{\mathcal{M}} = \frac{1}{2}\|\delta\|_{\mathcal{M}\to\mathcal{M}}$.*

Theorem 5.3.3 and Theorem 5.3.4 imply the following.

Corollary 5.3.5 ([134, Chapter IV, Section 1, 4.1.7]). *Let $\mathcal{A}$ be a C^*-subalgebra in $\mathcal{B}(H)$ and let δ be a derivation on $\mathcal{A}$. Then there exists an element $a \in \overline{\mathcal{A}}^{wo}$, such that $\delta(x) = [a, x]$ for all $x \in \mathcal{A}$.*

Next, we prove that any noninner derivation $\delta : LS(\mathcal{M}) \to LS(\mathcal{M})$ can not have its image of $\mathcal{M}$ to be contained in $\mathcal{M}$.

Proposition 5.3.6. *Let $\mathcal{M}$ be a von Neumann algebra and let $\mathcal{A}$ be a subalgebra of $LS(\mathcal{M})$, such that $\mathcal{M} \subset \mathcal{A}$. If $\delta : \mathcal{A} \to \mathcal{A}$ is a derivation, such that $\delta(\mathcal{M}) \subset \mathcal{M}$, then δ is inner.*

Proof. Since $\delta : \mathcal{A} \to \mathcal{A}$ is a derivation, $\mathcal{M} \subset \mathcal{A}$ and $\delta(\mathcal{M}) \subset \mathcal{M}$, it follows that the restriction $\delta|_{\mathcal{M}}$ is a derivation on $\mathcal{M}$. By Theorem 5.3.4, there exists $a \in \mathcal{M}$, such that

$$\delta(x) = [a, x], \quad x \in \mathcal{M}.$$

Define a derivation $\widetilde{\delta} : \mathcal{A} \to LS(\mathcal{M})$ by setting

$$\widetilde{\delta}(x) = [a, x], \quad x \in \mathcal{A}.$$

Then $\widetilde{\delta}|_{\mathcal{M}} = \delta|_{\mathcal{M}}$ and, therefore, Lemma 5.2.5 implies that $\delta = \widetilde{\delta}$, showing that δ is inner. □

We now move onto the study of derivations with values in ideals of a von Neumann algebra $\mathcal{M}$. The commutator estimates established in Theorem 3.6.11 allow us to show that any derivation from $\mathcal{M}$ into any (two-sided) ideal is necessarily inner. We emphasize that the following theorem does not assume any topological structure on the ideal.

Theorem 5.3.7. *Let $\mathcal{M}$ be an arbitrary von Neumann algebra and let $\mathcal{I}$ be an arbitrary (two-sided) ideal in $\mathcal{M}$. Then every derivation $\delta : \mathcal{M} \to \mathcal{I}$ is inner, i. e., there exists $a \in \mathcal{I}$ such that $\delta(x) = [a, x]$ for every $x \in \mathcal{M}$.*

Proof. Let $\delta : \mathcal{M} \to \mathcal{I}$ be a derivation. Considering δ as a derivation on $\mathcal{M}$, Theorem 5.3.4 implies that there exists $a \in \mathcal{M}$ such that $\delta(x) = [a, x]$ for every $x \in \mathcal{M}$.

By assumption, $[a, \mathcal{M}] = \delta(\mathcal{M}) \subset \mathcal{I}$. Since any two-sided ideal of $\mathcal{M}$ is necessarily a $*$-ideal (see, e. g., [93, Proposition 6.8.9]), it follows that $[a^*, x] = -[a, x^*]^* \in \mathcal{I}$ for any $x \in \mathcal{M}$. Hence, $[\mathrm{Re}(a), x] = [a + a^*, x]/2 \in \mathcal{I}$ and $[\mathrm{Im}(a), x] = [a - a^*, x]/2i \in \mathcal{I}$ for all $x \in \mathcal{M}$.

Taking $\varepsilon = 1/2$ in Theorem 3.6.11, we obtain that there exist $c_1, c_2 \in \mathcal{Z}(LS(\mathcal{M}))$ and unitary operators $u_1, u_2 \in \mathcal{M}$, such that

$$2\big|[a_i, u_i]\big| \geq |a_i - c_i|, \quad i = 1, 2.$$

By Corollary 3.3.9, there exist $b_i \in \mathcal{M}$ with $\|b_i\|_{\mathcal{M}} \leq 1$, such that $|a_i - c_i| = 2b_i^* |[a_i, u_i]| b_i$. Since $[a_i, u_i] \in \mathcal{I}$ and $\mathcal{I}$ is an ideal, it follows that $d_i := a_i - c_i \in \mathcal{I}$, $i = 1, 2$. Therefore, $d = d_1 + id_2 \in \mathcal{I}$. Since c_1, c_2 are central elements from $LS(\mathcal{M})$, it follows that $\delta(x) = [a, x] = [d, x]$ for all $x \in \mathcal{M}$. Thus, δ is inner. □

Our next step is derivations $\delta : \mathcal{I} \to \mathcal{I}$ for a Banach ideal $\mathcal{I}$ in $\mathcal{M}$ and our primary objective here is to show that any such derivation is necessarily spatial.

Definition 5.3.8. Let $\mathcal{I}$ be a (two-sided) ideal in $\mathcal{M}$ equipped with a complete norm $\|\cdot\|_{\mathcal{I}}$. We say that $(\mathcal{I}, \|\cdot\|_{\mathcal{I}})$ is a *Banach ideal* if $\|axb\|_{\mathcal{I}} \leq \|a\|_{\mathcal{M}} \|b\|_{\mathcal{M}} \|x\|_{\mathcal{I}}$ for any $x \in \mathcal{I}$ and $a, b \in \mathcal{M}$.

Note that any Banach ideal $\mathcal{M}$ is necessarily a normed $\mathcal{M}$-bimodule (see Definition 3.1.9). In particular, Proposition 3.1.10 holds for a Banach ideal $(\mathcal{I}, \|\cdot\|_{\mathcal{I}})$.

We start with some lemmas. The first lemma complements Theorem 5.3.1.

Lemma 5.3.9. *Let $(\mathcal{I}, \|\cdot\|_{\mathcal{I}})$ be a Banach ideal in a von Neumann algebra $\mathcal{M}$. Then every derivation $\delta : \mathcal{I} \to \mathcal{I}$ is continuous in the norm $\|\cdot\|_{\mathcal{I}}$.*

Proof. Without loss of generality, we can assume that $\delta^* = \delta$. It suffices to show that the graph of δ is closed.

Suppose the contrary that the graph of δ is not closed. We claim that there exists a projection $r \in \mathcal{M}$, such that δ is well-defined on $r\mathcal{M}r$.

Since the graph of δ is not closed, there exist a sequence $\{x_n\}_{n\in\mathbb{N}} \subset \mathcal{I}$ of self-adjoint elements and a self-adjoint element $0 \neq x \in \mathcal{I}$, such that $x_n \to 0$ and $\delta(x_n) \to x$ in $\mathcal{I}$ as $n \to \infty$.

Let x_+ and x_- be the positive and negative parts of x, respectively. Without loss of generality, we can assume that $x_+ \neq 0$, since otherwise we can consider the sequence $\{-x_n\}_{n\in\mathbb{N}}$. Let $p = E_x(0,+\infty)$. Since $\|x_n p\|_{\mathcal{I}} \le \|x_n\|_{\mathcal{I}} \to 0$, it follows that $x_n p \to 0$ in $\mathcal{I}$. Furthermore,

$$\delta(x_n p) = \delta(x_n)p + x_n\delta(p) = (\delta(x_n) - x)p + xp + x_n\delta(p) \to xp = x_+$$

in $\mathcal{I}$ as $n \to \infty$. Replacing the sequence $\{x_n\}_{n\in\mathbb{N}}$ with the sequence $\{x_n p\}_{n\in\mathbb{N}}$, we may assume without loss of generality that $x \ge 0$.

Choose $\lambda > 0$ such that $r = E_x(\lambda, +\infty)$ is nonzero. By spectral theory, for the function $f(t) = \frac{1}{t^{1/2}}\chi_{(\lambda,\infty)}(t)$, $t \in \mathbb{R}$, we have that $y := f(x) \in \mathcal{M}$ and $r = y^* x y$. Since $x \in \mathcal{I}$, it follows that $r \in \mathcal{I}$ and, therefore, $y = yr \in \mathcal{I}$. Since $y^* x_n y \to 0$ and

$$\delta(y^* x_n y) = \delta(y^*)x_n y + y^*\delta(x_n)y + y^* x_n \delta(y) \to y^* x y = r$$

in $\mathcal{I}$ as $n \to \infty$, we can replace the sequence $\{x_n\}_{n\in\mathbb{N}}$ with the sequence $\{y^* x_n y\}_{n\in\mathbb{N}}$, and assume without loss of generality that $x = r$ and that $x_n = r x_n r$.

Since $r \in \mathcal{I}$, the derivation δ is well-defined on $r\mathcal{M}r$. Consider the von Neumann algebra $r\mathcal{M}r$ and the derivation $\delta^{(r)} : r\mathcal{M}r \to r\mathcal{M}r$ defined by

$$\delta^{(r)}(a) = r\delta(rar)r, \quad a \in r\mathcal{M}r.$$

Since $r\mathcal{M}r \subset \mathcal{I}$, such a derivation is well-defined.

By Theorem 5.3.4, there exists $d \in r\mathcal{M}r$ such that

$$\delta^{(r)}(a) = [d, a], \quad a \in r\mathcal{M}r.$$

We have

$$\|\delta^{(r)}(x_n)\|_{\mathcal{I}} = \|dx_n - x_n d\|_{\mathcal{I}} \le 2\|d\|_{\mathcal{M}}\|x_n\|_{\mathcal{I}},$$

and so $\|\delta^{(r)}(x_n)\|_{\mathcal{I}} \to 0$. On the other hand,

$$\|\delta^{(r)}(x_n) - r\|_{\mathcal{I}} = \|r\delta(x_n)r - r\|_{\mathcal{I}} \le \|r\|_{\mathcal{M}}^2 \|\delta(x_n) - r\|_{\mathcal{I}} \to 0.$$

Since $r \neq 0$, this is a contradiction. Therefore, $\delta : \mathcal{I} \to \mathcal{I}$ is norm continuous. □

In the following lemma, we show that any derivation on a Banach ideal is necessarily continuous in the norm $\|\cdot\|_{\mathcal{M}}$.

Lemma 5.3.10. *Let $\mathcal{I}$ be a Banach ideal in a von Neumann algebra $\mathcal{M}$. Every derivation $\delta : (\mathcal{I}, \|\cdot\|_{\mathcal{M}}) \longrightarrow (\mathcal{I}, \|\cdot\|_{\mathcal{M}})$ is bounded and*

$$\|\delta\|_{(\mathcal{I},\|\cdot\|_{\mathcal{M}})\longrightarrow(\mathcal{I},\|\cdot\|_{\mathcal{M}})} \le 2\|\delta\|_{\mathcal{I}\to\mathcal{I}}.$$

Proof. By Lemma 5.3.9, $\delta : (\mathcal{I}, \|\cdot\|_{\mathcal{I}}) \longrightarrow (\mathcal{I}, \|\cdot\|_{\mathcal{I}})$ is bounded. Without loss of generality, we can assume that

$$\|\delta(y)\|_{\mathcal{I}} \leq \|y\|_{\mathcal{I}}, \quad y \in \mathcal{I}. \tag{5.14}$$

Let $x \in \mathcal{I}$ be such that $\delta(x) \neq 0$. Fix $0 < \varepsilon < 1$ and set

$$p = E_{|\delta(x)|}((1-\varepsilon)\|\delta(x)\|_{\mathcal{M}}, \|\delta(x)\|_{\mathcal{M}}].$$

By spectral theory, we have

$$0 \leq (1-\varepsilon)\|\delta(x)\|_{\mathcal{M}} p \leq |\delta(x)|p. \tag{5.15}$$

Since $\delta(x) \in \mathcal{I}$ and $p \in \mathcal{M}$, Proposition 3.1.6(i) implies that $|\delta(x)|p \in \mathcal{I}$. Inequality (5.15) combined with Proposition 3.3.10 implies that $p \in \mathcal{I}$ and

$$(1-\varepsilon)\|\delta(x)\|_{\mathcal{M}}\|p\|_{\mathcal{I}} \leq \||\delta(x)|p\|_{\mathcal{I}} = \|\delta(x)p\|_{\mathcal{I}}. \tag{5.16}$$

The Leibniz rule and (5.14) imply that

$$\begin{aligned}\|\delta(x)p\|_{\mathcal{I}} &= \|\delta(xp) - x\delta(p)\|_{\mathcal{I}} \leq \|\delta(xp)\|_{\mathcal{I}} + \|x\delta(p)\|_{\mathcal{I}} \\ &\leq \|xp\|_{\mathcal{I}} + \|x\|_{\mathcal{M}}\|p\|_{\mathcal{I}} \leq 2\|x\|_{\mathcal{M}}\|p\|_{\mathcal{I}}.\end{aligned}$$

Combining with (5.16), we infer that

$$(1-\varepsilon)\|\delta(x)\|_{\mathcal{M}}\|p\|_{\mathcal{I}} \leq 2\|x\|_{\mathcal{M}}\|p\|_{\mathcal{I}}.$$

Since $p \neq 0$, it follows that

$$(1-\varepsilon)\|\delta(x)\|_{\mathcal{M}} \leq 2\|x\|_{\mathcal{M}}.$$

Since $\varepsilon > 0$ is arbitrarily small, it follows that

$$\|\delta(x)\|_{\mathcal{M}} \leq 2\|x\|_{\mathcal{M}}.$$

Thus, $\delta : (\mathcal{I}, \|\cdot\|_{\mathcal{M}}) \longrightarrow (\mathcal{I}, \|\cdot\|_{\mathcal{M}})$ is bounded and

$$\|\delta\|_{(\mathcal{I},\|\cdot\|_{\mathcal{M}})\longrightarrow(\mathcal{I},\|\cdot\|_{\mathcal{M}})} \leq 2\|\delta\|_{\mathcal{I}\to\mathcal{I}},$$

as required. □

We now conclude this section by showing that any derivation on a Banach ideal is necessarily spatial.

Theorem 5.3.11. *Let $\mathcal{I}$ be a Banach $*$-ideal in the von Neumann algebra $\mathcal{M}$ and let $\delta : \mathcal{I} \to \mathcal{I}$ be a derivation. There exists an element d in the ultraweak closure of $\mathcal{I}$, such that $\delta(x) = [d, x]$ for all $x \in \mathcal{I}$. Furthermore, d can be chosen so that*

$$\|d\|_{\mathcal{M}} \le \|\delta\|_{\mathcal{I}\to\mathcal{I}}.$$

Proof. Denote by $\mathcal{I}_1$ and $\mathcal{I}_2$ the closure of the ideal $\mathcal{I}$ with respect to the uniform and ultraweak operator topology, respectively. Clearly, $\mathcal{I}_1$ is a C^*-subalgebra in $\mathcal{M}$ and $\mathcal{I}_2$ is a von Neumann subalgebra in $\mathcal{M}$ such that $\mathcal{I} \subset \mathcal{I}_1 \subset \mathcal{I}_2$.

By Lemma 5.3.10, δ extends to a linear mapping (also denoted by δ) from $\mathcal{I}_1$ to $\mathcal{I}_1$ continuous in the uniform norm. For $x, y \in \mathcal{I}_1$, fix sequences $\{x_n\}_{n\in\mathbb{N}}, \{y_n\}_{n\in\mathbb{N}} \subset \mathcal{I}$, such that $x_n \to x$ and $y_n \to y$ in the uniform norm as $n \to \infty$. Clearly, $x_n y_n \to xy$ in the uniform norm as $n \to \infty$. Since δ is continuous in the uniform norm, it follows that

$$\delta(xy) = \lim_{n\to\infty} \delta(x_n y_n) = \lim_{n\to\infty} \delta(x_n) y_n + x_n \delta(y_n).$$

Since δ is continuous in the uniform norm, it follows that $\delta(x_n)y_n \to \delta(x)y$ and $x_n\delta(y_n) \to x\delta(y)$ in the uniform norm. Thus,

$$\delta(xy) = \delta(x)y + x\delta(y).$$

Since $x, y \in \mathcal{I}_1$ are arbitrary, it follows that $\delta : \mathcal{I}_1 \to \mathcal{I}_1$ is a derivation.

By Theorem 5.3.3, δ extends to a derivation $\delta : \mathcal{I}_2 \to \mathcal{I}_2$. Since $\mathcal{I}_2$ is a von Neumann algebra, it follows from Theorem 5.3.4 that there exists $d \in \mathcal{I}_2$, such that

$$\delta(x) = [d, x], \quad x \in \mathcal{I}_2.$$

In particular, the latter formula holds for $x \in \mathcal{I}$.

Furthermore, since δ is inner, it is automatically norm continuous in $\mathcal{I}_2$, and again appealing to Theorem 5.3.4, one can choose d such that

$$\|d\|_{\mathcal{M}} = \frac{1}{2}\|\delta\|_{\mathcal{I}_2\to\mathcal{I}_2} = \frac{1}{2}\|\delta\|_{\mathcal{I}_1\to\mathcal{I}_1} \le \|\delta\|_{\mathcal{I}\to\mathcal{I}},$$

where the last inequality follows from Lemma 5.3.10. □

5.4 Optimal λ-elements associated with a self-adjoint derivation

The main objective in this section is to introduce the so-called optimal λ-elements for a self-adjoint derivation $\delta : LS(\mathcal{M}) \to LS(\mathcal{M})$. These optimal λ-elements comprise the main technical tool in our study of continuous derivations in Section 5.5 below.

As we will show in Theorem 5.4.9 below, an optimal λ-element of a self-adjoint derivation δ give rise to a special projection q, such that δ reduced by its orthocomplement $\mathbf{1} - q$ is a derivation, which maps $\mathcal{M}$ into itself. Thus, roughly speaking, optimal

λ-elements give us control on how many bounded operators from $\mathcal{M}$ are mapped into bounded operators under δ. As we have seen in Proposition 5.3.6, if δ is a derivation on $LS(\mathcal{M})$, such that $\delta(\mathcal{M}) \subset \mathcal{M}$, then δ is necessarily inner. Thus, optimal λ-elements are the key, which allows us to prove that a given derivation is inner. As we will see in Section 5.5, for a derivation δ on $LS(\mathcal{M})$, which is continuous in the local measure topology, there exists sufficiently many optimal λ-elements.

In the present section, we shall introduce optimal λ-elements of self-adjoint derivations on any absolutely solid $*$-subalgebra of $LS(\mathcal{M})$. We will also establish their essential property mentioned above.

We start this section by introducing the sum of uniformly bounded family of pairwise disjoint elements in elements in $\mathcal{M}$. Recall that $x, y \in \mathcal{M}$ are said to be disjoint if $xy = yx = 0$.

Lemma 5.4.1. *Let $\{x_n\}_{n\in\mathbb{N}}$ be a uniformly bounded family of pairwise disjoint self-adjoint elements from $\mathcal{M}$. Then there exists unique element $x \in \mathcal{M}$, denoted by $\sum_{n=1}^{\infty} x_n$, such that $xs(x_n) = x_n$ for all $n \in \mathbb{N}$ and $\sup_{n\in\mathbb{N}} s(x_n) = s(x)$. In this case, $\|x\|_{\mathcal{M}} \le \sup_{n\in\mathbb{N}} \|x_n\|_{\mathcal{M}}$.*

Proof. We claim that the series $\sum_{n=1}^{\infty} x_n$ converges in the strong operator topology. Let $\xi \in H$. Since $x_n, n \in \mathbb{N}$, are pairwise disjoint and self-adjoint, we have that $\{s(x_n)\}_{n\in\mathbb{N}}$ is a family of pairwise orthogonal projections. Furthermore, $\langle s(x_n)\xi, s(x_m)\xi\rangle = \langle \xi, s(x_n)s(x_m)\xi\rangle = 0$ for all $n \neq m$, i. e., the vectors $s(x_n)\xi, n \in \mathbb{N}$, (and so the vectors $x_n\xi, n \in \mathbb{N}$) are pairwise orthogonal.

Therefore, for any $k \in \mathbb{N}$, we have

$$\begin{aligned}\Big\|\sum_{n=1}^{k} x_n\xi\Big\|_H^2 &= \sum_{n=1}^{k} \|x_n s(x_n)\xi\|_H^2 \le \sup_{n=1,\dots,k} \|x_n\|_{\mathcal{M}}^2 \cdot \sum_{n=1}^{k} \|s(x_n)\xi\|_H^2 \\ &\le \sup_{n\in\mathbb{N}} \|x_n\|_{\mathcal{M}}^2 \Big\|\sum_{n=1}^{k} s(x_n)\xi\Big\|_H^2 = \sup_{n\in\mathbb{N}} \|x_n\|_{\mathcal{M}}^2 \Big\|\sup_{n=1,\dots,k} s(x_n)\xi\Big\|_H^2 \\ &\le \sup_{n\in\mathbb{N}} \|x_n\|_{\mathcal{M}}^2 \|\xi\|_H^2.\end{aligned}$$

Thus, the series $x := \sum_{n=1}^{\infty} x_n$ converges in the strong operator topology and $\|x\|_{\mathcal{M}} \le \sup_{n\in\mathbb{N}} \|x_n\|_{\mathcal{M}}$. It follows from the definition that $xs(x_n) = x_n$ and $s(x) = \sup_{n\in\mathbb{N}} s(x_n)$.

It remains to show that the operator x satisfying these properties is unique. Assume that y is another self-adjoint operator in $\mathcal{M}$ satisfying $ys(x_n) = x_n$ for all $n \in \mathbb{N}$ and $\sup_{n\in\mathbb{N}} s(x_n) = s(y)$. Then $(x-y)s(x_n) = x_n - x_n = 0$ for any $n \in \mathbb{N}$ and, therefore, $s(x) = \sup_{n\in\mathbb{N}} s(x_n) \le (s_r(x-y))^{\perp}$. Thus,

$$x - y = xs(x) - ys(y) = xs(x) - ys(x) = (x-y)s(x) = 0,$$

which proves uniqueness of x. □

In view of Lemma 5.4.1, the following definition makes sense.

Definition 5.4.2. Let $\{x_n\}_{n\in\mathbb{N}}$ be a uniformly bounded family of pairwise disjoint self-adjoint elements from $\mathcal{M}$. The *disjoint sum* of pairwise disjoint family $\{x_n\}$ is the unique element $x = \sum_{n=1}^{\infty} x_n$, such that $xs(x_n) = x_n$ for all $n \in \mathbb{N}$ and $\sup_{n\in\mathbb{N}} s(x_n) = s(x)$.

Remark 5.4.3. *We note that the assumption that the family $\{x_n\}$ is uniformly bounded can not be omitted, in general. Indeed, let $\mathcal{M} = \mathcal{B}(H)$ and let $\{p_n\}_{n\in\mathbb{N}}$ be a family of pairwise orthogonal projections in H and let $x_n = np_n$. Then $\{x_n\}_{n\in\mathbb{N}}$ is a pairwise disjoint family of self-adjoint elements, but the series $\sum_{n=1}^{\infty} x_n$ does not converge in $\mathcal{M}$. Note that, in this case, $LS(\mathcal{B}(H)) = \mathcal{B}(H)$, and hence, the notion of disjoint sum cannot be defined for an arbitrary pairwise disjoint family even in terms of locally measurable operators. We note, however, that for a commutative von Neumann algebra $\mathcal{M}$, the disjoint sum of a pairwise disjoint family is well-defined (see Section* 6.1).

We will repeatedly use the following properties of the disjoint sum of uniformly bounded family of pairwise disjoint operators.

Remark 5.4.4. *Let $a \in LS(\mathcal{M})$ and let $\{x_n\}_{n\in\mathbb{N}}$ be a uniformly bounded family of pairwise disjoint self-adjoint operators in $\mathcal{M}$. Let $x = \sum_{n=1}^{\infty} x_n$ be the disjoint sum of $\{x_n\}$.*
(i) *If $ax_n = 0$ for all $n \in \mathbb{N}$, except for $n = k$, then $ax = ax_k$;*
(ii) *If $ax_n = x_n a \geq 0$ for all $n \in \mathbb{N}$, then $ax = xa \geq 0$.*

We now introduce the notion of optimal λ-elements for a self-adjoint derivation (see Definition 5.1.3 for the notion of self-adjoint derivations). We note that for a self-adjoint derivation δ and a self-adjoint operator $x \in LS(\mathcal{M})$ (in the domain of δ), the operator $\delta(x)$ is self-adjoint, too. In particular, the spectral measure $E_{\delta(x)}$ is well-defined.

Definition 5.4.5. Let $\mathcal{M}$ be a von Neumann algebra with the unit ball $\mathcal{M}_1$ and let $\delta : \mathcal{A} \to \mathcal{A}$ be a self-adjoint derivation on an absolutely solid $*$-subalgebra $\mathcal{A}$ of $LS(\mathcal{M})$. For a given scalar $\lambda > 0$, we say that:
(i) A self-adjoint element $x \in \mathcal{A} \cap \mathcal{M}_1$ is a *λ-element* (corresponding to δ) if $s(x) \leq E_{\delta(x)}[\lambda, \infty)$. The projection $s_\delta(x) := s(x) \vee s(\delta(x)) \vee s(\delta^2(x))$ is called the *δ-support* of the λ-element x;
(ii) Two λ-elements x and y are said to be *δ-disjoint* if

$$xy = x\delta(y) = \delta(x)\delta(y) = 0;$$

(iii) A λ-element x is called *optimal* if there is no nonzero λ-element $y \in \mathcal{A}$, which is δ-disjoint to x.

We will show in Proposition 5.5.3 below that continuity of a self-adjoint derivation δ (in local measure topology or measure topology) is a sufficient condition for existence of λ-elements for sufficiently large $\lambda > 0$. In this section, our only aim is to establish essential properties of λ-elements.

We note that if $\{x_n\}_{n\in\mathbb{N}}$ are δ-disjoint λ-elements (for some $\lambda > 0$), then in particular, we have that $x_n x_m = 0$, $n \neq m$, so that the family $\{x_n\}_{n\in\mathbb{N}}$ consists of pairwise disjoint operators. Furthermore, since x_n belongs to the unit ball of $\mathcal{M}$, it follows that $\sup_{n\in\mathbb{N}} \|x_n\|_{\mathcal{M}} \leq 1$, and so the family $\{x_n\}_{n\in\mathbb{N}}$ is uniformly bounded. Thus, if $\{x_n\}_{n\in\mathbb{N}}$ are δ-disjoint λ-elements for some $\lambda > 0$, then the disjoint sum $\sum_{n=1}^{\infty} x_n$ is well-defined and belongs to the unit ball of $\mathcal{M}$.

We also note that for any family $\{x_n\}_{n\in\mathbb{N}}$ of δ-disjoint λ-elements, the operators x_n, $\delta(x_n)$ are necessarily self-adjoint. Therefore,

$$s(x_n)s(x_m) = s(x_n)s(\delta(x_m)) = s(\delta(x_n))s(\delta(x_m)) = 0$$

for all $n \neq m$.

Throughout this section, we assume that $\mathcal{M}$ is an arbitrary von Neumann algebra, $\mathcal{A}$ is an absolutely solid $*$-subalgebra of $LS(\mathcal{M})$, and δ is a self-adjoint derivation on $\mathcal{A}$.

In the following proposition, we show that the disjoint sum of δ-disjoint λ-elements $\{x_n\}_{n\in\mathbb{N}} \subset \mathcal{A}$ is again a λ-element, provided that the disjoint sum $x = \sum_{n=1}^{\infty} x_n$ belongs to the algebra $\mathcal{A}$.

Proposition 5.4.6. *Let $\mathcal{A}$ be an absolutely solid $*$-subalgebra in $LS(\mathcal{M})$ and let $\delta : \mathcal{A} \to \mathcal{A}$ be a self-adjoint derivation. Assume that $\{x_n\}_{n\in\mathbb{N}}$ is a family of δ-disjoint λ-elements for some fixed $\lambda > 0$. If the disjoint sum $x = \sum_{n=1}^{\infty} x_n$ belongs to the algebra $\mathcal{A}$, then x is a λ-element, too.*

Proof. Since x_n is a λ-element for every $n \in \mathbb{N}$, there exists $p_n \in P(\mathcal{M})$, such that $p_n \sim s(x_n)$ and $p_n \leq E_{\delta(x_n)}[\lambda, \infty)$. By assumption, the λ-elements $\{x_n\}_{n\in\mathbb{N}}$ are δ-disjoint. Therefore, since $p_n \leq E_{\delta(x_n)}[\lambda, \infty) \leq s(\delta(x_n))$, it follows that for all $n \neq m$,

$$\begin{aligned} p_n p_m &= p_n s(\delta(x_n))s(\delta(x_m))p_m = 0, \\ p_n x_m &= p_n s(\delta(x_n))x_m = 0. \end{aligned} \tag{5.17}$$

In particular, it follows from Remark 5.4.4(i) that

$$p_n x = p_n x_n, \quad n \in \mathbb{N}. \tag{5.18}$$

We denote

$$p = \sup_{n\in\mathbb{N}} p_n.$$

Since $p_n \sim s(x_n)$ by assumption and $s(x) = \sup_{n\in\mathbb{N}} s(x_n)$ by definition of the disjoint sum, it follows from Theorem 1.9.4(vi) that

$$s(x) = \sup_{n\in\mathbb{N}} s(x_n) \sim \sup_{n\in\mathbb{N}} p_n = p.$$

Hence, to show that x is a λ-element, it is sufficient to show that $p \preceq E_{\delta(x)}[\lambda,\infty)$. By Lemma 3.3.5, this would follow from the inequality

$$p\delta(x)p \geq \lambda p,$$

which we prove next.

Note that the equation

$$\delta(x_n)\delta(x_m) = 0, \quad n \neq m$$

implies that

$$s(\delta(x_n))\delta(x_m) = 0, \quad n \neq m,$$

and, therefore, we have that

$$p_n\delta(x_m) = p_n s(\delta(x_n))\delta(x_m) = 0, \quad n \neq m. \tag{5.19}$$

Since $p = \sup_{n\in\mathbb{N}} p_n$, it follows that

$$p\delta(x_n) = p_n\delta(x_n), \quad \delta(x_n)p = \delta(x_n)p, \quad n \in \mathbb{N}. \tag{5.20}$$

Moreover, combining equalities (5.19) with (5.17) and using the Leibniz rule, we infer that

$$\delta(p_n)x_m = \delta(p_n x_m) - p_n\delta(x_m) = 0 \quad \text{for all } n \neq m.$$

Thus, by Remark 5.4.4(i) with $a = \delta(p_n)$, for every fixed $n \in \mathbb{N}$, we have that

$$\delta(p_n)x = \delta(p_n)x_n. \tag{5.21}$$

Employing the Leibniz rule repeatedly, we write

$$\begin{aligned} p_n \cdot p\delta(x)p = p_n\delta(x)p &= \delta(p_n x)p - \delta(p_n)xp \\ &\overset{(5.18)}{=} \delta(p_n x_n)p - \delta(p_n)xp \\ &\overset{(5.21)}{=} \delta(p_n x_n)p - \delta(p_n)x_n p \\ &= p_n\delta(x_n)p \\ &\overset{(5.19)}{=} p_n\delta(x_n)p_n \end{aligned}$$

Therefore, since δ is a self-adjoint derivation, we infer that

$$p\delta(x)p \cdot p_n = (p_n \cdot p\delta(x)p)^* = p_n\delta(x_n)p_n \geq \lambda p_n.$$

Hence, applying Remark 5.4.4(ii) for the family $\{p_n\}_{n\in\mathbb{N}}$ and the element $a = p\delta(x)p - \lambda\mathbf{1}$, we conclude that

$$p\delta(x)p \geq \lambda p,$$

as required. □

Before establishing the main result of this section, Theorem 5.4.9 below, we prove some auxiliary lemmas.

Lemma 5.4.7. *Let $\delta : \mathcal{A} \to \mathcal{A}$ be a self-adjoint derivation on an absolutely solid $*$-subalgebra $\mathcal{A}$ of $LS(\mathcal{M})$. Let $p, q \in P(\mathcal{M})$ be such that $p, q \in \mathcal{A}$. If $p\delta(q)p \geq \lambda p$ for some $\lambda > 0$, then for the projection $e = s_l(qp)$ we have that $e \in \mathcal{A}$ and*

$$e \preceq E_{\delta(e)}[\lambda, \infty).$$

Proof. For brevity, we introduce the notation

$$f = s_r(qp).$$

Since $e \leq q$ and $f \leq p$ and since the algebra $\mathcal{A}$ is an absolutely solid $*$-subalgebra, it follows that $e, f \in \mathcal{A}$.

We have that

$$ef = (eq)(pf) = e(qp)f = s_l(qp)qps_r(qp) = qp.$$

Multiplying by f on the right, we obtain

$$ef = (qp)f = q(pf) = qf$$

and, therefore, $fe = fq$.

By the Leibniz rule, we have

$$\begin{aligned} f\delta(e)f &= f(f\delta(e))f = f(\delta(fe) - \delta(f)e)f \\ &= f(\delta(fq) - \delta(f)q)f = f(f\delta(q))f = f\delta(q)f. \end{aligned}$$

Since $f \leq p$, it follows that

$$f\delta(e)f = f \cdot p\delta(q)p \cdot f \geq f \cdot \lambda p \cdot f = \lambda f.$$

Appealing Lemma 3.3.5, we infer that

$$f \preceq E_{\delta(e)}[\lambda, \infty).$$

Since $f = s_r(qp) \sim s_l(qp) = e$, we obtain that $e \preceq E_{\delta(e)}[\lambda, \infty)$, as required. □

Before we proceed to the next result, recall (see Section 5.1) that for a projection $0 \neq e \in \mathcal{M}$ the reduced derivation $\delta^{(e)} : e\mathcal{A}e \to e\mathcal{A}e$ is defined by setting

$$\delta^{(e)}(x) = e\delta(x)e \quad \text{for all } x \in e\mathcal{A}e.$$

Lemma 5.4.8. *Let $\delta : \mathcal{A} \to \mathcal{A}$ be a self-adjoint derivation on an absolutely solid $*$-subalgebra $\mathcal{A}$ of $LS(\mathcal{M})$. Let $x \in \mathcal{A}$ be an optimal λ-element for some $\lambda > 0$ with the δ-support $s_\delta(x)$. Let $g = \mathbf{1} - s_\delta(x)$. If $q \in P(\mathcal{M})$ is such that $q \in \mathcal{A}$ and $q \leq g$, then $\delta^{(g)}(q) \in \mathcal{M}$ and $\|\delta^{(g)}(q)\|_{\mathcal{M}} \leq \lambda$.*

Proof. Since δ is a self-adjoint derivation, then so is $\delta^{(g)} : g\mathcal{A}g \to g\mathcal{A}g$. In particular, $\delta^{(g)}(q)$ is self-adjoint. Let p_+ and p_- be the spectral projections for $\delta^{(g)}(q)$ corresponding to the interval $(\lambda, +\infty)$ and $(-\infty, -\lambda)$, respectively.

By Remark 3.3.16, we have that $p_\pm \in \mathcal{A}$. To prove the assertion, it is sufficient to show that $p_\pm = 0$. Assume, by contradiction, that $p_+ \neq 0$. Since $p_+ = E_{\delta^{(g)}(q)}(\lambda, +\infty) \leq s(\delta^{(g)}(q)) \leq g$, the functional calculus and the definition of a reduced derivation imply that

$$p_+\delta(q)p_+ = p_+g\delta(q)gp_+ = p_+\delta^{(g)}(q)p_+ \geq \lambda p_+ \tag{5.22}$$

Using the Leibniz rule, we write

$$p_+\delta(q)p_+ = \delta(p_+q)p_+ - \delta(p_+) \cdot (p_+q)^*.$$

If $p_+q = 0$, then $p_+\delta(q)p_+ = 0$ and, therefore, inequality (5.22) implies that $p_+ = 0$, which is a contradiction. Therefore, $p_+q \neq 0$.

We set

$$e = s_l(qp_+) \neq 0.$$

Combining inequality (5.22) with Lemma 5.4.7, we have that $e \leq E_{\delta(e)}[\lambda, \infty)$. Thus, e is a λ-element.

We claim that e is a λ-element δ-disjoint to the λ-element x. Since $e = s_l(qp_+) \leq q \leq g$ and, by definition, g is orthogonal to $s_\delta(x) = s(x) \vee s(\delta(x)) \vee s(\delta^2(x))$, it follows that

$$xe = x \cdot s(x)g \cdot e = 0, \quad \delta(x)e = \delta(x) \cdot s(\delta(x))g \cdot e = 0$$

and

$$\delta^2(x)e = \delta^2(x) \cdot s(\delta^2(x))g \cdot e = 0.$$

By the Leibniz rule, we have

$$x\delta(e) = \delta(xe) - \delta(x)e = 0, \quad \delta(x)\delta(e) = \delta(\delta(x)e) - \delta^2(x)e = 0.$$

Thus, e is a λ-element δ-disjoint to the λ-element x. However, by the assumption, the λ-element x is optimal. Hence, $e = 0$. In other words, our assumption that $p_+ \neq 0$ leads to a contradiction. Thus, $E_{\delta^{(g)}(q)}(\lambda, +\infty) = p_+ = 0$.

Repeating the same argument for $p_- = 0$ with q replaced by $g - q$, we obtain that $E_{\delta^{(g)}(q)}(-\infty, -\lambda) = 0$ too. Hence, $\delta^{(g)}(q) \in \mathcal{M}$ and $\|\delta^{(g)}(q)\|_{\mathcal{M}} \leq \lambda$, as required. □

We now establish the main property of optimal λ-elements. This property shows that for a self-adjoint derivation δ, the reduction of δ by the orthogonal complement of the δ-support of an optimal λ-element x maps $\mathcal{M} \cap \mathcal{A}$ into itself.

Theorem 5.4.9. *Let $\delta : \mathcal{A} \to \mathcal{A}$ be a self-adjoint derivation on an absolutely solid $*$-subalgebra $\mathcal{A}$ of $LS(\mathcal{M})$. Let $x \in \mathcal{A}$ be an optimal λ-element with δ-support $s_\delta(x)$. For the projection $g = \mathbf{1} - s_\delta(x)$, we have that*

$$\delta^{(g)}(g(\mathcal{M} \cap \mathcal{A})g) \subset g(\mathcal{M} \cap \mathcal{A})g.$$

Proof. It is sufficient to show that

$$\|\delta^{(g)}(y)\|_{\mathcal{M}} \leq 3\lambda \|y\|_{\mathcal{M}}, \quad y = y^* \in g(\mathcal{M} \cap \mathcal{A})g. \tag{5.23}$$

Let $y = y^* \in g(\mathcal{M} \cap \mathcal{A})g$. Without loss of generality, we can assume that $-\mathbf{1} \leq y \leq \mathbf{1}$. We set

$$r = E_{\delta^{(g)}(y)}(3\lambda, \infty).$$

By assumption, we have that $y \in \mathcal{A}$ and, therefore, $\delta(y) \in \mathcal{A}$, too. Remark 3.3.16 implies that $r \in \mathcal{A}$. Noting that $\delta^{(g)}(y) \in g\mathcal{A}g$, we infer $r \leq g$, so that $r \in g(\mathcal{M} \cap \mathcal{A})g$.

Suppose that $r \neq 0$. We claim that ryr is a λ-element. By the Leibniz rule, we have that

$$r\delta(ryr)r = r\delta(y)r + (r\delta(r)yr + ry\delta(r)r). \tag{5.24}$$

Since $r = E_{\delta^{(g)}(y)}(3\lambda, \infty)$, the functional calculus implies that the first term on the right-hand side can be estimated as follows:

$$r\delta(y)r = rg\delta(y)gr = r\delta^{(g)}(y)r \geq 3\lambda r. \tag{5.25}$$

To estimate the second term on the right-hand side of (5.24), we note first that since $y \in g\mathcal{M}g$ and $r \leq g$, we have that

$$r\delta(r)yr + ry\delta(r)r = rg\delta(r)gyr + ryg\delta(r)gr = r\delta^{(g)}(r)yr + ry\delta^{(g)}(r)r.$$

Since $r \leq q$, Lemma 5.4.8 implies that $\|\delta^{(g)}(r)\|_{\mathcal{M}} \leq \lambda$. Therefore,

$$\|r\delta(r)yr + ry\delta(r)r\|_{\mathcal{M}} \leq 2\|\delta^{(g)}(r)\|_{\mathcal{M}} \leq 2\lambda.$$

In particular,

$$r\delta(r)yr + ry\delta(r)r \geq -2\lambda r.$$

Hence, combining this estimate with (5.24) and (5.25), we infer that

$$r\delta(ryr)r \geq 3\lambda r - 2\lambda r = \lambda r. \tag{5.26}$$

By Lemma 3.3.5, we have

$$r \preceq E_{\delta(ryr)}[\lambda, \infty).$$

Hence, since $s(ryr) \leq r$, we conclude that $s(ryr) \preceq E_{\delta(ryr)}[\lambda, \infty)$, i. e., ryr is indeed a λ-element.

Next, we show that ryr is λ-element, which is δ-disjoint to the λ-element x. Since $r \leq g$, and g is orthogonal to $s_\delta(x) = s(x) \vee s(\delta(x)) \vee s(\delta^2(x))$, it follows that

$$\begin{aligned} x \cdot ryr &= x \cdot s(x)g \cdot ryr = 0, \\ \delta(x) \cdot ryr &= \delta(x) \cdot s(\delta(x))g \cdot ryr = 0, \\ \delta^2(x) \cdot ryr &= \delta^2(x) \cdot s(\delta^2(x))g \cdot ryr = 0. \end{aligned}$$

Hence,

$$x \cdot \delta(ryr) = \delta(x \cdot ryr) - \delta(x) \cdot ryr = 0,$$

and

$$\delta(x)\delta(ryr) = \delta(\delta(x) \cdot ryr) - \delta^2(x) \cdot ryr = 0.$$

Thus, ryr is a λ-element, which is δ-disjoint to the λ-element x. However, by assumption the λ-element x is optimal. Hence, $ryr = 0$ and, therefore, by (5.26), we conclude that $r = 0$. In other words, our assumption that $r \neq 0$ leads to a contradiction. Hence, $E_{\delta^{(g)}(y)}(3\lambda, \infty) = r = 0$. Similarly, considering $-y$ instead of y, we conclude that $E_{\delta^{(g)}(y)}(-\infty, -3\lambda) = 0$. This proves (5.23), and hence, concludes the proof. □

5.5 Continuous derivations are inner

In this section, we show that any derivation on the algebra $LS(\mathcal{M})$ (resp., $S_0(\mathcal{M})$), which is continuous in the local measure topology is necessarily inner (resp., spatial). We first show that continuity in the local measure topology of a self-adjoint derivation is a sufficient condition for existence of λ-elements (see Definition 5.4.5) for sufficiently large $\lambda > 0$. Following that, for an appropriate sequence of optimal λ-elements, we shall construct a sequence of projections $\{q_n\} \subset P(\mathcal{M})$ and a sequence $\{d_n\}_{n\in\mathbb{N}} \subset \mathcal{M}$ such that

$\delta^{(q_n)} = \delta_{d_n}$ on $q_n \mathcal{M} q_n$ with $q_n \to \mathbf{1}$ locally in measure. Gluing the sequence $\{d_n\}$ in a way similar to martingale differences, we obtain an element d that generates the derivation δ on the whole algebra $LS(\mathcal{M})$ (and resp., on $S_0(\mathcal{M})$).

If the von Neumann algebra $\mathcal{M}$ is equipped with a faithful normal semifinite trace τ, we prove a similar result for derivations on the algebra $S(\mathcal{M}, \tau)$ (and resp., on $S_0(\mathcal{M}, \tau)$), which are continuous in the measure topology.

We recall that φ denotes a $*$-isomorphism between $\mathcal{Z}(\mathcal{M})$ and $L_\infty(\Omega, \Sigma, \mu)$ (see Remark 1.8.11) for a probability space (Ω, Σ, μ), and ψ denotes a normal state on $\mathcal{Z}(\mathcal{M})$ corresponding to the integration with respect to μ (Example 1.10.12). The dimension function on $\mathcal{M}$ is denoted by $\mathcal{D}$.

We start with several auxiliary results. In the first of these auxiliary results, we give some basic properties of linear mappings, which are continuous in the local measure topology (resp., measure topology).

Lemma 5.5.1.

(i) *Let X, Y be linear subspaces in $LS(\mathcal{M})$ and let $T : X \to Y$ be a linear mapping continuous in the local measure topology. Then for any $\varepsilon > 0$ there exists $N_\varepsilon > 0$, such that for any $x \in X$ with $\|x\|_\mathcal{M} \le 1$ there exists $z \in P(\mathcal{Z}(\mathcal{M}))$ satisfying $\psi(z^\perp) \le \varepsilon$ and*

$$\mathcal{D}\big(zE_{|T(x)|}(N_\varepsilon, \infty)\big) \le \varepsilon \varphi(z);$$

(ii) *Assume that $\mathcal{M}$ is equipped with a faithful normal semifinite trace τ and let X, Y be linear subspaces in $S(\mathcal{M}, \tau)$ and let $T : X \to Y$ be a linear mapping continuous in the measure topology t_τ. Then for any $\varepsilon > 0$ there exists $N_\varepsilon > 0$, such that*

$$\tau\big(E_{|T(x)|}(N_\varepsilon, \infty)\big) \le \varepsilon$$

for any $x \in X$ with $\|x\|_\mathcal{M} \le 1$.

Proof. (i) Let $\varepsilon > 0$ be fixed. By Proposition 4.3.10, the sets $\{V(\cdot,\cdot,\cdot)\}$ form a base of the local measure topology. Since T is continuous in the local measure topology, there exists $\lambda > 0$, such that

$$T(V(\lambda, \lambda, \lambda) \cap X) \subset V(1, \varepsilon, \varepsilon).$$

Then

$$\begin{aligned} T(X \cap \mathcal{M}_1) &\subset T\big(V(1, \lambda, \lambda) \cap X\big) \\ &= T(\lambda^{-1} V(\lambda, \lambda, \lambda) \cap X) \subset \lambda^{-1} V(1, \varepsilon, \varepsilon) = V(\lambda^{-1}, \varepsilon, \varepsilon). \end{aligned}$$

Setting $N_\varepsilon := \lambda^{-1}$, we have that

$$T(x) \in V(N_\varepsilon, \varepsilon, \varepsilon)$$

for any fixed $x \in X$ with $\|x\|_{\mathcal{M}} \le 1$. By definition of the base of neighborhoods $\{V(\cdot,\cdot,\cdot)\}$ (see Proposition 4.3.10), the latter inclusion implies that there exists $z \in P(\mathcal{Z}(\mathcal{M}))$, such that $\psi(z^{\perp}) \le \varepsilon$ and

$$\mathcal{D}(zE_{|T(x)|}(N_\varepsilon, \infty)) \le \mathcal{D}(zp) \le \varepsilon\varphi(z),$$

as required.

(ii) Recall (see Proposition 4.2.5) that the sets $\{U(\cdot,\cdot)\}$ form a base of the measure topology. Since T is continuous in measure, for every $\varepsilon > 0$ there exists $\delta > 0$, such that

$$T(U(\delta,\delta) \cap X) \subset U(\varepsilon,\varepsilon).$$

Then

$$T(X \cap \mathcal{M}_1) \subset T(U(1,\delta) \cap X) = T(\delta^{-1}U(\delta,\delta) \cap X) \subset \delta^{-1}U(\varepsilon,\varepsilon) = U(\varepsilon/\delta,\varepsilon).$$

Setting $N_\varepsilon = \frac{\varepsilon}{\delta}$, and referring to the definition of the sets $U(\cdot,\cdot)$ (see Proposition 4.2.5), we conclude the proof. □

Next, we estimate the δ-support of a λ-element of a self-adjoint derivation.

Lemma 5.5.2. *Let $\delta : \mathcal{A} \to \mathcal{A}$ be a self-adjoint derivation on an absolutely solid $*$-subalgebra $\mathcal{A}$ of $LS(\mathcal{M})$. Let x be a λ-element for δ for some $\lambda > 0$.*

(i) *For $\beta = \mathcal{D}(E_{\delta(x)}[\lambda,\infty)) \in L^+(\Omega,\Sigma,\mu)$, we have*

$$\mathcal{D}(s(x)) \le \beta, \quad \mathcal{D}(s(\delta(x))) \le 2\beta, \quad \mathcal{D}(s(\delta^2(x))) \le 4\beta.$$

In particular, for the δ-support $s_\delta(x)$ of x, we have $\mathcal{D}(s_\delta(x)) \le 7\beta$;

(ii) *Assume that $\mathcal{M}$ is a semifinite von Neumann algebra equipped with a faithful normal semifinite trace τ. If $\tau(E_{\delta(x)}[\lambda,\infty)) \le \alpha$ for some $\alpha > 0$, then*

$$\tau(s(x)) \le \alpha, \quad \tau(s(\delta(x))) \le 2\alpha, \quad \tau(s(\delta^2(x))) \le 4\alpha.$$

In particular, for the δ-support $s_\delta(x)$ of x we have $\tau(s_\delta(x)) \le 7\alpha$.

Proof. By the definition of a λ-element, we have

$$s(x) \prec E_{\delta(x)}[\lambda,\infty). \tag{5.27}$$

By the Leibniz rule, we have

$$\delta(x) = \delta(s(x)x) = \delta(s(x))x + s(x)\delta(x)$$

and, therefore,

$$s(\delta(x)) \leq s_l(\delta(s(x))x) \vee s_l(s(x)\delta(x)) \leq s_l(\delta(s(x))x) \vee s(x). \tag{5.28}$$

Since $s_r(y) \sim s_l(y)$ for any $y \in LS(\mathcal{M})$, Proposition 1.11.6(i) and inequality (5.28) imply that

$$\begin{aligned}\mathcal{D}(s(\delta(x))) &\leq \mathcal{D}(s_l(\delta(s(x))x)) + \mathcal{D}(s(x))\\ &= \mathcal{D}(s_r(\delta(s(x))x)) + \mathcal{D}(s(x)) \leq 2\mathcal{D}(s(x)).\end{aligned}$$

Applying this inequality twice, we infer that

$$\mathcal{D}(s(\delta^2(x))) \leq 2\mathcal{D}(s(\delta(x))) \leq 4\mathcal{D}(s(x)).$$

Combining Proposition 1.11.6(iv) with inequality (5.27), we infer that $\mathcal{D}(s(x)) \leq \mathcal{D}(E_{\delta(x)}[\lambda,\infty))$, and so

$$\mathcal{D}(s_\delta(x)) \leq \mathcal{D}(s(x)) + \mathcal{D}(s(\delta(x))) + \mathcal{D}(s(\delta^2(x))) \leq 7\mathcal{D}(E_{\delta(x)}[\lambda,\infty)),$$

concluding the proof of part (i).

To prove part (ii), assume now that τ is a faithful semifinite normal trace τ on $\mathcal{M}$. Inequality (5.27) implies that $\tau(s(x)) \leq \tau(E_{\delta(x)}[\lambda,\infty)) \leq \alpha$. Furthermore, inequality (5.28) implies that

$$\tau(s(\delta(x))) \leq \tau(s(x)) + \tau(s_r(s(x)\delta(x))).$$

Since

$$s_r(s(x)\delta(x)) \sim s_l(s(x)\delta(x)) \leq s(x),$$

it follows that

$$\tau(s(\delta(x))) \leq 2\tau(s(x))$$

and

$$\tau(s(\delta^2(x))) \leq 2\tau(s(\delta(x))) \leq 4\tau(s(x)),$$

as required. □

Now, we show that for self-adjoint derivations δ on $S(\mathcal{M},\tau)$ (or on $S_0(\mathcal{M},\tau)$) that are continuous in the measure topology, there is a plethora of optimal λ-elements as long as $\lambda > 0$ is sufficiently large. For self-adjoint derivations on $LS(\mathcal{M})$ (or on $S_0(\mathcal{M})$), which are continuous in the local measure topology, a similar result holds. We recall that φ denotes a $*$-isomorphism between $\mathcal{Z}(\mathcal{M})$ and $L_\infty(\Omega,\Sigma,\mu)$ (see Remark 1.8.11), and ψ denotes a normal state on $\mathcal{Z}(\mathcal{M})$ corresponding to the integration with respect to μ (Example 1.10.12). As before, we assume that (Ω,Σ,μ) is a probability space.

Proposition 5.5.3. *Let $\mathcal{M}$ be a von Neumann algebra.*

(i) *Suppose that $\mathcal{M}$ is equipped with a faithful normal semifinite trace τ and let δ be a self-adjoint derivation on $S(\mathcal{M},\tau)$ (resp., on $S_0(\mathcal{M},\tau)$), which is continuous in the measure topology t_τ. Then for any $\varepsilon > 0$ there exist a scalar $\lambda > 0$ and a (possibly zero) optimal λ-element $x \in S(\mathcal{M},\tau)$ (resp., $x \in S_0(\mathcal{M},\tau)$), such that*

$$\tau(E_{\delta(x)}[\lambda,\infty)) \le \varepsilon; \tag{5.29}$$

(ii) *Let δ be a self-adjoint derivation on $LS(\mathcal{M})$ (resp., on $S_0(\mathcal{M})$), which is continuous in the local measure topology. Then for any $\varepsilon > 0$ there exist a projection $z_\varepsilon \in P(\mathcal{Z}(\mathcal{M}))$, a scalar $\lambda > 0$ and an optimal λ-element $x \in z_\varepsilon LS(\mathcal{M})$ (resp., $x \in z_\varepsilon S_0(\mathcal{M})$) for the derivation $\delta^{(z_\varepsilon)}$ on the algebra $z_\varepsilon LS(\mathcal{M})$ (resp., $z_\varepsilon S_0(\mathcal{M})$), such that $\psi(z_\varepsilon^\perp) \le \varepsilon$ and*

$$\mathcal{D}(E_{\delta(z_\varepsilon x)}[\lambda,\infty)) \le \varepsilon\varphi(z_\varepsilon). \tag{5.30}$$

Proof. (i) We will prove both cases of derivations on $S(\mathcal{M},\tau)$ and $S_0(\mathcal{M},\tau)$ simultaneously. For this purpose, we denote by $\mathcal{A}$ either the algebra $S(\mathcal{M},\tau)$ or the algebra $S_0(\mathcal{M},\tau)$. Let $\delta : \mathcal{A} \to \mathcal{A}$ be a self-adjoint derivation continuous in the measure topology t_τ.

Fix $\varepsilon > 0$. Since $\delta : \mathcal{A} \to \mathcal{A}$ is continuous in the measure topology t_τ, Lemma 5.5.1(i) implies that there exists $N_\varepsilon > 0$, such that $\tau(E_{|\delta(x)|}(N_\varepsilon,\infty)) \le \varepsilon$ for all $x \in \mathcal{A} \cap \mathcal{M}_1$. Choose $\lambda > 0$ satisfying $\lambda > N_\varepsilon$. Then

$$\tau(E_{\delta(x)}[\lambda,\infty)) \le \tau(E_{|\delta(x)|}(N_\varepsilon,\infty)) \le \varepsilon, \quad x \in \mathcal{A} \cap \mathcal{M}_1. \tag{5.31}$$

Let S be the set of all families (finite or infinite) of all pairwise δ-disjoint λ-elements $\{x_i\}_{i\in I}$. If S is empty, then 0 is the desired optimal λ-element. In what follows, we assume that $S \ne \emptyset$. We supply S with a partial order by inclusion. By Zorn's lemma, there exists a maximal element in S. Let $\{x_i\}_{i\in I}$ be the maximal element in S. Since $\{x_i\}_{i\in I}$ are pairwise δ-disjoint λ-elements, it follows that, in particular, $\{x_i\}_{i\in I}$ are pairwise disjoint and uniformly bounded. Since x_i are pairwise disjoint and self-adjoint, it follows that the projections $s(x_i)$ are pairwise orthogonal. Separability of the Hilbert space H implies that the family $\{x_i\}_{i\in I}$ is at most countable. Hence, the disjoint sum

$$x = \sum_{i\in I} x_i,$$

is well-defined.

We claim that x is the required λ-element. If $\mathcal{A} = S(\mathcal{M},\tau)$, then $x \in \mathcal{M} \subset \mathcal{A}$ and, therefore, Proposition 5.4.6 implies that x is a λ-element.

For the case, when $\mathcal{A} = S_0(\mathcal{M},\tau)$, to conclude that x is a λ-element we first need to show that $x \in S_0(\mathcal{M},\tau)$. Choose an arbitrary finite subset $I_0 \subset I$ and let $x_0 = \sum_{i\in I_0} x_i$. Since $x_i \in S_0(\mathcal{M},\tau)$ for all $i \in I$ and x_0 is a finite sum of x_i, it follows that $x_0 \in S_0(\mathcal{M},\tau)$.

Therefore, by Proposition 5.4.6, x_0 is a λ-element. In particular, $x_0 \in S_0(\mathcal{M}, \tau) \cap \mathcal{M}_1$ and $s(x_0) \preceq E_{\delta(x_0)}[\lambda, \infty)$. Therefore, inequality (5.31) implies that

$$\tau(s(x_0)) \leq \tau(E_{\delta(x_0)}(\lambda, \infty)) \leq \varepsilon.$$

Since the finite subset I_0 is arbitrary, it follows that

$$\tau(s(x)) = \tau\Big(\sup_{i\in I} s(x_i)\Big) = \sum_{i\in I} \tau(s(x_i)) \leq \varepsilon.$$

Therefore, by Remark 2.6.3 we infer that $x \in S_0(\mathcal{M}, \tau)$. Referring to Proposition 5.4.6 once again, we conclude that x is a λ-element.

Thus, in both cases of $\mathcal{A} = S(\mathcal{M}, \tau)$ and $\mathcal{A} = S_0(\mathcal{M}, \tau)$, the constructed x is a λ-element. Since, in particular, this implies that $x \in \mathcal{A} \cap \mathcal{M}_1$, the estimate $\tau(E_{\delta(x)}[\lambda, \infty)) \leq \varepsilon$ follows from (5.31).

Thus, it remains to show that x is an optimal λ-element. For this purpose, suppose that $y \in \mathcal{A}$ is a nonzero λ-element, which is δ-disjoint to x. By definition, we have that

$$xy = x\delta(y) = \delta(x)\delta(y) = 0.$$

Therefore,

$$x_i y = s(x_i)xy = 0, \quad x_i\delta(y) = s(x_i)x\delta(y) = 0 \quad \text{for all } i \in I.$$

We claim that $\delta(x_i)\delta(y) = 0$ for all $i \in I$, too. By the Leibniz rule, we have that

$$x\delta^2(y) = \delta(x\delta(y)) - \delta(x)\delta(y) = 0.$$

Therefore, we obtain

$$x_i\delta^2(y) = s(x_i)x\delta^2(y) = 0 \quad \text{for all } i \in I.$$

Hence, another application of the Leibniz rule yields

$$\delta(x_i)\delta(y) = \delta(x_i\delta(y)) - x_i\delta^2(y) = 0 \quad \text{for all } i \in I.$$

Thus,

$$x_i y = x_i\delta(y) = y\delta(x_i) = \delta(x_i)\delta(y) \quad \text{for all } i \in I,$$

which means that y is a λ-element, which is δ-disjoint to all $\{x_i\}_{i\in I}$. Hence, $\{y\} \cup \{x_i\}_{i\in I} \in S$. However, this contradicts with the maximality of $\{x_i\}_{i\in I}$ in S. Thus, there are no δ-disjoint λ-elements to x, and so x is an optimal λ-element, as required.

(ii) We will prove both cases of derivations on $LS(\mathcal{M})$ and $S_0(\mathcal{M})$ simultaneously. For this purpose, we denote by $\mathcal{A}$ either the algebra $LS(\mathcal{M})$ or the algebra $S_0(\mathcal{M})$. Let $\delta : \mathcal{A} \to \mathcal{A}$ be a self-adjoint derivation continuous in the local measure topology.

Let $\varepsilon > 0$ be fixed. By Lemma 5.5.1, there exists $N_\varepsilon > 0$ such that for any $x \in \mathcal{A} \cap \mathcal{M}$ with $\|x\|_{\mathcal{M}} \le 1$, there exists $z \in P(\mathcal{Z}(\mathcal{M}))$, such that $\psi(z^\perp) \le \varepsilon$ and

$$\mathcal{D}(zE_{|\delta(x)|}(N_\varepsilon, \infty)) \le \varepsilon\varphi(z).$$

We fix an arbitrary $\lambda > N_\varepsilon$. Then, for any $x \in \mathcal{A} \cap \mathcal{M}_1$, there exists $z \in P(\mathcal{Z}(\mathcal{M}))$ satisfying $\psi(z^\perp) \le \varepsilon$ and

$$\mathcal{D}(zE_{\delta(x)}[\lambda, \infty)) \le \mathcal{D}(zE_{|\delta(x)|}(N_\varepsilon, \infty)) \le \varepsilon\varphi(z). \tag{5.32}$$

As in part (i), we let S be the set of all families (finite or infinite) of pairwise δ-disjoint λ-elements $\{x_i\}_{i\in I} \subset \mathcal{A}$. If S is empty, then 0 is the desired optimal λ-element with $z_\varepsilon = \mathbf{1}$. In what follows, we assume that $S \neq \emptyset$. We supply S with a partial order by inclusion. By Zorn's lemma, there exists a maximal element in S. Let $\{x_i\}_{i\in I}$ be the maximal element in S. Since $\{x_i\}_{i\in I}$ are pairwise δ-disjoint λ-elements, it follows that, in particular, $\{x_i\}_{i\in I}$ are pairwise disjoint and uniformly bounded. Since x_i are pairwise disjoint and self-adjoint, it follows that the projections $s(x_i)$ are pairwise orthogonal. Separability of the Hilbert space H implies that $\{x_i\}_{i\in I}$ is at most countable family. We define

$$x = \sum_{i\in I} x_i,$$

as the disjoint sum of $\{x_i\}_{i\in I}$.

In the case, when $\mathcal{A} = LS(\mathcal{M})$, we have that $x \in \mathcal{M} \subset LS(\mathcal{M})$ and, therefore, by Proposition 5.4.6, x is a λ-element for the derivation δ. By (5.32), there exists $z_\varepsilon \in P(\mathcal{Z}(\mathcal{M}))$, such that $\psi(z_\varepsilon^\perp) \le \varepsilon$ and

$$\mathcal{D}(z_\varepsilon E_{\delta(x)}[\lambda, \infty)) \le \varepsilon\varphi(z_\varepsilon).$$

Next, we consider the case when $\mathcal{A} = S_0(\mathcal{M})$ and prove that there exists $z_\varepsilon \in P(\mathcal{Z}(\mathcal{M}))$, such that $z_\varepsilon x$ is a λ-element for the derivation $\delta^{(z_\varepsilon)}$ on $z_\varepsilon S_0(\mathcal{M})$. We first construct the required projection $z_\varepsilon \in P(\mathcal{Z}(\mathcal{M}))$.

Let $n \subset \mathbb{N}$. We set

$$x_{(n)} = \sum_{i=1}^{n} x_i.$$

Then $x_{(n)} \in S_0(\mathcal{M})$. Therefore, Proposition 5.4.6 implies that $x_{(n)}$ is a λ-element in $S_0(\mathcal{M})$ for the derivation δ. We define $z_{(n)} = \varphi^{-1}(\chi_{[\mathcal{D}(s(x_{(n)}))\le\varepsilon]})$.

We have that

$$\begin{aligned}\mathcal{D}(z_{(n)}s(x_{(n)})) &= \varphi(z_{(n)})\mathcal{D}(s(x_{(n)})) \\ &= \chi_{[\mathcal{D}(s(x_{(n)}))\le\varepsilon]}\mathcal{D}(s(x_{(n)})) \le \varepsilon\varphi(z_{(n)}).\end{aligned} \tag{5.33}$$

Since $x_{(n)} \in S_0(\mathcal{M}) \cap \mathcal{M}_1$, (5.32) implies that there exists $z \in P(\mathcal{Z}(\mathcal{M}))$, such that $\psi(z^\perp) \le \varepsilon$ and $\mathcal{D}(zE_{\delta(x_{(n)})}[\lambda,\infty)) \le \varepsilon\varphi(z)$.

Note that $z \le z_{(n)}$. Indeed, if $\varphi(z)(\omega) = 1$ for some $\omega \in \Omega$, then

$$\begin{aligned}\mathcal{D}(s(x_{(n)}))(\omega) &\le \mathcal{D}(E_{\delta(x_{(n)})}[\lambda,\infty))(\omega) = (\varphi(z)(\omega))\mathcal{D}(E_{\delta(x_{(n)})}[\lambda,\infty))(\omega) \\ &= \mathcal{D}(zE_{\delta(x_{(n)})}[\lambda,\infty))(\omega) \le \varepsilon\varphi(z)(\omega) = \varepsilon,\end{aligned}$$

where in the first inequality we used the fact that $s(x_{(n)}) \le E_{\delta(x)}[\lambda,\infty)$. Therefore, $\omega \in [\mathcal{D}(s(x_{(n)})) \le \varepsilon]$, and so $\varphi(z_{(n)})(\omega) = 1$. Thus, $\varphi(z) \le \varphi(z_{(n)})$, proving that $z \le z_{(n)}$. In particular, $\psi(z_{(n)}^\perp) \le \psi(z^\perp) \le \varepsilon$.

If $n < m$, then $\mathcal{D}(s(x_{(n)})) \le \mathcal{D}(s(x_{(m)}))$, and so $[\mathcal{D}(s(x_{(m)})) \le \varepsilon] \subset [\mathcal{D}(s(x_{(n)})) \le \varepsilon]$, which implies that the sequence $\{z_{(n)}\}$ is decreasing. We define

$$z_\varepsilon := \inf_{n\in\mathbb{N}}\{z_{(n)}\}.$$

Since ψ is normal, it follows that

$$\psi(z_\varepsilon^\perp) = \psi\Big(\sup_{n\in\mathbb{N}}\{z_{(n)}^\perp\}\Big) = \sup_{n\in\mathbb{N}}\{\psi(z_{(n)}^\perp)\} \le \varepsilon.$$

Next, we show that $z_\varepsilon x \in z_\varepsilon S_0(\mathcal{M})$. By (5.33), we have that

$$\mathcal{D}(s(z_\varepsilon x_{(n)})) = \varphi(z_\varepsilon)\mathcal{D}(s(z_{(n)}x_{(n)})) \le \varepsilon\varphi(z_\varepsilon)\varphi(z_{(n)}) = \varepsilon\varphi(z_\varepsilon)$$

for any $n \in \mathbb{N}$. Therefore,

$$\mathcal{D}(s(z_\varepsilon x)) = \mathcal{D}\Big(\sup_{n\in\mathbb{N}} s(z_\varepsilon x_{(n)})\Big) \le \sup_{n\in\mathbb{N}}\Big(\mathcal{D}(z_\varepsilon E_{\delta(x_{(n)})}[\lambda,\infty))\Big) \le \varepsilon\varphi(z_\varepsilon), \tag{5.34}$$

i. e., $\mathcal{D}(s(z_\varepsilon x))$ is finite a. e. Definition 1.11.1(ii) implies that the projection $s(z_\varepsilon x)$ is finite and, therefore, $z_\varepsilon x \in z_\varepsilon S_0(\mathcal{M})$ (see Remark 2.6.3). Appealing to Proposition 5.4.6, we conclude that $z_\varepsilon x$ is a λ-element for the reduced derivation $\delta^{(z_\varepsilon)}$ on $z_\varepsilon S_0(\mathcal{M})$.

Thus, in both cases when $\mathcal{A} = LS(\mathcal{M})$ and $\mathcal{A} = S_0(\mathcal{M})$, the operator $z_\varepsilon x$ is a λ-element for the reduced derivation $\delta^{(z_\varepsilon)}$. Arguing as in part (i), one can show that $z_\varepsilon x$ is an optimal λ-element for the reduced derivation $\delta^{(z_\varepsilon)}$.

It remains to show that the estimate (5.30) holds.

To this end, without loss of generality, we can assume that $z_\varepsilon = \mathbf{1}$. Since $x = \sum_{i\in I} x_i$ is so-limit of the sequence $x_{(n)} = \sum_{i=1}^n x_i$ and $x_{(n)}x_{(m)} = x_{(n)}^2$ for $n < m$, it follows that $xx_{(n)} = x_{(n)}x = x_{(n)}^2$ for any $n \in \mathbb{N}$. Therefore, $|x - x_{(n)}| = |x| - |x_{(n)}|$. It is clear that the sequence

$|x_{(n)}| = \sum_{i=1}^{n} |x_i|$ increases and *so*-converges to $|x|$ as $n \to \infty$. Therefore, the sequence $|x - x_{(n)}|$ decreases and *so*-converges to 0 as $n \to \infty$. Consequently, $|x - x_{(n)}| \downarrow 0$. Since $|x - x_{(n)}|$ is majorized by a compact operator x, Theorem 4.3.26 implies that $|x - x_{(n)}| \to 0$ locally in measure as $n \to \infty$. Since the derivation δ is continuous locally in measure, it follows that $\delta(x) - \delta(\sum_{i=1}^{n} x_i) = \delta(x - x_{(n)}) \to 0$ locally in measure. Thus, for the disjoint sum $x = \sum_{i \in I} x_i$, we have that $\delta(x) = \sum_{i \in I} \delta(x_i)$, where the series converges locally in measure.

Furthermore, since $\{x_i\}$ are δ-disjoint λ-elements, it follows that the family $\{s(\delta(x_i))\}_{i \in I}$ is pairwise orthogonal. Therefore,

$$\begin{aligned} \mathcal{D}(E_{\delta(x)}[\lambda, \infty)) &= \mathcal{D}(E_{\sum_{i\in I} \delta(x_i)}[\lambda, \infty)) \\ &= \mathcal{D}\Big(\sup_{i \in I} E_{\delta(x_i)}[\lambda, \infty)\Big) \overset{(5.34)}{\le} \varepsilon\varphi(\mathbf{1}), \end{aligned}$$

as required. □

Corollary 5.5.4. *Let $\mathcal{M}$ be a von Neumann algebra.*

(i) *Suppose that $\mathcal{M}$ is equipped with a faithful normal semifinite trace τ and let δ be a self-adjoint derivation on $S(\mathcal{M}, \tau)$ (resp., on $S_0(\mathcal{M}, \tau)$), which is continuous in the measure topology t_τ. Then there exists an increasing sequence $\{q_n\}_{n \in \mathbb{N}} \subset P(\mathcal{M})$, such that $q_n \to \mathbf{1}$ in the measure topology t_τ and*

$$\delta^{(q_n)}(q_n \mathcal{M} q_n) \subset q_n \mathcal{M} q_n,$$

and respectively,

$$\delta^{(q_n)}(q_n \mathcal{M} q_n \cap S_0(\mathcal{M}, \tau)) \subset q_n \mathcal{M} q_n \cap S_0(\mathcal{M}, \tau);$$

(ii) *Let δ be a self-adjoint derivation on $LS(\mathcal{M})$ (resp., on $S_0(\mathcal{M})$), which is continuous in the local measure topology. Then there exists an increasing sequence $\{q_n\}_{n \in \mathbb{N}} \subset P(\mathcal{M})$, such that $q_n \to \mathbf{1}$ in the local measure topology $t(\mathcal{M})$ and*

$$\delta^{(q_n)}(q_n \mathcal{M} q_n) \subset q_n \mathcal{M} q_n,$$

and respectively,

$$\delta^{(q_n)}(q_n \mathcal{M} q_n \cap S_0(\mathcal{M})) \subset q_n \mathcal{M} q_n \cap S_0(\mathcal{M}).$$

Proof. (i) As in the proof of Proposition 5.5.3(i), we denote by $\mathcal{A}$ either the algebra $S(\mathcal{M}, \tau)$ or the algebra $S_0(\mathcal{M}, \tau)$. By Proposition 5.5.3(i), for any $n \in \mathbb{N}$, there exist scalar $\lambda_n > 0$ and an optimal λ_n-element $x_n \in \mathcal{A}$ (possibly zero), such that

$$\tau(E_{\delta(x_n)}[\lambda_n, \infty)) \le 2^{-n}. \tag{5.35}$$

We set

$$e_n := \mathbf{1} - s_\delta(x_n), \quad n \in \mathbb{N}.$$

Combining (5.35) with Lemma 5.5.2(ii), we infer that $\tau(e_n^\perp) \leq 7 \cdot 2^{-n}$.

We now define

$$q_n := \inf_{k \geq n} e_k, \quad n \in \mathbb{N}.$$

It is clear that $\{q_n\}_{n \in \mathbb{N}}$ is an increasing sequence. Furthermore, for any $n \in \mathbb{N}$, we have that

$$\tau(q_n^\perp) = \tau\Big(\sup_{k \geq n} e_k^\perp\Big) = \sum_{k=n}^{\infty} \tau(e_k^\perp) \leq \sum_{k=n}^{\infty} 7 \cdot 2^{-k} \leq 2^{4-n}$$

and, therefore, $\tau(q_n^\perp) \to 0$ as $n \to \infty$. By Corollary 4.2.10, we have that $q_n \to \mathbf{1}$ in measure.

By Theorem 5.4.9, we have that

$$\delta^{(e_n)}(e_n \mathcal{M} e_n \cap \mathcal{A}) \subset e_n \mathcal{M} e_n \cap \mathcal{A}.$$

Since $q_n \leq e_n$, it follows that

$$\delta^{(q_n)}(q_n \mathcal{M} q_n \cap \mathcal{A}) \subset q_n \mathcal{M} q_n \cap \mathcal{A},$$

as required.

(ii) We denote by $\mathcal{A}$ either the algebra $LS(\mathcal{M})$ or the algebra $S_0(\mathcal{M})$. By Proposition 5.5.3(ii) for every $n \in \mathbb{N}$, there exist a projection $z_n \in P(\mathcal{Z}(\mathcal{M}))$, a scalar $\lambda_n > 0$, and an optimal λ_n-element $x_n \in \mathcal{A} z_n$ (possibly, zero), such that $\psi(z_n^\perp) \leq 2^{-n}$ and

$$\mathcal{D}(E_{\delta(x_n)}[\lambda_n, \infty)) \leq 2^{-n} \varphi(z_n). \tag{5.36}$$

We note that since

$$\mu([\mathcal{D}(z_n^\perp) > 2^{-n}]) \leq \mu([\mathcal{D}(z_n^\perp) \neq 0]) = \mu([\varphi(z_n^\perp) \neq 0]) = \psi(z_n^\perp) \leq 2^{-n},$$

equality (4.10) implies that

$$\|z_n^\perp\|_{LS(\mathcal{M})} = \inf_{\lambda > 0} \max\{\lambda, \mu([\mathcal{D}(z_n^\perp) > \lambda])\} \leq 2^{-n}, \quad n \in \mathbb{N}. \tag{5.37}$$

Combining (5.36) with Lemma 5.5.2(i), we infer that $\mathcal{D}(z_n s_\delta(x_n)) \leq 7 \cdot 2^{-n} \varphi(z_n)$. Therefore,

$$\mu([\mathcal{D}(z_n s_\delta(x_n)) > 7 \cdot 2^{-n}]) \leq \mu([\varphi(z_n) > 1]) = 0,$$

and hence, referring to (4.10) once again, we obtain that

$$\|z_n s_\delta(x_n)\|_{LS(\mathcal{M})} = \inf_{\lambda>0} \max\{\lambda, \mu([\mathcal{D}(z_n s_\delta(x_n)) > \lambda])\} \le 7 \cdot 2^{-n}. \tag{5.38}$$

Setting

$$e_n = z_n(\mathbf{1} - s_\delta(x_n)), \quad n \in \mathbb{N},$$

and appealing to (5.37) and (5.38), we have that

$$\begin{aligned}\|e_n^\perp\|_{LS(\mathcal{M})} &= \|z_n s_\delta(x_n) + z_n^\perp\|_{LS(\mathcal{M})} \le \|z_n s_\delta(x_n)\|_{LS(\mathcal{M})} + \|z_n^\perp\|_{LS(\mathcal{M})} \\ &\le 7 \cdot 2^{-n} + 2^{-n} = 2^{3-n}.\end{aligned}$$

As in part (i), we now define

$$q_n = \inf_{k \ge n} e_k, \quad n \in \mathbb{N}.$$

It is clear that $\{q_n\}_{n\in\mathbb{N}}$ is an increasing sequence. Furthermore, Proposition 4.3.13(v) implies that

$$\|q_n^\perp\|_{LS(\mathcal{M})} = \Big\|\sup_{k\ge n} e_k^\perp\Big\|_{LS(\mathcal{M})} \le \sum_{k=n}^{\infty} \|e_k^\perp\|_{LS(\mathcal{M})} \le \sum_{k=n}^{\infty} 2^{3-k} = 2^{4-n},$$

i. e., $q_n \to \mathbf{1}$ locally in measure.

Finally, Theorem 5.4.9 implies that

$$\delta^{(e_n)}(e_n \mathcal{M} e_n \cap \mathcal{A}) \subset e_n \mathcal{M} e_n \cap \mathcal{A}.$$

Since $q_n \le e_n$, it follows that

$$\delta^{(q_n)}(q_n \mathcal{M} q_n \cap \mathcal{A}) \subset q_n \mathcal{M} q_n \cap \mathcal{A},$$

as required. □

Lemma 5.5.5. *Let $\{q_n\}_{n\in\mathbb{N}} \subset \mathcal{M}$ be an increasing sequence of projections such that $q_n \to \mathbf{1}$ in measure (resp., locally in measure). If $\{c_n\}_{n\in\mathbb{N}} \subset \mathcal{M}$ is such that:*
(i) $c_n \in q_n \mathcal{M} q_n$ *for all* $n \in \mathbb{N}$;
(ii) $c_n - q_n c_m q_n \in q_n \mathcal{Z}(\mathcal{M})$ *for all* $m \ge n$,

then there exists $d \in S(\mathcal{M}, \tau)$ (resp., $d \in LS(\mathcal{M})$), such that

$$c_n - q_n d q_n \in q_n \mathcal{Z}(\mathcal{M}) \quad \textit{for all } n \in \mathbb{N}.$$

Proof. We claim that there exists a sequence $\{d_n\}_{n\in\mathbb{N}} \subset \mathcal{M}$, such that

$$\begin{aligned} d_n - c_n &\in q_n \mathcal{Z}(\mathcal{M}), \quad n \in \mathbb{N}, \\ d_n &= q_n d_m q_n, \quad m \geq n. \end{aligned} \tag{5.39}$$

We construct this sequence recursively. Set $d_1 = c_1$ and assume that the elements $d_1, \ldots, d_n$ are already constructed. By assumption, we have that $c_n - q_n c_{n+1} q_n \in q_n \mathcal{Z}(\mathcal{M})$. In addition, by the inductive assumption, $d_n - c_n \in q_n \mathcal{Z}(\mathcal{M})$. Hence,

$$d_n - q_n c_{n+1} q_n = (d_n - c_n) + (c_n - q_n c_{n+1} q_n) \in q_n \mathcal{Z}(\mathcal{M}).$$

Let $Z_{n+1} \in \mathcal{Z}(\mathcal{M})$ be such that

$$d_n - q_n c_{n+1} q_n = q_n Z_{n+1}.$$

We set $d_{n+1} = c_{n+1} + q_{n+1} Z_{n+1}$. Since the sequence $\{q_n\}$ is increasing, it follows that

$$\begin{aligned} q_n d_{n+1} q_n &= q_n c_{n+1} q_n + q_n q_{n+1} Z_{n+1} q_{n+1} q_n \\ &= q_n c_{n+1} q_n + q_n Z_{n+1} = q_n c_{n+1} q_n + d_n - q_n c_{n+1} q_n \\ &= d_n. \end{aligned}$$

For $m > n$, we let $m = n + k$ and using again the fact that $\{q_n\}_{n\in\mathbb{N}}$ is increasing, we write

$$\begin{aligned} q_n d_m q_n &= q_n \cdot q_{n+k-1} d_{n+k} q_{n+k-1} \cdot q_n = q_n \cdot d_{n+k-1} \cdot q_n \\ &= q_n \cdot q_{n+k-2} d_{n+k-1} q_{n+k-2} \cdot q_n = q_n \cdot d_{n+k-2} \cdot q_n \\ &= \cdots = q_n d_{n+1} q_n = d_n, \end{aligned}$$

which implies that the sequence $\{d_n\}_{n\in\mathbb{N}}$ satisfying the required properties (5.39) exists.

For future purposes, we note that for any $m \geq n$, we have that

$$d_m - d_n = d_m - q_n d_m q_n = (\mathbf{1} - q_n) d_m + q_n d_m (\mathbf{1} - q_n)$$

and, therefore,

$$s_r(d_m - d_n) \leq s_r((\mathbf{1} - q_n) d_m) \vee s_r(q_n d_m (\mathbf{1} - q_n)) \leq s_r((\mathbf{1} - q_n) d_m) \vee (\mathbf{1} - q_n). \tag{5.40}$$

Consider now the case when $q_n \to \mathbf{1}$ in measure. We claim that $\{d_n\}_{n\in\mathbb{N}}$ is a Cauchy sequence in $(S(\mathcal{M}, \tau), t_\tau)$. Since $s_l(x) \sim s_r(x)$ for any $x \in LS(\mathcal{M})$, inequality (5.40) implies that

$$\begin{aligned} \tau(s_r(d_m - d_n)) &\leq \tau(s_r((\mathbf{1} - q_n) d_m)) + \tau(\mathbf{1} - q_n) \\ &= \tau(s_l((\mathbf{1} - q_n) d_m)) + \tau(\mathbf{1} - q_n) \leq 2\tau(\mathbf{1} - q_n). \end{aligned}$$

Since $q_n \to \mathbf{1}$ in measure, Corollary 4.2.10 implies that $\tau(\mathbf{1} - q_n) \to 0$, and so $\tau(s_r(d_m - d_n)) \to 0$ as $n, m \to \infty$. By Corollary 4.2.11, $\{d_n\}_{n\in\mathbb{N}}$ is a Cauchy sequence in measure. Since $(S(\mathcal{M},\tau), t_\tau)$ is a complete topological $*$-algebra (see Theorem 4.2.8), it follows that there exists $d \in S(\mathcal{M},\tau)$ such that $d_n \to d$ in measure. For every fixed $n \in \mathbb{N}$, we have

$$d_n = q_n d_m q_n \to q_n d q_n$$

locally in measure as $m \to \infty$. That is, $d_n = q_n d q_n$ for all $n \in \mathbb{N}$. By construction of the sequence $\{d_n\}_{n\in\mathbb{N}}$ (see (5.39)), this implies that $q_n d q_n - c_n \in q_n \mathcal{Z}(\mathcal{M})$, which concludes the proof for the case when $q_n \to \mathbf{1}$ in measure as $n \to \infty$.

Assume now that $q_n \to \mathbf{1}$ locally in measure as $n \to \infty$. We claim that $\{d_n\}_{n\in\mathbb{N}}$ is a Cauchy sequence in $(LS(\mathcal{M}), t(\mathcal{M}))$. Combining inequality (5.40) with Proposition 4.3.13(iv) and (v), we infer that

$$\begin{aligned}\|s_r(d_m - d_n)\|_{LS(\mathcal{M})} &\le \|s_r((\mathbf{1} - q_n)d_m)\|_{LS(\mathcal{M})} + \|\mathbf{1} - q_n\|_{LS(\mathcal{M})}\\ &= \|s_l((\mathbf{1} - q_n)d_m)\|_{LS(\mathcal{M})} + \|\mathbf{1} - q_n\|_{LS(\mathcal{M})}\\ &\le 2\|\mathbf{1} - q_n\|_{LS(\mathcal{M})} \to 0.\end{aligned}$$

Referring to Corollary 4.3.17, we infer that

$$\|d_m - d_n\|_{LS(\mathcal{M})} \to 0$$

as $n, m \to \infty$.

Thus, $\{d_n\}_{n\in\mathbb{N}}$ is a Cauchy sequence in $(LS(\mathcal{M}), t(\mathcal{M}))$. Since $(LS(\mathcal{M}), \|\cdot\|_{LS(\mathcal{M})})$ is an F-space (see Theorem 4.3.14), it follows that there exists $d \in LS(\mathcal{M})$ such that $d_n \to d$ in locally measure as $n \to \infty$. The proof of the inclusion $q_n d q_n - c_n \in q_n \mathcal{Z}(\mathcal{M})$ is a verbatim repetition of the case of measure convergence and is therefore omitted. □

We are now ready to prove the main result of this section that continuity of a derivation in the appropriate topology necessarily implies that the derivation is inner/spatial. The proof combines Corollary 5.5.4 with Proposition 5.3.6 as well as with numerous results established in Chapter 2 and Chapter 4.

Theorem 5.5.6. *Suppose that $\mathcal{M}$ is a von Neumann algebra. Then:*

(i) *A derivation $\delta : LS(\mathcal{M}) \to LS(\mathcal{M})$ is continuous in the local measure topology if and only if it is inner, i. e., there exists $d \in LS(\mathcal{M})$, such that $\delta = \delta_d$;*

(ii) *A derivation $\delta : S_0(\mathcal{M}) \to S_0(\mathcal{M})$ is continuous in the local measure topology if and only if it is spatial, i. e., there exists $d \in LS(\mathcal{M})$, such that $\delta = \delta_d$.*

If, in addition, $\mathcal{M}$ is a semifinite von Neumann algebra equipped with a faithful semifinite trace τ, then:

(iii) *A derivation $\delta : S(\mathcal{M},\tau) \to S(\mathcal{M},\tau)$ is continuous in the measure topology if and only if it is inner, i. e., there exists $d \in S(\mathcal{M},\tau)$, such that $\delta = \delta_d$;*

(iv) *A derivation $\delta : S_0(\mathcal{M},\tau) \to S_0(\mathcal{M},\tau)$ is continuous in the measure topology if and only if it is spatial, i. e., there exists $d \in S(\mathcal{M},\tau)$, such that $\delta = \delta_d$.*

Proof. The proofs for all four cases are similar. For this reason, we introduce the following notation: $\mathcal{A}$, $\hat{\mathcal{A}}$, $\|\cdot\|_{\mathcal{A}}$, $t_{\mathcal{A}}$. If $\mathcal{A} = LS(\mathcal{M})$ or $\mathcal{A} = S_0(\mathcal{M})$, then $\hat{\mathcal{A}} = LS(\mathcal{M})$ and $\|\cdot\|_{\mathcal{A}} = \|\cdot\|_{LS(\mathcal{M})}$, $t_{\mathcal{A}} = t(\mathcal{M})$. If $\mathcal{A} = S(\mathcal{M},\tau)$ or $\mathcal{A} = S_0(\mathcal{M},\tau)$, then $\hat{\mathcal{A}} = S(\mathcal{M},\tau)$ and $\|\cdot\|_{\mathcal{A}} = \|\cdot\|_{S(\mathcal{M},\tau)}$, $t_{\mathcal{A}} = t_\tau$.

Since $(\mathcal{A}, \|\cdot\|_{\mathcal{A}})$ is a topological algebra, it follows that if δ is inner/spatial, then it is necessarily continuous with respect to the topology $t_{\mathcal{A}}$. Thus, it is sufficient to show that if $\delta : \mathcal{A} \to \mathcal{A}$ is a derivation continuous in $t_{\mathcal{A}}$, then there exists $d \in \hat{\mathcal{A}}$, such that $\delta = \delta_d$.

Let $\delta : \mathcal{A} \to \mathcal{A}$ be a $t_{\mathcal{A}}$-continuous derivation. By Proposition 5.1.4 without loss of generality, we can assume that δ is a self-adjoint derivation. By Corollary 5.5.4, there exists an increasing sequence $\{q_n\}_{n\in\mathbb{N}} \subset P(\mathcal{M})$, such that $q_n \to \mathbf{1}$ in the topology $t_{\mathcal{A}}$ and

$$\delta^{(q_n)}(q_n\mathcal{M}q_n \cap \mathcal{A}) \subset q_n\mathcal{M}q_n \cap \mathcal{A}$$

for any $n \in \mathbb{N}$.

If $\mathcal{A} = LS(\mathcal{M})$ or $\mathcal{A} = S(\mathcal{M},\tau)$, then $q_n\mathcal{M}q_n \cap \mathcal{A} = q_n\mathcal{M}q_n$ and, therefore, $\delta^{(q_n)}(q_n\mathcal{M}q_n) \subset q_n\mathcal{M}q_n$. Hence, Theorem 5.3.4 implies that $\delta^{(q_n)}|_{q_n\mathcal{M}q_n} = [c_n,\cdot]$ for some $c_n \in q_n\mathcal{M}q_n$.

If $\mathcal{A} = S_0(\mathcal{M})$ or $\mathcal{A} = S_0(\mathcal{M},\tau)$, then without loss of generality we can assume that $\mathcal{M}$ is a semifinite von Neumann algebra. By Corollary 4.3.24 and Corollary 4.2.15, respectively, the space $q_n\mathcal{M}q_n \cap \mathcal{A}$ is a Banach ideal in $q_n\mathcal{M}q_n$ (with respect to the operator norm). Furthermore, by Proposition 2.6.7, the strong operator closure of $q_n\mathcal{M}q_n \cap \mathcal{A}$ is $q_n\mathcal{M}q_n$. Hence, Theorem 5.3.11 implies that $\delta^{(q_n)}|_{S_0(q_n\mathcal{M}q_n)\cap q_n\mathcal{M}q_n} = [c_n,\cdot]$ for some $c_n \in q_n\mathcal{M}q_n$.

Thus, for all four cases, we have that for every $n \in \mathbb{N}$ there exists $c_n \in q_n\mathcal{M}q_n$, such that

$$\delta^{(q_n)}(x) = [c_n, x] \quad \text{for all } x \in q_n(\mathcal{M} \cap \mathcal{A})q_n. \tag{5.41}$$

We claim that the sequences $\{q_n\}_{n\in\mathbb{N}} \subset P(\mathcal{M})$ and $\{c_n\}_{n\in\mathbb{N}} \subset \mathcal{M}$ satisfy the assumption of Lemma 5.5.5. As already noted, the sequence $\{q_n\}_{n\in\mathbb{N}}$ is increasing, and converges to the identity in the topology $t_{\mathcal{A}}$, and $c_n \in q_n\mathcal{M}q_n$. Thus, it is sufficient to show that $c_n - q_nc_mq_n \in q_m\mathcal{Z}(\mathcal{M})$ for all $m \geq n$.

Let $m \geq n$ and let $x \in q_n(\mathcal{M}\cap\mathcal{A})q_n$ be arbitrary. Since $q_nx = xq_n = x$, $c_nq_n = q_nc_n = c_n$ and $\{q_n\}$ is increasing, we have

$$\begin{aligned}[q_nc_mq_n - c_n, x] &= q_n[c_m, x]q_n - [c_n, x] \\ &= q_n \cdot q_m[c_m, x]q_m \cdot q_n - q_n[c_n, x]q_n = q_n\delta^{(q_m)}(x)q_n - \delta^{(q_n)}(x)\end{aligned}$$

$$= q_n \cdot q_m \delta(x) q_m \cdot q_n - q_n \delta(x) q_n = 0.$$

As discussed above, the strong operator closure of $q_n(\mathcal{M} \cap \mathcal{A})q_n$ coincides with $q_n \mathcal{M} q_n$ and, therefore,

$$[q_n c_m q_n - c_n, x] = 0, \quad \forall x \in q_n \mathcal{M} q_n.$$

By [134, Proposition 2.2.11], we have that

$$q_n c_m q_n - c_n \in \mathcal{Z}(q_n \mathcal{M} q_n) = q_n \mathcal{Z}(\mathcal{M})$$

for all $m \geq n$. That is, indeed, the sequences $\{q_n\}_{n \in \mathbb{N}} \subset P(\mathcal{M})$ and $\{c_n\}_{n \in \mathbb{N}} \subset \mathcal{M}$ satisfy the assumption of Lemma 5.5.5.

By Lemma 5.5.5, there exists $d \in \hat{\mathcal{A}}$, such that $c_n - q_n d q_n \in q_n \mathcal{Z}(\mathcal{M})$ for all $n \in \mathbb{N}$. Combining this with equality (5.41), we obtain that for any $x \in q_n(\mathcal{M} \cap \mathcal{A})q_n$, we have

$$\delta^{(q_n)}(x) = [c_n, x] = [c_n - q_n d q_n, x] + [q_n d q_n, x] = [q_n d q_n, x]. \tag{5.42}$$

Define a derivation $\widetilde{\delta} : \mathcal{A} \to \mathcal{A}$ by setting

$$\widetilde{\delta}(x) = [d, x], \quad x \in \mathcal{A}.$$

We claim that $\delta = \widetilde{\delta}$ on $q_n(\mathcal{M} \cap \mathcal{A})q_n$ for every $n \in \mathbb{N}$.

Fix $n \in \mathbb{N}$. For all $m \geq n$, we have

$$q_m \delta(x) q_m = [q_m d q_m, x] = q_m \widetilde{\delta}(x) q_m, \quad x \in q_n(\mathcal{M} \cap \mathcal{A})q_n.$$

Since $q_m \to \mathbf{1}$ in the topology $t_{\mathcal{A}}$ as $m \to \infty$, passing $m \to \infty$, we obtain that

$$q_m \delta(x) q_m \to \delta(x), \quad q_m \widetilde{\delta}(x) q_m \to \widetilde{\delta}(x), \quad x \in q_n(\mathcal{M} \cap \mathcal{A})q_n,$$

in the topology $t_{\mathcal{A}}$. Therefore,

$$\delta(x) = \widetilde{\delta}(x), \quad x \in q_n(\mathcal{M} \cap \mathcal{A})q_n.$$

If $x \in \mathcal{M}$, then $q_n x q_n \to x$ in topology $t_{\mathcal{A}}$. Hence, the union

$$\bigcup_{n \in \mathbb{N}} q_n(\mathcal{M} \cap \mathcal{A})q_n$$

is dense in the topology $t_{\mathcal{A}}$ in $\mathcal{M} \cap \mathcal{A}$, and hence, in $\mathcal{A}$. Since δ and $\widetilde{\delta}$ are $t_{\mathcal{A}}$-continuous derivations, it follows that

$$\delta(x) = [d, x] \quad \text{for all } x \in \mathcal{A},$$

as required. □

Bibliographical notes

Derivations on algebras of bounded operators

The study of derivations on Banach algebras is divided into two parts: the bounded and the unbounded ones. The study of bounded derivations started from the mid-1940s and continues up to the present day. As a result, a plethora of excellent results have been obtained, which can offer important tools to the investigation of unbounded derivations. The motivation for the study of unbounded derivations was given by the problem of constructing the dynamics in statistical mechanics. Moreover, according to S. Sakai, the necessity of studying unbounded derivations came from some observations of Kaplansky [98], in 1958, on two apparently unrelated papers, one by Šilov [142], in 1947, having to do with differentiation, and another one by Wielandt [161], in 1949, related to quantum mechanics. We refer the reader to the book by S. Sakai [135], where a rich literature on derivations can also be found, together with a very informative preface, interesting comments, and historical remarks in all sections. Historical remarks are also contained in the introduction and in the main body of the book by O. Bratteli [46], where the main initiators and contributors to the theory of unbounded derivations are listed.

Theorem 5.3.2 was proved by Singer (see [89, p. 351] and [97, Lemma 15]).

Theorem 5.3.1 was first established by Kaplansky for the particular case of derivations on type I von Neumann algebras in [97]. There, he conjectured that any derivation on a C^*-algebra is automatically continuous. This conjecture was proved a few years later by Sakai in [132]. In the same paper [97], Kaplansky proved that any derivation on type I von Neumann algebra is inner (see [97, Theorem 10]). Note that Kaplansky proved these results for an even more general class of operator algebras, so-called AW^*-algebras of type I. Nowadays, many authors refer to this class of $*$-algebras as *Kaplansky algebras*; see [59, 69, 97, 119].

In 1966, Kadison in [88, Theorem 4] and Sakai in [133, Theorem 1 and 2] proved that each derivation of a C^*-algebra acting on a Hilbert space H extends to a derivation of the strong-operator closure of that algebra, a von Neumann algebra, and that each derivation of a von Neumann algebra is inner.

Theorem 5.3.7 was proved in [37] and [36]. The full proof of Theorem 5.3.11 showing that any derivation $\delta: \mathcal{I} \to \mathcal{I}$ on a Banach ideal $(\mathcal{I}, \|\cdot\|_{\mathcal{I}})$ in the von Neumann algebra $\mathcal{M}$ has the form $\delta = \delta_a$ for some a from the ultraweak closure of $\mathcal{I}$, was obtained in [25]. The topic of derivations with values in ideals of von Neumann algebras is very extensive and is beyond the scope of the present manuscript. We list here just a few closest to the authors' interests [30, 31, 58, 82, 86, 94, 123, 124].

The first article about derivations of unbounded operator algebras appears due to C. Brödel and G. Lassner [52], who proved that every derivation of a so-called complete O^*-algebra $\mathcal{A}$ is spatial, and is the generator of a one-parameter automorphism group of $\mathcal{A}$. Other results concerning derivations of nonnormed topological $*$-algebras can be found in [83, 84].

In 1975, Bratelli and Robinson obtained a characterization of unbounded derivations of a C^*-algebra $\mathcal{A}$, which are the infinitesimal generators of strongly continuous one-parameter groups of $*$-automorphisms of $\mathcal{A}$ [47, Theorem 1]. Following this paper, investigations of unbounded derivations on operator algebras became very popular (see, e. g., [48, 87, 104, 119, 125]).

Derivations on algebras of locally measurable operators

Theorem 5.2.6 on the extension of derivations from some $*$-subalgebras of $LS(\mathcal{M})$ to the $*$-algebra $LS(\mathcal{M})$ was proved in [23].

Theorem 5.5.6, showing that continuity of a given derivation in the local measure topology (resp., measure topology) is a necessary and sufficient condition for that derivation to be inner, was proved in several stages.

In 2008, Albeverio, Ayupov, and Kudaybergenov demonstrated in [5] that any continuous (in measure topology) derivation on the algebra of all τ-compact operators affiliated with $\mathcal{M}$ is spatial, under the additional assumption that $\mathcal{M}$ is a type I von Neumann algebra. Under the same assumption of a type I von Neumann algebra, in 2010, Ayupov and Kudaybergenov proved that any continuous derivation on $S(\mathcal{M}, \tau)$ is inner [12]. In 2011, a similar result was proved in passing by Ber, de Pagter, and Sukochev in [28] for operator-valued functions taking their values in the algebra $\mathcal{B}(H)$ of all bounded linear operators on a Hilbert space H.

The first result without any assumptions on the von Neumann algebra was proved by Ber in [21]. Namely, it was shown that any derivation on the algebra $S(\mathcal{M}, \tau)$ which is continuous in measure is necessarily inner.

In 2013, the necessity and sufficiency of continuity (in the locally measure topology) of a derivation on $LS(\mathcal{M})$ for that derivation to be inner in the case of finite von Neumann algebras was proved by Ayupov and Kudaybergenov in [14]; see also [15]. This result was extended in 2014 by Ber, Chilin, and Sukochev in [24] to the case of a general von Neumann algebra. The idea of this proof was based on the earlier result of Ber for derivations on $S(\mathcal{M}, \tau)$ [21].

Theorem 5.5.6 concerning derivations on $S_0(\mathcal{M})$ for an arbitrary von Neumann algebra is a new result. Theorem 5.5.6 for derivations on $S_0(\mathcal{M}, \tau)$ was proved by Albeverio, Ayupov, and Kudaybergenov for type I algebras in [5] and in the general case by Ayupov and Kudaybergenov in [15].

It is noteworthy that in Corollary 5.19 in [28] for operator-valued functions taking their values in the algebra $\mathcal{B}(H)$ it is proved that *any* derivation of the algebras $LS(\mathcal{M})$, $S(\mathcal{M})$, and $S(\mathcal{M}, \tau)$ is inner when $\mathcal{M}$ is the tensor product algebra $L_\infty(0,1)\bar{\otimes}\mathcal{B}_h(H)$. The proofs in [28] rely on original ideas linking the theory of derivations on operator algebras with the theory of operator-valued functions.

6 Derivations on the algebras of measurable operators affiliated with a type I_{fin} von Neumann algebra

In this chapter, we consider derivations on the algebra $LS(\mathcal{M}) = S(\mathcal{M})$ of measurable operators affiliated with a finite type I von Neumann algebra and establish a criterion for the existence of noninner (equivalently, discontinuous) derivations on the algebra $S(\mathcal{M})$. As will be shown in Chapter 7, this is the only type of von Neumann algebra that admits noninner derivations on $LS(\mathcal{M})$.

As we will show in Section 6.6, any derivation D on the algebra $S(\mathcal{M})$ associated with a type I_{fin} algebra $\mathcal{M}$ can be uniquely written as a sum of an inner derivation $D_x, x \in S(\mathcal{M})$, and a so-called centrally induced derivation D_δ lifted from a derivation δ on $S(\mathcal{Z}(\mathcal{M}))$. Furthermore, a centrally induced derivation D_δ is inner if and only if the derivation δ (and hence D_δ itself) is trivial. Equivalently, any derivation on the algebra $S(\mathcal{M})$ associated with a type I_{fin} algebra $\mathcal{M}$ is inner if and only if any derivation on $S(\mathcal{Z}(\mathcal{M}))$ is trivial. This necessitates prior study of the derivations on the algebra $S(\mathcal{M})$ for a commutative von Neumann algebra $\mathcal{M}$, or equivalently, derivations on the algebra $L_0(\Omega, \Sigma, \mu)$ of all (classes of a. e. equal) measurable functions on a σ-finite measure space (Ω, Σ, μ).

Denoting by $\mathbb{C}_c(P(\Omega))$, the algebra of all simple complex-valued functions on (Ω, Σ, μ), it is not difficult to see that any derivation $\delta : \mathbb{C}_c(P(\Omega)) \to L_0(\Omega, \Sigma, \mu)$ must be trivial (see Proposition 6.4.2 below). Furthermore, the algebra $\mathbb{C}_c(P(\Omega))$ coincides with the algebra $L_0(\Omega, \Sigma, \mu)$ if and only if the measure space (Ω, Σ, μ) is atomic. Therefore, if the measure space (Ω, Σ, μ) is atomic, then any derivation on $L_0(\Omega, \Sigma, \mu)$ is necessarily trivial. For a nonatomic measure space (Ω, Σ, μ), we aim to construct many distinct (hence, nonzero) extensions of the trivial derivation on $\mathbb{C}_c(P(\Omega))$. Hence, the principal step in the solution of this problem consists of the construction of an extension of any (nonexpansive) derivation $\delta : \mathcal{B} \to L_0(\Omega, \Sigma, \mu)$, defined on a subalgebra $\mathcal{B}$ in $L_0(\Omega, \Sigma, \mu)$, up to a derivation on the whole algebra $L_0(\Omega, \Sigma, \mu)$.

The construction of an extension of a nonexpansive derivation $\delta : \mathcal{B} \to L_0(\Omega, \Sigma, \mu)$ up to a derivation defined on the whole algebra $L_0(\Omega, \Sigma, \mu)$ proceeds in several stages. The first step is constructing an extension of δ to the minimal subalgebra containing $\mathcal{B}$ and the Boolean algebra $P(\mathcal{M})$, followed by extending it (in the second step) to the minimal regular subalgebra containing $\mathcal{B}$ and $P(\mathcal{M})$. The third step extending a derivation defined on the subalgebra $\mathcal{B}$, which contains $P(\mathcal{M})$, to a derivation on the closure $\overline{\mathcal{B}}$ of the subalgebra $\mathcal{B}$ in $(L_0(\Omega, \Sigma, \mu), \rho_\mu)$, where ρ_μ is a special complete metric on $L_0(\Omega, \Sigma, \mu)$ generated by the measure μ. The fourth step is constructing an extension of a derivation δ, defined on a closed regular subalgebra $\mathcal{B}$ containing $P(\mathcal{M})$ to a derivation on the subalgebra $\mathcal{B}(a)$ generated by $\mathcal{B}$ and an integral element $a \in L_0(\Omega, \Sigma, \mu)$ with respect to $\mathcal{B}$.

https://doi.org/10.1515/9783111599687-007

Finally, in the fifth step, we shall construct an extension of a derivation δ to the subalgebra $\mathcal{B}(a)$ in the case when a is not necessarily an integral element with respect to $\mathcal{B}$. By completing these five steps and applying Zorn's lemma, we construct an extension of a nonexpansive derivation $\delta: \mathcal{B} \to L_0(\Omega, \Sigma, \mu)$ to a derivation $\hat{\delta}$ defined on the entire algebra $L_0(\Omega, \Sigma, \mu)$. The algebraic and topological properties of the subalgebra $\mathcal{B}$ will determine whether δ admits a unique extension $\hat{\delta}$ or infinitely many distinct extensions to a derivation on $L_0(\Omega, \Sigma, \mu)$. Employing this scheme for the trivial derivation $\delta : \mathbb{C}_c(P(\Omega)) \to L_0(\Omega, \Sigma, \mu)$, we will obtain a criterion for the existence of a nonzero derivation on $L_0(\Omega, \Sigma, \mu)$ (see Theorem 6.4.9).

A prominent place in the present chapter is held for the classical derivation ∂ defined on the algebra of all (classes of a. e. equal) differentiable functions. In Section 6.5, we prove that the algebra of all classes $f \in L_0[a, b]$ for which an almost everywhere finite approximate derivative f'_{ap} exists is the maximal algebra to which the classical derivation ∂ extends uniquely. Any larger subalgebra of $L_0[a, b]$ admits a nonunique extension of this derivation. In particular, there are infinitely many extensions of the classical derivation to a derivation defined on the entire subalgebra $L_0[a, b]$. This answers the question posed by Khinčin, mentioned in the introduction.

6.1 Algebraic and topological properties of subalgebras of measurable functions

In this section, we recall some important algebraic and topological properties of the algebra $S(\mathcal{M})$ in the special case when the von Neumann algebra $\mathcal{M}$ is commutative. In this special case, the von Neumann $\mathcal{M}$ is $*$-isomorphic to the algebra $L_\infty(\Omega, \Sigma, \mu)$ for some σ-finite measure space (Ω, Σ, μ) (see Theorem 1.8.10) and the algebra $S(\mathcal{M})$ is isomorphic to the algebra $L_0(\Omega, \Sigma, \mu)$ (see Theorem 2.2.15). Thus, throughout this chapter, for a commutative von Neumann algebra $\mathcal{M}$, we simply assume that $\mathcal{M} = L_\infty(\Omega, \Sigma, \mu)$ and $S(\mathcal{M}) = L_0(\Omega, \Sigma, \mu)$ for some σ-finite measure space (Ω, Σ, μ). Two particularly interesting properties of the algebra $L_0(\Omega, \Sigma, \mu)$ (and subalgebras of $L_0(\Omega, \Sigma, \mu)$) are the regularity and completeness of $L_0(\Omega, \Sigma, \mu)$ with respect to a special metric on $L_0(\Omega, \Sigma, \mu)$. These two properties, along with another algebraic property referred to as integral completeness (introduced in the following section), are crucial for our study of derivations. As we demonstrate in Section 6.3, these three properties of a subalgebra of $L_0(\Omega, \Sigma, \mu)$ completely determine whether a derivation $\delta : \mathcal{B} \to L_0(\Omega, \Sigma, \mu)$ on a subalgebra $\mathcal{B} \subset L_0(\Omega, \Sigma, \mu)$ can be uniquely extended to a derivation on $L_0(\Omega, \Sigma, \mu)$.

Throughout this chapter, we assume that (Ω, Σ, μ) is a σ-finite measure space and denote by $L_0(\Omega, \Sigma, \mu)$ the $*$-algebra of classes of almost everywhere equal measurable complex-valued functions. If necessary, we shall distinguish between the class $[f]$ and its representative f. To shorten notation, we denote the algebra $L_0(\Omega, \Sigma, \mu)$ by $L_0(\Omega)$ and the algebra $L_\infty(\Omega, \Sigma, \mu)$ by $L_\infty(\Omega)$, when the underlying measure space (Ω, Σ, μ) is clear from the context.

The lattice of all projections in $L_\infty(\Omega)$ is denoted by $P(\Omega)$. Any projection p in $L_0(\Omega)$ is in fact the class $[\chi_A]$ for some measurable set $A \in \Sigma$. We note in this commutative setting the lattice operations for projections in $L_\infty(\Omega)$ can be described as follows:

$$p \vee q = p + q - qp, \quad p \wedge q = pq, \quad p, q \in P(\Omega). \tag{6.1}$$

A partition of unity $\{p_n\}_{n\in\mathbb{N}}$ can be equivalently defined by a partition $\{\Omega_n\}_{n\in\mathbb{N}}$ of the measure space (Ω, Σ, μ). Namely, $\{\Omega_n\}_{n\in\mathbb{N}}$ is a partition of Ω if $\Omega_n \cap \Omega_m = \emptyset$ for $n \neq m$ and $\mu(\bigcup_{n\in\mathbb{N}} \Omega_n) = \mu(\Omega)$. If $\{\Omega_n\}_{n\in\mathbb{N}}$ is a partition of Ω, then $\{[\chi_{\Omega_n}]\}_{n\in\mathbb{N}}$ is a partition of unity in $L_\infty(\Omega)$.

For the equivalence class $[f] \in L_0(\Omega)$, the support projection $s([f])$ is defined as the equivalence class $[\chi_{G(f)}]$, where $G(f) = \{\omega \in \Omega : f(\omega) \neq 0\} \in \Sigma$.

In the commutative setting, we can also define the so-called weak inverse. If f is a measurable function, then the function $i(f)$ defined by

$$(i(f))(\omega) = \begin{cases} \frac{1}{f(\omega)}, & f(\omega) \neq 0, \\ 0, & \text{otherwise}, \end{cases} \tag{6.2}$$

is also measurable. Hence, the following definition makes sense.

Definition 6.1.1. For an equivalence class $a = [f] \in L_0(\Omega)$, the *weak inverse* of $[f]$ is defined as $i(a) := [i(f)] \in L_0(\Omega)$ with $i(f)$ defined by (6.2).

It is clear that $i(ab) = i(a)i(b)$, $i(i(a)) = a$, and $ai(a) = s(a)$ for any $a \in L_0(\Omega)$. In addition,

$$i(L_0(\Omega)) := \{i(a), a \in L_0(\Omega)\} = L_0(\Omega).$$

In the following proposition, we list some essential properties of supports and weak inverses. The proof of this proposition follows directly from the definition of support $s(a)$ and weak inverse $i(a)$ for $a \in L_0(\Omega)$.

Proposition 6.1.2. *If $a, b \in L_0(\Omega)$, $p, q \in P(\Omega)$, $0 \neq \lambda \in \mathbb{C}$, then:*

(i) $s(i(a)) = s(a)$, $s(\lambda a) = s(a)$ *and* $s(a^n) = s(a)$ *for all* $n \in \mathbb{N}$*;*

(ii) $i(ab) = i(a)i(b)$ *and* $s(ab) = s(a)s(b)$*;*

(iii) $ab = 0 \Leftrightarrow s(a)s(b) = 0$*;*

(iv) *If* $ab = 0$*, then* $i(a + b) = i(a) + i(b)$, $s(a + b) = s(a) + s(b)$*;*

(v) $s(a + b) \leq s(a) \vee s(b)$*;*

(vi) $s(b - a) \geq s(b) \vartriangle s(a)$, $s(p - q) = p \vartriangle q = (p - q)^2$*, where* $p \vartriangle q$ *is the symmetric difference of* p *and* q*;*

(vii) *An element* $a \in L_0(\Omega)$ *is invertible in* $L_0(\Omega)$ *if and only if* $s(a) = \mathbf{1}$*, in this case* $a^{-1} = i(a)$.

Using the notion of weak inverse, we can define the first algebraic property of subalgebras of $L_0(\Omega)$, which will be utilized in the subsequent discussion on the extension of derivations.

Definition 6.1.3. A subalgebra $\mathcal{B}$ in the algebra $L_0(\Omega)$ is said to be a *regular subalgebra* (in the sense of von Neumann), if

$$i(\mathcal{B}) := \{i(b), b \in \mathcal{B}\} = \mathcal{B}.$$

For a given subalgebra $\mathcal{B}$, such that $P(\Omega) \subset \mathcal{B} \subset L_0(\Omega)$, we aim to describe the least regular subalgebra containing $\mathcal{B}$. As one may expect, this algebra consists of all elements of the form $ai(b), a, b \in \mathcal{B}$. We first prove a technical lemma to describe the sum of two elements of this form.

Lemma 6.1.4. *Suppose that $\mathcal{B}$ is a subalgebra of $L_0(\Omega)$, which contains all projections in $L_0(\Omega)$, and let $d_1 = a_1 i(b_1)$, $d_2 = a_2 i(b_2)$ for some $a_1, a_2, b_1, b_2 \in \mathcal{B}$. Then the element $d_1 + d_2$ can be written as $u \cdot i(v)$ with $u, v \in \mathcal{B}$ defined by*

$$u = a_1 e + a_2 f + (a_1 b_2 + a_2 b_1) g, \quad v = b_1 e + b_2 f + b_1 b_2 g,$$

with

$$e = s(d_1) - s(d_1 d_2), \quad f = s(d_2) - s(d_1 d_2), \quad g = s(d_1 d_2).$$

Proof. Since $s(d_k) = s(a_k)s(b_k)$, and $s(a_k) = s(i(a_k)), s(b_k) = s(i(b_k))$ for $k = 1, 2$ it follows that $ef = 0, eg = 0, fg = 0$. Hence,

$$\begin{aligned}
u \cdot i(v) &= (a_1 e + a_2 f + (a_1 b_2 + a_2 b_1) g) \cdot i(b_1 e + b_2 f + b_1 b_2 g) \\
&= (a_1 e + a_2 f + (a_1 b_2 + a_2 b_1) g) \cdot (i(b_1) e + i(b_2) f + i(b_1) i(b_2) g) \\
&= a_1 i(b_1) e + a_2 i(b_2) f + (a_1 b_2 + a_2 b_1) i(b_1) i(b_2) g \\
&= a_1 i(b_1) e + a_2 i(b_2) f + a_1 i(b_1) g + a_2 i(b_2) g \\
&= a_1 i(b_1)(e + g) + a_2 i(b_2)(f + g) = d_1 + d_2,
\end{aligned}$$

as required. □

Proposition 6.1.5. *If $\mathcal{B}$ is a subalgebra of $L_0(\Omega)$, such that $P(\Omega) \subset \mathcal{B}$, then*

$$\mathcal{R}(\mathcal{B}) := \{a \cdot i(b) : a, b \in \mathcal{B}\} \tag{6.3}$$

is the smallest regular subalgebra in $L_0(\Omega)$ containing $\mathcal{B}$.

Proof. By Proposition 6.1.2(i), for any $b \in \mathcal{B}$, we have that $i(s(b)) = s(b) \in P(\Omega) \subset \mathcal{B}$. Therefore, we can write that $b = bs(b) = bi(s(b)) \in \mathcal{R}(\mathcal{B})$. Thus, $\mathcal{R}(\mathcal{B})$ contains $\mathcal{B}$.

Let us show that $\mathcal{R}(\mathcal{B})$ is a subalgebra of $L_0(\Omega)$. It is clear that

$$\alpha \cdot \mathcal{R}(\mathcal{B}) \subset \mathcal{R}(\mathcal{B})$$

for all $\alpha \in \mathbb{C}$. In addition, for any $a_k, b_k \in \mathcal{B}$, we have that

$$a_1 i(b_1) \cdot a_2 i(b_2) = a_1 a_2 i(b_1 b_2).$$

That is,

$$\mathcal{R}(\mathcal{B}) \cdot \mathcal{R}(\mathcal{B}) \subset \mathcal{R}(\mathcal{B}).$$

By Lemma 6.1.4, $\mathcal{R}(\mathcal{B})$ is closed under addition. Thus, $\mathcal{R}(\mathcal{B})$ is a subalgebra in $L_0(\Omega)$.

The equations

$$i(ab) = i(a)i(b), \quad i(i(a)) = a$$

imply that $i(a \cdot i(b)) = i(a) \cdot b$, $a, b \in \mathcal{B}$. Therefore, $\mathcal{R}(\mathcal{B})$ is a regular subalgebra in $L_0(\Omega)$.

The fact that $\mathcal{R}(\mathcal{B})$ is the smallest subalgebra among all regular subalgebras in $\mathcal{A}$ containing $\mathcal{B}$ follows immediately from the definition of $\mathcal{R}(\mathcal{B})$. □

To study the topological properties of the algebra $L_0(\Omega)$, we first need to recall another algebraic property of $L_0(\Omega)$.

Recall, that two projections $p, q \in P(\Omega)$ are disjoint if $p \wedge q = 0$. Since any $p, q \in P(\Omega)$ commute, it follows that $p \wedge q = pq$ (see Section 1.1). In particular, two projections $p, q \in P(\Omega)$ are disjoint if and only if they are orthogonal to one another.

Two elements a and b of the algebra $L_0(\Omega)$ are said to be *disjoint*, if $ab = 0$, or equivalently $s(a)s(b) = 0$. Since the measure space (Ω, Σ, μ) is σ-finite, it follows that a family $\{a_i\}_{i \in I}$ of pairwise disjoint elements in $L_0(\Omega)$ is at most countable. It is clear that for any family $\{a_n\}_{n \in \mathbb{N}}$ of pairwise disjoint measurable functions the sum $\sum_{n=1}^{\infty} a_n$ is also measurable.

Definition 6.1.6. A subalgebra $\mathcal{B}$ in $L_0(\Omega)$ is said to be *disjointly complete*, if for every family $\{a_n\}_{n \in \mathbb{N}}$ of pairwise disjoint elements of $\mathcal{B}$ the sum $\sum_{n=1}^{\infty} a_n$ belongs to $\mathcal{B}$.

As noted above, the algebra $L_0(\Omega)$ is disjointly complete.

Remark 6.1.7. *Suppose that $\{a_n\}_{n \in \mathbb{N}}$ is a family of pairwise disjoint elements in $L_0(\Omega)$. It is clear that the element $a = \sum_{n=1}^{\infty} a_n$ satisfies*

$$s(a) = \sup_{n \in \mathbb{N}} s(a_n), \quad as(a_n) = a_n.$$

Furthermore, a is the unique element in $L_0(\Omega)$, which satisfies these two equalities.

Now we introduce a special topology on the algebra $L_0(\Omega)$, which plays a crucial role in the construction of extensions of derivations. This topology is closely related to the disjoint completeness of the algebra $L_0(\Omega)$. In fact, a subalgebra $\mathcal{B}$ of $L_0(\Omega)$ is complete with respect to this topology if and only if it is disjointly complete (see Theorem 6.1.13 below).

By assumption, the measure space (Ω, Σ, μ) is σ-finite. Hence, there exists a strictly positive finite measure μ_f on $P(\Omega)$, which is equivalent to μ (see Remark 1.8.11). The mapping

$$d(e, q) := \mu_f(e \triangle q), \quad e, q \in P(\Omega),$$

defines a metric on $P(\Omega)$, and the pair $(P(\Omega), d)$ is a complete metric space (see, e. g., [155, Chapter III, Section 5]).

We now define a special metric on $L_0(\Omega)$, which extends the metric d on $P(\Omega)$.

Definition 6.1.8. Let (Ω, Σ, μ) be a σ-finite measure space and let μ_f be a strictly positive finite measure μ_f on $P(\Omega)$ which is equivalent to μ. We define

$$\rho_\mu(a, b) = \mu_f(s(a - b)), \quad a, b \in L_0(\Omega).$$

Proposition 6.1.9. *ρ_μ is a metric on $L_0(\Omega)$, such that*

$$\rho_\mu(i(a), i(b)) = \rho_\mu(a, b), \quad \rho_\mu(\lambda a, \lambda b) = \rho_\mu(a, b), \quad a, b \in L_0(\Omega), 0 \neq \lambda \in \mathbb{C}.$$

Moreover, the algebraic operations $a + b$, ab, are jointly continuous and the operations $a \mapsto (-a)$ is continuous in $(L_0(\Omega), \rho_\mu)$, i. e., $L_0(\Omega)$ is a topological ring in the topology given by ρ_μ.

Proof. We first show that ρ_μ is indeed a metric on $L_0(\Omega)$.

Since the measure μ_f is strictly positive, we clearly have that

$$\rho_\mu(a, b) = 0 \Leftrightarrow a = b.$$

Since $s(a - b) = s(b - a)$ for all $a, b \in L_0(\Omega)$, it follows that $\rho_\mu(a, b) = \rho_\mu(b, a)$. Finally, since $a - b = (a - c) + (c - b)$, $c \in L_0(\Omega)$, it follows that $s(a - b) \leq s(a - c) \vee s(c - b)$ (Proposition 6.1.2(v)) and, therefore, $\rho_\mu(a, b) \leq \rho_\mu(a, c) + \rho_\mu(c, b)$. Hence, ρ_μ is a metric on $L_0(\Omega)$.

To show the equality $\rho_\mu(i(a), i(b)) = \rho_\mu(a, b)$, suppose that a, $b \in L_0(\Omega)$. Setting $e = \mathbf{1} - s(a-b)$, we have that $(a-b)e = 0$, i. e., $a\,e = b\,e$. Hence, $i(a)e = i(a\,e) = i(b\,e) = i(b)e$ and, therefore, $(i(a) - i(b))e = 0$. Consequently, $e \leq \mathbf{1} - s(i(a) - i(b))$, i. e., $s(i(a) - i(b)) \leq s(a-b)$. Replacing in the argument above a by $i(a)$, and b by $i(b)$, we get that $s(a-b) \leq s(i(a) - i(b))$. Thus, $s(a - b) = s(i(a) - i(b))$ and, therefore, $\rho_\mu(i(a), i(b)) = \rho_\mu(a, b)$.

Thus, in order to complete the proof, we only need to show the continuity of the algebraic operations in $L_0(\Omega)$. For the continuity of taking the (additive) inverse, it is sufficient to note that

$$\rho_\mu(-a,-b) = \mu_f(s(-a+b)) = \mu_f(s(a-b)) = \rho_\mu(a,b), \quad a,b \in L_0(\Omega).$$

The continuity of the other two operations follows from the estimates:

$$\begin{aligned}\rho_\mu(a+b,c+d) &= \mu_f(s((a+b)-(c+d))) \le \mu_f(s(a-c) \vee s(b-d)) \\ &\le \mu_f(s(a-c)+s(b-d)) \le \rho_\mu(a,c)+\rho_\mu(b,d)\end{aligned}$$

and

$$\begin{aligned}\rho_\mu(ab,cd) &= \mu_f(s((ab-cb)+(cb-cd))) \le \mu_f(s(ab-cb) \vee s(cb-cd)) \\ &\le \mu_f(s(a-c)) + \mu_f(s(b-d)) = \rho_\mu(a,c)+\rho_\mu(b,d)\end{aligned}$$

for arbitrary $a,b,c,d \in L_0(\Omega)$. □

Remark 6.1.10. *Since* $s(e-f) = e \triangle f$ *(see Proposition* 6.1.2(iv)*), it follows that the metric* ρ_μ *extends the metric* $d(e,q) = \mu_f(e \triangle q)$, $e,g \in P(\Omega)$ *to a metric on* $L_0(\Omega)$. *Hence, Proposition* 6.1.9 *implies that* $P(\Omega)$ *is a closed subset in* $(L_0(\Omega),\rho_\mu)$.

As we show in the following proposition, the disjoint completeness of the algebra $L_0(\Omega)$ guarantees that the series $\sum_{n=1}^{\infty} a_n$ of pairwise disjoint elements always converges with respect to the metric ρ_μ.

Proposition 6.1.11. *Let* $\{a_n\}_{n\in\mathbb{N}} \in L_0(\Omega)$ *be a sequence of pairwise disjoint elements in* $L_0(\Omega)$. *Then the series* $\sum_{n=1}^{\infty} a_n$ *converges in* $(L_0(\Omega),\rho_\mu)$.

Proof. Denote by $a = \sum_{n=1}^{\infty} a_n$. Since $s(a) = \sup_{n\in\mathbb{N}} s(a_n)$, it follows that

$$\mu_f(s(a)) = \lim_{n\to\infty} \mu_f\Big(\sup_{1\le k\le n} s(a_k)\Big).$$

For any $n \in \mathbb{N}$, we have that $a-\sum_{k=1}^{n} a_k = \sum_{k=n+1}^{\infty} a_k$ and so $s(a-\sum_{k=1}^{n} a_k) = \sup_{k\ge n+1} s(a_k)$. Therefore,

$$\rho_\mu\left(\sum_{k=1}^{n} a_k, a\right) = \mu_f\left(s\left(a - \sum_{k=1}^{n} a_k\right)\right) = \mu_f\Big(\sup_{k\ge n+1} s(a_k)\Big) \to 0,$$

as $n \to \infty$. Thus, the series $\sum_{n=1}^{\infty} a_n$ converges with respect to ρ_μ to a. □

We now introduce a special topological property of a subalgebra $\mathcal{B}$ of $L_0(\Omega)$, which we use exclusively in this chapter.

Definition 6.1.12. For a subalgebra $\mathcal{B}$ of $L_0(\Omega)$, we denote by $\overline{\mathcal{B}}^{\rho_\mu}$ the closure of $\mathcal{B}$ in $L_0(\Omega)$ with respect to ρ_μ.

In the following theorem, we present a necessary and sufficient condition for a subalgebra $\mathcal{B} \subset L_0(\Omega)$ to be disjointly complete.

Theorem 6.1.13. *Let $\mathcal{B}$ be a (not necessarily proper) subalgebra of $L_0(\Omega)$, which contains all projections of $L_0(\Omega)$. Then the following conditions are equivalent:*
(i) *The algebra $\mathcal{B}$ is complete with respect to the metric ρ_μ;*
(ii) *The algebra $\mathcal{B}$ is disjointly complete.*

Proof. (i)⇒(ii). Suppose that $\{a_n\}_{n\in\mathbb{N}}$ is a family of pairwise disjoint functions. By Proposition 6.1.11, the series $\sum_{n=1}^{\infty} a_n$ converges in $(\mathcal{B}, \rho_\mu)$. Since $(\mathcal{B}, \rho_\mu)$ is complete, it follows that $\sum_{n=1}^{\infty} a_n \in \mathcal{B}$, proving that $\mathcal{B}$ is disjointly complete.

(ii)⇒(i). Suppose that $\mathcal{B}$ is disjointly complete and let $\{a_n\}_{n\in\mathbb{N}}$ be a Cauchy sequence in $(\mathcal{B}, \rho_\mu)$. Passing to a subsequence if necessary, we may assume that

$$\rho_\mu(a_{n+1}, a_n) = \mu_f(s(a_{n+1} - a_n)) < \frac{1}{2^n} \quad \text{for all } n \in \mathbb{N}.$$

For each fixed $k \in \mathbb{N}_0$, we set

$$q_0 = 0, \quad q_k := \bigwedge_{n=k}^{\infty} (\mathbf{1} - s(a_{n+1} - a_n)).$$

Clearly, $q_k \le q_{k+1}$ and $\mathbf{1} - q_k = \bigvee_{n=k}^{\infty} s(a_{n+1} - a_n)$, $k \in \mathbb{N}_0$. Thus,

$$\mu_f(\mathbf{1} - q_k) = \mu_f\left(\bigvee_{n=k}^{\infty} s(a_{n+1} - a_n)\right) \le \sum_{n=k}^{\infty} \mu_f(s(a_{n+1} - a_n))$$
$$\le \sum_{n=k}^{\infty} \frac{1}{2^n} = \frac{1}{2^{k-1}} \to 0 \quad \text{as } k \to \infty.$$

Thus, $q_k \uparrow \mathbf{1}$ as $k \to \infty$.

For every $n \ge k$, we have that $s(a_{n+1} - a_n)q_k = 0$, and so $a_{n+1}q_k = a_n q_k$. Hence, by induction we infer that

$$a_n q_k = a_k q_k, \quad n \ge k. \tag{6.4}$$

Since $q_n \uparrow \mathbf{1}$, the family $\{a_n(q_n - q_{n-1})\}_{n\in\mathbb{N}}$ is pairwise disjoint. Since $\mathcal{B}$ contains all projections of $L_0(\Omega)$, we also have that $\{a_n(q_n - q_{n-1})\}_{n\in\mathbb{N}} \subset \mathcal{B}$. Since $\mathcal{B}$ is disjointly complete, we have that $a = \sum_{n=1}^{\infty} a_n(q_n - q_{n-1}) \in \mathcal{B}$. Furthermore, by Proposition 6.1.11, we have that

$$a = \sum_{n=1}^{\infty} a_n(q_n - q_{n-1}) = \lim_{k\to\infty} \sum_{n=1}^{k} a_n(q_n - q_{n-1}) \in \mathcal{B}, \tag{6.5}$$

where the limit is taken with respect to the metric ρ_μ.

We claim that a is the limit of a_n with respect to ρ_μ. Fix $\varepsilon > 0$. Since

$$\rho_\mu(a_m, a_m q_k) = \mu_f(s(a_m(\mathbf{1} - q_k))) \leq \mu_f(\mathbf{1} - q_k) \to 0 \quad \text{as } k \to \infty,$$

there exists k_1 such that

$$\rho_\mu(a_m, a_m q_k) < \varepsilon, \quad m \in \mathbb{N},\ k \geq k_1. \tag{6.6}$$

By (6.5), there exists $k_2 \geq k_1$ such that

$$\rho_\mu\left(a, \sum_{n=1}^{k} a_n(q_n - q_{n-1})\right) < \varepsilon \quad \text{for all } k \geq k_2. \tag{6.7}$$

Furthermore, by (6.4) for every $m \geq k_2$, we have that

$$\rho_\mu\left(\sum_{n=1}^{k_2} a_n(q_n - q_{n-1}), \sum_{n=1}^{k_2} a_m(q_n - q_{n-1})\right) = 0. \tag{6.8}$$

Hence, combining (6.6), (6.7), and (6.8), for every $m \geq k_2$ we obtain that

$$\begin{aligned}
\rho_\mu(a, a_m) &\leq \rho_\mu\left(a, \sum_{n=1}^{k_2} a_n(q_n - q_{n-1})\right) \\
&\quad + \rho_\mu\left(\sum_{n=1}^{k_2} a_n(q_n - q_{n-1}), \sum_{n=1}^{k_2} a_m(q_n - q_{n-1})\right) \\
&\quad + \rho_\mu\left(\sum_{n=1}^{k_2} a_m(q_n - q_{n-1}), a_m\right) \\
&< \varepsilon + \rho_\mu(a_m q_{k_2}, a_m) < 2\varepsilon.
\end{aligned}$$

Thus, $\rho_\mu(a, a_m) \to 0$ as $m \to \infty$, i. e., $(\mathcal{B}, \rho_\mu)$ is a complete metric space. □

Since the algebra $L_0(\Omega)$ is disjointly complete, as an immediate corollary of Theorem 6.1.13, we obtain the following result.

Corollary 6.1.14. *The metric space $(L_0(\Omega), \rho_\mu)$ is complete.*

We also prove several assertions for the topological closure $\overline{\mathcal{B}}^{\rho_\mu}$ of a subalgebra $\mathcal{B}$. In particular, we show that $\overline{\mathcal{B}}^{\rho_\mu}$ is also a subalgebra of $L_0(\Omega)$ and the regularity of the subalgebra $\mathcal{B}$ is preserved under taking the topological closure.

Proposition 6.1.15. *Suppose that $\mathcal{B}$ is a subalgebra of $L_0(\Omega)$. Then the closure $\overline{\mathcal{B}}^{\rho_\mu}$ of the algebra $\mathcal{B}$ in $(L_0(\Omega), \rho_\mu)$ is a subalgebra of $L_0(\Omega)$. In addition, if the subalgebra $\mathcal{B}$ is regular, then $\overline{\mathcal{B}}^{\rho_\mu}$ is regular, too.*

Proof. It follows from Proposition 6.1.9 that $\overline{\mathcal{B}}^{\rho_\mu}$ is a subring in $L_0(\Omega)$. Let $\lambda \in \mathbb{C}$, $a \in \overline{\mathcal{B}}^{\rho_\mu}$, $a_n \in \mathcal{B}$, $\rho_\mu(a_n, a) \to 0$. If $\lambda \neq 0$, then

$$\rho_\mu(\lambda a_n, \lambda a) = \rho(a_n, a) \to 0,$$

i. e., $\lambda a \in \overline{\mathcal{B}}^{\rho_\mu}$. If $\lambda = 0$, then $\lambda a = 0 \in \mathcal{B} \subset \overline{\mathcal{B}}^{\rho_\mu}$. Consequently, $\overline{\mathcal{B}}^{\rho_\mu}$ is a subalgebra in $L_0(\Omega)$.

Suppose now that $\mathcal{B}$ is a regular subalgebra in $L_0(\Omega)$. Suppose that $a \in \overline{\mathcal{B}}^{\rho_\mu}$ and let $a_n \in \mathcal{B}$, $n \in \mathbb{N}$, be such that $\rho_\mu(a_n, a) \to 0$ as $n \to \infty$. By Proposition 6.1.9, we have that $\rho_\mu(i(a_n), i(a)) = \rho_\mu(a_n, a) \to 0$ as $n \to \infty$. Therefore, since $i(a_n) \in \mathcal{B}$ by assumption, we conclude that $i(a) \in \overline{\mathcal{B}}^{\rho_\mu}$. Thus, the subalgebra $\overline{\mathcal{B}}^{\rho_\mu}$ is also a regular subalgebra in $L_0(\Omega)$. ☐

In conclusion of this section, we provide an auxiliary characterization of topological closures of subalgebras of $L_0(\Omega)$.

Proposition 6.1.16. *Suppose that $\mathcal{B}$ is a subalgebra in $L_0(\Omega)$ containing all projections from $L_0(\Omega)$. Then, for every element $b \in \overline{\mathcal{B}}^{\rho_\mu}$ there exists a sequence $\{b_n\}_{n\in\mathbb{N}} \subset \mathcal{B}$, such that $s(b_n) \uparrow s(b)$ and $\rho_\mu(b_n, b) \to 0$ as $n \to \infty$.*

Proof. Since $b \in \overline{\mathcal{B}}^{\rho_\mu}$, there exists a sequence $\{c_n\}_{n\in\mathbb{N}} \subset \mathcal{B}$, such that $\rho_\mu(c_n, b) \to 0$ as $n \to \infty$. By Proposition 6.1.9, we have that $s(c_n) = c_n i(c_n) \to bi(b) = s(b)$ in $(L_0(\Omega), \rho_\mu)$. Similarly, for $d_n = c_n s(b)$, we have $\rho_\mu(d_n, b) \to 0$ and $\rho_\mu(s(d_n), s(b)) \to 0$ as $n \to \infty$, in addition, $s(d_n) \leq s(b)$.

Let

$$e_n = \sup_{1\leq k\leq n} s(d_k), \quad e = \sup_{n\in\mathbb{N}} e_n = \sup_{k\in\mathbb{N}} s(d_k).$$

Since $\rho_\mu(s(d_n), s(b)) \to 0$ and $s(d_n) \leq s(b)$, it follows that $s(b) = e$ and $e_n \uparrow s(b)$. In particular, $\rho_\mu(e_n, s(b)) \to 0$.

We define

$$b_n = d_n + (e_n - s(d_n)), \quad n \in \mathbb{N}.$$

Using the inclusion $P(\Omega) \subset \mathcal{B}$ and the convergence $\rho_\mu(s(d_n), s(b)) \to 0$, we have $\{b_n\}_{n\in\mathbb{N}} \subset \mathcal{B}$ and

$$\begin{aligned}\rho_\mu(b_n, b) &= \rho_\mu(d_n - b, -(e_n - s(d_n))) \\ &\leq \rho_\mu(d_n - b, 0) + \rho_\mu((e_n - s(d_n)), 0) \to 0 \quad \text{as } n \to \infty.\end{aligned}$$

Noting that, $s(b_n) = e_n \uparrow s(b)$, we conclude the proof. ☐

6.2 Integral closure of subalgebras of measurable functions

In this section, we introduce the third crucial property of subalgebras of $L_0(\Omega)$. This property is the integral closedness of a subalgebra $\mathcal{B}$ in $L_0(\Omega)$. The main purpose of this section is to introduce integral and weakly transcendental elements with respect to a subalgebra $\mathcal{B}$ of $L_0(\Omega)$ and study their properties. Defining then integrally closed subalgebras in $L_0(\Omega)$ as subalgebras containing all its integral elements, the main result of this section (see Theorem 6.2.11 below) shows that for any subalgebra $\mathcal{B}$ of $L_0(\Omega)$ there exists the least regular, integrally closed, and disjointly complete subalgebra $E(\mathcal{B})$, containing $\mathcal{B}$ and $P(\Omega)$. We shall give an exact construction of $E(\mathcal{B})$.

The algebra $E(\mathcal{B})$ plays a pivotal role in the construction of the extension of a derivation $\delta : \mathcal{B} \to L_0(\Omega)$. In fact, as we show in Section 6.3, this algebra $E(\mathcal{B})$ is the largest subalgebra of $L_0(\Omega)$ containing $\mathcal{B}$ and admitting the unique extension of the derivation $\delta : \mathcal{B} \to L_0(\Omega)$.

We start this section with the definition of elements integral with respect to a given subalgebra of $L_0(\Omega)$ and then present their properties. After that, we introduce the class of integrally closed subalgebras of $L_0(\Omega)$ and show how the property of integral closedness is connected with regularity and disjoint completeness. With these preparations at hand, we prove the main result of the section, Theorem 6.2.11.

We shall conclude this section by showing that if, for a given subalgebra $\mathcal{B}$ of $L_0(\Omega)$, the algebra $E(\mathcal{B})$ is proper, then there exist sufficiently many weakly transcendental elements with respect to $E(\mathcal{B})$.

As before, we assume that (Ω, Σ, μ) is a σ-finite measure space and denote by $L_0(\Omega)$ the algebra of all (classes of) complex-valued measurable functions.

Let $\mathcal{B}$ be an arbitrary subalgebra of $L_0(\Omega)$. Denote by $\mathcal{B}[x]$ the algebra of all polynomials with coefficient from $\mathcal{B}$, i. e.,

$$\mathcal{B}[x] = \left\{ \sum_{k=0}^{n} a_k x^{n-k} : a_k \in \mathcal{B},\ k = 0, 1, \ldots, n,\ n \in \mathbb{N} \right\},$$

where, as usual, $a_n x^0 = a_n$.

If $p(x) = \sum_{k=0}^{n} a_k x^{n-k} \in \mathcal{B}[x]$, then the greatest integer $n-k$ such that $a_k \neq 0$, is called the *degree* of the polynomial $p(x)$ and is denoted by $\deg p$, and in this case the coefficient a_k is called the *leading coefficient* of the polynomial $p(x)$. If the leading coefficient of the polynomial $p(x)$ is equal to $\mathbf{1}$, then $p(x)$ is called a *monic* polynomial. The set of all monic polynomials from $\mathcal{B}[x]$ will be denoted by $\mathcal{B}_u[x]$.

In what follows, we identify the algebra $\mathcal{B}[x]$ with a subalgebra of the algebra $(L_0(\Omega))[x]$.

Using the algebra $\mathcal{B}[x]$ of polynomials with coefficients from $\mathcal{B}$, for every $a \in L_0(\Omega)$, we can define the subalgebra

$$\mathcal{B}(a) = \{p(a) : p \in \mathcal{B}[x]\} \tag{6.9}$$

of $L_0(\Omega)$ generated by $\mathcal{B}$ and the element a. It is clear that $\mathcal{B} \subset \mathcal{B}(a)$, and if $a \in \mathcal{B}$, then $\mathcal{B}(a) = \mathcal{B}$.

Definition 6.2.1. Let $\mathcal{B}$ be a subalgebra of $L_0(\Omega)$. An element $a \in L_0(\Omega)$ is said to be:
(i) *algebraic* with respect to the subalgebra $\mathcal{B}$, if there exists a nonzero polynomial $p \in \mathcal{B}[x]$, such that $p(a) = 0$;
(ii) *integral* with respect to the subalgebra $\mathcal{B}$ (or simply a *$\mathcal{B}$-integral element*), if there exists a monic polynomial $p \in \mathcal{B}_u[x]$, such that $p(a) = 0$. In this case, we denote

$$\deg_{\mathcal{B}}(a) := \min\{\deg p : p \in B_u[x],\ p(a) = 0\}.$$

Note that for any $\mathcal{B}$-integral element $a \in L_0(\Omega)$ we have that $\deg_{e\mathcal{B}}(ea) \leq \deg_{\mathcal{B}}(a)$, $e \in P(\Omega)$. However, as the following example shows, the inequality can be strict in general.

Example 6.2.2. *Consider the algebra $L_0[0,1]$ with the Lebesgue measure on $[0,1]$ and let $\mathcal{B}$ be the subalgebra generated by all polynomials and projections from $L_0[0,1]$. Define*

$$a = \begin{cases} t, & t \in [0, \frac{1}{2}), \\ \sqrt{t}, & t \in [\frac{1}{2}, 1] \end{cases}$$

and $e = [\chi_{[0,1/2]}]$. It is clear that $\deg_{e\mathcal{B}}(ea) = 1$ and $\deg_{\mathcal{B}}(a) = 2$.

We now distinguish the class of $\mathcal{B}$-integral elements, such that for all $e \in P(\Omega)$ with $e \leq s(a)$, the inequality $\deg_{e\mathcal{B}}(ea) \leq \deg_{\mathcal{B}}(a)$ is, in fact, an equality.

Definition 6.2.3. Let $\mathcal{B}$ be a subalgebra of $L_0(\Omega)$ and let $a \in L_0(\Omega)$ be a $\mathcal{B}$-integral element. We say that a is a *proper $\mathcal{B}$-integral* element if, for any nonzero $e \leq s(a)$, it holds that $\deg_{e\mathcal{B}}(ea) = \deg_{\mathcal{B}}(a)$.

As shown in the following proposition, a proper $\mathcal{B}$-integral element has an important property that, if it is invertible, then its inverse is also a $\mathcal{B}$-integral element.

Proposition 6.2.4. *Assume that $\mathcal{B}$ is a regular subalgebra of the algebra $L_0(\Omega)$ with $P(\Omega) \subset \mathcal{B}$. Let $a \in L_0(\Omega)$ be a proper $\mathcal{B}$-integral element with $s(a) = \mathbf{1}$. Then the inverse element a^{-1} is a $\mathcal{B}$-integral element, too.*

Proof. By assumption, there exists a monic polynomial

$$p(x) = x^n + a_1 x^{n-1} + \cdots + a_{n-1}x + a_n \in \mathcal{B}_u[x],$$

such that $p(a) = 0$ and $n = \deg(p) = \deg_{\mathcal{B}}(a)$. Furthermore, by the definition of proper integral elements, for any $0 \neq e \leq s(a) = \mathbf{1}$, we have that $\deg_{\mathcal{B}}(a) = \deg_{\mathcal{B}e}(ae)$.

Let us show first that the free coefficient a_n of the polynomial p is invertible in $L_0(\Omega)$. It is sufficient to show that $s(a_n) = \mathbf{1}$. Suppose that $e = \mathbf{1} - s(a_n) \neq 0$ and consider the monic polynomial

$$q(x) = x^{n-1} + ea_1x^{n-2} + \cdots + ea_{n-2}x + ea_{n-1}.$$

Since $P(\Omega) \subset \mathcal{B}$, it follows that $q \in \mathcal{B}[x]$. In addition, since $ea_n = 0$ by the definition of e, it follows that

$$q(ea)a = ea^n + ea_1a^{n-1} + \cdots + ea_{n-2}a^2 + ea_{n-1}a + ea_n = ep(a) = 0.$$

By assumption $s(a) = \mathbf{1}$ and, therefore, there exists the inverse element a^{-1}. Multiplying the equality $q(ea)a = 0$ by a^{-1}, we obtain that $q(ea) = 0$. Hence,

$$n = \deg_{\mathcal{B}}(a) = \deg_{\mathcal{B}e}(ae) \leq \deg(q) = n - 1.$$

The obtained contradiction implies that $e = 0$, i. e., $s(a_n) = \mathbf{1}$ and, therefore, a_n is invertible in $L_0(\Omega)$.

By the assumption, the algebra $\mathcal{B}$ is regular and $a_n \in \mathcal{B}$. Therefore, $a_n^{-1} = i(a_n) \in \mathcal{B}$. Consider now the monic polynomial

$$q(x) = x^n + a_{n-1}a_n^{-1}x^{n-1} + \cdots + a_1a_n^{-1}x + a_n^{-1} \in \mathcal{B}[x].$$

We have that

$$\begin{aligned} q(a^{-1})a_na^n &= (a^{-n} + a_{n-1}a_n^{-1}a^{-n+1} + \cdots + a_1a_n^{-1}a^{-1} + a_n^{-1})a_na^n \\ &= a_n + a_{n-1}a + \cdots + a_1a^{n-1} + a^n = p(a) = 0. \end{aligned}$$

Since both a and a_n are invertible, multiplying the above equality by a_n^{-1} and a^{-n}, we conclude that $q(a^{-1}) = 0$. Thus, a^{-1} is a $\mathcal{B}$-integral element, as required. □

For future purpose, we also prove a technical lemma, which shows that for an invertible proper $\mathcal{B}$-integral element a, any polynomial $q \in \mathcal{B}[x]$ for which $q(a) = 0$, necessarily belongs to the principal ideal in $\mathcal{B}[x]$ generated by a monic polynomial $p \in \mathcal{B}_u[x]$ whose degree is equal to $\deg_{\mathcal{B}}(a)$.

Lemma 6.2.5. *Let $\mathcal{B}$ be a regular subalgebra of an algebra $L_0(\Omega)$, $P(\Omega) \subset \mathcal{B}$, and let $a \in L_0(\Omega)$ be an invertible proper $\mathcal{B}$-integral element. Let*

$$p(x) = x^n + a_1x^{n-1} + \cdots + a_{n-1}x + a_n \in \mathcal{B}_u[x]$$

be a monic polynomial, such that $n = \deg(p) = \deg_{\mathcal{B}}(a)$ and $p(a) = 0$. If $q \in \mathcal{B}[x]$ and $q(a) = 0$, then $q(x) = u(x)p(x)$ for some $u \in \mathcal{B}[x]$.

Proof. By [154, Chapter 3, Section 14], we have that $q(x) = u(x)p(x) + r(x)$ for some $u, r \in \mathcal{B}[x]$ with $\deg r < \deg p = n$. By assumption, we have that $p(a) = 0 = q(a)$ and, therefore, $r(a) = q(a) - u(a)p(a) = 0$. Suppose that $r \neq 0$ and write

$$r(x) = b_0x^m + b_1x^{m-1} + \cdots + b_{m-1}x + b_m, \quad b_i \in \mathcal{B},\ i = 0, 1, \ldots, m,\ m < n,$$

where $b_0 \neq 0$, i. e., $s(b_0) \neq 0$. Since $\mathcal{B}$ is a regular algebra, we have that $i(b_0) \in \mathcal{B}$ and, therefore, the polynomial

$$i(b_0)r(x) = s(b_0)x^m + i(b_0)b_1x^{m-1} + \cdots + i(b_0)b_{m-1}x + i(b_0)b_m$$

is also contained in the algebra $\mathcal{B}[x]$. For the monic polynomial,

$$\begin{aligned} q_1(x) &= x^m + i(b_0)r(x) - s(b_0)x^m \\ &= x^m + i(b_0)b_1x^{m-1} + \cdots + i(b_0)b_{m-1}x + i(b_0)b_m \in \mathcal{B}_u[x] \end{aligned}$$

we have that

$$q_1(s(b_0)a) = i(b_0)r(s(b_0)a) = i(b_0)s(b_0)r(a) = 0.$$

Consequently, using assumption $\deg_{\mathcal{B}e}(ae) = \deg_{\mathcal{B}}(a)$ for every nonzero $e \in P(\Omega)$, we have

$$n = \deg_{\mathcal{B}}(a) = \deg_{\mathcal{B}s(b_0)s(a)}(s(b_0)a) \leq \deg q_1 = m = \deg r < n.$$

The obtained contradiction implies that $r(x) = 0$ and $q(x) = u(x)p(x)$. □

In the following proposition, we show that for any integral element $a \in L_0(\Omega)$ with respect to a given regular disjointly complete subalgebra $\mathcal{B}$, there exists finitely many disjoint sets $\{A_k\}_{k=1}^m \subset \Sigma$, $m \in \mathbb{N}$, which on one hand exhaust the support of a, and on the other hand every $a[\chi_{A_k}]$ is properly $\chi_{A_k}\mathcal{B}$-integral.

Proposition 6.2.6. *Assume that $\mathcal{B}$ is a regular disjointly complete subalgebra in $L_0(\Omega)$ with $P(\Omega) \subset \mathcal{B}$ and let $a \in L_0(\Omega)$ be a nonzero $\mathcal{B}$-integral element. Then there exist integers $n_1 < \cdots < n_m \leq \deg_{\mathcal{B}}(a)$ and nonzero pairwise disjoint projections $e_1, \ldots, e_m \in P(\Omega)$, such that:*
(i) $s(a) = \sum_{k=1}^m e_k$;
(ii) *every ae_k is a proper $\mathcal{B}e_k$-integral element in $L_0(\Omega)e_k$, such that $n_k = \deg_{\mathcal{B}e_k}(ae_k)$.*

Proof. For every nonzero $e \leq s(a)$, we set

$$n(e) := \min\{\deg p : p \in \mathcal{B}_u[x],\ p(ae) = 0\}.$$

For the positive integer n_1 given by

$$n_1 = n_1(a) := \min\{n(e) : 0 \neq e \leq s(a)\},$$

let us consider the set of idempotents from $P(\Omega)$ defined by

$$J_1 := \{e \in P(\Omega) : n(e) = n_1,\ 0 \neq e \leq s(a)\} \cup \{0\}.$$

Consider two monic polynomials $p_j(x) = x^{n_1} + a_{n_1-1}^{(j)}x^{n_1-1} + \cdots + a_0^{(j)} \in \mathcal{B}_u[x]$, such that $p_j(ae) = 0$, $0 \neq e \in J_1$, $j = 1, 2$. Let us show that

$$a_k^{(1)}e = a_k^{(2)}e \quad \text{for all } k = 0, \ldots, n_1 - 1.$$

If this is not the case, then there exists the greatest integer $k_0 \in \{0, \ldots, n_1 - 1\}$, so that $a_{k_0}^{(1)}e \neq a_{k_0}^{(2)}e$. In this case,

$$(a_{k_0}^{(1)}e - a_{k_0}^{(2)}e)(ae)^{k_0} + \cdots + (a_0^{(1)}e - a_0^{(2)}e) = p_1(ae) - p_2(ae) = 0. \tag{6.10}$$

Multiplying both parts of equality (6.10) by $i(a_{k_0}^{(1)}e - a_{k_0}^{(2)}e)$, we obtain that

$$(as(a_{k_0}^{(1)}e - a_{k_0}^{(2)}e))^{k_0} + (a_{k_0-1}^{(1)}e - a_{k_0-1}^{(2)}e)i(a_{k_0}^{(1)}e - a_{k_0}^{(2)}e)(as(a_{k_0}^{(1)}e - a_{k_0}^{(2)}e))^{k_0-1}$$
$$+ \cdots + (a_0^{(1)}e - a_0^{(2)}e)i(a_{k_0}^{(1)}e - a_{k_0}^{(2)}e) = 0. \tag{6.11}$$

Since $\mathcal{B}$ is a regular subalgebra in $L_0(\Omega)$, the polynomial on the left side of equality (6.11) is a monic polynomial from $\mathcal{B}_u[x]$. The inequalities

$$0 \neq s(a_{k_0}^{(1)}e - a_{k_0}^{(2)}e) \leq e \leq s(a)$$

imply that $n(s(a_{k_0}^{(1)}e - a_{k_0}^{(2)}e)) \leq k_0 < n_1$, which contradicts the choice of n_1. Hence, $a_k^{(1)}e = a_k^{(2)}e$ for all $k = 0, \ldots, n_1 - 1$.

It means that for every nonzero idempotent $e \in J_1$ there exists a unique polynomial $p_e(x) = x^{n_1} + a_{n_1-1}^{(e)}x^{n_1-1} + \cdots + a_0^{(e)} \in \mathcal{B}_u[x]$, such that $p_e(ae) = 0$ and $s(a_k^{(e)}) \leq e$, $k = 0, \ldots, n_1 - 1$, where $n_1 = n_1(a)$.

Let

$$0 \neq e \in J_1, \quad f \in P(\Omega), \quad 0 \neq f \leq e, \quad p \in \mathcal{B}_u[x], \quad p(x) = x^{n_1} + b_1x^{n_1-1} + \cdots + b_{n_1},$$

and $p(ae) = 0$. Then

$$(af)^{n_1} + b_1(af)^{n_1-1} + \cdots + fb_{n_1} = fp(ae) = 0,$$

therefore, $f \in J_1$.

By [101, Section 1.1.6], there exists at most a countable set $G = \{g_i\} \subset J_1$ consisting of pairwise disjoint elements, such that $\sup_i g_i = e_1$.

For every $g_i \in G$, there exists a unique polynomial

$$p_{g_i}(x) = x^{n_1} + a_{n_1-1}^{(g_i)}x^{n_1-1} + \cdots + a_0^{(g_i)} \in \mathcal{B}_u[x],$$

such that

$$p_{g_i}(ag_i) = 0 \quad \text{and} \quad s(a_k^{(g_i)}) \leq g_i, \quad k = 0, \ldots, n_1 - 1, \text{where } n_1 = n_1(a). \tag{6.12}$$

For every $k \in \{0,\dots,n_1-1\}$, we define $a_k \in L_0(\Omega)$, such that $a_k = \sum_{i=1}^{\infty} a_k^{(g_i)}$. Since $\mathcal{B}$ is a disjointly complete subalgebra, it follows that $a_k \in \mathcal{B}$ for all $k = 0,\dots,n_1-1$.

Summing (6.12) and using Proposition 6.1.11, we obtain that

$$(ae_1)^{n_1} + a_{n_1-1}(ae_1)^{n_1-1} + \cdots + a_0 = 0,$$

i. e., $e_1 \in J_1$ and, therefore, $J_1 = e_1 P(\Omega)$. In particular, ae_1 is a $\mathcal{B}e_1$-integral element from the algebra $L_0(\Omega)e_1$ and $n_1 = \deg_{\mathcal{B}e_1}(ae_1) = \deg_{\mathcal{B}e_1}(ae)$ for all nonzero idempotents $e \le e_1$.

If $e_1 = s(a)$, then Proposition 6.2.6 is proved and $m = 1$.

Suppose that $e_1 \neq s(a)$. From the definition of the set J_1, it follows that for every nonzero idempotent $e \le s(a) - e_1$ the inequality $n_1 < n(e)$ holds.

Set

$$n_2 = n_2(a) := \min\{n(e) : e \in P(\Omega),\ 0 \neq e \le s(a) - e_1\}$$

and consider the set

$$J_2 := \{e \in P(\Omega) : n(e) = n_2,\ 0 \neq e \le s(a) - e_1\} \cup \{0\}.$$

It is clear from the definition of $n(e)$ that $n_1 < n_2$. Repeating the same argument, we obtain that

$$e_2 = \sup\{e : e \in J_2\} \in J_2.$$

In addition, $e_2 \le s(a) - e_1$ and $J_2 = e_2 P(\Omega)$. Moreover, ae_2 is a $\mathcal{B}e_2$-integral element from $L_0(\Omega)e_2$ and $n_2 = \deg_{\mathcal{B}e_2}(ae_2) = \deg_{\mathcal{B}e_2}(ae)$ for all nonzero idempotents $e \le e_2$.

If $e_1 + e_2 = s(a)$, then the proof of Proposition 6.2.6 is completed with $m = 2$. If, however, $e_1 + e_2 \neq s(a)$, then we continue the same procedure. Since a is a $\mathcal{B}$-integral element with respect to $\mathcal{B}$, there exists a monic polynomial $q(x) = x^n + b_{n-1}x^{n-1} + \cdots + b_0 \in \mathcal{B}_u[x]$ of degree $n = \deg_{\mathcal{B}}(a)$, such that $q(a) = 0$. Consequently, $n(e) \le n$ for all nonzero idempotents $e \le s(a)$, since $(ae)^n + eb_{n-1}(ae)^{n-1} + \cdots + eb_0 = eq(a) = 0$. Therefore, the outlined procedure necessarily terminates for some $m \in \mathbb{N}$ and $n_m \le \deg_{\mathcal{B}}(a)$, i. e., we obtain that $e_1 + \cdots + e_m = s(a)$ and $n_1 < \cdots < n_m \le \deg_{\mathcal{B}}(a)$. □

Now we define the integral closure of subalgebras of $L_0(\Omega)$. Let $\mathcal{B}$ be a subalgebra of $L_0(\Omega)$. Denote by

$$I(\mathcal{B}) = \{a \in L_0(\Omega) : a \text{ is } \mathcal{B}\text{-integral}\}, \tag{6.13}$$

i. e., $I(\mathcal{B})$ is the set of all $\mathcal{B}$-integral elements. It is well known [45, Chapter V, Section 1, n.2] that $I(\mathcal{B})$ is a subring in the algebra $L_0(\Omega)$.

Since every element $b \in \mathcal{B}$ is the root of the monic polynomial $p(x) = x - b$ of $\mathcal{B}[x]$, we have that $\mathcal{B} \subset I(\mathcal{B})$. In addition, if $a \in I(\mathcal{B})$, $\lambda \in \mathbb{C}$, and $p(x) = x^n + a_1x^{n-1} + \cdots + a_n \in \mathcal{B}_u[x]$ are such that $p(a) = 0$, then

$$0 = \lambda^n p(a) = (\lambda a)^n + a_1 \lambda (\lambda a)^{n-1} + \cdots + \lambda^n a_n = q(\lambda a),$$

where $q(x) = x^n + \lambda a_1 x^{n-1} + \cdots + \lambda^n a_n \in \mathcal{B}_u[x]$. Consequently, $\lambda a \in I(\mathcal{B})$ for all $\lambda \in \mathbb{C}$, $a \in I(\mathcal{B})$. Thus, $I(\mathcal{B})$ is a subalgebra in $L_0(\Omega)$ containing $\mathcal{B}$. In addition, by [45, Chapter V, Section 1, n.1, Proposition 6], we have that $I(I(\mathcal{B})) = I(\mathcal{B})$.

We now introduce the notion of integral closedness (see, e. g., [45, Chapter V, Section 1, n.2]).

Definition 6.2.7. The subalgebra $I(\mathcal{B})$, defined by (6.13), is called the *integral closure* of a subalgebra $\mathcal{B}$ in the algebra $L_0(\Omega)$. If $I(\mathcal{B}) = \mathcal{B}$, then the subalgebra $\mathcal{B}$ is said to be an *integrally closed* subalgebra in $L_0(\Omega)$.

Next, we study how the three properties of regularity, disjoint completeness, and integral closedness are connected to one another. In the following proposition, we establish a connection between the integral closure and closure with respect to the metric ρ_μ of a subalgebra in $L_0(\Omega)$ (see Definition 6.1.8). Recall that the symbol $\overline{\mathcal{B}}^{\rho_\mu}$ stands for the topological closure of $\mathcal{B}$ (i. e., the closure of $\mathcal{B}$ in $(L_0(\Omega), \rho_\mu)$).

Proposition 6.2.8. *Let $\mathcal{B}$ be an arbitrary subalgebra of $L_0(\Omega)$, containing all projections from $L_0(\Omega)$. Then $I(\overline{\mathcal{B}}^{\rho_\mu}) \subset \overline{I(\mathcal{B})}^{\rho_\mu}$ and $I(\overline{I(\mathcal{B})}^{\rho_\mu}) = \overline{I(\mathcal{B})}^{\rho_\mu}$.*

Proof. To show the first inclusion, assume that $a \in I(\overline{\mathcal{B}}^{\rho_\mu})$ and let $p(x) = x^n + a_1 x^{n-1} + \cdots + a_{n-1}x + a_n \in \overline{\mathcal{B}}_u^{\rho_\mu}[x]$ be such that $p(a) = 0$. By Proposition 6.1.16 for every $a_m \in \overline{\mathcal{B}}^{\rho_\mu}$, there exists a sequence $\{b_k^{(m)}\}_{k\in\mathbb{N}} \subset \mathcal{B}$ such that

$$\rho_\mu(b_k^{(m)}, a_m) = \mu_f(s(a_m - b_k^{(m)})) \to 0, \quad \text{as } k \to \infty, \text{for all } m = 1, \ldots, n,$$

and $s(b_k^{(m)}) \uparrow s(a_m)$ as $k \to \infty$.

For every $k \in \mathbb{N}$, let us define the projection

$$e_k := \left(\mathbf{1} - (s(a_1) - s(b_k^{(1)}))\right) \cdots \left(\mathbf{1} - (s(a_n) - s(b_k^{(n)}))\right) \in P(\Omega).$$

Since $s(b_k^{(m)}) \uparrow s(a_m)$, it follows that $\mathbf{1} - (s(a_m) - s(b_k^{(m)})) \uparrow \mathbf{1}$ as $k \to \infty$ for all $m = 1, \ldots, n$. Hence, $e_k \uparrow \mathbf{1}$ as $k \to \infty$.

Furthermore, since

$$(a_m - b_k^{(m)})e_k = (a_m - b_k^{(m)})(\mathbf{1} - s(a_m - b_k^{(m)}))e_k = 0,$$

it follows that $a_m e_k = b_k^{(m)} e_k$. Therefore, using the inclusions $b_k^{(m)} \subset \mathcal{B}$ and $P(\Omega) \subset \mathcal{B}$, we infer that $a_m e_k \in \mathcal{B}$ for all $m = 1, \ldots, n$ and $k \in \mathbb{N}$.

Fix k and consider the polynomial

$$q(x) := x^n + a_1 e_k x^{n-1} + \cdots + a_{n-1} e_k x + a_n e_k \in \mathcal{B}_u[x].$$

Since

$$q(ae_k) = a^n e_k + a_1 a^{n-1} e_k + \cdots + a_{n-1} a e_k + a_n e_k = p(a)e_k = 0,$$

it follows that $ae_k \in I(\mathcal{B})$.

Now, using the convergence $e_k \uparrow \mathbf{1}$, we conclude that

$$\rho_\mu(a, ae_k) = \mu_f(s(a - ae_k)) \leq \mu_f(\mathbf{1} - e_k) \to 0 \quad \text{as } k \to \infty.$$

Thus, $a \in \overline{I(\mathcal{B})}^{\rho_\mu}$, which implies the inclusion $I(\overline{\mathcal{B}}^{\rho_\mu}) \subset \overline{I(\mathcal{B})}^{\rho_\mu}$.

To show the equality $I(\overline{I(\mathcal{B})}^{\rho_\mu}) = \overline{I(\mathcal{B})}^{\rho_\mu}$, we note that $I(\mathcal{B})$ is also an algebra containing $P(\Omega)$. Therefore, by the already proven inclusion applied to the algebra $I(\mathcal{B})$, we obtain that

$$I(\overline{I(\mathcal{B})}^{\rho_\mu}) \subset \overline{I(I(\mathcal{B}))}^{\rho_\mu} = \overline{I(\mathcal{B})}^{\rho_\mu} \subset I(\overline{I(\mathcal{B})}^{\rho_\mu}),$$

i. e., $I(\overline{I(\mathcal{B})}^{\rho_\mu}) = \overline{I(\mathcal{B})}^{\rho_\mu}$. □

Now, we show that the integral closure of every regular disjointly complete subalgebra $\mathcal{B}$ is again a regular subalgebra.

Proposition 6.2.9. *Let $\mathcal{B}$ be a regular disjointly complete subalgebra of $L_0(\Omega)$. Then the subalgebra $I(\mathcal{B})$ is regular in $L_0(\Omega)$, too.*

Proof. By Proposition 6.2.6 for every nonzero element $a \in I(\mathcal{B})$, there exist natural numbers $n_1 < \cdots < n_m$ and nonzero pairwise disjoint idempotents $e_1, \ldots, e_m$, such that $s(a) = \sum_{k=1}^{n} e_k$, and for every $k = 1, \ldots, m$, the element ae_k is a proper integral element with respect to the algebra $\mathcal{B}e_k$ with $n_k = \deg_{\mathcal{B}e_k}(ae_k)$. In particular, $ae_k \in I(\mathcal{B}e_k)$ for all $k = 1, \ldots, m$.

By Proposition 6.2.4 applied to the algebra $L_0(\Omega)e_k$ and the disjointly complete regular subalgebra $\mathcal{B}e_k$, we obtain that the inverse element $i(ae_k) = i(a)e_k$ of the element ae_k in $L_0(\Omega)e_k$ is a $\mathcal{B}e_k$-integral element. Hence, $i(a)e_k \in I(\mathcal{B})$ for all $k = 1, \ldots, m$, which implies the inclusion

$$i(a) = i(a)s(a) = i(a)e_1 + \cdots + i(a)e_m \in I(\mathcal{B}).$$

Thus, the subalgebra $I(\mathcal{B})$ is regular in $L_0(\Omega)$. □

Before we proceed to the main result of this section, we introduce the following notation.

Definition 6.2.10. Let $\mathcal{B}$ be an arbitrary subalgebra of $L_0(\Omega)$. We denote by $\mathbb{C}(\mathcal{B}, P(\Omega))$ the least subalgebra of $L_0(\Omega)$ containing both $\mathcal{B}$ and $P(\Omega)$, i. e., the algebra of all elements in $L_0(\Omega)$ of the form

$$\sum_{l=1}^{m} b_l f_l + \sum_{k=1}^{n} \lambda_k e_k,$$

where $f_l, e_k \in P(\Omega)$, $b_l \in \mathcal{B}$, $\lambda_k \in \mathbb{C}$, and $l = 1, \ldots, m$, $k = 1, \ldots, n$ for some fixed $m, n \in \mathbb{N}$.

We are now ready to prove the main result of this section. We show that for an arbitrary subalgebra $\mathcal{B}$ of the algebra $L_0(\Omega)$, there exists the least regular, integrally closed, and disjointly complete subalgebra of $L_0(\Omega)$ that contains both $\mathcal{B}$ and $P(\Omega)$.

Theorem 6.2.11. *Let $\mathcal{B}$ be an arbitrary subalgebra of $L_0(\Omega)$. Then the subalgebra*

$$E(\mathcal{B}) = \overline{I(\overline{\mathcal{R}(\mathbb{C}(\mathcal{B}, P(\Omega)))}^{\rho_\mu})}^{\rho_\mu} \tag{6.14}$$

is the least regular, integrally closed, and disjointly complete subalgebra of $L_0(\Omega)$, which contains the subalgebra $\mathcal{B}$ and all projections from $L_0(\Omega)$.

Proof. Let $\mathbb{C}(\mathcal{B}, P(\Omega))$ be the least subalgebra of $L_0(\Omega)$, containing $\mathcal{B}$ and $P(\Omega)$, as defined in Definition 6.2.10. By Proposition 6.1.5, the algebra $\mathcal{R}(\mathbb{C}(\mathcal{B}, P(\Omega)))$ is the least regular subalgebra of $L_0(\Omega)$, containing the subalgebra $\mathbb{C}(\mathcal{B}, P(\Omega))$. Referring to Proposition 6.1.15, we conclude that the topological closure $\overline{\mathcal{R}(\mathbb{C}(\mathcal{B}, P(\Omega)))}^{\rho_\mu}$ of $\mathcal{R}(\mathbb{C}(\mathcal{B}, P(\Omega)))$ is a regular subalgebra of $L_0(\Omega)$.

Hence, using Proposition 6.2.9 for the algebra $\overline{\mathcal{R}(\mathbb{C}(\mathcal{B}, P(\Omega)))}^{\rho_\mu}$ and then again Proposition 6.1.15, we obtain that the algebra

$$E(\mathcal{B}) := \overline{I(\overline{\mathcal{R}(\mathbb{C}(\mathcal{B}, P(\Omega)))}^{\rho_\mu})}^{\rho_\mu}$$

is regular. By Proposition 6.2.8, we have

$$I(E(\mathcal{B})) = I(\overline{I(\overline{\mathcal{R}(\mathbb{C}(\mathcal{B}, P(\Omega)))}^{\rho_\mu})}^{\rho_\mu}) = \overline{I(\overline{\mathcal{R}(\mathbb{C}(\mathcal{B}, P(\Omega)))}^{\rho_\mu})}^{\rho_\mu} = E(\mathcal{B}).$$

Furthermore, by Theorem 6.1.13, the algebra $E(\mathcal{B})$ is disjointly complete, since it is closed in the metric ρ_μ. Thus, the subalgebra $E(\mathcal{B})$ is a regular, integrally closed, and disjointly complete subalgebra of $L_0(\Omega)$, which contains both $\mathcal{B}$ and $P(\Omega)$.

We claim that $E(\mathcal{B})$ is in fact the least subalgebra of $L_0(\Omega)$ satisfying these conditions.

Let $\mathcal{B}_0$ be an arbitrary regular, integrally closed, and disjointly complete subalgebra of $L_0(\Omega)$, containing $P(\Omega)$ and $\mathcal{B}$. Then $\mathbb{C}(\mathcal{B}, P(\Omega)) \subset \mathcal{B}_0$ and, therefore, $\mathcal{R}(\mathbb{C}(\mathcal{B}, P(\Omega))) \subset \mathcal{R}(\mathcal{B}_0) = \mathcal{B}_0$. Consequently,

$$I(\overline{\mathcal{R}(\mathbb{C}(\mathcal{B}, P(\Omega)))}^{\rho_\mu}) \subset I(\overline{\mathcal{B}_0}^{\rho_\mu}) = I(\mathcal{B}_0) = \mathcal{B}_0,$$

which implies the inclusion

$$E(\mathcal{B}) = \overline{I(\overline{R(\mathbb{C}(\mathcal{B}, P(\Omega)))}^{\rho_\mu})}^{\rho_\mu} \subset \overline{\mathcal{B}_0}^{\rho_\mu} = \mathcal{B}_0. \qquad \square$$

Let $\mathcal{B}$ be an arbitrary subalgebra of $L_0(\Omega)$. In order to describe properties of elements, which do not belong to the least regular, integrally closed, and disjointly complete subalgebra of $L_0(\Omega)$, that contains the subalgebra $\mathcal{B}$ and all projections from $L_0(\Omega)$, we introduce the notion of weakly transcendental elements with respect to the subalgebra $\mathcal{B}$.

Definition 6.2.12. An element $a \in L_0(\Omega)$ is called a *transcendental element* with respect to $\mathcal{B}$, if a is not algebraic with respect to $\mathcal{B}$. We say that a nonzero element $a \in L_0(\Omega)$ is a *weakly transcendental element* with respect to $\mathcal{B}$, if for every nonzero $e \leq s(a)$ the element ae is not a $\mathcal{B}$-integral element.

The technical lemma below will later provide an instrument for the extension of derivations. It shows that there are sufficiently many weakly transcendental elements with respect to a proper, regular, integrally closed, and disjointly complete subalgebra $\mathcal{B}$ of $L_0(\Omega)$.

Lemma 6.2.13. *Let $\mathcal{B}$ be a proper, regular, integrally closed, and disjointly complete subalgebra of $L_0(\Omega)$, containing all projections from $L_0(\Omega)$. Then for every $a \in L_0(\Omega) \setminus \mathcal{B}$ there exists $0 \neq e(a) \leq s(a)$, such that $e(a) \neq s(a)$, $ae(a) \in \mathcal{B}$, and the element $a(\mathbf{1} - e(a))$ is weakly transcendental with respect to $\mathcal{B}$.*

Proof. Fix an element $a \in L_0(\Omega) \setminus \mathcal{B}$ and consider the set

$$F = \{q \in P(\Omega) : aq \in \mathcal{B}\}.$$

Let $e(a) = \sup F$. Let $q \in F$ and $p \in P(\Omega)$ be such that $p \leq q$. Since $P(\Omega) \subset \mathcal{B}$, we have that $ap = aq \cdot p \in \mathcal{B}$, and so $p \in F$. By [101, Section 1.1.6] there exists at most a countable disjoint set $G = \{e_i\} \subset F$, such that $\sup_i e_i = e$. It is clear that $e_i = s(b_i)$ for $b_i = ae_i \in \mathcal{B}$. By assumption, $\mathcal{B}$ is disjointly complete. Since $\{e_i\}$ are pairwise disjoint, it follows that there exists $b_0 = \sum_i b_i \in \mathcal{B}$. By the definition of b_i, we have that

$$b_0 = \sum_i ae_i = a \sum_i e_i = ae(a).$$

Consider the element $c = (\mathbf{1} - e(a))a$. Using the inequality $e(a) \leq s(a)$, we have

$$s(c) = (s(a) - e(a))s(a) \quad \text{and} \quad c = (s(a) - e(a))a = s(c)a.$$

Since $a \in L_0(\Omega) \setminus \mathcal{B}$ and $ae(a) = b_0 \in \mathcal{B}$, it follows $e(a) \neq s(a)$.

Let us show that the element c is weakly transcendental with respect to the subalgebra $\mathcal{B}$. If it is not the case, then there exists a nonzero idempotent $q \leq s(c) \leq s(a)$, such that cq is integral with respect to $\mathcal{B}$. Since $\mathcal{B}$ is integrally closed, it follows that $cq \in \mathcal{B}$. The equality $c = as(c)$ implies that $aq = as(c)q = cq \in \mathcal{B}$. Thus, $q \in F$ and $q \leq e(a)$. Combining this with the inequality $q \leq s(c) \leq \mathbf{1} - e(a)$, we conclude that $q = 0$, which contradicts the choice of q. Thus, the element c is weakly transcendental with respect to $\mathcal{B}$. □

6.3 Extension of derivations on algebras of measurable functions

In this section, we construct an extension of a derivation $\delta : \mathcal{B} \to L_0(\Omega)$ on a subalgebra $\mathcal{B}$ of $L_0(\Omega)$. The construction of an extension of any derivation δ, defined on a subalgebra $\mathcal{B}$ on $L_0(\Omega)$, to a derivation on the whole algebra $L_0(\Omega)$, proceeds in several stages. This process enables us to extend any nonexpansive derivation δ on an arbitrary subalgebra $\mathcal{B}$ of $L_0(\Omega)$.

First, we construct an extension of δ up to a minimal subalgebra, containing the subalgebra $\mathcal{B} \subset L_0(\Omega)$ and the Boolean algebra $P(\Omega)$ of all projections in $L_0(\Omega)$. At the second step, we can extend the derivation δ up to a derivation on the least regular subalgebra of $L_0(\Omega)$, containing both $\mathcal{B}$ and $P(\Omega)$ and, subsequently, to the topological closure with respect to the metric ρ_μ (see Definition 6.1.8) of this least regular algebra containing $\mathcal{B}$ and $P(\Omega)$.

With these first three relatively easy stages completed, we can work in the setting of a derivation δ, defined on a regular and disjointly complete algebra containing all projections from $L_0(\Omega)$. The next stage is building an extension of the derivation up to a derivation on the subalgebra $\mathcal{B}(a)$ generated by $\mathcal{B}$ and an integral (resp., weakly transcendental) element $a \in L_0(\Omega)$ with respect to $\mathcal{B}$. A simple procedure involving Zorn's lemma then yields the first main result of this section (and in fact of this chapter), which establishes that any nonexpansive derivation defined on a subalgebra $\mathcal{B}$ of $L_0(\Omega)$ can be extended up to a derivation on the whole algebra $L_0(\Omega)$.

In this section, we also shall prove the second main result of this chapter. For this result, we are looking for the largest subalgebra $\mathcal{A}$ of $L_0(\Omega)$, which contains $\mathcal{B}$ and admits a unique extension of a derivation δ on an arbitrary subalgebra $\mathcal{B}$ of $L_0(\Omega)$. As we show, this largest algebra is, in fact, the least regular, integrally closed, and disjointly complete subalgebra $E(\mathcal{B})$ of $L_0(\Omega)$, containing $\mathcal{B}$ and all projections from $L_0(\Omega)$ (this algebra exists by Theorem 6.2.11). In particular, we show that if $E(\mathcal{B})$ is a proper subalgebra of $L_0(\Omega)$, then the derivation δ has infinitely many distinct extensions up to a derivation on the algebra $L_0(\Omega)$.

Let $\mathcal{B}$ be an arbitrary subalgebra of $L_0(\Omega)$, and let $\delta : \mathcal{B} \to L_0(\Omega)$ be a derivation. The commutativity of the algebra $L_0(\Omega)$ yields several additional properties of derivations in $L_0(\Omega)$, to the properties we already discussed in Chapter 5. We list these simple properties of derivations in the following proposition.

Proposition 6.3.1. *Let $\mathcal{B}$ be a subalgebra in $L_0(\Omega)$ and let $P(\Omega)$ be the Boolean algebra of all projections from $L_0(\Omega)$. If $\delta : \mathcal{B} \to L_0(\Omega)$ is a derivation, then the following hold for all $b \in \mathcal{B}$ and $e \in \mathcal{B} \cap P(\Omega)$:*
(i) $\delta(b^n) = nb^{n-1}\delta(b)$, $n \in \mathbb{N}$;
(ii) $\delta(e) = 0$;
(iii) $\delta(be) = \delta(b)e$;
(iv) *if* $s(b) \in \mathcal{B}$, *then* $s(\delta(b)) \le s(b)$;
(v) *if* $i(b) \in \mathcal{B}$, *then* $\delta(i(b)) = -\delta(b)i(b^2)$.

Proof. The assertion (i) easily follows from the Leibniz rule and commutativity of $L_0(\Omega)$, while (ii) and (iii) are already established in Proposition 5.1.8.

(iv) If $s(b) \in \mathcal{B}$, then appealing to (iii) we obtain that

$$\delta(b) = \delta(s(b)b) = s(b)\delta(b),$$

which immediately implies that $s(\delta(b)) \leq s(b)$.

(v) Suppose that $i(b) \in \mathcal{B}$. We have that $s(b) = b \cdot i(b) \in \mathcal{B}$ and by (iv), we have the equalities $\delta(s(b)) = 0$. Hence, using the Leibniz rule, we obtain

$$0 = \delta(s(b)) = \delta(b \cdot i(b)) = \delta(b)i(b) + \delta(i(b))b,$$

that is $\delta(i(b))b = -\delta(b)i(b)$. Thus, referring to part (iii), we conclude that

$$\begin{aligned} \delta(i(b)) &= \delta(i(b)s(b)) = \delta(i(b))s(b) = \delta(i(b))b \cdot i(b) \\ &= -\delta(b)i(b)i(b) = -\delta(b)i(b^2), \end{aligned}$$

as required. □

Next, we introduce the notion of nonexpansive derivations.

Definition 6.3.2. Let $\mathcal{B}$ be a subalgebra of $L_0(\Omega)$. A derivation $\delta : \mathcal{B} \to L_0(\Omega)$ is called *nonexpansive* if $s(\delta(a)) \leq s(a)$ for all $a \in \mathcal{B}$.

We note the following property of nonexpansive derivations.

Proposition 6.3.3. *If $\mathcal{B}$ is a subalgebra of $L_0(\Omega)$ and $\delta : \mathcal{B} \to L_0(\Omega)$ is a nonexpansive derivation, then $\delta : \mathcal{B} \to L_0(\Omega)$ is continuous with respect to the metric ρ_μ on $L_0(\Omega)$, introduced in Definition* 6.1.8.

Proof. Since the derivation δ is nonexpansive, for any $a \in \mathcal{B}$ we have that $s(\delta(a)) \leq s(a)$. Hence, for all $a, b \in \mathcal{B}$ we have that

$$\rho_\mu(\delta(a), \delta(b)) = \mu_f(s(\delta(a-b))) \leq \mu_f(s(a-b)) = \rho_\mu(a,b),$$

which implies that δ is uniformly continuous in the metric ρ_μ. □

Proposition 6.3.1(iv) implies also the following.

Corollary 6.3.4. *If $\mathcal{B}$ is a subalgebra of $L_0(\Omega)$, which contains all projections from $L_0(\Omega)$, then every derivation $\delta : \mathcal{B} \to L_0(\Omega)$ is nonexpansive and, therefore, is continuous with respect to the metric ρ_μ on $L_0(\Omega)$.*

Proof. By assumption, every $e \in P(\Omega)$ is contained in $\mathcal{B}$. In particular, $s(a) \in \mathcal{B}$ for any $a \in \mathcal{B}$. Hence, by Proposition 6.3.1(iv), we have that $\delta(s(a)) \leq s(a)$ for any $a \in \mathcal{B}$, that is δ is nonexpansive. The second assertion follows from Proposition 6.3.3. □

At the same time, there are examples of derivations, which are not nonexpansive.

Example 6.3.5. *Let $(\Omega, \Sigma, \mu) = ([0,2], \Sigma, m)$ with Lebesgue measure m. Consider the element $a = a(t) = t\chi_{[0,1]}(t)$ from $L_\infty([0,2], m)$ and set $\mathcal{B} = a \cdot L_\infty([0,2], m)$. It is clear that for the subalgebra*

$$\mathcal{B}^2 = \mathcal{B} \cdot \mathcal{B} = \left\{ \sum_{i=1}^{n} f_i g_i : f_i, g_i \in \mathcal{B},\ i = 1, \ldots, n,\ n \in \mathbb{N} \right\}$$

the equality $\mathcal{B}^2 = \{b(t) = t^2\chi_{[0,1]}c(t) : c(t) \in L_\infty([0,2], m)\}$ holds.

Let us show that $\mathbb{C}a \cap \mathcal{B}^2 = \{0\}$. Indeed, if this is not the case, then there exists a nonzero number $\lambda \in \mathbb{C}$ and $0 \neq c \in L_\infty([0,2], m)$, such that $0 \neq \lambda t = t^2 c(t)$ for all t from some Lebesgue measurable set $E \subset (0,1]$ with $m(E) = 1$ and $|c(t)| \leq \|c\|_\infty$ for all $t \in E$. Consequently, $0 \neq \lambda = tc(t)$ for all $t \in E$. Since $m(E) = 1$, it follows that there exists a sequence $t_n \to 0$ in E. Hence, $t_n c(t_n) \to 0$ and, therefore, $\lambda = 0$, which contradicts with the choice of λ. Hence, $\mathbb{C}a \cap \mathcal{B}^2 = \{0\}$.

This means that every element b from the algebra $\mathcal{C} = \mathbb{C}a + \mathcal{B}^2$ can be uniquely written in the form $b = f(b)a + b^{(2)}$, where $f(b) \in \mathbb{C}$, $b^{(2)} \in \mathcal{B}^2$. We set

$$\delta(b) = f(b)\chi_{[1,2]} \in L_0(\Omega).$$

Linearity of the mapping $\delta : \mathcal{C} \to L_0(\Omega)$ follows from the linearity of the functional f. Since $\delta(\mathcal{C}^2) \subset \delta(\mathcal{B}^2) = \{0\}$, $\mathcal{C} \subset \chi_{[0,1]}L_0(\Omega)$, and $\delta(\mathcal{C}) \subset \chi_{[1,2]}L_0(\Omega)$, we have that $\delta(xy) = 0 = \delta(x)y + x\delta(y)$ for all $x, y \in \mathcal{C}$. Consequently, δ is a derivation on $\mathcal{C}$ and $s(\delta(a))s(a) = 0$, $s(\delta(a)) \neq 0 \neq s(a)$, i. e., the derivation δ is not nonexpansive.

Now, we proceed with the first step in our extension procedure, i. e., we extend a derivation $\delta : \mathcal{B} \to L_0(\Omega)$ defined on an arbitrary subalgebra $\mathcal{B}$ of $L_0(\Omega)$, to a derivation defined on the least subalgebra of $L_0(\Omega)$ containing $\mathcal{B}$ and $P(\Omega)$.

Recall (see Definition 6.2.10), that the least subalgebra in $L_0(\Omega)$ containing $\mathcal{B}$ and $P(\Omega)$ is the subalgebra $\mathbb{C}(\mathcal{B}, P(\Omega))$, containing all elements of the form

$$\sum_{i=1}^{n} b_i e_i + \sum_{j=1}^{m} a_j f_j,$$

where $b_i \in \mathcal{B}$, $e_i, f_j \in P(\Omega)$, $a_j \in \mathbb{C}$, $i = 1, \ldots, n$, $j = 1, \ldots, m$, for some fixed $n, m \in \mathbb{N}$.

Lemma 6.3.6. *Suppose that $\delta : \mathcal{B} \to L_0(\Omega)$ is a nonexpansive derivation on a subalgebra $\mathcal{B}$ of $L_0(\Omega)$ and let $a \in \mathbb{C}(\mathcal{B}, P(\Omega))$ with a representation $a = \sum_{i=1}^{n} b_i e_i + \sum_{j=1}^{m} a_j f_j \in \mathbb{C}(\mathcal{B}, P(\Omega))$. The mapping $\delta_1 : \mathbb{C}(\mathcal{B}, P(\Omega)) \to L_0(\Omega)$, defined by*

$$\delta_1(a) = \sum_{i=1}^{n} \delta(b_i) e_i \tag{6.15}$$

is well-defined.

Proof. Let

$$a = \sum_{i=1}^{n_1} b_i^{(1)} e_i^{(1)} + \sum_{j=1}^{m_1} a_j^{(1)} f_j^{(1)} = \sum_{i=1}^{n_2} b_i^{(2)} e_i^{(2)} + \sum_{j=1}^{m_2} a_j^{(2)} f_j^{(2)} \tag{6.16}$$

be two possible representations of $a \in \mathbb{C}(\mathcal{B}, P(\Omega))$.

Denote by ∇ the finite Boolean subalgebra of $P(\Omega)$, generated by

$$e_1^{(1)}, \dots, e_{n_1}^{(1)}, \quad e_1^{(2)}, \dots, e_{n_2}^{(2)}, \quad f_1^{(1)}, \dots, f_{m_1}^{(1)}, \quad f_1^{(2)}, \dots, f_{m_2}^{(2)}.$$

Let $q_1, \dots, q_l$ be the list of all atoms in ∇.

Fix an arbitrary number $k \in \{1, \dots, l\}$. Since $q_k e = q_k$ for any $e \in \nabla$ with $e \geq q_k$, and $q_k e = 0$ for any $e \in \nabla$ with $e \leq q_k$, equality (6.16) implies that

$$q_k \sum_{i:e_i^{(1)} \geq q_k} b_i^{(1)} + q_k \sum_{j:f_j^{(1)} \geq q_k} a_j^{(1)} = q_k \sum_{i:e_i^{(2)} \geq q_k} b_i^{(2)} + q_k \sum_{j:f_j^{(2)} \geq q_k} a_j^{(2)}. \tag{6.17}$$

Setting

$$c_k := \sum_{i:e_i^{(1)} \geq q_k} b_i^{(1)} - \sum_{i:e_i^{(2)} \geq q_k} b_i^{(2)}, \quad \beta_k = \sum_{j:f_j^{(1)} \geq q_k} a_j^{(1)} - \sum_{j:f_j^{(2)} \geq q_k} a_j^{(2)},$$

equality (6.17) can be written as $c_k q_k + \beta_k q_k = 0$. We claim that

$$q_k \delta(c_k) = 0.$$

If $\beta_k = 0$, then $q_k c_k = 0$ and, therefore, by Proposition 6.1.2(iii), $q_k s(c_k) = 0$. Since δ is a nonexpansive derivation, we have $q_k s(\delta(c_k)) = 0$ that implies $q_k \delta(c_k) = 0$, as required.

Assume now that $\beta_k \neq 0$. Setting $d_k = -\beta_k^{-1} c_k$, the equality $c_k q_k + \beta_k q_k = 0$ implies that $q_k d_k = q_k$, in particular, $q_k d_k = q_k = q_k^2 = q_k d_k^2$. Thus, $q_k(d_k^2 - d_k) = 0$ and by Proposition 6.1.2(iii), we have $q_k s(d_k^2 - d_k) = 0$. Since the derivation δ is nonexpansive, it follows that $q_k s(\delta(d_k^2 - d_k)) = 0$, and hence, $q_k \delta(d_k^2 - d_k) = 0$.

Now, using the equality $q_k d_k = q_k$ and Proposition 6.3.1(i), we infer that

$$q_k \delta(d_k) = q_k \delta(d_k^2) = 2 q_k \delta(d_k),$$

i. e., $q_k \delta(d_k) = 0$. Consequently,

$$q_k \delta(c_k) = -\beta_k q_k \delta(d_k) = 0,$$

which proves the claim.

Since k is arbitrary, it follows that $q_k \delta(c_k) = 0$ for all $k \in \{1, \dots, l\}$. This implies that

$$\sum_{i=1}^{n_1}\delta(b_i^{(1)})e_i^{(1)} = \sum_{k=1}^{l} q_k \sum_{i:e_i^{(1)}\geq q_k}\delta(b_i^{(1)})$$
$$= \sum_{k=1}^{l} q_k \sum_{i:e_i^{(2)}\geq q_k}\delta(b_i^{(2)}) = \sum_{i=1}^{n_2}\delta(b_i^{(2)})e_i^{(2)},$$

proving that the mapping δ_1 is well-defined on $\mathbb{C}(\mathcal{B}, P(\Omega))$. □

We now show that the mapping $\delta_1 : \mathbb{C}(\mathcal{B}, P(\Omega)) \to L_0(\Omega)$, defined in the above lemma, is the unique extension of a nonexpansive derivation $\delta : \mathcal{B} \to L_0(\Omega)$ up to the algebra $\mathbb{C}(\mathcal{B}, P(\Omega))$.

Proposition 6.3.7. *Let $\mathcal{B}$ be an arbitrary subalgebra of $L_0(\Omega)$ and let $\delta : \mathcal{B} \to L_0(\Omega)$ be a nonexpansive derivation. Then $\delta_1 : \mathbb{C}(\mathcal{B}, P(\Omega)) \to L_0(\Omega)$, defined by* (6.15), *is a derivation, which uniquely extends δ up to the algebra $\mathbb{C}(\mathcal{B}, P(\Omega))$.*

Proof. The linearity of the mapping δ_1 and the equality $\delta_1(b) = \delta(b)$ for all $b \in \mathcal{B}$ follow immediately from the definition of the mapping δ_1 (see (6.15)).

Let us show that δ_1 is a derivation on $\mathbb{C}(\mathcal{B}, P(\Omega))$. If $a, b \in \mathcal{B}$, $e, f \in P(\Omega)$, then

$$\delta_1((ae)(bf)) = \delta_1(abef) = \delta(ab)ef = (\delta(a)b + a\delta(b))ef$$
$$= \delta_1(ae)bf + ae\delta_1(bf)$$

and

$$\delta_1(aef) = \delta(a)ef = \delta_1(ae)f.$$

For arbitrary elements,

$$c = \sum_{i=1}^{k} b_i e_i + \sum_{j=1}^{m} a_j f_j, \quad d = \sum_{s=1}^{l} b'_s e'_s + \sum_{p=1}^{t} a'_p f'_p,$$

from $\mathbb{C}(\mathcal{B}, P(\Omega))$, we have

$$\delta_1(cd) = \delta_1\Big(\sum_{i,s} b_i e_i b'_s e'_s + \sum_{i,p} b_i e_i a'_p f'_p + \sum_{j,s} b'_s e'_s a_j f_j + \sum_{j,p} a_j a'_p f_j f_p\Big)$$
$$= \sum_{i,s}(\delta_1(b_i e_i)b'_s e'_s + b_i e_i \delta_1(b'_s e'_s)) + \sum_{i,p}\delta_1(b_i e_i)a'_p f'_p + \sum_{j,s}\delta_1(b'_s e'_s)a_j f_j$$
$$= \delta_1(c)d + c\delta_1(d).$$

Thus, $\delta_1 : \mathbb{C}(\mathcal{B}, P(\Omega)) \to L_0(\Omega)$ is a derivation.

To conclude the proof, assume that $\delta_2 : \mathbb{C}(\mathcal{B}, P(\Omega)) \to L_0(\Omega)$ is another derivation such that $\delta_2(b) = \delta(b)$ for all $b \in \mathcal{B}$. For any $b_i \in \mathcal{B}$, $e_i, f_j \in P(\Omega)$, $a_j \in \mathbb{C}$, Proposition 6.3.1(ii) and (iii) imply the equalities

$$\delta_2\left(\sum_{i=1}^{n} b_i e_i + \sum_{j=1}^{m} a_j f_j\right) = \sum_{i=1}^{n} \delta_2(b_i)e_i = \sum_{i=1}^{n} \delta(b_i)e_i$$
$$= \delta_1\left(\sum_{i=1}^{n} b_i e_i + \sum_{j=1}^{m} a_j f_j\right).$$

That is, $\delta_1 = \delta_2$, showing the uniqueness of the derivation δ_1. □

Thus, we can always uniquely extend a nonexpansive derivation $\delta : \mathcal{B} \to L_0(\Omega)$ on a subalgebra $\mathcal{B}$, to a subalgebra $\mathbb{C}(\mathcal{B}, P(\Omega))$, containing $\mathcal{B}$ and $P(\Omega)$. Hence, in the following, we can assume that $\mathcal{B}$ contains all projections from $L_0(\Omega)$. In this case, Corollary 6.3.4 implies that any derivation $\delta : \mathcal{B} \to L_0(\Omega)$ is nonexpansive.

Our second step is extending a derivation $\delta : \mathcal{B} \to L_0(\Omega)$, with $P(\Omega) \subset \mathcal{B}$, to a minimal regular subalgebra $\mathcal{R}(\mathcal{B})$, containing $\mathcal{B}$. As for the previous step, we introduce the intended extension $\delta_1 : \mathcal{R}(\mathcal{B}) \to L_0(\Omega)$ in a separate lemma. We recall that, by Proposition 6.1.5, the least regular subalgebra $\mathcal{R}(\mathcal{B})$ containing $\mathcal{B}$ can be described as

$$\mathcal{R}(\mathcal{B}) = \{ai(b) : a, b \in \mathcal{B}\}.$$

Lemma 6.3.8. *Let $\mathcal{B}$ be an arbitrary subalgebra of $L_0(\Omega)$, containing the Boolean algebra $P(\Omega)$ and let $\delta : \mathcal{B} \to L_0(\Omega)$ be a derivation. For any $c \in \mathcal{R}(\mathcal{B})$ with a representation $c = ai(b)$, we define*

$$\delta_1(ai(b)) = \delta(a)i(b) - a\delta(b)i(b^2). \tag{6.18}$$

Then $\delta_1 : \mathcal{R}(\mathcal{B}) \to L_0(\Omega)$ is a well-defined linear mapping.

Proof. We first show that δ_1 is well-defined.

Suppose that $a_1 i(b_1) = a_2 i(b_2)$, $a_k, b_k \in \mathcal{B}$, $k = 1, 2$. Then

$$s(b_1)s(b_2)(a_1 b_2 - a_2 b_1) = a_1 i(b_1) \cdot b_1 b_2 - a_2 i(b_2) \cdot b_2 b_1$$
$$= (a_2 i(b_2) - a_2 i(b_2)) \cdot b_1 b_2 = 0.$$

Hence, defining

$$h := s(a_1)s(b_1) = s(a_1 i(b_1)) = s(a_2 i(b_2)) = s(a_2)s(b_2),$$

we note that $h \le s(b_1) \wedge s(b_2) = s(b_1) \cdot s(b_2)$ and, therefore,

$$h \cdot (a_1 b_2 - a_2 b_1) = 0.$$

Since $P(\Omega) \subset \mathcal{B}$, the above equality, together with Proposition 6.3.1(iii) implies that

$$h \cdot \delta(a_1 b_2 - a_2 b_1) = \delta(h \cdot (a_1 b_2 - a_2 b_1)) = 0.$$

Using now the inclusion $P(\Omega) \subset \mathcal{B}$ and equalities

$$s(i(b_1^2b_2^2)) = s(b_1^2)s(b_2^2) = s(b_1)s(b_2) \geq h,$$

we obtain that

$$\begin{aligned}
&[\delta(a_1)i(b_1) - a_1\delta(b_1)i(b_1{}^2)] - [\delta(a_2)i(b_2) - a_2\delta(b_2)i(b_2{}^2)] \\
&\quad = h([\delta(a_1)i(b_1) - a_1\delta(b_1)i(b_1{}^2)] - [\delta(a_2)i(b_2) - a_2\delta(b_2)i(b_2{}^2)]) \\
&\quad = hi(b_1^2)i(b_2^2)b_1^2b_2^2([\delta(a_1)i(b_1) - a_1\delta(b_1)i(b_1{}^2)] \\
&\qquad - [\delta(a_2)i(b_2) - a_2\delta(b_2)i(b_2{}^2)]) \\
&\quad = h\,i(b_1{}^2b_2{}^2)(\delta(a_1)b_1b_2{}^2 - a_1\delta(b_1)b_2{}^2 - \delta(a_2)b_2b_1{}^2 + a_2\delta(b_2)b_1{}^2) \\
&\quad = h\,i(b_1{}^2b_2{}^2)(\delta(a_1)b_1b_2{}^2 - a_2b_1b_2\delta(b_1) - \delta(a_2)b_2b_1{}^2 + a_1b_1b_2\delta(b_2)) \\
&\quad = h\,i(b_1b_2)(\delta(a_1)b_2 - a_2\delta(b_1) - \delta(a_2)b_1 + a_1\delta(b_2)) \\
&\quad = hi(b_1b_2)\delta(a_1b_2 - a_2b_1) = 0.
\end{aligned}$$

Thus, the mapping δ_1 is well-defined.

Next, we show that $\delta_1 : \mathcal{R}(\mathcal{B}) \to L_0(\Omega)$ is linear. It is clear that $\delta_1(\lambda a\, i(b)) = \lambda\, \delta_1(a\, i(b))$ for every $\lambda \in \mathbb{C}$. Thus, to show that δ_1 is a linear map, we need only to establish that $\delta_1(d_1 + d_2) = \delta_1(d_1) + \delta_1(d_2)$, $d_1, d_2 \in \mathcal{R}(\mathcal{B})$.

Let $d_k = a_k i(b_k)$, $a_k, b_k \in \mathcal{B}$, $k = 1, 2$. By Lemma 6.1.4, we can write $d_1 + d_2 = u \cdot i(v)$, where

$$u = a_1e + a_2f + (a_1b_2 + b_2a_1)g, \quad v = b_1e + b_2f + b_1b_2g,$$

and pairwise disjoint projections e, f, g are given by

$$e = s(d_1) - s(d_1d_2), \quad f = s(d_2) - s(d_1d_2), \quad g = s(d_1d_2).$$

Since for all $t \in P(\Omega)$, $a, b \in \mathcal{B}$, the equalities

$$\delta_1(t\, a\, i(b)) = \delta(t\, a)i(b) - t\, a\delta(b)i(b^2) = t\delta_1(a\, i(b))$$

hold, it follows that

$$\begin{aligned}
e\delta_1(d_1 + d_2) &= \delta_1(e(d_1 + d_2)) = \delta_1(eu \cdot i(ev)) = \delta_1(a_1i(b_1)e) \\
&= e(\delta(a_1)i(b_1) - a_1\delta(b_1)i(b_1{}^2)) \\
&= e\delta_1(d_1),
\end{aligned}$$

and

$$\begin{aligned} f\delta_1(d_1+d_2) &= \delta_1(f(d_1+d_2)) = \delta_1(fu\cdot i(ev)) = \delta_1(a_2 i(b_2)f) \\ &= f(\delta(a_2)i(b_2) - a\delta(b_2)i({b_2}^2)) \\ &= f\delta_1(d_2). \end{aligned}$$

Furthermore, since $g(d_1+d_2) = gu\cdot i(gv) = (a_1b_2 + a_2b_1)i(b_1b_2)g$, we also have that

$$\begin{aligned} g\delta_1(d_1+d_2) &= \delta_1(g(d_1+d_2)) \\ &= g\delta(a_1b_2+a_2b_1)i(b_1b_2) - (a_1b_2+a_2b_1)\delta(b_1b_2)i(b_1^2b_2^2) \\ &= g(\delta(a_1)b_2 + a_1\delta(b_2) + \delta(a_2)b_1 + a_2\delta(b_1))i(b_1b_2) \\ &\quad - (a_1b_2+a_2b_1)(\delta(b_1)b_2 + b_1\delta(b_2))\cdot i({b_1}^2{b_2}^2) \\ &= g(\delta(a_1)i(b_1) - a_1\delta(b_1)i({b_1}^2) + \delta(a_2)i(b_2) - a_2\delta(b_2)i({b_2}^2)) \\ &= g(\delta_1(d_1) + \delta_1(d_2)). \end{aligned}$$

Since $s(u) \le s(d_1) \vee s(d_2) = e + f + g$, $s(v) \le e + f + g$, we arrive at

$$s(\delta_1(d_1+d_2)) \le e + f + g.$$

In addition, since $P(\Omega) \subset \mathcal{B}$, the derivation δ is nonexpansive (see Proposition 6.3.1(iv)), we have that

$$s(\delta_1(d_1)) \le s(d_1) = e + g, \quad s(\delta_1(d_2)) \le s(d_2) = f + g.$$

Thus,

$$\begin{aligned} \delta_1(d_1+d_2) &= \delta_1(d_1+d_2)(e+f+g) \\ &= e\delta(d_1) + f\delta_1(d_2) + g(\delta_1(d_1) + \delta_1(d_2)) \\ &= \delta_1(d_1) + \delta_1(d_2), \end{aligned}$$

proving that δ_1 is a linear mapping. □

The next proposition is the second step in the extension of derivations. It shows that we can always extend uniquely a derivation $\delta : \mathcal{B} \to L_0(\Omega)$, $P(\Omega) \subset \mathcal{B}$, up to derivation on the least regular subalgebra of $L_0(\Omega)$, containing $\mathcal{B}$.

Proposition 6.3.9. *Let $\mathcal{B}$ be an arbitrary subalgebra in an algebra $L_0(\Omega)$ with $P(\Omega) \subset \mathcal{B}$ and let $\delta : \mathcal{B} \to L_0(\Omega)$ be a derivation. Then the mapping $\delta_1 : \mathcal{R}(\mathcal{B}) \to L_0(\Omega)$, defined by* (6.18)*, is a derivation, which uniquely extends δ up to $\mathcal{R}(\mathcal{B})$.*

Proof. By Lemma 6.3.8, the mapping $\delta_1 : \mathcal{R}(\mathcal{B}) \to L_0(\Omega)$ defined by setting

$$\delta_1(a\, i(b)) = \delta(a)i(b) - a\delta(b)i(b^2),$$

is well-defined and linear.

Let us show that δ_1 is a derivation. For all $d_k = a_k i(b_k)$, $a_k, b_k \in \mathcal{B}$, $k = 1, 2$, we have

$$\begin{aligned}
\delta_1(d_1 d_2) &= \delta_1(a_1 a_2 i(b_1 b_2)) \\
&\overset{(6.18)}{=} (\delta(a_1)a_2 + a_1\delta(a_2))i(b_1 b_2) - a_1 a_2(\delta(b_1)b_2 + b_1\delta(b_2))i({b_1}^2{b_2}^2) \\
&= (\delta(a_1)i(b_1) - a_1\delta(b_1)i({b_1}^2))a_2 i(b_2) \\
&\quad + (\delta(a_2)i(b_2) - a_2\delta(b_2)i({b_2}^2))a_1 i(b_1) \\
&= \delta_1(d_1)d_2 + \delta_1(d_2)d_1.
\end{aligned}$$

That is, δ_1 is a derivation from $\mathcal{R}(\mathcal{B})$ into $L_0(\Omega)$. In addition, if $b \in \mathcal{B}$, then

$$\delta_1(b) = \delta_1(b\, i(\mathbf{1})) = \delta(b)i(\mathbf{1}) - b\delta(\mathbf{1})i(\mathbf{1}) = \delta(b).$$

Thus, δ_1 is an extension of the derivation δ onto the algebra $\mathcal{R}(\mathcal{B})$.

Let us show that δ_1 is a unique extension. Assume that $\delta_2 : \mathcal{R}(\mathcal{B}) \to L_0(\Omega)$ is another derivation, such that $\delta_2(b) = \delta(b)$ for all $b \in \mathcal{B}$. By Proposition 6.3.1(v), for all $a, b \in \mathcal{B}$, we have that

$$\delta_2(a\, i(b)) = \delta_2(a)i(b) + a\delta_2(i(b)) = \delta_2(a)i(b) - a\delta_2(b)i(b^2) = \delta_1(a\, i(b)).$$

Since $\mathcal{R}(\mathcal{B})$ consists of elements of the form $a\, i(b)$, where $a, b \in \mathcal{B}$, we infer that $\delta_2 = \delta_1$, as required. □

The third step is to extend a derivation δ, defined on a subalgebra $\mathcal{B}$, up to a derivation on the topological closure $\overline{\mathcal{B}}^{\rho_\mu}$ of the subalgebra $\mathcal{B}$ in $(L_0(\Omega), \rho_\mu)$, where ρ_μ is the metric defined in Definition 6.1.8.

Proposition 6.3.10. *Let $\mathcal{B}$ be a subalgebra of $L_0(\Omega)$ with $P(\Omega) \subset \mathcal{B}$. Then for every derivation $\delta : \mathcal{B} \to L_0(\Omega)$, there exists a unique derivation $\overline{\delta} : \overline{\mathcal{B}}^{\rho_\mu} \to L_0(\Omega)$, such that $\overline{\delta}(b) = \delta(b)$ for all $b \in \mathcal{B}$.*

Proof. Since $P(\Omega) \subset \mathcal{B}$, Corollary 6.3.4 implies that δ is uniformly continuous in the metric ρ_μ. By Proposition 6.1.9, $(L_0(\Omega), \rho_\mu)$ is a topological ring, and so we can uniquely extend (by continuity) the derivation $\delta : \mathcal{B} \to L_0(\Omega)$ up to a derivation $\overline{\delta} : \overline{\mathcal{B}}^{\rho_\mu} \to L_0(\Omega)$. □

The fourth step in our program is extending a derivation $\delta : \mathcal{B} \to L_0(\Omega)$ onto the algebra $\mathcal{B}(a) = \{p(a) : p \in \mathcal{B}[x]\}$ generated by the subalgebra $\mathcal{B}$ and an integral element $a \in I(\mathcal{B})$ (see Definition 6.2.1). For the realization of this step, we need to introduce two types of derivatives of polynomials with coefficients in $\mathcal{B}$.

The first derivation, p^δ is an extension of a derivation $\delta : \mathcal{B} \to L_0(\Omega)$ up to a derivation on $\mathcal{B}[x]$ (with values in $(L_0(\Omega))[x]$).

Definition 6.3.11. Let $\mathcal{B}$ be a subalgebra of $L_0(\Omega)$ and $\delta : \mathcal{B} \to L_0(\Omega)$ be a derivation. For every polynomial $p(x) = (\sum_{k=0}^{n} a_k x^{n-k}) \in \mathcal{B}[x]$, we set

$$p^{\delta}(x) = \sum_{k=0}^{n} \delta(a_k)x^{n-k}. \tag{6.19}$$

Lemma 6.3.12. *The mapping $p \to p^{\delta}$ is a derivation from the algebra $\mathcal{B}[x]$ into $(L_0(\Omega))[x]$.*

Proof. It is clear that the mapping $p \mapsto p^{\delta}$ is linear. Hence, it is sufficient to verify the Leibniz rule for monomials. Let $a, b \in \mathcal{B}$ and $n, m \in \mathbb{N}$. We have

$$\begin{aligned}((ax^n)(bx^m))^{\delta} &= (abx^{n+m})^{\delta} = \delta(ab)x^{n+m} \\ &= (\delta(a)x^n)(bx^m) + (ax^n)(\delta(b)x^m) \\ &= (ax^n)^{\delta}(bx^m) + (ax^n)(bx^m)^{\delta},\end{aligned}$$

as required. □

We also introduce another derivation on $\mathcal{B}[x]$, which mimics the classical derivation on polynomials.

Definition 6.3.13. Let $p(x) = (\sum_{k=0}^{n} a_k x^k) \in \mathcal{B}[x]$. We define polynomial $p'(x)$ by setting

$$p'(x) = \left(\sum_{k=0}^{n} a_k x^k\right)' = \sum_{k=1}^{n} k a_k x^{k-1}. \tag{6.20}$$

The mapping $p \to p'$ is clearly a derivation on $\mathcal{B}[x]$ (with values in $\mathcal{B}[x]$).

In the next lemma, we show that if a is a properly $\mathcal{B}$-integral element (see Definition 6.2.3) with respect to a regular subalgebra containing the Boolean algebra $P(\Omega)$, then for the monic polynomial $p \in \mathcal{B}[x]$, such that $n = \deg(p) = \deg_{\mathcal{B}}(a)$ and $p(a) = 0$, the derivative p' of p evaluated at a is an invertible element in $L_0(\Omega)$.

Lemma 6.3.14. *Let $\mathcal{B}$ be a regular subalgebra of $L_0(\Omega)$, $P(\Omega) \subset \mathcal{B}$, and let $a \in L_0(\Omega)$ be a proper invertible $\mathcal{B}$-integral element. Let $p(x) \in \mathcal{B}[x]$ be a monic polynomial, such that $n = \deg(p) = \deg_{\mathcal{B}}(a)$ and $p(a) = 0$. Then $p'(a)$ is invertible in $\mathcal{A}$.*

Proof. If $\deg_{\mathcal{B}}(a) = 1$, then $p'(a) = \mathbf{1}$, and hence, $p'(a)$ is invertible. Suppose that $\deg_{\mathcal{B}}(a) \geq 2$. It is sufficient to show that $s(p'(a)) = \mathbf{1}$.

Note that the polynomial $q(x) = (1/n)p'(x) \in \mathcal{B}[x]$ is a monic polynomial and $s(q(a)) = s(p'(a))$. Suppose that $e := \mathbf{1} - s(q(a)) \neq 0$. In this case, the polynomial $q_1(x) = x^{n-1} + eq(x) - ex^{n-1}$ is a monic polynomial, moreover, $q_1(x) \in \mathcal{B}[x]$, since $P(\Omega) \subset \mathcal{B}$. Furthermore,

$$\deg q_1 = n - 1 < \deg_{\mathcal{B}}(a) \quad \text{and} \quad q_1(ae) = eq(ae) = eq(a) = 0.$$

Thus, $\deg_{\mathcal{B}e}(ae) \leq n - 1$. However, the element ae is a proper $\mathcal{B}$-integral element and, therefore, by definition, $\deg_{\mathcal{B}e}(ae) = \deg_{\mathcal{B}}(a) = n$, which is a contradiction. Hence, $e = 0$, i. e., $\mathbf{1} = s(q(a)) = s(p'(a))$, as required. □

Next, we extend the derivation $\delta : \mathcal{B} \to L_0(\Omega)$ to a derivation $\delta_1 : \mathcal{B}(a) \to L_0(\Omega)$, where a is a proper $\mathcal{B}$-integral element. We divide this step into several separate lemmas.

Lemma 6.3.15. *Let $\mathcal{B}$ be a regular subalgebra of $L_0(\Omega)$ with $P(\Omega) \subset \mathcal{B}$ and let a be a proper $\mathcal{B}$-integral element with $s(a) = \mathbf{1}$. Let $p(x) \in \mathcal{B}[x]$ be a monic polynomial, such that $n = \deg(p) = \deg_{\mathcal{B}}(a)$ and $p(a) = 0$. For an arbitrary element $y = q(a) \in \mathcal{B}(a)$, where $q(x) = \sum_{k=0}^{m} b_k x^{m-k} \in \mathcal{B}[x]$, we define*

$$\delta_1(y) = q^{\delta}(a) - q'(a)p^{\delta}(a)(p'(a))^{-1}. \tag{6.21}$$

Then $\delta_1(y)$ does not depend on the choice of the polynomial $q \in \mathcal{B}[x]$, such that $y = q(a)$.

Proof. By Lemma 6.3.14, the element $p'(a)$ is invertible.

Suppose that $y = q_1(a) = q_2(a)$, where $q_1, q_2 \in \mathcal{B}[x]$. Consider the polynomial $q_3(x) := q_1(x) - q_2(x) \in \mathcal{B}[x]$. We have that $q_3(a) = 0$. Therefore, since a is a proper $\mathcal{B}$-integral element, by Lemma 6.2.5 we have that $q_3(x) = u(x)p(x)$ for some $u \in \mathcal{B}[x]$. Hence, we can write

$$\begin{aligned}
&[q_1^{\delta}(a) - q_1'(a)p^{\delta}(a)(p'(a))^{-1}] - [q_2^{\delta}(a) - q_2'(a)p^{\delta}(a)(p'(a))^{-1}] \\
&\quad = q_3^{\delta}(a) - q_3'(a)p^{\delta}(a)(p'(a))^{-1} \\
&\quad = (up)^{\delta}(a) - (up)'(a)p^{\delta}(a)(p'(a))^{-1}
\end{aligned}$$

Since both $p \mapsto p^{\delta}$ and $p \mapsto p'$ are derivations on $\mathcal{B}[x]$ (see Lemma 6.3.12 and Definition 6.3.13, resp.) and $p(a) = 0$ by assumption, we conclude that

$$\begin{aligned}
&[q_1^{\delta}(a) - q_1'(a)p^{\delta}(a)(p'(a))^{-1}] - [q_2^{\delta}(a) - q_2'(a)p^{\delta}(a)(p'(a))^{-1}] \\
&\quad = u^{\delta}(a)p(a) + u(a)p^{\delta}(a) - u'(a)p(a)p^{\delta}(a)(p'(a))^{-1} \\
&\qquad - u(a)p'(a)p^{\delta}(a)(p'(a))^{-1} \\
&\quad = u(a)p^{\delta}(a) - u(a)p'(a)p^{\delta}(a)(p'(a))^{-1} \\
&\quad = u(a)p^{\delta}(a) - u(a)p^{\delta}(a) = 0.
\end{aligned}$$

Thus, $\delta_1(y)$ does not depend on the choice of the polynomial $q \in \mathcal{B}[x]$, such that $y = q(a)$. □

Next, we show that the mapping defined in Lemma 6.3.15 is in fact a derivation on $\mathcal{B}(a)$.

Lemma 6.3.16. *Let $\mathcal{B}$ be a regular subalgebra of $L_0(\Omega)$ with $P(\Omega) \subset \mathcal{B}$ and let a be an invertible proper $\mathcal{B}$-integral element. Assume that $\delta : \mathcal{B} \to L_0(\Omega)$ is a derivation. The mapping $\delta_1 : \mathcal{B}(a) \to L_0(\Omega)$, defined by (6.21), is a derivation, which uniquely extends δ.*

Proof. Let $p(x) = x^n + a_1x^{n-1} + \cdots + a_{n-1}x + a_n \in \mathcal{B}_u[x]$ be a monic polynomial, such that $p(a) = 0$ and $n = \deg_{\mathcal{B}}(a)$. By Lemma 6.2.5, we have that there exists the inverse element $(p'(a))^{-1}$.

Let $y \in \mathcal{B}(a)$ and let $q \in \mathcal{B}[x]$ be such that $y = q(a)$. By Lemma 6.3.15, the mapping

$$\delta_1(y) = q^{\delta}(a) - q'(a)p^{\delta}(a)(p'(a))^{-1},$$

where $q^{\delta}(x) = \sum_{k=0}^{m} \delta(b_k)x^{m-k}$ and $q'(x) = \sum_{k=0}^{m-1}(m-k)b_k x^{m-k-1}$ do not depend on the choice of $q \in \mathcal{B}[x]$ satisfying $y = q(a)$.

We claim that δ_1 is a derivation.

Recall that the mappings $p(x) \to p'(x)$ and $p(x) \to p^{\delta}(x)$ are derivations from $\mathcal{B}[x]$ into $L_0(\Omega)[x]$ (see Definition 6.3.13 and Lemma 6.3.12, resp.). Hence, for every $y_i = q_i(a)$, $q_i \in \mathcal{B}[x]$, $\lambda_i \in \mathbb{C}$, $i = 1, 2$, we have that

$$\begin{aligned}
&\delta_1(\lambda_1 y_1 + \lambda_2 y_2)\\
&\quad = (\lambda_1 q_1 + \lambda_2 q_2)^{\delta}(a) - (\lambda_1 q_1 + \lambda_2 q_2)'(a)p^{\delta}(a)(p'(a))^{-1}\\
&\quad = (\lambda_1 q_1^{\delta}(a) - \lambda_1 q_1'(a)p^{\delta}(a)(p'(a))^{-1})\\
&\qquad + (\lambda_2 q_2^{\delta}(a) - \lambda_2 q_2'(a)p^{\delta}(a)(p'(a))^{-1})\\
&\quad = \lambda_1\delta_1(y_1) + \lambda_2\delta_1(y_2),
\end{aligned}$$

i. e., δ_1 is linear. For the Leibniz rule, we have

$$\begin{aligned}
\delta_1(y_1y_2) &= (q_1q_2)^{\delta}(a) - (q_1q_2)'(a)p^{\delta}(a)(p'(a))^{-1}\\
&= q_1^{\delta}(a)q_2(a) + q_1(a)q_2^{\delta}(a) - (q_1'(a)q_2(a) + q_1(a)q_2'(a))p^{\delta}(a)(p'(a))^{-1}\\
&= (q_1^{\delta}(a) - q_1'(a)p^{\delta}(a)(p'(a))^{-1})q_2(a)\\
&\quad + q_1(a)(q_2^{\delta}(a) - q_2'(a)p^{\delta}(a)(p'(a))^{-1})\\
&= \delta_1(y_1)y_2 + y_1\delta_1(y_2).
\end{aligned}$$

Thus, δ_1 is a derivation on $\mathcal{B}(a)$.

To show that δ_1 is an extension of δ, assume that $b \in \mathcal{B}$. Then $b = q(a)$ for the polynomial $q(x) \equiv b \in \mathcal{B}[x]$ and, therefore,

$$\delta_1(b) = q^{\delta}(a) - q'(a)p^{\delta}(a)(p'(a))^{-1} = \delta(b) - 0 \cdot p^{\delta}(a)(p'(a))^{-1} = \delta(b),$$

which implies that δ_1 is indeed an extension of δ.

Assume now that $\delta_2 : \mathcal{B}(a) \to L_0(\Omega)$ is another derivation, such that $\delta_2(b) = \delta(b)$ for all $b \in \mathcal{B}$. Using the equalities,

$$p^{\delta}(a) + p'(a)\delta_2(a) = p^{\delta_2}(a) + p'(a)\delta_2(a) = \delta_2(p(a)) = 0,$$

we obtain that

$$\delta_2(a) = -p^{\delta}(a)(p'(a))^{-1} = \delta_1(a).$$

Consequently,

$$\delta_2(q(a)) = q^{\delta_2}(a) + q'(a)\delta_2(a) = q^{\delta}(a) + q'(a)\delta_1(a) = \delta_1(q(a))$$

for all $q \in \mathcal{B}[x]$. Thus, δ_1 is a unique extension of δ. □

With the extension by proper $\mathcal{B}$-integral elements concluded, we can now show that any derivation $\delta : \mathcal{B} \to L_0(\Omega)$ on a regular disjointly complete subalgebra of $L_0(\Omega)$ that contains the Boolean algebra $P(\Omega)$ can be uniquely extended to a derivation $\delta_1 : \mathcal{B}(a) \to L_0(\Omega)$, where $\mathcal{B}(a)$ is the subalgebra of $L_0(\Omega)$ generated by $\mathcal{B}$ and an arbitrary $\mathcal{B}$-integral element $a \in L_0(\Omega)$.

Proposition 6.3.17. *Let $\mathcal{B}$ be a regular disjointly complete subalgebra of $L_0(\Omega)$ with $P(\Omega) \subset \mathcal{B}$ and let a be an arbitrary nonzero $\mathcal{B}$-integral element. Then, for every derivation $\delta : \mathcal{B} \to L_0(\Omega)$, there exists a unique derivation $\tilde{\delta} : \mathcal{B}(a) \to L_0(\Omega)$, such that $\tilde{\delta}(b) = \delta(b)$ for all $b \in \mathcal{B}$.*

Proof. By Proposition 6.2.6, there exist natural numbers $n_1 < n_2 < \cdots < n_m$ and nonzero pairwise disjoint projections $e_1, e_2, \ldots, e_m \in P(\Omega)$, such that $\sum_{k=1}^{m} e_k = s(a)$, and every ae_k is a proper $\mathcal{B}e_k$-integral element. In addition, $n_k = \deg_{\mathcal{B}e_k}(ae_k)$.

We set

$$e_0 = \mathbf{1} - s(a), \quad \mathcal{B}_k := \mathcal{B}e_k, \quad k = 0, 1, 2, \ldots, m.$$

It is clear that $\mathcal{B}(a) \cap L_0(\Omega)e_k = \mathcal{B}_k(ae_k)$ and

$$\delta_k(xe_k) := \delta(xe_k) = \delta(x)e_k$$

is a derivation from $\mathcal{B}_k$ into $L_0(\Omega)e_k$ for every $k = 0, 1, 2, \ldots, m$.

For every $k = 1, \ldots, m$, the element ae_k is a proper $\mathcal{B}_k$-integral element and $s(ae_k) = e_k = \mathbf{1}_{\mathcal{B}_k}$. Hence, by Lemma 6.3.16, there exists a unique derivation

$$\delta_k^{(1)} : \mathcal{B}_k(ae_k) \to L_0(\Omega)e_k,$$

such that $\delta_k^{(1)}(b) = \delta_k(b)$ for all $b \in \mathcal{B}_k$. In addition, since $e_0 s(a) = 0$, it follows that $\mathcal{B}_0(ae_0) = \mathcal{B}_0(0) = \mathcal{B}_0$. Therefore

$$\mathcal{B}(a) = \mathcal{B}_0 + \sum_{k=1}^{m} \mathcal{B}_k(ae_k).$$

For every $y \in \mathcal{B}(a)$, we set

$$\tilde{\delta}(y) = \delta_0(ye_0) + \delta_1^{(1)}(ye_1) + \delta_2^{(1)}(ye_2) + \cdots + \delta_m^{(1)}(ye_m).$$

We claim that δ is a derivation from $\mathcal{B}(a)$ into $L_0(\Omega)$. The linearity of the mapping $\tilde{\delta}$ follows from linearity of the derivations δ_0 and $\delta_k^{(1)}$, $k = 1, \ldots, m$.

To prove the Leibniz rule, let $y_1, y_2 \in \mathcal{B}(a)$. Since δ_0 is a derivation on $\mathcal{B}(0)$ and every $\delta_k^{(1)}$ is a derivation on $\mathcal{B}_k(ae_k)$, $k = 1, \ldots, m$, it follows that

$$\begin{aligned}
\tilde{\delta}(y_1y_2) &= \delta_0(y_1y_2e_0) + \delta_1^{(1)}(y_1y_2e_1) + \cdots + \delta_m^{(1)}(y_1y_2e_m) \\
&= (\delta_0(y_1e_0)y_2e_0 + y_1e_0\delta_0(y_2e_0)) + (\delta_1^{(1)}(y_1e_1)y_2e_1 + y_1e_1\delta_1^{(1)}(y_2e_1)) \\
&\quad + \cdots + (\delta_m^{(1)}(y_1e_m)y_2e_m + y_1e_m\delta_m^{(1)}(y_2e_m)) \\
&= (\delta_0(y_1e_0) + \delta_1^{(1)}(y_1e_1) + \cdots + \delta_m^{(1)}(y_1e_m)) \cdot y_2 \\
&\quad + y_1 \cdot (\delta_0(y_2e_0) + \delta_1^{(1)}(y_2e_1) + \cdots + \delta_m^{(1)}(y_2e_m)) \\
&= \tilde{\delta}(y_1)y_2 + y_1\tilde{\delta}(y_2).
\end{aligned}$$

Thus, $\tilde{\delta} : \mathcal{B}(a) \to L_0(\Omega)$ is a derivation. It is clear that $\tilde{\delta}$ is an extension of δ.

Let $\delta_1 : \mathcal{B}(a) \to L_0(\Omega)$ be another derivation, such that $\delta_1(b) = \delta_0(b)$ for all $b \in \mathcal{B}$. Then $\delta_1(be_k) = \delta_0(be_k)$ and by uniqueness of the extensions $\delta_k^{(1)}$ (see Lemma 6.3.16), we have that $\delta_1(ye_k) = \delta_k^{(1)}(ye_k)$ for all $y \in \mathcal{B}(a)$, $k = 1, \ldots, m$. Thus, for every $y \in \mathcal{B}(a)$, the equalities

$$\begin{aligned}
\delta_1(y) &= \delta_1(ye_0) + \delta_1(ye_1) + \cdots + \delta_1(ye_m) \\
&= \delta_0(ye_0) + \delta_1^{(1)}(ye_1) + \delta_2^{(1)}(ye_2) + \cdots + \delta_m^{(1)}(ye_m) = \tilde{\delta}(y)
\end{aligned}$$

hold. In other words, $\tilde{\delta}$ is a unique extension of δ. □

The fifth step in our program is extending a derivation $\delta : \mathcal{B} \to L_0(\Omega)$ onto the algebra $\mathcal{B}(a)$, generated by the subalgebra $\mathcal{B}$ and a weakly transcendental with respect to $\mathcal{B}$ element a. We start with an auxiliary lemma. We recall that the derivations $p \mapsto p^\delta$ and $p \mapsto p'$ on the algebra $\mathcal{B}[x]$ are defined in Definition 6.3.11 and Definition 6.3.13, respectively.

Lemma 6.3.18. *Assume that $\mathcal{B}$ is a regular subalgebra of $L_0(\Omega)$ with $P(\Omega) \subset \mathcal{B}$ and $a \in L_0(\Omega)$ is a weakly transcendental $\mathcal{B}$-integral element. Let $\delta : \mathcal{B} \to L_0(\Omega)$ be a derivation. If $f, g \in \mathcal{B}[x]$ are such that $f(a) = g(a)$, then*

$$f^\delta(a) = g^\delta(a), \quad f'(a)s(a) = g'(a)s(a).$$

Proof. Introduce the polynomial $h(x) = f(x) - g(x) \in \mathcal{B}[x]$. It is sufficient to prove that

$$h^\delta(a) = 0, \quad h'(a)s(a) = 0.$$

Since $f(a) = g(a)$, it follows that $h(a) = 0$. Let us write

$$h(x) = \sum_{j=0}^{n} a_j x^{n-j},$$

for some $a_j \in \mathcal{B}, j = 0, \dots, n, n \in \mathbb{N}$.

Since $s(a_0) = a_0 i(a_0)$, we have

$$\begin{aligned} 0 = i(a_0)h(a) &= i(a_0) \sum_{j=0}^{n} a_j a^{n-j} \\ &= s(a_0)a^n + a_1 i(a_0)a^{n-1} + \dots + i(a_0)a_n \\ &= (s(a_0)a)^n + a_1 i(a_0)(s(a_0)a)^{n-1} + \dots + i(a_0)a_n. \end{aligned}$$

Hence, since $i(a_0)a_j \in \mathcal{B}, j = 1, \dots, n$, it follows that $s(a_0)a = s(a_0)s(a)a$ is a $\mathcal{B}$-integral element. Since $s(a_0)s(a) \le s(a)$ and by the assumption the element a is weakly transcendental with respect to $\mathcal{B}$ (see Definition 6.2.12), it follows that $s(a_0)a = 0$. In particular, $a_0 a^n = 0$ and, therefore, for the polynomial

$$h_1(x) := \sum_{j=1}^{n} a_j x^{n-j}$$

from $\mathcal{B}[x]$ the equality $h_1(a) = 0$ holds.

Repeating this procedure n times, we obtain that

$$s(a_j)a = 0, \quad j = 0, 1, \dots, n-1, \quad a_n = 0.$$

Hence, $\delta(a_j)a^{n-j} = \delta(a_j)s(a_j)a^{n-j} = 0$ for every $j = 0, \dots, n-1$, and $\delta(a_n) = 0$. Therefore,

$$h^{\delta}(a) = \sum_{j=0}^{n} \delta(a_j)a^{n-j} = 0,$$

$$h'(a)s(a) = \sum_{j=0}^{n-1} (n-j)a_j a^{n-j-1} s(a) = 0,$$

as required. □

In the following proposition, we extend a derivation $\mathcal{B} \to L_0(\Omega)$ up to a derivation $\delta_1 : \mathcal{B}(a) \to L_0(\Omega)$, where $\mathcal{B}(a)$ is the algebra generated by $\mathcal{B}$ and a weakly transcendental element with respect to $\mathcal{B}$. We note that here we already obtain infinitely many extensions of δ.

Proposition 6.3.19. *Assume that $\mathcal{B}$ is a regular subalgebra of $L_0(\Omega)$ with $P(\Omega) \subset \mathcal{B}$ and $a \in L_0(\Omega)$ is a weakly transcendental $\mathcal{B}$-integral element. Then, for every derivation δ :*

$\mathcal{B} \to L_0(\Omega)$ and for every $c \in L_0(\Omega)$ with $s(c) \leq s(a)$, there exists a unique derivation $\delta_1 : \mathcal{B}(a) \to L_0(\Omega)$, such that $\delta_1(a) = c$ and $\delta_1(b) = \delta(b)$ for all $b \in \mathcal{B}$.

Proof. For every $f(a) \in \mathcal{B}(a), f \in \mathcal{B}[x]$, define a mapping $\delta_1 : \mathcal{B}(a) \to L_0(\Omega)$ by setting

$$\delta_1(f(a)) := f^{\delta}(a) + f'(a)c.$$

Since $s(c) \leq s(a)$, Lemma 6.3.18 guarantees that the mapping $\delta_1 : \mathcal{B}(a) \to L_0(\Omega)$ is well-defined.

Taking into account the mappings $f \to f', f \to f^{\delta}$ are derivations from $\mathcal{B}[x]$ into $L_0(\Omega)[x]$ (see Definition 6.3.13 and Lemma 6.3.12, resp.), we obtain that the mapping $\delta_1 : \mathcal{B}(a) \to L_0(\Omega)$ is a derivation from $\mathcal{B}(a)$ into $L_0(\Omega)$ and $\delta_1(b) = \delta(b)$ for all $b \in \mathcal{B}$.

In addition, taking $f(x) = x$, we have $f(a) = a, f'(a) = \mathbf{1}, f^{\delta}(a) = 0$ and, therefore, $\delta_1(a) = c$.

Thus, δ_1 is indeed an extension of δ satisfying the required properties.

Let $\delta_2 : \mathcal{B}(a) \to L_0(\Omega)$ be another derivation, such that $\delta_2(b) = \delta(b)$ for all $b \in \mathcal{B}$ and $\delta_2(a) = c$. Then for every polynomial $f \in \mathcal{B}[x]$, we have

$$\delta_2(f(a)) = f^{\delta}(a) + f'(a)\delta_2(a) = f^{\delta}(a) + f'(a)c = \delta_1(f(a)).$$

Thus, δ_1 is a unique extension with the required properties. □

Now we collect all five steps together in the following theorem, which is the first main result of this section.

Theorem 6.3.20. *Let $\mathcal{B}$ be an arbitrary subalgebra of the algebra $L_0(\Omega)$. Then for every nonexpansive derivation $\delta : \mathcal{B} \to L_0(\Omega)$, there exists a derivation $\widetilde{\delta} : L_0(\Omega) \to L_0(\Omega)$, such that $\widetilde{\delta}(b) = \delta(b)$ for all $b \in \mathcal{B}$.*

Proof. Consider the set X of all pairs $(\mathcal{A}_i, \delta_i)$, where $\mathcal{A}_i$ is a subalgebra of $\mathcal{A}$, $\mathcal{B} \subset \mathcal{A}_i$ and $\delta_i : \mathcal{A}_i \to L_0(\Omega)$ is a derivation so that $\delta_i(b) = \delta(b)$ for all $b \in \mathcal{B}$ and $s(\delta_i(a)) \leq s(a)$ for all $a \in \mathcal{A}_i$. The set X is nonempty since $(\mathcal{B}, \delta) \in X$. We define a partial order on X by setting $(\mathcal{A}_i, \delta_i) \leq (\mathcal{A}_j, \delta_j)$, if and only if $\mathcal{A}_i \subset \mathcal{A}_j$, $\delta_j(a) = \delta_i(a)$ for all $a \in \mathcal{A}_i$. Let $Y := \{(\mathcal{A}_i, \delta_i)\}_{i \in I}$ be an arbitrary linearly ordered subset of X. Set $\mathcal{A}_I = \cup_{i \in I} \mathcal{A}_i$. It is clear that $\mathcal{A}_I$ is a subalgebra of $L_0(\Omega)$, $\mathcal{B} \subset \mathcal{A}_I$, and the derivation $\delta_I : \mathcal{A}_I \to L_0(\Omega)$, where $\delta_I(a) = \delta_i(a)$ for all $a \in \mathcal{A}_i, i \in I$ is correctly defined, furthermore, $s(\delta_I(a)) \leq s(a)$ for all $a \in \mathcal{A}_I$. Consequently, $(\mathcal{A}_I, \delta_I) \in X$ and $(\mathcal{A}_i, \delta_i) \leq (\mathcal{A}_I, \delta_I)$ for all $i \in I$.

By Zorn's lemma, there exists the maximal element $(\mathcal{A}_0, \delta_0) \in X$. From Theorem 6.2.11, maximality of $(\mathcal{A}_0, \delta_0)$ and Proposition 6.3.7, Proposition 6.3.9, Proposition 6.3.10, and Proposition 6.3.17, we infer that $E(\mathcal{A}_0) = \mathcal{A}_0$.

Suppose that $\mathcal{A}_0 \neq L_0(\Omega)$. Then by Lemma 6.2.13 there exists a weakly transcendental element c with respect to $\mathcal{A}_0$. Proposition 6.3.19 implies that the derivation δ_0 extends up to a derivation $\delta_1 : \mathcal{A}_0(c) \to L_0(\Omega)$. This means that $(\mathcal{A}_0(c), \delta_1) \in X$, in addition, $\mathcal{A}_0(c) \neq \mathcal{A}_0$, which contradicts with maximality of $(\mathcal{A}_0, \delta_0)$ in X. Hence, $\mathcal{A}_0 = L_0(\Omega)$

and $\widetilde{\delta} := \delta_0$ is the required extension of the derivation δ up to a derivation defined on the whole algebra $L_0(\Omega)$. □

With this general extension theorem, we can now prove that for any subalgebra $\mathcal{B}$ and a nonexpansive derivation $\delta : \mathcal{B} \to L_0(\Omega)$, the least regular, integrally closed, and disjointly complete subalgebra $E(\mathcal{B})$ containing $\mathcal{B}$ and $P(\Omega)$ is the largest subalgebra of $L_0(\Omega)$, which contains $\mathcal{B}$ and admits a unique extension of the derivation δ.

We first prove that any derivation defined on a regular subalgebra $\mathcal{B}$ of the algebra $L_0(\Omega)$ with $P(\Omega) \subset \mathcal{B}$ can be uniquely extended up to a derivation defined on the integral closure $I(\mathcal{B})$ of the subalgebra $\mathcal{B}$.

Proposition 6.3.21. *Let $\mathcal{B}$ be a regular subalgebra of an algebra $L_0(\Omega)$, containing the Boolean algebra $P(\Omega)$. Then for every derivation $\delta : \mathcal{B} \to L_0(\Omega)$ there exists a unique derivation $\delta_1 : I(\mathcal{B}) \to L_0(\Omega)$, such that $\delta_1(b) = \delta(b)$ for all $b \in \mathcal{B}$.*

Proof. By Corollary 6.3.4, the derivation δ is nonexpansive. Therefore, by Theorem 6.3.20, there exists a derivation $\delta_1 : I(\mathcal{B}) \to L_0(\Omega)$, such that $\delta_1(b) = \delta(b)$ for all $b \in \mathcal{B}$. Suppose that $\delta_2 : I(\mathcal{B}) \to L_0(\Omega)$ is another derivation such that $\delta_2(b) = \delta(b)$ for all $b \in \mathcal{B}$. By Proposition 6.3.17, for every nonzero $a \in I(\mathcal{B})$, the derivations δ_1 and δ_2 coincide on the subalgebra $\mathcal{B}(a)$ generated by $\mathcal{B}$ and the integral element $a \in I(\mathcal{B})$. Therefore, $\delta_1(a) = \delta_2(a)$ for every $a \in I(\mathcal{B})$. □

Let $\mathcal{B}$ be an arbitrary subalgebra in $L_0(\Omega)$. By Theorem 6.3.20, every nonexpansive derivation $\delta : \mathcal{B} \to L_0(\Omega)$ extends up to a derivation $\widetilde{\delta}$, defined on the whole algebra $L_0(\Omega)$. It means that the derivation δ has an extension on an arbitrary subalgebra $\mathcal{C}$ of $L_0(\Omega)$, which contains the subalgebra $\mathcal{B}$. We now present the necessary and sufficient conditions on the subalgebra $\mathcal{C}$, which guarantee uniqueness of the extension of the derivation δ. The following theorem is the second main result of this section.

Theorem 6.3.22. *Let $\mathcal{B}$ be an arbitrary subalgebra of $L_0(\Omega)$ and let $E(\mathcal{B})$ be the least regular, integrally closed, and disjointly complete subalgebra $E(\mathcal{B})$ containing $\mathcal{B}$ and $P(\Omega)$. Assume that $\delta : \mathcal{B} \to L_0(\Omega)$ is a nonexpansive derivation. Then $E(\mathcal{B})$ is the largest subalgebra of $L_0(\Omega)$ containing $\mathcal{B}$ and admitting a unique extension of the derivation δ.*

Proof. By Theorem 6.2.11, the algebra $E(\mathcal{B})$ is equal to

$$E(\mathcal{B}) = \overline{I(\overline{\mathcal{R}(\mathbb{C}(\mathcal{B}, P(\Omega)))}^{\rho_\mu})}^{\rho_\mu}.$$

A combination of Proposition 6.3.7 and Proposition 6.3.9 guarantees that there exists a unique extension $\bar{\delta}$ of δ up to a derivation on $\mathcal{R}(\mathbb{C}(\mathcal{B}, P(\Omega)))$. By Proposition 6.3.10 and Proposition 6.3.21, $\bar{\delta}$ (and hence δ) can be uniquely extended up to a derivation on $I(\overline{\mathcal{R}(\mathbb{C}(\mathcal{B}, P(\Omega)))}^{\rho_\mu})$. Referring to Proposition 6.3.10 once again, there exists a unique extension $\hat{\delta}$ onto the algebra $E(\mathcal{B}) = \overline{I(\overline{\mathcal{R}(\mathbb{C}(\mathcal{B}, P(\Omega)))}^{\rho_\mu})}^{\rho_\mu}$.

Assume now that there exists a subalgebra $\mathcal{C}$ on $L_0(\Omega)$, such that $\mathcal{B} \subset \mathcal{C}$, the derivation $\delta : \mathcal{B} \to L_0(\Omega)$ extends uniquely up to a derivation $\tilde{\delta} : \mathcal{C} \to L_0(\Omega)$, but $\mathcal{C} \not\subset E(\mathcal{B})$.

By Lemma 6.2.13, for an element $a_0 \in \mathcal{C} \setminus E(\mathcal{B})$ there exists a projection e, such that $ae_0 \in E(\mathcal{B})$, and $a = a_0 - a_0 e$ is a weakly transcendental element with respect to the subalgebra $E(\mathcal{B})$. By Proposition 6.3.19, the subalgebra $(E(\mathcal{B}))(a)$ admits two derivations δ_1 and δ_2, extending $\widehat{\delta}$, so that $\delta_1(a) = 0$ and $\delta_2(a) = a$.

It is clear that $a_0 = a + a_0 e \in (E(\mathcal{B}))(a)$, and

$$\begin{aligned}\delta_1(a_0) &= \delta_1(a + a_0 e) = \delta_1(a_0 e) = \delta_2(a_0 e) = \delta_2(a) + \delta_2(a_0 e) - a \\ &= \delta_2(a + a_0 e) - a = \delta_2(a_0) - a \neq \delta_2(a_0).\end{aligned}$$

By Theorem 6.3.20, we can extend the derivation $\delta_1 : (E(\mathcal{B}))(a) \to L_0(\Omega)$ (resp., $\delta_2 : (E(\mathcal{B}))(a) \to L_0(\Omega)$) up to derivation $\widehat{\delta}_1 : L_0(\Omega) \to L_0(\Omega)$ (resp., $\widehat{\delta}_2 : L_0(\Omega) \to L_0(\Omega)$) the derivation from $L_0(\Omega)$ into $L_0(\Omega)$, extending the derivation $\delta_1 : (E(\mathcal{B}))(a) \to L_0(\Omega)$ (resp., $\delta_2 : (E(\mathcal{B}))(a) \to L_0(\Omega)$) (see Theorem 6.3.20). It is clear that the restrictions $\widehat{\delta}_1|_{\mathcal{C}}$ and $\widehat{\delta}_2|_{\mathcal{C}}$ are two different nonexpansive derivations on the subalgebra $\mathcal{C}$, which coincide on the subalgebra $\mathcal{B}$, which is a contradiction. Thus, $\mathcal{C} \subset E(\mathcal{B})$ and, therefore, $E(\mathcal{B})$ is the largest subalgebra of $L_0(\Omega)$ containing $\mathcal{B}$ and admitting a unique extension of the derivation δ. □

The following proposition establishes the existence of infinitely many distinct extensions of a nonexpansive derivation $\delta : \mathcal{B} \to L_0(\Omega)$ in the case when $E(\mathcal{B}) \neq L_0(\Omega)$.

Proposition 6.3.23. *Suppose that $\mathcal{B}$ is a subalgebra of $L_0(\Omega)$ and assume that the least regular, integrally closed, and disjointly complete subalgebra $E(\mathcal{B})$ of $L_0(\Omega)$ is a proper subalgebra of $L_0(\Omega)$. Then there exist uncountably many distinct (in particular, nonzero) extensions of the derivation δ up to a derivation from $L_0(\Omega)$ into $L_0(\Omega)$.*

Proof. Appealing to Lemma 6.2.13, we can find a weakly transcendental element $a \in L_0(\Omega) \setminus E(\mathcal{B})$ with respect to the subalgebra $E(\mathcal{B})$. Using Proposition 6.3.19, for every $\lambda \in \mathbb{C}$ we construct a derivation $\delta_\lambda : (E(\mathcal{B}))(a) \to L_0(\Omega)$, such that $\delta_\lambda|_{\mathcal{B}} = \delta$ and $\delta_\lambda(a) = \lambda a$. By Theorem 6.3.20, for every $\lambda \in \mathbb{C}$, there exists a derivation $\hat{\delta}_\lambda : L_0(\Omega) \to L_0(\Omega)$, such that $\hat{\delta}_\lambda|_{\mathcal{B}} = \delta$ and $\hat{\delta}_\lambda(a) = \lambda a$. Since the set $\{\lambda a : \lambda \in \mathbb{C}\}$ is an uncountable set of pairwise distinct elements from $L_0(\Omega)$, it follows that $\{\hat{\delta}_\lambda\}_{\lambda \in \mathbb{C}}$ is also an uncountable family of pairwise distinct derivations on $L_0(\Omega)$, each of which is an extension of the derivation δ. □

6.4 Existence of nonzero derivations on the algebra of measurable functions

The central purpose of this section is presenting the necessary and sufficient condition for the existence of nonzero derivations $\delta : L_0(\Omega) \to L_0(\Omega)$. The starting point here is the algebra $\mathbb{C}_c(P(\Omega))$ of all simple elements in $L_0(\Omega)$ (see Definition 6.4.1). As we will show, this algebra is a regular, integrally closed, and disjointly complete subalgebra of $L_0(\Omega)$. Furthermore, any derivation on $\mathbb{C}_c(P(\Omega))$ is necessarily trivial (see Proposition 6.4.2).

Since $\mathbb{C}_c(P(\Omega))$ is a regular, integrally closed, and disjointly complete subalgebra of $L_0(\Omega)$, Proposition 6.3.23 implies that there exist infinitely many nonzero derivations on $L_0(\Omega)$ when $\mathbb{C}_c(P(\Omega))$ is a proper subalgebra on $L_0(\Omega)$. On the other hand, if $\mathbb{C}_c(P(\Omega))$ coincides with $L_0(\Omega)$, then Theorem 6.3.22 guarantees that all derivations on $L_0(\Omega)$ are zero, as the unique extension of the zero derivation on $\mathbb{C}_c(P(\Omega))$.

Thus, to establish the necessary and sufficient conditions for the existence of nonzero derivations $\delta : L_0(\Omega) \to L_0(\Omega)$, we study topological and algebraic properties of $\mathbb{C}_c(P(\Omega))$ as well as necessary and sufficient conditions for this algebra to be a proper subalgebra of $L_0(\Omega)$.

Definition 6.4.1. We denote by $\mathbb{C}_c(P(\Omega))$ the set of all simple elements in $L_0(\Omega)$, i. e., elements of the form

$$\sum_{n=1}^{\infty} \lambda_n [\chi_{A_n}],$$

for some $\lambda_n \in \mathbb{C}$, $n \in \mathbb{N}$, and pairwise disjoint family $\{A_n\}_{n\in\mathbb{N}} \subset \Sigma$.

We note that by Proposition 6.1.11 the series $\sum_{n=1}^{\infty} \lambda_n [\chi_{A_n}]$ for any disjoint family $\{A_n\}_{n\in\mathbb{N}}$ converges with respect to the metric ρ_μ.

We now show that any derivation on $\mathbb{C}_c(P(\Omega))$ is trivial.

Proposition 6.4.2. *Let $\delta : \mathbb{C}_c(P(\Omega)) \to L_0(\Omega)$ be a derivation. Then $\delta = 0$.*

Proof. By Proposition 6.3.1(ii) we have that $\delta([\chi_A]) = 0$ for any $A \in \Sigma$. Therefore, referring to Proposition 6.3.10 and Proposition 6.1.11, we have that for any $a = \sum_{n=1}^{\infty} \lambda_n [\chi_{A_n}] \in \mathbb{C}_c(P(\Omega))$ with $\{\lambda_n\}_{n\in\mathbb{N}} \subset \mathbb{C}$ and any pairwise disjoint family $\{A_n\}_{n\in\mathbb{N}} \subset \Sigma$,

$$\delta(a) = \sum_{n=1}^{\infty} \lambda_n \delta([\chi_{A_n}]) = 0.$$

Thus, any derivation on $\mathbb{C}_c(P(\Omega))$ is trivial. □

Next, we turn to proving algebraic and topological properties of the algebra $\mathbb{C}_c(P(\Omega))$. We start with regularity and disjoint completeness.

Proposition 6.4.3. *The algebra $\mathbb{C}_c(P(\Omega))$ is a regular, disjointly complete subalgebra of $L_0(\Omega)$, which contains all projections from $L_0(\Omega)$.*

Proof. Clearly, $P(\Omega) \subset \mathbb{C}_c(P(\Omega))$ and $\mathbb{C}_c(P(\Omega))$ are disjointly complete.

It is clear that for a simple element $a = \sum_{n=1}^{\infty} \lambda_n [\chi_{A_n}] \in \mathbb{C}_c(P(\Omega))$, its weak inverse has the form

$$i(a) = \sum_{n:\lambda_n \neq 0} \lambda_n^{-1} [\chi_{A_n}] \in \mathbb{C}_c(P(\Omega)).$$

Thus, $\mathbb{C}_c(P(\Omega))$ is a regular subalgebra of $L_0(\Omega)$. □

We establish now that the algebra $\mathbb{C}_c(P(\Omega))$ is also integrally closed.

Proposition 6.4.4. *The subalgebra $\mathbb{C}_c(P(\Omega))$ is integrally closed.*

Proof. Suppose that a is an integral element with respect to $\mathbb{C}_c(P(\Omega))$, i. e., there exists a monic polynomial $p(x) = x^n + a_n x^{n-1} + \cdots + a_0$ with $a_0, \dots, a_n \in \mathbb{C}_c(P(\Omega))$, such that $p(a) = 0$.

Since every $a_0, \dots, a_n$ is a simple element, we can find an at-most countable family $\nabla \subset \Sigma$, such that for every $A \in \nabla$ and $i = 0, \dots, n$, we have the equality

$$a_i[\chi_A] = \alpha[\chi_A]$$

for some $\alpha \in \mathbb{C}$.

Fix $A \in \nabla$ and denote by λ_i the numbers satisfying $a_i[\chi_A] = \lambda_i[\chi_A]$. It is sufficient to show that $a[\chi_A]$ is a simple element.

Introduce the monic polynomial $q(x) = x^n + \lambda_{n-1}x^{n-1} + \cdots + \lambda_0 \in \mathbb{C}[x]$. Since the field $\mathbb{C}$ is algebraically closed, there exist $\alpha_1, \dots, \alpha_n$ from $\mathbb{C}$, such that $q(x) = (x - \alpha_1) \cdots (x - \alpha_n)$. Hence, using the equalities $a_i[\chi_A] = \lambda_i[\chi_A]$, $i = 0, \dots, n-1$, we obtain that

$$\begin{aligned} 0 = p(a)[\chi_A] &= (a^n + a_{n-1}a^{n-1} + \cdots + a_0) \cdot [\chi_A] \\ &= (a[\chi_A])^n + \lambda_{n-1}(a[\chi_A])^{n-1} + \cdots + \lambda_0[\chi_A] = q(a[\chi_A])[\chi_A] \\ &= (a[\chi_A] - \alpha_1[\chi_A]) \cdot \cdots \cdot (a[\chi_A] - \alpha_n[\chi_A]) = 0. \end{aligned}$$

The latter equality implies that

$$\inf_{k=0,\dots,n} s(a[\chi_A] - \alpha_k[\chi_A]) = s(a[\chi_A] - \alpha_1[\chi_A]) \cdots s(a[\chi_A] - \alpha_n[\chi_A]) = 0. \tag{6.22}$$

Let us denote by B_k measurable sets, such that $B_k \subset A$ and $[\chi_{B_k}] = s(a[\chi_A] - \alpha_k[\chi_A])$, $k = 0, \dots, n$. By (6.22), we have that $\mu(\bigcap_{k=0}^n B_k) = 0$. Therefore, up to a set of measure zero, we have the equality $A = \bigcup_{k=0}^n (A \setminus B_k)$. Choose a (finite) family $\{A_i\}_{i \in N}, N \in \mathbb{N}$ of pairwise disjoint sets, such that $A_i \subset A \setminus B_{k(i)}$ for every $i = 1, \dots, N$ and $A = \bigcup_{i=1}^n A_i$ (up to measure zero set).

Since $A_i \subset A \setminus B_{k(i)}$, the definition of B_k implies that $[\chi_{A_i}] \leq [\chi_{A \setminus B_{k(i)}}] = [\chi_A] - [\chi_{B_{k(i)}}]$. Therefore, using the equality $[\chi_{B_k}] = s(a[\chi_A] - \alpha_k[\chi_A]) \leq [\chi_A]$, $k = 0, \dots, n$, we have that

$$\begin{aligned} a[\chi_{B_{k(i)}}] - \alpha_{k(i)}[\chi_{B_{k(i)}}] &= (a[\chi_A] - \alpha_{k(i)}[\chi_A])[\chi_{B_{k(i)}}] \\ &= (a[\chi_A] - \alpha_{k(i)}[\chi_A]) \cdot s((a[\chi_A] - \alpha_{k(i)}[\chi_A])) = a[\chi_A] - \alpha_{k(i)}[\chi_A], \end{aligned}$$

or, equivalently,

$$a[\chi_{A\setminus B_{k(i)}}] = \alpha_{k(i)}[\chi_{A\setminus B_{k(i)}}].$$

The choice of A_i implies that

$$a[\chi_{A_i}] = \alpha_{k(i)}[\chi_{A_i}],$$

i. e., $a[\chi_{A_i}]$ is a simple element for every $i = 1, \dots, N$. Since, up to measure zero, we have the equality $A = \bigcup_{i=1}^{N} A_i$, and the sets A_i are pairwise disjoint, we infer that $a[\chi_A]$ is a simple element too. This suffices to conclude that the subalgebra $\mathbb{C}_c(P(\Omega))$ is integrally closed. □

Combining this result with Proposition 6.4.3, we obtain the following.

Corollary 6.4.5. *The $*$-algebra $\mathbb{C}_c(P(\Omega))$ is a regular, integrally closed, and disjointly complete subalgebra of $L_0(\Omega)$, which contains all projections from $L_0(\Omega)$.*

We now establish a necessary and sufficient condition for the algebra $\mathbb{C}_c(P(\Omega))$ to be a proper subalgebra of $L_0(\Omega)$.

Theorem 6.4.6. *Assume that (Ω, Σ, μ) is a σ-finite measure space. The following conditions are equivalent:*
(i) $L_0(\Omega) = \mathbb{C}_c(P(\Omega))$;
(ii) *The measure space (Ω, Σ, μ) is atomic.*

Proof. (i)⇒(ii). Assume that $L_0(\Omega) = \mathbb{C}_c(P(\Omega))$ and that the measure space (Ω, Σ, μ) is not atomic. In this case, there exists $A \in \Sigma$, such that the measure space $(A, \Sigma|_A, \mu|_A)$ has no atoms, and, by assumption, $L_0(A) = \mathbb{C}_c(P(A))$.

By [143, Theorem 3.5.2], there exists a $*$-isomorphism Φ from the von Neumann algebra $L_\infty[0,1]$ onto the commutative von Neumann algebra $L_\infty(A, \Sigma|_A, \mu|_A)$. Let $x \in L_\infty[0,1]$, where $x(t) = t$. Suppose that $y \in \mathbb{C}_c(P(A))$ is such that $\rho_\mu(\Phi(x), y) < 1$. Then there exists $0 < \alpha < 1$, such that

$$\rho_\mu(\Phi(x), y) = \mu_f(s(\Phi(x)), s(y)) = \mathbf{1} - \alpha$$

and, therefore, for some measurable set B with $\mu_f(B) > 0$, we have that $\Phi(x)[\chi_B] = \beta[\chi_B]$ where $\beta \in \mathbb{C}$. Hence, on a measurable set of nonzero measure x is constant. This contradiction implies that $\rho_\mu(\Phi(x), y) = 1$ for any $y \in \mathbb{C}_c(P(A))$. In particular, since $\mathbb{C}_c(P(A))$ is disjointly complete, and so closed with respect to the metric ρ_μ (see Theorem 6.1.13), it follows that $0 \neq x \in L_0(A) \setminus \mathbb{C}_c(P(A))$. This contradiction implies that the measure space (Ω, Σ, μ) is atomic.

(ii)⇒(i). Suppose that the measure space (Ω, Σ, μ) is atomic. Denote by Δ the set of all atoms from $P(\Omega)$ (which is countable). In this case, the von Neumann algebra $L_\infty(\Omega)$ coincides with the $*$-algebra

$$l_\infty(\Delta) = \left\{\{\alpha_q\}_{q\in\Delta} : \alpha_q \in \mathbb{C},\ \sup_{q\in\Delta} |\alpha_q| < \infty\right\},$$

and the $*$-algebra $L_0(\Omega)$ can be identified with the $*$-algebra $\mathbb{C}^\Delta$ of all complex-valued functions on Δ with respect to the pointwise algebraic operations. Hence, the equality $L_0(\Omega) = \mathbb{C}_c(P(\Omega))$ clearly holds. □

We are now ready to present the main result of this section. The following theorem gives an important criterion for the existence of a nonzero derivation on the algebra $L_0(\Omega)$.

Theorem 6.4.7. *Let (Ω, Σ, μ) be a σ-finite measure space. The following conditions are equivalent:*
(i) *The measure space (Ω, Σ, μ) is not atomic;*
(ii) *There exists a nonzero derivation $\delta : L_0(\Omega) \to L_0(\Omega)$.*

Proof. (i)⇒(ii). Since the measure space is not atomic, Theorem 6.4.6 implies that $L_0(\Omega) \neq \mathbb{C}_c(P(\Omega))$. By Corollary 6.4.5, the algebra $\mathbb{C}_c(P(\Omega))$ is a regular, integrally closed, and disjointly complete subalgebra of $L_0(\Omega)$, which contains all projections from $L_0(\Omega)$, i. e., $\mathbb{C}_c(P(\Omega)) = E(\mathbb{C}_c(P(\Omega))) \neq L_0(\Omega)$. Hence, by Proposition 6.3.23 there exists uncountably many nonzero derivations on $L_0(\Omega)$, extending the trivial derivation on $\mathbb{C}_c(P(\Omega))$.

(ii)⇒(i). Assume, on the contrary, that the measure space (Ω, Σ, μ) is atomic. By Theorem 6.4.6, we have that $L_0(\Omega) = \mathbb{C}_c(P(\Omega))$. By Proposition 6.4.2, any derivation on $L_0(\Omega) = \mathbb{C}_c(P(\Omega))$ is trivial, which is a contradiction. Therefore, the measure space (Ω, Σ, μ) is not atomic. □

Remark 6.4.8. *As noted in the proof of Theorem 6.4.7, if the measure space (Ω, Σ, μ) is not atomic, then there exists uncountably many distinct derivations on $L_0(\Omega, \Sigma, \mu)$.*

For convenience, we also present a form of the latter theorem, formulated in terms of commutative von Neumann algebras.

Theorem 6.4.9. *Let $\mathcal{M}$ be an arbitrary commutative von Neumann algebra, $S(\mathcal{M})$ be the $*$-algebra of all measurable operators affiliated with $\mathcal{M}$. Then the following conditions are equivalent:*
(i) *The Boolean algebra $P(\mathcal{M})$ of all projections in $\mathcal{M}$ is atomic;*
(ii) *Every derivation on the algebra $S(\mathcal{M})$ is identically zero.*

Remark 6.4.10. *Suppose that the Boolean algebra $P(\mathcal{M})$ of all projections in a commutative von Neumann algebra is not atomic. Then Theorem 6.4.9 guarantees that there exists a nonzero derivation $\delta : S(\mathcal{M}) \to S(\mathcal{M})$. In particular, there exists a nonzero derivation $\delta : \mathcal{M} \to S(\mathcal{M})$. In contrast, by Theorem 5.3.4 any derivation $\delta_1 : \mathcal{M} \to \mathcal{M}$ is inner, and hence, zero due to commutativity of $\mathcal{M}$. For an example of a nonzero derivation on $L_0[a, b]$, see the following section.*

6.5 Maximal unique extension of the classical derivation

In this section, we focus on extensions of the classical derivation ∂ defined on the algebra $D[a,b], a,b \in \mathbb{R}, a < b$, of (classes of a.e. equal) almost everywhere differentiable functions. The primary objective of this section is to show that the largest subalgebra of $L_0[a,b]$ admitting a unique extension of ∂ is the algebra $AD[a,b]$ of all (classes of a. e. equal) approximately differentiable functions on $[a,b]$.

As shown in Theorem 6.3.22, the largest subalgebra of $L_0(\Omega)$ admitting a unique extension of a nonexpansive derivation $\delta : \mathcal{B} \to L_0(\Omega)$ on a subalgebra $\mathcal{B} \subset L_0(\Omega)$ coincides with the least regular, integrally closed, and disjointly complete algebra $E(\mathcal{B})$, which contains $\mathcal{B}$ and $P(\Omega)$. Thus, the bulk of this section is devoted to showing that $E(D[a,b])$ is exactly the algebra $AD[a,b]$ of approximately differentiable functions.

Throughout this section, we assume that $[a,b]$ is an arbitrary closed interval $[a,b] \subset \mathbb{R}$, $a < b$, with the σ-algebra Σ of all Lebesgue measurable subsets and the classical Lebesgue measure m, defined on Σ. As before, $L_0[a,b]$ denotes the algebra of classes of almost everywhere equal measurable functions. In this section, we make the distinction between a function f on $[a,b]$ and the class $[f]$ of almost everywhere equal functions.

Note that for any differentiable function f on $[a,b]$, its derivative f' is also a measurable function as a pointwise limit of measurable functions. We now show that for a class $[f] \in L_0[a,b]$ we can talk about the equivalence class $[f'] \in L_0[a,b]$.

Lemma 6.5.1. *Suppose that f and g are almost everywhere differentiable functions on $[a,b]$, such that $f = g$ almost everywhere. Then f' and g' are measurable and $f' = g'$ almost everywhere.*

Proof. Let A be the set of all points $t \in [a,b]$, such that $f(t) = g(t)$ and both derivatives f', g' exist and are finite. By [71, Theorem 3.1.4], f' and g' are measurable on A. The function $h = f - g$ has everywhere defined derivative $f' - g'$ on A. Since $h(t) = 0$ for all $t \in A$, the equality $h'(t) = 0$ holds on A. The proof is complete since the latter set has full measure. □

The above lemma allows us to define the classical derivation ∂ defined on the algebra of almost everywhere differentiable functions.

Definition 6.5.2. Let $D[a,b]$ be the subalgebra of $L_0[a,b]$, consisting of all classes $[f] \in L_0[a,b]$ of almost everywhere equal functions, such that there exists the almost everywhere finite classical derivative $t \mapsto f'(t), t \in [a,b]$. Define the derivation $\partial : D[a,b] \to L_0[a,b]$ by setting

$$\partial[f] = [f'], \quad t \in [a,b].$$

We first show that the derivation $\partial : D[a,b] \to L_0[a,b]$ is nonexpansive so that the results of Section 6.3 are applicable to ∂. To this end, we recall definition of density points of a measurable set and Lebesgue's density theorem.

Assume that $E \in \Sigma$. An interior point $t \in (a,b)$ is called a *density point* of the set E, if

$$\lim_{\varepsilon \downarrow 0} \frac{m(E \cap [t-\varepsilon, t+\varepsilon])}{2\varepsilon} = 1.$$

Similarly, the endpoints a and b are called *density points* of the set E, if

$$\lim_{\varepsilon \downarrow 0} \frac{m(E \cap [a, a+\varepsilon])}{\varepsilon} = 1, \quad \lim_{\varepsilon \downarrow 0} \frac{m(E \cap [b-\varepsilon, b])}{\varepsilon} = 1.$$

Theorem 6.5.3 (Lebesgue's density theorem, [71]). *Let $A \subset \Sigma$ and let $D(A)$ be the set of all density points of A. Then $m(A \triangle D(A)) = 0$.*

We now prove that $\partial : D[a,b] \to L_0[a,b]$ is a nonexpansive derivation.

Proposition 6.5.4. *The derivation $\partial : D[a,b] \to L_0[a,b]$ is nonexpansive.*

Proof. Let f be almost everywhere differentiable on $[a,b]$ and suppose that the set $N(f) := \{t \in [a,b] : f(t) = 0\}$ has nonzero measure. If t is a density point of $N(f)$, and at this point there is a derivative of f, then we have $f'(t) = 0$. Thus, $N(f)$ is a subset of the set $N(f') := \{t \in [a,b] : f'(t) = 0\}$. This means that $s(\partial(f)) = s([f']) \leq s([f])$, as required. □

As shown in Section 6.3, the extension of a given derivation $\delta : \mathcal{B} \to L_0(\Omega)$ from a subalgebra $\mathcal{B}$ of $L_0(\Omega)$ starts with the extension of δ to a subalgebra generated by $\mathcal{B}$ and all projections in $L_0(\Omega)$. For this reason, we show that the algebra $D[a,b]$ does not contain all the projections in $L_0[a,b]$. Recall that the fat Cantor set (also known as the Smith–Volterra–Cantor set) is created by iteratively removing subintervals of width $\frac{1}{2^{2n}}$ of the remaining intervals (see, e. g., [8, Section 15]).

Proposition 6.5.5. *Let E be the fat Cantor set in $[0,1]$. Then $\chi_E \notin D[0,1]$. In particular, the algebra $D[0,1]$ does not contain all projections from $L_0[0,1]$.*

Proof. The set E is a closed nowhere dense subset of $[a,b]$ with $m(E) > 0$. Denote by F the set of all points from E, which are density points for E. By the Lebesgue density theorem, we have that $m(F) = m(E) > 0$. Since the set E is nowhere dense, it follows that in every neighborhood of a point $t \in F$ there exist points, which do not belong to E. It means that finite derivative $(\chi_E)'(t)$ does not exist at any point $t \in F$. Consequently, $[\chi_E] \notin D[a,b]$, as required. □

Next, we recall the notion of approximate derivative. To define the approximate derivative of a function on $[a,b]$, we first need to introduce the approximate limit.

Assume that $f : [a,b] \to \mathbb{C}$ is a measurable function and let $t_0 \in [a,b]$. If $E \in \Sigma$ is such that t_0 is a density point for E, then as shown in [71], (if it exists) the limit $\lim_{\substack{E \ni t \to t_0 \\ t \neq t_0}} f(t)$ does not depend on the choice of the measurable set $E \in \Sigma$ for which t_0 is a density point. Hence, one can define the *approximate limit* of f at a point $t_0 \in [a,b]$ (denoted by $\operatorname{ap}\lim_{t\to t_0} f(t)$) by

$$\operatorname{ap}\lim_{t\to t_0} f(t) = \lim_{\substack{E\ni t\to t_0\\ t\neq t_0}} f(t).$$

Definition 6.5.6. A function $f : [a,b] \to \mathbb{C}$ is said to be *approximately differentiable* at a point $t_0 \in [a,b]$, if there exists the approximate limit

$$\operatorname{ap}\lim_{t\to t_0} \frac{f(t)-f(t_0)}{t-t_0}.$$

This approximate limit is called the *approximate derivative* of the function f at the point t_0 and is denoted by $f'_{\mathrm{ap}}(t_0)$.

The approximate derivative of a function is a generalization of the classical derivations. Indeed, assume that for a function $f : [a,b] \to \mathbb{C}$ there exists a classical derivative $f'(t_0) = \lim_{t\to t_0} \frac{f(t)-f(t_0)}{t-t_0}$ at a point $t_0 \in [a,b]$. Then, for the measurable set $E = [a,b] \in \Sigma$, the point t_0 is clearly a density point for E and, therefore,

$$f'_{\mathrm{ap}}(t_0) = \operatorname{ap}\lim_{t\to t_0} \frac{f(t)-f(t_0)}{t-t_0} = \lim_{t\to t_0} \frac{f(t)-f(t_0)}{t-t_0} = f'(t_0). \tag{6.23}$$

By the Lebesgue density Theorem 6.5.3, almost every point of a measurable set $E \in \Sigma$ is a density point of E. Therefore, the following definition makes sense.

Definition 6.5.7. Denote by $AD[a,b]$ the set of all classes $[f] \in L_0[a,b]$, such that f is approximately differentiable almost everywhere.

Proposition 6.5.8. *The set $AD[a,b]$ is a proper $*$-subalgebra of $L_0[a,b]$ containing $D[a,b]$ and all projections in $L_0[a,b]$.*

Proof. The inclusion $D[a,b] \subset AD[a,b]$ follows directly from the equation (6.23).

Since a density point of two subsets $E, F \in \Sigma$ is a density point for their intersection $E \cap F$, it follows that for a. e. approximately differentiable functions f and g on $[a,b]$ the functions $f+g, fg, \bar{f}$, and λf, $\lambda \in \mathbb{C}$ are also a. e. approximately differentiable on $[a,b]$. Therefore, $AD[a,b]$ is a $*$-subalgebra of $L_0[a,b]$.

Let us show that $AD[a,b]$ contains all projections from $L_\infty[a,b]$. For a measurable subset A in $[a,b]$ consider the subset $A_0 \subseteq A$ of all points of density of A. By the Lebesgue density Theorem 6.5.3, we know that the Lebesgue measure of the set $A \setminus A_0$ vanishes. Since the characteristic function χ_{A_0} has an approximate derivative equal to zero almost everywhere in A_0, it follows that the class containing the function χ_A belongs to $AD[a,b]$. Hence, $AD[a,b]$ contains all projections from $L_\infty[a,b]$.

Finally, to show that $AD[a,b]$ is a proper subalgebra of $L_0[a,b]$, let f be a continuous function on $[a,b]$, for which the equality $|f'_{\mathrm{ap}}(t)| = +\infty$ holds a. e. (for an example of such function see [85]). Since the function f is continuous, we have that $[f] \in L_0[a,b]$. We claim that $[f] \notin AD[a,b]$.

If $g \in [f]$ and $E = \{t \in [a,b] : f(t) = g(t)\}$, then $m(E) = b - a$, and therefore, every point $t_0 \in [a,b]$ is a density point for E. Hence,

$$\frac{f(t) - f(t_0)}{t - t_0} = \frac{g(t) - g(t_0)}{t - t_0}$$

for all $t, t_0 \in E$, which implies the equality $f'_{\mathrm{ap}}(t_0) = g'_{\mathrm{ap}}(t_0)$ for all points $t_0 \in E$, in which the finite approximate derivative $g'_{\mathrm{ap}}(t_0)$ exists. By the choice of function f, we have that $|f'_{\mathrm{ap}}(t)| = +\infty$ a. e. This means that the function $g(t)$ does not have finite approximate derivative as well. Consequently, $[f] \notin AD[a,b]$ and, therefore, $AD[a,b] \neq L_0[a,b]$. □

Recall that by Lusin's theorem, for a measurable function $f : [a,b] \to \mathbb{C}$, there always exists a compact $E \subset \Sigma$ of sufficiently large measure, such that f restricted onto E is continuous. A similar property holds for an approximately differentiable function f with restriction $f|_E$ being a continuously differentiable function (see, e. g., [71, Chapter 3, Section 3.1, Theorem 3.1.16]). For our purpose of extension of the classical derivation, we adapt this characterization of approximately differentiable functions for classes of a. e. equal functions. We denote by $C^1[a,b]$ the algebra of all complex-valued continuously differentiable functions on $[a,b]$.

Proposition 6.5.9. *For an element $[f] \in L_0[a,b]$, the following conditions are equivalent:*
(i) $[f] \in AD[a,b]$;
(ii)

$$f = \sum_{n=1}^{\infty} \chi_{A_n} g_n \tag{6.24}$$

for some partition $\{A_n\}_{n\in\mathbb{N}} \subset \Sigma$ of $[a,b]$ and a family $\{g_n\}_{n\in\mathbb{N}} \subset C^1[a,b]$.

Proof. (i)⇒(ii). Let a function $f : [a,b] \to \mathbb{C}$ have the finite approximate derivative $f'_{\mathrm{ap}}(t)$ at all points $t \in E \in \Sigma$ with $m(E) = b - a$. By [71, Chapter 3, Section 3.1, Theorem 3.1.16], for every $n \in \mathbb{N}$ there exists a function $g_n \in C^1[a,b]$, such that $m(E \setminus \{t \in [a,b] : f(t) = g_n(t)\}) < 1/n$.

Set

$$F_n = \{t \in [a,b] : f(t) = g_n(t)\},$$

and define

$$A_1 = F_1, \quad A_{n+1} = F_{n+1} \setminus \bigcup_{k=1}^{n} A_k, \quad n \in \mathbb{N}.$$

It is easy to see that

$$m\left([a,b] \setminus \bigcup_{k=1}^{n} A_k\right) = m([a,b] \setminus F_n) < 1/n$$

and $A_n \cap A_k = \emptyset$ for $n \neq k$. Hence, $m(\bigcup_{n=1}^{\infty} A_n) = b - a$, i. e., $\{A_n\}_{n\in\mathbb{N}}$ is a partition of $[a,b]$. Since, in addition, $\chi_{A_n} \cdot f = \chi_{A_n} \cdot g_n$ for all $n \in \mathbb{N}$, we conclude the proof of the implication (i)⇒(ii).

(ii)⇒(i). By the Lebesgue density Theorem 6.5.3, we can assume that $A_n, n \in \mathbb{N}$ coincides with the set of its density points. Since $g_n \in C^1[a,b]$, it follows that at every point $t \in A_n$, the finite approximate derivative $f'_{\mathrm{ap}}(t) = g'_n(t)$ exists. Consequently, the finite approximate derivative $f'_{\mathrm{ap}}(t)$ exists on the whole set $\bigcup_{n=1}^{\infty} A_n$, which implies that $[f] \in AD[a,b]$. □

Now, in order to prove that the algebra $AD[a,b]$ is the largest algebra, which admits a unique extension of the derivation ∂ on $D[a,b]$, we first need to establish some topological and algebraic properties of the algebra $AD[a,b]$. We first show that $AD[a,b]$ is the smallest disjointly complete $*$-subalgebra of $L_0[a,b]$ containing $D[a,b]$ and all projections from $L_0[a,b]$.

Proposition 6.5.10. *The algebra $AD[a,b]$ is the smallest disjointly complete $*$-subalgebra of $L_0[a,b]$ containing $D[a,b]$ and all projections from $L_0[a,b]$.*

Proof. By Proposition 6.5.8, the $*$-algebra $AD[a,b]$ contains $D[a,b]$ and all projections from $L_0[a,b]$. Disjoint completeness of $AD[a,b]$ immediately follow from Proposition 6.5.9.

Let $\mathcal{A} \subset L_0[a,b]$ be a disjointly complete $*$-subalgebra of $L_0[a,b]$ containing $D[a,b]$ and all projections in $L_0[a,b]$. Let $[f] \in AD[a,b]$. By Proposition 6.5.9, we can write $f = \sum_{n=1}^{\infty} \chi_{A_n} g_n$ for some partition $\{A_n\}_{n\in\mathbb{N}} \subset \Sigma$ of $[a,b]$ and family $\{g_n\}_{n\in\mathbb{N}} \subset C^1[a,b]$. For every $k \in \mathbb{N}$, the partial sums $\sum_{n=1}^{k} \chi_{A_n} g_n$ are contained in $\mathcal{A}$. By Proposition 6.1.11, the series $f = \sum_{n=1}^{\infty} \chi_{A_n} g_n$ converges with respect to the metric ρ_m. By Theorem 6.1.13, we have that $\mathcal{A}$ is closed with respect to the metric ρ_m. Hence, $[f] \in \mathcal{A}$, which means $\mathcal{A} \subseteq AD[a,b]$. □

Next, we deal with algebraic properties of the algebra $AD[a,b]$. We start with regularity of this algebra.

Proposition 6.5.11. *$AD[a,b]$ is a regular subalgebra of $L_0[a,b]$.*

Proof. Recall that the weak inverse $i(g)$ to a measurable function g is defined in Definition 6.1.1 and to prove that $AD[a,b]$ is regular, it is sufficient to show that $[i(g)] \in AD[a,b]$ for any $[g] \in AD[a,b]$.

We show first that $[i(g)] \in AD[a,b]$ for all $g \in C^1[a,b]$. If $g \in C^1[a,b]$, $t_0 \in [a,b]$, $g(t_0) \neq 0$, then there exists a neighborhood V of the point t_0, such that $g(t) \neq 0$ and $(i(g))(t) = \frac{1}{g(t)}$ for all $t \in V$. Hence, there exists a finite derivative $(i(g))'(t) = -\frac{g'(t)}{(g(t))^2}$ for all $t \in V$.

If $E_g = \{t \in [a,b] : g(t) = 0\}$ and $m(E_g) = 0$, then $(i(g))'(t)$ exists a. e. and, therefore, $[i(g)] \in D[a,b] \subset AD[a,b]$. If $m(E_g) > 0$, then for every point $t_0 \in E_g$, which is a density point of the set E_g, we have that

$$(i(g))'(t_0) = \lim_{E_g\setminus\{t_0\}\ni t\to t_0} \frac{(i(g))(t) - (i(g))(t_0)}{t - t_0} = 0.$$

Consequently, the approximate derivative $(i(g))'_{\mathrm{ap}}(t)$ always exists for almost all $t \in [a,b]$ in the case when $g \in C^1[a,b]$. Thus, $i([g]) \in AD[a,b]$ for all $g \in C^1[a,b]$.

Let now $[f] \in AD[a,b]$ be arbitrary. By Proposition 6.5.9, we can write $f = \sum_{n=1}^{\infty} \chi_{A_n} g_n$ for a partition $\{A_n\}_{n\in\mathbb{N}}$ of $[a,b]$ and a family $\{g_n\} \subset C^1[a,b]$. Since

$$[\chi_{A_n}]i([f]) = i([\chi_{A_n} f]) = i([\chi_{A_n} g_n]) = \chi_{A_n} i([g_n]) \in AD[a,b],$$

it follows that

$$i([f]) = \sum_{n=1}^{\infty} [\chi_{A_n}]i([f]) = \sum_{n=1}^{\infty} [\chi_{A_n}]i([g_n]) \in \overline{AD[a,b]}^{\rho_\mu} = AD[a,b].$$

Thus, $AD[a,b]$ is a regular subalgebra of $L_0[a,b]$. □

Next, we show that the algebra $AD[a,b]$ is integrally closed, i. e., $AD[a,b]$ contains all elements, which are integral with respect to $AD[a,b]$.

Proposition 6.5.12. *The algebra $AD[a,b]$ is integrally closed.*

Proof. Let $[f]$ be an integral element with respect to $AD[a,b]$. By Proposition 6.2.6, any integral element can be written as the sum of disjointly supported proper integral elements (on their respective supports). Hence, without loss of generality, we can assume that $[f] \in L_0[a,b]$ is a proper $s([f])AD[a,b]$-integral element.

Let $[f] \in L_0[a,b]$ be a proper $s([f])AD[a,b]$-integral element, i. e., f is a root of a monic polynomial

$$p(x) = x^n + a_1 x^{n-1} + \cdots + a_n \tag{6.25}$$

with $a_k \in AD[a,b]$, $k = 1,\dots,n$, and $\deg_{[\chi_A]AD[a,b]}([\chi_A f]) = \deg_{AD[a,b]}([f])$ for any $A \subset s(f)$.

Assume first that $a_1,\dots,a_n \in D[a,b]$. If $n = 1$, then $[f]$ is a root of monic polynomial $p(x) = x + a_1$ and, therefore, $[f]$ is an almost everywhere differentiable function, and so $[f] \in AD[a,b]$. Suppose now that $n > 1$.

For almost all points $t \in s(f)$, the equality

$$p(f(t)) = f(t)^n + a_1(t)f(t)^{n-1} + \cdots + a_{n-1}(t)f(t) + a_n(t) = 0$$

holds. Thus, for every $t \in s(f)$, the number $f(t)$ is a root of the polynomial

$$p_t(x) = x^n + a_1(t)x^{n-1} + \cdots + a_n(t).$$

Denote by A the set of all $t \in s(f)$, such that $f(t)$ is not a simple root of $p_t(x)$. For any $t \in A$, we have that $f(t)$ is a root of the polynomial

$$(p_t)'(x) = nx^{n-1} + (n-1)a_1(t)x^{n-2} + \cdots + a_{n-1}(t).$$

In particular, $A \in \Sigma$. If $m(A) > 0$, then $[\chi_A][f]$ is a root of the monic polynomial

$$n^{-1}p'(x) = x^{n-1} + (n-1)n^{-1}g_1x^{n-2} + \cdots + n^{-1}g_{n-1},$$

where $g_k = [\chi_A][a_k]$, $k = 1, \ldots, n-1$. This contradicts with the equality

$$\deg_{[\chi_A]AD[a,b]}([\chi_A f]) = \deg_{AD[a,b]}([f]).$$

Hence, $m(A) = 0$.

Thus, for almost all $t \in s([f])$, the number $f(t)$ is a simple root of a complex polynomial $p_t(x)$. Let us fix one such point $t_0 \in s([f]) \setminus A$ and set $z_0 = (a_1(t_0), \ldots, a_m(t_0), f(t_0)) \in \mathbb{C}^{m+1}$. Consider the function F on $\mathbb{C}^{m+1}$ defined by

$$F(\xi_1, \ldots, \xi_m, y) = y^m + \xi_1 y^{m-1} + \cdots + \xi_m.$$

It is differentiable on $\mathbb{C}^{m+1}$, moreover,

$$F(z_0) = 0 \quad \text{and} \quad F'_y = my^{m-1} + (m-1)\xi_1 y^{m-2} + \cdots + \xi_{m-1}.$$

Since $f(t_0)$ is a simple root of $p_t(x)$, we have that $F'_y(z_0) \neq 0$. Since F'_y is continuous, there is a neighborhood $V(z_0) \subset \mathbb{C}^{m+1}$ of z_0, such that for any $z \in V(z_0)$ we have $F'_y(z) \neq 0$. Moreover, all other partial derivatives that are $F'_{\xi_k} = y^{m-k}$ are continuous. Hence, by the implicit function theorem (see, e. g., [76, Section I.B]), there exists a neighborhood $W \subset \mathbb{C}^m$ of $(a_1(t_0), \ldots, a_m(t_0))$, such that $W \subset \pi(V(z_0))$ (here a projection $\pi : \mathbb{C}^{m+1} \to \mathbb{C}^m$ defined as $\pi(\xi_1, \ldots, \xi_m, \xi_{m+1}) = (\xi_1, \ldots, \xi_m)$) and there is a unique differentiable function $G : W \to \mathbb{C}$, such that

$$G(a_1(t_0), \ldots, a_m(t_0)) = f(t_0) \quad \text{and} \quad F(w, G(w)) = 0 \quad \text{for all } w \in W.$$

Take $\varepsilon > 0$, such that $(a_1(t), \ldots, a_m(t)) \in W$ for almost all $t \in (t_0 - \varepsilon, t_0 + \varepsilon)$. Then $g(t) = G(a_1(t), \ldots, a_m(t))$ is almost everywhere differentiable on $(t_0 - \varepsilon, t_0 + \varepsilon)$. Since $F(w, G(w)) = 0$ for all $w \in W$ and $(a_1(t), \ldots, a_m(t)) \in W$ for almost all $t \in (t_0 - \varepsilon, t_0 + \varepsilon)$, it follows that $p(g(t)) = 0$ for almost all $t \in (t_0 - \varepsilon, t_0 + \varepsilon)$. Thus, $\chi_B g$ is a root of the polynomial p, where $B = (t_0 - \varepsilon, t_0 + \varepsilon)$. Since f is also root of the polynomial p, it follows that $\chi_B f$ is a root of the polynomial $\frac{p(x)-p(g(t))}{x-g(t)}$ whose degree is strictly less than m. Hence,

$$m(\{t : f(t) \neq g(t)\} \cap (t_0 - \varepsilon, t_0 + \varepsilon)) = 0,$$

i. e., $f(t)$ and $g(t)$ coincide almost everywhere in $(t_0-\varepsilon, t_0+\varepsilon)\cap s(f)$. Since the set $s(f)\setminus A$ is measurable, it is a countable union of compact sets, and so there exists a countable cover of $s(f)\setminus A$ of the intervals of the form $(t_0-\varepsilon, t_0+\varepsilon)$, $t_0 \in s(f)\setminus A$. On each of these intervals, f coincides with some a. e. differentiable function. Since $m(A) = 0$, it follows that f is an almost everywhere differentiable function. This completes the proof in the case when the coefficients $a_1, \dots, a_n$ of the polynomial in (6.25) are a. e. differentiable on $[a, b]$.

Now we shall consider the general case $a_1, \dots, a_n \in AD[a, b]$. By Proposition 6.5.9, we can write $a_i = \sum_{n=1}^{\infty} \chi_{A_{i,n}} g_{i,n}$, where $A_{i,n} \cap A_{i,k} = \emptyset$ for $n \neq k$, $m(\bigcup_n A_{i,n}) = 1$ and $g_{i,n}$ is an almost everywhere differentiable function on $[a, b]$ for all $n, i \in \mathbb{N}$. Further, consider a partition $\{B_k\}_{k\in\mathbb{N}}$ of $[a, b]$ consisting of subsets of the form $\bigcap_{i=1}^{m} A_{i,n_i}$, $n_1, \dots, n_m \in \mathbb{N}$. By the result proven above, the function $f\chi_{B_k}$ is approximately differentiable for every $k \in \mathbb{N}$. Since $[f] = \sum_{k=1}^{\infty} [f\chi_{B_k}]$ and $AD[a, b]$ are disjointly complete, it follows that $[f]$ is approximately differentiable, too. Thus, the algebra $AD[a, b]$ is integrally closed, as required. □

Now, we have established all the necessary topological and algebraical properties of the algebra $AD[a, b]$ and its connection with the algebra $D[a, b]$, which we formulate in the following corollary.

Corollary 6.5.13. *The subalgebra $AD[a, b]$ is the least regular, integrally closed, and disjointly complete $*$-subalgebra of $L_0[a, b]$, which contains the subalgebra $D[a, b]$ and all projections from $L_0[a, b]$.*

Proof. By Proposition 6.5.10, we have that $AD[a, b]$ is the least disjointly complete subalgebra of $L_0[a, b]$, which contains the subalgebra $D[a, b]$ and all projections from $L_0[a, b]$. By Proposition 6.5.11 and Proposition 6.5.12, the algebra $AD[a, b]$ is also regular and integrally closed. Hence, the claim follows. □

We now define the extension $\partial_{\text{ap}} : AD[a, b] \to L_0[a, b]$ of the classical derivation ∂. We recall that the approximate derivative f'_{ap} for a. e. approximately differentiable function $f : [a, b] \to \mathbb{C}$ is a measurable function on $([a, b], \Sigma, m)$ [136, Chapter IX, Section 11]. In particular, the following definition makes sense.

Definition 6.5.14. For $[f] \in AD[a, b]$, we define the mapping ∂_{ap} by setting

$$\partial_{\text{ap}}([f]) = [f'_{\text{ap}}]. \tag{6.26}$$

Proposition 6.5.15. *∂_{ap} is a nonexpansive derivation on $AD[a, b]$, which extends the derivation ∂ on $D[a, b]$.*

Proof. The fact that $\partial_{\text{ap}} : AD[a, b] \to L_0[a, b]$ extends $\partial : D[a, b] \to L_0[a, b]$ follows immediately from equality (6.23).

As we have already noted, a common density point of two sets $E, F \in \Sigma$ is a density point for their intersection $E \cap F$. Therefore, for two arbitrary a. e. approximately differentiable functions f and g, the equalities $(f+g)'_{\mathrm{ap}} = f'_{\mathrm{ap}} + g'_{\mathrm{ap}}$ and $(fg)'_{\mathrm{ap}} = f'_{\mathrm{ap}} g + f g'_{\mathrm{ap}}$ hold a. e. Using also the equality $(af)'_{\mathrm{ap}}(t) = af'_{\mathrm{ap}}(t)$, we obtain that the mapping ∂_{ap} is a derivation from $AD[a,b]$ into $L_0[a,b]$. By Proposition 6.5.8, $AD[a,b]$ contains all projections from $L_0[a,b]$. Therefore, Proposition 6.3.1(iv) guarantees that the mapping ∂_{ap} is a nonexpansive derivation. □

To obtain the main result of this section, we only need to show that $AD[a,b]$ is the largest subalgebra of $L_0[a,b]$ containing $D[a,b]$ with the property that the derivation $\partial : D[a,b] \to L_0[a,b]$ extends uniquely up to the derivation $\partial_{ad} : AD[a,b] \to L_0[a,b]$.

Theorem 6.5.16. *The algebra $AD[a,b]$ is the largest subalgebra of the algebra $L_0[a,b]$, such that $D[a,b] \subset AD[a,b]$ and the classical derivation ∂ on $D[a,b]$ extends uniquely up to a derivation on $AD[a,b]$. In addition, this unique extension of ∂ coincides with ∂_{ad} defined in Definition* 6.5.14.

Proof. By Corollary 6.5.13, the subalgebra $AD[a,b]$ is the least regular, integrally closed, and disjointly complete subalgebra of $L_0[a,b]$, which contains the subalgebra $D[a,b]$ and all projections in $L_0[a,b]$. The claim now follows from Theorem 6.3.22. □

Corollary 6.5.17. *There exist infinitely many extensions of the derivation $\partial : D[a,b] \to L_0[a,b]$ up to a derivation $\delta : L_0[a,b] \to L_0[a,b]$.*

Proof. By Proposition 6.5.8, the algebra $AD[a,b]$ is a proper subalgebra of $L_0[a,b]$. Since $AD[a,b]$ is the least regular, integrally closed, and disjointly complete subalgebra of $L_0[a,b]$, which contains the subalgebra $D[a,b]$ and all projections from $L_0[a,b]$, Proposition 6.3.23 implies that there are infinitely many derivations on the algebra $L_0[a,b]$, which extend ∂. □

As we show next, the derivation $\partial_{\mathrm{ap}} : AD[a,b] \to L_0[a,b]$, which extends the classical derivation $\partial : D[a,b] \to L_0[a,b]$, is, in fact, surjective.

Proposition 6.5.18. *We have that $\partial_{\mathrm{ap}}(AD[a,b]) = L_0[a,b]$.*

Proof. Suppose that $[f] \in L_0[a,b]$. Define $A_n = \{t \in [a,b] : n-1 \leq |f(t)| < n\} \in \Sigma$. It is clear that $[f_n] := [f\chi_{A_n}] \in L_\infty[a,b]$. In addition, since $\sum_{k=n}^{\infty} m(A_k) \to 0$ as $n \to \infty$, it follows that $[f] = \sum_{n=1}^{\infty} [f_n]$ with respect to the metric ρ_m.

Since $[f_n] \in L_\infty[a,b]$, it follows that $[f_n]$ is integrable. Hence, by the Lebesgue differentiation theorem, there exists $[g_n] \in D[a,b] \subset AD[a,b]$ such that $[g'_n] = [f_n]$ for all $n \in \mathbb{N}$.

Since the algebra $AD[a,b]$ contains all projections from $L_0[a,b]$, we can define

$$[h_n] = [g_n][\chi_{A_n}] \in AD[a,b].$$

In addition, since $AD[a,b]$ is a disjointly complete algebra, it follows that $[h] = \sum_{n=1}^{\infty}[h_n] \in AD[a,b]$. By Proposition 6.1.11, we have that $[h] = \sum_{n=1}^{\infty}[h_n]$ with respect to the metric ρ_m.

Using also the fact that ∂_{ad} is continuous with respect to ρ_m, we conclude that

$$\begin{aligned}\partial_{ad}([h]) &= \sum_{n=1}^{\infty}\partial_{ad}([h_n]) = \sum_{n=1}^{\infty}[\chi_{A_n}]\partial_{ad}([g_n]) \\ &= \sum_{n=1}^{\infty}[\chi_{A_n}]\partial([g_n]) = \sum_{n=1}^{\infty}[\chi_{A_n}]([f_n]) = [f].\end{aligned}$$

□

In conclusion of this section, we connect our results to the classical Lebesgue results. Denote by $AC[a,b]$ the subalgebras of $L_0[a,b]$ of classes $[f]$, with representative f being absolutely continuous functions.

Remark 6.5.19. *We note that although $AC[a,b]$ is an algebra of classes of almost everywhere equal functions, it is actually isomorphic to the algebra of all absolutely continuous functions on $[a,b]$. Indeed, assume that f and g are absolutely continuous functions on $[a,b]$ and $[f] = [g]$. In this case, $f = g$ almost everywhere and, therefore, they are equal on some dense subset of $[a,b]$. Since the functions are continuous on $[a,b]$, we obtain that $f = g$ on $[a,b]$, as required.*

The Lebesgue theorem guarantees that $AC[a,b] \subset D[a,b]$. Therefore, the classical derivative ∂ is a derivation on $AC[a,b]$ (with values in $L_0[a,b]$). In addition, the derivation ∂_{ap} on $AC[a,b]$ extends uniquely to the derivation $\partial : AC[a,b] \to L_0[a,b]$ up to a derivation on $AD[a,b]$. We show now that, as for derivation ∂ defined on the algebra $D[a,b]$, the algebra $AD[a,b]$ is the largest subalgebra of $L_0[a,b]$, which admits a unique extension of ∂, defined on $AC[a,b]$.

Theorem 6.5.20. *The algebra $AD[a,b]$ is the largest subalgebra of $L_0[a,b]$, containing $AC[a,b]$, which admits a unique extension of the derivation $\partial : AC[a,b] \to L_0[a,b]$.*

Proof. By Theorem 6.3.22, it is sufficient to show that the least regular, integrally closed, and disjointly complete subalgebra $E(AC[a,b])$ of $L_0[a,b]$ containing $AC[a,b]$ and all projections from $L_0[a,b]$ is equal to $AD[a,b]$.

Since $AC[a,b] \subset D[a,b]$, Corollary 6.5.13 implies that $E(AC[a,b]) \subset AD[a,b]$. To show the converse inclusion, assume that $[f] \in AD[a,b]$. By Proposition 6.5.9, there exist a partition $\{A_n\}_{n\in\mathbb{N}}$ of $[a,b]$ and a sequence $\{[g_n]\}_{n\in\mathbb{N}}$ with $g_n \in C^1[a,b]$, such that $[f] = \sum_{n=1}^{\infty}[\chi_{A_n}g_n]$. Proposition 6.1.11 implies that

$$[f] = \sum_{n=1}^{\infty}[\chi_{A_n}][g_n] \tag{6.27}$$

with respect to the metric ρ_m. Since $g_n \in C^1[a,b]$, it follows that $[g_n] \in AC[a,b]$. Since the algebra $E(AC[a,b])$ is disjointly complete and contains both $AC[a,b]$ and all projections from $L_0[a,b]$, decomposition (6.27) implies that $[f] \in E(AC[a,b])$, which concludes the proof. □

6.6 Derivation on algebras of measurable operators associated with a general type I_{fin} von Neumann algebra

In the final section of this chapter, we will establish necessary and sufficient conditions for the existence of noninner (equivalently, $t(\mathcal{M})$-discontinuous) derivations on the algebra $S(\mathcal{M})$ associated with a type I_{fin} von Neumann algebra. The nature of derivations on $S(\mathcal{M})$ is completely determined by the nature of derivations on the algebra $S(\mathcal{Z}(\mathcal{M}))$, i. e., there are noninner (equivalently, $t(\mathcal{M})$-discontinuous) derivations on $S(\mathcal{M})$ if and only if there are noninner derivations (equivalently, $t(\mathcal{Z}(\mathcal{M}))$-discontinuous) on $S(\mathcal{Z}(\mathcal{M}))$. With the necessary and sufficient condition of the existence of noninner derivations on $S(\mathcal{Z}(\mathcal{M}))$ established in Theorem 6.4.9, we will obtain a necessary and sufficient condition of existence of noninner derivations on $S(\mathcal{M})$.

Our approach consists of writing any derivation on $S(\mathcal{M})$ for a type I_{fin} as the sum of a unique inner derivation and a unique centrally induced derivation built from a derivation on $S(\mathcal{Z}(\mathcal{M}))$. We will start with the construction of a centrally induced derivation for a type I_n, $n \in \mathbb{N}$ algebra $\mathcal{M}$ and will move on to a general type I_{fin} later.

We now fix some $n \in \mathbb{N}$ and assume that $\mathcal{M}$ is a type I_n von Neumann algebra. Let (Ω, Σ, μ) be a σ-finite measure space, such that $\mathcal{Z}(\mathcal{M}) = L_\infty(\Omega, \Sigma, \mu)$ (up to a $*$-isomorphism). Recall (see Theorem 2.2.15) that (up to a $*$-isomorphism) we have that $S(\mathcal{Z}(\mathcal{M})) = L_0(\Omega)$.

As it is shown in Proposition 1.9.16(ii), up to a $*$-isomorphism, we can assume that

$$\mathcal{M} = M_{n\times n}(L_\infty(\Omega)), \quad \mathcal{Z}(\mathcal{M}) = \{aI, a \in L_\infty(\Omega)\}, \tag{6.28}$$

where I denotes the identity $n \times n$-matrix. In this case, by Proposition 2.2.16 (up to a $*$-isomorphism), we have that

$$S(\mathcal{M}) = M_{n\times n}(L_0(\Omega)), \quad \mathcal{Z}(S(\mathcal{M})) = \{aI, a \in L_0(\Omega)\}.$$

By e_{ij}, $i,j = 1,\dots,n$, we denote the matrix units.

Our first aim is to establish the existence of centrally induced derivations on $M_{n\times n}(L_0(\Omega))$ as derivations lifted from a derivation on $L_0(\Omega)$.

Let δ be a derivation on the algebra $L_0(\Omega)$. We define the lifted mapping D_δ on the algebra $M_{n\times n}(L_0(\Omega))$ by componentwise action of δ, i. e.,

$$D_\delta(x) = \big(\delta(x_{ij})\big)_{i,j=1}^n, \quad x = (x_{ij})_{i,j=1}^n \in M_{n\times n}(L_0(\Omega)). \tag{6.29}$$

In the following proposition, we show that D_δ is indeed a derivation on the algebra $M_{n\times n}(L_0(\Omega))$.

Proposition 6.6.1. *Let δ be a derivation on $L_0(\Omega)$. The mapping D_δ defined by* (6.29) *is a derivation on the algebra $M_{n\times n}(L_0(\Omega))$. In addition,*

$$D_\delta(xI) = \delta(x)I, \quad x \in L_0(\Omega). \tag{6.30}$$

Proof. It is clear that D_δ is a linear mapping. The equality $D_\delta(xI) = \delta(x)I$, $x \in L_0(\Omega)$ follows directly from the definition of D_δ. Therefore, we only need to prove that the Leibniz rule holds for D_δ.

Let $x = (x_{ij})_{i,j=1}^n$ and $y = (y_{ij})_{i,j=1}^n$ be in $M_{n\times n}(L_0(\Omega))$. We have that

$$xy = \left(\sum_{k=1}^n x_{ik}y_{kj}\right)_{i,j=1}^n .$$

Hence, using the Leibniz rule for the derivation δ, we write

$$\begin{aligned} D_\delta(xy) &= \left(\delta\left(\sum_{k=1}^n x_{ik}y_{kj}\right)\right)_{i,j=1}^n = \left(\sum_{k=1}^n \delta(x_{ik}y_{kj})\right)_{i,j=1}^n \\ &= \left(\sum_{k=1}^n \delta(x_{ik})\cdot y_{kj}\right)_{i,j=1}^n + \left(\sum_{k=1}^n x_{ik}\cdot\delta(y_{kj})\right)_{i,j=1}^n \\ &= (\delta(x_{ij}))_{i,j=1}^n\cdot(y_{ij})_{i,j=1}^n + (x_{ij})_{i,j=1}\cdot(\delta(y_{ij}))_{i,j=1} \\ &= D_\delta(x)y + xD_\delta(y). \end{aligned}$$

Thus, D_δ is a derivation on $M_{n\times n}(L_0(\Omega))$, as required. □

Definition 6.6.2. Let δ be a derivation on $L_0(\Omega)$. The derivation D_δ on the algebra $M_{n\times n}(L_0(\Omega))$ defined by (6.29) is called the *centrally induced* derivation implemented by δ.

In particular, if we have a nonzero derivation δ on the algebra $L_0(\Omega)$, then the centrally induced derivation D_δ is a nonzero derivation on $M_{n\times n}(L_0(\Omega))$, which does not vanish on the center $\{xI, x \in L_0(\Omega)\}$.

As we show next, any derivation D on the algebra $M_{n\times n}(L_0(\Omega))$ implements a derivation δ_D on $L_0(\Omega)$ and, moreover, on the center of the algebra $M_{n\times n}(L_0(\Omega))$ the centrally induced derivation D_{δ_D} implemented by δ_D coincides with the original derivation D.

Let D be a derivation on $M_{n\times n}(L_0(\Omega))$. By Lemma 5.1.5, D is a derivation on the center of $M_{n\times n}(L_0(\Omega))$. Since the center of $M_{n\times n}(L_0(\Omega))$ is $\{xI, x \in L_0(\Omega)\}$, it follows that we can define a mapping δ_D on $L_0(\Omega)$, such that

$$D(xI) = \delta_D(x)I, \quad x \in L_0(\Omega). \tag{6.31}$$

As we show in the following lemma, δ_D is a derivation on the algebra $L_0(\Omega)$.

Lemma 6.6.3. *Assume that D is a derivation on the algebra $M_{n\times n}(L_0(\Omega))$. The mapping δ_D, defined by (6.31), is a derivation on the algebra $L_0(\Omega)$. In addition, if D_{δ_D} is the centrally induced derivation on $M_{n\times n}(L_0(\Omega))$ implemented by δ_D, then*

$$D_{\delta_D}(xI) = D(xI), \quad x \in L_0(\Omega).$$

Proof. By the definition of δ_D for any $x, y \in L_0(\Omega)$, we have

$$\delta_D(xy)I = D((xI)(yI)) = D(xI)(yI) + (xI)D(yI) = (\delta_D(x)y + x\delta_D(y))I.$$

Hence,

$$\delta_D(xy) = \delta_D(x)y + x\delta_D(y), \quad x, y \in L_0(\Omega),$$

and so δ_D is a derivation on $L_0(\Omega)$. In addition, using equality (6.30), we conclude that

$$D_{\delta_D}(xI) = \delta_D(x)I = D(xI), \quad x \in L_0(\Omega),$$

as required. □

An immediate corollary of Lemma 6.6.3 shows that only the trivial derivation can be simultaneously centrally induced and inner. Therefore, the classes of inner derivations and (nonzero) centrally induced derivations complement one another.

Proposition 6.6.4. *If D is a derivation on $M_{n\times n}(L_0(\Omega))$, which is simultaneously centrally induced and inner, then $D = 0$.*

Proof. Since D is centrally induced, it follows that there is a derivation δ on $L_0(\Omega)$, such that

$$D((x_{ij})_{i,j=1}^n) = (\delta(x_{ij}))_{i,j=1}^n, \quad (x_{ij})_{i,j=1}^n \in M_{n\times n}(L_0(\Omega)).$$

In particular, by Lemma 6.6.3, we have that

$$D(xI) = \delta(x)I, \quad x \in L_0(\Omega).$$

Since D is inner, it must vanish on the center of $M_{n\times n}(L_0(\Omega))$. Therefore,

$$D(xI) = 0, \quad x \in L_0(\Omega).$$

Comparing the latter formulae, we obtain that

$$\delta(x) = 0, \quad x \in L_0(\Omega).$$

Thus, we conclude that

$$D\big((x_{ij})_{i,j=1}^n\big) = \big(\delta(x_{ij})\big)_{i,j=1}^n = 0, \quad (x_{ij})_{i,j=1}^n \in M_{n\times n}\big(L_0(\Omega)\big).$$ □

Suppose now that we have an inner derivation $D_x, x \in M_{n\times n}(L_0(\Omega))$ on the algebra $M_{n\times n}(L_0(\Omega))$. Then D_x clearly vanishes on the center $\mathcal{Z}(S(\mathcal{M})) = \{xI, x \in L_0(\Omega)\}$. As the next step of our approach, we show that any derivation vanishing on the center of $M_{n\times n}(L_0(\Omega))$ is necessarily inner. We start with an auxiliary lemma.

Lemma 6.6.5. *For a derivation $D : M_{n\times n}(L_0(\Omega)) \to M_{n\times n}(L_0(\Omega))$, set*

$$x_D = \sum_{i=1}^n e_{i1} D(e_{1i}). \tag{6.32}$$

We have

$$D(e_{kl}) = [e_{kl}, x_D], \quad 1 \le k,l \le n. \tag{6.33}$$

Proof. By Proposition 5.1.8, we have that $D(I) = 0$. Using the Leibniz rule, we write

$$\begin{aligned} x_D &= \sum_{i=1}^n \big(D(e_{i1}e_{1i}) - D(e_{i1})e_{1i}\big) \\ &= \sum_{i=1}^n \big(D(e_{ii}) - D(e_{i1})e_{1i}\big) = D(I) - \sum_{i=1}^n D(e_{i1})e_{1i} \\ &= -\sum_{i=1}^n D(e_{i1})e_{1i}. \end{aligned} \tag{6.34}$$

Therefore, for every $k, l = 1, \ldots, n$, we have

$$\begin{aligned} [e_{kl}, x_D] &= e_{kl}x_D - x_D e_{kl} \overset{(6.34)}{=} e_{kl}\sum_{i=1}^n e_{i1}D(e_{1i}) + \sum_{i=1}^n D(e_{i1})e_{1i}e_{kl} \\ &= e_{k1}D(e_{1l}) + D(e_{k1})e_{1l} \\ &= D(e_{k1}e_{1l}) = D(e_{kl}), \end{aligned} \tag{6.35}$$

as required. □

Proposition 6.6.6. *For a derivation $D : M_{n\times n}(L_0(\Omega)) \to M_{n\times n}(L_0(\Omega))$, the following conditions are equivalent:*
(i) *The derivation D is inner;*
(ii) *D vanishes on the center of $M_{n\times n}(L_0(\Omega))$, i. e., $D(xI) = 0$ for every $x \in L_0(\Omega)$.*

Proof. The implication (i)⇒(ii) is trivial. To prove the converse implication, consider the matrix x_D defined by (6.32). We claim that x_D implements D, i. e.,

$$D(x) = [x, x_D], \quad x \in M_{n\times n}\big(L_0(\Omega)\big).$$

For every $x = (x_{ij})_{i,j=1}^n \in M_{n\times n}(L_0(\Omega))$, writing $x = \sum_{i,j=1}^n x_{ij}e_{ij}$ and using the Leibniz rule, we obtain

$$D(x) = \sum_{i,j=1}^n D(x_{ij}I \cdot e_{ij}) = \sum_{i,j=1}^n D(x_{ij}I)e_{ij} + \sum_{i,j=1}^n (x_{ij}I)D(e_{ij}).$$

Since D is a trivial derivation on the center of $M_{n\times n}(L_0(\Omega))$, it follows that the first sum on the right-hand side vanishes. Hence,

$$\begin{aligned} D(x) &= \sum_{i,j=1}^n (x_{ij}I)D(e_{ij}) \overset{(6.33)}{=} \sum_{i,j=1}^n x_{ij}I \cdot [e_{ij}, x_D] \\ &= \left[\sum_{i,j=1}^n x_{ij}e_{ij}, x_D\right] = [x, x_D]. \end{aligned}$$

Thus, D is an inner derivation implemented by x_D, as required. □

So far, we have showed that there exist distinct classes of derivations on the algebra $M_{n\times n}(L_0(\Omega))$, centrally induced and inner derivations. We now show that any derivation D on the algebra $M_{n\times n}(L_0(\Omega))$ can be uniquely written as a sum of an inner derivation and a centrally induced one.

Theorem 6.6.7. *Assume that $\mathcal{M}$ is a type I_n algebra for some $n \in \mathbb{N}$ (identified with $M_{n\times n}(L_\infty(\Omega))$). Assume that D is a derivation on the algebra $S(\mathcal{M})$ (identified with $M_{n\times n}(L_0(\Omega))$). There exists a unique centrally induced derivation D_1 and a unique inner derivation D_2 on the algebra $S(\mathcal{M})$, such that $D = D_1 + D_2$.*

Proof. Let x_D be as in Lemma 6.6.5 and let δ_D be defined by (6.31). Let D_1 be the centrally induced derivation on the algebra $M_{n\times n}(L_0(\Omega))$ implemented by δ_D and let D_2 be an inner derivation on the algebra $M_{n\times n}(L_0(\Omega))$ implemented by x_D. We claim that $D = D_1 + D_2$.

Let $y = (y_{ij})_{i,j=1}^n \in M_{n\times n}(L_0(\Omega))$ be arbitrary. Writing $y = \sum_{i,j=1}^n y_{ij}I \cdot e_{i,j}$ and using the Leibniz rule, we have

$$D(y) = \sum_{i,j=1}^n D(y_{ij}I \cdot e_{i,j}) = \sum_{i,j=1}^n D(y_{ij}I)e_{ij} + \sum_{i,j=1}^n y_{ij}I \cdot D(e_{ij}).$$

By Lemma 6.6.3, we have

$$D(y_{ij}I) = \delta_D(y_{ij})I, \quad 1 \le i,j \le n.$$

By Lemma 6.6.5, we have

$$\sum_{i,j=1}^n y_{ij}I \cdot D(e_{ij}) = \sum_{i,j=1}^n y_{ij}I[e_{ij}, x_D] = [y, x_D].$$

Therefore, we infer that

$$D(y) = \sum_{i,j=1}^{n} \delta_D(y_{ij})e_{ij} + \sum_{i,j=1}^{n} y_{ij}I \cdot [e_{ij}, x_D]$$
$$= D_1(y) + [y, x_D] = D_1(y) + D_2(y),$$

proving that $D = D_1 + D_2$.

To prove uniqueness, assume that D_3 is a centrally induced derivation on $M_{n\times n}(L_0(\Omega))$ and D_4 is an inner derivation on $M_{n\times n}(L_0(\Omega))$, such that $D_1 + D_2 = D = D_3 + D_4$. Then $D_1 - D_3 = D_4 - D_2$. Since D_2 and D_4 are inner derivations, $D_4 - D_2$ is also an inner derivation. Since D_1 and D_3 are centrally induced derivations, it is clear that $D_3 - D_1$ is also a centrally induced derivation. Thus, the derivation $D_1 - D_3 = D_4 - D_2$ is simultaneously inner and centrally induced. Referring to Proposition 6.6.4, we conclude that $D_1 - D_3 = D_4 - D_2 = 0$, i. e., $D_1 = D_3$ and $D_2 = D_4$, as required. □

We now move on to the general type I_{fin} algebra $\mathcal{M}$. We first establish a general form of derivations on the algebra $S(\mathcal{M})$ similar to Theorem 6.6.7.

Assume that $\mathcal{M}$ is a type I_{fin} algebra. By Theorem 1.9.17, there exists a central partition of unity $\{z_n\}_{n\in\mathbb{N}}$ (some of the z_n may be zero), such that the algebra $z_n\mathcal{M}$ is $*$-isomorphic to the algebra $M_{n\times n}(L_\infty(\Omega_n))$ for some σ-finite measure space $(\Omega_n, \Sigma_n, \mu_n)$, such that $\bigoplus_{n\in\mathbb{N}}(L_\infty(\Omega_n)) = L_\infty(\Omega) \cong \mathcal{Z}(\mathcal{M})$. Thus, without loss of generality, we can write

$$\mathcal{M} = \bigoplus_{n\in\mathbb{N}} M_{n\times n}(L_\infty(\Omega_n)).$$

First, we define a centrally induced derivation. Let δ be a derivation on $\mathcal{Z}(S(\mathcal{M})) = L_\infty(\Omega, \Sigma, \mu)$. By Proposition 5.1.8, the restriction δ_n of δ onto $z_n\mathcal{Z}(S(\mathcal{M}))$ is a derivation on $z_n\mathcal{Z}(S(\mathcal{M}))$. Let D_{δ_n} be a centrally induced derivation on $z_nS(\mathcal{M})$ as defined in Definition 6.6.2. Since $\{z_n\}_{n\in\mathbb{N}}$ are pairwise orthogonal, it follows that for every $y \in S(\mathcal{M})$, we can define

$$D_\delta(y) = \sum_{n=1}^{\infty} D_{\delta_n}(z_ny), \tag{6.36}$$

where the series converges locally in measure. It is clear that D_δ is a derivation on $S(\mathcal{M})$ and, therefore, the following definition makes sense.

Definition 6.6.8. Let δ be a derivation on $\mathcal{Z}(S(\mathcal{M}))$ and let δ_n be restriction of δ onto $z_n\mathcal{Z}(S(\mathcal{M}))$. Then the derivation D_δ defined by (6.36) is called the *centrally induced derivation* on $S(\mathcal{M})$ implemented by δ.

Theorem 6.6.9. *Let $\mathcal{M}$ be a von Neumann algebra of type I_{fin}. For every derivation D of $S(\mathcal{M})$, there exists a unique centrally induced derivation D_1 and a unique inner derivation D_2 on the algebra $S(\mathcal{M})$, such that $D = D_1 + D_2$.*

Proof. Let $\{z_n\}_{n\in\mathbb{N}}$ be a central partition of unity as above.

Let D be an arbitrary derivation on the algebra $S(\mathcal{M})$. By Proposition 5.1.8, the restriction $D_n = D|_{z_nS(\mathcal{M})}$ of the derivation D to the $*$-subalgebra $z_nS(\mathcal{M}) = S(z_n\mathcal{M})$ is a derivation on the algebra $S(z_n\mathcal{M})$.

Since the algebra $z_n\mathcal{M}$ is a type I_n algebra, Theorem 6.6.7 implies that there exist a unique derivation $\delta_n: S(z_n\mathcal{Z}(\mathcal{M})) \to S(z_n\mathcal{Z}(\mathcal{M}))$ and an inner derivation $[\cdot, x_n], x_n \in S(z_n\mathcal{M})$ on the algebra $S(z_n\mathcal{M})$, such that

$$D_n = D_{\delta_n} + [\cdot, x_n], \quad n \in \mathbb{N}. \tag{6.37}$$

Since $x_n \in z_nS(\mathcal{M})$ and $\{z_n\}_{n\in\mathbb{N}}$ are pairwise orthogonal, with $\sum_{n=1}^{\infty} z_n = \mathbf{1}$, Proposition 2.3.10 and Theorem 2.5.6 guarantees that there exists

$$x = \sum_{n=1}^{\infty} x_n \in S(\mathcal{M}), \tag{6.38}$$

such that $z_nx = x_n$ for every $n \in \mathbb{N}$ and $x \in S(\mathcal{M})$.

Note also, that for any $y \in S(\mathcal{M})$, we have that

$$z_n[y, x] = [z_ny, z_nx] = [z_ny, x_n], \quad n \in \mathbb{N}. \tag{6.39}$$

Similarly, since δ_n is a derivation on the algebra $z_n\mathcal{Z}(S(\mathcal{M}))$, we can define

$$\delta(y) = \sum_{n=1}^{\infty} \delta_n(z_ny), \quad y \in S(\mathcal{Z}(\mathcal{M})). \tag{6.40}$$

It is clear that the mapping δ is a derivation on the algebra $\mathcal{Z}(S(\mathcal{M}))$. In addition, $\delta(z_ny) = \delta_n(z_ny)$ for all $y \in S(\mathcal{M}), n \in \mathbb{N}$.

Let D_δ be the centrally induced derivation on $S(\mathcal{M})$ implemented by δ, i. e.,

$$D_\delta(y) = \sum_{n=1}^{\infty} D_{\delta_n}(z_nx), \quad x \in S(\mathcal{M}). \tag{6.41}$$

Note that the derivation D_δ satisfies

$$z_nD_\delta(z_ny) = D_{\delta_n}(z_ny), \quad y \in S(\mathcal{M}).$$

We now show that D_δ and $[\cdot, x]$ defined above are the unique centrally induced and inner derivations on $S(\mathcal{M})$ with $D = D_\delta + [\cdot, x]$.

Note that $D(z_ny) = z_nD(z_ny) = D_n(z_ny)$ for every $y \in S(\mathcal{M})$ and $n \in \mathbb{N}$. Therefore, by (6.37) and the definition of x and δ (see (6.38) and (6.40)), we have that

$$\begin{aligned} D(z_ny) &= z_nD_n(z_ny) = D_{\delta_n}(z_ny) + [z_ny, x_n] \\ &= z_nD_\delta(z_ny) + z_n[y, x] \\ &= D_\delta(z_ny) + [z_ny, x]. \end{aligned}$$

Hence, since $\sum_{n=1}^{\infty} z_n = \mathbf{1}$, it follows that

$$D = D_\delta + [\cdot, x].$$

To conclude the proof, we only need to show that δ and $[\cdot, x]$ in this decomposition are unique.

Assume that δ_1 (resp., $[\cdot, w], w \in S(\mathcal{M})$) is another derivation on $\mathcal{Z}(S(\mathcal{M}))$ (resp., inner derivation on $S(\mathcal{M})$), such that $D = D_{\delta_1} + [\cdot, w]$.

Fix any $n \in \mathbb{N}$, such that $z_n\mathcal{M}$ is not trivial. On one hand, for any $y \in S(\mathcal{M})$, we have that

$$D_n(z_n y) = D(z_n y) = z_n D_\delta(z_n y) + z_n[z_n y, w] = D_{\delta_n}(z_n y) + D_{x_n}(z_n y).$$

On the other hand,

$$\begin{aligned} D_n(z_n y) &= z_n D_{\delta_1}(z_n y) + z_n[z_n y, w] \\ &= D_{\delta_1^{(z_n)}}(z_n y) + [z_n y, z_n w]. \end{aligned}$$

Thus, the derivation D_n on the algebra $z_n S(\mathcal{M})$ can be written as

$$D_{\delta_n} + [, \cdot, x_n] = D_n = D_{\delta_1^{(z_n)}} + [\cdot, z_n w].$$

Since the algebra $z_n\mathcal{M}$ is type I_n algebra, Theorem 6.6.7 implies that

$$D_{\delta_n} = D_{\delta_1^{(z_n)}}, \quad [\cdot, x_n] = [\cdot, z_n w], \quad n \in \mathbb{N}.$$

Since $\sum_{n=1}^{\infty} z_n = \mathbf{1}$, it follows that $D_\delta = D_{\delta_1}$ and $[\cdot, x] = [\cdot, w]$, which proves the uniqueness of the decomposition. □

Finally, we prove the main result of this section, a necessary and sufficient condition for the existence of noninner derivations on the algebra $S(\mathcal{M})$ for a type I_{fin} algebra $\mathcal{M}$. We first prove an auxiliary lemma with such a necessary and sufficient condition for a type I_n algebra $\mathcal{M}$.

Lemma 6.6.10. *For a von Neumann algebra $\mathcal{M}$ of type I_n, $n \in \mathbb{N}$, the following conditions are equivalent:*
(i) *Every derivation on the algebra $S(\mathcal{M})$ is inner;*
(ii) *The Boolean algebra $P(\mathcal{Z}(\mathcal{M}))$ of all central projections from $\mathcal{M}$ is atomic.*

Proof. Without loss of generality, $\mathcal{M} = M_{n\times n}(L_0(\Omega))$.

(i)⇒(ii). Suppose by contradiction that Ω is not atomic. By Theorem 6.4.7, there exists a nonzero derivation $\delta : L_0(\Omega) \to L_0(\Omega)$. Let D be the centrally induced derivation on the algebra $M_{n\times n}(L_0(\Omega))$ corresponding to δ (see Definition 6.6.2). The derivation D does not vanish on the center of $S(\mathcal{M})$, and by Proposition 6.6.6, it is not inner. Thus, there exists a noninner derivation, which is a contradiction. Hence, Ω is atomic.

(ii)⇒(i). Since Ω is atomic, Theorem 6.4.7 guarantees that any derivation on $L_0(\Omega)$ is trivial. Therefore, any centrally induced derivation on $M_{n\times n}(L_0(\Omega))$ is trivial. By Theorem 6.6.7, any derivation on $M_{n\times n}(L_0(\Omega))$ is inner. □

Theorem 6.6.11. *For a von Neumann algebra $\mathcal{M}$ of type I_{fin}, the following conditions are equivalent:*
(i) *Every derivation on the algebra $S(\mathcal{M})$ is inner;*
(ii) *The Boolean algebra $P(\mathcal{Z}(\mathcal{M}))$ of all central projections from $\mathcal{M}$ is atomic.*

Proof. By Theorem 1.9.17, there exists a central partition of unity $\{z_n\}_{n\in\mathbb{N}}$, such that $z_n\mathcal{M}$ (is either trivial or) is of type I_n for every $n \in \mathbb{N}$.

(i)⇒(ii). Let $n \in \mathbb{N}$ be such that $z_n \neq 0$ and let δ be a derivation on the algebra $S(z_n\mathcal{M})$. Define a derivation D on the algebra $S(\mathcal{M})$ by setting

$$D(x) = \delta(xz_n), \quad x \in S(\mathcal{M}).$$

By assumption, D is inner. Hence, there exists $x_n \in S(\mathcal{M})$, such that

$$D(x) = [x, x_n], \quad x \in S(\mathcal{M}).$$

Since $D(x) = D(xz_n)$, it follows that

$$\delta(x) = D(x) = [x, x_n z_n], \quad x \in S(z_n\mathcal{M}).$$

Hence, every derivation on the algebra $S(z_n\mathcal{M})$ is inner. By Lemma 6.6.10, the Boolean algebra $P(\mathcal{Z}(z_n\mathcal{M}))$ is atomic. Since $n \in \mathbb{N}$ is arbitrary, it follows that the Boolean algebra $P(\mathcal{Z}(\mathcal{M}))$ is atomic.

(ii)⇒(i). Let D be a derivation on the algebra $S(\mathcal{M})$. Let $n \in \mathbb{N}$, such that $z_n \neq 0$. Define a derivation D_n on $S(z_n\mathcal{M})$ by setting

$$D_n(x) = D(xz_n), \quad x \in S(z_n\mathcal{M}).$$

By assumption, the Boolean algebra $P(\mathcal{Z}(z_n\mathcal{M}))$ is atomic. By Lemma 6.6.10, there exists $y_n \in S(z_n\mathcal{M})$, such that

$$D_n(x) = [x, y_n], \quad x \in S(z_n\mathcal{M}).$$

We have

$$D(x) = \sum_{n=1}^{\infty} D(x)z_n = \sum_{n=1}^{\infty} D(xz_n) = \sum_{n=1}^{\infty} [xz_n, y_n], \quad x \in S(\mathcal{M}).$$

Since $y_n \in S(z_n\mathcal{M}) = z_nS(\mathcal{M})$ and $\{z_n\}_{n\in\mathbb{N}}$ is a central partition of unity, by Proposition 2.3.10, there exists unique $y \in S(\mathcal{M})$, such that $yz_n = y_n$, $n \in \mathbb{N}$. Since

$$z_nD(x) = [xz_n, y_n] = [x, y]z_n$$

for any $n \in \mathbb{N}$, and $x \in S(\mathcal{M})$, referring to Lemma 2.3.7, we conclude that

$$D(x) = [x, y], \quad x \in S(\mathcal{M}).$$

Hence, every derivation on $S(\mathcal{M})$ is inner. □

By Theorem 5.5.6(i), a derivation on $S(\mathcal{M}) = LS(\mathcal{M})$ is inner if and only if it is $t(\mathcal{M})$-continuous. Therefore, we can also give the following statement of the above result.

Corollary 6.6.12. *Let $\mathcal{M}$ be type I_{fin} algebra. There are $t(\mathcal{M})$-discontinuous derivations on $S(\mathcal{M})$ if and only if the Boolean algebra $P(\mathcal{Z}(\mathcal{M}))$ is not atomic.*

Bibliographical notes

The notion of the metric ρ_μ (see Definition 6.1.8) is due to von Neumann [157, 159, 160], who defined this notion in the general setting of regular rings [157], [160, p. 161, Definition 3] and who proved in [159] that $*$-algebras $S(\mathcal{R})$ and $\mathcal{R}_\infty$ are not $*$-isomorphic; here, $\mathcal{R}$ is a hyperfinite factor of type II_1 and $\mathcal{R}_\infty$ is the ρ-closure of an increasing sequence matrix $*$-subalgebras of $\mathcal{R}$, and ρ is the rank metric on $\mathcal{R}$ (see [160, p. 188, Theorem E]). The metric ρ_μ was also later reintroduced by Ciach [56].

In 2000, Sh. Ayupov proved that any derivation on the algebra $L_0(\Omega, \Sigma, \mu)$ vanishes (and so, is inner) provided that the measure space is atomic [9].

Theorem 6.4.7 (equivalently, Theorem 6.4.9) with necessary and sufficient conditions for the existence of nontrivial derivations on the algebra of measurable operators affiliated with a commutative von Neumann algebra is due to three present co-authors (Ber, Chilin, and Sukochev). It was firstly announced in [38] with full proofs appearing in [22]. In fact, these results constitute a possible answer to Khinčin's question discussed in the introduction. Via the methods drawn from Boolean-valued analysis a similar result was obtained by A. Kusraev [102, 103] for commutative AW*-algebras. Later, in 2024, the main results of [22, 38] have been extended to arbitrary disjointly complete commutative regular algebras (see [26, Theorem 1.1]).

The construction of the extension of derivations by algebraic and transcendental elements is presented in [57, 161]. The notion of a weakly transcendental element with respect to a subalgebra was invented in [22]. Extension of derivations by a weakly transcendental element was also introduced in [22].

Approximately differentiable functions are studied in great detail in the monographs [136, Chapter IX, Section 11] and [71, Chapter 3, Section 3.1]. The proof of the assertion that the maximal subalgebra in $L_0[a, b]$ admitting a unique extension of the classical derivative coincides with the algebra of all approximately differentiable functions appeared in [33]. This result was first announced in [18]. Note that any extension of the classical derivation ∂ up to a derivation on the algebra $L_0[a, b]$ does not commute with translations [29].

A complete description of derivations on the $*$-algebra $S(\mathcal{M})$ of measurable operators affiliated with a finite von Neumann algebra $\mathcal{M}$ of type I (Section 6.6) is due to S. Albeverio, Sh. Ayupov, and K. Kudaybergenov [4, 6], and [7].

For further research in this direction, see [4, 6, 7, 10, 16, 20, 29].

7 Complete description of derivations on algebras of locally measurable operators

In this chapter, we present the full resolution of Ayupov-Kadison-Liu problem stated in the introduction. Specifically, we establish necessary and sufficient conditions on the von Neumann algebra $\mathcal{M}$ which guarantee that all derivations on the algebra $LS(\mathcal{M})$ (and respectively, on the algebra $S(\mathcal{M},\tau)$) are inner.

As it has been already noted in Chapter 6, if $\mathcal{M}$ is a von Neumann algebra of type I_{fin}, then the $*$-algebra $LS(\mathcal{M})$ may support a noninner derivation. In the present chapter, we show that if the algebra $\mathcal{M}$ does not have a type I_{fin} direct summand, then the $*$-algebras $LS(\mathcal{M})$, $S(\mathcal{M},\tau)$ (resp., $S_0(\mathcal{M})$ and $S_0(\mathcal{M},\tau)$) have only inner (resp., spatial) derivations. Thus, the first cohomology group of each of these algebras is trivial.

As shown in Theorem 5.5.6, a derivation δ on the algebra $LS(\mathcal{M})$ (resp., on $S_0(\mathcal{M})$) is inner (resp., spatial) if and only if δ is continuous with respect to the topology of convergence locally in measure. Similar results hold for derivations on the algebras $S(\mathcal{M},\tau)$ and $S_0(\mathcal{M},\tau)$ when the local measure topology is replaced with the measure topology t_τ. Hence, it is sufficient to show that every derivation on the algebra $LS(\mathcal{M})$ (or $S_0(\mathcal{M})$, $S(\mathcal{M},\tau)$, $S_0(\mathcal{M},\tau)$) is continuous in the respective topology.

Assuming that a von Neumann algebra $\mathcal{M}$ has no type I_{fin} direct summand, we can write $\mathcal{M}$ as a direct sum of a type II_1 algebra, a type III algebra, and a properly infinite semifinite algebra. We prove automatic continuity of derivations assuming that $\mathcal{M}$ is of either of these types arguing by contradiction. We start with derivations on $LS(\mathcal{M})$ for a type II_1 von Neumann algebra $\mathcal{M}$ in Section 7.1, and then tackle the type III and semifinite properly infinite case in Section 7.3 and Section 7.4, respectively. In all three cases, we first establish rather technical results concerning discontinuous derivations on $LS(\mathcal{M})$. The main result yielding a necessary and sufficient condition for the existence of noninner derivations on $LS(\mathcal{M})$ (and other algebras) for a general von Neumann algebra $\mathcal{M}$ is given in Section 7.6.

In Section 7.7, we complete the exposition by showing that any derivation with values in a quasi-normed bimodule over $\mathcal{M}$ is necessarily inner.

7.1 Automatic continuity of derivations in the case of type II_1 algebras

In this section, we show that for a type II_1 algebra $\mathcal{M}$ any derivation δ on $LS(\mathcal{M}) = S(\mathcal{M})$ is necessarily continuous in the local measure topology.

Let us start with the following technical construction for an arbitrary von Neumann algebra with no abelian projections. We recall that a type II_1 algebra is necessarily an algebra with no abelian projection, and so the result of Lemma 7.1.1 applies to any type II_1 algebra.

https://doi.org/10.1515/9783111599687-008

Lemma 7.1.1. *Let $\mathcal{M}$ be a von Neumann algebra with no abelian projections. Then there exists a system $\{e_{i,j}^{(n)} : i,j = 1,\ldots,2^n,\ n \in \mathbb{N}\}$ of partial isometries from $\mathcal{M}$, such that:*

(i) $(e_{i,j}^{(n)})^* = e_{j,i}^{(n)}$, $e_{i,j}^{(n)} e_{k,l}^{(n)} = \delta_{jk} e_{i,l}^{(n)}$, $i,j,k,l = 1,\ldots,2^n$ *and* $\sum_{i=1}^{2^n} e_{i,i}^{(n)} = \mathbf{1}$ *(i. e. $\{e_{i,j}^{(n)} : i,j = 1,\ldots,2^n\}$ is a system of matrix units in $\mathcal{M}$);*

(ii) $e_{2i-1,2j-1}^{(n+1)} + e_{2i,2j}^{(n+1)} = e_{i,j}^{(n)}$, $i,j = 1,\ldots,2^n$, $n \in \mathbb{N}$.

Proof. We construct the system $\{e_{i,j}^{(n)} : i,j = 1,\ldots,2^n,\ n \in \mathbb{N}\}$ by induction on $n \in \mathbb{N}$. From [134, Proposition 2.2.13], it follows that $\mathbf{1}$ can be written as $\mathbf{1} = e_{1,1}^{(1)} + e_{2,2}^{(1)}$ for some mutually orthogonal and equivalent projections $e_{1,1}^{(1)}$ and $e_{2,2}^{(1)}$. Then there exists a partial isometry $e_{1,2}^{(1)}$ that $(e_{1,2}^{(1)})^* e_{1,2}^{(1)} = e_{2,2}^{(1)}$, $e_{1,2}^{(1)}(e_{1,2}^{(1)})^* = e_{1,1}^{(1)}$. We set $e_{2,1}^{(1)} = (e_{1,2}^{(1)})^*$.

Assume that the $\{e_{i,j}^{(k)} : i,j = 1,\ldots,2^k,\ k \le n\}$ are already constructed. We will now define $\{e_{i,j}^{(n+1)} : i,j = 1,\ldots,2^{n+1}\}$. Again from [134, Proposition 2.2.13], it follows that $e_{1,1}^{(n)}$ can be written as $e_{1,1}^{(n)} = e_{1,1}^{(n+1)} + e_{2,2}^{(n+1)}$ for some mutually orthogonal and equivalent projections $e_{1,1}^{(n+1)}, e_{2,2}^{(n+1)}$. Then there exists a partial isometry $e_{1,2}^{(n+1)}$, such that $(e_{1,2}^{(n+1)})^* e_{1,2}^{(n+1)} = e_{2,2}^{(n+1)}$, $e_{1,2}^{(n+1)}(e_{1,2}^{(n+1)})^* = e_{1,1}^{(n+1)}$. We set $e_{2,1}^{(n+1)} = (e_{1,2}^{(n+1)})^*$ and for $i,j = 1,\ldots,2^n$, $s,t \in \{0,1\}$, we set $e_{2i-s,2j-t}^{(n+1)} = e_{i,1}^{(n)} e_{2-s,2-t}^{(n+1)} e_{1,j}^{(n)}$.

We have that

$$
\begin{aligned}
e_{2i_1-s_1,2j_1-t_1}^{(n+1)} e_{2i_2-s_2,2j_2-t_2}^{(n+1)} &= e_{i_1,1}^{(n)} e_{2-s_1,2-t_1}^{(n+1)} e_{1,j_1}^{(n)} e_{i_2,1}^{(n)} e_{2-s_2,2-t_2}^{(n+1)} e_{1,j_2}^{(n)} \\
&= \delta_{j_1 i_2} e_{i_1,1}^{(n)} e_{2-s_1,2-t_1}^{(n+1)} e_{2-s_2,2-t_2}^{(n+1)} e_{1,j_2}^{(n)} \\
&= \delta_{j_1 i_2} \delta_{t_1 s_2} e_{i_1,1}^{(n)} e_{2-s_1,2-t_2}^{(n+1)} e_{1,j_2}^{(n)} \\
&= \delta_{2j_1-t_1,2i_2-s_2} e_{2i_1-s_1,2j_2-t_2}^{(n+1)}, \\
(e_{2i-s,2j-t}^{(n+1)})^* &= e_{j,1}^{(n)} e_{2-t,2-s}^{(n+1)} e_{1,i}^{(n)} = e_{2j-t,2i-s}^{(n+1)}, \\
e_{2i-1,2j-1}^{(n+1)} + e_{2i,2j}^{(n+1)} &= e_{i,1}^{(n)} e_{1,1}^{(n+1)} e_{1,j}^{(n)} + e_{i,1}^{(n)} e_{2,2}^{(n+1)} e_{1,j}^{(n)} \\
&= e_{i,1}^{(n)} e_{1,j}^{(n)} = e_{i,j}^{(n)}
\end{aligned} \tag{7.1}
$$

for any $i, i_1, i_2, j, j_1, j_2 = 1,\ldots,2^n$, $s, s_1, s_2, t, t_1, t_2 \in \{0,1\}$.

Equality (7.1) also implies that

$$
\sum_{k=1}^{2^{n+1}} e_{k,k}^{(n+1)} = \sum_{i=1}^{2^n} (e_{2i-1,2i-1}^{(n+1)} + e_{2i,2i}^{(n+1)}) = \sum_{i=1}^{2^n} e_{i,i}^{(n)} = \mathbf{1}.
$$

Thus, $\{e_{i,j}^{(n)} : i,j = 1,\ldots,2^n,\ n \in \mathbb{N}\}$ is the required system. □

Assume now that $\mathcal{M}$ is a type II_1 algebra equipped with a faithful normal finite trace τ, in which case $LS(\mathcal{M}) = S(\mathcal{M})$. Our aim is to show that any derivation on $S(\mathcal{M})$ is continuous in the local measure topology. Since the trace τ is finite, Theorem 4.4.3 guarantees that the topological algebras $(S(\mathcal{M}), t(\mathcal{M}))$ and $(S(\mathcal{M},\tau), t_\tau)$ coincide. By Proposition 5.1.15, a derivation $\delta : S(\mathcal{M}) \to S(\mathcal{M})$ is continuous in measure if and only if δ is continuous as a mapping from $(\mathcal{M}, \|\cdot\|_{\mathcal{M}})$ to $(S(\mathcal{M}), t_\tau)$. In the following two lemmas, we give an auxiliary construction for a derivation $\delta : S(\mathcal{M}) \to S(\mathcal{M})$, that is discontinuous as a mapping from $(\mathcal{M}, \|\cdot\|_{\mathcal{M}})$ into $(S(\mathcal{M}), t_\tau)$. First, we show that for some nonzero central projection $z \in \mathcal{M}$, every element in the reduced algebra $z\mathcal{M}$ is necessarily a point of discontinuity of the derivation $\delta : \mathcal{M} \to S(\mathcal{M})$.

Lemma 7.1.2. *Let $\mathcal{M}$ be a von Neumann algebra equipped with a faithful finite normal trace τ. Let δ be a self-adjoint discontinuous derivation from $(\mathcal{M}, \|\cdot\|_{\mathcal{M}})$ into $(S(\mathcal{M}), t_\tau)$. Then there exists a nonzero $z \in P(\mathcal{Z}(\mathcal{M}))$, such that for any $y \in z\mathcal{M}$, there exists $\{a_n\}_{n\in\mathbb{N}} \subset \mathcal{M}$, such that $a_n \to 0$ in the uniform norm and $\delta(a_n) \to y$ in measure.*

Proof. Let us define

$$\mathcal{I} = \{y \in S(\mathcal{M}) : \exists\{x_n\} \subset \mathcal{M},\ \|x_n\|_{\mathcal{M}} \to 0,\ \delta(x_n) \xrightarrow{t_\tau} y\}.$$

Clearly, $\mathcal{I}$ is a linear subspace in $S(\mathcal{M})$. Since $(S(\mathcal{M}), t_\tau)$ is an F-space (see Theorem 4.2.8) and δ is discontinuous, the closed graph theorem (see, e. g., [129, Chapter 2, Theorem 2.15]) implies that $\mathcal{I}$ is a nontrivial subspace of $S(\mathcal{M})$. We claim that $\mathcal{I} \cap \mathcal{M} \neq \{0\}$. Let $0 \neq y \in S(\mathcal{M})$ and $\{x_n\} \subset \mathcal{M}$ be such that $x_n \to 0$ in the uniform norm and $\delta(x_n) \to y$ in measure. Since δ is self-adjoint, we can assume that x_n and y are self-adjoint. Without loss of generality, $y_+ \neq 0$ (otherwise, consider $-x_n$ instead of x_n). Let $\varepsilon > 0$ be such that $E_y(\varepsilon, \infty) \neq 0$. Then, by the functional calculus, the operator $a = \max\{y, \varepsilon \cdot \mathbf{1}\}^{-1/2} E_y(\varepsilon, \infty)$ is bounded. Furthermore, $ax_n a \to 0$ in the uniform norm as $n \to \infty$, and by the Leibniz rule

$$\delta(ax_n a) = \delta(a)x_n a + a\delta(x_n)a + ax_n\delta(a) \to aya = E_y(\varepsilon, \infty),$$

in measure, i. e., $E_y(\varepsilon, \infty) \in \mathcal{I}$ guaranteeing that $\mathcal{I} \cap \mathcal{M} \neq \{0\}$.

Next, we claim that $\mathcal{I}$ is closed in the measure topology. Let $\{y_n\}_{n\in\mathbb{N}} \subset \mathcal{I}$ be such that $y_n \to y \in S(\mathcal{M})$ in measure. Since $y_n \in \mathcal{I}$, $n \in \mathbb{N}$, there exists $\{x_{m,n}\}_{m\in\mathbb{N}} \in \mathcal{M}$, such that $\|x_{m,n}\|_{\mathcal{M}} \to 0$ and $y_n = \lim_{m\to\infty} \delta(x_{m,n})$ in measure for every fixed $n \in \mathbb{N}$. For any $n \in \mathbb{N}$, there exists m_n, such that $\|x_{m_n,n}\|_{\mathcal{M}} < \frac{1}{n}$ and $\|y_n - \delta(x_{m_n,n})\|_{S(\mathcal{M},\tau)} < \frac{1}{n}$. Setting $a_n = x_{m_n,n}$, we have $\|a_n\|_{\mathcal{M}} \to 0$ and

$$\begin{aligned}\|y - \delta(a_n)\|_{S(\mathcal{M},\tau)} &\leq \|y - y_n\|_{S(\mathcal{M},\tau)} + \|y_n - \delta(a_n)\|_{S(\mathcal{M},\tau)} \\ &\leq \|y - y_n\|_{S(\mathcal{M},\tau)} + 1/n \to 0,\end{aligned}$$

i. e., $y \in \mathcal{I}$, proving that $\mathcal{I}$ is closed in the measure topology.

We define $\mathcal{J} = \mathcal{I} \cap \mathcal{M}$ and first show that $\mathcal{J}$ is a two-sided ideal in $\mathcal{M}$. If $y \in \mathcal{J}$, $x \in \mathcal{M}$, then there exists a sequence $\{x_n\}_{n\in\mathbb{N}} \subset \mathcal{M}$, such that $y = \lim_{n\to\infty} \delta(x_n)$ and $\lim_{n\to\infty} \|x_n\|_{\mathcal{M}} = 0$. Therefore, $\lim_{n\to\infty} \|xx_n\|_{\mathcal{M}} = 0$ and by the Leibniz rule

$$xy = \lim_{n\to\infty}(x\delta(x_n)) = \lim_{n\to\infty}(\delta(xx_n) - \delta(x)x_n) = \lim_{n\to\infty}\delta(xx_n)$$

in measure, i. e., $xy \in \mathcal{J}$. Similarly, $yx \in \mathcal{J}$, proving that $\mathcal{J}$ is a two-sided ideal in $\mathcal{M}$.

As mentioned above, the ideal $\mathcal{J}$ is nontrivial. Our next aim is to show that $\mathcal{J}$ is closed in the ultrastrong operator topology. Let $\{b_k\}_{k\in\mathbb{N}} \subset \mathcal{J}$ be such that $b_k \to b \in \mathcal{M}$ in the ultrastrong operator topology. We claim that $b_k \to b$ in measure as $k \to \infty$.

Since

$$|\langle |b_k - b|^2\xi, \eta\rangle| = |\langle (b_k - b)\xi, (b_k - b)\eta\rangle| \le \|(b_k - b)\xi\|\|(b_k - b)\eta\|$$

for all $\xi, \eta \in H$, it follows that $|b_k - b|^2$ converges to 0 in the ultraweak operator topology. Therefore, $\tau(|b_k - b|^2) \to 0$ as $k \to \infty$ (see Theorem 1.8.7). By spectral theory, for any $x \in \mathcal{M}$ and any $t > 0$, we have that $t^2 E_{|x|}(t,\infty) \le |x|^2 E_{|x|}(t,\infty)$ and, therefore, $\tau(E_{|x|}(t,\infty)) \le t^{-2}\tau(|x|^2) < \infty$, where the last inequality holds since the trace τ is finite. Taking $t = \tau(|x|^2)^{1/4}$ in the definition of the F-norm $\|\cdot\|_{S(\mathcal{M},\tau)}$ (see Definition 4.2.2), we have that

$$\|x\|_{S(\mathcal{M},\tau)} \le \tau(|x|^2)^{1/4} + \tau(|x|^2)^{1/2}, \quad x \in S(\mathcal{M},\tau).$$

Since $\tau(|b_k - b|^2) \to 0$ as $k \to \infty$, we conclude that $b_k \to b$ in measure as $k \to \infty$, as claimed. Since $\mathcal{I}$ is closed in measure, it follows that $b \in \mathcal{I}$ ensuring that $b \in \mathcal{J}$. Thus, the ideal $\mathcal{J}$ is closed in the ultrastrong operator topology.

Clearly, $\mathcal{J}$ is a $*$-algebra, which is closed in the ultrastrong operator topology. By [61, Theorem 2, p. 45], it is ultraweakly closed. Hence, by [61, Corollary 3, p. 46], there exists $0 \ne z \in P(\mathcal{Z}(\mathcal{M}))$, such that $\mathcal{J} = z\mathcal{M}$. The assertion now follows. □

Lemma 7.1.3. *Let $\mathcal{M}$ be a type II_1 von Neumann algebra equipped with a faithful finite normal trace τ and let $\{e_{i,j}^{(n)} : i,j = 1,\dots,2^n,\ n \in \mathbb{N}\}$ be the system of partial isometries from $\mathcal{M}$ constructed Lemma 7.1.1. Assume that δ is a discontinuous self-adjoint derivation from $(\mathcal{M}, \|\cdot\|_{\mathcal{M}})$ to $(S(\mathcal{M}), t_\tau)$, such that the central projection z in Lemma 7.1.2 is* **1**. *Then for every $k \in \mathbb{N}$ and every $\varepsilon > 0$, there exists $y \in (\{e_{i,j}^{(k)},\ i,j = 1,\dots,2^k\})' \cap \mathcal{M}$, such that*

$$\|y\|_{\mathcal{M}} < \varepsilon, \quad \|\delta(y) - \mathbf{1}\|_{S(\mathcal{M},\tau)} < \varepsilon.$$

Proof. We fix $k \in \mathbb{N}$ and $\varepsilon > 0$.

Since $\|\lambda x\|_{S(\mathcal{M},\tau)} \to 0$ as $\lambda \to 0$ for any $x \in S(\mathcal{M},\tau)$, we can find $0 < \lambda < \varepsilon$, such that

$$\sum_{i=1}^{2^k}\|\lambda\,\delta(e_{i,1}^{(k)})\|_{S(\mathcal{M},\tau)} + \sum_{i=1}^{2^k}\|\lambda\,\delta(e_{1,i}^{(k)})\|_{S(\mathcal{M},\tau)} + 2^k\lambda < \varepsilon. \tag{7.2}$$

By Lemma 7.1.2, there exists $w \in \mathcal{M}$, such that

$$\|w\|_{\mathcal{M}} < \lambda, \quad \|\delta(w) - e_{1,1}^{(k)}\|_{S(\mathcal{M},\tau)} < \lambda.$$

We set

$$y = \sum_{i=1}^{2^k} e_{i,1}^{(k)} w e_{1,i}^{(k)} \in \mathcal{M}.$$

It is clear that $\|y\|_{\mathcal{M}} < \lambda < \varepsilon$.

To show that $y \in (\{e_{i,j}^{(k)},\ i,j = 1,\dots,2^k\})'$, it is sufficient to show that $ye_{m_1,m_2}^{(k)} = e_{m_1,m_2}^{(k)}y$ for any $m_1, m_2 = 1,\dots,2^k$. By constructuion, we have

$$\begin{aligned}
e_{m_1,m_2}^{(k)} \cdot y &= \sum_{i=1}^{2^k} e_{m_1,m_2}^{(k)} e_{i,1}^{(k)} w e_{1,i}^{(k)} = e_{m_1,m_2}^{(k)} e_{m_2,1}^{(k)} w e_{1,m_2}^{(k)} \\
&= e_{m_1,1}^{(k)} w e_{1,m_2}^{(k)} \\
&= e_{m_1,1}^{(k)} w e_{1,m_1}^{(k)} e_{m_1,m_2}^{(k)} = \sum_{i=1}^{2^k} e_{i,1}^{(k)} w e_{1,i}^{(k)} e_{m_1,m_2}^{(k)} \\
&= y \cdot e_{m_1,m_2}^{(k)},
\end{aligned}$$

as required.

It remains to show that $\|\delta(y) - \mathbf{1}\|_{S(\mathcal{M},\tau)} < \varepsilon$. Introducing

$$A := \sum_{i=1}^{2^k} \delta(e_{i,1}^{(k)}) \cdot w e_{1,i}^{(k)}, \quad B := \sum_{i=1}^{2^k} e_{i,1}^{(k)} \cdot \delta(w) \cdot e_{1,i}^{(k)}, \quad C := \sum_{i=1}^{2^k} e_{i,1}^{(k)} w \cdot \delta(e_{1,i}^{(k)})$$

and using the Leibniz rule we can estimate

$$\begin{aligned}
\|\delta(y) - \mathbf{1}\|_{S(\mathcal{M},\tau)} &= \Big\|\sum_{i=1}^{2^k} \delta(e_{i,1}^{(k)} w e_{1,i}^{(k)}) - \mathbf{1}\Big\|_{S(\mathcal{M},\tau)} \\
&= \|A + (B - \mathbf{1}) + C\|_{S(\mathcal{M},\tau)} \\
&\le \|A\|_{S(\mathcal{M},\tau)} + \|B - \mathbf{1}\|_{S(\mathcal{M},\tau)} + \|C\|_{S(\mathcal{M},\tau)}.
\end{aligned} \tag{7.3}$$

Appealing to Proposition 4.2.7(iii) and (i), we can estimate each term as follows:

$$\begin{aligned}
\|A\|_{S(\mathcal{M},\tau)} &\le \sum_{i=1}^{2^k} \|\delta(e_{i,1}^{(k)}) \cdot w e_{1,i}^{(k)}\|_{S(\mathcal{M},\tau)} \\
&\le \sum_{i=1}^{2^k} \|\|w\|_{\mathcal{M}} \delta(e_{i,1}^{(k)})\|_{S(\mathcal{M},\tau)} \le \sum_{i=1}^{2^k} \|\lambda \delta(e_{i,1}^{(k)})\|_{S(\mathcal{M},\tau)}; \\
\|B - \mathbf{1}\|_{S(\mathcal{M},\tau)} &\le \sum_{i=1}^{2^k} \|e_{i,1}^{(k)} \cdot (\delta(w) - e_{1,1}^{(k)}) \cdot e_{1,i}^{(k)}\|_{S(\mathcal{M},\tau)} \\
&\le \sum_{i=1}^{2^k} \|\delta(w) - e_{1,1}^{(k)}\|_{S(\mathcal{M},\tau)} \le 2^k \lambda;
\end{aligned}$$

$$\|C\|_{S(\mathcal{M},\tau)} \le \sum_{i=1}^{2^k} \|w \cdot \delta(e_{i,1}^{(k)})\|_{S(\mathcal{M},\tau)} \le \sum_{i=1}^{2^k} \| \|w\|_{\mathcal{M}} \delta(e_{i,1}^{(k)})\|_{S(\mathcal{M},\tau)}$$
$$\le \sum_{i=1}^{2^k} \|\lambda \cdot \delta(e_{1,i}^{(k)})\|_{S(\mathcal{M},\tau)}.$$

Therefore, by (7.3), we infer that

$$\|\delta(y) - \mathbf{1}\|_{S(\mathcal{M},\tau)} \le \sum_{i=1}^{2^k} \|\lambda\delta(e_{i,1}^{(k)})\|_{S(\mathcal{M},\tau)} + \sum_{i=1}^{2^k} \|\lambda\delta(e_{1,i}^{(k)})\|_{S(\mathcal{M},\tau)} + 2^k\lambda.$$

The choice of λ (see (7.2)) guarantees that $\|\delta(y) - \mathbf{1}\|_{S(\mathcal{M},\tau)} < \varepsilon$, as required. □

Before we proceed to the proof of automatic continuity of derivations on $S(\mathcal{M})$ for a type II_1 von Neumann algebra $\mathcal{M}$, we state several auxiliary estimates for the F-norm $\|\cdot\|_{S(\mathcal{M},\tau)}$.

Lemma 7.1.4. *Let $\mathcal{M}$ be a type II_1 von Neumann algebra equipped with a faithful finite normal trace τ and let $\delta : \mathcal{M} \to S(\mathcal{M})$ be a derivation. If $a, b \in \mathcal{M}$ and $u \in U(\mathcal{M})$ are such that $a - b$ commutes with u, then for any $\lambda \ge 1$ we have the estimate*

$$\|\lambda^{-1}[\delta(a), u]\|_{S(\mathcal{M},\tau)} \le 2\|\lambda^{-1}\delta(b)\|_{S(\mathcal{M},\tau)} + \| \|a - b\|_{\mathcal{M}} \delta(u)\|_{S(\mathcal{M},\tau)}$$
$$+ \| \|a - b\|_{\mathcal{M}} \delta(u^*)\|_{S(\mathcal{M},\tau)}.$$

Proof. Since $a - b$ commutes with u, the Leibniz rule implies that

$$\begin{aligned} u\delta(a-b)u^* &= \delta(u(a-b)u^*) - \delta(u)\cdot(a-b)u^* - u(a-b)\cdot\delta(u^*) \\ &= \delta(a-b) - \delta(u)\cdot(a-b)u^* - u(a-b)\cdot\delta(u^*). \end{aligned}$$

By linearity, the latter equality can be rewritten as

$$\delta(a) - u\delta(a)u^* = (\delta(b) - u\delta(b)u^*) + \delta(u)\cdot(a-b)u^* + u(a-b)\cdot\delta(u^*).$$

Using the triangle inequality and the fact that $\lambda \ge 1$, we infer

$$\begin{aligned} &\|\lambda^{-1}\cdot(\delta(a) - u\delta(a)u^*)\|_{S(\mathcal{M},\tau)} \\ &\quad \le \|\lambda^{-1}\delta(b)\|_{S(\mathcal{M},\tau)} + \|\lambda^{-1}u\delta(b)u^*\|_{S(\mathcal{M},\tau)} \\ &\qquad + \|\lambda^{-1}\cdot\delta(u)\cdot(a-b)u^*\|_{S(\mathcal{M},\tau)} + \|\lambda^{-1}\cdot u(a-b)\cdot\delta(u^*)\|_{S(\mathcal{M},\tau)} \\ &\quad \le \|\lambda^{-1}\delta(b)\|_{S(\mathcal{M},\tau)} + \|\lambda^{-1}u\delta(b)u^*\|_{S(\mathcal{M},\tau)} \\ &\qquad + \|\delta(u)\cdot(a-b)u^*\|_{S(\mathcal{M},\tau)} + \|u(a-b)\cdot\delta(u^*)\|_{S(\mathcal{M},\tau)}. \end{aligned}$$

Since u is unitary, Proposition 4.2.7(iii) implies that

$$\begin{aligned}\|\lambda^{-1}[\delta(a),u]\|_{S(\mathcal{M},\tau)} &= \|\lambda^{-1}(\delta(a)-u\delta(a)u^*)\cdot u\|_{S(\mathcal{M},\tau)}\\ &\le \|\lambda^{-1}(\delta(a)-u\delta(a)u^*)\|_{S(\mathcal{M},\tau)}\\ &\le 2\|\lambda^{-1}\delta(b)\|_{S(\mathcal{M},\tau)} + \|\|a-b\|_{\mathcal{M}}\delta(u)\|_{S(\mathcal{M},\tau)}\\ &\quad + \|\|a-b\|_{\mathcal{M}}\delta(u^*)\|_{S(\mathcal{M},\tau)},\end{aligned}$$

as required. □

Lemma 7.1.5. *Let $\mathcal{M}$ be a type II_1 von Neumann algebra equipped with a faithful finite normal trace τ and let $\delta : \mathcal{M} \to S(\mathcal{M})$ be a derivation. If $y \in \mathcal{M}$ and $u, v \in U(\mathcal{M})$, then for any $\lambda \ge 1$ we have the estimate*

$$\begin{aligned}&\|\lambda^{-1}[\delta(vy),u] - \lambda^{-1}[v,u]\|_{S(\mathcal{M},\tau)}\\ &\quad\le 2\|\lambda^{-1}\|y\|_{\mathcal{M}}\delta(v)\|_{S(\mathcal{M},\tau)} + 2\|\delta(y)-\mathbf{1}\|_{S(\mathcal{M},\tau)}.\end{aligned}$$

Proof. Using the Leibniz rule, we write

$$[\delta(vy),u] = [\delta(v)y,u] + [v\delta(y),u] = [\delta(v)y,u] + [v(\delta(y)-\mathbf{1}),u] + [v,u].$$

Therefore, we have

$$\lambda^{-1}[\delta(vy),u] - \lambda^{-1}[v,u] = \lambda^{-1}[\delta(v)y,u] + \lambda^{-1}[v(\delta(y)-\mathbf{1}),u].$$

Since $\lambda \ge 1$, it follows that

$$\begin{aligned}&\|\lambda^{-1}[\delta(vy),u] - \lambda^{-1}[v,u]\|_{S(\mathcal{M},\tau)}\\ &\quad\le \|\lambda^{-1}[\delta(v)y,u]\|_{S(\mathcal{M},\tau)} + \|\lambda^{-1}[v(\delta(y)-\mathbf{1}),u]\|_{S(\mathcal{M},\tau)}\\ &\quad\le \|\lambda^{-1}[\delta(v)y,u]\|_{S(\mathcal{M},\tau)} + \|[v(\delta(y)-\mathbf{1}),u]\|_{S(\mathcal{M},\tau)}\\ &\quad\le \|\lambda^{-1}\delta(v)yu\|_{S(\mathcal{M},\tau)} + \|\lambda^{-1}u\delta(v)y\|_{S(\mathcal{M},\tau)}\\ &\qquad + \|v(\delta(y)-\mathbf{1})u\|_{S(\mathcal{M},\tau)} + \|uv(\delta(y)-\mathbf{1})\|_{S(\mathcal{M},\tau)}.\end{aligned}$$

Since u and v are unitaries, Proposition 4.2.7(iii) implies that

$$\begin{aligned}&\|\lambda^{-1}[\delta(vy),u] - \lambda^{-1}[v,u]\|_{S(\mathcal{M},\tau)}\\ &\quad\le 2\|\lambda^{-1}\|y\|_{\mathcal{M}}\delta(v)\|_{S(\mathcal{M},\tau)} + 2\|\delta(y)-\mathbf{1}\|_{S(\mathcal{M},\tau)},\end{aligned}$$

which completes the proof. □

We now prove the automatic continuity of derivations on the algebra $S(\mathcal{M})$ for a type II_1 von Neumann algebra $\mathcal{M}$ under the additional assumption that $\mathcal{M}$ is equipped with a faithful normal finite trace τ. For an $n \times n$ matrix $A = (a_{ij})_{i,j=1}^n$ and $m \times m$-matrix

$B = (b_{kl})_{k,l=1}^{m}$, we denote by $A \otimes B$ the tensor product of A, and B, which is an $nm \times nm$ matrix with $[k + (i-1)m, l + (j-1)m]$-th element equal to $a_{ij}b_{kl}$ (see, e. g., [79]).

Proposition 7.1.6. *Let $\mathcal{M}$ be a type II_1 von Neumann algebra equipped with a faithful finite normal trace τ. Any derivation $\delta : S(\mathcal{M}) \to S(\mathcal{M})$ is necessarily continuous in measure.*

Proof. Let $\delta : S(\mathcal{M}) \to S(\mathcal{M})$ be a derivation. By Proposition 5.1.4, we can assume that δ is self-adjoint. By Proposition 5.1.15, it is sufficient to show that $\delta : \mathcal{M} \to S(\mathcal{M})$ is continuous as a mapping from $(\mathcal{M}, \|\cdot\|_{\mathcal{M}})$ to $(S(\mathcal{M}), t_\tau)$.

Suppose the contrary and let $0 \neq z \in P(\mathcal{Z}(\mathcal{M}))$ be as in Lemma 7.1.2. Without loss of generality, we can assume that $z = \mathbf{1}$. Let $\{e_{i,j}^{(n)} : i,j = 1, \dots, 2^n,\ n \in \mathbb{N}\}$ be the system of partial isometries from $\mathcal{M}$ constructed in Lemma 7.1.1.

Denote by $M_{2^n \times 2^n}(\mathbb{C})$ the $*$-algebra of $2^n \times 2^n$-matrices with complex entries. For every $n \in \mathbb{N}$, we define a mapping $F_n : M_{2^n \times 2^n}(\mathbb{C}) \to \mathcal{M}$, by setting

$$F_n\big((a_{i,j})_{i,j=1}^{2^n}\big) = \sum_{i,j=1}^{2^n} a_{i,j} e_{i,j}^{(n)}, \quad (a_{i,j})_{i,j=1}^{2^n} \in M_{2^n \times 2^n}(\mathbb{C}).$$

It is clear that F_n is an injective unital $*$-homomorphism of the algebra $M_{2^n \times 2^n}(\mathbb{C})$ to $\mathcal{M}$.

We set $\mathcal{R}_n = F_n(M_{2^n \times 2^n}(\mathbb{C}))$ for any $n \in \mathbb{N}$. By the construction of $\{e_{i,j}^{(n)} : i,j = 1, \dots, 2^n,\ n \in \mathbb{N}\}$ (see Lemma 7.1.1(ii)) for any $A = (a_{i,j})_{i,j=1}^{2^n} \in M_{2^n \times 2^n}(\mathbb{C})$, we have that

$$F_n(A) = \sum_{i,j=1}^{2^n} \sum_{s,t=1}^{2} a_{i,j} \delta_{s,t} e_{2(i-1)+s, 2(j-1)+t}^{(n+1)} = F_{n+1}(A \otimes I_2),$$

where I_n is $n \times n$ unit matrix.

Let $A, B \in M_{2\times 2}(\mathbb{C})$, and $n < m$. Then

$$\begin{aligned} F_n(I_{2^{n-1}} \otimes A) F_m(I_{2^{m-1}} \otimes B) &= F_m(I_{2^{n-1}} \otimes A \otimes I_{2^{m-n}}) F_m(I_{2^{m-1}} \otimes B) \\ &= F_m(I_{2^{n-1}} \otimes A \otimes I_{2^{m-n-1}} \otimes B) \\ &= F_m(I_{2^{m-1}} \otimes B) F_m(I_{2^{n-1}} \otimes A \otimes I_{2^{m-n}}) \\ &= F_m(I_{2^{m-1}} \otimes B) F_n(I_{2^{n-1}} \otimes A). \end{aligned} \tag{7.4}$$

Let σ_1 and σ_2 denote the Pauli matrices, i. e.,

$$\sigma_1 = \begin{pmatrix} 0 & 1 \\ 1 & 0 \end{pmatrix}, \quad \sigma_2 = \begin{pmatrix} 0 & i \\ -i & 0 \end{pmatrix}.$$

Note that $\sigma_1\sigma_2 = -\sigma_2\sigma_1$.

For $n \in \mathbb{N}$, we define

$$u_n = F_n(I_{2^{n-1}} \otimes \sigma_1), \quad v_n = F_n(I_{2^{n-1}} \otimes \sigma_2). \tag{7.5}$$

It is clear that for every $n \in \mathbb{N}$, u_n, and v_n are unitaries and, furthermore, by (7.4) we have that

$$u_n v_n = -v_n u_n, \quad u_n v_k = v_k u_n, \quad k \neq n. \tag{7.6}$$

We aim to obtain a contradiction by showing that

$$\|2 \cdot \mathbf{1}\|_{S(\mathcal{M},\tau)} = \|2 v_n u_n\|_{S(\mathcal{M},\tau)} < \|2 \cdot \mathbf{1}\|_{S(\mathcal{M},\tau)}, \quad n \in \mathbb{N}.$$

To this end, we construct a sequence $x_n \in \mathcal{M}$, such that

$$\|3^{-n}[\delta(x_n), u_n] - 2v_n u_n\|_{S(\mathcal{M},\tau)} < \|2 \cdot \mathbf{1}\|_{S(\mathcal{M},\tau)} \quad \text{and} \quad \|3^{-n}\delta[x_n, u_n]\|_{S(\mathcal{M},\tau)} \to 0$$

as $n \to \infty$.

Since $\|\lambda y\|_{S(\mathcal{M},\tau)} \to 0$ as $\lambda \to 0, y \in S(\mathcal{M},\tau)$, for every $n \in \mathbb{N}$, we can find λ_n, such that

$$\|\lambda_n \delta(u_n)\|_{S(\mathcal{M},\tau)} \le 2^{-n-2}, \quad \|\lambda_n \delta(u_n^*)\|_{S(\mathcal{M},\tau)} \le 2^{-n-2}$$

and $\|\lambda_n \delta(v_n)\|_{S(\mathcal{M},\tau)} \le 2^{-n-2}$. Setting $\varepsilon_1 = \min\{\frac{1}{4} \cdot \|2 \cdot \mathbf{1}\|_{S(\mathcal{M},\tau)}, \lambda_1\}$ and defining the sequence $\{\varepsilon_n\}_{n \in \mathbb{N}}$ inductively as

$$\varepsilon_n = \min\left\{\lambda_n, \frac{1}{2}\varepsilon_{n-1}\right\}, \quad \|\varepsilon_n \delta(v_n)\|_{S(\mathcal{M},\tau)} \le \frac{\|2 \cdot \mathbf{1}\|_{S(\mathcal{M},\tau)}}{2^{n+1}}, \quad n \ge 2,$$

we have that

$$\sum_{k=1}^{\infty} \varepsilon_k \le \frac{1}{2} \cdot \|2 \cdot \mathbf{1}\|_{S(\mathcal{M},\tau)}, \quad \sum_{k=n+1}^{\infty} \varepsilon_k \le \varepsilon_n, \quad n \in \mathbb{N}, \tag{7.7}$$

$$\sum_{k=1}^{\infty} \|\varepsilon_k \delta(v_k)\|_{S(\mathcal{M},\tau)} \le \frac{1}{2} \cdot \|2 \cdot \mathbf{1}\|_{S(\mathcal{M},\tau)}, \tag{7.8}$$

and

$$\|\varepsilon_n \delta(u_n)\|_{S(\mathcal{M},\tau)} + \|\varepsilon_n \delta(u_n^*)\|_{S(\mathcal{M},\tau)} \to 0, \quad n \to \infty. \tag{7.9}$$

By Lemma 7.1.3 for every $k \in \mathbb{N}$, there exists $y_k \in \mathcal{R}_k' \cap \mathcal{M}$, such that

$$\|y_k\|_{\mathcal{M}} < \frac{\varepsilon_k}{3^k}, \quad \|\delta(y_k) - \mathbf{1}\|_{S(\mathcal{M},\tau)} < \frac{\varepsilon_k}{3^k}. \tag{7.10}$$

Note that if $k > n$, then $\mathcal{R}_n \subset \mathcal{R}_k$ so that $\mathcal{R}_k' \subset \mathcal{R}_n'$. Therefore, $y_k \in \mathcal{R}_n'$ for all $n < k$.

We define

$$x = \sum_{k=1}^{\infty} 3^k v_k y_k, \quad x_n = \sum_{k=1}^{n} 3^k v_k y_k,$$

where the series converges in the uniform norm by (7.10) and (7.7).

Since $x_n - x = \sum_{k=n+1}^{\infty} 3^k v_k y_k$, $y_k \in \mathcal{R}'_n$ for all $k > n$, and $v_k u_n = u_n v_k$ for $k \neq n$ (see (7.6)), it follows that for every $n \in \mathbb{N}$ the operator $x - x_n$ commutes with u_n. Therefore, using Lemma 7.1.4 (with $a = x_n$, $b = x$, $u = u_n$ and $\lambda = 3^n$), we obtain the estimate

$$\begin{aligned}
&\|3^{-n}[\delta(x_n), u_n]\|_{S(\mathcal{M},\tau)} \\
&\quad \leq 2\|3^{-n}\delta(x)\|_{S(\mathcal{M},\tau)} + \big\|\|x - x_n\|_{\mathcal{M}}\delta(u_n)\big\|_{S(\mathcal{M},\tau)} \\
&\qquad + \big\|\|x - x_n\|_{\mathcal{M}}\delta(u_n^*)\big\|_{S(\mathcal{M},\tau)} \\
&\quad \overset{(7.10)}{\leq} 2\|3^{-n}\delta(x)\|_{S(\mathcal{M},\tau)} + \left\|\left(\sum_{k=n+1}^{\infty} \varepsilon_k\right)\delta(u_n)\right\|_{S(\mathcal{M},\tau)} \\
&\qquad + \left\|\left(\sum_{k=n+1}^{\infty} \varepsilon_k\right)\delta(u_n^*)\right\|_{S(\mathcal{M},\tau)} \\
&\quad \overset{(7.7)}{\leq} 2\|3^{-n}\delta(x)\|_{S(\mathcal{M},\tau)} + \|\varepsilon_n\delta(u_n)\|_{S(\mathcal{M},\tau)} \\
&\qquad + \|\varepsilon_n\delta(u_n^*)\|_{S(\mathcal{M},\tau)}.
\end{aligned} \tag{7.11}$$

We now estimate $\|3^{-n}[\delta(x_n), u_n] - 2v_n u_n\|_{S(\mathcal{M},\tau)}$, $n \in \mathbb{N}$. Note that using the commutation relation (7.6) we can write

$$\sum_{k=1}^{n} 3^{k-n}[v_k, u_n] = [v_n, u_n] = 2v_n u_n.$$

Therefore, using the triangle inequality, we can estimate

$$\begin{aligned}
&\|3^{-n}[\delta(x_n), u_n] - 2v_n u_n\|_{S(\mathcal{M},\tau)} \\
&\quad = \left\|\sum_{k=1}^{n} 3^{k-n}[\delta(v_k y_n), u_n] - \sum_{k=1}^{n} 3^{k-n}[v_k, u_n]\right\|_{S(\mathcal{M},\tau)} \\
&\quad \leq \sum_{k=1}^{n} \|3^{k-n}[\delta(v_k y_k), u_n] - 3^{k-n}[v_k, u_n]\|_{S(\mathcal{M},\tau)}.
\end{aligned}$$

By Lemma 7.1.5 (with $y = y_k$, $v = v_k$, $u = u_n$ and $\lambda = 3^{n-k}$), we have

$$\begin{aligned}
&\|3^{k-n}[\delta(v_k y_k), u_n] - 3^{k-n}[v_k, u_n]\|_{S(\mathcal{M},\tau)} \\
&\quad \leq 2\|3^{k-n}\|y_k\|_{\mathcal{M}}\delta(v_k)\|_{S(\mathcal{M},\tau)} + 2\|\delta(y_k) - \mathbf{1}\|_{S(\mathcal{M},\tau)}.
\end{aligned}$$

Since

$$3^{k-n}\|y_k\|_{\mathcal{M}} \leq 3^{-n}\varepsilon_k \leq \varepsilon_k,$$

we conclude that

$$\begin{aligned}&\|3^{-n}[\delta(x_n),u_n]-2v_nu_n\|_{S(\mathcal{M},\tau)}\\&\quad\le 2\sum_{k=1}^{n}\big(\|\varepsilon_k\delta(v_k)\|_{S(\mathcal{M},\tau)}+\|\delta(y_k)-\mathbf{1}\|_{S(\mathcal{M},\tau)}\big).\end{aligned}\tag{7.12}$$

Combining (7.11) and (7.12) and using the triangle inequality, we infer

$$\begin{aligned}\|2\cdot\mathbf{1}\|_{S(\mathcal{M},\tau)}&=\|\,|2v_nu_n|\,\|_{S(\mathcal{M},\tau)}=\|2v_nu_n\|_{S(\mathcal{M},\tau)}\\&\le\|3^{-n}[\delta(x_n),u_n]-2v_nu_n\|_{S(\mathcal{M},\tau)}+\|3^{-n}[\delta(x_n),u_n]\|_{S(\mathcal{M},\tau)}\\&\le 2\|3^{-n}\delta(x)\|_{S(\mathcal{M},\tau)}+\|\varepsilon_n\delta(u_n)\|_{S(\mathcal{M},\tau)}+\|\varepsilon_n\delta(u_n^*)\|_{S(\mathcal{M},\tau)}\\&\quad+2\sum_{k=1}^{n}\big(\|\varepsilon_k\delta(v_k)\|_{S(\mathcal{M},\tau)}+\|\delta(y_k)-\mathbf{1}\|_{S(\mathcal{M},\tau)}\big).\end{aligned}$$

By (7.9) and properties of the F-norm $\|\cdot\|_{S(\mathcal{M},\tau)}$, the first three summands tend to 0 as $n\to\infty$. Passing $n\to\infty$, and referring to (7.8) and (7.10) we conclude that

$$\|2\cdot\mathbf{1}\|_{S(\mathcal{M},\tau)}\le 2\sum_{k=1}^{\infty}\big(\|\varepsilon_k\delta(v_k)\|_{S(\mathcal{M},\tau)}+\|\delta(y_k)-\mathbf{1}\|_{S(\mathcal{M},\tau)}\big)<\|2\cdot\mathbf{1}\|_{S(\mathcal{M},\tau)}.$$

The obtained contradiction concludes the argument. □

We can now present the main result of this section. We emphasize that the following results hold for an arbitrary type II_1 algebra without the additional assumption of existence of a finite trace on $\mathcal{M}$.

Theorem 7.1.7. *Let $\mathcal{M}$ be a type II_1 von Neumann algebra. Then any derivation on $LS(\mathcal{M})=S(\mathcal{M})$ is continuous locally in measure.*

Proof. Let $\mathcal{M}$ be an arbitrary type II_1 von Neumann algebra with a faithful normal semifinite trace τ. Let $\{z_n\}_{n\in\mathbb{N}}$ be the central partition of unity such that $\tau(z_n)<+\infty$ for every $n\in\mathbb{N}$ (see Corollary 1.10.16). By Corollary 3.4.7, the algebra $S(\mathcal{M})$ is $*$-isomorphic to the algebra $\prod_{n\in\mathbb{N}}S(z_n\mathcal{M})$.

Let now δ be a derivation on $S(\mathcal{M})$. By Proposition 5.1.10, $\delta|_{z_nS(\mathcal{M})}$ is a derivation on $z_nS(\mathcal{M})$. Since $\tau(z_n)<+\infty$, it follows from Proposition 7.1.6 that $\delta|_{z_nS(\mathcal{M})}$ is $t_{\tau|_{z_n\mathcal{M}}}$-continuous. In particular, $\delta|_{z_nS(\mathcal{M})}$ is $t(z_n\mathcal{M})$-continuous. Referring to Proposition 5.1.11, we conclude that δ is $t(\mathcal{M})$-continuous. □

In conclusion of this section, we show that for a type II_1 von Neumann algebra $\mathcal{M}$ equipped with a faithful normal semifinite trace τ any derivation on $S_0(\mathcal{M},\tau)$ is also continuous in measure.

Theorem 7.1.8. *Let $\mathcal{M}$ be a type II_1 von Neumann algebra equipped with a faithful normal semifinite trace τ. Then any derivation on $S_0(\mathcal{M},\tau)$ is continuous in measure.*

Proof. If τ is finite, then $S_0(\mathcal{M},\tau)=S(\mathcal{M},\tau)=S(\mathcal{M})$, and so, by Proposition 7.1.6, δ is continuous in measure.

Suppose now that τ is infinite and assume that δ is t_τ-discontinuous. Since $(S_0(\mathcal{M},\tau), t_\tau)$ is an F-space (see Proposition 4.2.14), the closed graph theorem implies that there exists a sequence $\{a_n\}_{n\in\mathbb{N}} \subset S_0(\mathcal{M},\tau)$, such that $a_n \to 0$ and $\delta(a_n) \to b \neq 0$ in measure.

By Corollary 1.10.16, there exists a central partition of unity $\{z_k\}_{k\in\mathbb{N}}$ such that $\tau(z_k) < +\infty$ for every $k \in \mathbb{N}$. Since $b \neq 0$, there exists $z \in \{z_k\}_{k\in\mathbb{N}}$, such that $bz \neq 0$. It is clear that $zS_0(\mathcal{M},\tau) = S_0(z\mathcal{M},\tau|_{z\mathcal{M}})$ and by Proposition 5.1.9 we have that $\delta(S_0(z\mathcal{M},\tau|_{z\mathcal{M}})) \subset S_0(z\mathcal{M},\tau|_{z\mathcal{M}})$. Since $\tau|_{z\mathcal{M}}$ is finite, by the proven above $\delta|_{S_0(z\mathcal{M},\tau|_{z\mathcal{M}})}$ is t_τ continuous. Therefore, since $za_n \to 0$ in measure, on one hand we have that $z\delta(a_n) = \delta(za_n) \to 0$ in measure, and on the other hand $z\delta(a_n) \to zb$ in measure. This implies that $zb = 0$, which is a contradiction. Thus, δ is t_τ-continuous. □

7.2 Some properties of discontinuous derivations

In the proof of automatic continuity of derivations on the algebras $LS(\mathcal{M})$, $S_0(\mathcal{M})$, $S(\mathcal{M},\tau)$, and $S_0(\mathcal{M},\tau)$ for a properly infinite von Neumann algebra $\mathcal{M}$, we also argue by contradiction. The purpose of this section is to collect some immediate properties of $t(\mathcal{M})$-discontinuous derivations on $LS(\mathcal{M})$ for an arbitrary properly infinite von Neumann algebra $\mathcal{M}$ as well as t_τ-discontinuous derivations on $S(\mathcal{M},\tau)$. The key idea here is the notion of a so-called paving projection, which is the technical backbone of our approach in the proof of automatic continuity of derivations in Section 7.3, Section 7.4, and Section 7.5.

Definition 7.2.1. A projection $p \in P(\mathcal{M})$ is called *paving* (for $\mathcal{M}$) if there exists an isometry $v \in \mathcal{M}$ such that:

(i) the projections $p_n := v^n p (v^n)^*$, $n \in \mathbb{N}_0$, are pairwise orthogonal;

(ii) $\sum_{n=0}^{\infty} p_n = \sup_{n\in\mathbb{N}_0} p_n = \mathbf{1}$.

As we show in the following proposition, a properly infinite von Neumann algebra always contains a paving projection.

Proposition 7.2.2. *Let $\mathcal{M}$ be a properly infinite von Neumann algebra. If $r \in P(\mathcal{M})$ is such that the central support of r equals $\mathbf{1}$, then there exists a paving projection $p \in P(\mathcal{M})$, such that $r \sim p$.*

Proof. By Theorem 1.9.9(vi), there exists a central projection $z \in P(\mathcal{Z}(\mathcal{M}))$, such that rz is finite and $r(\mathbf{1}-z)$ is properly infinite. Since equivalence of projections is preserved by multiplication by central projections (see Theorem 1.9.4(iii)), it suffices to prove the assertion separately for the projection rz in the algebra $z\mathcal{M}$ and for the projection $r(\mathbf{1}-z)$ in the algebra $(\mathbf{1}-z)\mathcal{M}$. Therefore, we may assume without loss of generality that either $z = 0$ or $z = \mathbf{1}$.

Suppose first that $z = 0$. In this case, r is a properly infinite projection whose central support equals $\mathbf{1}$. Since we assume that $\mathcal{M}$ acts on a separable Hilbert space, Theorem 1.9.9(viii) implies that $r \sim \mathbf{1}$. By Theorem 1.9.9(v), there exists a sequence $\{p_n\}_{n\in\mathbb{N}_0}$

of pairwise orthogonal projections, such that $p_n \sim \mathbf{1}$ for all $n \in \mathbb{N}_0$ and $\sup_{n\in\mathbb{N}_0} p_n = \mathbf{1}$. Since the projections p_n, $n \in \mathbb{N}_0$, are equivalent, for every $n \in \mathbb{N}_0$ there exists a partial isometry $v_n \in \mathcal{M}$, such that $v_n^* v_n = p_n$ and $v_n v_n^* = p_{n+1}$.

Define an element $v \in \mathcal{M}$ by setting

$$v = \sum_{n=0}^{\infty} v_n,$$

where the series converges in the strong operator topology. We have that

$$v^* v = \sup_{n\in\mathbb{N}_0} p_n = \mathbf{1},$$

and

$$vv^* = \sup_{n\in\mathbb{N}_0} p_{n+1} = \mathbf{1} - p_0.$$

In particular, v is an isometry.

Note that

$$vp_n v^* = \sum_{k,l} v_k p_n v_l^* = v_n p_n v_n^* = v_n v_n^* = p_{n+1}, \quad n \in \mathbb{N}_0.$$

Hence,

$$v^n p_0 (v^n)^* = p_n, \quad n \in \mathbb{N}_0.$$

In other words, we have that $\{v^n p_0 (v^n)^*\}_{n\in\mathbb{N}_0}$ is a sequence of pairwise orthogonal projections with $\sup_{n\in\mathbb{N}_0} v^n p_0 (v^n)^* = \mathbf{1}$. Thus, the projection p_0 is paving. Since, in addition, $r \sim \mathbf{1}$ and $p_0 \sim \mathbf{1}$, we conclude that $r \sim p_0$, as required.

Suppose now that $z = \mathbf{1}$. In this case, the projection r is finite in a properly infinite von Neumann algebra $\mathcal{M}$. Since we assume that $\mathcal{M}$ acts on a separable Hilbert space, the projection $\mathbf{1}$ is a countably decomposable projection. Hence, by [93, Proposition 6.3.12], there exists a sequence $\{p_n\}_{n\in\mathbb{N}_0}$ of pairwise orthogonal projections, such that $p_n \sim r$ and $\sup_{n\in\mathbb{N}_0} p_n = \mathbf{1}$. Repeating the first part of the proof, we conclude that p_0 is a paving projection equivalent to r. □

The following proposition is crucial in our approach. It shows that for a self-adjoint $t(\mathcal{M})$-discontinuous derivation on $LS(\mathcal{M})$ for a properly infinite algebra $\mathcal{M}$ we can always find a sequence of self-adjoint operators $a_n \in LS(\mathcal{M})$ such that $a_n \to 0$ and $\delta(a_n) \to p$ locally in measure, where p is a paving projection in the algebra $\mathcal{M}$. A similar result holds for derivations on the algebras $S_0(\mathcal{M})$, $S(\mathcal{M},\tau)$, and $S_0(\mathcal{M},\tau)$. For this reason, we formulate the following proposition for a general setting.

Proposition 7.2.3. *Let $\mathcal{M}$ be a properly infinite von Neumann algebra and let t be the local measure topology $t(\mathcal{M})$ or the measure topology t_τ (in which case $\mathcal{M}$ is assumed to be equipped with a faithful normal semifinite trace τ). Assume that $\mathcal{A}$ is an absolutely solid $*$-subalgebra of $LS(\mathcal{M})$, which is closed in the topology t and δ is a self-adjoint derivation on $\mathcal{A}$, which is discontinuous in the topology t. Then:*

(i) *there exists a nonzero central projection $z \in P(\mathcal{Z}(\mathcal{M}))$ and a paving projection p in the algebra $z\mathcal{M}$ such that $p \in \mathcal{A}$;*

(ii) *there exists a sequence $\{a_n\}_{n\in\mathbb{N}_0} \in z\mathcal{A}$, such that $a_n = a_n^*$ and $a_n p = a_n$ for all $n \in \mathbb{N}_0$;*

(iii) *$a_n \to 0$ and $\delta(a_n) \to p$ as $n \to \infty$ with respect to the topology t.*

Proof. For convenience, we let $\|\cdot\|_{\mathcal{A}}$ denote either $\|\cdot\|_{LS(\mathcal{M})}$ (if $\mathcal{A}$ is closed in the local measure topology) or $\|\cdot\|_{S(\mathcal{M},\tau)}$ (if $\mathcal{A}$ is closed with respect to the measure topology). Since both $(LS(\mathcal{M}), \|\cdot\|_{LS(\mathcal{M})})$ and $(S(\mathcal{M},\tau), \|\cdot\|_{S(\mathcal{M},\tau)})$ are F-spaces (see Theorem 4.2.8 and Theorem 4.3.14) and $\mathcal{A}$ is closed with respect to $\|\cdot\|_{\mathcal{A}}$, it follows that $(\mathcal{A}, \|\cdot\|_{\mathcal{A}})$ is also an F-space.

Since δ is a $\|\cdot\|_{\mathcal{A}}$-discontinuous derivation on $\mathcal{A}$, the closed graph theorem (see, e. g., [129, Chapter 2, Theorem 2.15]) implies that there exists a sequence $\{a_n\}_{n\in\mathbb{N}_0} \subset \mathcal{A}$ and an element $0 \neq a \in \mathcal{A}$, such that $\|a_n\|_{\mathcal{A}} \to 0$ and $\|\delta(a_n) - a\|_{\mathcal{A}} \to 0$ as $n \to \infty$. Since $(\mathcal{A}, \|\cdot\|_{\mathcal{A}})$ is a topological $*$-algebra and δ is self-adjoint, we can assume that $a_n, n \in \mathbb{N}_0$, and a are self-adjoint. Without loss of generality, $a_+ \neq 0$ (otherwise, consider $-a_n$ instead of a_n).

Since $a_+ \neq 0$, there exists $\varepsilon > 0$, such that the projection $r = E_a(\varepsilon, \infty)$ is nonzero. By the functional calculus, the operator $\max\{a, \varepsilon\mathbf{1}\}^{-\frac{1}{2}}$ is bounded and, therefore, the operator

$$b = \max\{a, \varepsilon\mathbf{1}\}^{-\frac{1}{2}} r$$

is bounded, too, and, moreover,

$$bab = r. \tag{7.13}$$

Since $\mathcal{A}$ is an absolutely solid $*$-subalgebra, Proposition 3.3.13 implies that $r \in \mathcal{A}$.

Since $\|a_n\|_{\mathcal{A}} \to 0$ as $n \to \infty$, it follows that $\|ba_n b\|_{\mathcal{A}} \to 0$ as $n \to \infty$ (see Proposition 4.2.7(v) for the case $\|\cdot\|_{\mathcal{A}} = \|\cdot\|_{S(\mathcal{M},\tau)}$ and Proposition 4.3.13(vi) for the case $\|\cdot\|_{\mathcal{A}} = \|\cdot\|_{LS(\mathcal{M})}$). In addition, by the Leibniz rule, we have

$$\begin{aligned}\delta(ba_n b) &= \delta(b)\cdot a_n \cdot b + b\cdot\delta(a_n)\cdot b + b\cdot a_n\cdot\delta(b)\\ &\to \delta(b)\cdot 0\cdot b + b\cdot a\cdot b + b\cdot 0\cdot\delta(b)\\ &\overset{(7.13)}{=} r\end{aligned}$$

with respect to $\|\cdot\|_{\mathcal{A}}$. Therefore, replacing the sequence $\{a_n\}_{n\in\mathbb{N}_0}$ by the sequence $\{ba_nb\}_{n\in\mathbb{N}_0}$, we may assume, without loss of generality, that $\|a_n\|_{\mathcal{A}} \to 0$ and $\|\delta(a_n) - r\|_{\mathcal{A}} \to 0$ as $n \to \infty$.

Now, let z be the central support of r. By Proposition 7.2.2, there exists a paving projection $p \in z\mathcal{M}$, such that $p \sim r$. Let w be a partial isometry, such that $w^*w = p$, $ww^* = r$. Then $w^*a_nw \in \mathcal{A}$, $\|w^*a_nw\|_{\mathcal{A}} \to 0$ as $n \to \infty$ and

$$\begin{aligned}\delta(w^*a_nw) &= \delta(w^*)\cdot a_n\cdot w + w^*\cdot\delta(a_n)\cdot w + w^*\cdot a_n\cdot\delta(w)\\ &\to \delta(w^*)\cdot 0\cdot w + w^*\cdot r\cdot w + w^*\cdot 0\cdot\delta(w) = w^*rw = p\end{aligned}$$

with respect to the F-norm $\|\cdot\|_{\mathcal{A}}$. Therefore, without loss of generality, we can assume that $r = p$.

Thus, there exists a sequence $\{a_n\}_{n\in\mathbb{N}_0} \subset \mathcal{A}$ of self-adjoint operators, a non-zero central projection $z \in \mathcal{M}$ and a paving projection $p \in z\mathcal{M}$, such that $p \in \mathcal{A}$, $\|a_n\|_{\mathcal{A}} \to 0$, and $\|\delta(a_n) - p\|_{\mathcal{A}} \to 0$ as $n \to \infty$. Replacing a_n with pa_np, we finally obtain the required sequence $\{a_n\}_{n\in\mathbb{N}_0} \subset \mathcal{A}$ of self-adjoint elements and a paving projection $p \in z\mathcal{M} \cap \mathcal{A}$, such that $a_np = a_n$, $\|a_n\|_{\mathcal{A}} \to 0$ and $\|\delta(a_n) - p\|_{\mathcal{A}} \to 0$ as $n \to \infty$, as required. □

Since $S(\mathcal{M},\tau)$ is an absolutely solid $*$-subalgebra of $LS(\mathcal{M})$ and an F-space (see Corollary 3.3.15 and Theorem 4.2.8), for a t_τ-discontinuous derivation δ on $S(\mathcal{M},\tau)$, Proposition 7.2.3 provides the first building block for the proof of automatic continuity of derivations on $S(\mathcal{M},\tau)$ for a properly infinite semifinite von Neumann algebra $\mathcal{M}$.

The pair $(LS(\mathcal{M}), \|\cdot\|_{LS(\mathcal{M})})$ is also an F-space, and so Proposition 7.2.3 is applicable to $t(\mathcal{M})$-discontinuous derivations on $LS(\mathcal{M})$. However, for a $t(\mathcal{M})$-discontinuous derivation δ on the algebra $LS(\mathcal{M})$, Proposition 7.2.3 gives a sequence $\{a_n\}_{n\in\mathbb{N}_0}$, which converges to zero only in the local measure topology. In the following two results, we use Egorov type theorems (see Section 4.4) to show that the sequence $\{a_n\}_{n\in\mathbb{N}_0}$ can be chosen to converge in a stronger topology.

Corollary 7.2.4. *Let $\mathcal{M}$ be a properly infinite von Neumann algebra with faithful normal semifinite trace τ and let δ be a $t(\mathcal{M})$-discontinuous self-adjoint derivation on the algebra $LS(\mathcal{M})$. Then:*

(i) *there exists a nonzero central projection $z \in P(\mathcal{Z}(\mathcal{M}))$ and a paving projection p in the algebra $z\mathcal{M}$ with the corresponding isometry v satisfying $\delta(v)z \in zS(\mathcal{M},\tau)$;*

(ii) *there exists a sequence $\{a_n\}_{n\in\mathbb{N}_0} \subset zS(\mathcal{M},\tau)$, such that $a_n = a_n^*$, $a_np = a_n$ and $\delta(a_n) \subset zS(\mathcal{M},\tau)$ for all $n \in \mathbb{N}_0$;*

(iii) *$a_n \to 0$ and $\delta(a_n) \to p$ in measure as $n \to \infty$.*

Proof. As mentioned above $(LS(\mathcal{M}), \|\cdot\|_{LS(\mathcal{M})})$ satisfies the assumptions of Proposition 7.2.3. By Proposition 7.2.3 (with $\mathcal{A} = LS(\mathcal{M})$ and $t = t(\mathcal{M})$), there exist a (nonzero) central projection $z_0 \in P(\mathcal{Z}(\mathcal{M}))$, a paving projection $p \in z_0\mathcal{M}$, and a sequence $\{a_n\}_{n\in\mathbb{N}_0} \subset z_0LS(\mathcal{M})$, such that $a_n = a_n^*$, $a_np = a_n$ and $a_n \to 0$ and $\delta(a_n) \to p$ locally in measure as $n \to \infty$.

Passing to a subsequence if required and using Theorem 4.4.7 twice, we find a nonzero central projection $z_1 \le z_0$, such that

$$\{a_n z_1\}_{n\in\mathbb{N}_0}, \quad \{\delta(a_n z_1)\}_{n\in\mathbb{N}_0} \subset S(z_1\mathcal{M},\tau), \quad a_n z_1 \to 0, \quad \delta(a_n z_1) \to p z_1,$$

where the convergences hold in the measure topology. Let v be the isometry corresponding to the paving projection p. By Lemma 4.4.5, there exists a nonzero central projection $z \le z_1$, such that $\delta(v)z \in S(z\mathcal{M},\tau)$.

Thus, we have that $\{a_n z\}_{n\in\mathbb{N}_0}$, with $(a_n z)^* = a_n z$, $a_n zp = a_n z$, and $\{\delta(a_n z)\}_{n\in\mathbb{N}_0} \subset S(z\mathcal{M},\tau)$, $a_n z \to 0$, and $\delta(a_n z) \to pz$ in measure. The observation that the projection pz is paving in the von Neumann algebra $z\mathcal{M}$ concludes the proof. □

Corollary 7.2.5. *Let $\mathcal{M}$ be a type III von Neumann algebra and let δ be a $t(\mathcal{M})$-discontinuous self-adjoint derivation on the algebra $LS(\mathcal{M})$. Then:*

(i) *there exists a nonzero central projection $z \in P(\mathcal{Z}(\mathcal{M}))$ and a paving projection p in the algebra $z\mathcal{M}$ with the corresponding isometry v satisfying $\delta(v)z \in z\mathcal{M}$;*

(ii) *there exists a sequence $\{a_n\}_{n\in\mathbb{N}_0} \subset z\mathcal{M}$ such that $a_n = a_n^*$, $a_n p = a_n$, and $\delta(a_n) \subset z\mathcal{M}$ for all $n \in \mathbb{N}_0$;*

(iii) *$a_n \to 0$ and $\delta(a_n) \to p$ with respect to the uniform norm as $n \to \infty$.*

Proof. By Theorem 4.3.14, the pair $(LS(\mathcal{M}), \|\cdot\|_{LS(\mathcal{M})})$ satisfies the assumptions of Proposition 7.2.3. By Proposition 7.2.3 (with $\mathcal{A} = LS(\mathcal{M})$ and $t = t(\mathcal{M})$), there exists a nonzero central projection $z_0 \in P(\mathcal{Z}(\mathcal{M}))$ and a paving projection p in the algebra $z_0\mathcal{M}$, a sequence $\{a_n\}_{n\in\mathbb{N}_0} \in z_0 LS(\mathcal{M})$, such that $a_n = a_n^*$ and $a_n p = a_n$ and $a_n \to 0$ and $\delta(a_n) \to p$ locally in measure. Passing to a subsequence if needed and using Theorem 4.4.9, we find a nonzero central projection $z_1 \le z_0$, such that $\{a_n z_1\}_{n\in\mathbb{N}_0}, \{\delta(a_n z_1)\}_{n\in\mathbb{N}_0} \subset \mathcal{M}z_1$, $a_n z_1 \to 0$, and $\delta(a_n z_1) \to pz_1$ in the uniform norm. Furthermore, using Lemma 4.4.8, we can choose a nonzero central projection $z \le z_1$, such that $\delta(v)z \in \mathcal{M}z$. We still have that $\{a_n z\}_{n\in\mathbb{N}_0}, \{\delta(a_n z)\}_{n\in\mathbb{N}_0} \subset z\mathcal{M}$, $a_n z \to 0$, and $\delta(a_n z) \to pz_2$ with respect to the uniform norm. Noting that the projection pz is paving in the algebra $z\mathcal{M}$, we conclude the proof. □

7.3 Automatic continuity of derivations in the case of type *III* von Neumann algebras

In this section, we show that in the case when $\mathcal{M}$ is a type *III* von Neumann algebra, every derivation $\delta : LS(\mathcal{M}) \to LS(\mathcal{M})$ is necessarily continuous locally in measure. We prove this by obtaining a contradiction under the assumption that there is a $t(\mathcal{M})$-discontinuous derivation δ on $LS(\mathcal{M})$. The contradiction is achieved by the construction of an element $b \in \mathcal{M}$ and a sequence of pairwise orthogonal equivalent nonzero projections $\{e_k\}_{k\in\mathbb{N}_0} \subset \mathcal{M}$, such that

$$e_k\delta(b)e_k \geq \frac{k}{8}e_k, \quad k \in \mathbb{N}_0.$$

As it is shown in Proposition 3.3.17, this cannot happen for a locally measurable operator $\delta(b)$ and nonzero projections $\{e_k\}_{k\in\mathbb{N}_0} \subset \mathcal{M}$.

We start by constructing an auxiliary sequence $\{b_k\}_{n\in\mathbb{N}_0}$ of bounded operators, such that $\|b_k\|_{\mathcal{M}}$ vanishes significantly fast, but $\delta(b_k) \geq \frac{1}{4}\cdot\mathbf{1}$ for all $k \in \mathbb{N}_0$. For convenience, we start with a technical assumption.

Assumption 7.3.1. *Suppose that $\mathcal{M}$ is a type III von Neumann algebra and let $\delta : LS(\mathcal{M}) \to LS(\mathcal{M})$ be a self-adjoint derivation. Assume also that:*
(i) *there exists a paving projection $p \in P(\mathcal{M})$ with the corresponding isometry $v \in \mathcal{M}$ satisfying $\delta(v) \in \mathcal{M}$;*
(ii) *there exists a sequence $\{a_n\}_{n\in\mathbb{N}_0} \subset \mathcal{M}$, such that $\delta(a_n) \in \mathcal{M}$, $a_n \to 0$, and $\delta(a_n) \to p$ in the uniform norm $\|\cdot\|_{\mathcal{M}}$ as $n \to \infty$;*
(iii) *$a_n = a_n^*$ and $a_np = a_n$ for all $n \in \mathbb{N}_0$.*

Notation 7.3.2. *Suppose Assumption 7.3.1 holds. For every $n \in \mathbb{N}_0$, we set*

$$y_n = \left(\mathbf{1} + \sum_{m=0}^{n} |\delta(v^m)|^2\right)^{\frac{1}{2}} \in \mathcal{M}.$$

Remark 7.3.3.
(i) *We note that for any $k, m \in \mathbb{N}_0$, we have that $|\delta(v^m)|^2 \leq y_{m+k}^2$. Therefore, $y_{m+k}^{-1}|\delta(v^m)|^2 y_{m+k}^{-1} \leq \mathbf{1}$, and so*

$$\left\||\delta(v^m)|y_{m+k}^{-1}\right\|_{\mathcal{M}}^2 = \left\|y_{m+k}^{-1}|\delta(v^m)|^2 y_{m+k}^{-1}\right\|_{\mathcal{M}} \leq 1.$$

Using the polar decomposition for $\delta(v^m)$, we conclude that

$$\|\delta(v^m)y_{m+k}^{-1}\|_{\mathcal{M}} \leq 1, \quad m, k \in \mathbb{N}_0. \tag{7.14}$$

(ii) *Since $a_n \to 0$ and $\delta(a_n) \to p$ with respect to the uniform norm, passing to a subsequence if necessary, we can assume without loss of generality that*

$$\|a_n\|_{\mathcal{M}}, \|\delta(a_n) - p\|_{\mathcal{M}}, \|y_n a_n\|_{\mathcal{M}} \leq 2^{-n}, \quad n \in \mathbb{N}_0. \tag{7.15}$$

In the next proposition, we construct an initial sequence $\{b_k\} \in \mathcal{M}$, gluing which we shall construct the element b mentioned at the beginning of this section.

Lemma 7.3.4. *Suppose Assumption 7.3.1 holds. For every fixed $k \in \mathbb{N}_0$, the series*

$$b_k = \sum_{m=0}^{\infty} v^m a_{m+k} (v^m)^*$$

converges in the uniform norm and

$$\|b_k\|_{\mathcal{M}} \le 2^{-k}, \quad k \in \mathbb{N}_0.$$

Proof. By (7.15), we have that $\|a_{m+k}\|_{\mathcal{M}} \le 2^{-m-k}$. Since v is an isometry, we have that

$$\|v^m a_{m+k}(v^m)^*\|_{\mathcal{M}} \le 2^{-m-k}, \quad m \in \mathbb{N}_0.$$

Thus, the series $b_k = \sum_{m=1}^{\infty} v^m a_{m+k}(v^m)^*$ converges in the uniform norm. Furthermore, we have

$$v^m a_{m+k}(v^m)^* = (v^m p(v^m)^*) \cdot v^m a_{m+k}(v^m)^* \cdot (v^m p(v^m)^*),$$

Since p is a paving projection, the projections $v^m p(v^m)^*, m \in \mathbb{N}_0$, are pairwise orthogonal. Therefore,

$$\|b_k\|_{\mathcal{M}} = \sup_{m\in\mathbb{N}_0} \|v^m a_{m+k}(v^m)^*\|_{\mathcal{M}} \le \sup_{m\in\mathbb{N}_0} 2^{-m-k} = 2^{-k},$$

as required. □

Next, we want to study how the derivation δ is acting on each b_k, constructed above. In order to do this, we first construct an auxiliary sequence $\{d_k\}_{k\in\mathbb{N}_0}$, which as we show in the proof of Proposition 7.3.6 below, is the image of $\{b_k\}_{k\in\mathbb{N}_0}$ under δ.

Lemma 7.3.5. *Suppose Assumption* 7.3.1 *holds. For every $m, k \in \mathbb{N}_0$, the operator $\delta(v^m a_{m+k}(v^m)^*)$ is bounded and for every fixed $k \in \mathbb{N}_0$, the series*

$$d_k := \sum_{m=0}^{\infty} \delta(v^m a_{m+k}(v^m)^*),$$

converges in the strong operator topology. Furthermore,

$$\|d_k - \mathbf{1}\|_{\mathcal{M}} \le 6 \cdot 2^{-k}, \quad k \in \mathbb{N}_0.$$

Proof. By the Leibniz rule, for every $m, k \in \mathbb{N}_0$, we have that

$$\begin{aligned} \delta(v^m a_{m+k}(v^m)^*) &= \delta(v^m)a_{m+k}(v^m)^* + v^m a_{m+k}\delta((v^m)^*) \\ &\quad + v^m \cdot (\delta(a_{m+k}) - p) + v^m p(v^m)^*. \end{aligned} \tag{7.16}$$

Therefore, since v is an isometry, inequalities (7.14) and (7.15) imply that

$$\begin{aligned} \|\delta(v^m a_{m+k}(v^m)^*) - v^m p(v^m)^*\|_{\mathcal{M}} &\le 2\|\delta(v^m)a_{m+k}\|_{\mathcal{M}} + \|\delta(a_{m+k}) - p\|_{\mathcal{M}} \\ &\le 2\|\delta(v^m)y_{m+k}^{-1}\|_{\mathcal{M}}\|y_{m+k}a_{m+k}\|_{\mathcal{M}} + 2^{-m-k} \\ &\le 3 \cdot 2^{-m-k}. \end{aligned} \tag{7.17}$$

In particular, equality (7.16) implies that

$$\|\delta(v^m a_{m+k}(v^m)^*)\|_{\mathcal{M}} \leq 3 \cdot 2^{-m-k} + 1$$

and, therefore, $\delta(v^m a_{m+k}(v^m)^*)$ is bounded for all $m, k \in \mathbb{N}_0$.

Next, by (7.17), the series

$$\sum_{m=0}^{\infty} (\delta(v^m a_{m+k}(v^m)^*) - v^m p(v^m)^*)$$

converges in the uniform norm. Since p is a paving projection, we also have that $\sum_{m=0}^{\infty} v^m p(v^m)^* = \mathbf{1}$ with respect to the strong operator topology. Hence,

$$d_k = \sum_{m=0}^{\infty} \delta(v^m a_{m+k}(v^m)^*),$$

converges in the strong operator topology. Moreover, we have that

$$\begin{aligned} \|d_k - \mathbf{1}\|_{\mathcal{M}} &= \left\| \sum_{m=0}^{\infty} \delta(v^m a_{m+k}(v^m)^*) - \sum_{m=0}^{\infty} v^m p(v^m)^* \right\|_{\mathcal{M}} \\ &\leq \sum_{m=0}^{\infty} \|\delta(v^m a_{m+k}(v^m)^*) - v^m p(v^m)^* \|_{\mathcal{M}} \\ &\overset{(7.17)}{\leq} \sum_{m=0}^{\infty} 3 \cdot 2^{-m-k} = 6 \cdot 2^{-k}, \end{aligned}$$

which concludes the proof. □

With Lemma 7.3.4 and Lemma 7.3.5 at hand, we are ready to prove the existence of the auxiliary sequence $\{b_k\}$ for any $t(\mathcal{M})$-discontinuous derivation on $LS(\mathcal{M})$ mentioned before Assumption 7.3.1.

Proposition 7.3.6. *Let $\mathcal{M}$ be a type III von Neumann algebra and let $\delta : LS(\mathcal{M}) \to LS(\mathcal{M})$ be a $t(\mathcal{M})$-discontinuous self-adjoint derivation. Then there exists a nonzero projection $z \in P(\mathcal{Z}(\mathcal{M}))$, such that in the reduced algebra $z\mathcal{M}$ there exists $\{b_k\}_{k\in\mathbb{N}_0} \subset z\mathcal{M}$ with $\|b_k\|_{\mathcal{M}} \leq 2^{-k}$ and $\delta(b_k) \geq \frac{1}{4} \cdot \mathbf{1}$ for all $k \geq 3$.*

Proof. Since $\mathcal{M}$ is a type *III* von Neumann algebra and δ is a $t(\mathcal{M})$-discontinuous derivation on $LS(\mathcal{M})$, it follows from Corollary 7.2.5 that there exists a nonzero central projection $z \in P(\mathcal{Z}(\mathcal{M}))$, such that the reduced derivation $\delta^{(z)}$ satisfies Assumption 7.3.1 in the algebra $z\mathcal{M}$. Hence, by Lemma 7.3.4, there exists $b_k \in z\mathcal{M}$, such that $\|b_k\|_{\mathcal{M}} \leq 2^{-k}$ for all $k \in \mathbb{N}_0$. It remains to show that $\delta(b_k) = \delta^{(z)}(b_k) \geq \frac{1}{4} \cdot \mathbf{1}$ for all $k \in \mathbb{N}_0$.

We first show that $\delta(b_k) = d_k$, where $d_k \in z\mathcal{M}$ is defined in Lemma 7.3.5. We set

$$r_m = \sup_{k\leq m} p_k,$$

where $p_m = v^m p(v^m)^*$, $m \in \mathbb{N}_0$. Since p is a paving projection, it follows that $r_m \uparrow \mathbf{1}$ as $m \to \infty$, and so by Proposition 3.3.7(iii) it is sufficient to show that

$$r_m \delta(b_k) r_m = r_m d_k r_m, \quad \forall m \in \mathbb{N}_0.$$

Since $r_m = \sum_{k \le m} p_k$, the latter equality would follow from the equality

$$p_{m_1} \delta(b_k) p_{m_2} = p_{m_1} d_k p_{m_2}, \tag{7.18}$$

for all $m_1, m_2 \in \mathbb{N}_0$. Therefore, we now concentrate on the proof of equation (7.18).

Fix arbitrary $m_1, m_2 \in \mathbb{N}_0$. Assume first that $m_1 \neq m_2$. For the left-hand side of (7.18), the Leibniz equality implies that

$$p_{m_1} \delta(b_k) p_{m_2} = \delta(p_{m_1} b_k p_{m_2}) - \delta(p_{m_1}) b_k p_{m_2} - p_{m_1} b_k \delta(p_{m_2}). \tag{7.19}$$

Since $b_k = \sum_{m=0}^{\infty} p_m v^m a_{m+k} (v^m)^* p_m$, it follows that the first summand on the right-hand side of (7.19) vanishes, and we have

$$p_{m_1} \delta(b_k) p_{m_2} = -\delta(p_{m_1}) v^{m_2} a_{m_2+k} (v^{m_2})^* - v^{m_1} a_{m_1+k} (v^{m_1})^* \delta(p_{m_2}).$$

Again using the Leibniz rule, we obtain

$$\begin{aligned} p_{m_1} \delta(b_k) p_{m_2} &= -\delta(p_{m_1} v^{m_2} a_{m_2+k} (v^{m_2})^*) - \delta(v^{m_1} a_{m_1+k} (v^{m_1})^* p_{m_2}) \\ &\quad + p_{m_1} \delta(v^{m_2} a_{m_2+k} (v^{m_2})^*) + \delta(v^{m_1} a_{m_1+k} (v^{m_1})^*) p_{m_2} \\ &= -\delta(p_{m_1} p_{m_2} v^{m_2} a_{m_2+k} (v^{m_2})^*) - \delta(v^{m_1} a_{m_1+k} (v^{m_1})^* p_{m_1} p_{m_2}) \\ &\quad + p_{m_1} \delta(v^{m_2} a_{m_2+k} (v^{m_2})^*) + \delta(v^{m_1} a_{m_1+k} (v^{m_1})^*) p_{m_2} \\ &= p_{m_1} \delta(v^{m_2} a_{m_2+k} (v^{m_2})^*) + \delta(v^{m_1} a_{m_1+k} (v^{m_1})^*) p_{m_2}. \end{aligned} \tag{7.20}$$

On the other hand, for the right-hand side of (7.18), by the definition of d_k we have

$$p_{m_1} d_k p_{m_2} = \sum_{m=0}^{\infty} p_{m_1} \delta(v^m a_{m+k} (v^m)^*) p_{m_2}.$$

Since $a_{m+k} p = p a_{m+k}$ and v is the isometry corresponding to the paving projection p, we can write

$$p_{m_1} d_k p_{m_2} = \sum_{m=0}^{\infty} p_{m_1} \delta(p_m v^m a_{m+k} (v^m)^* p_m) p_{m_2},$$

and, therefore, using the Leibniz rule repeatedly together with pairwise orthogonality of $\{p_m\}_{m \in \mathbb{N}_0}$ we infer that

$$\begin{aligned}
&p_{m_1} d_k p_{m_2} \\
&= \sum_{m=0}^{\infty} \big(\delta(p_{m_1} p_m v^m a_{m+k} (v^m)^* p_m) p_{m_2} - \delta(p_{m_1}) p_m v^m a_{m+k} (v^m)^* p_m p_{m_2}\big) \\
&= \delta(p_{m_1} v^{m_1} a_{m_1+k} (v^{m_1})^* p_{m_1}) p_{m_2} - \delta(p_{m_1}) p_{m_2} v^{m_2} a_{m_2+k} (v^{m_2})^* p_{m_2} \\
&= \delta(p_{m_1} v^{m_1} a_{m_1+k} (v^{m_1})^* p_{m_1}) p_{m_2} - \delta(p_{m_1} p_{m_2} v^{m_2} a_{m_2+k} (v^{m_2})^* p_{m_2}) \\
&\quad + p_{m_1} \delta(p_{m_2} v^{m_2} a_{m_2+k} (v^{m_2})^* p_{m_2}) \\
&= \delta(p_{m_1} v^{m_1} a_{m_1+k} (v^{m_1})^* p_{m_1}) p_{m_2} + p_{m_1} \delta(p_{m_2} v^{m_2} a_{m_2+k} (v^{m_2})^* p_{m_2}).
\end{aligned}$$

Thus, combining this equality with (7.20), we conclude that $p_{m_1}\delta(b_k)p_{m_1} = p_{m_1} d_k p_{m_1}$ for $m_1 \neq m_2$. Similarly, for $m_1 = m_2$, we have $p_{m_1}\delta(b_k)p_{m_1} = p_{m_1} d_k p_{m_1}$.

Thus, $\delta(b_k) = d_k$ for all $k \in \mathbb{N}_0$. By Lemma 7.3.5, we have that $\|d_k - \mathbf{1}\|_{\mathcal{M}} \leq 6 \cdot 2^{-k}$. Therefore,

$$\delta(b_k) = d_k = \mathbf{1} - (\mathbf{1} - d_k) \geq \mathbf{1} - 6 \cdot 2^{-k} \cdot \mathbf{1} \geq \frac{1}{4} \cdot \mathbf{1},$$

for all $k \geq 3$, which proves the required estimate. □

Now, having constructed the sequence $\{b_k\}$, we are ready to prove the main result of this section, that any derivation on the algebra $LS(\mathcal{M})$ for a type *III* von Neumann algebra $\mathcal{M}$ is necessarily $t(\mathcal{M})$-continuous.

Theorem 7.3.7. *Let $\mathcal{M}$ be a type III von Neumann algebra. Every derivation $\delta : LS(\mathcal{M}) \to LS(\mathcal{M})$ is continuous locally in measure.*

Proof. Assume to the contrary that a derivation $\delta : LS(\mathcal{M}) \to LS(\mathcal{M})$ is $t(\mathcal{M})$-discontinuous. By Proposition 5.1.4, without loss of generality, we can assume that δ is self-adjoint. By Proposition 7.3.6, there exists a nonzero projection $z \in P(\mathcal{Z}(\mathcal{M}))$, such that in the reduced algebra $z\mathcal{M}$ there exists $\{b_k\}_{k\in\mathbb{N}_0} \subset z\mathcal{M}$, such that $\|b_k\|_{\mathcal{M}} \leq 2^{-k}$ and $\delta(b_k) \geq \frac{1}{4} \cdot \mathbf{1}$ for all $k \geq 3$.

Since $\mathcal{M}$ is a type *III* von Neumann algebra, it follows that z is a properly infinite projection. Hence, by Theorem 1.9.9(v), there exists a sequence $\{e_k\}_{k\in\mathbb{N}_0}$ of pairwise orthogonal properly infinite equivalent projections such that $z = \sum_{k=0}^{\infty} e_k$. By Lemma 4.4.8, we can choose a nonzero central projection $z_1 \leq z$, such that $\delta(e_k z_1) = \delta(e_k) z_1 \in \mathcal{M}$ for all $k \in \mathbb{N}_0$. Without loss of generality, we can assume that $z_1 = \mathbf{1}$.

Since $\delta(e_k)$ is bounded, we can choose a sequence $\{n_k\}_{k\in\mathbb{N}_0} \subset \mathbb{N}$, such that each

$$n_k \geq k + \|\delta(e_k)\|_{\mathcal{M}}, \; 2n_k \cdot 2^{-n_k} \leq \frac{1}{8}, \quad k \in \mathbb{N}_0.$$

Since

$$\|k e_k b_{n_k} e_k\|_{\mathcal{M}} \leq k \|b_{n_k}\|_{\mathcal{M}} \leq k \cdot 2^{-n_k} \leq k \cdot 2^{-k},$$

it follows that the series

$$b := \sum_{k\geq 8} k e_k b_{n_k} e_k$$

converges in the uniform norm. Therefore,

$$\delta(b) \in LS(\mathcal{M}). \tag{7.21}$$

For every $k \geq 3$, the repeated application of the Leibniz rule and pairwise orthogonality of $\{e_k\}_{k\in\mathbb{N}_0}$ imply that

$$\begin{aligned} e_k\delta(b)e_k &= \delta(e_k b e_k) - \delta(e_k)be_k - e_k b\delta(e_k) \\ &= k\cdot(\delta(e_k b_{n_k} e_k) - \delta(e_k)e_k b_{n_k} e_k - e_k b_{n_k} e_k \delta(e_k)) \\ &= k\cdot(e_k\delta(b_{n_k})e_k + \delta(e_k)(\mathbf{1}-e_k)b_{n_k}e_k + e_k b_{n_k}(\mathbf{1}-e_k)\delta(e_k)). \end{aligned}$$

Since $\|b_{n_k}\|_{\mathcal{M}} \leq 2^{-n_k}$ and $\|\delta(e_k)\|_{\mathcal{M}} \leq n_k$, we have that

$$\begin{aligned} &\big\|e_k\delta(e_k)\cdot(\mathbf{1}-e_k)\cdot b_{n_k}e_k + e_k b_{n_k}\cdot(\mathbf{1}-e_k)\cdot\delta(e_k)e_k\big\|_{\mathcal{M}} \\ &\quad \leq 2\big\|\delta(e_k)\big\|_{\mathcal{M}}\cdot\|b_{n_k}\|_{\mathcal{M}} \leq 2n_k\cdot 2^{-n_k} \leq \frac{1}{8}, \end{aligned}$$

in particular,

$$e_k\delta(e_k)\cdot(\mathbf{1}-e_k)\cdot b_{n_k}e_k + e_k b_{n_k}\cdot(\mathbf{1}-e_k)\cdot\delta(e_k)e_k \geq -\frac{1}{8}e_k.$$

Hence, since $\delta(b_{n_k}) \geq \frac{1}{4}\cdot\mathbf{1}$, we infer that

$$e_k\delta(b)e_k \geq k\cdot\left(e_k\delta(b_{n_k})e_k - \frac{1}{8}e_k\right) \geq \frac{k}{8}e_k.$$

Since $\{e_k\}_{k\in\mathbb{N}_0}$ are pairwise equivalent projections and $\delta(d) \in LS(\mathcal{M})$, Proposition 3.3.17 implies that $e_k = 0$ for all $k \in \mathbb{N}_0$, which is a contradiction. Thus, our initial assumption that δ is $t(\mathcal{M})$-discontinuous is false. Hence, δ must be continuous locally in measure. □

7.4 Automatic continuity of derivations in the case of properly infinite semifinite von Neumann algebra

In this section, we show that for a semifinite properly infinite von Neumann algebra $\mathcal{M}$, any derivation on $LS(\mathcal{M})$ is necessarily continuous locally in measure. We also show that any derivation on the algebra $S(\mathcal{M},\tau)$ for a faithful normal semifinite trace τ on $\mathcal{M}$ is continuous in measure. The proofs in both cases are similar and will be presented simultaneously.

7.4.1 Preliminary construction

In this subsection, we present a construction of an auxiliary sequence $\{b_k\}_{k\in\mathbb{N}_0}$ and a projection q, such that the uniform norm of $b_k q$ decreases diadically, but the derivation δ of b_k reduced by q remains larger than q. The main result of this subsection is Theorem 7.4.14 and is similar to Proposition 7.3.6 for a type *III* von Neumann algebra. However, this construction for a semifinite properly infinite von Neumann algebra is technically more complicated since it involves the measure topology, rather than the uniform norm topology as it was in the type *III* case.

We begin with the setting that we use throughout this subsection.

Assumption 7.4.1. *Suppose that $\mathcal{M}$ is a semifinite properly infinite von Neumann algebra equipped with a faithful normal semifinite trace τ. Let $\delta : LS(\mathcal{M}) \to LS(\mathcal{M})$ (or, $\delta : S(\mathcal{M},\tau) \to S(\mathcal{M},\tau)$) be a self-adjoint derivation. Assume also that:*
(i) *there exists a paving projection $p \in P(\mathcal{M})$ with the corresponding isometry $v \in \mathcal{M}$ satisfying $\delta(v) \in S(\mathcal{M},\tau)$;*
(ii) *there exists a sequence $\{x_n\}_{n\in\mathbb{N}_0} \subset S(\mathcal{M},\tau)$, such that $\delta(x_n) \in S(\mathcal{M},\tau)$, $x_n \to 0$, and $\delta(x_n) \to p$ in measure;*
(iii) *the sequence $\{x_n\}_{n\in\mathbb{N}_0}$ satisfies in addition that $x_n = x_n^*$ and $x_n p = x_n$ for all $n \in \mathbb{N}_0$.*

As before, for the paving projection p and the associated isometry v, we use the notation $p_n = v^n p (v^n)^*$, $n \in \mathbb{N}_0$ (see Definition 7.2.1). First, we construct an auxiliary sequence $\{y_n\}_{n\in\mathbb{N}_0}$.

Notation 7.4.2. *Suppose Assumption* 7.4.1 *holds. For every $n \in \mathbb{N}_0$, we set*

$$y_n := \left(\mathbf{1} + \sum_{m=0}^{n} |\delta(v^m)|^2\right)^{\frac{1}{2}}.$$

Remark 7.4.3.
(i) *By Assumption* 7.4.1(i), *we have that $\delta(v) \in S(\mathcal{M},\tau)$ and, therefore, by the Leibniz rule $\delta(v^m) \in S(\mathcal{M},\tau)$ for all $m \in \mathbb{N}_0$. Since the algebra $S(\mathcal{M},\tau)$ is invariant under taking the square root (see Theorem* 3.2.1(iii)), *we that $y_n \in S(\mathcal{M},\tau)$ for every $n \in \mathbb{N}_0$. We also note that $y_n \geq \mathbf{1}$ and, therefore, by the functional calculus, $y_n^{-1} \in \mathcal{M}$ and $\|y_n^{-1}\|_{\mathcal{M}} \leq 1$ for any $n \in \mathbb{N}_0$. Furthermore, repeating the argument of Remark* 7.3.3(i), *one can show that*

$$\|\delta(v^m) y_{m+k}^{-1}\|_{\mathcal{M}} \leq 1, \quad m,k \in \mathbb{N}_0. \tag{7.22}$$

(ii) *Since $x_n \to 0$ and $\delta(x_n) \to p$ in measure, passing to a subsequence, we can assume that $\|x_n\|_{S(\mathcal{M},\tau)}, \|\delta(x_n) - p\|_{S(\mathcal{M},\tau)} \leq 2^{-n}$ for every $n \in \mathbb{N}_0$. Using also the inclusion $\{y_n\}_{n\in\mathbb{N}_0} \in S(\mathcal{M},\tau)$ and Proposition* 4.2.12(v), *we can assume without loss of generality that*

$$\|x_n\|_{S(\mathcal{M},\tau)}, \|\delta(x_n) - p\|_{S(\mathcal{M},\tau)}, \|y_n x_n\|_{S(\mathcal{M},\tau)} \le 2^{-n}, \quad n \in \mathbb{N}_0. \tag{7.23}$$

Having defined the auxiliary sequence $\{y_n\}_{n\in\mathbb{N}_0}$, we also define an auxiliary sequence $\{f_k\}_{k\in\mathbb{N}_0} \in P(\mathcal{M})$ of projections. Shifting this sequence by powers of the isometry v and gluing them together, we will construct the projection q, mentioned above.

Notation 7.4.4. *Suppose Assumption* 7.4.1 *holds and let* $\{y_n\}_{n\in\mathbb{N}_0}$ *be as in Notation* 7.4.2. *We set*

$$f_k := E_{|y_k x_k|}[0, 2^{-k}] \wedge E_{|\delta(x_k)-p|}[0, 2^{-k}] \wedge p, \quad k \in \mathbb{N}_0.$$

Lemma 7.4.5. *Suppose Assumption* 7.4.1 *holds and let* $\{f_k\}_{k\in\mathbb{N}_0}$ *be as introduced in Notation* 7.4.4. *For every* $k \in \mathbb{N}_0$, *the projection* $p - f_k$ *is* τ*-finite and*

$$\tau(p - f_k) \le 2^{1-k}, \quad k \in \mathbb{N}_0.$$

Proof. By the definition of the projections f_k and Kaplansky's identity (see Theorem 1.9.4(v)) for every $k \in \mathbb{N}_0$, we have that

$$\begin{aligned} p - f_k &= p - p \wedge (E_{|y_k x_k|}[0, 2^{-k}] \wedge E_{|\delta(x_k)-p|}[0, 2^{-k}]) \\ &\sim p \vee (E_{|y_k x_k|}[0, 2^{-k}] \wedge E_{|\delta(x_k)-p|}[0, 2^{-k}]) \\ &\quad - (E_{|y_k x_k|}[0, 2^{-k}] \wedge E_{|\delta(x_k)-p|}[0, 2^{-k}]) \\ &\le \mathbf{1} - (E_{|y_k x_k|}[0, 2^{-k}] \wedge E_{|\delta(x_k)-p|}[0, 2^{-k}]) \\ &= E_{|y_k x_k|}(2^{-k}, \infty) \vee E_{|\delta(x_k)-p|}(2^{-k}, \infty). \end{aligned}$$

By (7.23), we have that

$$\|y_k x_k\|_{S(\mathcal{M},\tau)}, \|\delta(x_k) - p\|_{S(\mathcal{M},\tau)} \le 2^{-k}, \quad k \in \mathbb{N}_0.$$

Corollary 4.2.6 ensures that

$$\tau(E_{|y_k x_k|}(2^{-k}, \infty)) \le 2^{-k}, \quad \tau(E_{|\delta(x_k)-p|}(2^{-k}, \infty)) \le 2^{-k}, \quad k \in \mathbb{N}_0.$$

Therefore, for any $k \in \mathbb{N}_0$ we infer that

$$\begin{aligned} \tau(p - f_k) &\le \tau(E_{|y_k x_k|}(2^{-k}, \infty)) + \tau(E_{|\delta(x_k)-p|}(2^{-k}, \infty)) \\ &\le 2^{-k} + 2^{-k} = 2^{1-k}, \end{aligned}$$

as required. □

We can now define the projection q, which will give us the necessary reduction.

Notation 7.4.6. *Suppose Assumption 7.4.1 holds and let f_k, $k \in \mathbb{N}_0$ be as in Notation 7.4.4. We set*

$$q_k = \sup_{m\in\mathbb{N}_0} v^m f_{m+k}(v^m)^*$$

and

$$q = \inf_{k\geq 3} q_k.$$

Remark 7.4.7. *Since $f_k \leq p$ by definition and since v is an isometry, we have that*

$$v^m f_{m+k}(v^m)^* = v^m p(v^m)^* \cdot v^m f_{m+k}(v^m)^* \cdot v^m p(v^m)^* = p_m v^m f_{m+k}(v^m)^* p_m.$$

By the definition of the paving projection p, the projections p_m, $m \in \mathbb{N}_0$ are pairwise orthogonal. Hence, we can write

$$q_k = \sum_{m=0}^{\infty} v^m f_{m+k}(v^m)^* = \sum_{m=0}^{\infty} p_m v^m f_{m+k}(v^m)^* p_m, \tag{7.24}$$

where the series converge in the strong operator topology.

Proposition 7.4.8. *Suppose Assumption 7.4.1 holds and let q be as in Notation 7.4.6. The projection $\mathbf{1} - q$ is τ-finite and*

$$\tau(\mathbf{1} - q) \leq 1.$$

Proof. By the definition of q, we have that $q = \inf_{k\geq 3} q_k$. Therefore, we first work with projections q_k, $k \in \mathbb{N}_0$.

By Assumption 7.4.1(i), p is a paving projection. Therefore,

$$\sum_{m=0}^{\infty} v^m p(v^m)^* = \mathbf{1}.$$

By Remark 7.4.7, we also have that

$$q_k = \sum_{m=0}^{\infty} v^m f_{m+k}(v^m)^*.$$

Thus,

$$\mathbf{1} - q_k = \sum_{m=0}^{\infty} (v^m p(v^m)^* - v^m f_{m+k}(v^m)^*) = \sum_{m=0}^{\infty} v^m (p - f_{m+k})(v^m)^*.$$

Since v is an isometry, it follows that $(v^m)^*v^m = \mathbf{1}$ for all $m \in \mathbb{N}_0$. In addition, by Lemma 7.4.5, for every $m \in \mathbb{N}_0$ the projection $p - f_m$ is τ-finite. Hence, the cyclicity of the trace τ implies that

$$\tau(v^m(p - f_{m+k})(v^m)^*) = \tau((p - f_{m+k}) \cdot (v^m)^* v^m) = \tau(p - f_{m+k}).$$

Therefore,

$$\tau(\mathbf{1} - q_k) = \sum_{m=0}^{\infty} \tau(v^m(p - f_{m+k})(v^m)^*) = \sum_{m=0}^{\infty} \tau(p - f_{m+k}) = \sum_{m=k}^{\infty} \tau(p - f_m).$$

By Lemma 7.4.5, we have that $\tau(p - f_m) \le 2^{1-m}$, for every $m \in \mathbb{N}_0$. Therefore,

$$\tau(\mathbf{1} - q_k) = \sum_{m=k}^{\infty} \tau(p - f_m) \le \sum_{m=k}^{\infty} 2^{1-m} = 2^{2-k}. \tag{7.25}$$

Thus, equality $q = \inf_{k \ge 3} q_k$ combined with estimate (7.25) implies that

$$\tau(\mathbf{1} - q) = \tau\Big(\sup_{k \ge 3}(\mathbf{1} - q_k)\Big) \le \sum_{k=3}^{\infty} \tau(\mathbf{1} - q_k) \le \sum_{k=3}^{\infty} 2^{2-k} = 1,$$

as required. □

Having constructed the projection q, we now construct the sequence $\{b_k\}_{k \in \mathbb{N}_0}$, such that the uniform norm of $b_k q$ decreases as $k \to \infty$, but $\delta(b_k)$ reduced by q remains large.

Notation 7.4.9. *Suppose Assumption* 7.4.1 *holds. We set*

$$b_k := \sum_{m=0}^{\infty} v^m x_{m+k} (v^m)^*, \quad k \in \mathbb{N}_0.$$

Remark 7.4.10. *By the definition of the projections* $p_n = v^n p (v^n)^*, n \in \mathbb{N}_0$ *(see Definition* 7.2.1*), we have that*

$$b_k = \sum_{m=0}^{\infty} v^m x_{m+k} (v^m)^* = \sum_{m=0}^{\infty} p_m v^m x_{m+k} (v^m)^* p_m.$$

In the following lemma, we show that each b_k, $k \in \mathbb{N}_0$ is well-defined as a series convergent in the measure topology. We also estimate the uniform norm of the operators $b_k q_k$, $k \in \mathbb{N}_0$.

Lemma 7.4.11. *Suppose Assumption* 7.4.1 *holds. For every* $k \in \mathbb{N}_0$, *the series* b_k *defined in Notation* 7.4.9 *converges in measure. Moreover, with the projection* q_k *defined in Notation* 7.4.6, *for every* $k \in \mathbb{N}_0$ *the operator* $b_k q_k$ *is bounded and*

$$\|b_k q_k\|_{\mathcal{M}} \le 2^{-k}, \quad k \in \mathbb{N}_0.$$

Proof. We first prove that the series

$$b_k = \sum_{m=0}^{\infty} v^m x_{m+k} (v^m)^*$$

converges in measure for every $k \in \mathbb{N}_0$.

Fix $k \in \mathbb{N}_0$. Since $\|x_k\|_{S(\mathcal{M},\tau)} \le 2^{-k}$ (see (7.23)), Proposition 4.2.7(iii) implies that

$$\|v^m x_{m+k} (v^m)^*\|_{S(\mathcal{M},\tau)} \le \|x_{m+k}\|_{S(\mathcal{M},\tau)} \le 2^{-m-k}, \quad m \in \mathbb{N}_0.$$

Therefore, for all $n_1, n_2 \in \mathbb{N}_0$ with $n_1 > n_2$ we have

$$\begin{aligned} \Big\| \sum_{m=n_1}^{n_2} v^m x_{m+k} (v^m)^* \Big\|_{S(\mathcal{M},\tau)} &\le \sum_{m=n_1}^{n_2} \|v^m x_{m+k} (v^m)^*\|_{S(\mathcal{M},\tau)} \\ &\le \sum_{m=n_1}^{n_2} 2^{-m-k} \to 0, \quad n_1, n_2 \to \infty. \end{aligned}$$

Since $(S(\mathcal{M},\tau), \|\cdot\|_{S(\mathcal{M},\tau)})$ is complete (see Theorem 4.2.8), it follows that the series $b_k = \sum_{m=0}^{\infty} v^m x_{m+k} (v^m)^*$ converges in measure and

$$\|b_k\|_{S(\mathcal{M},\tau)} \le \sum_{m=0}^{\infty} \|v^m x_{m+k} (v^m)^*\|_{S(\mathcal{M},\tau)} \le \sum_{m=0}^{\infty} 2^{-m-k} = 2^{1-k}.$$

To prove the second assertion of this lemma, we fix $k \in \mathbb{N}_0$ and recall (see Remark 7.4.7) that q_k can be written as

$$q_k = \sum_{m=0}^{\infty} v^m f_{m+k} (v^m)^*,$$

where the series converges in the strong operator topology. Therefore, we have

$$\begin{aligned} b_k q_k &= \Big(\sum_{m_1=0}^{\infty} v^{m_1} x_{m_1+k} (v^{m_1})^* \Big) \cdot \Big(\sum_{m_2=0}^{\infty} v^{m_2} f_{m_2+k} (v^{m_2})^* \Big) \\ &= \sum_{m_1,m_2=0}^{\infty} v^{m_1} x_{m_1+k} (v^{m_1})^* v^{m_2} f_{m_2+k} (v^{m_2})^*. \end{aligned}$$

By Assumption 7.4.1, we have that $x_n p = p x_n = x_n$ for every $n \in \mathbb{N}_0$. Since the projections $p_m, m \in \mathbb{N}_0$, are pairwise orthogonal, we can write

$$\begin{aligned} b_k q_k &= \sum_{m_1,m_2=0}^{\infty} v^{m_1} x_{m_1+k} (v^{m_1})^* \cdot p_{m_1} p_{m_2} \cdot v^{m_2} f_{m_2+k} (v^{m_2})^* \\ &= \sum_{m=0}^{\infty} v^m x_{m+k} (v^m)^* \cdot v^m f_{m+k} (v^m)^* = \sum_{m=0}^{\infty} v^m x_{m+k} f_{m+k} (v^m)^*, \end{aligned}$$

where the series converges in the measure topology.

By definition of the projection f_k, we have that $f_k \leq E_{|y_k x_k|}[0, 2^{-k}]$ and, therefore,

$$\|y_k x_k f_k\|_{\mathcal{M}} \leq 2^{-k}. \tag{7.26}$$

By Remark 7.4.3(i), the operator y_k has bounded inverse with $\|y_k^{-1}\|_{\mathcal{M}} \leq 1$. Therefore,

$$\|x_k f_k\|_{\mathcal{M}} = \|y_k^{-1} y_k x_k f_k\|_{\mathcal{M}} \leq \|y_k x_k f_k\|_{\mathcal{M}} \overset{(7.26)}{\leq} 2^{-k}.$$

Since, in addition, the family $\{v^m x_{m+k} f_{m+k} (v^m)^*\}_{m \in \mathbb{N}_0}$ is pairwise disjoint, it follows that $b_k q_k = \sum_{m=0}^{\infty} v^m x_{m+k} f_{m+k} (v^m)^*$ as the disjoint sum of a uniformly bounded family of pairwise disjoint operators (see Definition 5.4.2).

By Lemma 5.4.1, for any $k \in \mathbb{N}_0$, we have that

$$\begin{aligned} \|b_k q_k\|_{\mathcal{M}} &= \sup_{m \in \mathbb{N}_0} \|v^m x_{m+k} f_{m+k} (v^m)^*\|_{\mathcal{M}} \\ &\leq \sup_{m \in \mathbb{N}_0} \|x_{m+k} f_{m+k}\|_{\mathcal{M}} = \sup_{m \geq k} \|x_m f_m\|_{\mathcal{M}} \leq 2^{-k}, \end{aligned}$$

as required. □

As the following step of our construction, we study how the derivation δ acts on every b_k, $k \in \mathbb{N}_0$. To this end, similar to Lemma 7.3.5, we first introduce an auxiliary element d_k, which as we show in the Lemma 7.4.13 below, is $\delta(b_k)$ reduced by q_k.

Lemma 7.4.12. *Suppose Assumption* 7.4.1 *holds and let* $q_k \in P(\mathcal{M})$, $k \in \mathbb{N}_0$, *be as in Notation* 7.4.6.

(i) *For every* $k, m \in \mathbb{N}_0$, *the operator* $q_k \delta(v^m x_{m+k} (v^m)^*) q_k$ *is bounded;*

(ii) *For every* $k \in \mathbb{N}_0$, *the series*

$$d_k := \sum_{m=0}^{\infty} q_k \delta(v^m x_{m+k} (v^m)^*) q_k, \tag{7.27}$$

converges in the strong operator topology and

$$\|d_k - q_k\|_{\mathcal{M}} \leq 6 \cdot 2^{-k}.$$

Proof. (i) Fix $m, k \in \mathbb{N}_0$. For convenience, we define

$$\begin{aligned} A_{m,k} &:= q_k \delta(v^m) x_{m+k} (v^m)^* q_k, \\ B_{m,k} &:= q_k v^m \cdot (\delta(x_{m+k}) - p) \cdot (v^m)^* q_k, \\ C_{m,k} &:= q_k (v^m p (v^m)^*) q_k. \end{aligned}$$

Note that since δ is a self-adjoint derivation, we have that

$$A_{m,k}^* = q_k v^m x_{m+k} \delta((v^m)^*) q_k.$$

By the Leibniz rule, we can write

$$\begin{aligned} &q_k \delta(v^m x_{m+k} (v^m)^*) q_k \\ &\quad = q_k \delta(v^m) x_{m+k} (v^m)^* q_k + q_k v^m x_{m+k} \delta((v^m)^*) q_k \\ &\quad\quad + q_k v^m \delta(x_{m+k}) (v^m)^* q_k \\ &\quad = q_k \delta(v^m) x_{m+k} (v^m)^* q_k + q_k v^m x_{m+k} \delta((v^m)^*) q_k \\ &\quad\quad + q_k v^m \cdot (\delta(x_{m+k}) - p) \cdot (v^m)^* q_k + q_k (v^m p (v^m)^*) q_k \\ &\quad = A_{m,k} + A_{m,k}^* + B_{m,k} + C_{m,k}. \end{aligned} \tag{7.28}$$

It is clear that the operator $C_{m,k}$ is bounded. Hence, it is sufficient to show that the operators $A_{m,k}$ and $B_{m,k}$ are bounded.

Let $\{y_k\}_{k\in\mathbb{N}_0}$ and $\{f_k\}_{k\in\mathbb{N}_0}$ be as in Notation 7.4.2 and Notation 7.4.4, respectively. Recall also that $p_n = v^n p (v^n)^*$, $n \in \mathbb{N}_0$ (see Definition 7.2.1).

By Assumption 7.4.1, we have that $x_{m+k} p = x_{m+k}$. Since, in addition, v is an isometry, we can write

$$\begin{aligned} A_{m,k} &= q_k \delta(v^m) x_{m+k} \cdot (v^m)^* v^m \cdot p (v^m)^* q_k \\ &= q_k \delta(v^m) x_{m+k} (v^m)^* \cdot p_m \cdot q_k. \end{aligned} \tag{7.29}$$

By definition of paving projections, the projections p_n, $n \in \mathbb{N}_0$, are pairwise orthogonal. Since $f_k \le p$ by definition of f_k, we note that

$$\begin{aligned} p_m q_k &\overset{(7.24)}{=} \sum_{l=0}^{\infty} p_m p_l v^l f_{l+k} (v^l)^* p_l \\ &= p_m v^m f_{m+k} (v^m)^* p_m = v^m f_{m+k} (v^m)^*. \end{aligned}$$

Hence, by (7.29), we have

$$\begin{aligned} A_{m,k} &= q_k \delta(v^m) x_{m+k} (v^m)^* \cdot v^m f_{m+k} (v^m)^* = q_k \delta(v^m) x_{m+k} f_{m+k} (v^m)^* \\ &= q_k \cdot \delta(v^m) y_{m+k}^{-1} \cdot y_{m+k} x_{m+k} f_{m+k} \cdot (v^m)^*. \end{aligned}$$

By (7.22), we have that $\|\delta(v^m) y_{m+k}^{-1}\|_{\mathcal{M}} \le 1$. In addition, by (7.26), we have $\|y_{m+k} x_{m+k} f_{m+k}\|_{\mathcal{M}} \le 2^{-m-k}$. Hence, the operator $A_{m,k}$ is bounded and

$$\begin{aligned} \|A_{m,k}\|_{\mathcal{M}} &\le \|\delta(v^m) y_{m+k}^{-1} \cdot y_{m+k} x_{m+k} f_{m+k}\|_{\mathcal{M}} \\ &\le \|\delta(v^m) y_{m+k}^{-1}\|_{\mathcal{M}} \cdot \|y_{m+k} x_{m+k} f_{m+k}\|_{\mathcal{M}} \le 2^{-m-k}. \end{aligned}$$

Thus,

$$\|A_{m,k}\|_{\mathcal{M}} \leq 2^{-m-k}. \tag{7.30}$$

To estimate the uniform norm of the operator $B_{m,k}$, we recall that $q_k = \sup_{m\in\mathbb{N}_0} v^m f_{m+k}(v^m)^*$, and $f_k \leq E_{|\delta(x_k)-p|}[0,2^{-k}]$. Hence,

$$\begin{aligned}
B_{m,k} &= q_k v^m \cdot (\delta(x_{m+k}) - p)(v^m)^* \cdot q_k \\
&= q_k v^m \cdot (\delta(x_{m+k}) - p)(v^m)^* \cdot v^m f_{m+k}(v^m)^* \cdot q_k \\
&= q_k v^m \cdot (\delta(x_{m+k}) - p) f_{m+k}(v^m)^* q_k \\
&= q_k v^m \cdot (\delta(x_{m+k}) - p) E_{|\delta(x_{m+k})-p|}[0,2^{-k-m}] \cdot f_{m+k}(v^m)^* q_k
\end{aligned}$$

and, therefore,

$$\|B_{m,k}\|_{\mathcal{M}} \leq \|(\delta(x_{m+k}) - p)E_{|\delta(x_{m+k})-p|}[0,2^{-k-m}]\|_{\mathcal{M}} \leq 2^{-k-m}. \tag{7.31}$$

Combining (7.28), (7.30), and (7.31), we conclude the proof of part (i).

(ii) Combining (7.30) and (7.31), we infer that the series

$$\sum_{m=0}^{\infty} A_{m,k} + A_{m,k}^* + B_{m,k}$$

converges in the uniform norm for every fixed $k \in \mathbb{N}_0$. In addition, since p is a paving projection, it follows that $\sum_{m=0}^{\infty} v^m p(v^m)^* = \mathbf{1}$ with respect to the strong operator topology. Therefore, we have that

$$\sum_{m=0}^{\infty} C_{m,k} = \sum_{m=0}^{\infty} q_k(v^m p(v^m)^*)q_k = q_k$$

with respect to the strong operator topology.

Thus, equality (7.28) implies that the series

$$d_k = \sum_{m=0}^{\infty} q_k \delta(v^m x_{m+k}(v^m)^*)q_k = \sum_{m=0}^{\infty}(A_{m,k} + A_{m,k}^* + B_{m,k} + C_{m,k})$$

converges in the strong operator topology. In particular, equalities (7.30) and (7.31) imply that

$$\begin{aligned}
\|d_k - q_k\|_{\mathcal{M}} &= \left\|\sum_{m=0}^{\infty}(A_{m,k} + A_{m,k}^* + B_{m,k} + C_{m,k}) - \sum_{m=0}^{\infty} C_{m,k}\right\|_{\mathcal{M}} \\
&= \left\|\sum_{m=0}^{\infty}(A_{m,k} + A_{m,k}^* + B_{m,k})\right\|_{\mathcal{M}}
\end{aligned}$$

$$\leq \sum_{m=0}^{\infty} 2\|A_{m,k}\|_{\mathcal{M}} + \|B_{m,k}\|_{\mathcal{M}} \leq \sum_{m=0}^{\infty} 3 \cdot 2^{-m-k}$$
$$\leq 6 \cdot 2^{-k},$$

as required. □

Lemma 7.4.13. *Suppose Assumption* 7.4.1 *holds and let b_k, q_k, $k \in \mathbb{N}_0$ be as in Notations* 7.4.9 *and* 7.4.6 *and d_k, $k \in \mathbb{N}_0$ be as in* (7.27). *For every $k \in \mathbb{N}_0$, we have*

$$q_k \delta(b_k) q_k = d_k.$$

Proof. Let $k \in \mathbb{N}_0$ be fixed. To shorten notation, we introduce

$$r_m = \sup_{l \leq m} p_l = \sum_{l=0}^{m} p_l.$$

By definition of the paving projection p, we have that $\sup_{m \in \mathbb{N}_0} p_m = \mathbf{1}$ and, therefore, $r_m \uparrow \mathbf{1}$ as $m \to \infty$. Hence, by Proposition 3.3.7(iii), it is sufficient to show that

$$r_m \cdot q_k \delta(b_k) q_k \cdot r_m = r_m \cdot d_k \cdot r_m, \tag{7.32}$$

for every $m \in \mathbb{N}_0$. Fixing $m \in \mathbb{N}_0$, we consider the left-hand side and right-hand side of (7.32) separately.

Since $p_m, m \in \mathbb{N}_0$ are pairwise orthogonal, equality (7.24) implies that

$$r_m q_k = \sum_{\substack{l_1 \leq m \\ l_2 \in \mathbb{N}_0}} p_{l_1} p_{l_2} v^{l_2} f_{l_2+k} (v^{l_2})^* p_{l_2} = \sum_{l \leq m} p_l v^l f_{l+k} (v^l)^* p_l = q_k r_m. \tag{7.33}$$

Since (see Remark 7.4.10)

$$b_k = \sum_{m=0}^{\infty} p_m v^m x_{m+k} (v^m)^* p_m,$$

it follows that

$$(\mathbf{1} - r_m) b_k r_m = r_m b_k (\mathbf{1} - r_m) = 0$$

and, therefore, we can write

$$b_k = r_m b_k r_m + (\mathbf{1} - r_m) b_k (\mathbf{1} - r_m). \tag{7.34}$$

By the Leibniz rule, we have

$$\begin{aligned} &r_m \cdot \delta((\mathbf{1} - r_m) b_k (\mathbf{1} - r_m)) \cdot r_m \\ &\quad = \delta(r_m \cdot (\mathbf{1} - r_m) b_k (\mathbf{1} - r_m)) \cdot r_m - \delta(r_m) \cdot (\mathbf{1} - r_m) b_k (\mathbf{1} - r_m) \cdot r_m = 0. \end{aligned}$$

Combining the latter equality with (7.33) and (7.34), we obtain that

$$\begin{aligned} r_m \cdot q_k \delta(b_k) q_k \cdot r_m &\overset{(7.33)}{=} q_k \cdot r_m \delta(b_k) r_m \cdot q_k \\ &\overset{(7.34)}{=} q_k \cdot r_m \delta(r_m b_k r_m) r_m \cdot q_k + q_k \cdot r_m \delta((\mathbf{1} - r_m) b_k (\mathbf{1} - r_m)) r_m \cdot q_k \\ &= q_k \cdot r_m \delta(r_m b_k r_m) r_m \cdot q_k. \end{aligned}$$

Since $r_m b_k r_m = \sum_{l=0}^{m} v^l x_{l+k} (v^l)^*$, we conclude that the left-hand side of (7.32) can be written as

$$\begin{aligned} r_m \cdot q_k \delta(b_k) q_k \cdot r_m &= r_m q_k \delta\left(\sum_{l=0}^{m} v^l x_{l+k} (v^l)^* \right) q_k r_m \\ &= r_m q_k \left(\sum_{l=0}^{m} \delta(v^l x_{l+k} (v^l)^*) \right) q_k \cdot r_m \\ &= r_m \left(\sum_{l=0}^{m} q_k \delta(v^l x_{l+k} (v^l)^*) q_k \right) r_m. \end{aligned} \tag{7.35}$$

On the other hand, for the right-hand side of (7.32), we have

$$r_m d_k r_m = \sum_{l=0}^{\infty} r_m q_k \delta(v^l x_{l+k} (v^l)^*) q_k r_m \overset{(7.33)}{=} \sum_{l=0}^{\infty} q_k r_m \delta(v^l x_{l+k} (v^l)^*) r_m q_k. \tag{7.36}$$

Since $v^l x_{l+k} (v^l)^* = p_l v^l x_{l+k} (v^l)^* p_l$ for all $l \in \mathbb{N}_0$, we have that $(\mathbf{1} - r_m) v^l x_{l+k} (v^l)^* (\mathbf{1} - r_m) = v^l x_{l+k} (v^l)^*$ for all $l > m$. Therefore, for all $l > m$, the Leibniz rule implies that

$$\begin{aligned} r_m \delta(v^l x_{l+k} (v^l)^*) r_m &= r_m \delta((\mathbf{1} - r_m) v^l x_{l+k} (v^l)^* (\mathbf{1} - r_m)) r_m \\ &= r_m \delta(\mathbf{1} - r_m) \cdot v^l x_{l+k} (v^l)^* \cdot (\mathbf{1} - r_m) r_m \\ &\quad + r_m (\mathbf{1} - r_m) \delta(v^l x_{l+k} (v^l)^*) (\mathbf{1} - r_m) r_m \\ &\quad + r_m (\mathbf{1} - r_m) \cdot v^l x_{l+k} (v^l)^* \cdot \delta(\mathbf{1} - r_m) r_m \\ &= 0. \end{aligned}$$

Hence, combining this equality with (7.36), we obtain that

$$\begin{aligned} r_m d_k r_m &= \sum_{l=0}^{m} q_k r_m \delta(v^l x_{l+k} (v^l)^*) r_m q_k \\ &\overset{(7.33)}{=} r_m \left(\sum_{l=0}^{m} q_k \delta(v^l x_{l+k} (v^l)^*) q_k \right) r_m. \end{aligned}$$

Recalling equality (7.35), we obtain that

$$r_m \cdot q_k \delta(b_k) q_k \cdot r_m = r_m \left(\sum_{l=0}^{m} q_k \delta(v^l x_{l+k} (v^l)^*) q_k \right) r_m = r_m d_k r_m,$$

which proves (7.32) and hence concludes the proof. □

Now, we have completed our preliminary construction and are ready to formulate the main result of this subsection, which gives an auxiliary sequence b_k and a projection q we are looking for.

Theorem 7.4.14. *Suppose Assumption* 7.4.1 *holds. Then there exists a projection* $q \in P(\mathcal{M})$ *and a sequence* $\{b_k\}_{n\in\mathbb{N}_0} \subset S(\mathcal{M}, \tau)$*, such that:*
(i) $\tau(\mathbf{1} - q) \leq 1$;
(ii) $\|b_k q\|_{\mathcal{M}} \leq 2^{-k}$ *for all* $k \geq 3$;
(iii) $q\delta(b_k)q \geq \frac{1}{4}q$ *for all* $k \geq 3$.

Proof. Let q be as in Notation 7.4.6 and let $\{b_k\}_{k\in\mathbb{N}_0}$ be as in Notation 7.4.9.

Proposition 7.4.8 guarantees that the first assertion of theorem holds. By Lemma 7.4.11, we have that $\|b_k q_k\|_{\mathcal{M}} \leq 2^{-k}$ for all $k \in \mathbb{N}_0$. Since $q = \inf_{k\geq 3} q_k$, we obtain that

$$\|b_k q\|_{\mathcal{M}} = \|b_k q_k q\|_{\mathcal{M}} \leq \|b_k q_k\|_{\mathcal{M}} \leq 2^{-k}, \quad k \geq 3,$$

and, therefore, the second assertion also holds. Thus, it remains to show that $q\delta(b_k)q \geq \frac{1}{4}q$ for all $k \geq 3$.

By Lemma 7.4.13, we have that

$$q_k \delta(b_k) q_k = q_k d_k q_k = q_k - q_k(q_k - d_k) q_k.$$

By Lemma 7.4.12(ii), we have that $\|q_k - d_k\|_{\mathcal{M}} \leq 6 \cdot 2^{-k}$ and, therefore,

$$q_k(q_k - d_k)q_k \leq 6 \cdot 2^{-k} q_k.$$

Thus, we infer that

$$q_k \delta(b_k) q_k = q_k - q_k(q_k - d_k)q_k \geq q_k - 6 \cdot 2^{-k} q_k \geq \frac{1}{4} q_k, \quad k \geq 3.$$

Since $q = \inf_{k\geq 3} q_k$, Proposition 3.3.4(ii) implies that

$$q\delta(b_k)q \geq \frac{1}{4}q, \quad k \geq 3,$$

concluding the proof. □

7.4.2 Automatic continuity of derivations on the algebras of locally measurable operators for semifinite properly infinite algebras

With the technical construction of Theorem 7.4.14 completed, we can proceed to the proof of automatic continuity of derivations on $LS(\mathcal{M})$. To this end we require an additional auxiliary lemma.

Lemma 7.4.15. *Suppose that $\{e_k\}_{k\in\mathbb{N}_0}$ is a sequence of pairwise orthogonal equivalent properly infinite projections in $\mathcal{M}$ and let δ be a self-adjoint derivation on $LS(\mathcal{M})$ (or on $S(\mathcal{M},\tau)$), such that $\delta(e_k) \in S(\mathcal{M},\tau)$. Then there exists a sequence $\{f_k\}_{k\in\mathbb{N}_0}$ of pairwise orthogonal and equivalent properly infinite projections, such that $f_k \leq e_k$ and $f_k\delta(e_k) \in \mathcal{M}$ for every $k \in \mathbb{N}_0$.*

Proof. By assumption, the derivation δ is self-adjoint, and so the operator $e_k\delta(e_k)^2e_k$ is positive for any $k \in \mathbb{N}_0$. Since $\delta(e_k) \in S(\mathcal{M},\tau)$, it follows that $e_k\delta(e_k)^2e_k \in S(\mathcal{M},\tau)$. By Proposition 2.4.2(iv), there exists $n_k \in \mathbb{N}$, such that the projection $E_{e_k\delta(e_k)^2e_k}(n_k,\infty)$ is τ-finite, in particular finite.

For every $k \in \mathbb{N}_0$, we set

$$f_k = e_k \wedge E_{e_k\delta(e_k)^2e_k}[0,n_k].$$

Then $f_k \leq e_k$ and $f_ke_k\delta(e_k)^2e_kf_k \in \mathcal{M}$. Hence, $f_k\delta(e_k) = f_ke_k\delta(e_k)$ is bounded, too. By Kaplansky's identity (see Theorem 1.9.4(v)), we have that

$$\begin{aligned} e_k - f_k = e_k - \left(e_k \wedge E_{e_k\delta(e_k)^2e_k}[0,n_k]\right) &\sim \left(e_k \vee E_{e_k\delta(e_k)^2e_k}[0,n_k]\right) - E_{e_k\delta(e_k)^2e_k}[0,n_k] \\ &\leq \mathbf{1} - E_{e_k\delta(e_k)^2e_k}[0,n_k] = E_{e_k\delta(e_k)^2e_k}(n_k,\infty). \end{aligned}$$

Therefore, the projection $e_k - f_k$ is finite for every $k \in \mathbb{N}_0$. Since e_k is a properly infinite projection for every $k \in \mathbb{N}_0$, applying Proposition 1.9.10(i) to the algebra $e_k\mathcal{M}e_k$, we conclude that $f_k = e_k - (e_k - f_k)$ is a properly infinite projection with central support equal to $s_c(e_k)$. Since the projections $\{e_k\}_{k\in\mathbb{N}_0}$ are pairwise equivalent, Theorem 1.9.4(ii) implies that $s_c(e_k)$ are all equal for all $k \in \mathbb{N}_0$. Hence, the projections f_k, $k \in \mathbb{N}_0$ are properly infinite projections with equal central supports, and so the projections f_k, $k \in \mathbb{N}_0$, are all equivalent (see Theorem 1.9.9(viii)). □

We now prove the main result of this section of automatic continuity of derivations on $LS(\mathcal{M})$ and $S(\mathcal{M},\tau)$ for a properly infinite semifinite von Neumann algebra $\mathcal{M}$.

Theorem 7.4.16. *Let $\mathcal{M}$ be a semifinite properly infinite von Neumann algebra.*

(i) *Every derivation $\delta : LS(\mathcal{M}) \to LS(\mathcal{M})$ is continuous locally in measure;*

(ii) *If $\mathcal{M}$ is equipped with a faithful normal semifinite trace τ, then every derivation $\delta : S(\mathcal{M},\tau) \to S(\mathcal{M},\tau)$ is continuous in measure.*

Proof. The proof for both the cases of algebras $LS(\mathcal{M})$ and $S(\mathcal{M},\tau)$ follows the same lines. Therefore, we let $(\mathcal{A}, \|\cdot\|_{\mathcal{A}})$ be either $(LS(\mathcal{M}), \|\cdot\|_{LS(\mathcal{M})})$ or $(S(\mathcal{M},\tau), \|\cdot\|_{S(\mathcal{M},\tau)})$ and show that any derivation $\delta : \mathcal{A} \to \mathcal{A}$ is continuous with respect to $\|\cdot\|_{\mathcal{A}}$.

Assume to the contrary that the derivation $\delta : (\mathcal{A}, \|\cdot\|_{\mathcal{A}}) \to (\mathcal{A}, \|\cdot\|_{\mathcal{A}})$ is not continuous. By Proposition 5.1.4, without loss of generality, we can assume that δ is a self-adjoint derivation.

By Corollary 7.2.4 in the case $(\mathcal{A}, \|\cdot\|_{\mathcal{A}}) = (LS(\mathcal{M}), \|\cdot\|_{LS(\mathcal{M})})$ and by Proposition 7.2.3 in the case $(\mathcal{A}, \|\cdot\|_{\mathcal{A}}) = (S(\mathcal{M},\tau), \|\cdot\|_{S(\mathcal{M},\tau)})$, there exists a (nonzero) central projection $z \in P(\mathcal{Z}(\mathcal{M}))$ and a paving projection $p \in z\mathcal{M} \subset S(\mathcal{M},\tau)$ (with the corresponding isometry v satisfying $z\delta(v) \in zS(\mathcal{M},\tau)$) and a sequence $\{x_n\}_{n\in\mathbb{N}_0} \subset zS(\mathcal{M},\tau)$, such that $x_n = x_n^*$, $x_n p = x_n$ and $x_n \to 0$, and $\delta(x_n) \to p$ in measure. Hence, for the derivation $\delta^{(z)}$ considering within the reduced algebra $z\mathcal{M}$, Assumption 7.4.1 holds. By Theorem 7.4.14, there exists a projection $q \in P(z\mathcal{M})$ and a sequence $\{q_k\}_{k\in\mathbb{N}_0} \subset S(z\mathcal{M},\tau)$, such that $\tau(z-q) \le 1$, $\|b_k q\|_{\mathcal{M}} \le 2^{-k}$, and $q\delta^{(z)}(b_k)q \ge \frac{1}{4}q$ for all $k \ge 3$. Without loss of generality, we assume that $z = \mathbf{1}$.

Since $\mathbf{1} - q$ is finite, Proposition 1.9.10(i) implies that the projection q is properly infinite. Therefore, by Theorem 1.9.9(viii), there exists a sequence $\{e_k\}_{k\in\mathbb{N}_0}$ of pairwise orthogonal properly infinite and pairwise equivalent projections such that $q = \sum_{k=0}^{\infty} e_k$. By Lemma 4.4.5, there exists a nonzero central projection $z_0 \in \mathcal{M}$, such that $\delta(e_k z_0) = \delta(e_k)z_0 \in S(\mathcal{M},\tau)$ for all $k \in \mathbb{N}_0$. Without loss of generality, we can assume that $z_0 = \mathbf{1}$.

By Lemma 7.4.15, there exists a sequence $\{f_k\}_{k\in\mathbb{N}_0}$ of pairwise orthogonal equivalent properly infinite projections, such that $f_k \le e_k$ and $f_k\delta(e_k) \in \mathcal{M}$. Since $f_k\delta(e_k)$ is bounded for every $k \in \mathbb{N}_0$, we can choose a sequence $\{n_k\}_{k\in\mathbb{N}_0} \subset \mathbb{N}$, such that

$$n_k \ge k + \|f_k\delta(e_k)\|_{\mathcal{M}}, \quad k \in \mathbb{N}_0.$$

We set

$$b = \sum_{k=8}^{\infty} k e_k b_{n_k} e_k.$$

Since $e_k \le q$ and $\|b_k q\|_{\mathcal{M}} \le 2^{-k}$, it follows that

$$\|k e_k b_{n_k} e_k\|_{\mathcal{M}} \le k\|b_{n_k} q\|_{\mathcal{M}} \le k \cdot 2^{-n_k} \le k \cdot 2^{-k}$$

and, therefore, the series $b = \sum_{k=8}^{\infty} k e_k b_{n_k} e_k$ converges with respect to the uniform norm. That is, $b \in \mathcal{M}$ and, therefore,

$$\delta(b) \in \mathcal{A} \subset LS(\mathcal{M}). \tag{7.37}$$

We claim that

$$f_k\delta(b)f_k \ge \frac{k}{8} f_k$$

for all $k \ge 8$.

Note first that since $\{e_k\}_{k\in\mathbb{N}_0}$ are pairwise orthogonal, repeated application of the Leibniz rule yields

$$
\begin{aligned}
e_k\delta(b)e_k &= \delta(e_kbe_k) - \delta(e_k)be_k - e_kb\delta(e_k)\\
&= k\cdot(\delta(e_kb_{n_k}e_k) - \delta(e_k)e_kb_{n_k}e_k - e_kb_{n_k}e_k\delta(e_k))\\
&= k\cdot(e_k\delta(b_{n_k})e_k + \delta(e_k)(\mathbf{1}-e_k)b_{n_k}e_k + e_kb_{n_k}(\mathbf{1}-e_k)\delta(e_k)).
\end{aligned}
$$

Hence, since $f_k \le e_k$, it follows that

$$
\begin{aligned}
f_k\delta(b)f_k = k\cdot\Big(&f_k\delta(b_{n_k})f_k + f_k\delta(e_k)\cdot(\mathbf{1}-e_k)\cdot b_{n_k}f_k\\
&+ f_kb_{n_k}\cdot(\mathbf{1}-e_k)\cdot\delta(e_k)f_k\Big).
\end{aligned}
\tag{7.38}
$$

The choice of $\{n_k\}$ and the estimate $\|b_kq\|_{\mathcal{M}} \le 2^{-k}$, $k \ge 3$ implies the following estimate on the last two terms on the right-hand side of (7.38):

$$
\begin{aligned}
&\|f_k\delta(e_k)\cdot(\mathbf{1}-e_k)\cdot b_{n_k}f_k + f_kb_{n_k}\cdot(\mathbf{1}-e_k)\cdot\delta(e_k)f_k\|_{\mathcal{M}}\\
&\quad\le 2\|f_k\delta(e_k)\|_{\mathcal{M}}\cdot\|b_{n_k}q\|_{\mathcal{M}} \le 2n_k\cdot 2^{-n_k} \le \frac{1}{8}
\end{aligned}
$$

for all $k \ge 8$. In particular,

$$
f_k\delta(e_k)\cdot(\mathbf{1}-e_k)\cdot b_{n_k}f_k + f_kb_{n_k}\cdot(\mathbf{1}-e_k)\cdot\delta(e_k)f_k \ge -\frac{1}{8}f_k, \quad k \ge 8.
$$

Hence, combining this estimate with (7.38), we infer that

$$
f_k\delta(b)f_k \ge k\Big(f_k\delta(b_{n_k})f_k - \frac{1}{8}f_k\Big).
$$

Since, in addition, $f_k \le e_k \le q$ and $q\delta(b_k)q \ge \frac{1}{4}q$, we conclude that

$$
f_k\delta(b)f_k \ge k\cdot\Big(f_kq\delta(b_{n_k})qf_k - \frac{1}{8}f_k\Big) \ge k\Big(\frac{1}{4}f_k - \frac{1}{8}f_k\Big) = \frac{k}{8}f_k.
$$

Since the projections $\{f_k\}_{k\in\mathbb{N}_0}$ are pairwise equivalent and $\delta(b) \in LS(\mathcal{M})$ (see (7.37)), Proposition 3.3.17 implies that $f_k = 0$ for all $k \in \mathbb{N}_0$, which is a contradiction. Hence, $\delta : \mathcal{A} \to \mathcal{A}$ is continuous with respect to $\|\cdot\|_{\mathcal{A}}$, as required. □

7.5 Automatic continuity of derivations on the algebras of compact and τ-compact operators

In this section, we show that any derivation on the algebra $S_0(\mathcal{M})$ of all compact operators and the algebra $S_0(\mathcal{M})$ of τ-compact operators $S_0(\mathcal{M},\tau)$ associated with a semifinite properly infinite von Neumann algebra is automatically continuous with respect to the measure topology (in the latter case $\mathcal{M}$ is assumed to be equipped with a faithful normal semifinite trace τ).

We first prove an auxiliary lemma, which shall provide the contradiction for our argument in the proof of Theorem 7.5.2.

Lemma 7.5.1. *Suppose that $\mathcal{M}$ is a properly infinite von Neumann algebra. If $x \in LS(\mathcal{M})$ and $\{p_m\}_{m\in\mathbb{N}} \subset \mathcal{M}$ is a sequence of nonzero pairwise orthogonal and pairwise equivalent projections, such that the sequence $\{p_m x p_m - p_m\}_{m\in\mathbb{N}}$ converges to zero locally in measure, then $x \notin S_0(\mathcal{M})$.*

Proof. Since $(LS(\mathcal{M}), t(\mathcal{M}))$ is a topological $*$-algebra (see Theorem 4.3.14), without loss of generality, we can assume that x is self-adjoint. Furthermore, passing to a subsequence if necessary, we can assume that

$$\|p_m x p_m - p_m\|_{LS(\mathcal{M})} < 2^{-m-1}, \quad m \in \mathbb{N}.$$

By assumption, the sequence $\{p_m\}_{m\in\mathbb{N}}$ consists of pairwise equivalent projections. Denote by $\{v_m\}_{m\in\mathbb{N}} \subset \mathcal{M}$ a sequence of partial isometries, such that

$$v_m v_m^* = p_1, \quad v_m^* v_m = p_m, \quad m \in \mathbb{N}.$$

Proposition 4.3.4 guarantees that there exists $r_m \in P_{\text{fin}}(\mathcal{M})$, $z_m \in P(\mathcal{Z}(\mathcal{M}))$, such that

$$\|(p_m x p_m - p_m)(\mathbf{1} - r_m)\|_{\mathcal{M}} < 2^{-m}, \quad \mathcal{D}(r_m z_m) \le 2^{-m-1}\varphi(z_m),$$

and $\psi(\mathbf{1} - z_m) \le 2^{-m-1}$ for all $m \in \mathbb{N}$. Here, φ denotes a $*$-isomorphism between $\mathcal{Z}(\mathcal{M})$ and $L_\infty(\Omega, \Sigma, \mu)$, and ψ denotes a normal state on $\mathcal{Z}(\mathcal{M})$ corresponding to the integration in (Ω, Σ, μ).

Let $f_m = s_l(p_m r_m)$, $m \in \mathbb{N}$. Then $f_m \preceq r_m$ and $f_m \le p_m$. Since $p_m r_m p_m = f_m p_m r_m p_m f_m \le f_m p_m f_m = f_m$, it follows that $p_m - p_m r_m p_m \ge p_m - f_m$ and

$$\begin{aligned}\|(p_m x p_m - p_m)(p_m - f_m)\|_{\mathcal{M}} &= \|(p_m x p_m - p_m)(\mathbf{1} - f_m)\|_{\mathcal{M}} \\ &\le \|(p_m x p_m - p_m)(\mathbf{1} - r_m)\|_{\mathcal{M}} < 2^{-m}\end{aligned}$$

for all $m \in \mathbb{N}$. In addition,

$$\mathcal{D}(f_m z_m) \le \mathcal{D}(r_m z_m) \le 2^{-m-1}\varphi(z_m), \quad m \in \mathbb{N}.$$

Hence, without loss of generality, we can assume that $r_m = f_m$, which guarantees, in particular, that $r_m \le p_m$ for all $m \in \mathbb{N}$.

For every $m \in \mathbb{N}$, we define

$$\lambda_m := \|(p_m x p_m - p_m)(p_m - r_m)\|_{\mathcal{M}}, \quad e_m := E_{|p_m x p_m - p_m|}[0, \lambda_m],$$

where the spectral measure $E_{|p_m x p_m - p_m|}$ is considered in the algebra $p_m \mathcal{M} p_m$, implying, in particular, that $e_m \le p_m$ for all $m \in \mathbb{N}$.

Lemma 4.3.1(i), applied to the algebra $p_m \mathcal{M} p_m$, implies that

$$\mathcal{D}(e_m^{\perp}) = \mathcal{D}(E_{|p_m x p_m - p_m|}(\lambda_m, \infty)) \le \mathcal{D}(r_m).$$

Therefore, $\mathcal{D}(e_m^{\perp} z_m) \le \mathcal{D}(r_m z_m) \le 2^{-m-1}\varphi(z_m)$.

We set

$$z := \inf_{m\in\mathbb{N}} z_m \in P(\mathcal{Z}(\mathcal{M})), \quad r := z \cdot \inf_{m\in\mathbb{N}} v_m e_m v_m^* \in P(\mathcal{M}), \quad m \in \mathbb{N}.$$

Since ψ is a normal state on $\mathcal{Z}(\mathcal{M})$ and $\psi(\mathbf{1} - z_m) \le 2^{-m-1}$, it follows that

$$\psi(\mathbf{1} - z) = \psi\Big(\sup_{m\in\mathbb{N}}(\mathbf{1} - z_m)\Big) = \sup_{m\in\mathbb{N}} \psi(\mathbf{1} - z_m) < \frac{1}{2},$$

implying that $\psi(z) > \frac{1}{2}$. In particular, $z \ne 0$. Furthermore, since $z \ge r$ and it follows that $\mathcal{D}(z) \ge \mathcal{D}(r)$.

We have that

$$\begin{aligned}
\mathcal{D}(z - r) &= \mathcal{D}\Big(z - z \cdot \inf_{m\in\mathbb{N}} v_m e_m v_m^*\Big) = \mathcal{D}\Big(\sup_{m\in\mathbb{N}}(z - z v_m e_m v_m^*)\Big) \\
&\le \sup_{m\in\mathbb{N}} \mathcal{D}(z - z v_m e_m v_m^*) = \sup_{m\in\mathbb{N}}(\mathcal{D}(z) - \mathcal{D}(z v_m e_m v_m^*)) \\
&= \sup_{m\in\mathbb{N}}(\mathcal{D}(z) - \mathcal{D}(z e_m v_m^* v_m)) = \sup_{m\in\mathbb{N}}(\mathcal{D}(z) - \mathcal{D}(z e_m p_m)) \\
&= \sup_{m\in\mathbb{N}}(\mathcal{D}(z) - \mathcal{D}(z e_m)) = \sup_{m\in\mathbb{N}} \mathcal{D}(z e_m^{\perp}) \\
&\le \sup_{m\in\mathbb{N}} \mathcal{D}(z_m e_m^{\perp}) \le \frac{1}{2} \cdot \mathbf{1}.
\end{aligned}$$

Multiplying both sides of this inequality by $\varphi(z)$, we infer that

$$\mathcal{D}(z - r) = \mathcal{D}(z(z - r)) = \varphi(z)\mathcal{D}(z - r) \le \frac{1}{2} \cdot \varphi(z).$$

Thus,

$$\mathcal{D}(r) \ge \mathcal{D}(z) - \frac{1}{2} \cdot \varphi(z) = \varphi(z)\Big(\mathcal{D}(\mathbf{1}) - \frac{1}{2} \cdot \mathbf{1}\Big). \tag{7.39}$$

Since $\mathcal{M}$ is a properly infinite von Neumann algebra, it follows that the function $\mathcal{D}(\mathbf{1})$ takes infinite value almost everywhere (see Remark 1.11.5). Therefore, the right-hand side of (7.39) takes infinite value almost everywhere on the support of $\varphi(z)$. This implies that $r \ne 0$.

Next, we define

$$q_m := v_m^* r v_m, \quad m \in \mathbb{N}.$$

The definition of r immediately implies that $q_m \le e_m$ for all $m \in \mathbb{N}$. Since $e_m = E_{|p_m x p_m - p_m|}[0, \lambda_m] \le p_m$, it follows that

$$\begin{aligned}\|q_m x q_m - q_m\|_{\mathcal{M}} &= \|q_m p_m x p_m e_m q_m - q_m p_m e_m q_m\|_{\mathcal{M}} \\ &\le \|(p_m x p_m - p_m) e_m\|_{\mathcal{M}} \le \lambda_m < 2^{-m}, \quad m \in \mathbb{N}.\end{aligned}$$

Hence, $q_m x q_m - q_m \ge -\frac{1}{2} \cdot \mathbf{1}$ for all $m \in \mathbb{N}$. Multiplying the latter inequality by q_m from both sides, we obtain that $q_m x q_m - q_m \ge -\frac{1}{2} q_m$, so that

$$q_m x q_m \ge \frac{1}{2} q_m, \quad m \in \mathbb{N}. \tag{7.40}$$

By definition of the projections $\{q_m\}_{m \in \mathbb{N}}$, we have that $q_m \sim r$ for all $m \in \mathbb{N}$. Since, in addition, $q_m \le p_m$, $m \in \mathbb{N}$, and the projections p_m, $m \in \mathbb{N}$ are pairwise orthogonal, it follows that the projections q_m are also pairwise orthogonal. Thus, the sequence $\{q_m\}_{m \in \mathbb{N}}$ consists of pairwise orthogonal and pairwise equivalent projections in the algebra $z\mathcal{M}$. In particular, the projection $\sup_{m \in \mathbb{N}} q_m$ is an infinite projection in the algebra $z\mathcal{M}$.

Now, assume by contradiction that $x \in S_0(\mathcal{M})$. Since $S_0(\mathcal{M})$ is an absolutely solid $*$-subalgebra in $LS(\mathcal{M})$ (see Proposition 2.6.4(i)), Proposition 3.3.13 implies that $q_m x q_m \in S_0(\mathcal{M})$ for all $m \in \mathbb{N}$. Inequality (7.40) implies that $q_m \in S_0(\mathcal{M})$ for all $m \in \mathbb{N}$, i. e., each projection q_m, $m \in \mathbb{N}$ is a finite projection. Since $q_m \sim r$, it follows that r is also a finite projection, and so $\mathcal{D}(r)$ is finite almost everywhere. Inequality (7.39) implies that $\mathcal{D}(r) + \frac{1}{2}\varphi(z) \ge \mathcal{D}(z)$, implying that $\mathcal{D}(z)$ is also finite almost everywhere. Definition 1.11.1(ii) implies that the projection z is finite. This is a contradiction since $\mathcal{M}$ is a properly infinite von Neumann algebra. The obtained contradiction implies that $x \notin S_0(\mathcal{M})$. □

We now prove automatic continuity of derivations on the algebras of compact and τ-compact operators associated with a semifinite properly infinite algebra $\mathcal{M}$.

Theorem 7.5.2. *Let $\mathcal{M}$ be a semifinite properly infinite von Neumann algebra, equipped with a faithful normal semifinite trace τ.*
(i) *Every derivation $\delta : S_0(\mathcal{M}) \to S_0(\mathcal{M})$ is continuous locally in measure;*
(ii) *Every derivation $\delta : S_0(\mathcal{M}, \tau) \to S_0(\mathcal{M}, \tau)$ is continuous in measure.*

Proof. We will prove both cases (i) and (ii) simultaneously. For this end, we denote by $(\mathcal{A}, \|\cdot\|_{\mathcal{A}})$ either the pair $(S_0(\mathcal{M}), t(\mathcal{M}))$ (for case (i)) or the pair $(S_0(\mathcal{M}, \tau), \|\cdot\|_{S(\mathcal{M},\tau)})$ (for case (ii)). By Corollary 3.3.15 and Proposition 4.2.14, the pair $(S_0(\mathcal{M}, \tau), \|\cdot\|_{S(\mathcal{M},\tau)})$ is an absolutely solid $*$-subalgebra of $LS(\mathcal{M})$ and an F-space. Similarly, by Corollary 3.3.15 and Theorem 4.3.23, the pair $(S_0(\mathcal{M}), \|\cdot\|_{LS(\mathcal{M})})$ is an absolutely solid $*$-subalgebra of $LS(\mathcal{M})$ and an F-space. Thus, in both cases $(\mathcal{A}, \|\cdot\|_{\mathcal{A}})$ is an absolutely solid $*$-subalgebra of $LS(\mathcal{M})$ and an F-space. In particular, $(\mathcal{A}, \|\cdot\|_{\mathcal{A}})$ satisfies the assumptions of Proposition 7.2.3.

Let δ be a derivation on $\mathcal{A}$. By Proposition 5.1.4, without loss of generality, we can assume that δ is self-adjoint.

By Proposition 7.2.3, we can find a paving projection $p \in z\mathcal{A}$ and sequence $\{a_n\}_{n\in\mathbb{N}_0} \subset \mathcal{A}$, such that $\|a_n\|_{\mathcal{A}} \to 0$ and $\|\delta(a_n) - p\|_{\mathcal{A}} \to 0$ as $n \to \infty$ with $a_n = a_n^*$ and $a_n p = a_n$ for all $n \in \mathbb{N}_0$.

Let v be the isometry corresponding to the paving projection p (see Definition 7.2.1). As before, we denote $p_m = v^m p(v^m)^*$. By definition of paving projections, the projections $p_m, m \in \mathbb{N}_0$ are pairwise orthogonal and $\sup_{m\in\mathbb{N}_0} p_m = \mathbf{1}$.

For all $m, n \in \mathbb{N}_0$, we set

$$a_{m,n} = v^m a_n (v^m)^* = (v^m p) a_n (v^m p)^* = p_m v^m a_n (v^m)^* p_m.$$

Then we have

$$\|a_{m,n}\|_{\mathcal{A}} \le \big\| \|p_m v^m\|_{\mathcal{M}} \|(v^m)^* p_m\|_{\mathcal{M}} a_n \big\|_{\mathcal{A}} \le \|a_n\|_{\mathcal{A}} \to 0.$$

Furthermore, since $\|\delta(a_n) - p\|_{\mathcal{A}} \to 0$ as $n \to \infty$, Proposition 4.2.7 for the case $(\mathcal{A}, \|\cdot\|_{\mathcal{A}}) = (S_0(\mathcal{M},\tau), t_\tau)$ and Proposition 4.3.13(vi) for the case $(\mathcal{A}, \|\cdot\|_{\mathcal{A}}) = (S_0(\mathcal{M}), t(\mathcal{M}))$ implies that

$$\begin{aligned}\|\delta(a_{m,n}) - p_m\|_{\mathcal{A}} \le \|\delta(v^m)\cdot a_n \cdot (v^m)^*\|_{\mathcal{A}} + \|v^m \delta(a_n)(v^m)^* - p_m\|_{\mathcal{A}} \\ + \|v^m a_n \delta(v^m)^*\|_{\mathcal{A}} \to 0\end{aligned}$$

as $n \to \infty$.

We can choose $n(m)$ such that

$$\|a_{m,n(m)}\|_{\mathcal{A}} < 2^{-m-1}, \quad \|\delta(a_{m,n(m)}) - p_m\|_{\mathcal{A}} < 2^{-m-1}, \quad \|a_{m,n(m)}\delta(p_m)\|_{\mathcal{A}} < 2^{-m-1},$$

for all $m \in \mathbb{N}_0$. Setting

$$a = \sum_{m=0}^{\infty} a_{m,n(m)}$$

we have that

$$\|a\|_{\mathcal{A}} \le \sum_{m=0}^{\infty} \|a_{m,n(m)}\|_{\mathcal{A}} \le \sum_{m=0}^{\infty} 2^{-m-1} < \infty.$$

Since $a_{m,n(m)} \in \mathcal{A}$ and $(\mathcal{A}, \|\cdot\|_{\mathcal{A}})$ is complete, it follows that $a \in \mathcal{A}$. Thus, $\delta(a) \in \mathcal{A}$.

Since $\{p_m\}_{m\in\mathbb{N}_0}$ are pairwise orthogonal, it follows from the Leibniz rule and the definition of a that

$$p_m \cdot \delta(a) \cdot p_m = p_m \cdot (\delta(a p_m) - a\delta(p_m)) = p_m \delta(a_{m,n(m)}) - a_{m,n(m)}\delta(p_m).$$

By the choice of $n(m)$, we obtain that

$$\begin{aligned} &\|p_m\delta(a)p_m - p_m\|_{\mathcal{A}} \\ &\le \|p_m\delta(a_{m,n(m)}) - p_m\|_{\mathcal{A}} + \|a_{m,n(m)}\delta(p_m)\|_{\mathcal{A}} \\ &< 2^{-m-1} + 2^{-m-1} = 2^{-m} \to 0, \quad m \to \infty. \end{aligned} \tag{7.41}$$

If $(\mathcal{A}, \|\cdot\|_{\mathcal{A}}) = (S_0(\mathcal{M}), \|\cdot\|_{LS(\mathcal{M})})$, then by Lemma 7.5.1, convergence (7.41) implies that $\delta(a) \notin S_0(\mathcal{M})$, which is a contradiction. If $(\mathcal{A}, \|\cdot\|_{\mathcal{A}}) = (S_0(\mathcal{M}, \tau), \|\cdot\|_{S(\mathcal{M},\tau)})$, then since the topology of convergence in measure is stronger than the topology of convergence locally in measure (see Proposition 4.4.1), convergence (7.41) implies that $\|p_m\delta(a)p_m - p_m\|_{LS(\mathcal{M})} \to 0$ as $m \to \infty$. Referring to Lemma 7.5.1, we obtain in this case that $\delta(a) \notin S_0(\mathcal{M})$. Since $\delta(a) \in S_0(\mathcal{M}, \tau) \subset S_0(\mathcal{M})$, this is again a contradiction. Thus, in both the case of $(\mathcal{A}, \|\cdot\|_{\mathcal{A}}) = (S_0(\mathcal{M}), \|\cdot\|_{LS(\mathcal{M})})$ or $(\mathcal{A}, \|\cdot\|_{\mathcal{A}}) = (S_0(\mathcal{M}, \tau), \|\cdot\|_{S(\mathcal{M},\tau)})$, we conclude that the derivation δ is continuous in the respective topology. □

7.6 Vanishing first cohomology group

In this section, we combine the results of the previous sections to prove the main result of the present chapter. Namely, we show that all derivations on algebras of locally measurable and τ-measurable operators are inner, provided that the algebra $\mathcal{M}$ does not have a type I_{fin} direct summand. For the algebras of compact and τ-compact operators, such an assumption on the von Neumann algebra $\mathcal{M}$ is a sufficient condition for all derivations to be spatial.

Theorem 7.6.1. *Let $\mathcal{M}$ be a von Neumann algebra with no type I_{fin} direct summand. Then:*
(i) *any derivation δ on $LS(\mathcal{M})$ is inner;*
(ii) *any derivation δ on $S(\mathcal{M})$ is inner;*
(iii) *any derivation δ on $S_0(\mathcal{M})$ is spatial, i. e., there exists $d \in LS(\mathcal{M})$, such that $\delta = \delta_d$.*

If, in addition, $\mathcal{M}$ is equipped with a faithful normal semifinite trace τ then:
(iv) *any derivation δ on $S(\mathcal{M}, \tau)$ is inner;*
(v) *any derivation δ on $S_0(\mathcal{M}, \tau)$ is spatial, i. e., there exists $d \in S(\mathcal{M}, \tau)$, such that $\delta = \delta_d$.*

Proof. Let $\mathcal{M}$ be a von Neumann algebra with no type I_{fin} direct summand. By Theorem 1.9.14, there exist pairwise orthogonal central projections z_1, z_2, z_3 such that $z_1 + z_2 + z_3 = \mathbf{1}$ and $z_1\mathcal{M}$ is a type II_1 algebra, $z_2\mathcal{M}$ is a semifinite properly infinite algebra, and $z_3\mathcal{M}$ is a type III algebra.

(i) Assume first that δ is a derivation on $LS(\mathcal{M})$. By definition of reduced derivations (see Definition 5.1.7), for every $i = 1, 2, 3$, the mapping $\delta^{(z_i)}$ is a derivation on $LS(z_i\mathcal{M})$. Since $z_1\mathcal{M}$ is a type II_1 algebra, Theorem 7.1.7 implies that the derivation $\delta^{(z_1)}$ is $t(z_1\mathcal{M})$-continuous. Similarly, Theorem 7.3.7 guarantees that $\delta^{(z_3)}$ is $t(z_3\mathcal{M})$-continuous on $LS(z_3\mathcal{M})$, while Theorem 7.4.16(i)) guarantees that $\delta^{(z_2)}$ is $t(z_2\mathcal{M})$-continuous on

$LS(z_2\mathcal{M})$. Referring to Proposition 5.1.11, we obtain that δ is $t(\mathcal{M})$-continuous derivation on $LS(\mathcal{M})$. Hence, by Theorem 5.5.6(i). we have that δ is inner.

(ii) Since $S(\mathcal{M})$ is an absolutely solid $*$-subalgebra of $LS(\mathcal{M})$ (see Corollary 3.3.15), containing $\mathcal{M}$, the assertion follows from part (i) and Theorem 5.2.7(ii).

(iii) Since $z_3 S_0(\mathcal{M}) = 0$, without loss of generality, we can assume that $z_3 = 0$. Since $z_1\mathcal{M}$ is a type II_1 von Neumann algebra, we have that $S_0(\mathcal{M})z_1 = S(\mathcal{M}z_1)$. By (ii), $\delta^{(z_1)} = \delta_{d_1}$ for some $d_1 \in S(\mathcal{M}z_1)$.

Since $z_2\mathcal{M}$ is a semifinite properly infinite algebra, Theorem 7.5.2(i) implies that $\delta^{(z_2)}$ is continuous locally in measure. By Theorem 5.5.6(ii) there exists $d_2 \in LS(\mathcal{M}z_2)$, such that $\delta^{(z_2)} = \delta_{d_2}$ on $S_0(z_2\mathcal{M})$. Setting $d = d_1 + d_2 \in LS(\mathcal{M})$, Proposition 5.1.9 implies that for any $x \in \mathcal{M}$, we have

$$\begin{aligned}\delta(x) &= \delta(z_1 x + z_2 x) = z_1\delta(z_1 x) + z_2\delta(z_2 x) = \delta^{(z_1)}(z_1 x) + \delta^{(z_2)}(z_2 x)\\ &= [d_1, z_1 x] + [d_2, z_2 x] = [d, x],\end{aligned}$$

proving that δ is spatial.

Next, assume that $\mathcal{M}$ is a semifinite von Neumann algebra equipped with a faithful normal semifinite trace τ. In this case, $z_3 = 0$.

(iv) Since $S(\mathcal{M}, \tau)$ is an absolutely solid $*$-subalgebra of $LS(\mathcal{M})$ (see Corollary 3.3.15), containing $\mathcal{M}$, the assertion follows from part (i) and Theorem 5.2.7(ii).

(v) Since $z_1\mathcal{M}$ is type II_1, Theorem 7.1.8 implies that $\delta^{(z_1)}$ is continuous in measure. Since $z_2\mathcal{M}$ is properly infinite, Theorem 7.5.2(ii) implies that $\delta^{(z_2)}$ is continuous is measure. By Theorem 5.5.6(iii), there exists $d_i \in S(z_i\mathcal{M}, \tau)$, $i = 1, 2$, such that $\delta^{(z_i)} = \delta_{d_i}$ on $S_0(z_i\mathcal{M}, \tau|_{z_i\mathcal{M}})$. Setting $d = d_1 + d_2 \in S(\mathcal{M}, \tau)$, Proposition 5.1.9 implies that

$$\begin{aligned}\delta(x) &= \delta(z_1 x + z_2 x) = z_1\delta(z_1 x) + z_2\delta(z_2 x) = \delta^{(z_1)}(z_1 x) + \delta^{(z_2)}(z_2 x)\\ &= [d_1, z_1 x] + [d_2, z_2 x] = [d, x],\end{aligned}$$

proving that δ is spatial. □

Combining the previous theorem with the necessary and sufficient condition for existence on noninner derivations on $LS(\mathcal{M})$ for a type I_{fin} algebra established in Chapter 6 we infer the following necessary and sufficient condition for the existence of noninner derivations on $LS(\mathcal{M})$ for an arbitrary von Neumann algebra $\mathcal{M}$.

Theorem 7.6.2. *Let $\mathcal{M}$ be an arbitrary von Neumann algebra. The following conditions are equivalent:*

(i) *The type I_{fin} summand of $\mathcal{M}$ is atomic;*

(ii) *All derivations on $LS(\mathcal{M})$ (resp., on $S(\mathcal{M})$) are inner;*

(iii) *All derivations on $S_0(\mathcal{M})$ are spatial, i. e., there exists $d \in LS(\mathcal{M})$, such that $\delta = \delta_d$;*

If, in addition, $\mathcal{M}$ is equipped with a faithful normal semifinite trace τ then the above is equivalent to any of the following conditions:

(iv) *All derivations on* $S(\mathcal{M},\tau)$ *are inner;*

(v) *All derivations on* $S_0(\mathcal{M},\tau)$ *are spatial, i. e., there exists* $d \in S(\mathcal{M},\tau)$*, such that* $\delta = \delta_d$.

Proof. By Theorem 1.9.14, there exist pairwise orthogonal central projections z_1 and z_2, such that $z_1 + z_2 = \mathbf{1}$ and $z_1\mathcal{M}$ has no direct summand of type I_{fin} and $z_2\mathcal{M}$ is a type I_{fin} algebra.

We first prove that (i) is equivalent to each of (ii) and (iii).

(i)⇔(ii). By Proposition 5.1.11, a derivation δ on $LS(\mathcal{M})$ (resp., on $S(\mathcal{M})$) is inner if and only if $\delta^{(z_i)}$ is inner on $LS(z_i\mathcal{M})$ (resp., on $S(z_i\mathcal{M})$) for $i = 1, 2$. Since $z_1\mathcal{M}$ has no direct summand of type I_{fin}, Theorem 7.6.1 implies that $\delta^{(z_1)}$ is always inner (on both $LS(z_1\mathcal{M})$ and $S(z_1\mathcal{M})$). By Theorem 6.6.11, the derivation $\delta^{(z_2)}$ is inner on $LS(z_2\mathcal{M}) = S(z_2\mathcal{M})$ if and only if the algebra $z_2\mathcal{M}$ is atomic.

(i)⇒(iii). Let δ be a derivation on $S_0(\mathcal{M})$. By Proposition 5.1.9, we have that $\delta = \delta^{(z_1)} + \delta^{(z_2)}$. Since $\delta^{(z_1)}$ is a derivation on the algebra $S_0(z_1\mathcal{M})$ and the algebra $z_1\mathcal{M}$ has no direct summand of type I_{fin}, Theorem 7.6.1 implies that $\delta^{(z_1)}$ is always spatial, i. e., there exists $d_1 \in LS(\mathcal{M}z_1)$, such that $\delta|_{z_1S_0(\mathcal{M})} = \delta_{d_1}$. Since $z_2\mathcal{M}$ is a finite von Neumann algebra, we have that $z_2S_0(\mathcal{M}) = S_0(z_2\mathcal{M}) = S(z_2\mathcal{M})$. Since $z_2\mathcal{M}$ is atomic, Theorem 6.6.11 guarantees that the derivation $\delta^{(z_2)}$ is inner on $S(z_2\mathcal{M})$ (i. e., there exists $d_2 \in S(z_2\mathcal{M})$ such that $\delta^{(z_2)} = \delta_{d_2}$). Hence, $\delta = \delta_{d_1+d_2}$ is a spatial derivation implemented by $d_1 + d_2 \in LS(\mathcal{M})$.

(iii)⇒(i). Let δ be a derivation on $S(z_2\mathcal{M})$. The mapping $\tilde{\delta}$, defined by $\tilde{\delta}(x) = \delta(z_2x)$, $x \in S_0(\mathcal{M})$ is a derivation on $S_0(\mathcal{M})$. By assumption, $\tilde{\delta}$ is spatial, so that there exists $d \in LS(\mathcal{M})$, such that $\tilde{\delta} = \delta_d$. Therefore, for any $x \in S(z_2\mathcal{M})$ we have that

$$\delta(x) = \delta(z_2x) = \tilde{\delta}(z_2x) = [d, z_2x] = [z_2d, x].$$

Since $z_2d \in S(z_2\mathcal{M})$, it follows that the derivation δ is an inner derivation on $S(z_2\mathcal{M})$. Thus, any derivation on $S(z_2\mathcal{M})$ is inner. Referring to Theorem 6.6.11 we conclude that $z_2\mathcal{M}$ is atomic, as required.

Next, to prove the equivalence of (i) with each of (iv) and (v), we assume that $\mathcal{M}$ is a semifinite von Neumann algebra equipped with a faithful normal semifinite trace τ.

(i)⇒(iv). It follows from (ii) that all derivations on $LS(\mathcal{M})$ are inner. Since $S(\mathcal{M},\tau)$ is an absolutely solid $*$-subalgebra of $LS(\mathcal{M})$ containing $\mathcal{M}$ (see Corollary 3.3.15), Theorem 5.2.7(ii) implies that all derivations on $S(\mathcal{M},\tau)$ are inner.

(iv)⇒(v). Let δ be a derivation on $S_0(\mathcal{M},\tau)$. Proposition 5.1.9 implies that $\delta(S_0(\mathcal{M}z_i,\tau)) \subset S_0(\mathcal{M}z_i,\tau)$, $i = 1, 2$. In particular, $\delta_i = \delta|_{S_0(\mathcal{M}z_i,\tau)}$, $i = 1, 2$ is a derivation on $S_0(\mathcal{M}z_i,\tau)$. Furthermore, $\delta = \delta_1 + \delta_2$.

Since $z_1\mathcal{M}$ has no direct summand of type I_{fin}, Theorem 7.6.1(v) implies that there exists $d_1 \in S(z_1\mathcal{M}, \tau|_{z_1\mathcal{M}})$, such that $\delta_1 = [\cdot, d_1]$. Since $\delta = \delta_1 + \delta_2$, it is sufficient to show that δ_2 is implemented by an element from $S(z_2\mathcal{M}, \tau|_{z_2\mathcal{M}})$. Without loss of generality, we can assume that $z_2 = \mathbf{1}$ (in which case $\delta_2 = \delta$).

By Theorem 5.5.6(iv), $\delta : S_0(\mathcal{M}, \tau) \to S_0(\mathcal{M}, \tau)$ is a spatial derivation, if and only if $\delta : S_0(\mathcal{M}, \tau) \to S_0(\mathcal{M}, \tau)$ is continuous in measure. Assume the contrary. Since the pair $(S_0(\mathcal{M}, \tau), \|\cdot\|_{S(\mathcal{M},\tau)})$ is an F-space (see Proposition 4.2.14), by the closed graph theorem there exists a sequence $\{x_n\}_{n\in\mathbb{N}} \subset S_0(\mathcal{M}, \tau)$, and $0 \neq x \in S_0(\mathcal{M}, \tau)$, such that $x_n \to 0$ and $\delta(x_n) \to x$ in measure as $n \to \infty$.

Since $\mathcal{M}$ is a type I_{fin} algebra, Corollary 1.10.16 implies that there exists a central partition $\{w_n\}_{n\in\mathbb{N}}$ of unity, such that w_n is τ-finite for all $n \in \mathbb{N}$. Since $x \neq 0$, there exists $k \in \mathbb{N}$, such that $w_k x \neq 0$. For this $k \in \mathbb{N}$, we have that $w_k x_n \to 0$, and

$$\delta(w_k x_n) = w_k \delta(x_n) \to w_k x \neq 0, \tag{7.42}$$

in measure as $n \to \infty$.

Proposition 5.1.9 implies that $\delta(S_0(\mathcal{M}w_k, \tau)) \subset S_0(\mathcal{M}w_k, \tau)$. Since the projection w_k is τ-finite, it follows that $S_0(\mathcal{M}w_k, \tau) = S(\mathcal{M}w_k, \tau)$ and, therefore, $\delta(S(\mathcal{M}w_k, \tau)) \subset S(\mathcal{M}w_k, \tau)$, i. e., $\delta|_{S(w_k\mathcal{M},\tau)}$ is a derivation on $S(w_k\mathcal{M}, \tau)$. Define a derivation $\delta_0 : S(\mathcal{M}, \tau) \to S(\mathcal{M}, \tau)$, by setting $\delta_0(x) = \delta(w_k x)$, $x \in S(\mathcal{M}, \tau)$. By assumption, δ_0 is inner and, therefore, is continuous in measure. In particular, $\delta(w_k x_n) = \delta_0(x_n) \to 0$ in measure. This contradicts with (7.42). Therefore, $\delta : S_0(\mathcal{M}, \tau) \to S_0(\mathcal{M}, \tau)$ is continuous in measure. By Theorem 5.5.6(iv), the derivation δ is spatial, as required.

(v)⇒(i). Suppose on the contrary that a type I_{fin} direct summand of $\mathcal{M}$ is not atomic. By Corollary 1.10.16, there exists $z \in P(\mathcal{Z}(\mathcal{M}))$, such that $\tau(z) < \infty$ and $\mathcal{M}z$ is not an atomic von Neumann algebra of type I_{fin}. Then $S_0(\mathcal{M}, \tau)z = S_0(\mathcal{M}z, \tau) = S(\mathcal{M}z, \tau) = S(\mathcal{M}z)$.

By Theorem 6.6.11, there exists a noninner derivation δ_0 on $S(\mathcal{M}z)$. Let δ be a derivation on $S_0(\mathcal{M}, \tau)$, defined by setting $\delta(x) = \delta_0(xz)$, $x \in S_0(\mathcal{M}, \tau)$. By assumption, all derivations on $S_0(\mathcal{M}, \tau)$ are spatial, and so there exists $d \in LS(\mathcal{M})$, such that $\delta = [d, \cdot]$. Then for any $x \in S(\mathcal{M}z)$, we have

$$\delta_0(x) = \delta(x) = [d, x] = [dz, x],$$

which contradicts with the choice of δ_0 since $dz \in S(\mathcal{M}z)$. Therefore, a type I_{fin} direct summand of $\mathcal{M}$ is atomic. □

Theorem 5.2.7 combined with Theorem 7.6.2 implies also the following result.

Corollary 7.6.3. *Let $\mathcal{M}$ be an arbitrary von Neumann algebra with an atomic type I_{fin} direct summand and let $\mathcal{A}$ be an absolutely solid $*$-subalgebra, such that $\mathcal{M} \subset \mathcal{A} \subset LS(\mathcal{M})$. Then any derivation on $\mathcal{A}$ is inner.*

Proof. By Theorem 7.6.2, all derivations on $LS(\mathcal{M})$ are inner. Hence, Theorem 5.2.7(ii) implies that all derivations on $\mathcal{A}$ are inner, too. □

7.7 Derivations with values in quasi-normed bimodules of locally measurable operators

In this section, we study derivations acting on a von Neumann algebra $\mathcal{M}$ with values in a (quasi-normed) bimodule over $\mathcal{M}$. Recall that bimodules over a given von Neumann algebra are defined in Definition 3.1.3.

Definition 7.7.1. Let $\mathcal{E}$ be an $\mathcal{M}$-bimodule of locally measurable operators over a given von Neumann algebra $\mathcal{M}$. A linear mapping $\delta : \mathcal{M} \to \mathcal{E}$ is called a *derivation* (with values in $\mathcal{E}$), if $\delta(ab) = \delta(a)b + a\delta(b)$ for all $a, b \in \mathcal{M}$. A derivation $\delta : \mathcal{M} \to \mathcal{E}$ is called *inner*, if there exists an element $d \in \mathcal{E}$, such that $\delta(x) = [d, x] = dx - xd$ for all $x \in \mathcal{M}$.

When $\mathcal{M}$ has an atomic direct summand of type I_{fin} we can immediately describe derivations from $\mathcal{M}$ into an $\mathcal{M}$-bimodule $\mathcal{E}$. This result follows from a combination of Theorem 7.6.2 and the commutator estimates established in Theorem 3.6.11 without any assumption of a topological structure on $\mathcal{E}$.

Theorem 7.7.2. *Let $\mathcal{M}$ be a von Neumann algebra with atomic direct summand of type I_{fin} and let $\mathcal{E}$ be an $\mathcal{M}$-bimodule of locally measurable operators. Then any derivation $\delta : \mathcal{M} \to \mathcal{E}$ is inner.*

Proof. By Theorem 5.2.6, there exists a derivation $\widetilde{\delta} : LS(\mathcal{M}) \to LS(\mathcal{M})$, such that $\widetilde{\delta}(x) = \delta(x)$ for all $x \in \mathcal{M}$. By Theorem 7.6.2, there exists an element $a \in LS(\mathcal{M})$, such that $\widetilde{\delta}(x) = [a, x]$ for all $x \in LS(\mathcal{M})$. It is clear that $[a, \mathcal{M}] = \widetilde{\delta}(\mathcal{M}) = \delta(\mathcal{M}) \subset \mathcal{E}$.

Let $a_1 = \mathrm{Re}(a)$, $a_2 = \mathrm{Im}(a)$. Since $[a^*, x] = -[a, x^*]^* \in \mathcal{E}$ for any $x \in \mathcal{M}$, it follows that $[a_1, x] = [a + a^*, x]/2 \in \mathcal{E}$ and $[a_2, x] = [a - a^*, x]/2i \in \mathcal{E}$ for all $x \in \mathcal{M}$.

Taking $\varepsilon = 1/2$ in Theorem 3.6.11, we obtain that there exist self-adjoint $c_1, c_2 \in \mathcal{Z}(LS(\mathcal{M}))$, and unitary operators $u_1, u_2 \in \mathcal{M}$, such that

$$2\big|[a_i, u_i]\big| \geq |a_i - c_i|, \quad i = 1, 2.$$

Since $[a_i, u_i] \in \mathcal{E}$ and $\mathcal{E}$ is an $\mathcal{M}$-bimodule, Proposition 3.3.10 implies that $d_i := a_i - c_i \in \mathcal{E}$, $i = 1, 2$. Therefore, $d = d_1 + id_2 \in \mathcal{E}$. Since c_1, c_2 are central elements from $LS(\mathcal{M})$, it follows that $\delta(x) = [a, x] = [d, x]$ for all $x \in \mathcal{M}$, as required. □

When $\mathcal{M}$ has a not atomic direct summand of type I_{fin}, then a derivation from $\mathcal{M}$ into an $\mathcal{M}$-bimodule $\mathcal{E}$ may be noninner. Indeed, let $\mathcal{M}$ be a nonatomic von Neumann algebra of type I_{fin} and let $\mathcal{E} = S(\mathcal{M})$. As shown in Theorem 6.4.9, there are noninner derivations from $S(\mathcal{M})$ into $S(\mathcal{M})$. In particular, there are noninner derivations from $\mathcal{M}$ into the $\mathcal{M}$-bimodule $S(\mathcal{M})$. Note that in this case $S(\mathcal{M}) = LS(\mathcal{M})$ is also an $\mathcal{M}$-bimodule equipped with an F-norm $\|\cdot\|_{LS(\mathcal{M})}$. Therefore, for all derivations from $\mathcal{M}$ into an $\mathcal{M}$-bimodule $\mathcal{E}$ to be inner, the bimodule $\mathcal{E}$ must be equipped with a more robust topology.

The first result for derivations with values in $\mathcal{M}$-bimodules is Ringrose's theorem, which shows that any derivation on a von Neumann algebra $\mathcal{M}$ with values in a quasi-normed $\mathcal{M}$-bimodule $\mathcal{E}$ is automatically continuous as a mapping from $(\mathcal{M}, \|\cdot\|_{\mathcal{M}})$ to $(\mathcal{E}, \|\cdot\|_{\mathcal{E}})$. We start with an auxiliary lemma.

Lemma 7.7.3. *If $\mathcal{I}$ is a closed two-sided ideal in a C^*-algebra $\mathcal{A}$ equipped with norm $\|\cdot\|_{\mathcal{A}}$, $a \in \mathcal{I}$, $b \in \mathcal{I}^+$, $\|b\|_{\mathcal{A}} \leq 1$, and $aa^* \leq b^4$, then $a = bc$ for some $c \in \mathcal{I}$ with $\|c\|_{\mathcal{A}} \leq 1$.*

Proof. Without loss of generality, we may assume that $\mathcal{A}$ has an identity $\mathbf{1}$. For every $t > 0$, we define $c_t \in \mathcal{I}$, by setting $c_t = (b + t\mathbf{1})^{-1}a$. Since $\sigma(b) \subset [0,1]$, the functional calculus and Proposition 1.7.6(ii) implies that

$$c_t c_t^* = (b + t\mathbf{1})^{-1} aa^* (b + t\mathbf{1})^{-1} \leq (b + t\mathbf{1})^{-1} b^4 (b + t\mathbf{1})^{-1} \leq \mathbf{1}.$$

Hence, $\|c_t\|_{\mathcal{A}} \leq 1$. Moreover, by the resolvent identity we have

$$c_s - c_t = (t - s)(b + s\mathbf{1})^{-1}(b + t\mathbf{1})^{-1}a,$$

and so

$$\begin{aligned}(c_s - c_t)(c_s - c_t)^* &= |t - s|^2 (b + s\mathbf{1})^{-1}(b + t\mathbf{1})^{-1} aa^* (b + t\mathbf{1})^{-1}(b + s\mathbf{1})^{-1} \\ &\leq |t - s|^2 (b + s\mathbf{1})^{-1}(b + t\mathbf{1})^{-1} b^4 (b + t\mathbf{1})^{-1}(b + s\mathbf{1})^{-1} \\ &\leq |t - s|^2 \mathbf{1}.\end{aligned}$$

Hence, $\|c_s - c_t\|_{\mathcal{A}} \leq |t - s|$ whenever $s, t > 0$.

Thus, $t \mapsto c_t$ is a uniformly bounded norm-continuous mapping. Therefore, c_t converges in norm, as $t \to 0$, to some c in $\mathcal{A}$. Since

$$c_t \in \mathcal{I}, \quad \|c_t\|_{\mathcal{A}} \leq 1, \quad (b + t\mathbf{1})c_t = a,$$

we conclude that $c \in \mathcal{I}$, $\|c\|_{\mathcal{A}} \leq 1$, and $bc = a$, as required. □

Theorem 7.7.4 (Ringrose's theorem). *Suppose that $\mathcal{M}$ is a von Neumann algebra and $(\mathcal{E}, \|\cdot\|_{\mathcal{E}})$ is a quasi-normed $\mathcal{M}$-bimodule. Then any derivation $\delta : \mathcal{M} \to \mathcal{E}$ is a continuous mapping from $(\mathcal{M}, \|\cdot\|_{\mathcal{M}})$ into $(\mathcal{E}, \|\cdot\|_{\mathcal{E}})$.*

Proof. Suppose that $\delta : \mathcal{M} \to \mathcal{E}$ is a derivation. For any $a \in \mathcal{M}$, we define

$$\varphi_a : \mathcal{M} \to \mathcal{E} : \varphi_a(x) = \delta(ax),$$

and let

$$\mathcal{I} = \{a \in \mathcal{M} : \varphi_a \text{ is continuous from } \mathcal{M} \text{ into } \mathcal{E}\}.$$

It is clear that $\mathcal{I}$ is a right ideal in $\mathcal{M}$ and the relation

$$\varphi_{ba}(x) = \delta(bax) = b\delta(ax) + \delta(b)ax, \quad a \in \mathcal{I},\ b, x \in \mathcal{M},$$

shows that $\mathcal{I}$ is also a left ideal. Since

$$\varphi_a(x) = \delta(ax) = \delta(a)x + a\delta(x), \quad a, x \in \mathcal{M},$$

it follows that $\mathcal{I}$ is the set of all $a \in \mathcal{M}$ for which the mapping

$$\psi_a : \mathcal{M} \to \mathcal{E} : \psi_a(x) = a\delta(x),$$

is continuous.

We claim that $\mathcal{I}$ is a closed ideal in $\mathcal{M}$. Suppose that $\{a_n\}_{n\in\mathbb{N}} \subset \mathcal{I}$ and $a \in \mathcal{M}$ are such that $\|a - a_n\|_{\mathcal{M}} \to 0$. Then each ψ_{a_n} is a bounded linear operator, and

$$\psi_a(x) = \lim_{n\to\infty} \psi_{a_n}(x), \quad x \in \mathcal{M}.$$

The uniform boundedness principle (see, e. g., [129, Theorem 2.6]) implies that ψ_a is norm continuous, and so $a \in \mathcal{I}$. Thus, $\mathcal{I}$ is a closed two-sided ideal in $\mathcal{M}$.

We claim that the restriction $\delta|_{\mathcal{I}}$ is continuous. Assume the contrary. Then there exists $\{a_n\}_{n\in\mathbb{N}} \subset \mathcal{I}$, such that

$$\sum_{n=1}^{\infty} \|a_n\|_{\mathcal{M}}^2 \le 1, \quad \|\delta(a_n)\|_{\mathcal{E}} \to \infty, \quad n \to \infty.$$

Define b by setting $b = (\sum_{n=1}^{\infty} a_n a_n^*)^{\frac{1}{4}}$, where the series converges in the uniform norm. Since $\mathcal{I}$ is norm closed, it follows that $b \in \mathcal{I}$. Furthermore, $b \in \mathcal{I}_+$, $\|b\| \le 1$, and $a_n a_n^* \le b^4$. Lemma 7.7.3 implies that $a_n = bc_n$ for some $c_n \in \mathcal{I}$ with $\|c_n\|_{\mathcal{M}} \le 1$. We have that

$$\|\delta(bc_n)\|_{\mathcal{E}} = \|\delta(a_n)\|_{\mathcal{E}} \to \infty,$$

i. e., the mapping φ_b is unbounded. This contradicts with the inclusion $b \in \mathcal{I}$. Thus, the restriction $\delta|_{\mathcal{I}}$ is a continuous mapping.

Next, we claim that the factor algebra $\mathcal{M}/\mathcal{I}$ is finite-dimensional. Suppose the contrary. Then there exists an infinite-dimensional closed commutative $*$-subalgebra $\mathcal{M}_0$ in $\mathcal{M}/\mathcal{I}$ [118]. Since the carrier space X of $\mathcal{M}_0$ is infinite, the isomorphism between $\mathcal{M}_0$ and $C_0(X)$ implies that there exists a positive operator h in $\mathcal{M}_0$ whose spectrum $\sigma(h)$ is infinite. Let $f_1, f_2, \ldots$ be non-negative continuous functions defined on the positive real axis, such that

$$f_j f_k = 0, \quad j \ne k, \quad f_j(h) \ne 0, \quad j \in \mathbb{N}.$$

Denote by γ the natural mapping from $\mathcal{M}$ onto $\mathcal{M}/\mathcal{I}$ and let $d \in \mathcal{M}$ be a positive element, such that $\gamma(d) = h$. Setting $x_j = f_j(d)$, $j \in \mathbb{N}$, we have that $x_j \in \mathcal{M}$ and

$$\gamma(x_j^2) = \gamma(f_j(d))^2 = [f_j(\gamma(d))]^2 = [f_j(h)]^2 \neq 0.$$

Thus

$$x_j \in \mathcal{M}, \quad x_j^2 \notin \mathcal{I}; \quad x_j x_k = 0, \quad j \neq k.$$

Multiplying x_j by an appropriate scalar, if necessary, we may suppose also that $\|x_j\|_{\mathcal{M}} \leq 1$.

Since $x_j^2 \notin \mathcal{I}$, the mapping $\varphi_{x_j^2}$ is unbounded. Choose $y_1, y_2, \dots \in \mathcal{M}$, such that

$$\|y_j\|_{\mathcal{M}} \leq 2^{-j}, \quad \|\delta(x_j^2 y_j)\|_{\mathcal{E}} \geq C_{\mathcal{E}}(\|\delta(x_j)\|_{\mathcal{E}} + j).$$

For $c := \sum_{j=1}^{\infty} x_j y_j \in \mathcal{M}$, we have $\|c\|_{\mathcal{M}} \leq 1$ and $x_j c = x_j^2 y_j$. Hence,

$$\begin{aligned}\|x_j \delta(c)\|_{\mathcal{E}} &= \|\delta(x_j c) - \delta(x_j)c\|_{\mathcal{E}} \geq C_{\mathcal{E}}^{-1}\|\delta(x_j^2 y_j)\|_{\mathcal{E}} - \|\delta(x_j)c\|_{\mathcal{E}} \\ &\geq \|\delta(x_j)\|_{\mathcal{E}} + j - \|\delta(x_j)\|_{\mathcal{E}}\|c\|_{\mathcal{M}} \geq j,\end{aligned}$$

which is a contradiction, since $\|x_j\|_{\mathcal{M}} \leq 1$ and the mapping $x \to x\delta(c)$ is bounded. Thus, $\mathcal{M}/\mathcal{I}$ is finite-dimensional.

Let $a_1, \dots, a_n \in \mathcal{M}$ be such that $\gamma(a_1), \dots, \gamma(a_n)$ is a basis in $\mathcal{M}/\mathcal{I}$. Let $\{x_m\}_{m\in\mathbb{N}} \subset \mathcal{M}$ be such that $\|x_m\|_{\mathcal{M}} \to 0$. Let $\alpha_{im} \in \mathbb{C}$ be such that

$$\gamma(x_m) = \alpha_{1m}\gamma(a_1) + \dots + \alpha_{nm}\gamma(a_n),$$

for every $m \in \mathbb{N}$. Then

$$x_m = \alpha_{1m}a_1 + \dots + \alpha_{nm}a_n + x_m'$$

for some $x_m' \in \mathcal{I}$. Since $\gamma(x_m) \to 0$, it follows that $\alpha_{im} \to 0$ for all $i = 1, \dots, n$ as $m \to 0$, and so $\|x_m'\|_{\mathcal{M}} \to 0$. Since $x_m' \in \mathcal{I}$ and the restriction $\delta|_{\mathcal{I}}$ is continuous, it follows that $\|\delta(x_m')\|_{\mathcal{E}} \to 0$ as $m \to \infty$. Thus, since $C_{\mathcal{E}} \geq 1$, we conclude that

$$\|\delta(x_m)\|_{\mathcal{E}} \leq C_{\mathcal{E}}^n(|\alpha_{1m}|\|\delta(a_1)\|_{\mathcal{E}} + \dots + |\alpha_{nm}|\|\delta(a_n)\|_{\mathcal{E}} + \|\delta(x_m')\|_{\mathcal{E}}) \to 0,$$

as $m \to \infty$.

Since $\{x_m\}_{m\in\mathbb{N}} \subset \mathcal{M}$ is arbitrary, it follows that $\delta : (\mathcal{M}, \|\cdot\|_{\mathcal{M}}) \to (\mathcal{E}, \|\cdot\|_{\mathcal{E}})$ is continuous. □

The following lemma shows that a quasi-normed bimodule $(\mathcal{E}, \|\cdot\|_{\mathcal{E}})$ of locally measurable operator over a finite von Neumann algebra $\mathcal{M}$ is continuously embedded into $(LS(\mathcal{M}), t(\mathcal{M}))$.

Lemma 7.7.5. *Assume that $\mathcal{M}$ is a finite von Neumann algebra and let $(\mathcal{E}, \|\cdot\|_{\mathcal{E}})$ be a quasi-normed $\mathcal{M}$-bimodule. If $\{a_n\}_{n\in\mathbb{N}} \subset \mathcal{E}$ and $\|a_n\|_{\mathcal{E}} \to 0$, then $a_n \to 0$ locally in measure.*

Proof. Since the von Neumann algebra $\mathcal{M}$ is finite, there exists a faithful normal finite trace τ on $\mathcal{M}$ (see Theorem 1.10.13). By Proposition 4.4.1, it is sufficient to show that $a_n \xrightarrow{t_\tau} 0$.

Suppose to the contrary that a_n does not converge to zero in the measure topology t_τ. Passing to a subsequence, if necessary, and referring to Proposition 4.2.9(iv), we may choose $\varepsilon, \delta > 0$, such that

$$\tau(E_{|a_n|}(\varepsilon, \infty)) > \delta, \quad \|a_n\|_{\mathcal{E}} < (2C_{\mathcal{E}})^{-n}\varepsilon$$

for all $n \in \mathbb{N}$.

We set

$$p_n = E_{|a_n|}(\varepsilon, \infty), \quad q_n = \sup_{m\geq n} p_m, \quad q = \inf_{n\geq 1} q_n.$$

Since τ is a normal finite trace and $\tau(p_n) > \delta$, we have that

$$\tau(q) \geq \delta. \tag{7.43}$$

Since $0 \leq \varepsilon p_n \leq |a_n|$, Proposition 3.3.10 implies that

$$\|p_n\|_{\mathcal{E}} \leq \varepsilon^{-1}\|a_n\|_{\mathcal{E}} < (2C_{\mathcal{E}})^{-n}.$$

Set

$$r_{n,s} = \left(\bigvee_{m=n}^{n+s} p_m\right) \wedge q.$$

It is clear that $r_{n,s} \leq r_{n,s+1}$. Since $\tau(\mathbf{1}) < \infty$ the set $P(\mathcal{M})$ is an orthocomplemented complete modular lattice and, therefore, $P(\mathcal{M})$ is a continuous geometry [96]. In particular,

$$\sup_{s\geq 1} r_{n,s} = \left(\bigvee_{m=n}^{\infty} p_m\right) \wedge q = q_n \wedge q = q.$$

Consequently, for all $n \in \mathbb{N}$, there exists an integer s_n, such that $\tau(q - r_{n,s_n}) < 2^{-n}$. For every $n \in \mathbb{N}$ we set $f_n = \inf_{m\geq n} r_{m,s_m}$. The sequence of projections $\{f_n\}_{n\in\mathbb{N}}$ is increasing, moreover, $f_n \leq q$ and

$$\tau(q - f_n) = \tau\Big(q - \inf_{m\geq n} r_{m,s_m}\Big) = \tau\Big(\sup_{m\geq n}(q - r_{m,s_m})\Big) \leq 2^{-(n-1)}.$$

Therefore, $f_n \uparrow q$ as $n \to \infty$.

By Proposition 3.3.11(ii), we have that

$$\|f_{n-1}\|_{\mathcal{E}} \le \|f_n\|_{\mathcal{E}} \le \|r_{n,s_n}\|_{\mathcal{E}} \le \left\| \bigvee_{k=n}^{n+s_n} p_k \right\|_{\mathcal{E}}$$
$$\le \sum_{k=n}^{n+s} C_{\mathcal{E}}^k \|p_k\|_{\mathcal{E}} < \sum_{k=n}^{n+s} 2^{-k} < 2^{-(n-1)},$$

which implies the equality $f_n = 0$. Since $f_n \uparrow q$ as $n \to \infty$, we obtain $q = 0$, which contradicts (7.43). Thus, $a_n \to 0$ in measure, as required. □

Next, we show that any derivation with values in a quasi-normed bimodule over a finite von Neumann algebra is necessarily continuous in the local measure topology.

Lemma 7.7.6. *Let $\mathcal{M}$ be a finite von Neumann algebra and let $(\mathcal{E}, \|\cdot\|_{\mathcal{E}})$ be a quasi-normed $\mathcal{M}$-bimodule. Every derivation $\delta : LS(\mathcal{M}) \to LS(\mathcal{M})$ with $\delta(\mathcal{M}) \subset \mathcal{E}$ is continuous in the local measure topology.*

Proof. Since $(LS(\mathcal{M}), t(\mathcal{M}))$ is an F-space (see Theorem 4.3.14), it is sufficient to show that the graph of the linear operator δ is closed. Suppose the contrary. Then there exist a sequence $\{a_n\}_{n\in\mathbb{N}} \subset LS(\mathcal{M})$ and $0 \neq b \in LS(\mathcal{M})$, such that $\|a_n\|_{LS(\mathcal{M})} \to 0$ and $\|\delta(a_n) - b\|_{LS(\mathcal{M})} \to 0$ as $n \to \infty$.

By Proposition 4.4.10 and passing, if necessary, to a subsequence, we may assume that $a_n = b_n + c_n$, where $b_n \in \mathcal{M}$, $c_n \in LS(\mathcal{M})$, $n \in \mathbb{N}$, $\|b_n\|_{\mathcal{M}} \to 0$, and $\|s(|c_n|)\|_{LS(\mathcal{M})} \to 0$ as $n \to \infty$.

Since the restriction $\delta|_{\mathcal{M}}$ of the derivation δ to the von Neumann algebra $\mathcal{M}$ is a derivation from $\mathcal{M}$ into the quasi-normed $\mathcal{M}$-bimodule $\mathcal{E}$, Ringrose's Theorem 7.7.4 implies that $\|\delta(b_n)\|_{\mathcal{E}} \to 0$. Referring to Lemma 7.7.5, we infer that $\delta(b_n) \to 0$ locally in measure.

From the equalities,

$$\delta(c_n) = \delta(c_n s(|c_n|)) = \delta(c_n)s(|c_n|) + c_n\delta(s(|c_n|)),$$

we have that

$$s(\delta(c_n)) \le s_l\big(\delta(c_n)s(|c_n|)\big) \vee s_r\big(\delta(c_n)s(|c_n|)\big)$$
$$\vee\, s_l\big(c_n\delta(s(|c_n|))\big) \vee s_r\big(c_n\delta(s(|c_n|))\big).$$

Note that

$$s_l(c_n) \sim s_r(c_n) = s(|c_n|),\ s_l(\delta(c_n)s(|c_n|)) \sim s_r(\delta(c_n)s(|c_n|)) \le s(|c_n|),$$

and

$$s_r\big(c_n\delta(s(|c_n|))\big) \sim s_l\big(c_n\delta(s(|c_n|))\big) \le s_l(c_n) \precsim s(|c_n|).$$

Therefore, by Proposition 4.3.13(iv) and (v), we have that

$$\|s(\delta(c_n))\|_{LS(\mathcal{M})} \le 4\|s(|c_n|)\|_{LS(\mathcal{M})}.$$

Since $\|s(|c_n|)\|_{LS(\mathcal{M})} \to 0$, it follows that $\|s(\delta(c_n))\|_{LS(\mathcal{M})} \to 0$. Hence, referring to Corollary 4.3.17, we conclude that $\delta(c_n) \to 0$ locally in measure.

Thus,

$$0 \neq b \xleftarrow{t(\mathcal{M})} \delta(a_n) = \delta(b_n) + \delta(c_n) \xrightarrow{t(\mathcal{M})} 0.$$

The obtained contradiction implies that δ is $t(\mathcal{M})$-continuous. □

Now, we present the main result of this section, which shows that any derivation from $\mathcal{M}$ into a quasi-normed bimodule over $\mathcal{M}$ is necessarily inner.

Theorem 7.7.7. *Let $\mathcal{M}$ be an arbitrary von Neumann algebra and let $\mathcal{E}$ be a quasi-normed $\mathcal{M}$-bimodule of locally measurable operators. Then any derivation $\delta : \mathcal{M} \to \mathcal{E}$ is inner. Furthermore, there exists $d \in \mathcal{E}$ such that $\delta(x) = [d, x]$ for all $x \in \mathcal{M}$ and $\|d\|_{\mathcal{E}} \le 2C_{\mathcal{E}}\|\delta\|_{\mathcal{M}\to\mathcal{E}}$. If $\delta^* = \delta$ or $\delta^* = -\delta$, then d may be chosen so that $\|d\|_{\mathcal{E}} \le \|\delta\|_{\mathcal{M}\to\mathcal{E}}$.*

Proof. By Theorem 5.2.6, there exists a unique derivation $\overline{\delta} : LS(\mathcal{M}) \to LS(\mathcal{M})$, such that $\overline{\delta}(x) = \delta(x)$ for all $x \in \mathcal{M}$.

By Theorem 1.9.14, there exists a central projection $z \in P(\mathcal{Z}(\mathcal{M}))$, such that $z\mathcal{M}$ is a finite von Neumann algebra and $(\mathbf{1} - z)\mathcal{M}$ is a properly infinite von Neumann algebra. By Theorem 7.6.1(i), the derivation $\overline{\delta}^{(\mathbf{1}-z)} : LS((\mathbf{1} - z)\mathcal{M}) \to LS((\mathbf{1} - z)\mathcal{M})$ is $t((\mathbf{1}-z)\mathcal{M})$-continuous. Since the von Neumann algebra $z\mathcal{M}$ is finite and for the derivation $\overline{\delta}^{(z)} : LS(z\mathcal{M}) \to LS(z\mathcal{M})$ the inclusion $\overline{\delta}^{(z)}(z\mathcal{M}) \subset z\mathcal{E}$ holds, Lemma 7.7.6 implies that $\overline{\delta}^{(z)}$ is also $t(z\mathcal{M})$-continuous for all $j \in J$. Therefore, Proposition 5.1.11 implies that the derivation $\overline{\delta}$ is $t(\mathcal{M})$-continuous. Theorem 5.5.6(i) guarantees that $\overline{\delta}$ is inner, i. e., there exists $a \in LS(\mathcal{M})$, such that $\overline{\delta}(x) = [a, x]$ for all $x \in LS(\mathcal{M})$. It is clear that $[a, \mathcal{M}] = \overline{\delta}(\mathcal{M}) = \delta(\mathcal{M}) \subset \mathcal{E}$.

Let $a_1 = \mathrm{Re}(a)$, $a_2 = \mathrm{Im}(a)$. Since $[a^*, x] = -[a, x^*]^* \in \mathcal{E}$ for any $x \in \mathcal{M}$, it follows that $[a_1, x] = [a + a^*, x]/2 \in \mathcal{E}$ and $[a_2, x] = [a - a^*, x]/2i \in \mathcal{E}$ for all $x \in \mathcal{M}$.

Taking $\varepsilon = 1/2$ in Theorem 3.6.11, we obtain that there exist self-adjoint central elements $c_1, c_2 \in \mathcal{Z}(LS(\mathcal{M}))$, and unitary operators $u_1, u_2 \in \mathcal{M}$, such that

$$2|[a_i, u_i]| \ge |a_i - c_i|, \quad i = 1, 2.$$

Since $[a_i, u_i] \in \mathcal{E}$ and $\mathcal{E}$ is an $\mathcal{M}$-bimodule, Proposition 3.3.10 implies that $d_i := a_i - c_i \in \mathcal{E}$, $i = 1, 2$. Therefore, $d = d_1 + id_2 \in \mathcal{E}$. Since c_1, c_2 are central elements from $LS(\mathcal{M})$, it follows that $\delta(x) = [a, x] = [d, x]$ for all $x \in \mathcal{M}$, proving that δ is inner.

Now, suppose that $\delta^* = \delta$. In this case,

$$[d + d^*, x] = [d, x] - [d, x^*]^* = \delta(x) - (\delta(x^*))^* = \delta(x) - \delta^*(x) = 0$$

for any $x \in \mathcal{M}$. Consequently, the operator $\mathrm{Re}(d) = (d + d^*)/2$ commutes with every element from $\mathcal{M}$. This implies that

$$\delta(x) = [d, x] = [\mathrm{Re}(d), x] + [i\,\mathrm{Im}(d), x] = [i\,\mathrm{Im}(d), x], \quad x \in \mathcal{M}.$$

Therefore, we may assume that $\delta(x) = [d, x]$, $x \in \mathcal{M}$, where $d = ia$ for some self-adjoint $a \in \mathcal{E}$. By Theorem 3.6.11, there exist $c = c^*$ from the center of the algebra $LS(\mathcal{M})$ and a family $\{u_\varepsilon\}_{\varepsilon>0}$ of unitary operators from $\mathcal{M}$ such that

$$\big|[a, u_\varepsilon]\big| \geq (1 - \varepsilon)|a - c|. \tag{7.44}$$

For $b = ia - ic$ and $\varepsilon = 1/2$, we have

$$|b| = |a - c| \leq 2\big|[a, u_{1/2}]\big| = 2\big|[-id, u_{1/2}]\big| = 2\big|[d, u_{1/2}]\big|.$$

Since $[d, u_{1/2}] = \delta(u_{1/2}) \in \mathcal{E}$, Proposition 3.1.6(i) and Proposition 3.3.10 imply that $b \in \mathcal{E}$. Moreover,

$$\delta(x) = [d, x] = [ia, x] = [b, x]$$

for all $x \in \mathcal{M}$. Appealing to (7.44), we have that

$$(1 - \varepsilon)|b| = (1 - \varepsilon)|a - c| \leq \big|[a, u_\varepsilon]\big| = \big|[d, u_\varepsilon]\big| = \big|\delta(u_\varepsilon)\big|$$

and, therefore, it follows from Proposition 3.3.10 that

$$(1 - \varepsilon)\|b\|_{\mathcal{E}} \leq \big\|\delta(u_\varepsilon)\big\|_{\mathcal{E}} \leq \|\delta\|_{\mathcal{M}\to\mathcal{E}}$$

for all $\varepsilon > 0$. Since $\varepsilon > 0$ is arbitrary, we conclude that $\|b\|_{\mathcal{E}} \leq \|\delta\|_{\mathcal{M}\to\mathcal{E}}$.

If $\delta^* = -\delta$, then taking $\mathrm{Im}(d)$ instead of $\mathrm{Re}(d)$ and repeating the preceding argument, we obtain that $\delta(x) = [b, x]$, where $b \in \mathcal{E}$ and $\|b\|_{\mathcal{E}} \leq \|\delta\|_{\mathcal{M}\to\mathcal{E}}$.

Now, suppose that $\delta \neq \delta^*$ and $\delta \neq -\delta^*$. Proposition 3.1.10 implies that

$$\begin{aligned}\|\delta^*\|_{\mathcal{M}\to\mathcal{E}} &= \sup\{\|\delta(x^*)^*\|_{\mathcal{E}} : \|x\|_{\mathcal{M}} \leq 1\} \\ &= \sup\{\|\delta(x)\|_{\mathcal{E}} : \|x\|_{\mathcal{M}} \leq 1\} = \|\delta\|_{\mathcal{M}\to\mathcal{E}}.\end{aligned}$$

Consequently,

$$\|\mathrm{Re}(\delta)\|_{\mathcal{M}\to\mathcal{E}} = 2^{-1}\|\delta + \delta^*\|_{\mathcal{M}\to\mathcal{E}} \leq \|\delta\|_{\mathcal{M}\to\mathcal{E}}.$$

Similarly, $\|\mathrm{Im}(\delta)\|_{\mathcal{M}\to\mathcal{E}} \leq \|\delta\|_{\mathcal{M}\to\mathcal{E}}$. Since $(\mathrm{Re}(\delta))^* = \mathrm{Re}(\delta)$, $(\mathrm{Im}(\delta))^* = \mathrm{Im}(\delta)$, there exist $d_1, d_2 \in \mathcal{E}$, such that $\mathrm{Re}(\delta)(x) = [d_1, x]$, $\mathrm{Im}(\delta)(x) = [d_2, x]$ for all $x \in \mathcal{M}$ and $\|d_i\|_{\mathcal{E}} \leq \|\delta\|_{\mathcal{M}\to\mathcal{E}}$, $i = 1, 2$. Taking $d = d_1 + id_2$, we have that $d \in \mathcal{E}$,

$$\delta(x) = (\mathrm{Re}(\delta) + i \cdot \mathrm{Im}(\delta))(x) = [d_1, x] + i[d_2, x] = [d, x], \quad x \in \mathcal{M},$$

and $\|d\|_{\mathcal{E}} \leq 2C_{\mathcal{E}}(\|d_1\|_{\mathcal{E}} + \|d_2\|_{\mathcal{E}}) \leq 2C_{\mathcal{E}}\|\delta\|_{\mathcal{M}\to\mathcal{E}}$. □

Bibliographical notes

The main result of this chapter provides the full resolution of the Ayupov–Kadison–Liu derivation problem, described in the introduction. Namely, it gives the answer to the following question, posed by Sh. Ayupov in 2000 (and restated later in 2014 by Kadison and Liu): Is any derivation on the algebra of locally measurable operators affiliated with an arbitrary von Neumann algebra continuous in the local measure topology (resp., inner)?

In 2000, in his paper [9], Sh. Ayupov established that any derivation on the $*$-algebra $S(M)$ is continuous in the C-topology, which is generated by the metric introduced in Definition 6.1.8.

In 2008, Albeverio, Ayupov, Kudaybergenov resolved the Ayupov–Kadison–Liu derivation problem for type I von Neumann algebras [4, 7]. In 2008, Gutman, Kusraev, and Kutateladze used a Boolean-valued approach to resolve the derivation problem for type I von Neumann algebras [77]. In 2010, Ber, de Pagter, and Sukochev in [27, 28] employed an operator-valued functions approach for treating the derivation problem for infinite type I von Neumann algebras. In 2012, Ayupov and Kudaybergenov resolved the derivation problem for $LS(M)$ and their subalgebras for type I and III von Neumann algebras [13]. In 2011, Ber proved the automatic continuity in the measure topology of derivations on the algebra $S(\mathcal{M}, \tau)$ affiliated with a properly infinite semifinite von Neumann algebra $\mathcal{M}$ [19]. In 2013, Ber, Chilin, Sukochev showed the automatic continuity in the local measure topology of derivations on the algebra $LS(\mathcal{M})$ for a properly infinite von Neumann algebra $\mathcal{M}$ [23].

In 2019, Ber, Kudaybergenov, Sukochev announced in [32] the complete solution of the Ayupov–Kadison–Liu derivation problem by establishing the continuity of derivations of $S(\mathcal{M})$ for type II_1 von Neumann algebras. The complete proof of this result was published in [34]. Ring derivations on Murray–von Neumann algebras associated with finite type II von Neumann algebras were treated in [80].

For a description of derivations on Arens algebras and their properties, see, e. g., [1–3, 11]. For an overview of problems related to the Ayupov–Kadison–Liu problem, we refer the reader to [100].

Theorem 7.7.4 was proved by Ringrose in 1972 [128] for a general case of Banach bimodules over a C^*-algebra. It was extended to the case of quasi-normed bimodules in [35]. Theorem 7.7.7 was proved in [24] for Banach bimodules over a given von Neumann algebra and extended to quasi-normed case in [35].

All results of the present chapter remain valid in nonseparable Hilbert spaces. The proofs presented here require only slight modifications in the nonseparable case. For the complete proofs in the nonseparable case, we refer the reader to aforementioned papers.

Bibliography

[1] S. Albeverio, Sh. Ayupov, and R. Abdullaev. Arens spaces associated with von Neumann algebras and normal states. *Positivity*, 14(1):105–121, 2010.

[2] S. Albeverio, Sh. Ayupov, R. Abdullaev, and K. Kudaybergenov. Additive derivations on generalized Arens algebras. *Lobachevskii J. Math.*, 32(3):194–202, 2011.

[3] S. Albeverio, Sh. Ayupov, and K. Kudaybergenov. Non-commutative Arens algebras and their derivations. *J. Funct. Anal.*, 253(1):287–302, 2007.

[4] S. Albeverio, Sh. Ayupov, and K. Kudaybergenov. Derivations on the algebra of measurable operators affiliated with a type I von Neumann algebra. *Sib. Adv. Math.*, 18(2):86–94, 2008.

[5] S. Albeverio, Sh. Ayupov, and K. Kudaybergenov. Derivations on the algebra of τ-compact operators affiliated with a type I von Neumann algebra. *Positivity*, 12(2):375–386, 2008.

[6] S. Albeverio, Sh. Ayupov, and K. Kudaybergenov. Description of derivations on locally measurable operator algebras of type I. *Extr. Math.*, 24(1):1–15, 2009.

[7] S. Albeverio, Sh. Ayupov, and K. Kudaybergenov. Structure of derivations on various algebras of measurable operators for type I von Neumann algebras. *J. Funct. Anal.*, 256(9):2917–2943, 2009.

[8] C. Aliprantis and O. Burkinshaw. *Principles of real analysis*. Academic Press, Inc., Boston, MA, second edition, 1990.

[9] Sh. Ayupov. Derivations in algebras of measurable operators. *Akad. Nauk Respub. Uzbek.*, 56(3):14–17, 2000.

[10] Sh. Ayupov, K. Karimov, and K. Kudaybergenov. Isomorphism between the algebra of measurable functions and its subalgebra of approximately differentiable functions. *Vladikavkaz. Mat. Zh.*, 25(2):25–37, 2023.

[11] Sh. Ayupov and K. Kudaybergenov. Derivations of noncommutative Arens algebras. *Funkc. Anal. Prilozh.*, 41(4):70–72, 2007.

[12] Sh. Ayupov and K. Kudaybergenov. Derivations on algebras of measurable operators. *Infin. Dimens. Anal. Quantum Probab. Relat. Top.*, 13(2):305–337, 2010.

[13] Sh. Ayupov and K. Kudaybergenov. Additive derivations on algebras of measurable operators. *J. Oper. Theory*, 67(2):495–510, 2012.

[14] Sh. Ayupov and K. Kudaybergenov. Innerness of continuous derivations on algebras of measurable operators affiliated with finite von Neumann algebras. *J. Math. Anal. Appl.*, 408(1):256–267, 2013.

[15] Sh. Ayupov and K. Kudaybergenov. Spatiality of derivations on the algebra of τ-compact operators. *Integral Equ. Oper. Theory*, 77(4):581–598, 2013.

[16] Sh. Ayupov, K. Kudaybergenov, and K. Karimov. On dimension of the space of derivations on commutative regular algebras. *J. Math. Sci. (N. Y.)*, 271(6):694–699, 2023.

[17] E. Azoff. Spectrum and direct integral. *Trans. Am. Math. Soc.*, 197:211–223, 1974.

[18] A. Ber. Maximal unique extension of differentiation in commutative regular algebras. In *16-th Crimean Autumn Mathematical School—Symposium*, volume 16, pages 47–50, 2006.

[19] A. Ber. Continuity of derivations on properly infinite $*$-algebras of τ-measurable operators. *Mat. Zametki*, 90(5):776–780, 2011.

[20] A. Ber. Derivations in commutative regular algebras. *Sib. Adv. Math.*, 21(3):161–169, 2011.

[21] A. Ber. Continuous derivations on $*$-algebras of τ-measurable operators are inner. *Math. Notes*, 93(5):654–659, 2013.

[22] A. Ber, V. Chilin, and F. Sukochev. Non-trivial derivations on commutative regular algebras. *Extr. Math.*, 21(2):107–147, 2006.

[23] A. Ber, V. Chilin, and F. Sukochev. Continuity of derivations of algebras of locally measurable operators. *Integral Equ. Oper. Theory*, 75(4):527–557, 2013.

[24] A. Ber, V. Chilin, and F. Sukochev. Continuous derivations on algebras of locally measurable operators are inner. *Proc. Lond. Math. Soc. (3)*, 109(1):65–89, 2014.

https://doi.org/10.1515/9783111599687-009

[25] A. Ber, V. Chilin, and F. Sukochev. Derivations on Banach $*$-ideals in von Neumann algebras. *Vladikavkaz. Mat. Zh.*, 20(2):23–28, 2018.

[26] A. Ber, V. Chilin, and F. Sukochev. Derivations in disjointly complete commutative regular algebras. *Quaest. Math.*, 47:S23–S86, 2024.

[27] A. Ber, B. de Pagter, and F. Sukochev. Some remarks on derivations in algebras of measurable operators. *Mat. Zametki*, 87(4):502–513, 2010.

[28] A. Ber, B. de Pagter, and F. Sukochev. Derivations in algebras of operator-valued functions. *J. Oper. Theory*, 66(2):261–300, 2011.

[29] A. Ber, J. Huang, K. Kudaybergenov, and F. Sukochev. Non-existence of translation-invariant derivations on algebras of measurable functions. *Quaest. Math.*, 46(5):909–926, 2023.

[30] A. Ber, J. Huang, G. Levitina, and F. Sukochev. Derivations with values in ideals of semifinite von Neumann algebras. *J. Funct. Anal.*, 272(12):4984–4997, 2017.

[31] A. Ber, J. Huang, G. Levitina, and F. Sukochev. Derivations with values in the ideal of τ-compact operators affiliated with a semifinite von Neumann algebra. *Commun. Math. Phys.*, 390(2):577–616, 2022.

[32] A. Ber, K. Kudaybergenov, and F. Sukochev. Derivation on Murray–von Neumann algebras. *Usp. Mat. Nauk*, 74(5(449)):183–184, 2019.

[33] A. Ber, K. Kudaybergenov, and F. Sukochev. Notes on derivations of Murray–von Neumann algebras. *J. Funct. Anal.*, 279(5):108589, 26, 2020.

[34] A. Ber, K. Kudaybergenov, and F. Sukochev. Derivations of Murray–von Neumann algebras. *J. Reine Angew. Math.*, 791:283–301, 2022.

[35] A. Ber, G. Levitina, and V. Chilin. Derivations with values in quasinormable bimodules of locally measurable operators. *Mat. Tr.*, 17(1):3–18, 2014.

[36] A. Ber and F. Sukochev. Commutator estimates in W^*-algebras. *J. Funct. Anal.*, 262(2):537–568, 2012.

[37] A. Ber and F. Sukochev. Commutator estimates in W^*-factors. *Trans. Am. Math. Soc.*, 364(10):5571–5587, 2012.

[38] A. Ber, F. Sukochev, and V. Chilin. Derivations in commutative regular algebras. *Mat. Zametki*, 75(3):453–454, 2004.

[39] S. Berberian. *Baer *-rings. Die Grundlehren der mathematischen Wissenschaften in Einzeldarstellungen.* Band 195. Springer, New York–Berlin, 1972.

[40] A. Bikchentaev. Majorization for products of measurable operators. In *Proceedings of the International Quantum Structures Association 1996 (Berlin)*, volume 37, pages 571–576, 1998.

[41] A. Bikchentaev. Local convergence in measure on semifinite von Neumamn algebras. *Tr. Mat. Inst. Steklova*, 255:41–54, 2006.

[42] A. Bikchentaev. Local convergence in measure on semifinite von Neumann algebras. II. *Mat. Zametki*, 82(5):783–786, 2007.

[43] M. Birman and M. Solomjak. *Spectral theory of selfadjoint operators in Hilbert space*. Mathematics and its Applications (Soviet Series). D. Reidel Publishing Co., Dordrecht, 1987. Translated from the 1980 Russian original by S. Khrushchëv and V. Peller.

[44] B. Blackadar. *Operator algebras*, volume 122 of *Encyclopaedia of Mathematical Sciences*. Springer-Verlag, Berlin, 2006. Theory of C^*-algebras and von Neumann algebras, Operator Algebras and Non-commutative Geometry, III.

[45] N. Bourbaki. *Commutative algebra. Chapters 1–7.* Elements of Mathematics (Berlin). Springer-Verlag, Berlin, 1998. Translated from the French, Reprint of the 1989 English translation.

[46] O. Bratteli. *Derivations, dissipations and group actions on C^*-algebras*, volume 1229 of *Lecture Notes in Mathematics*. Springer-Verlag, Berlin, 1986.

[47] O. Bratteli and D. Robinson. Unbounded derivations of C^*-algebras. *Commun. Math. Phys.*, 42:253–268, 1975.

[48] O. Bratteli and D. Robinson. Unbounded derivations of C^*-algebras. II. *Commun. Math. Phys.*, 46(1):11–30, 1976.

[49] O. Bratteli and D. Robinson. *Operator algebras and quantum statistical mechanics. Vol. 1.* Texts and Monographs in Physics. Springer-Verlag, New York, 1979. C^*- and W^*-algebras, algebras, symmetry groups, decomposition of states.

[50] O. Bratteli and D. Robinson. *Operator algebras and quantum statistical mechanics. 2.* Texts and Monographs in Physics. Springer-Verlag, Berlin, second edition, 1997. Equilibrium states. Models in quantum statistical mechanics.

[51] M. Breuer. Fredholm theories in von Neumann algebras. I. *Math. Ann.*, 178:243–254, 1968.

[52] C. Brödel and G. Lassner. Derivationen auf gewissen Op*-Algebren. *Math. Nachr.*, 67:53–58, 1975.

[53] V. Chilin and M. Muratov. Comparison of topologies on $*$-algebras of locally measurable operators. *Positivity*, 17(1):111–132, 2013.

[54] T. Chow. A spectral theory for direct integrals of operators. *Math. Ann.*, 188:285–303, 1970.

[55] T. Chow and F. Gilfeather. Functions of direct integrals of operators. *Proc. Am. Math. Soc.*, 29:325–330, 1971.

[56] L. Ciach. Linear-topological spaces of operators affiliated with a von Neumann algebra. *Bull. Pol. Acad. Sci., Math.*, 31(3–4):161–166, 1983.

[57] H. Dales. *Banach algebras and automatic continuity*, volume 24 of *London Mathematical Society Monographs. New Series.* The Clarendon Press Oxford University Press, New York, 2000. Oxford Science Publications.

[58] K. Davidson. *Nest algebras*, volume 191 of *Pitman Research Notes in Mathematics Series*. Longman Scientific & Technical, Harlow; copublished in the United States with John Wiley & Sons, Inc., New York, 1988. Triangular forms for operator algebras on Hilbert space.

[59] J. Deel. Derivations of AW^*-algebras. *Proc. Am. Math. Soc.*, 42:85–95, 1974.

[60] J. Dixmier. *C^*-algebras*, volume 15 of *North-Holland Mathematical Library*. North-Holland Publishing Co., Amsterdam, 1977. Translated from the French by Francis Jellett.

[61] J. Dixmier. *von Neumann algebras*, volume 27 of *North-Holland Mathematical Library*. North-Holland Publishing Co., Amsterdam, 1981. With a preface by E. C. Lance, Translated from the second French edition by F. Jellett.

[62] P. Dixon. Generalized B^*-algebras. *Proc. Lond. Math. Soc. (3)*, 21:693–715, 1970.

[63] P. Dixon. Unbounded operator algebras. *Proc. Lond. Math. Soc. (3)*, 23:53–69, 1971.

[64] P. Dodds, B. de Pagter, and F. Sukochev. *Noncommutative integration and operator theory*, volume 349 of *Progress in Mathematics*. Birkhäuser/Springer, Cham, 2023.

[65] P. Dodds, T. Dodds, P. Dowling, C. Lennard, and F. Sukochev. A uniform Kadec–Klee property for symmetric operator spaces. *Math. Proc. Camb. Philos. Soc.*, 118(3):487–502, 1995.

[66] N. Dunford and J. Schwartz. *Linear operators. Part I.* Wiley Classics Library. John Wiley & Sons, Inc., New York, 1988. General theory, With the assistance of William G. Bade and Robert G. Bartle, Reprint of the 1958 original, A Wiley-Interscience Publication.

[67] H. Dye. The Radon-Nikodým theorem for finite rings of operators. *Trans. Am. Math. Soc.*, 72:243–280, 1952.

[68] K. Dykema, J. Noles, F. Sukochev, and D. Zanin. On reduction theory and Brown measure for closed unbounded operators. *J. Funct. Anal.*, 271(12):3403–3422, 2016.

[69] G. Elliott. On derivations of AW^*-algebras. *Tohoku Math. J. (2)*, 30(2):263–276, 1978.

[70] T. Fack and H. Kosaki. Generalized s-numbers of τ-measurable operators. *Pac. J. Math.*, 123(2):269–300, 1986.

[71] H. Federer. *Geometric measure theory*. Die Grundlehren der mathematischen Wissenschaften, Band 153. Springer-Verlag New York Inc., New York, 1969.

[72] D. Fremlin. *Measure theory*. Number v. 1 in Measure Theory. Torres Fremlin, 2000.

[73] F. Gesztesy, A. Gomilko, F. Sukochev, and Y. Tomilov. On a question of A. E. Nussbaum on measurability of families of closed linear operators in a Hilbert space. *Isr. J. Math.*, 188:195–219, 2012.

[74] F. Gesztesy, Y. Latushkin, K. Makarov, F. Sukochev, and Y. Tomilov. The index formula and the spectral shift function for relatively trace class perturbations. *Adv. Math.*, 227(1):319–420, 2011.

[75] A. Grothendieck. Réarrangements de fonctions et inégalités de convexité dans les algèbres de von Neumann munies d'une trace. *Sémin. Bourbaki*, 3:127–139, 1954–1956.

[76] R. Gunning and H. Rossi. *Analytic functions of several complex variables*. AMS Chelsea Publishing Series. AMS Chelsea Pub., 2009.

[77] A. Gutman, A. Kusraev, and S. Kutateladze. The Wickstead problem. *Sib. Èlektron. Mat. Izv.*, 5:293–333, 2008.

[78] P. Halmos. *Measure theory*. D. Van Nostrand Company, Inc., New York, N. Y., 1950.

[79] E. Hewitt and K. Ross. *Abstract harmonic analysis*. Number v. 2 in Abstract Harmonic Analysis. Academic Press, 1963.

[80] J. Huang, K. Kudaybergenov, and F. Sukochev. Ring derivations of Murray–von Neumann algebras. *Linear Algebra Appl.*, 672:28–52, 2023.

[81] J. Huang and F. Sukochev. Interpolation between $L_0(\mathcal{M},\tau)$ and $L_\infty(\mathcal{M},\tau)$. *Math. Z.*, 293(3–4):1657–1672, 2019.

[82] J. Huang and F. Sukochev. Derivations with values in noncommutative symmetric spaces. *C. R. Math. Acad. Sci. Paris*, 361:1357–1365, 2023.

[83] A. Inoue and S. Ota. Derivations on algebras of unbounded operators. *Trans. Am. Math. Soc.*, 261(2):567–577, 1980.

[84] A. Inoue, S. Ota, and J. Tomiyama. Derivations of operator algebras into spaces of unbounded operators. *Pac. J. Math.*, 96(2):389–404, 1981.

[85] V. Jarni'k. Sur les fonctions de deux variables re'elles. *Fundam. Math.*, 27(1):147–150, 1936.

[86] B. Johnson and S. Parrott. Operators commuting with a von Neumann algebra modulo the set of compact operators. *J. Funct. Anal.*, 11:39–61, 1972.

[87] P. Jorgensen. Extensions of unbounded *-derivations in UHF C^*-algebras. *J. Funct. Anal.*, 45(3):341–356, 1982.

[88] R. Kadison. Derivations of operator algebras. *Ann. Math. (2)*, 83:280–293, 1966.

[89] R. Kadison. Which Singer is that? In *Surveys in differential geometry*, volume 7 of *Surv. Differ. Geom.*, pages 347–373. Int. Press, Somerville, MA, 2000.

[90] R. Kadison and Z. Liu. A note on derivations of Murray–von Neumann algebras. *Proc. Natl. Acad. Sci. USA*, 111(6):2087–2093, 2014.

[91] R. Kadison and J. Ringrose. Cohomology of operator algebras. I. Type I von Neumann algebras. *Acta Math.*, 126:227–243, 1971.

[92] R. Kadison and J. Ringrose. *Fundamentals of the theory of operator algebras. Vol. I*, volume 15 of *Graduate Studies in Mathematics*. American Mathematical Society, Providence, RI, 1997. Elementary theory, Reprint of the 1983 original.

[93] R. Kadison and J. Ringrose. *Fundamentals of the theory of operator algebras. Vol. II*, volume 16 of *Graduate Studies in Mathematics*. American Mathematical Society, Providence, RI, 1997. Advanced theory, Corrected reprint of the 1986 original.

[94] V. Kaftal and G. Weiss. Compact derivations relative to semifinite von Neumann algebras. *J. Funct. Anal.*, 62(2):202–220, 1985.

[95] N. Kalton, N. Peck, and J. Roberts. *An F-space sampler*. London Mathematical Society Lecture Note Series. Cambridge University Press, 1984.

[96] I. Kaplansky. Projections in Banach algebras. *Ann. Math. (2)*, 53:235–249, 1951.

[97] I. Kaplansky. Modules over operator algebras. *Am. J. Math.*, 75:839–858, 1953.

[98] I. Kaplansky. Functional analysis. In *Some aspects of analysis and probability*, volume 4 of *Surveys Appl. Math.*, pages 1–34. Wiley, New York, 1958.

[99] A. Khinčin. Recherches sur la structure des fonctions mesurables. *Rec. Math. Moscou*, 31:265–285, 377–433, 1923.

[100] K. Kudaybergenov. On solution of Ayupov's problem. *Bull. Inst. Math.*, 5:219–233, 2022.

[101] A. Kusraev. *Dominated operators*, volume 519 of *Mathematics and its Applications*. Kluwer Academic Publishers, Dordrecht, 2000. Translated from the 1999 Russian original by the author, Translation edited and with a foreword by S. Kutateladze.

[102] A. Kusraev. Derivations and automorphisms in the algebra of measurable complex-valued functions. *Vladikavkaz. Mat. Zh.*, 7(3):45–49 (electronic), 2005.
[103] A. Kusraev. Automorphisms and derivations in an extended complex f-algebra. *Sib. Mat. Zh.*, 47(1):97–107, 2006.
[104] C. Lance and A. Niknam. Unbounded derivations of group C^*-algebras. *Proc. Am. Math. Soc.*, 61(2):310–314 (1977), 1976.
[105] M. Lennon. Direct integrals of locally measurable operators. *Math. Scand.*, 32:123–132, 1973.
[106] M. Lennon. Direct integral decomposition of spectral operators. *Math. Ann.*, 207:257–268, 1974.
[107] M. Lennon. On sums and products of unbounded operators in Hilbert space. *Trans. Am. Math. Soc.*, 198:273–285, 1974.
[108] S. Lord, F. Sukochev, and D. Zanin. *Singular traces*, volume 46 of *de Gruyter Studies in Mathematics*. De Gruyter, Berlin, 2013. Theory and applications.
[109] M. Muratov and V. Chilin. Topological algebras of measurable and locally measurable operators. *Sovrem. Mat. Fundam. Napravl.*, 61:115–163, 2016.
[110] G. Murphy. *C^*-algebras and operator theory*. Academic Press Inc., Boston, MA, 1990.
[111] F. Murray and J. Von Neumann. On rings of operators. *Ann. Math. (2)*, 37(1):116–229, 1936.
[112] F. Murray and J. von Neumann. On rings of operators. II. *Trans. Am. Math. Soc.*, 41(2):208–248, 1937.
[113] F. Murray and J. von Neumann. On rings of operators. IV. *Ann. Math.*, 44(4):716–808, 1943.
[114] M. Naĭmark. *Normed rings*. P. Noordhoff N. V., Groningen, 1959. Translated from the first Russian edition by Leo F. Boron.
[115] S. Nayak. On Murray–von Neumann algebras—I: topological, order-theoretic and analytical aspects. *Banach J. Math. Anal.*, 15(3):Paper No. 45, 40, 2021.
[116] E. Nelson. Notes on non-commutative integration. *J. Funct. Anal.*, 15:103–116, 1974.
[117] A. Nussbaum. Reduction theory for unbounded closed operators in Hilbert space. *Duke Math. J.*, 31:33–44, 1964.
[118] T. Ogasawara. Finite-dimensionality of certain Banach algebras. *J. Sci. Hiroshima Univ., Ser. A Math. Phys. Chem.*, 17(3):359–364, 1954.
[119] D. Olesen. Derivations of AW^*-algebras are inner. *Pac. J. Math.*, 53:555–561, 1974.
[120] V. Ovčinnikov. The s-numbers of measurable operators. *Funkc. Anal. Prilozh.*, 4(3):78–85, 1970.
[121] V. Ovčinnikov. The completely continuous operators with respect to a von Neumann algebra. *Funkc. Anal. Prilož.*, 6(1):37–40, 1972.
[122] A. Padmanabhan. Convergence in measure and related results in finite rings of operators. *Trans. Am. Math. Soc.*, 128(3):359–378, 1967.
[123] S. Popa. The commutant modulo the set of compact operators of a von Neumann algebra. *J. Funct. Anal.*, 71(2):393–408, 1987.
[124] S. Popa and F. Rădulescu. Derivations of von Neumann algebras into the compact ideal space of a semifinite algebra. *Duke Math. J.*, 57(2):485–518, 1988.
[125] R. Powers and S. Sakai. Unbounded derivations in operator algebras. *J. Funct. Anal.*, 19:81–95, 1975.
[126] M. Reed and B. Simon. *Methods of modern mathematical physics. II. Fourier analysis, self-adjointness*. Academic Press [Harcourt Brace Jovanovich, Publishers], New York–London, 1975.
[127] M. Reed and B. Simon. *Methods of modern mathematical physics. I. Functional analysis*. Academic Press Inc. [Harcourt Brace Jovanovich Publishers], New York, second edition, 1980.
[128] J. Ringrose. Automatic continuity of derivations of operator algebras. *J. Lond. Math. Soc. (2)*, 5:432–438, 1972.
[129] W. Rudin. *Functional analysis*. International Series in Pure and Applied Mathematics. McGraw-Hill Inc., New York, second edition, 1991.
[130] K. Saitô. On the algebra of measurable operators for a general AW*-algebra. *Tôhoku Math. J. (2)*, 21:249–270, 1969.
[131] K. Saitô. On the algebra of measurable operators for a general AW^*-algebra. II. *Tôhoku Math. J. (2)*, 23:525–534, 1971.

[132] S. Sakai. On a conjecture of Kaplansky. *Tôhoku Math. J. (2)*, 12:31–33, 1960.
[133] S. Sakai. Derivations of W^*-algebras. *Ann. Math. (2)*, 83:273–279, 1966.
[134] S. Sakai. *C^*-algebras and W^*-algebras*. Ergebnisse der Mathematik und ihrer Grenzgebiete, Band 60. Springer-Verlag, New York, 1971.
[135] S. Sakai. *Operator algebras in dynamical systems*, volume 41 of *Encyclopedia of Mathematics and its Applications*. Cambridge University Press, Cambridge, 1991. The theory of unbounded derivations in C^*-algebras.
[136] S. Saks. *Theory of the integral*. Dover Publications Inc., New York, second revised edition, 1964. English translation by L. C. Young. With two additional notes by Stefan Banach.
[137] S. Sankaran. The *-algebra of unbounded operators. *J. Lond. Math. Soc.*, 34:337–344, 1959.
[138] S. Sankaran. Stochastic convergence for operators. *Quart. J. Math. Oxf. Ser. (2)*, 15:97–102, 1964.
[139] K. Schmüdgen. *Unbounded operator algebras and representation theory*, volume 37 of *Operator Theory: Advances and Applications*. Birkhäuser Verlag, Basel, 1990.
[140] K. Schmüdgen. *Unbounded self-adjoint operators on Hilbert space*, volume 265 of *Graduate Texts in Mathematics*. Springer, Dordrecht, 2012.
[141] I. Segal. A non-commutative extension of abstract integration. *Ann. Math. (2)*, 57:401–457, 1953.
[142] G. Šilov. On a property of rings of functions. *Doklady Akad. Nauk SSSR (N. S.)*, 58:985–988, 1947.
[143] A. Sinclair and R. Smith. *Finite von Neumann algebras and masas*, volume 351 of *London Mathematical Society Lecture Note Series*. Cambridge University Press, Cambridge, 2008.
[144] M. Sonis. A certain class of operators in von Neumann algebras with Segel's measure on the projectors. *Mat. Sb. (N. S.)*, 84(126):353–368, 1971.
[145] W. Stinespring. Integration theorems for gages and duality for unimodular groups. *Trans. Am. Math. Soc.*, 90:15–56, 1959.
[146] M. Stone. On unbounded operators in Hilbert space. *J. Indian Math. Soc. (N. S.)*, 15:155–192, 1951.
[147] S. Strătilă and L. Zsidó. *Lectures on von Neumann algebras*. Editura Academiei, Bucharest, 1979. Revision of the 1975 original, Translated from the Romanian by Silviu Teleman.
[148] A. Ströh and G. West. τ-compact operators affiliated to a semifinite von Neumann algebra. *Proc. Roy. Ir. Acad. Sect. A*, 93(1):73–86, 1993.
[149] M. Takesaki. *Theory of operator algebras. I*. Springer-Verlag, New York, 1979.
[150] M. Takesaki. *Theory of operator algebras. II*, volume 125 of *Encyclopaedia of Mathematical Sciences*. Springer-Verlag, Berlin, 2003. Operator Algebras and Non-commutative Geometry, 6.
[151] M. Takesaki. *Theory of operator algebras. III*, volume 127 of *Encyclopaedia of Mathematical Sciences*. Springer-Verlag, Berlin, 2003. Operator Algebras and Non-commutative Geometry, 8.
[152] M. Terp. l_p-spaces associated with von Neumann algebras, 1981.
[153] Y. Tseng. Spectral representation of self-adjoint functional transformations in a non-hilbertian space. *Sci. Rep. Nat. Tsinghua Univ.*, 3:113–125, 1935.
[154] B. van der Waerden. *Algebra. Vol. 1*. Frederick Ungar Publishing Co., New York, 1970. Translated by Fred Blum and John R. Schulenberger.
[155] D. Vladimirov. *Boolean algebras in analysis*, volume 540 of *Mathematics and its Applications*. Kluwer Academic Publishers, Dordrecht, 2002. Translated from the Russian manuscript, Foreword and appendix by S. S. Kutateladze.
[156] J. von Neumann. Zur algebra der funktionaloperationen und theorie der normalen operatoren. *Math. Ann.*, 102:370–427, 1930.
[157] J. von Neumann. Continuous rings and their arithmetics. *Proc. Natl. Acad. Sci.*, 23(6):341–349, 1937.
[158] J. von Neumann. On rings of operators. III. *Ann. Math. (2)*, 41:94–161, 1940.
[159] J. von Neumann. The non-isomorphism of certain continuous rings. *Ann. Math. (2)*, 67:485–496, 1958.
[160] J. von Neumann. *Collected works. Vol. IV: Continuous geometry and other topics*. Pergamon Press, Oxford–London–New York–Paris, 1962. General editor: A. H. Taub.
[161] H. Wielandt. Über die Unbeschränktheit der Operatoren der Quantenmechanik. *Math. Ann.*, 121:21, 1949.

[162] F. Yeadon. Convergence of measurable operators. *Proc. Camb. Philos. Soc.*, 74:257–268, 1973.
[163] F. Yeadon. On a result of P. G. Dixon. *J. Lond. Math. Soc. (2)*, 9:610–612, 1974/75.
[164] F. Yeadon. Non-commutative L^p-spaces. *Math. Proc. Camb. Philos. Soc.*, 77:91–102, 1975.
[165] B. Zakirov and V. Chilin. Abstract characterization of EW^*-algebras. *Funkc. Anal. Prilozh.*, 25(1):76–78, 1991.
[166] L. Zsidó. The norm of a derivation in a W^*-algebra. *Proc. Am. Math. Soc.*, 38(1):147–150, 1973.

Subject index

https://doi.org/10.1515/9783111599687-010

Notation index

https://doi.org/10.1515/9783111599687-011

www.ingramcontent.com/pod-product-compliance
Lightning Source LLC
Chambersburg PA
CBHW080651220326
41598CB00033B/5171